Space Sciences Series of ISSI

For other titles published in this series, go to
www.springer.com/series/6592

L. Bengtsson, S. Koumoutsaris, R.-M. Bonnet,
R.P. Allan, C. Fröhlich, J. Heintzenberg,
P. Ingmann, R. Kandel, N.G. Loeb, B. Soden,
K.E. Trenberth
Editors

Observing and Modeling Earth's Energy Flows

Previously published in *Surveys in Geophysics*, Volume 33, Issues 3–4, 2012

Springer

Editors
Lennart Bengtsson
International Space Science Institute
Hallerstrasse 6, 3012 Bern, Switzerland
bengtsson@issibern.ch

Symeon Koumoutsaris
Risk Management Solutions Ltd
Peninsular House Monument Street 30
London EC3R 8NB, United Kingdom
skoumoutsaris@rms.com

Roger-Maurice Bonnet
International Space Science Institute
Hallerstrasse 6, 3012 Bern, Switzerland
rmbonnet@issibern.ch

Richard P. Allan
University of Reading Department of Meteorology
Earley Gate, Reading RG6 6BB
United Kingdom
r.p.allan@reading.ac.uk

Claus Fröhlich
Physikalisch-Meteorologisches Observatorium Davos/
World Radiation Center Dorfstrasse 33
7260 Davos, Switzerland
claus.froehlich@pmodwrc.ch

Jost Heintzenberg
Leibniz Institute for Tropospheric Research
Permoserstrasse 15, 04318 Leipzig
Germany
jost.heintzenberg@tropos.de

Paul Ingmann
European Space Agency
Earth Observation Mission Science Division
Keplerlaan 1, 2201 AZ Noordwijk
Netherlands
paul.ingmann@esa.int

Robert Kandel
Ecole Polytechnique
Laboratoire de Météorologie Dynamique
Route de Saclay, 91128 Palaiseau, France
kandel@lmd.polytechnique.fr

Norman G. Loeb
National Aeronautics and Space Administration Langley Research Center
Langley Blvd, Hampton Virginia 23666
USA
Norman.G.Loeb@nasa.gov

Brian Soden
University of Miami Rosenstiel School of Marine & Atmospheric Science (RSMAS)
Div. MPO Rickenbacker Causeway 4600
Miami, Florida 33149, USA
bsoden@rsmas.miami.edu

Kevin E. Trenberth
National Center of Atmospheric Research
Climate & Global Dynamics
1850 Table Mesa Drive
Boulder, Colorado 80305, USA
trenbert@ucar.edu

ISBN: 978-94-017-7970-8
DOI: 10.1007/978-94-007-4327-4

Springer Dordrecht Heidelberg New York London

Softcover re-print of the Hardcover 1st edition 2012

Cover Illustration: Moon and Venus over Switzerland, Photographer: David Kaplan

Printed on acid-free paper

Springer is part of Springer Science+Business Media (www.springer.com)

Contents

Surv Geophys (2012) 33:333–336
DOI 10.1007/s10712-012-9194-y

Foreword: International Space Science Institute (ISSI) Workshop on Observing and Modeling Earth's Energy Flows

Lennart Bengtsson

Received: 27 April 2012 / Accepted: 30 April 2012 / Published online: 22 May 2012

The Earth's climate, as well as planetary climates in general, is broadly regulated by three fundamental parameters: the total solar irradiance, the planetary albedo and the planetary emissivity. Observations from series of different satellites during the last three decades indicate that these three quantities are generally very stable. The total solar irradiation of some 1,361 W/m^2 at 1 A.U. varies within 1 W/m^2 during the 11-year solar cycle (Fröhlich 2012). The albedo is close to 29 % with minute changes from year to year but with marked zonal differences (Stevens and Schwartz 2012). The only exception to the overall stability is a minor *decrease* in the planetary emissivity (the ratio between the radiation to space and the radiation from the surface of the Earth). This is a consequence of the increase in atmospheric greenhouse gas amounts making the atmosphere gradually more opaque to long-wave terrestrial radiation. As a consequence, radiation processes are slightly out of balance as less heat is leaving the Earth in the form of thermal radiation than the amount of heat from the incoming solar radiation. Present space-based systems cannot yet measure this imbalance, but the effect can be inferred from the increase in heat in the oceans where most of the heat accumulates. Minor amounts of heat are used to melt ice and to warm the atmosphere and the surface of the Earth.

The reverse is happening when there is a volcanic eruption that emits particles that reflect solar radiation. In this case less heat is entering the Earth's system than leaving it and a cooling take place, as the system is moving toward another equilibrium. In reality the Earth's system is never in equilibrium as it is continuously exposed to disturbances in its energy balance. In addition to external and anthropogenic effects there are natural processes in the Earth's system caused by variations in the cloud cover, in the exchange of heat between ocean and atmosphere, in the surface conditions such as snow and ice cover, and in changes in the three-dimensional distribution of water vapor. Other changes in the energy balance include land vegetation that affects both the surface albedo and the surface fluxes of heat and water.

Some of the processes now affecting the Earth's climate system are relatively well understood, and others less so. The radiative effect of the greenhouse gases added to the

L. Bengtsson (✉)
International Space Science Institute, Hallerstrasse 6, 3012 Bern, Switzerland
e-mail: lennart.bengtsson@zmaw.de

atmosphere since the beginning of the industrialization at the end of the eighteenth century is equivalent to a reduction in the outgoing radiation to space by some 2.8 W/m^2. Less well understood are the radiative effects of anthropogenic aerosols, but an overall cooling effect of the order of 1 W/m^2 is suggested. The Earth has partly adjusted to this by increasing its surface temperature by 0.6–0.8 °C during the same period, but the radiation balance has not yet been restored. Based on the ongoing warming of the oceans the radiation imbalance during the last decades is estimated to be of the order of 0.5 W/m^2 with generally higher values in later years. Other associated changes of importance for the Earth's climate follow a temperature increase, including an increasing amount of water vapor in the atmosphere. The water vapor amount adjusts rapidly to atmospheric temperature by about 6–7 % for each degree of warming.

The scientific papers in the present volume address different observational and modeling aspects of the Earth's energy flows. It represents the outcome of the second workshop held within the ISSI Earth Science Programme The workshop took place from 10 to 14 January, 2011, in Bern, Switzerland, with the objective of providing an in-depth overview of the Earth's energy flows. The participants in the workshop were experts in a wide range of disciplines and included solar physicists, experts on space observations, atmospheric radiation experts, meteorologists, oceanographers and climate modelers.

While the radiative forcing of the Earth has undergone many changes over time with consequences for the climate, the scientific interest and public concern during the last few decades have been focused on anthropogenic changes of the composition of the atmosphere and its effects on the climate. During the last 50 years, CO_2 amounts have increased by 75 ppm or by almost 25 %. The amounts of other important greenhouse gases, CH_4, N_2O and the CFCs, have also increased significantly. Such rapid increases were not previously encountered during the long history of our planet. During the same time, anthropogenic aerosols have also increased and affected the radiation balance by offsetting part of the greenhouse warming.

The ability of the greenhouse gases including water vapor to absorb long-wave radiation, such as the heat radiation from the Earth, has been known since the nineteenth century as described in the presentation by Kandel (2012). The main effect of the greenhouse gases is to warm the lower atmosphere, the troposphere, and to cool the lower stratosphere. Such a warming signature is presently observed, suggesting that the increasing amounts of greenhouse gases are the main cause of the observed global warming. A corresponding contribution by solar radiation would have had a different signature, as the warming in this case would encompass the full depth of the atmosphere. Furthermore, as documented in this issue, the best estimate of surface and lower tropospheric temperature suggests that the solar influence is small compared to the anthropogenic forcing. It appears that the modulation of low altitude clouds by galactic cosmic rays is unlikely to provide an explanation to the observed temperature increase (Lockwood 2012).

Three central issues were at the forefront of the Workshop: *Firstly*, how accurately can we estimate the present forcing of the Earth's system? *Secondly*, how well do we know the flow of energy in the Earth's system? *Thirdly*, how well can we determine the response to the forcing, including the internal energy feedbacks of the Earth's system?

The external radiative forcing of the system is the solar irradiation and its variation, anthropogenic greenhouse gases and aerosols, plus aerosols from volcanic eruptions.

We know that the Earth presently is receiving more heat than it emits, although present observations cannot yet measure this imbalance. However, as is shown in some of the papers in this issue, the imbalance can be estimated from the accumulation of heat in the Earth's system where some 90 % ends up in the oceans. For the period 1993–2008 Lyman

(2012) has estimated the heat flux in the upper 700 m of the ocean, normalized for the full area of the Earth, to be 0.64 ± 0.11 W/m^2. For shorter periods, variations in the trends are considerable, suggesting that the irregular warming of the lower atmosphere and the surface of the Earth is probably related to the irregular warming of the deeper parts of the ocean (Trenberth and Fasullo 2012).

The energy flow within the atmosphere and between the ocean, land and atmosphere and the energy fluxes in global and annual average are known to within 1–2 W/m^2 but with slightly larger values (say ± 5 W/m^2) for the atmospheric absorption of solar radiation and the down-welling long-wave radiation at the surface. The workshop identified the need to determine these factors more accurately.

Perhaps, the most difficult problem is to estimate or calculate the climate response to forcing perturbations; most of the presentations and discussions at the workshop were concerned with exploring a multitude of issues in this context that need to be better understood.

The direct effect of increasing the amounts of greenhouse gases in equilibrium is straightforward to calculate as it can in principle be obtained from the Stefan–Boltzmann equation. That readily shows that a relative change in radiative forcing, dF, is related to temperature change, dT, as $dF/F = 4dT/T$, where the forcing is measured in W/m^2 and the temperature in degrees Kelvin. A doubling of CO_2 is equivalent to $\sim$3.7 W/m^2 which means that the corresponding increase in temperature, dT, is about 1.1 K. This is the increased equilibrium temperature that the system will gradually approach.

What significantly complicate the situation are the feedback processes of the Earth's system that will either decrease or increase this value. Atmospheric water vapor follows temperature and will act as a positive feedback factor as water vapor in itself is a powerful greenhouse gas. Surface albedo is also likely to exercise a positive feedback as melting ice and snow at higher temperatures are likely to lead to reduced reflectivity of solar radiation. However, the largest problem is related to clouds. Clouds reflect solar radiation and consequently cool the Earth's system, but clouds also enhance the absorption of terrestrial radiation. The reflected amount dominates, so the combined effect indicates a net cooling by some 20 W/m^2. Of particular importance here are stratiform clouds over the oceans at low latitudes where they act as powerful regulators of the global temperature. Any changes in the amount of clouds in these areas will have a major influence on how the global temperature will evolve. The reason for this is the large difference in albedo of these clouds compared to the underlying ocean. The possible cloud feedback, that is how a warmer climate in turn will affect the clouds, is poorly understood. It is also the main reason that climate models such as those used within the Intergovernmental Panel for Climate Change (IPCC) provide such widely different results.

The reason for the comparatively slow and irregular warming of the Earth's system during the last century is still in many respects an open scientific issue. A possible explanation by Schwartz (2012) is the different response timescales between the atmosphere and upper ocean areas, on the one hand, and the deep ocean, on the other hand. The upper layers determine what we might call transient climate sensitivity while the deep ocean determines the equilibrium response. The timescales of the two compartments differ widely, from the order of a decade for the upper ocean layer to several 100 years when we include the lower ocean compartment. Schwartz's results suggest that transient processes with a modest warming might be more representative of the present century while the planet over longer times will gradually approach a higher surface temperature as it moves toward radiation equilibrium.

The present issue includes the majority of papers presented at the workshop and includes a summary of each of the different subsections. In spite of impressive scientific progress in recent years, the need for continuous space observations must be highlighted. Firstly, because of the complexity of the Earth's energy flows including their high variability in time and space (mainly because of clouds), that will require a comprehensive sampling that must satisfactorily cover the diurnal cycle in different parts of the world. Secondly, a long-term commitment to Earth's radiation measurements is essential because of low-frequency variations caused by both external (solar and volcanic) and anthropogenic effects, as well by different internal variations of the climate system.

References

Fröhlich C (2012) Total solar irradiance observations. Surv Geophys (this issue)

Kandel R (2012) Understanding and measuring earth's energy budget: from Fourier, Humboldt and Tyndall to CERES and beyond. Surv Geophys (this issue)

Lockwood M (2012) Solar influence on global and regional climates. Surv Geophys (this issue)

Lyman J (2012) Estimating global energy flow from the global upper ocean. Surv Geophys (this issue)

Schwartz S (2012) Determination of Earth's transient and equilibrium climate sensitivities from observations over the twentieth century: strong dependence on assumed forcing. Surv Geophys (this issue)

Stevens B, Schwartz S (2012) Observing and modeling earth's energy flows. Surv Geophys (this issue)

Trenberth K, Fasullo J (2012) Tracking Earth's energy: from El Nino to global warming. Surv Geophys (this issue)

Surv Geophys (2012) 33:337–350
DOI 10.1007/s10712-011-9162-y

Understanding and Measuring Earth's Energy Budget: From Fourier, Humboldt, and Tyndall to CERES and Beyond

Robert Kandel

Received: 2 November 2011 / Accepted: 14 November 2011 / Published online: 11 January 2012

Abstract This historical perspective on the determination of Earth's energy fluxes, beginning with the classical description of climate, outlines the establishment of the basic physics of the Earth climate system in the nineteenth century. After recalling the early twentieth century ground-based attempts to determine the Earth's energy budget, I review the growing contributions of observations from space to quantifying these exchanges. In particular, space observations have shown that variations of solar luminosity have been extremely small (of order 0.1%) over past decades and probably past centuries and that they play practically no role in present-day climate variations or variations that may be expected in coming decades. Overall geographical structure, diurnal and seasonal cycles, and some of the interannual and interdecadal variations of Earth's energy exchanges with the Sun and space are now quite well determined, but much remains to be done regarding, on the one hand, fluxes at the surface and, on the other hand, variations of clouds. Improvements are essential if scientific assessment of anthropogenic climate change risk is to keep up with the changes themselves.

Keywords Earth's energy flows · Satellite observations · Historical background

1 Historical Perspective

The Earth's energy budget is at the heart of physical climatology, and the principles of this science have been well established for over a century and a half, well before the development of any understanding of the physical structure of the Sun and stars, the nature of the physical processes accounting for their luminosities, and indeed the details of the transfer of radiation in the atmospheres of stars and planets. Only celestial mechanics is an older science.

R. Kandel (✉)
Laboratoire de Météorologie Dynamique, Institut Pierre Simon Laplace, Ecole Polytechnique, Palaiseau, France
e-mail: bobkandel@yahoo.com

1.1 Descriptive Climatology: From Ancient Greece to the Nineteenth Century

We humans find ourselves at the surface of the Earth, within its changing atmosphere, sometimes roaming the land and sailing the seas. More than two thousand years ago, it was already clear to natural philosophers that we could not begin to understand Earth's climate on the basis of either a flat Earth or a spherically symmetric model. At the very least, climate depends on the inclination of the Sun's rays, and thus on latitude. Already in 1693, Edmund Halley elaborated a first physical model for the latitudinal dependence of surface temperature, trade winds, and meridional heat transport (cf. e.g., Kutzbach 1996; Edwards 2010, Chap. 2). Beginning in the fifteenth century, with the globalization of exploration and trade, it became clear that, even at the same latitude, climate is not the same on the eastern and western shores of the great oceans.

1.2 Physical Climatology

Physical climatology is the oldest attempt to understand the functioning of a natural astronomical object—our planet. At the same time, like any living science, it is a young science, thanks to the remarkable post-World-War-2 developments in three broad areas. These are: (1) the new possibilities of both passive and active observation (from space at all wavelengths, on and under the sea surface, and on and in the ice); (2) the new tools for exploring the past (especially isotopic analyses for dating and paleo-thermometry); and (3) the methods and tools of computation, both for assimilating and analyzing observational data and for simulating the dynamics of the system.

1.3 Developments in the Nineteenth Century

In the first half of the nineteenth century, Horace de Saussure, Fourier (1827), and John Tyndall, among others, established the basic physical mechanisms of climate on our planet. Although their vocabulary was not the same as today's, and although they certainly did not have data for the entire globe, they established the essential lines of the global energy budget diagram that is found in the first chapter of most modern climatology texts. Solar radiative flux (now often denoted shortwave or SW flux), partly converted into heat, overwhelmingly dominates energy input, heat flux from the Earth's interior being negligible. Energy balance is secured by infrared ("heat") radiation (now often given as longwave or LW) to space through a partly absorbing atmosphere.

In his *Cosmos—Essai d'une description physique du Monde* (Physical Description of the World, edition in French published in 1846), Alexander von Humboldt treats climatology as a science of description, but he does discuss its physics. He defines the term *climatology* as "designating the ensemble of atmospheric variations that affect our senses: temperature, humidity, changes of barometric pressure, the calm of the atmosphere, winds, atmospheric electricity, the purity of the air or the presence of more or less deleterious miasmas (*what we today would call air pollution*), and finally the transparency and the serenity of the sky. This last term influences not only the effects of the heat radiated by the soil, the growth of vegetables and the ripening of fruit, but also our morale and the harmony of our faculties." (Humboldt 1846, pp. 377–378, transl. R. Kandel). Following Clausius and Clapeyron, Humboldt also noted that warmth is favorable to the existence of water in the gaseous state. Since water vapor is the major contributor to the infrared opacity of the atmosphere, this is the key to what is known today as water–vapor feedback.

For our planet, there is and always has been an abundance of data not available for any other celestial object. In the course of the nineteenth century, geologists showed that planet Earth had a long history of change. Thanks in particular to the work of Swiss naturalist Louis Agassiz, it became clear that Earth's recent history involved a succession of ice ages separated by interglacial periods. This implies substantial changes in the volume of water in the solid state on land, and dated sea-level changes (over 100 m since the last glacial maximum 18,000 years ago) are consistent with a substantially constant total water volume. Although there are some opposing views, most specialists judge that, over at least the past few million years, extraterrestrial volatile contributions to the planet's water budget can be neglected. Fourier had noted as early as 1827 that the latitudinal and seasonal distribution of solar energy input, fundamental factors of Earth's climates, would necessarily change in the event of changes in the Earth's orbit and/or the orientation of its axis of rotation. Adhémar in France and Croll in Great Britain hypothesized that such changes were important factors in the glacial–interglacial changes. These ideas were to be developed by Serbian mathematician Milutin Milankovitch early in the twentieth century.

2 Physical Climatology in the Twentieth Century: From Arrhenius to the IGY

Over a century ago, Swedish chemist Arrhenius (1896) formulated the hypothesis that continuing combustion of fossil fuels (then mostly coal) would ultimately increase atmospheric carbon dioxide (CO_2). Enhancing the natural greenhouse effect, this would change world climate. The systematic measurements begun at Mauna Loa in the International Geophysical Year (1957) led—within a few years—to the *discovery* that such anthropogenic alteration in the atmosphere was indeed under way. From 290 ppm (parts per million by volume, i.e., cm^3 of CO_2 per m^3 of air) in 1900, and 310 ppm in 1957, atmospheric CO_2 concentration has now (2011) passed 390 ppm.

Starting in 1910, Milankovitch (1920) used relatively simple one-dimensional climate models—including the basic physics of the greenhouse effect together with calculations of lunar and planetary gravitational perturbations of the Earth's orbit and rotation—to develop his theory that the changes in the latitudinal and seasonal distribution of astronomical insolation could account for the glacial–interglacial changes of Earth's climate. Although not accepted at first, this theory was later confirmed by the isotopic dating of these changes (cf. Imbrie and Imbrie 1979) and the more detailed calculations of Berger (1988) in Louvain-la-Neuve (Belgium).

Starting about 1900, Charles Greeley Abbot of the Smithsonian Institution began systematic measurements—in part from high-altitude sites—of the solar irradiance. He arrived at an average value of 1.946 cal cm^{-2} min^{-1} (1,357 Wm^{-2}). Over the next half century, Abbot claimed to have found variations of order several percent in the solar irradiance, corresponding to variations in solar activity. However, Sterne and Dieter (1958) showed that the variations detected by Abbot were not intrinsically solar but rather arose in the Earth's atmosphere. Sometime later, Foukal and Vernazza (1979) were able to show that one could detect the 27-day solar rotation period, governing the passage of large sunspot groups across the solar disk seen from Earth, in the Smithsonian record, but at a very weak level (10^{-4}). Although the ground-based observations could not reveal solar irradiance variations at the 1% level, they could not exclude them either. Early climate sensitivity estimates considered for the most part two cases that were as follows: sensitivity of climate (parameterized by global mean surface air temperature) to a 2% change in total solar irradiance (TSI) and sensitivity to doubling of atmospheric CO_2. Depending on the year of the estimate, this could be from 300 to 600 ppm, or from 330 to 660 ppm, at any rate, a radiative forcing at the tropopause of about 4 Wm^{-2} (Schneider and Dickinson 1974; Manabe and Wetherald 1975).

Abbot and Fowle (1908) were the first to put numbers into a global energy budget (cf. Hunt et al. 1986). They estimated Earth's Bond albedo at 34% (not very far from modern values), based on rough estimates of global cloud cover (based mostly on Northern Hemisphere data) together with measurements of reflectance of low clouds observed from Mount Wilson from which they gave mean cloud reflectance at 65%. Later, Aldrich (1920) revised these figures on the basis of balloon measurements (still of low clouds) giving mean reflectance as 78% and Bond albedo as 43%. The early estimates of the global radiation budget relied on still rougher estimates—in no sense observational—of the "effective height" of thermal infrared emission to space as controlled by water vapor and carbon dioxide absorption.

Using the Smithsonian solar irradiance data summarized in Abbot and Fowle (1908), Milankovitch assumed a constant Sun, although aware of Abbot's belief in intrinsic solar variations as high as 7%. Milankovitch (1920, pp. 327–329) also noted that the global mean albedo used depends in part on estimates of cloud cover, some made by Arrhenius, and that one can (and presumably should) also separate contributions from sea and land surfaces as well as from lower and higher clouds. He also estimated the solar radiation absorbed in the atmosphere and clouds, obtaining 22%, practically identical to modern values (e.g., 21% according to Trenberth et al. 2009). Although these estimates used reconstructions based in part on observations from balloons, none of them involved observations from space. Without waiting for Sputnik, however, indirect space observation was shown to be possible, using the lunar surface observed from Earth as the radiometer.

Beginning in Strasbourg, and later in Paris, André Danjon (1928) developed this method to estimate Earth's Bond albedo[1], measuring the earthlight or *lumière cendrée* (ashen light), that is, sunlight reflected by Earth illuminating the part of the lunar disk not directly illuminated by the Sun. Such estimates are neither truly global nor regional and are subject to strong bias due to the highly restricted angular and time sampling. Thus, as noted already by Danjon, measurements made from France with a waning crescent Moon, near the western horizon after sunset, depend strongly on cloud cover over the Atlantic, and with a waxing crescent near the eastern horizon before sunrise, on clouds over the Indian Ocean as well as central Asia. Danjon found somewhat high values close to 40%. Recent work (Qiu et al. 2003), using earthlight measurements at the Big Bear Solar Observatory (California), reconciled Danjon's results with modern satellite-based Bond albedo values close to 30%. However, the claim (Pallé et al. 2003, 2004) that recent earthlight measurements have given new information on the annual cycle and interannual variations of Earth's Bond albedo cannot be sustained (Wielicki et al. 2005).

3 Physical Climatology in the Space Age

With the successful October 1957 launch of Sputnik-1, and the increasing numbers of artificial satellites launched in following years, the ability to observe the Earth system from outside with global coverage provided more and more accurate measurements of the fundamental factors of physical climatology: net solar radiation input and outflow to space of thermal infrared radiation.

[1] The ratio of globally reflected solar radiation to the incident solar radiation intercepted by the planet.

3.1 The Solar Constant: A First Look

Abbot (1920) understood well the importance of observing the full solar spectrum from outside the atmosphere but, considering only rocket-borne instrumentation, he was doubtful regarding the possibility of attaining high accuracy, even leaving aside the question of continuous monitoring. *Systematic* monitoring of the TSI from artificial satellites only began in the late 1970s with Nimbus-7. Since then, partially overlapping series of measurements from different instruments on different satellites show clearly that there is an 11-year cycle in TSI, but only at the 0.1% level (Fröhlich and Lean 2004; Fröhlich 2009). However, absolute accuracy remains a problem, in part because it depends on accuracy of measurement of the instrument aperture area. At different times, based on different instruments, TSI (at 1 AU Sun–Earth distance) has been given as 1,391, 1,372, and 1,368 W m^{-2}. Loeb et al. (2009) have proposed to revise this value downward to 1,361 W m^{-2}. None of these changes is to be taken as an intrinsic change of TSI.

3.2 The Earth's Radiation Budget

With artificial satellites, it became possible to measure the terms of the Earth's radiation budget at the "top of the atmosphere" (TOA, arbitrarily defined at 30 km altitude in the Earth Radiation Budget Experiment): incident solar shortwave (SW) flux, reflected SW, and emitted terrestrial thermal longwave (LW) fluxes. Early results using simple detectors on the Explorer-7 satellite showed that Earth's Bond albedo was closer to 0.3 than to 0.4. Between 1962 and 1966 (Vonder Haar and Suomi 1971), data from the Tiros-IV (Television Infrared Observation Satellite Program) and ESSA-III (Environmental Science Services Administration) satellites made it possible to establish the annual cycle and zonal mean values of reflected (and, thus, of absorbed) solar SW and emitted thermal LW radiation. These observations then yielded the variation with latitude and season of zonal mean net radiative flux (Ellis et al. 1978, Fasullo and Trenberth 2008a, b, Stephens et al. 1981), driving (but also driven by) atmospheric and oceanic circulation. Zonal means of absorbed solar energy and emitted thermal LW reveal the role of clouds in the intertropical convergence zone near the equator. Of course, more recent satellite missions (Jacobowitz et al. 1984; House et al. 1986; Barkstrom et al. 1989; Ramanathan et al. 1989; Wielicki et al. 1995, 1996, 2002; Kandel et al. 1998; Duvel et al. 2001; Golovko et al. 2003; Kandel and Viollier 2005, 2010; Kato et al. 2008; Murphy et al. 2009; Trenberth et al. 2009, Lacis et al. 2010; Schmidt et al. 2010) have provided higher accuracy and much more detail than the early measurements, but the essentials of the picture have not changed.

3.3 The Parasol Factor: Clouds and Albedo

Although it was recognized early that clouds are major contributors to Earth's albedo, many (most) early attempts at climate models tended to focus instead on the contributions of ice and snow, implicitly represented in "energy balance" models (e.g., Budyko 1969; Sellers 1969) representing surface temperature as a function of latitude, the single dimension considered. Such models can have a built-in ice–albedo feedback. One must of course recognize that only daytime albedo matters and, at high latitudes, only summertime albedo; moreover, ice–albedo feedback is strongest in the absence of clouds, and relatively strong cloud (and atmospheric) reflectance for the higher solar zenith angles prevalent at high latitudes may limit the significance of such positive climate–radiation feedback. Other one-dimensional "radiative–convective" climate models considered only the vertical dimension in global mean, and in particular with an extremely crude representation of clouds. Moreover, even with 3-dimensional general circulation models (GCMs) with representation of clouds, it took sometime before diurnal variation and interactive clouds were considered. On the side of observation, studies of the role of clouds require a sufficiently narrow field of view (NFOV) (cf. Sect. 2.3, Kandel and Viollier 2005), but they also require that one be able to calculate with confidence the reflected SW flux on the basis of the measured reflected radiance from the cloud to the satellite. Beginning with the rotating scanner of the Nimbus-7/ERB instrument (Jacobowitz et al. 1984), observationally based bidirectional reflectance distribution functions were constructed for this purpose and later applied to the Earth Radiation Budget Experiment (ERBE) NFOV radiances as well as to ScaRaB (Scanner for Radiation Budget) and finally Clouds and the Earth's Radiant Energy System (CERES) measurements. Also, because certain types of cloud exhibit significant systematic diurnal variation, the time sampling issue cannot be neglected (cf. e.g., Kandel and Viollier 2005).

3.4 Terrestrial Thermal Infrared Radiation

Whereas earthlight did make possible indirect ground-based observations of Earth's reflected solar radiation, the planet's thermal infrared emission to space could only be observed from space. Nimbus-3 measurements in several spectral bands of the infrared made it possible to estimate outgoing broadband LW radiation; together with the broadband SW measurements, this yielded the geographical structure of Earth's radiation balance (Raschke et al. 1973). The observation of negative radiation balance over the Sahara was a key factor in the formulation of the Charney (1975) desertification mechanism. More generally, observation of TOA outgoing broadband LW, together with ground-based observations (or calculations using observed surface temperature and emittance) of upward LW flux, yields a parameter of the "greenhouse effect," for example, the normalized greenhouse factor defined by Raval and Ramanathan (1989). Observation of outgoing LW with adequate spatial resolution is essential to evaluate correctly the greenhouse role of clouds in the Earth's radiation budget, often neglected in early discussions and even in some ill-informed recent remarks. With the development of the International Satellite Cloud Climatology Program (Rossow and Schiffer 1991; Rossow et al. 1996), we now have a global cloud climatology (beginning in 1983) with regional spatial resolution together with resolution of the diurnal cycle, as revealed in outgoing LW radiation as well as in reflected SW. Although imperfect, the ISCCP products make it possible to calculate the terms of the Earth's radiation budget from the top of the atmosphere to the surface (Zhang et al. 2004). However, the ISCCP record is influenced by artefacts associated with satellite viewing geometry and, thus, it has not yet been possible to estimate a long-term trend (Wang et al. 2002; Kato 2009).

3.5 Cloud-Radiative Forcing

Important progress in climate model evaluation came with the definition of "cloud-radiative forcing" or CRF (Charlock and Ramanathan 1985; Cess et al. 1989; Fouquart et al. 1990). In models with representation of clouds, run to equilibrium, that is, with geographical structure and vertical profiles of temperature, humidity, and especially clouds, the « forcing » appears by taking this equilibrium structure with clouds and repeating the radiative transfer calculations with zero cloud opacity (SW or LW or both) but without allowing atmospheric structure to respond. Models that give the correct mean radiation budget (as measured by early Earth radiation budget missions) differ regarding the CRF. Can (SW and LW) CRF be determined observationally? By distinguishing clear-sky areas, one determines fluxes for these areas and assumes that this is equivalent to the zero-cloud-opacity flux in the radiative transfer calculation. Thus, even though the analogy is not perfect, the observationally derived CRF values became powerful tools for model *in*validation ("falsification" in the language of Popper).

Note that the CRF is an « all or nothing » forcing, not to be compared with the climate *change* forcing of +4 W m^{-2} by doubled CO_2. Thus, in the global mean, SW CRF is found to be close to −50 W m^{-2}, LW CRF close to +20 W m^{-2}, net CRF −30 W m^{-2}; the value for removing *all* CO_2 is of the same order, +32 W m^{-2}. The critical problem for climate sensitivity is cloud–radiation *feedback*. How will net CRF *change* as climate changes?

3.6 The Earth's Water Budget

Earth's *energy* (radiation) budget essentially involves the transformation of a fraction (close to 70%) of SW radiative energy flux coming from the Sun, *outside* the Earth–ocean–atmosphere system, into heat and then, after considerable interesting complex exchanges within the system, into thermal infrared radiation to space. In Earth's *water* budget, exchanges with outer space are generally (although not unanimously) considered to be negligible today and to have been so for at least the last few million years (cf. Kandel 2009a). However, the complex and interesting exchanges between ice, ocean, and atmosphere, between water vapor, clouds, and precipitation in the atmosphere, are major actors in the radiation budget, climate, and climate change. Certainly, the glacial–interglacial changes of the past few million years involved major changes in the water budget, not only in ice volume and extent and in sea level but also in atmospheric water vapor and clouds.

3.7 Surface Energy and Water Budgets

At the surface, the energy budget must not be assumed to be identical with the radiation budget, either locally or globally (cf. Möller 1963; Manabe and Wetherald 1967; Kandel 1981). Radiation balance (i.e. net radiation) at the surface is often but not always strongly influenced by the latent heat flux of evaporation, a major term of the water budget; sensible heat flux must also be considered, especially over dry surface areas. Direct measurement of the surface radiation and water fluxes is, in principle, possible everywhere but certainly not easy. Moreover, absolute calibration and inter-calibration of measurements made in different locations with different instruments raise serious difficulties. With adequate information (vertical profiles) on the state of the atmosphere (including clouds and non-water aerosols), satellite observations of outgoing radiation (broad- or narrowband SW and LW) can be and have been used to compute radiation budget terms at the surface (ISCCP: Zhang et al. 2004; CERES: Wielicki et al. 1998; Pinker et al. 2005; Trenberth et al. 2009). For surface stations with adequately calibrated radiation measurements, such satellite-based surface radiation budget terms can be "validated." However, it should be noted that the Baseline Surface Radiation Network (Ohmura et al. 1998) does not cover well, if at all, the sea surface, apart from a few island stations that should not be assumed to be representative of the open ocean.

4 Variations in the Earth's Radiation Budget and Climate Change

4.1 Solar Variations?

As already noted (Sect. 3.1 above), measurements from 1902 to 1962 of solar irradiance reaching the ground-based (but high-altitude) stations of the Smithsonian Institution raised the question whether there were significant (several percent) variations with solar activity of the total solar irradiance. The analyses by Sterne and Dieter (1958) and Hoyt (1978) indicated that intrinsic TSI variation was less than a few tenths of a percent.

Eddy (1976) noted, however, that the prolonged "Maunder minimum" period of low solar activity of the second half of the seventeenth century corresponded to the "Little Ice Age" and raised the question whether indeed reduced total solar irradiance (reduced by 0.3%? or by as much as 2%?) could account for the cooler climate.

Standard climate sensitivity calculations in the 1970s (e.g., Schneider and Dickinson 1974; Manabe and Wetherald 1975) considered 2% increase in TSI (i.e., 6.8 W m^{-2} after division by 4 to get the global mean) and doubling atmospheric CO_2 (from 300 to 600 ppm); both give tropopause radiative forcing of order 4–5 W m^{-2} yielding similar though not identical climate response, of the order of 2°C "global warming" expressed in terms of global annual mean surface air temperature. As already noted, observations from space have shown that—since 1978—intrinsic TSI variations are far smaller, at the 0.1% level, varying with the 11-year sunspot cycle. Ground-based solar irradiance observations hardly go back more than a century and cannot detect variations at this level. However, ground-based observations of solar *activity* go back to Galileo, and indirect determinations of solar activity by way of abundances of carbon-14 and other cosmogenic isotopes go back many thousands of years (Bard et al. 2000). It is understandable that some research on past climate variations has used (and some even today uses) indicators of past solar *activity* as indirect indicators (proxies) of past total solar irradiance. Some work may well have fallen into the trap of circular reasoning, choosing those indicators that correlate best with past climate changes, and then using the correlations between past TSI reconstructed in this way and observed climate variability as demonstration of solar variations driving climate.

In the 1970s, before well-calibrated series of TSI data from space existed, Maunder Minimum—Little Ice Age coincidence could legitimately suggest solar TSI forcing of climate (Wigley 1981). But today, invoking solar activity together with physically unjustified amplification (i.e., fudge factors) of very weak TSI variations to explain observed climate change is circular reasoning if not astrology. In absolute terms, the larger part of measured present-day relatively weak variations ($\sim$0.3 W m^{-2}, after division by 4 to get the global mean) of TSI are variations in the ultraviolet ($\sim$0.2 W m^{-2}, several percent in relative terms).

A number of questions should be raised.

1) Can measured weak (0.1%) variations of TSI with the (11- and 22-year) solar magnetic cycle be at all considered a valid guide to long-term (100- to 100,000-year) variations of TSI, as a function of variations of solar cycle *amplitude*?
2) What can physical models of solar structure and magnetism tell us about the likelihood of significant-amplitude (>0.3%) TSI variations on time scales of 100–100,000 years?
3) For the past, should not terrestrial proxies (isotope ratios…) of solar activity variations be used as guides to past solar ultraviolet rather than to TSI variations?
4) Why not forget TSI forcing for now?
5) Would not attempts to understand solar forcing of past climate changes be better advised to consider only relatively more significant solar UV irradiance changes, forcing changes in stratospheric physical and chemical processes, and thereby affecting processes in the troposphere?
6) Alternatively, consideration of past changes in particle fluxes revealed by cosmogenic isotopes as factors in climate change, notably by way of cloud freezing and cloud condensation nuclei, must consider how such atmospheric processes operate as a

function of altitude and (geomagnetic) latitude. On long timescales, such changes must depend not only on solar wind changes but also on extra-solar-system events.

4.2 Terrestrial Variations Seen from Space

4.2.1 Albedo Forcing

In modeling, forcing can be introduced arbitrarily. For example, for clouds (the locale of transition of atmospheric water from the gaseous to the liquid or solid state), albedo forcing can be introduced by assuming them transparent to solar radiation (shortwave cloud-radiative forcing, SWCRF: Charlock and Ramanathan 1985) while retaining (or not) their infrared opacity (LWCRF). Much work assumes that measured albedo differences between neighboring cloudy and cloud-free areas provide a comparable SWCRF (Cess et al. 1989). One can also assess the role of surface albedo forcing, for example, what might be the consequences of painting the Sahara black (Black and Tarmy 1963) or the roofs of houses white (Oleson et al. 2010). Purportedly realistic albedo forcing calculations have considered the effects of natural and anthropogenic aerosols added to the troposphere and stratosphere. For stratospheric aerosols, Minnis et al. (1993) showed that ERBE data revealed significant global-scale changes in albedo over more than a year following the violent eruption of Mount Pinatubo, treated as an example of radiative forcing (Stenchikov et al. 1998). This appears reasonable to the extent that climate is considered as mostly an affair of troposphere and surface and, in particular, with radiative forcing defined at the tropopause. With regard to tropospheric aerosols, whether natural or anthropogenic, additional albedo of added aerosols can be regarded as a forcing, and with availability of aerosol measurements independent of measurements of reflected SW flux, a direct aerosol albedo forcing can be derived for use in models.

Recent work has also considered "indirect" aerosol forcing, corresponding to observed and modeled changes of cloud properties in the presence of aerosols modifying cloud drop size distribution. Modelers in recent years take into account such an "indirect" forcing term, but one may question whether a cloud's albedo change in the presence of aerosols can usefully be assumed to occur without significant changes in the cloud's action in the water cycle. Other possible albedo forcing terms are essential in a number of geo-engineering schemes (e.g., Trenberth and Dai 2007; Jones et al. 2010) considered as a way to compensate for continued anthropogenic intensification of the greenhouse effect. Apart from such real or conjectured forced changes in albedo, measurements from space have revealed significant albedo changes—clearly apparent in regional and zonal means—linked to El Nino Southern Oscillation (ENSO) variations of cloud cover (Wielicki et al. 2002; Loeb et al. 2007). Geostationary satellite narrowband data have improved diurnal averaging of the broadband data. The broadband SW data from the Geostationary ERB (GERB) instrument on board Meteosat (Harries et al. 2005), although imperfect because of the multiple reflections, improve monitoring of rapid albedo changes. In the future, the Broad-Band Radiometer (BBR) on board the projected EarthCARE mission (Kandel et al. 2003; Capderou and Viollier 2006; Gelsthorpe et al. 2008) is to provide three nearly simultaneous views of each pixel along the satellite track, so as to reduce reliance on statistically based bidirectional reflectance distribution functions and to attain 10 W m^{-2} accuracy in deriving *instantaneous* reflected SW flux from the SW radiance measurements.

4.2.2 Greenhouse Forcing

Spectra of Earth's outgoing longwave radiation have revealed changes consistent with the measured increase in atmospheric methane concentrations (Harries et al. 2001). Advanced atmospheric sounders (e.g., IASI on board METOP) can be used to monitor local variations in atmospheric carbon dioxide and other gases. Useful as such measurements can be, they do not constitute measurement of atmospheric greenhouse forcing as defined by changes in the outgoing longwave radiation (OLR) at the tropopause driven by the measured (in situ) anthropogenic increases in CO_2 and other gases. Satellite measurements of changes in broadband OLR (as well as reflected broadband SW) yield both forcing and response, in what remains a non-equilibrium situation. Changes in the outgoing spectrum at wavelengths dominated by CO_2 include the reduced transmission of surface and low-atmosphere emission (the forcing), but they also include the effects of tropospheric temperature, humidity, and cloud changes, and in the stratosphere of increased CO_2 emittance combined with (and causing) reduced temperatures there. With regard to water vapor, atmospheric residence time of order 9 days is so short that anthropogenic emissions (e.g., the H_2O produced along with CO_2 when burning fossil hydrocarbons) have practically zero direct effect. Water vapor is not a well-mixed greenhouse gas. However, changes in atmospheric water vapor and cloud distributions inevitably modify the outgoing broadband LW radiation, and these changes act as feedbacks on the energy fluxes in the physical climate system.

Anthropogenic greenhouse forcing is recent and accelerating over the past few decades since the IGY and the start of the space age, so that the system is not in equilibrium. According to model calculations (Hansen et al. 2005; Murphy et al. 2009) and measurements of the heat accumulation in the oceans (Levitus et al. 2009), the global annual mean radiation balance or rather energy imbalance should today be of order $+0.5$ to $+1\ \mathrm{W\ m^{-2}}$. Observations from space have not yet produced sufficiently accurate estimates of global mean net radiation (cf. Loeb et al. 2009; Trenberth 2009). Closing the energy budget requires in particular enhanced absolute accuracy of total solar irradiance measurements, and NASA has projects in this direction.

5 Concluding Remarks: Feedbacks and the Future

We are engaged in a chase with climate change. Over the next few decades, climate change depends principally on the one hand on the evolution of anthropogenic changes in atmospheric composition and, on the other hand, on the strength of the various climate feedbacks. If human "decision-makers" are to bring into play a feedback of human intelligence, with organized rapid reduction of anthropogenic greenhouse gas emissions (Kandel 2009b), we as scientists need to reduce uncertainties regarding the *physical* climate feedbacks. These include in particular water–vapor feedback (Soden et al. 2002), cloud–radiation feedback, and ice–albedo feedback. Critical uncertainties remain regarding the cloud–radiation feedback, particularly with regard to low cloud over the tropical oceans (Bony and Dufresne 2005; Dufresne and Bony 2008). Better understanding of the various feedbacks, in particular those involving clouds, requires spatial resolution adequate to resolve cloud field heterogeneity, improved methods for measuring and taking into account real-time anisotropic reflectance, better time sampling, and of course still better radiometric accuracy. Some of these questions are well or better addressed by satellite missions with less than global coverage, as, for example, ERBS and TRMM in the past, Meteosat with GERB, and coming missions such as Megha-Tropiques (Desbois et al. 2007) and EarthCARE (Gelsthorpe et al. 2008). The question remains: *Can we measure and understand the changes in Earth's energy and water fluxes faster than human activities change them?*

Acknowledgments Lennart Bengtsson and Roger-Maurice Bonnet gave me the opportunity (or duty) to present this perspective at the ISSI workshop on energy flows in the Earth system. For the opportunity to continue in astro-geophysical research over the years since Sputnik-1, I thank in particular Jean-Claude Pecker and Roger Cayrel, and since the launch of Meteosat-1, Jacques Blamont, Pierre Morel, and André Berroir. Regarding the Earth's radiation budget, my research depended on the support of the Centre National d'Etudes Spatiales (CNES), and ERB studies have always been a team effort involving, for me, many scientists and engineers in France (Service d'Aéronomie, Laboratoire de Météorologie Dynamique, Laboratoire d'Optique Atmosphérique, CNES-Toulouse) and elsewhere, in particular in the United States at various university laboratories and at NASA (LaRC, GSFC, and GISS), in Canada (Ottawa, Montreal), Russia (Moscow, Obninsk, St. Petersburg), elsewhere in Europe (ESOC and ESTEC, Cologne and Geesthacht, Reading, Brussels, Valencia), and in Japan. But the errors and eccentricities in this perspective are my own. The author thanks Lennart Bengtsson and Kevin Trenberth for useful comments on a first version of this paper.

References

Abbot CG (1920) The larger opportunities for research on the relations of solar and terrestrial radiation. Proc Natl Acad Sci 6:82–95

Abbot CG, Fowle FE Jr (1908) Income and outgo of heat from the earth, and the dependence of its temperature thereon. Ann Astrophys Obs (Smithson Inst Wash DC) 2:159–176

Aldrich LB (1920) The reflecting power of clouds. Smithson Misc Collect 69(10):375–381

Arrhenius S (1896) On the influence of carbonic acid in the air upon the temperature of the ground. Philos Mag 5th series, 237–276

Bard E, Raisbeck G, Yiou F, Jouzel J (2000) Solar irradiance during the last 1200 yr based on cosmogenic nuclides. Tellus B 52:985–992

Barkstrom BR et al (1989) Earth radiation budget experiment (ERBE) archival and April 1985 results. Bull Am Meteorol Soc 70:1254–1262

Berger AL (1988) Milankovitch theory and climate. Rev Geophys 26:624–657

Black JF, Tarmy BL (1963) The use of asphalt coatings to increase rainfall. J Appl Meteorol 2:557–564. doi:10.1175/1520-0450(1963)002<0557:TUOACT>2.0.CO;2

Bony S, Dufresne J-L (2005) Marine boundary layer clouds at the heart of cloud feedback uncertainties in climate models. Geophys Res Lett 32:L20806. doi:10.1029/2005GL023851

Budyko MI (1969) The effect of solar radiation variations on the climate of the earth. Tellus 21:611–619

Capderou M, Viollier M (2006) True along-track scan to improve radiation budget estimations. J Atmos Ocean Technol 23:1093–1103. doi:10.1175/JTECH1907.1

Cess RD, Potter GL, Blanchet JP, Boer GJ, Ghan SJ, Kiehl JT, Le Treut H, Li Z-X, Liang X-Z, Morcrette J–J, Randall DA, Riches MR, Roeckner E, Schlese U, Slingo A, Taylor KE, Washington WM, Wetherald RT, Yanai I (1989) Interpretation of cloud climate feedback as produced by 14 atmospheric general circulation models. Science 245:513–516

Charlock T, Ramanathan V (1985) The albedo field and cloud radiative forcing produced by a general circulation model with internally generated cloud optics. J Atmos Sci 42:1408–1429

Charney J (1975) Dynamics of deserts and drought in the Sahel. Q J R Meteorol Soc 101:193–202

Danjon A (1928) Recherches sur la photométrie de la lumière cendrée et l'albedo de la terre. Ann Obs Strasbourg 2:165–180

Desbois M, Capderou M, Eymard L, Roca R, Viltard N, Viollier M, Karouche N (2007) Megha-Tropiques: un satellite hydrométéorologique franco-indien. La Météorologie N° 57:19–27

Dufresne J-L, Bony S (2008) An assessment of the primary sources of spread of global warming estimates from coupled atmosphere–ocean models. J Clim 21:5135–5144. doi:10.1175/2008JCLI2239.1

Duvel JP, Viollier M, Raberanto P, Kandel R, Haeffelin M, Pakhomov LA, Golovko VA, Mueller J, Stuhlmann R, the International ScaRaB Scientific Working Group (ISSWG) (2001) The ScaRaB-Resurs earth radiation budget dataset and first results. Bull Am Metereol Soc 82:1397–1408

Eddy JA (1976) The maunder minimum. Science 192:1189–1202

Edwards PN (2010) A vast machine. MIT Press, Cambridge
Ellis JS, Vonder Haar TH, Levitus S, Oort AH (1978) The annual variation in the global heat balance of the earth. J Geophys Res 83(C4):1958–1962
Fasullo JT, Trenberth KE (2008a) The annual cycle of the energy budget. Part I: global mean and land–ocean exchanges. J Clim 21:2297–2311. doi:10.1175/2007JCLI1935.1
Fasullo JT, Trenberth KE (2008b) The annual cycle of the energy budget. Part II: meridional structures and polar transports. J Clim 21:2313–2325. doi:10.1175/2007JCLI1936.1
Foukal P, Vernazza J (1979) The effect of magnetic fields on solar luminosity. Astrophys J 234:707–715
Fouquart Y, Buriez JC, Herman M, Kandel RS (1990) The influence of clouds on radiation: a climate modeling perspective. Rev Geophys 28:145–166
Fourier J (1827) Mémoire sur les Températures du Globe Terrestre et des Espaces Planétaires. Mémoires de l'Académie Royale des Sciences 7:569–604
Fröhlich C (2009) Evidence of a long-term trend in total solar irradiance. Astron Astrophys 501:L27–L30. doi:10.1051/0004-6361/200912318
Fröhlich C, Lean J (2004) Solar radiative output and its variability: evidence and mechanisms. Astron Astrophys Rev 12:273–320. doi:10.1007/s00159-004-0024-1
Gelsthorpe RV et al (2008) EarthCARE and its payload. Proc SPIE 7152:715207. doi:10.1117/12.804914
Golovko VA, Pakhomov LA, Uspensky AB (2003) Исследование поля уходящего излучения Земли с помощью сканирующего радиометра радиационного баланса на российских спутниках серии « Метеор » и « Ресурс » (Earth radiation budget investigations from "Meteor" and "Resurs" satellites.) *Электронный журнал « ИССЛЕДОВАНО В РОССИИ »* (Electronic journal "Investigated in Russia") 106: 1280–1299. http://zhurnal.ape.relarn.ru/articles/2003/106.pdf
Hansen J, Nazarenko L, Ruedy R, Sato Mki, Willis J, Del Genio A, Koch D, Lacis A, Lo K, Menon S, Novakov T, Perlwitz Ju, Russell G, Schmidt GA, Tausnev N (2005) Earth's energy imbalance: Confirmation and implications. Science 308:1431–1435. doi:10.1126/science.1110252
Harries JE, Brindley HE, Sagoo PJ, Bantges RJ (2001) Increases in greenhouse forcing inferred from the outgoing longwave radiation spectra of the Earth in 1970 and 1997. Nature 410:355–357
Harries JE et al (2005) The geostationary earth radiation budget project. Bull Am Meteorol Soc 86:945–960. doi:10.1175/BAMS-86-7-945
House FB, Gruber A, Hunt GE, Mecherikunnel AT (1986) History of satellite missions and measurements of the Earth radiation budget. Rev Geophys 24:357–377
Hoyt DV (1978) Variations of the solar constant measurements of the Smithsonian astrophysical observatory, 1923–1954. In: Proceeding 3rd conference on atmospheric radiation. American Meteorological Society, Boston
Humboldt A de (1846) Cosmos. Essai d'une description physique du Monde. Première partie. Gide et Cie, Paris
Hunt GE, Kandel RS, Mecherikunnel AT (1986) A history of presatellite investigations of the Earth's radiation budget. Rev Geophys 24:351–356
Imbrie J, Imbrie KP (1979) Ice ages. Solving the mystery. Enslow, Short Hills
Jacobowitz H, Soule HV, Kyle HL, House FB, the Nimbus-7 ERB Experiment Team (1984) The earth radiation budget (ERB) experiment: an overview. J Geophys Res 89:5021–5038
Jones A et al (2010) Geoengineering by stratospheric SO2 injection: results from the Met Office Hadgem2 climate model and comparison with the Goddard institute for space studies model. Atmos Chem Phys 10:5999–6006. doi:10.5194/acp-10-5999-2010
Kandel R (1981) Surface temperature sensitivity to increased atmospheric CO_2. Nature 293:634–636
Kandel R (2009a) Water and cosmos: from the big bang to the creation of water and earth. In: Tvedt T, Oestigaard T (eds) The ideas of water. I.B. Tauris, London, pp 39–65
Kandel R (2009b) Turning the tide on climate change. CEFIC, Brussels. http://sendit.cefic.org/DMZ/DownLoad.aspx?UID=e4a3fb41-082a-452b-8fc3-1fe70bef6ff5
Kandel R, Viollier M (2005) Planetary radiation budgets. Space Sci Rev 120:1–26
Kandel R, Viollier M (2010) Observation of the Earth's radiation budget from space. Comptes Rendus Geosci 342:286–300
Kandel R, Viollier M, Raberanto P, Duvel JP, Pakhomov LA, Golovko VA, Trishchenko AP, Mueller J, Raschke E, Stuhlmann R, the International ScaRaB Scientific Working Group (ISSWG) (1998) The ScaRaB earth radiation budget dataset. Bull Am Meteorol Soc 79:765–783
Kandel R, Bodas Salcedo A, Gimeno Ferrer JF, López-Baeza E (2003) Broad-band radiometry for Earth-CARE. In: Proceeding NASDA-CRL EarthCARE workshop, jointly held with 2nd international workshop on Spaceborne Cloud Profiling Radar/lidar, Tokyo
Kato S (2009) Interannual variability of global radiation budget. J Clim 22:4893–4907. doi:10.1175/2009JCLI2795.1

Kato S, Rose FG, Rutan DA, Charlock TP (2008) Cloud effects on the meridional atmospheric energy budget estimated from clouds and the Earth's Radiant Energy System (CERES) data. J Clim 21:4223–4241. doi:10.1175/2008JCLI1982.1

Kutzbach JE (1996) Steps in the evolution of climatology: from descriptive to analytic. In: Fleming JR (ed) Historical essays on meteorology 1919–1995. American Meteorological Society, Boston, pp 353–377

Lacis AC, Schmidt GA, Rind D, Ruedy RR (2010) Atmospheric CO_2. Principal control knob governing Earth's temperature. Science 330:356–359

Levitus S, Antonov J, Boyer T, Locarnini RA, Garcia HE, Mishonov AV (2009) Global ocean heat content 1955–2007 in light of recently revealed instrumentation problems. Geophys Res Lett 36:L07608. doi:10.1029/2008GL037155

Loeb NG, Wielicki BA, Rose FG, Doelling DR (2007) Variability in global top-of-atmosphere shortwave radiation between 2000 and 2005. Geophys Res Lett 34:L03704. doi:10.1029/2006GL028196

Loeb NG et al (2009) Towards optimal closure of the Earth's top-of-atmosphere atmospheric radiation budget. J Clim 22:748–766

London J (1957) A study of the atmospheric heat balance. Final report, AFCRC-TR-57-287 NTIS PB 115626), New York University, College of Engineering, New York

Manabe S, Wetherald RT (1967) Thermal equilibrium of the atmosphere with a given distribution of relative humidity. J Atmos Sci 24:241–259

Manabe S, Wetherald RT (1975) The effects of doubling the CO_2 concentration on the climate of a general circulation model. J Atmos Sci 32:3–15

Milankovitch M (1920) Théorie mathématique des phénomènes thermiques produits par la radiation solaire. Gauthier-Villars, Paris

Minnis P, Harrison EF, Stowe LL, Gibson GG, Denn FM, Doelling DR, Smith WL Jr (1993) Radiative climate forcing by the Mount Pinatubo eruption. Science 259:1411–1415

Möller F (1963) On the influence of changes in the CO_2 concentration in air on the radiation balance of the Earth's surface and on the climate. J Geophys Res 68:3877–3886

Murphy DM, Solomon S, Portmann RW, Rosenlof KH, Forster PM, Wong T (2009) An observationally based energy balance for the Earth since 1950. J Geophys Res 114:D17107. doi:10.1029/2009JD012105

Ohmura A et al (1998) Baseline surface radiation network (BSRN/WCRP): new precision radiometry for climate research. Bull Am Meteorol Soc 79:2115–2136

Oleson KW, Bonan GB, Feddema J (2010) The effects of white roofs on urban temperature in a global climate model. Geophys Res Lett 37:L03701. doi:10.1029/2009GL042194

Pallé E, Goode PR, Yurchyshyn V, Qiu J, Hickey J, Montanes-Rodrigues P, Chu M-C, Kolbe E, Brown CT, Koonin SE (2003) Earthshine and the Earth's albedo II: observations and simulations over three years. J Geophys Res 108(D22):4710. doi:10.1029/2003JD003611

Pallé E, Goode PR, Montañés-Rodriguez P, Koonin SE (2004) Changes in Earth's reflectance over the past two decades. Science 304:1299–1301

Pinker RT, Zhang B, Dutton EG (2005) Do satellites detect trends in surface solar radiation? Science 308:850–854

Qiu J et al (2003) Earthshine and the Earth's albedo: 1 Earthshine observations and measurements of the lunar phase function for accurate measurements of the Earth's bond albedo. J Geophys Res 108(D22):4709. doi:10.1029/2003JD003610

Ramanathan V, Cess RD, Harrison EF, Minnis P, Barkstrom BR, Ahmad E, Hartmann D (1989) Cloud radiative forcing and climate: results from the earth radiation budget experiment. Science 243:57–63

Raschke E, Vonder Haar TH, Bandeen WR, Pasternak M (1973) The annual radiation balance of the Earth-atmosphere system during 1969–1970 from Nimbus-3 measurements. J Atmos Sci 30:341–364

Raval A, Ramanathan V (1989) Observational determination of the greenhouse effect. Nature 342:758–761

Rossow WB, Schiffer RA (1991) ISCCP cloud data products. Bull Am Meteorol Soc 72:2–20

Rossow WB, Walker AW, Beuschel D, Roiter M (1996) International Satellite Cloud Climatology Project (ISCCP); Description of new cloud datasets. WMO/TD-No. 737, World Climate Research Program (ICSU and WMO), Geneva

Schmidt GA, Ruedy RA, Miller RL, Lacis AA (2010) Attribution of the present-day total greenhouse effect. J Geophys Res 115:D20106. doi:10.1029/2010JD014287

Schneider SH, Dickinson RE (1974) Climate modeling. Rev Geophys Space Phys 12:447–493

Sellers WD (1969) A global climatic model based on the energy balance of the earth–atmosphere system. J Appl Meteorol 8:392–400

Soden BJ et al (2002) Global cooling after the eruption of Mount Pinatubo: a test of climate feedback by water vapor. Science 296:727–730

Stenchikov GL et al (1998) Radiative forcing from the 1991 Mount Pinatubo volcanic eruption. J Geophys Res 103(D12):13837–13857
Stephens GL, Campbell GG, Vonder Haar TH (1981) Earth radiation budgets. J Geophys Res 86(C10): 9739–9760
Sterne TE, Dieter N (1958) The constancy of the solar constant. Smithson Contrib Astrophys 3:9–13
Trenberth KE (2009) An imperative for climate change planning: tracking Earth's global energy. Curr Opin Environ Sustain 1:19–27. doi:10.1016/j.cosust.2009.06.001
Trenberth KE, Dai A (2007) Effects of Mount Pinatubo volcanic eruption on the hydrological cycle as an analog of geoengineering. Geophys Res Lett 34:L15702. doi:10.1029/2007GL030524
Trenberth KE, Fasullo JT, Kiehl J (2009) Earth's global energy budget. Bull Am Meteorol Soc 90:311–323. doi:10.1175/2008BAMS2634.1
Vonder Haar T, Suomi V (1971) Measurements of the Earth's radiation budget from satellites during a five-year period. J Atmos Sci 28:305–314
Wang PH, Minnis P, Wielicki BA, Wong T, Vann LB (2002) Satellite observations of long-term changes in tropical cloud and outgoing long-wave radiation from 1985 to 1998. Geophys Res Lett 29(10):1397. doi:10.1029/2001GL014264
Wielicki BA et al (1995) Mission to planet Earth: role of clouds and radiation in climate. Bull Am Meteorol Soc 76:2125–2153
Wielicki BA et al (1996) Clouds and the Earth's radiant energy system (CERES): an Earth observing system experiment. Bull Am Meteorol Soc 77:853–868
Wielicki BA et al (1998) Clouds and the Earth's radiant energy system (CERES): algorithm overview. IEEE Trans Geosci Remote Sens 36:1127–1141
Wielicki BA et al (2002) Evidence for large decadal variation in the tropical mean radiative energy budget. Science 295:841–844
Wielicki BA, Wong T, Loeb N, Minnis P, Priestley K, Kandel R (2005) Changes in Earth's albedo measured by satellite. Science 308:825
Wigley TML (1981) Climate and paleoclimate: what can we learn about solar luminosity variations? Solar Phys 74:435–471
Zhang YC, Rossow WB, Lacis AA, Mishchenko MI, Oinas V (2004) Calculation of radiative fluxes from the surface to top-of-atmosphere based on ISCCP and other global datasets: refinements of the radiative transfer model and the input data. J Geophys Res 109:D19105. doi:10.1029/2003JD004457

Surv Geophys (2012) 33:351–357
DOI 10.1007/s10712-011-9165-8

Climate and Earth's Energy Flows

Matthew D. Palmer

Received: 24 June 2011 / Accepted: 30 November 2011 / Published online: 3 February 2012

Abstract Under equilibrium conditions, climate can be viewed in simple terms as the average energy pathways that incoming solar radiation takes before exiting the system in order to maintain overall energy balance. Similarly, future climate change will ultimately be determined by how the Earth's energy balance and average energy pathways change in response to external radiative forcings, such as anthropogenic greenhouse gases, and internal redistributions. Here, we give an overview of climate research in the context of Earth's energy flows and make the case for improved observations of total energy as a more physically robust metric of climate change than the commonly used surface temperature record.

Keywords Climate · Energy · Oceans · Ocean heat content · Radiation

1 Background

Energy is the currency of Earth's climate system. The non-uniform absorption of incoming solar energy gives rise to equator-to-pole surface temperature gradients that drive the planetary large-scale circulation of both the atmosphere and ocean. For an equilibrium climate, the long-term (e.g., century timescale) net top-of-atmosphere radiation (TOA) is zero. That is, the amount of incoming shortwave solar radiation is balanced by the reflected outgoing shortwave radiation and emitted longwave radiation. In some sense, an equilibrium climate state can be defined by the "average" pathways of the incoming solar energy, before it leaves the system (e.g., Trenberth and Fasullo 2011, this volume, Fig. 1).

Multi-century climate model simulations show that, about such an equilibrium state, the system generates internal variability—transient fluctuations in TOA and other important climate variables, such as surface temperature. At timescales of days or weeks, internal variability is dominated by weather systems and the chaotic nature of the atmospheric circulation. At timescales of years to decades, internal variability is associated with large-

M. D. Palmer (✉)
Met Office Hadley Centre, FitzRoy Road, Exeter EX1 3PB, UK
e-mail: matthew.palmer@metoffice.gov.uk

scale ocean–atmosphere modes, such as the El-Nino Southern Oscillation (McPhaden et al. 2006), the North Atlantic Oscillation (Hurrell et al. 2001), the Southern Annular Mode (Connolley 1997) and the Atlantic Multi-decadal Oscillation (Knight et al. 2005). All such modes of variability redistribute energy within the ocean and among other climate system components. This internally generated "noise" of the climate system is an important concept when interpreting observed or simulated climate changes.

At its simplest, climate change arises when an equilibrium climate is perturbed by an external forcing, so that the net TOA, and therefore the net convergence/divergence of energy into the system, is no longer zero. Changes in total solar irradiance, natural or anthropogenic aerosols and concentrations of atmospheric greenhouses can all contribute to "external" forcing of the climate system. The relative importance of these various factors in shaping Earth's climate record can be assessed under the framework of detection and attribution (Hegerl et al. 2007). This framework also provides a means of quantitatively assessing global climate model performance and has potential to reduce the large range of climate projections by scaling model projections using observational constraints (e.g., Stott and Forest 2007).

A standard metric used to interpret climate change projections is climate sensitivity, which is defined as the equilibrium response of global surface temperature to an imposed change in radiative forcing (e.g., a doubling of carbon dioxide concentrations, relative to pre-industrial levels, Meehl et al. 2011). Thus, models with high climate sensitivity will predict a greater surface temperature rise for a given CO_2 emissions scenario than those with low climate sensitivity. The climate response is determined by the strength of the climate feedbacks (Stocker et al. 2001), and how these affect the global energy balance. In reality, CO_2 emissions will continue to evolve and are unlikely to remain constant for any length of time. Therefore, additional terms, such as the rate of ocean heat uptake, are important for determining the rate of surface temperature rise (Raper et al. 2002). In essence then, climate change projections are ultimately dependent on how the energy pathways change in response to increased greenhouse gas concentrations (and other external forcings, such as anthropogenic aerosols).

2 Observing Earth's Energy Budget

There are fundamentally two ways in which we can estimate Earth's energy budget. The first is to measure the net energy flux (TOA) using direct satellite measurement of the incoming and outgoing radiation. The second is to estimate the energy storage terms, which on decadal and longer timescales is dominated by accumulation of heat within the global ocean (Bindoff et al. 2007; see also Trenberth and Fasullo 2011, this volume). The two approaches are highly complementary and make use of independent observational data sets. The ability to cross-reference these estimates provides an important means to diagnose the presence of systematic errors in the global climate observing system.

The first satellite measurements targeted at measuring Earth's radiation balance date back to 1975 and the launch of NASA's Earth Radiation Budget (ERB) instrument onboard the Nimbus-6 and Nimbus-7 satellites (Jacobowitz et al. 1979). This was followed in the 1980s by NASA's Earth Radiation Budget Experiment mission (ERBE; Barkstrom 1984). More recently, several international efforts have contributed measurements and improved understanding of Earth's radiation balance. These include the Clouds and Earth's Radiant Energy System (CERES; Wielicki et al. 1996); the Scanner for Radiation Budget (ScaRaB; Kandel et al. 1998); and the Geostationary Earth Radiation Budget Project (GERB; Harries et al. 2005). The great virtue of satellite observations is the spatial coverage they provide, although temporal sampling can be problematic for non-geostationary platforms. However, absolute calibration of satellite instruments makes estimating the residual of the globally integrated radiation terms a challenge. In addition, biases between different satellite sensors can introduce spurious climate signals unless great care is taken. Loeb et al. (2011, this volume) discuss the need to inter-calibrate successive satellite missions and document the progress being made toward development of a homogeneous climate record for the CERES data sets.

The accumulation of energy in the Earth System associated with anthropogenic global warming is manifest in numerous climate variables, including rising ocean heat content, global sea level, near surface temperature, reduced sea-ice coverage, ice-cap and glacial mass loss (e.g., Kennedy et al. 2010). However, in energetic terms, by far, the single largest term is associated with increased ocean heat storage (Bindoff et al. 2007). Recent attempts to quantify the rate of heat uptake using in situ observations in the 0–700 m layer represent an energy accumulation equivalent to a radiative forcing of 0.64 ± 0.11 W m^{-2} for the period 1993–2008 (Lyman et al. 2010). The largest single contribution to the uncertainty arises from bias correction of the expendable bathythermograph (XBT) measurements, which dominate the observational database for the 1970s to 2000s. As such, it may be possible to reduce this uncertainty if better agreement can be reached among a group of researchers actively developing new XBT corrections. Lyman (2011, this volume) explores to what extent ocean heat content changes are constrained by the in situ observations on shorter timescales.

3 Observing and Modeling Earth's Energy Flows

The preceding section has focused on Earth's energy budget. Of equal importance to our understanding of the climate system is the way in which the incoming solar radiation is redistributed and how this redistribution might change in response to anthropogenic climate change.

Our most comprehensive estimates of the atmospheric circulation over the twentieth century come from atmospheric reanalyses (e.g., Kalnay et al. 1996; Uppala et al. 2005). These are essentially the same data assimilation models used for weather forecasting, but often used at a lower spatial resolution in order to facilitate a continuous run over 40 years or more (e.g., Compo et al. 2006) to estimate the historical atmospheric state. The most sophisticated systems ingest a large suite of in situ and satellite measurements, but are fundamentally poorly constrained in areas with few historical observations, such as the polar regions.

One must be cautious in the application of atmospheric reanalysis data for studying Earth's energy flows and climate, in general, for a number of reasons. Reanalyses often use prescribed SSTs as the lower boundary condition, which provide an infinite source of heat and freshwater to the simulated atmosphere. Therefore, these models do not conserve heat or freshwater, and many have unrealistic TOA imbalances (Trenberth et al. 2009, 2011). In addition, inhomogeneous input observations and biases in the underlying physical models can limit the utility of reanalyses for studying climate and climate change, as discussed by Bengtson et al. (2007).

With careful consideration of the limitations described above, reanalyses are still a useful tool for informing us about part of Earth's energy flows. Trenberth et al. (2009) have presented an updated estimate of Earth's annual mean energy budget for the period

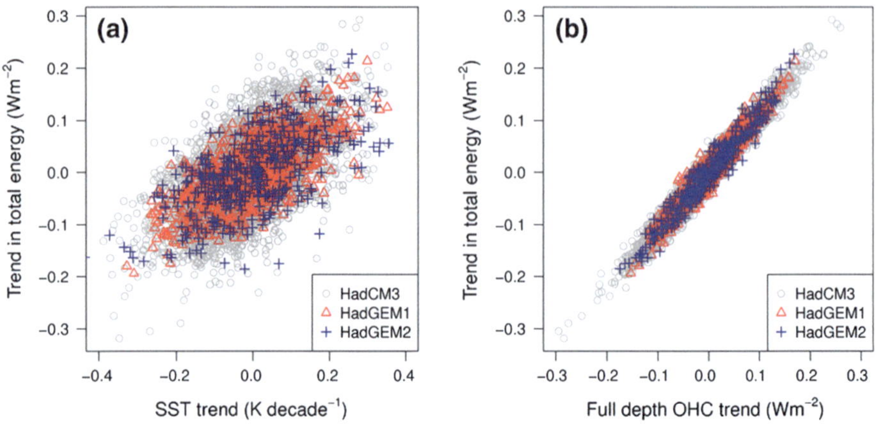

Fig. 1 Plot of linear decadal trends in total energy (W m^{-2}) regressed against: **a** decadal trends in globally averaged sea surface temperature (SST, K decade^{-1}) and **b** decadal trends in full-depth ocean heat content (OHC, W m^{-2}). Note that the trend in total energy is equivalent to the average top-of-atmosphere radiation balance (TOA) over the same period. Figure reproduced from Palmer et al. (2011)

2000–2005 based on new CERES measurements (see Trenberth and Fasullo 2011, Fig. 1, this volume) in combination with a wide range of observational and reanalysis data sources. However, there is not universal agreement among the research community on the values presented by Trenberth et al. (2009), and discussions have highlighted the need to work toward error bars for each of the energy flow terms presented by the authors (see Stevens and Schwartz 2011, this volume). In particular, Kato et al. (2011, this volume) suggest a total downwelling longwave radiation of 347 $\pm$ 7 W m^{-2} based on Cloudsat measurements (Stephens et al. 2002). This value is substantially larger than the 333 W m^{-2} put forward by Trenberth et al. (2009), but it remains unclear where the difference between these terms can be accounted for in the energy budget.

Contrary to atmospheric reanalyses, coupled climate models provide a physically consistent representation of the Earth System that conserves heat and freshwater. Since these tools are the primary means by which we make projections of future climate change, it is essential to evaluate their performance against current observations. Evaluating the top-of-atmosphere and surface energy budgets of climate models against observations provides a powerful means of diagnosing model errors. Trenberth and Fasullo (2010) have performed such an analysis using the CERES observations. The authors report that many of the models used in the AR4 show a similar systematic bias in surface downward shortwave radiation over the Southern Ocean that is also seen in atmospheric reanalyses. This bias in absorbed solar radiation arises mainly from a lack of mid-level cloud that is common feature of many state-of-the-art atmospheric models. The associated increase in sea surface temperature reduces poleward temperature gradients and leads to anomalously low southward ocean heat transport. These model deficiencies undermine projected changes in Southern Hemisphere cloud changes and migration of storm tracks.

4 Total Energy as a Metric for Climate Change

In recent years, there has been considerable controversy surrounding the reduced rate of surface temperature rise over the first decade of the twenty first century (e.g., Willet et al.

2010). A similar reduction in the rate of ocean heat uptake is also apparent in the estimate of Lyman et al. (2010), although it remains questionable whether this feature is statistically significant. The 0–700 m layer changes reported by Lyman et al. (2010) represent only the upper 15–20% of the full ocean depth, and this has motivated studies to look into the relationships between globally averaged temperature, TOA and full-depth ocean heat content in coupled climate model simulations. On decadal to multi-decadal timescales, the essential balance in the Earth System is between total ocean heat content, which is the primary energy storage term (Bindoff et al. 2007), and TOA, which is the total energy flux entering or leaving the system.

Trenberth and Fasullo (2011, this volume) and Katsman and van Oldenborgh (2011) demonstrate that in the presence of anthropogenic global warming, hiatuses in surface temperature rise are associated with increased heat export to the deeper ocean. These findings are supported by the work of Palmer et al. (2011), who find that decadal trends in surface temperature place a relatively weak constraint on TOA over the same period (Fig. 1). These studies highlight the ability of the ocean to re-arrange large quantities of heat on decadal timescales through internal variability. The implication is that the recent hiatus in surface temperature rise may not require us to invoke changes in external forcings as the primary explanation (although these may have also played a role). In addition, these studies point toward measurement of accumulated energy in the Earth System as a more robust metric of anthropogenic global warming than surface temperature.

5 Summary

A key goal of weather and climate research is to understand how internal variability and external forcings contribute to changes in TOA and the flow of energy within the Earth system on a range of timescales. This goal will ultimately be reached through a combination of improved Earth system observation and numerical model simulations. Central to these aims are the continuation of CERES-like TOA measurements (Loeb et al. 2011, this volume) and the development of a deep ocean observing array to better constrain future ocean heat content change (Palmer et al. 2010; Garzoli et al. 2010). These observations will help us to improve coupled climate models, understand their shortcomings and ultimately make better projections of changes in Earth's energy budget and how the energy redistribution pathways may change under anthropogenic climate change.

Acknowledgments I am grateful to Norman Loeb, Mark Ringer and Peter Stott for useful discussions on this work. Comments from Kevin Trenberth and an anonymous reviewer helped to improve the manuscript. This work was supported by the UK Joint DECC/Defra Met Office Hadley Centre Climate Programme (GA01101) and the European Community's 7th framework programme (FP7/2007-2013) under grant agreement No. GA212643 (THOR: "Thermohaline Overturning—at Risk," 2008–2012).

References

Barkstrom BR (1984) The earth radiation budget experiment (ERBE). Bull Am Meteorol Soc 65:1170–1185

Bengtson L, Arkin P, Berrisford P, Bougeault P, Folland CK, Gordon C, Haines K, Hodges KI, Jones P, Kallberg P, Rayner N, Simmons AJ, Stammer D, Thorne PW, Uppala S, Vose RS (2007) The need for a dynamical climate reanalysis. Bull Am Meteorol Soc 88:495–501

Bindoff NL, Willebrand J, Artale V, Cazenave A, Gregory J, Gulev S, Hanawa K, Le Quéré C, Levitus S, Nojiri Y, Shum CK, Talley LD, Unnikrishnan A (2007) Observations: oceanic climate change and sea level. In: Solomon S, Qin D, Manning M, Chen Z, Marquis M, Averyt KB, Tignor M, Miller HL (eds)

Climate change 2007: the physical science basis. Contribution of Working Group I to the Fourth Assessment Report of the Intergovernmental Panel on Climate Change. Cambridge University Press, Cambridge

Compo GP, Whitaker JS, Sardeshmukh PD (2006) Feasibility of a 100 year reanalysis using only surface pressure data. Bull Am Math Soc 87:175–190

Connolley WM (1997) Variability in annual mean circulation in southern high latitudes. Clim Dyn 13:745–756

Garzoli S, Boebel O, Bryden H, Fine R, Fukasawa M, Gladyshev S, Johnson G, Johnson M, MacDonald A, Meinen C, Mercier H, Orsi A, Piola A, Rintoul S, Speich S, Visbeck M, Wanninkhof R (2010) Progressing towards global sustained deep ocean observations. In: Hall J, Harrison DE, Stammer D (eds) Proceedings of OceanObs'09: sustained ocean observations and information for society, vol 2. ESA Publication WPP-306, Venice, Italy, 21–25 September 2009. doi:10.5270/OceanObs09.cwp.34

Harries JE et al (2005) The geostationary earth radiation budget project. Bull Am Meteorol Soc 86:945–960. doi:10.1175/BAMS-86-7-945

Hegerl GC, Zwiers FW, Braconnot P, Gillett NP, Luo Y, Marengo Orsini JA, Nicholls N, Penner JE, Stott PA (2007) Understanding and Attributing Climate Change. In: Solomon S et al (eds) Climate change 2007: the physical science basis. Contribution of Working Group I to the Fourth Assessment Report of the Intergovernmental Panel on Climate Change

Hurrell JW, Kushnir Y, Visbeck M (2001) The north atlantic oscillation. Science 291(5504):603–605

Jacobowitz H, Smith WL, Howell HB, Nagle FW, Hickey JR (1979) The first 18 months of planetary radiation budget measurements from the nimbus 6 ERB experiment. J Atmos Sci 36:501–507. doi:10.1175/1520-0469(1979)036<0501:TFMOPR>2.0.CO;2

Kalnay E et al (1996) The NCEP/NCAR 40-year reanalysis project. Bull Am Meteorol Soc 77:437–470

Kandel R et al (1998) The ScaRaB earth radiation budget dataset. Bull Am Meteorol Soc 79:765–783. doi:10.1175/1520-0477(1998)079<0765:TSERBD>2.0.CO;2

Kato S, Loeb NG, Rutan DA, Rose FG, Sun-Mack S, Miller WF, Chen Y (2011) Uncertainty in modeled surface irradiances using satellite-derived cloud properties (this volume)

Katsman CA, van Oldenborgh GJ (2011) Tracing the upper ocean's 'missing heat'. Geophys Res Lett 38:L14610. doi:10.1029/2011GL048417

Kennedy JJ, Thorne PW, Peterson TC, Ruedy R, Stott PA, Parker DE, Good SA, Titchner H, Willett K (2010) How do we know the world has warmed? In: Arndt DS et al (eds) State of the climate 2009. Bull Am Meteorol Soc 91(7):S26–S27. Cambridge University Press, Cambridge

Knight JR, Allan RJ, Folland CK, Vellinga M, Mann ME (2005) A signature of persistent natural thermohaline circulation cycles in observed climate. Geophys Res Lett 32:L20708. doi:10.1029/2005GL024233

Loeb N, Kato S, Su W, Wong T, Rose F, Doelling D, Norris J (2011) Advances in understanding top-of-atmosphere radiation variability from satellite observations (this issue)

Lyman JM (2011) Estimating global energy flow from the global upper ocean (this volume)

Lyman JM, Good SA, Gouretski VV, Ishii M, Johnson GC, Palmer MD, Smith DM, Willis JK (2010) Robust warming of the global upper ocean. Nature 465:334–337

McPhaden MJ, Zebiak SE, Glantz MH (2006) ENSO as an integrating concept in earth science. Science 314:1740–1745

Meehl GA, Arblaster JM, Fasullo JT, Hu A, Trenberth KE (2011) Model-based evidence of deep-ocean heat uptake during surface-temperature hiatus periods. Nat Clim Change 1:360–364. doi:10.1038/NCLIMATE1229

Palmer M, Antonov J, Barker P, Bindoff N, Boyer T, Carson M, Domingues C, Gille S, Gleckler P, Good S, Gouretski V, Guinehut S, Haines K, Harrison D, Ishii M, Johnson G, Levitus S, Lozier S, Lyman J, Meijers A, von Shuckmann K, Smith D, Wijffels S, Willis J (2010) Future observations for monitoring global ocean heat content. In: Hall J, Harrison DE, Stammer D (eds) Proceedings of OceanObs'09: sustained ocean observations and information for society, vol 2. ESA Publication WPP-306, Venice, Italy, 21–25 September 2009. doi:10.5270/OceanObs09.cwp.68

Palmer MD, McNeall DJ, Dunstone NJ (2011) Importance of the deep ocean for estimating decadal changes in Earth's radiation balance. Geophys Res Lett (in press)

Raper SCB, Gregory JM, Stouffer RJ (2002) The role of climate sensitivity and ocean heat uptake on AOGCM transient temperature response. J Clim 15:124–130

Stephens GL, Austin RT, Benedetti A, Boain R, Durden SL, Illingworth A, Mace G, Miller SD, Mitrescu C, O'Conner E, Rossow WB, Sassen K, Vane DG, Wang Z (2002) The CloudSat mission and the EPS constellation: a new dimension of space-based observations of clouds and precipitation. Bull Am Math Soc 83:1771–1790

Stevens B, Schwartz SE (2011) Observing and modeling earth's energy flows. doi:10.1007/s10712-012-9184-0

Stocker TF et al (2001) Chapter 7: physical climate processes and feedbacks. In: Houghton JT et al. (eds) Climate change 2001: the scientific basis. Contribution of Working Group I to the Third Assessment Report of the Intergovernmental Panel on Climate Change. Cambridge University Press, Cambridge, pp 417–470

Stott PA, Forest CE (2007) Ensemble climate predictions using climate models and observational constraints. Philos Trans R Soc Series A 365:2029–2052

Trenberth KE, Fasullo JT (2010) Simulation of present day and 21st century energy budgets of the southern oceans. J Clim 23(2):440–454

Trenberth KE, Fasullo JT (2011) Tracking earth's energy: from el nino to global warming (this volume)

Trenberth KE, Fasullo JT, Kiehl J (2009) Earth's global energy budget. Bull Am Math Soc 90:311–323

Trenberth KE, Fasullo JT, Mackaro J (2011) Atmospheric moisture transports from ocean to land and global energy flows in reanalyses. J Clim 24:4907–4924. doi:10.1175/2011JCLI4171.1

Uppala SM et al (2005) The ERA-40 re-analysis. Q J R Meteorol Soc 131(612):2961–3012

Wielicki BA, Barkstrom BR, Harrison EF, Lee RB, Smith GL, Cooper JE (1996) Clouds and the earth's radiant energy system (CERES): an earth observing system experiment. Bull Am Meteorol Soc 77:853–868

Willet KM, Alexander LV, Thorne PW (eds) (2010) Global climate [in "State of the Climate in 2009"]. Bull Am Meteorol Soc 91(7):S79–S106

Surv Geophys (2012) 33:359–385
DOI 10.1007/s10712-012-9175-1

Advances in Understanding Top-of-Atmosphere Radiation Variability from Satellite Observations

Norman G. Loeb · Seiji Kato · Wenying Su · Takmeng Wong · Fred G. Rose · David R. Doelling · Joel R. Norris · Xianglei Huang

Received: 23 May 2011 / Accepted: 3 January 2012 / Published online: 27 January 2012

Abstract This paper highlights how the emerging record of satellite observations from the Earth Observation System (EOS) and A-Train constellation are advancing our ability to more completely document and understand the underlying processes associated with variations in the Earth's top-of-atmosphere (TOA) radiation budget. Large-scale TOA radiation changes during the past decade are observed to be within 0.5 Wm^{-2} per decade based upon comparisons between Clouds and the Earth's Radiant Energy System (CERES) instruments aboard Terra and Aqua and other instruments. Tropical variations in emitted outgoing longwave (LW) radiation are found to closely track changes in the El Niño-Southern Oscillation (ENSO). During positive ENSO phase (El Niño), outgoing LW radiation increases, and decreases during the negative ENSO phase (La Niña). The coldest year during the last decade occurred in 2008, during which strong La Nina conditions persisted throughout most of the year. Atmospheric Infrared Sounder (AIRS) observations show that the lower temperatures extended throughout much of the troposphere for several months, resulting in a reduction in outgoing LW radiation and an increase in net incoming radiation. At the global scale, outgoing LW flux anomalies are partially compensated for by decreases in midlatitude cloud fraction and cloud height, as observed by Moderate Resolution Imaging Spectrometer and Multi-angle Imaging SpectroRadiometer, respectively. CERES data show that clouds have a net radiative warming influence during La Niña conditions and a net cooling influence during El Niño, but the magnitude of the anomalies varies greatly from one ENSO event to another. Regional cloud-radiation variations among several Terra and A-Train instruments show consistent patterns and

N. G. Loeb (✉) · S. Kato · T. Wong · D. R. Doelling
NASA Langley Research Center, Mail Stop 420, Hampton, VA 23681, USA
e-mail: norman.g.loeb@nasa.gov

W. Su · F. G. Rose
Science Systems and Applications, Inc., Hampton, VA 23666, USA

J. R. Norris
Scripps Institution of Oceanography, San Diego, CA 92037, USA

X. Huang
University of Michigan, Ann Arbor, MI 48109, USA

exhibit marked fluctuations at monthly timescales in response to tropical atmosphere-ocean dynamical processes associated with ENSO and Madden–Julian Oscillation.

Keywords Radiation budget · Clouds · Climate variability

1 Introduction

The top-of-atmosphere (TOA) Earth radiation budget (ERB) is determined from the difference between how much energy is absorbed and emitted by the planet. Climate forcing results in an imbalance in the TOA radiation budget that has direct implications for global climate, but the large natural variability in the Earth's radiation budget due to fluctuations in atmospheric and ocean dynamics complicates this picture. An illustration of the variability in TOA radiation is provided in Fig. 1, which shows a continuous 31-year record of tropical (20°S–20°N) TOA broadband outgoing longwave (LW) radiation (OLR) between 1979 and 2010 from non-scanner and scanner instruments. Figure 1a is an update to earlier

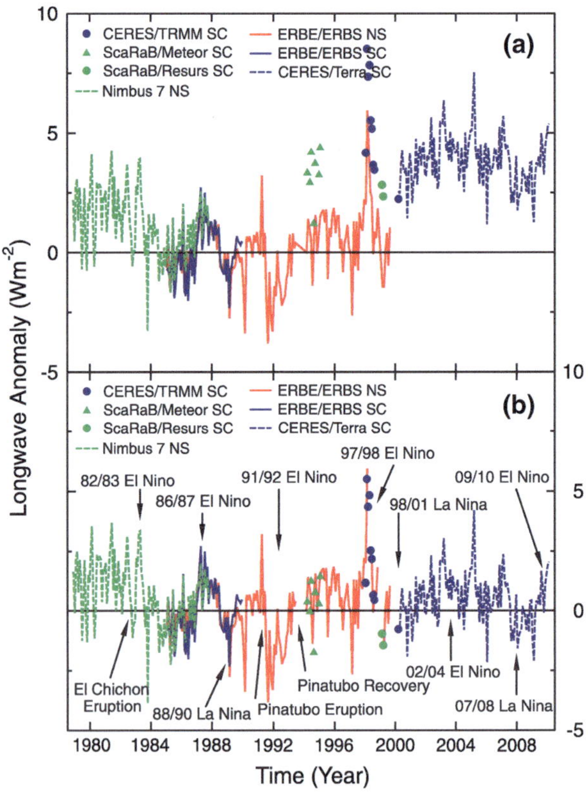

Fig. 1 LW TOA flux anomalies for 20°S–20°N from November 1978 to February 2010 **a** with no overlap correction, and **b** with overlap correction based upon ERBS Non-scanner WFOV Edition3_Rev1 (*red solid line*), Nimbus-7 Non-scanner (*green dashed line*), ERBS Scanner (*blue solid line*), CERES Terra crosstrack SSF1deg-lite_Ed2.5 (*blue dashed line*), CERES/TRMM Scanner Edition2 (*blue circle*), ScaRaB/Meteor Scanner (*green triangle*) and ScaRaB/Resurs Scanner (*green circle*). Anomalies are defined with respect to the 1985–1989 period

versions by Wielicki et al. (2002) and Wong et al. (2006). It shows rather marked jumps of up to 3 Wm^{-2} among the different satellites owing to absolute calibration differences, which are within measurement uncertainty. Because there is an overlap, the entire record can be placed on a common radiometric scale (Fig. 1b). Anomalies of up to 5 Wm^{-2} are observed during major El Niño-Southern Oscillation (ENSO; Philander 1990) events such as the 1997/98 El Niño. The record also provides quantitative data on the LW effect of the Pinatubo eruption and subsequent recovery.

With the availability of multiple years of data from new and improved passive instruments launched as part of the Earth Observing System (EOS) and active instruments belonging to the A-Train constellation (L'Ecuyer and Jiang 2010), a more complete observational record of ERB variations and the underlying processes is now possible. For the first time, simultaneous global observations of the ERB and a multitude of cloud, aerosol, and surface properties and atmospheric state data are available with a high degree of precision. These data are a far more comprehensive set for evaluating climate model simulations compared to ERB observations alone (Wielicki et al. 2002).

This study analyzes ERB observations from the Clouds and the Earth's Radiant Energy System (CERES) aboard Terra and Aqua together with data from several other instruments on Terra, Aqua and A-Train. We first examine how consistently independent satellite instruments observe large-scale (tropical and global) changes in radiation budget during the past decade. Next, we explore the relationship between large-scale outgoing SW and LW TOA radiative flux variations and ENSO and relate the TOA flux variations to cloud property variations. Finally, we illustrate how the new generation of satellite observations collectively enables a far more detailed description of the regional response of clouds and radiation to circulation changes associated with ENSO and Madden–Julian Oscillation (MJO; Madden and Julian 1971, 1994).

2 CERES Observations

This study uses 10 years (March 2000–February 2010) of CERES Terra and 7 years 8 months (July 2002–February 2010) of CERES Aqua regional monthly mean data from the CERES SSF1deg-lite_Ed2.5 data product. CERES SSF1deg-lite_Ed2.5 provides CERES TOA radiative fluxes and coincident cloud and aerosol properties from the Moderate Resolution Imaging Spectroradiometer (MODIS) (Salomonson et al. 1989; Barnes et al. 1998). Cloud retrievals are based upon Minnis et al. (2011), while aerosol data are from Remer et al. (2005). Each parameter is available at 1°-regional, zonal and global time–space scales. TOA fluxes are provided for clear and all-sky conditions in the longwave (LW), shortwave (SW), and window (WN) regions (Loeb et al. 2005). Net incoming TOA flux is determined from the difference between the observed TOA incoming solar irradiance and the sum of the outgoing reflected solar and emitted LW radiation. TOA solar irradiance is based upon time-varying observations from the Solar Radiation and Climate Experiment (SORCE; Woods et al. 2000) mission, which are updated daily (SORCE Level 3 Total Solar Irradiance Version 10 available from: http://lasp.colorado.edu/sorce/data/tsi_data.htm).

Regional mean TOA fluxes for 1° equal-area grid boxes in the CERES SSF1deg-Ed2.5 data product are interpolated using the assumption of constant meteorological conditions (termed non-geostationary or non-GEO) similar to the process used to average CERES ERBE-like data (Young et al. 1998). In the SW, TOA fluxes between observation times are interpolated, taking into account TOA solar irradiance changes throughout the day and the albedo dependence on solar zenith angle for the scene observed at the CERES observation time. To interpolate LW TOA flux for cloud-free scenes over land, CERES SSF1deg-lite applies a half-sine fit to approximate the diurnal cycle of OLR (Young et al. 1998). Otherwise, only the available daytime and nighttime LW observations are used to determine a daily mean. After interpolation, the time series is used to produce monthly means, calculated using the combination of observed and interpolated parameters from all days containing at least one CERES observation (Doelling et al. 2012).

A cloud mask (Minnis et al. 2011) is applied to MODIS 1-km pixels within CERES footprints to identify scenes that are clear and cloudy. Clear-sky TOA fluxes are determined from CERES footprints for which 99.9% or more of the MODIS pixels within the CERES footprint are identified as cloud-free. This study also uses clear-sky TOA fluxes from the recently released Energy Balanced and Filled (EBAF) Edition2.5A for March 2000–February 2010. Because of the coarse spatial resolution of CERES (20 km at nadir), clear-sky TOA fluxes in SSF1deg-lite_Ed2.5 only include contributions from cloud-free regions occurring over relatively large spatial scales and meteorological conditions and geographical regions where clouds occur less frequently. In EBAF, clear-sky TOA flux contributions at smaller spatial scales are recovered by inferring broadband TOA fluxes from MODIS radiances in clear portions of partly cloudy CERES footprints. The methodology, described in Loeb et al. (2009), involves using MODIS-CERES narrow-to-broadband regressions to convert MODIS narrowband radiances averaged over the clear portions of a footprint to broadband radiances. The "broadband" MODIS radiances are then converted to TOA radiative fluxes using CERES clear-sky Angular Distribution Models (Loeb et al. 2005).

Another advance in CERES SSF1deg-lite_Ed2.5 is that it includes the latest CERES instrument calibration improvements. The CERES team reanalyzed ground and in-flight calibration data and implemented a new algorithm to characterize the spectral degradation of the CERES optics with time that, if left uncorrected, leads to a decrease in SW radiances and a day/night inconsistency in LW radiances. A summary of the instrument calibration improvements required to overcome these problems is provided in Appendix 1.

3 Results

3.1 Comparison of TOA Radiation Variability from Different Instruments

Tropical (30°S–30°N) and global deseasonalized monthly anomalies of outgoing SW and LW and incoming net radiation from CERES Terra for March 2000–February 2010 and CERES Aqua for July 2002–February 2010 are provided in Fig. 2 and summarized in Table 1. A deseasonalized monthly anomaly is determined by differencing the average in a given month from the average of all years of the same month. As expected, variability is greater in the tropics than for the globe. Pronounced positive net TOA flux anomalies in 2008 are clearly evident in Fig. 2e, f. This period is characterized by 22 consecutive months of negative outgoing LW TOA flux anomalies in the tropics (between June 2007 and March 2009), and negative outgoing SW TOA flux anomalies in both the tropics and globally during much of 2008 and early 2009. The minimum tropical outgoing LW TOA flux anomaly occurs in January 2008, the coldest of all Januaries during the CERES Terra record. The peak net incoming TOA flux anomaly occurs in June 2009, shortly after the transition from La Nina to El Nino conditions, and reaches 2.5 Wm^{-2} in the tropics and 1.5 Wm^{-2} globally.

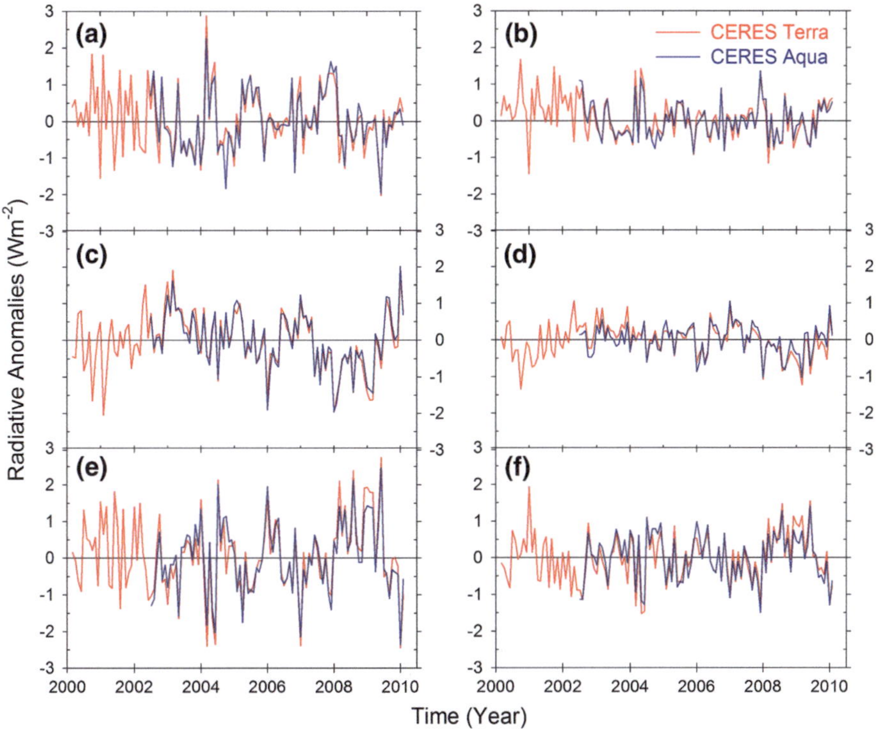

Fig. 2 CERES Terra and Aqua deseasonalized monthly for SW (**a** and **b**), LW (**c** and **d**), Net (**e** and **f**) TOA radiation. *Left column* corresponds to 30°S–30°N and *right column* corresponds to global. The CERES Terra record is from March 2000–February 2010, and the CERES Aqua record is from July 2002–February 2010. Both Terra and Aqua records use their respective means for July 2002–February 2010 to compute the anomalies

When CERES Terra and CERES Aqua are compared for July 2002 through February 2010 (Fig. 3; Table 1), differences are generally much smaller than the variability, as evident from the coefficient of determination (r^2) values, which exceed 0.92 in the tropics, and 0.81 globally (Table 1). Differences in SW trends between the two records are generally smaller than 0.3 Wm^{-2} per decade. For global LW, CERES Terra shows a 0.63 Wm^{-2} per decade steeper increase than CERES Aqua. The cause for the drift between Terra and Aqua LW TOA fluxes is due to an artificial discontinuity in the middle of the Aqua record that will be corrected for in the next release (see Appendix 2 for more details).

The CERES data record also shows excellent agreement with data records from other instruments. This is illustrated in Fig. 4a–c and Table 2. Figure 4a compares the stability of the CERES Terra SW observations with anomalies in Photosynthetically Active Radiation (PAR) from the Sea-Viewing Wide-Field-of-View Sensor (SeaWiFS; Hooker et al. 1992) (2009 reprocessing), launched in August 1997 onboard the SeaStar spacecraft. A detailed description of the methodology used in the comparison is provided in Loeb et al. (2007). Briefly, deseasonalized CERES SW TOA flux anomalies for all-sky ocean between 30°S and 30°N from the CERES Terra SSF1deg-lite_Ed2.5 data product are plotted against deseasonalized anomalies in SeaWIFS PAR. The SeaWIFS PAR anomalies are then multiplied by the slope of the regression line (-6.09 W m^{-2} per E m^{-2} day^{-1}) in order to

Table 1 Statistical summary tropical (30°S–30°N) and global CERES Terra deseasonalized monthly anomalies for SW, LW and net TOA radiation from July 2002–February 2010 for CERES Terra and CERES Aqua SSF1deg-lite_Ed2.5 and their difference ("Diff")

	30°S–30°N					Global				
	σ	m	ε	δ	r^2	σ	m	ε	δ	r^2
SW										
Terra	0.82	−0.03	0.94	–	–	0.52	0.00	0.51	–	–
Aqua	0.78	0.02	0.98	–	–	0.49	−0.11	0.53	–	–
Diff	0.23	0.04	0.27	0.27	0.92	0.18	−0.12	0.25	0.27	0.89
LW										
Terra	0.79	−1.30	1.27	–	–	0.43	−0.82	0.56	–	–
Aqua	0.78	−0.90	1.23	–	–	0.40	−0.19	0.49	–	–
Diff	0.18	0.39	0.21	0.45	0.95	0.19	0.63	0.36	0.72	0.81
Net										
Terra	1.06	1.00	1.36	–	–	0.67	0.55	0.77	–	–
Aqua	0.97	0.56	1.26	–	–	0.64	0.04	0.76	–	–
Diff	0.28	−0.44	0.37	0.58	0.93	0.23	−0.51	0.39	0.64	0.88
WN										
Terra	0.34	−0.59	0.61	–	–	0.17	−0.34	0.23	–	–
Aqua	0.32	−0.19	0.51	–	–	0.16	−0.04	0.22	–	–
Diff	0.13	0.4	0.31	0.5	0.86	0.09	0.3	0.27	0.4	0.68

σ, Standard deviation of monthly anomalies (Wm^{-2}); m, slope of linear regression fit to monthly anomalies (Wm^{-2} per decade); ε, 95% confidence in m (Wm^{-2} per decade); $\delta = (m^2 + \varepsilon^2)^{1/2}$ (Wm^{-2} per decade), r^2, coefficient of determination

place the two records on the same radiometric scale. Results show agreement in monthly anomalies to 0.29 Wm^{-2}, and agreement in the overall slope to <0.4 Wm^{-2} per decade at the 95% confidence level for all data through the end of 2008. In early 2008, SeaStar spacecraft anomalies caused significant reductions in sampling, resulting in a much noisier comparison.

A comparison of tropical CERES Terra SW TOA flux and MODIS Terra cloud fraction anomalies (Fig. 4b) shows a slightly negative change over the decade (Table 2) within their respective 95% confidence intervals, and the r^2 between these variables is 0.82. Figure 4c compares CERES Terra and Aqua LW TOA flux anomalies with Atmospheric Infrared Sounder (AIRS) outgoing LW flux (monthly AIRX3STM.005 product) anomalies for 30°S–30°N from September 2002 and June 2009. AIRS TOA fluxes from ascending and descending orbits are linearly averaged without any correction for diurnal cycle effects. CERES SSF1deg-lite applies a half-sine fit over land to approximate the diurnal cycle of OLR (Young et al. 1998). The r^2 value exceeds 0.9 in both the AIRS/CERES Terra and AIRS/CERES Aqua comparisons (Table 2). Also, LW TOA flux anomalies are negative from mid-2007 through March 2009, and steadily increase thereafter in all three data records. Susskind et al. (2012) provide a more extensive AIRS-CERES analysis comparing regional and zonal LW TOA fluxes and find excellent agreement in the pattern of OLR anomaly trends between AIRS and CERES.

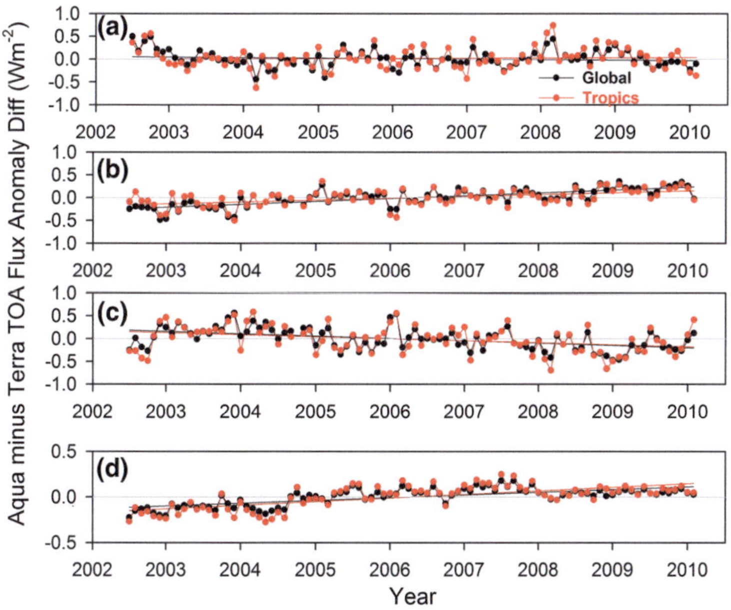

Fig. 3 Global and tropical (30°S–30°N) Aqua minus Terra CERES TOA flux anomaly difference for **a** SW, **b** LW, **c** Net and **d** WN for July 2002–February 2010

3.2 ENSO Influence on TOA Radiation Variability at Tropical and Global Scales

To explore the relationship between interannual variations in CERES TOA flux and ENSO, Fig. 5a shows outgoing LW TOA flux anomalies from CERES Terra with multivariate ENSO index (MEI; Wolter and Timlin 1998) for the tropics (30°S–30°N) and globe. The CERES TOA flux anomalies are smoothed with a 2-month running mean, consistent with MEI. In the tropics, there is a fairly good relationship between LW TOA flux anomalies and MEI, with an r^2 value of 0.56. Earlier studies (Allan and Singo 2002; Wong et al. 2006) showed a similar relationship between LW TOA flux and ENSO: for example, the 1997/98 El Niño resulted in monthly anomalies greater than 5 Wm^{-2} in the tropics. Positive MEI values (El Niño-like conditions) are associated with positive outgoing LW TOA flux anomalies, and negative MEI values (La Niña-like conditions) are associated with negative outgoing LW TOA flux anomalies. From mid-2007 through spring 2009, La Niña conditions dominate with two broad peaks in MEI occurring during the 2008 and 2009 winter seasons. Both of these events are associated with negative peaks in outgoing LW TOA flux anomalies, reaching -1.5 Wm^{-2}, and a pronounced reduction in tropospheric temperatures, as observed by AIRS (Fig. 6a). Anomalies in water vapor mixing ratio (Fig. 6b) are negative in the lower troposphere up to 600 mb through mid-2008, and become positive thereafter.

Globally, the relationship between outgoing LW TOA flux anomalies and MEI is much weaker ($r^2 = 0.25$), but there is still a tendency for negative LW TOA flux anomalies during extended periods of negative MEI. The pronounced minimum in tropical LW TOA flux anomaly in early 2008 is consistent with expectation: January 2008 was the coldest month in over a decade. Why then are the global average LW anomalies so much weaker than those in the tropics? It turns out that LW anomalies in the middle latitudes associated

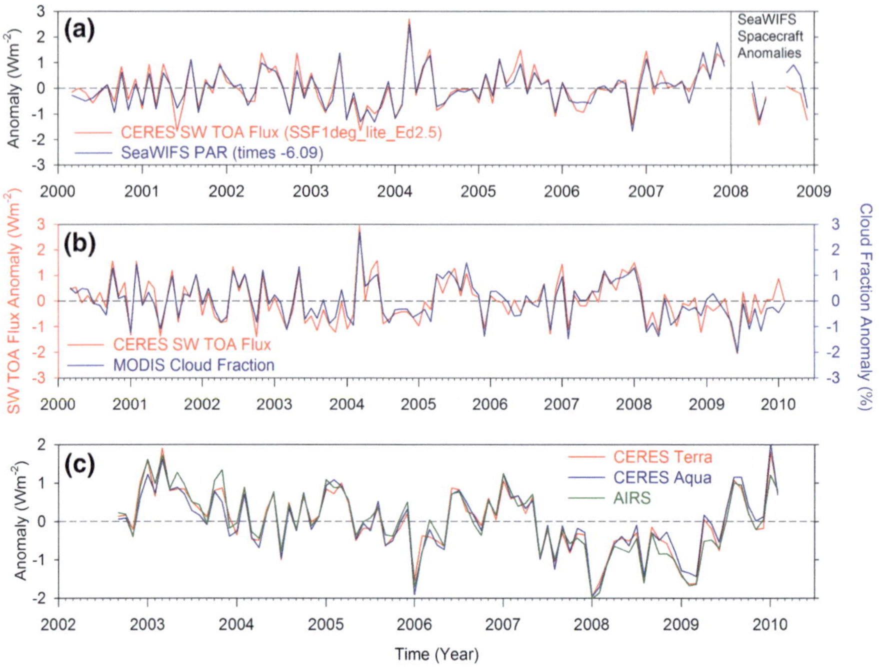

Fig. 4 Monthly anomalies in **a** CERES Terra SW TOA flux from SSF1deg-lite_Ed2.5 and SeaWiFS PAR scaled by a factor of −6.09 (corresponding to the slope of the regression line fit relating CERES SW TOA flux and SeaWiFS PAR anomalies) over ocean for 30°S–30°N from March 2000 to December 2009, **b** CERES Terra SW TOA flux and MODIS cloud fraction for 30°S–30°N between March 2000 and February 2010 and **c** global LW TOA flux from CERES Terra, CERES Aqua and AIRS Aqua for September 2002–February 2010

with large cloud property changes compensate for the decrease in outgoing LW radiation in the tropics. Figure 5b shows anomalies in midlatitude (30°S–60°S and 30°N–60°N) mean CERES LW TOA flux, MODIS cloud fraction, and Multi-angle Imaging SpectroRadiometer (MISR) cloud-top height. The MISR cloud-top height anomalies are determined from the MISR Cloud Fraction by Altitude product (Di Girolamo et al. 2010). During the 2007–2008 La Niña (yellow shaded region), midlatitude MODIS cloud fraction and MISR cloud-top height anomalies are both pronounced and negative. Since LW emission to space is greater in cloud-free regions and from clouds with lower tops, it follows that midlatitude LW TOA flux should increase, resulting in weaker global mean LW TOA flux anomalies. While it is plausible that the midlatitude cloud-radiation changes are related to circulation changes associated with the La Niña event in the tropics, other factors may also play a role.

To examine the role that clouds play in modulating tropical and global TOA flux variations, Fig. 7a–f compare all-sky and clear-sky outgoing SW and LW and net incoming TOA flux anomalies. Clear-sky TOA flux anomalies are based upon CERES TOA fluxes from cloud-free scenes, as identified from MODIS observations. We consider two sets of clear-sky TOA flux datasets. “Clear-sky (SSF1deg)” considers only CERES footprints in which 99.9% or more of the MODIS pixels within the CERES footprint are identified as cloud-free. “Clear-sky (EBAF)” is from the CERES Energy Balanced and Filled (CERES_EBAF_Edition2.5) Product (Loeb et al. 2009) and includes clear-sky TOA flux contributions from cloud-free regions in partly cloudy CERES footprints. In the SW,

Table 2 Same as Table 1 but for tropics (30°S–30°N)

	σ	m	ε	δ	r^2
CERES Terra and SeaWIFS (03/2000–12/2008)					
SW					
Terra	0.78	0.12	0.63	–	–
SeaWIFS	0.73	0.41	0.60	–	–
Diff	0.29	0.30	0.28	0.41	0.86
CERES Terra and MODIS (03/2000–02/2010)					
SW/f					
CERES SW	0.83	−0.25	0.55	–	–
MODIS f	0.77	−0.48	0.57	–	–
					0.82
CERES Terra and AIRS (09/2002–06/2009)					
LW					
CERES	0.79	−1.35	1.33	–	–
AIRS	0.83	−1.68	1.60	–	–
Diff	0.21	−0.33	0.22	0.40	0.93
CERES Aqua and AIRS (09/2002–06/2009)					
LW					
CERES	0.78	−0.93	1.29	–	–
AIRS	0.83	−1.68	1.60	–	–
Diff	0.25	−0.76	0.37	0.84	0.90

there is little correlation between clear and all-sky TOA fluxes. This is not surprising given the high correlation between CERES all-sky SW TOA flux and MODIS cloud fraction anomalies shown in Fig. 4b. However, in the LW, clear and all-sky TOA flux anomalies show very similar changes throughout the decade, both in the tropics and globally. During the 2008 La Niña, the magnitude of tropical clear-sky outgoing LW TOA flux anomalies is very similar to all-sky, consistent with the pronounced cooling throughout the troposphere during this period (Fig. 6a). In other parts of the LW record, clouds appear to introduce more high-frequency variations, as is evident from the abrupt spikes in the all-sky but not the clear-sky anomalies. As net incoming TOA flux is derived from both SW and LW contributions, the correlation between clear and all-sky net TOA flux is weaker, but nonetheless still apparent.

To compare the relative contribution of the surface, atmosphere and clouds on LW TOA flux variations, we use a plane-parallel radiative transfer code (Fu and Liou 1992, 1993) and perform a radiative perturbation analysis with observed variations in surface skin temperature (T_s) from the Global Modeling and Assimilation Office (GMAO)'s Goddard Earth Observing System (GEOS) Data Assimilation System (DAS) V4 (2003–2007) and V5 (2008–2009) products (Suarez 2005), atmospheric temperature ($T(z)$) and water vapor mixing ratio ($w(z)$) from AIRS, and effective cloud-top pressure (P_c), cloud fraction (f) and cloud optical depth (τ_c) from MODIS (Minnis et al. 2011). We set water vapor mixing ratio above 70 hPa to 10^{-6} g/g, and ozone is a fixed Midlatitude Summer standard atmosphere profile. The contribution to OLR variability by a given variable is computed from the difference between an OLR time series calculated using input variables specified by their monthly mean values for 84 months (January 2003–December 2009) and a second

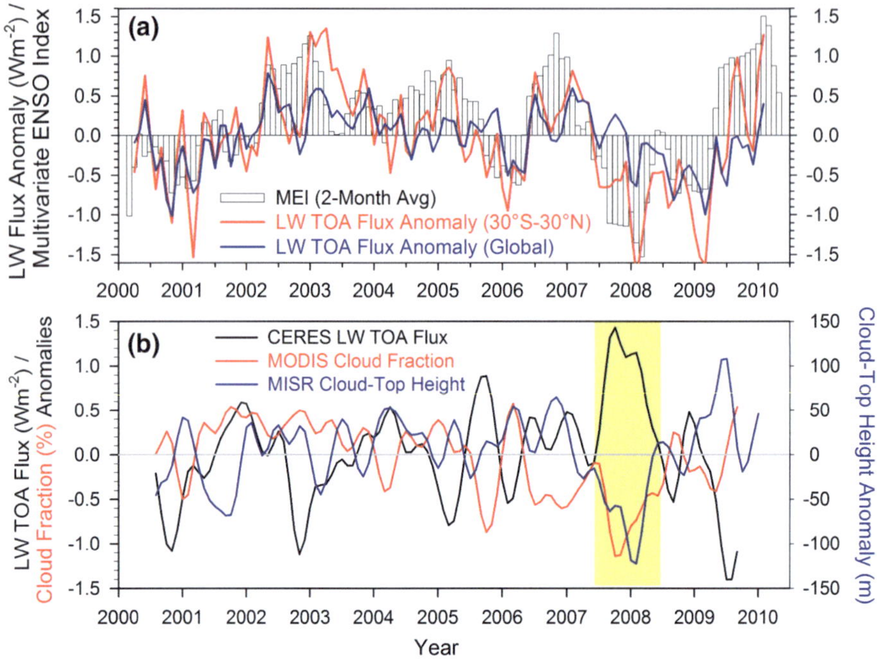

Fig. 5 **a** Deseasonalized anomalies in tropical (30°S–30°N) and global CERES LW TOA flux together with the multivariate ENSO Index (MEI). A 2-month running average is used to determine the LW TOA flux anomalies. **b** Deseasonalized anomalies in midlatitude (30°S–60°S and 30°N–60°N) mean CERES LW TOA flux, MODIS cloud fraction and MISR cloud-top height. An 11-point 6-month low-pass Lanczos filter is applied to the monthly anomalies

calculation identical to the first but with the parameter in question held fixed at its climatological monthly mean values. The climatological monthly mean values of each input variable are determined using data from all Januaries, Februaries, etc., for 2003–2009 in each $1° \times 1°$ deg region. In order to quantify the combined influence of variations in atmospheric temperature and humidity on OLR variability, the same approach is used but holding both variables fixed at their climatological values when calculating the difference. Similarly, the impact of clouds is determined by computing the combined contribution to OLR variability by P_c, f and τ_c. As a simplification, we assume single-layer clouds in the calculations, with cloud-top pressure given by the observed cloud fraction-weighted mean cloud-top pressure in a gridbox. With this simplification, it is not possible to determine whether a cloud fraction change corresponds to a change in low or high cloud, nor can we ascribe a change in cloud-top pressure to a change in low or high cloud altitude.

Figure 8a–g show the results of this analysis for all-sky (left-hand column) and clear-sky (right-hand column) conditions in the tropics. When variations from all parameters are included (Fig. 8a, b), calculated OLR anomalies (black line) agree well with CERES (red line). Atmospheric ($T(z)$ and ($w(z)$)) contributions explain most of the clear-sky LW TOA flux variations (Fig. 8d), and while the surface temperature contribution is correlated with that of $T(z)$, it is only appreciable in magnitude in early 2008, during the peak of the 2007–2008 La Niña event. Atmospheric temperature and water vapor mixing ratio contributions to LW TOA flux variability are negatively correlated (Fig. 8e, f), with lower (drier) temperature (water vapor) changes during the cold phase of ENSO reducing

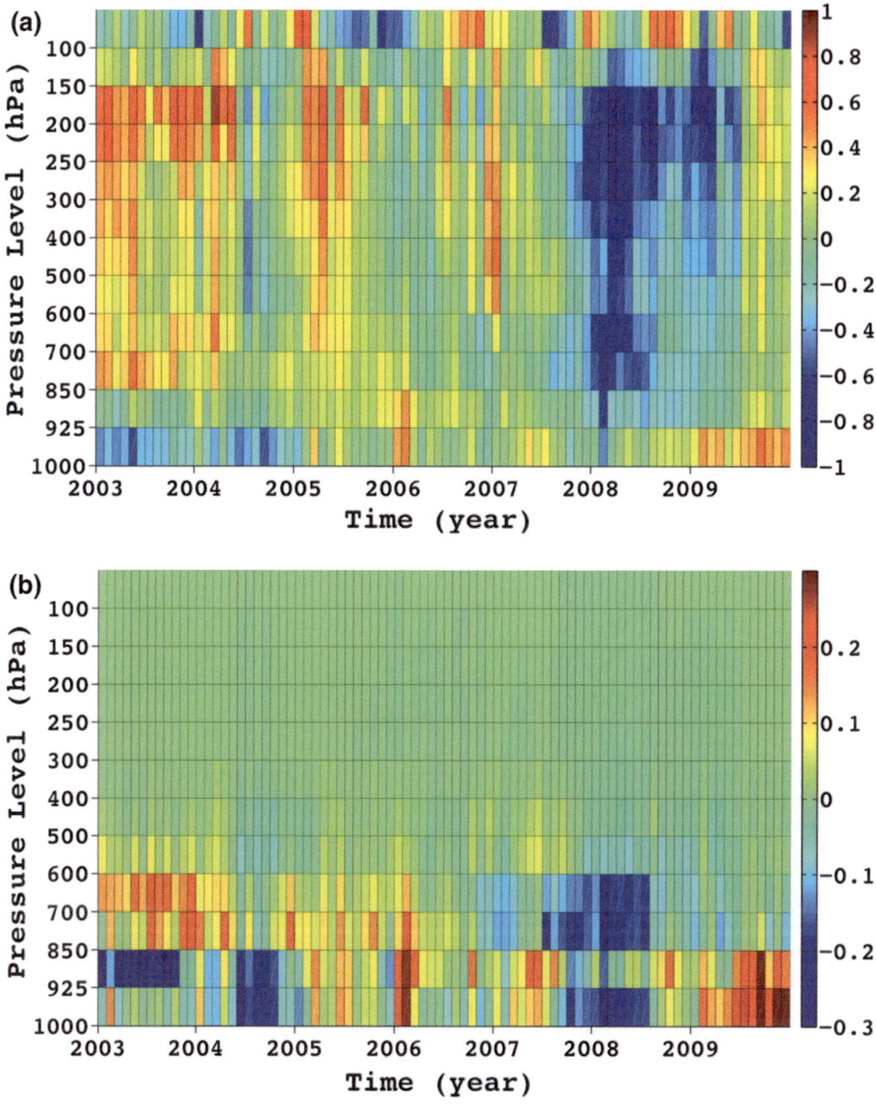

Fig. 6 AIRS low-latitude (30°S–30°N) temperature anomalies (K) as a function of pressure level

(enhancing) LW TOA flux. Overall, atmospheric temperature variations have a greater impact on LW TOA flux variability than water vapor changes. In all-sky conditions, atmosphere and cloud contributions generally track one another and contribute roughly equally to LW TOA flux variability. LW contributions by P_c and f are generally in phase with one another, particularly during stronger ENSO events.

A summary of tropical and global TOA flux anomalies averaged over intervals of positive and negative MEI is provided in Table 3. All tropical and most global mean LW Cloud Radiative Effect (CRE) anomalies are positive (enhanced radiative warming to the system) during La Niña events and negative (reduced radiative warming to the system) during El Niño events. The magnitude of LW cloud radiative anomalies in comparison to

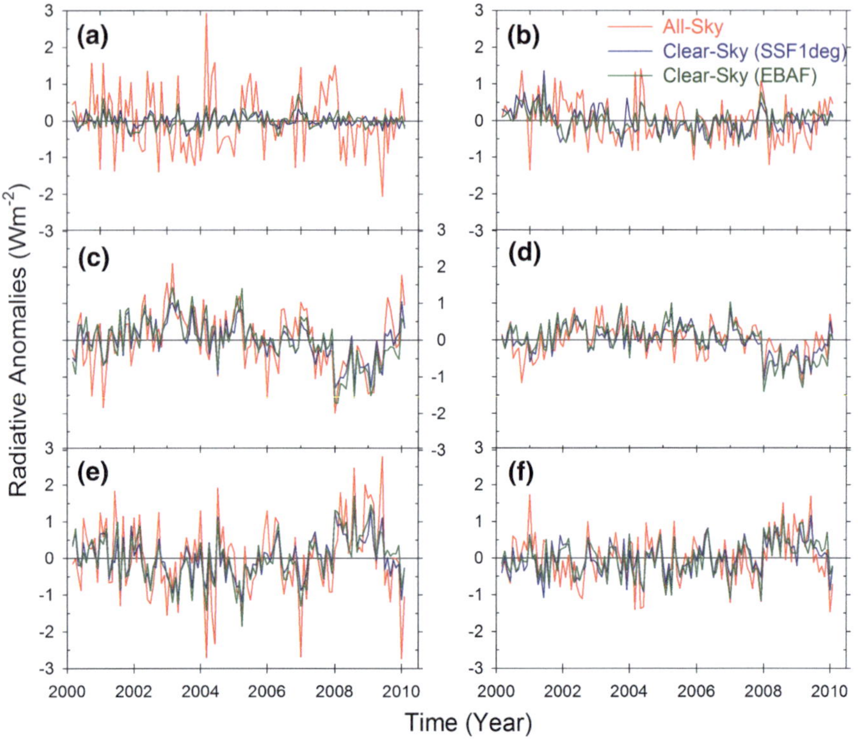

Fig. 7 CERES Terra deseasonalized monthly anomalies for SW (**a** and **b**), LW (**c** and **d**), Net (**e** and **f**) TOA radiation from March 2000–February 2010. *Left column* corresponds to 30°S–30°N and *right column* corresponds to global

clear-sky anomalies varies from one ENSO event to another, however. Average LW cloud radiative effects (CRE) are pronounced during the 03/00–03/02 La Niña event, as well as during the 05/06–05/07 and 05/09–02/10 El Niño events. In contrast, average LW CREs are relatively weak during the 06/07–04/09 La Niña and 04/02–09/05 El Niño events. Average clear-sky anomalies during the 06/07–04/09 La Niña are a factor of 2 larger than in any other ENSO interval during the decade.

Table 4 provides a summary of anomalies in cloud fraction, cloud-top pressure and visible cloud optical depth from MODIS. In two of the three El Niño events, mean cloud fraction anomalies are near-zero, and cloud-top pressure anomalies are positive. At the peak of the El Niño events (early 2003, 2005, 2007), LW flux contributions from cloud fraction and cloud-top pressure are both positive (Fig. 8g) (reducing LW CRE). During the 05/09–02/10 El Niño event, the decrease in LW CRE is associated with a sharp decline in cloud cover of −0.8%, which is partially compensated for by an increase in cloud-top height (Fig. 8g). Average cloud fraction anomalies are near-zero or positive during La Niña events (Table 4), and in 2 of the 3 events, moderate negative cloud-top pressure anomalies are observed.

In the tropics, SW CREs (Table 3) dominate over clear-sky radiative effects by at least a factor of 3, and SW CRE anomalies are positive in all 3 El Niño events and negative in 2 of the 3 La Niña events. The strongest El Niño event (05/09–02/10) is associated with the greatest change in both cloud fraction and cloud optical depth (both showing decreases)

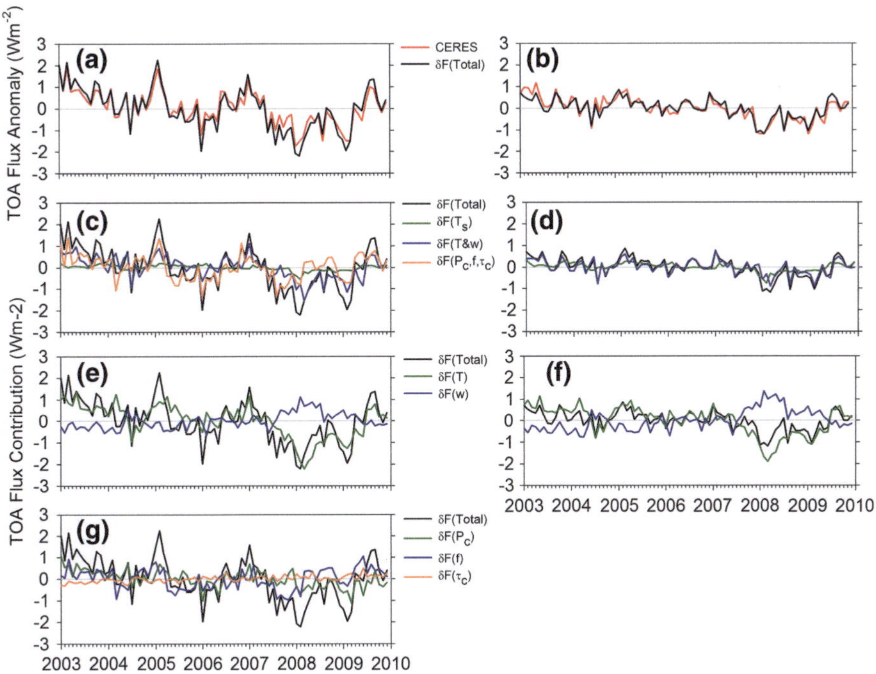

Fig. 8 CERES and calculated monthly anomalies in tropical mean LW TOA flux for **a** all-sky and **b** clear-sky; contribution to LW TOA flux from surface temperature (*dF*(*Ts*)), atmospheric temperature and humidity (*dF*(*T*&*w*)), clouds (P_c, *f*, t_c) for **c** all-sky and **d** clear-sky; contribution to LW TOA flux from *T* and *w* for **e** all-sky and **f** clear-sky; contribution to LW TOA flux from P_c, *f* and t_c

(Table 4). Average net CRE anomalies are near-zero in all but two ENSO events. During the 10/05–04/06 La Niña event, reduced SW cloud radiative cooling combined with enhanced LW CRE anomalies produce a net CRE of 0.39 Wm^{-2}. In contrast, during the 05/06–05/07 El Niño, the tropical net CRE anomaly reaches −0.34 Wm^{-2} due to a large negative LW CRE anomaly and a near-zero SW CRE. Thus, while the net radiative effect of clouds is that of warming (cooling) across the tropics during La Niña (El Niño) events, the magnitude is quite small and varies greatly from one event to another.

Results in Tables 3 and 4 are generally consistent with a recent study by Zelinka and Hartmann (2011) who used satellite observations to examine the relationship between tropical mean CRE and surface temperature anomalies. They find that LW CRE decreases with surface temperature, and SW CRE increases. This is consistent with negative (positive) LW CRE anomalies and positive (negative) SW CRE anomalies during El Nino (La Niña) (Table 3). Further, Zelinka and Hartmann (2011) show that while high clouds rise with warming, in agreement with the fixed anvil temperature (Hartmann and Larson 2002) and proportionately higher anvil temperature (Zelinka and Hartmann 2010) hypotheses, the increase is compensated by a decrease in high cloud amount. This is qualitatively consistent with a decrease (increase) in both cloud fraction and cloud-top pressure during El Niño (La Niña) (Table 4 and implied from Fig. 8g). Furthermore, the SW radiative cooling effect of clouds decreases in warmer conditions, owing to a decrease in tropical mean cloud fraction.

Results in the present study and Zelinka and Hartmann (2011) show a tendency for a negative LW and positive SW cloud feedback response to ENSO in the tropics. This is

Table 3 Summary of cloud and clear-sky radiative effect by MEI interval (Wm^{-2})

Date	MEI	SW (All)	SW (Clr)	SW CRE	LW (All)	LW (Clr)	LW CRE	NET (All)	NET (Clr)	NET CRE
Tropics (30°S–30°N)										
03/00–03/02	−0.25	**0.18**	−0.02	**−0.21**	**−0.21**	0.06	**0.27**	**0.22**	**0.16**	0.06
04/02–09/05	0.55	−0.04	0.02	0.06	**0.40**	**0.32**	−0.08	**−0.33**	**−0.32**	−0.01
10/05–04/06	−0.46	−0.15	0.03	**0.19**	**−0.38**	**−0.18**	**0.20**	**0.47**	0.08	**0.39**
05/06–05/07	0.61	−0.03	0.02	0.05	**0.40**	0.01	**−0.39**	**−0.44**	−0.10	**−0.34**
06/07–04/09	−0.67	0.06	−0.02	−0.08	**−0.78**	**−0.61**	0.16	**0.56**	**0.48**	0.08
05/09–02/10	0.96	**−0.28**	**−0.05**	**0.23**	**0.38**	0.04	**−0.34**	**−0.21**	−0.10	−0.11
Global										
03/00–03/02	−0.25	**0.32**	**0.24**	−0.09	**−0.18**	0.04	**0.21**	0.03	−0.10	0.13
04/02–09/05	0.55	−0.04	−0.02	0.02	**0.21**	**0.20**	−0.01	−0.15	−0.16	0.00
10/05–04/06	−0.46	−0.18	**−0.26**	−0.07	−0.07	0.13	0.20	0.20	0.07	0.13
05/06–05/07	0.61	−0.21	−0.16	0.04	**0.27**	0.11	−0.16	−0.14	−0.02	−0.11
06/07–04/09	−0.67	−0.14	−0.05	0.10	**−0.28**	**−0.40**	−0.12	**0.30**	**0.32**	−0.02
05/09–02/10	0.96	0.08	0.00	−0.08	−0.10	**−0.24**	−0.14	−0.07	0.14	−0.21

Each value corresponds to the mean anomaly for the indicated time period. *Bold values* exceed the standard error in the mean

Table 4 Summary of tropical mean (30°S–30°N) cloud property changes by MEI interval

Date	MEI	Cloud fraction (%)	Cloud-top pressure (hPa)	Cloud optical depth
03/00–03/02	−0.25	0.095	**1.4**	**0.041**
04/02–09/05	0.55	0.03	**1.5**	**0.026**
10/05–04/06	−0.46	**0.38**	**−4.2**	−0.013
05/06–05/07	0.61	0.13	**1.7**	**−0.034**
06/07–04/09	−0.67	0.00	**−3.1**	**−0.020**
05/09–02/10	0.96	**−0.80**	−1.8	**−0.11**

Each value corresponds to the mean anomaly for the indicated time period. *Bold values* exceed the standard error in the mean

consistent with Lin et al. (2002) based upon CERES Tropical Rainfall Measurement Mission (TRMM) data. Trenberth et al. (2010) find the overall feedback to be positive in the tropics. Globally, Dessler (2010) finds positive cloud feedbacks both for SW and LW after accounting for changes in clear-sky SW and LW properties masked by clouds. Positive SW cloud feedback is also observed in Clement et al. (2009) and Eitzen et al. (2011), who focused on marine boundary layer regimes.

3.3 ENSO Influence on TOA Radiation Variability Regional Scales

The regional pattern of TOA radiation, cloud property and skin temperature anomalies during the first 2 months of 2008 at the height of the 2008 La Niña event are provided in Figs. 9 and 10. During this period, negative SST anomalies persist across the equatorial Pacific basin, consistent with expectation during a La Niña event. Negative SST anomalies

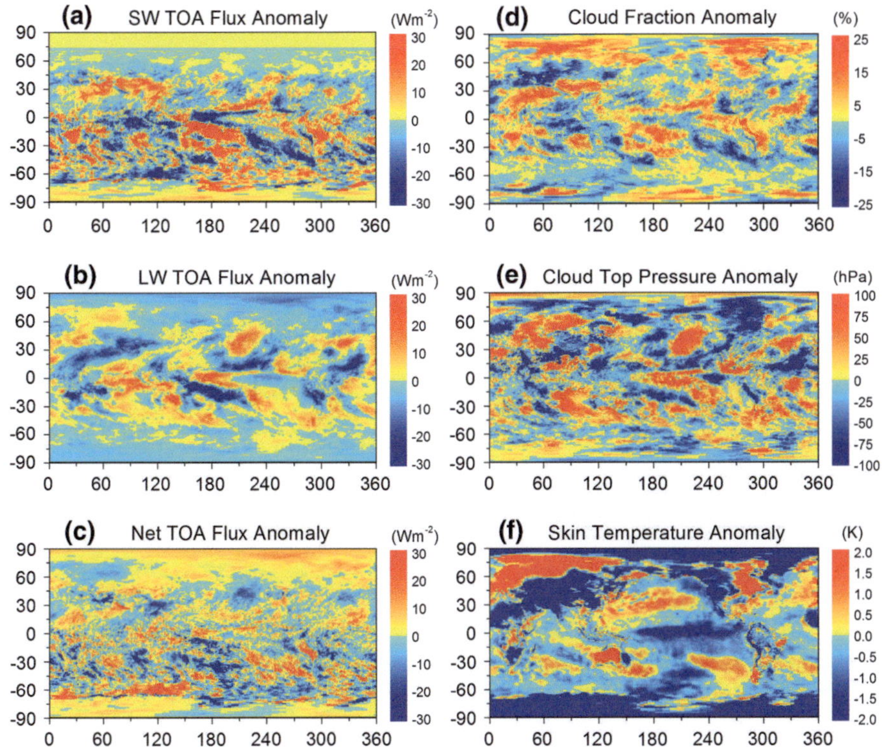

Fig. 9 Regional anomalies (relative to March 2000–February 2010 climatology) for January 2008. **a** SW TOA flux; **b** LW TOA flux; **c** net TOA flux; **d** cloud fraction; **e** cloud-top pressure; **f** skin temperature

are also found along the west coasts of the Americas, and positive anomalies occur in the west and central Pacific Ocean between 30°N and 45°N. This pattern is consistent with a negative Pacific Decadal Oscillation (PDO) pattern—in fact, 2008 saw the lowest PDO index since 1971 (Xue and Reynolds 2009). Regional patterns of CERES SW (Figs. 9a, 10a) and LW (Figs. 9b, 10b) TOA flux anomalies are closely linked to patterns of MODIS cloud fraction (Figs. 9d, 10d) and cloud-top pressure (Figs. 9e, 10e), respectively. Positive anomalies in cloud fraction and negative anomalies in cloud-top pressure likely due to stronger than normal low-level convergence in the South Pacific Convergence Zone (SPCZ) are associated with positive anomalies in reflected SW TOA flux and negative anomalies in LW TOA flux, respectively. In the North Pacific and Atlantic Oceans between 30°N and 60°N in January (Fig. 9), large areas of positive cloud-top pressure anomalies are closely tied to positive anomalies in LW TOA flux. Because SW TOA flux anomalies associated with these features are weak, these features show up quite clearly as areas of negative net TOA flux anomalies in Fig. 9c. Otherwise, with the exception of a few areas of extensive negative net TOA flux anomalies in the Southern Oceans associated with positive cloud fraction anomalies, most of the globe shows weak positive net TOA flux anomalies.

There is a striking difference between the January and February 2008 regional pattern of cloud property and TOA radiation anomalies. In February, a pronounced dipole pattern appears in which negative outgoing LW (positive SW) TOA flux anomalies occur over the Indian Ocean extending eastward to Indonesia and over the SPCZ, and positive outgoing

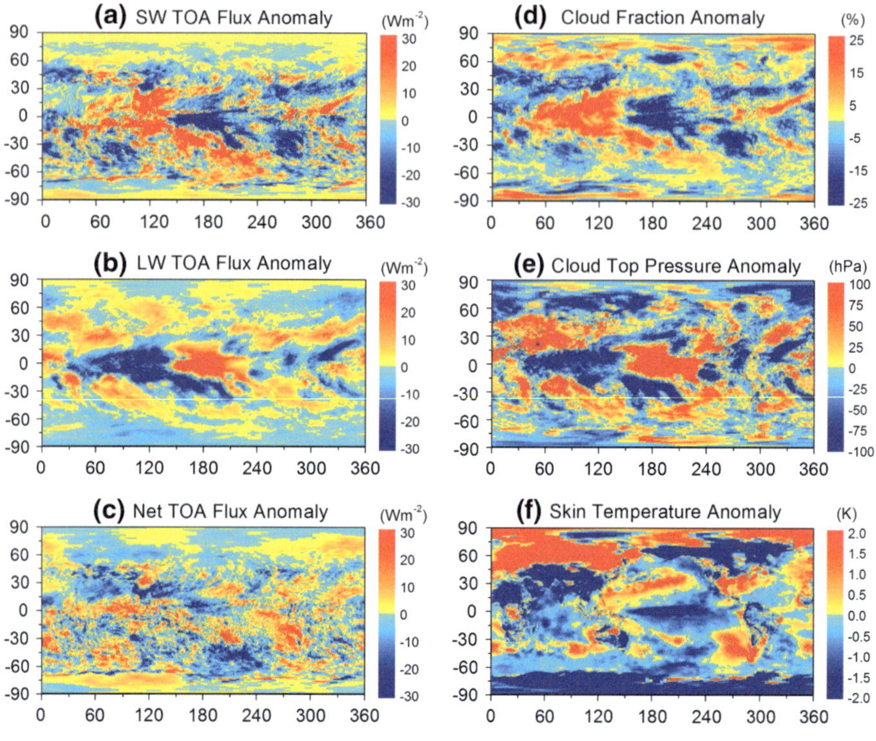

Fig. 10 Same as Fig. 9 but for February 2008

LW (negative SW) TOA flux anomalies occur over the Central equatorial Pacific Ocean. Consistent with this pattern, cloud fraction and cloud-top height increase to the west, while over the Central Pacific Ocean cloud fraction and cloud-top height decrease. The February 2008 pattern is consistent with expectation for a La Niña event. In January, the dipole pattern is much weaker. LW TOA flux anomalies over the Indian Ocean and Indonesia are positive, and over the Central Pacific Ocean, there is a mix of both weak positive and negative cloud fraction anomalies. This occurs in spite of the fact that January 2008 falls in the middle of the 06/2007–04/2009 La Niña event.

What happened to the convection over the Indian Ocean and Indonesia during January 2008? It turns out that, during January, this region corresponds to a period in which convection associated with the MJO was out of phase with La Niña-related convection. Gottschalk and Bell (2009) show that the 200-hPa velocity potential anomaly over this region is positive indicating upper-level convergence and suppressed convection. In contrast, during February, MJO and La Niña-related convection are in phase, resulting in intense convection. This interpretation and the CERES and MODIS results in Figs. 9 and 10 are corroborated by CALIPSO and Cloudsat observations of the difference between the cloud frequency of occurrence by altitude from 2008 and 2007 for January (Fig. 11a) and February (Fig. 11b) as a function of longitude across the Pacific Ocean between 0°S and 2.5°S. Results in Fig. 11 are from a merged satellite dataset that combines CALIPSO, Cloudsat, CERES and MODIS (C3M) (Kato et al. 2010). Negative anomalies in cloud frequency throughout the troposphere are seen between 90°E and 120°E in January 2008 compared to January 2007 (Fig. 11a). In February 2008, the convection reappears and

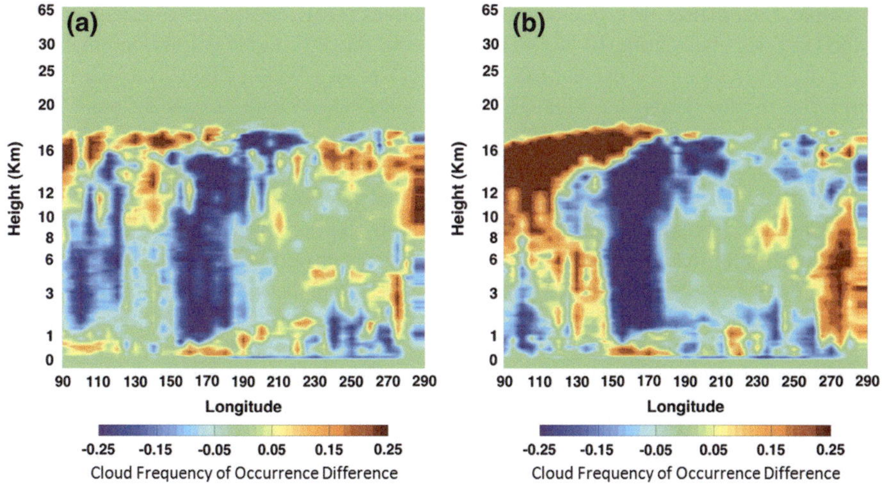

Fig. 11 Longitude-height cross-section of January 2008 minus January 2007 cloud frequency of occurrence difference for 0°S–2.5°S: **a** January 2008 minus January 2007; **b** February 2008 minus February 2007

covers a large area contributing to strong negative SW and positive LW TOA flux anomalies (Fig. 10a, b). In both January and February, the negative difference in cloud frequency between 150°E, and the dateline is associated with the westward displacement of convection that is typical of La Niña events. This area coincides with negative cloud fraction anomalies in Figs. 9d and 10d.

4 Summary and Conclusions

There are now over a decade of cloud, aerosol, radiation and atmospheric state observations from advanced satellite instruments aboard the Terra and Aqua spacecrafts, and close to 5 years of vertical cloud and aerosol profile data from CloudSat and CALIPSO. These data provide a rich resource for studying ERB variations and the underlying processes. With the latest CERES calibration improvements, large-scale top-of-atmosphere (TOA) radiation changes during the past decade are observed to within 0.5 Wm^{-2} per decade based upon comparisons between CERES Terra and Aqua, and between CERES and SeaWIFS, MODIS and AIRS. Monthly anomalies between the various datasets are highly correlated (coefficient of determination, r^2, is generally >0.85).

The first decade of the Terra record was characterized by relatively modest La Niña and El Niño conditions during the first 7 years followed by prolonged strong La Niña conditions between June 2007 and April 2009, and a strong El Niño event immediately after. CERES outgoing LW radiation closely tracks with the Multivariate ENSO Index (MEI). During El Niño conditions outgoing LW flux increases, and decreases during La Niña conditions. Net incoming TOA flux was positive during the 2007–2009 La Niña conditions, primarily due to a pronounced decrease in outgoing LW TOA flux in the tropics in both cloud-free and all-sky conditions. This period also coincides with strong cooling and drying throughout the troposphere as captured by AIRS temperature profile data. Global outgoing LW TOA flux anomalies in late 2007/early 2008 are weaker than in the tropics due to compensation by clouds at middle latitudes: MODIS cloud fraction and MISR cloud-top height anomalies are both negative, resulting in positive middle latitude LW TOA flux anomalies.

Tropical anomalies in LW cloud radiative effect (CRE) are positive during La Niña events and negative during El Niño conditions. In the SW, cloud effects dominate by a factor of 3 over anomalies in clear-sky TOA flux. The net effect of clouds is to warm (cool) across the tropics during La Niña (El Niño). A striking feature is how different ENSO events are from one another.

The Terra, Aqua and other A-Train data provide excellent coverage of regional TOA flux and cloud property changes in response to atmosphere-ocean dynamics associated with ENSO and MJO events. The spatial and vertical distribution of clouds and radiation over the Indian Ocean and Indonesia when MJO convection is out of phase with La Nina convection in January 2008, and in phase in February 2008 is clearly captured by CERES, MODIS, CALIPSO and Cloudsat observations. Such dramatic events captured so clearly by these advanced instruments provides a wealth of new data for climate model evaluation and improvement.

While the new satellite instruments discussed in this study have clearly advanced the state-of-the-art in cloud-radiation observational capabilities, there is a critical need to extend the length of these records over multiple decades and further improve their accuracy in order to quantify how clouds are changing in a warmer climate and how cloud changes impact the Earth's radiation budget. One key observational requirement to address long-term climate change is an improvement in instrument calibration, particularly for the imager and radiation budget instruments. While the estimated stability of the CERES TOA radiation record of roughly 0.5 Wm^{-2} per decade is a factor of 3–4 better than anticipated prior to the launch of CERES, there is a need for another factor of 2–3 improvement in order to constrain cloud feedback.

Acknowledgments We thank the CERES science, algorithm and data management teams and the NASA Science Mission Directorate for supporting this research. CERES SSF1deg-lite_Ed2.5 data were obtained from http://ceres.larc.nasa.gov/compare_products.php. The NASA Langley Atmospheric Sciences Data Center processed the instantaneous Single Scanner Footprint data used to produce SSF1deg-lite_Ed2.5. We thank Drs. Mark Zelinka and Dennis Hartmann for sharing their perspectives on short-term cloud-radiation variability and cloud feedback. We thank Dr. Dale Walikainen for providing the results used in Fig. 17.

Appendix 1: CERES Instrument Calibration Improvements

The primary purpose of an instrument's onboard calibration subsystem is to enable calibration scientists to detect, quantify and correct for changes in instrument sensitivity throughout the mission so that subtle changes in the climate system can be unambiguously detected. To ensure that ERB are accurate and consistent over the entire record, there is a need to periodically reprocess the entire record to account for instrument calibration drift too small to observe with a short record. The CERES team has recently conducted a major re-analysis of ground and in-flight calibration data and has implemented a new algorithm to characterize the spectral degradation of the CERES optics with time.

The two main quantities required in order to convert CERES detector output signals to unfiltered radiances are the instrument gain coefficients and spectral response functions. Both are determined prior to launch in a radiometric calibration facility. The primary in-flight calibration system used to detect drifts in the CERES sensor gains is the Internal Calibration Module (ICM). ICM calibrations are performed on-orbit weekly to determine monthly gains. To reduce noise, a 5-month running mean is used. Recently, Shankar et al. (2010) re-evaluated the ground calibration data collected prior to CERES Terra and Aqua launches and derived new pre-launch calibration data. Thomas et al. (2010) summarize results of a reanalysis of CERES gain changes through March 2010.

Detecting in-orbit changes in spectral response function is more challenging as there are no onboard calibration sources that cover the entire spectral range observed by CERES. This is problematic as it is well established that instrument UV exposure and molecular contamination cause a loss of measurement sensitivity with time, particularly in the blue end of the solar spectrum (Clark and Dibattista 1978). Indeed, early analyses of the CERES SW data revealed significant decreases in albedo that were more pronounced for clear ocean scenes than cloudy scenes. Further, direct comparison of coincident CERES nadir radiances from CERES instrument pairs on both platforms were found to depend strongly on the instrument scan mode (Matthews et al. 2005). The primary mode of operation of CERES instruments is the crosstrack mode (i.e., perpendicular to the groundtrack), but CERES instruments can also be operated in a rotating azimuth plane mode (RAP), where the scan head rotates in azimuth as it scans in elevation. The direct comparison analyses showed that the degradation was more pronounced when the instruments were operated in RAP mode. This feature was subsequently used to derive Edition2 user-applied revisions to the RAP instrument for clear ocean and all-sky scenes. Subsequently analyses (Loeb et al. 2007) of SW fluxes showed a significant overall improvement for CERES Terra, but CERES Aqua still appeared to exhibit an artificial drift. In addition, the CERES team also found systematic daytime–nighttime LW TOA flux inconsistency.

TOA fluxes in CERES SSF1deg-lite_Ed2.5 are derived using a more comprehensive approach to correct for temporally varying spectral changes in SW and LW radiance measurements. Based upon data collected by prior missions, Matthews (2009) assumed the following expression for the loss of transmission with wavelength, or spectral degradation, in the SW:

$$\mathrm{Tr}(\lambda) = 1 - Me^{-\alpha\lambda} \tag{1}$$

where M and α are coefficients of the fit to the data. Assuming Eq. 1 represents the shape of the change in spectral response, the direct comparison method can be used to infer wavelength-dependent changes in SW spectral response for each instrument pair on Terra and Aqua. We use $M = 1$ in Eq. 1, and derive an α value in each month of the Terra and Aqua missions by assuming spectral darkening in the SW only occurs for the instrument in RAP mode. The RAP instrument's α in a given month is selected from a set of candidate values to ensure that the ratio of unfiltered radiances (FM2/FM1 or FM3/FM4) for clear ocean scenes identified using the CERES cloud mask applied to MODIS radiances for 30°S and 30°N remains close to 1. The spectral darkening correction to the RAP instrument is thus only done relative to the crosstrack instrument. However, because modifying α changes the shape of the spectral response function (SRF), the current approach is wavelength dependent, thereby eliminating the need for separate scene-type-dependent correction factors. Spectral response functions (and associated spectral correction coefficients) for up to 83 α candidate values are considered in this procedure. Figure 12a–d show the spectral shape of the changes to the SRFs for each CERES instrument throughout the Terra and Aqua missions. Results for FM4 are available only from July 2002 through March 2005 due to an anomaly in the SW channel. FM2 and FM3 show the greatest spectral darkening because they have spent more time in RAP mode. Changes in spectral

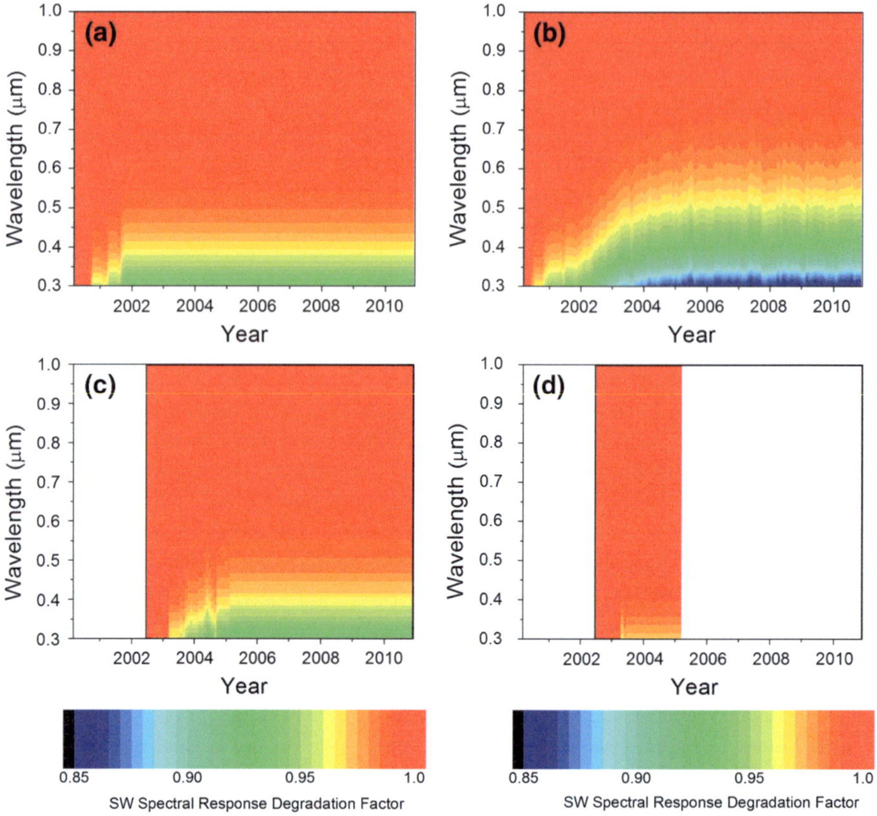

Fig. 12 SW spectral response degradation factor for **a** FM1, **b** FM2, **c** FM3, and **d** FM4

response reach 15% between 0.3 and 0.4 μm for FM2 and FM3, but remain <10% for FM1 and FM4.

Unfiltered CERES emitted LW radiances are determined directly from filtered total (TOT) channel measurements at night. Since the TOT channel radiance is comprised of both SW and LW contributions during daytime, unfiltered LW radiance is determined from TOT and SW filtered radiance measurements (Loeb et al. 2001). A drift in either the SW or LW portions of the TOT channel or in the SW channel can result in a drift in unfiltered LW radiance. Early validation of the CERES data showed a marked day–night bias with time in unfiltered LW radiance. This is illustrated in Fig. 13, which shows all-sky global mean daytime and nighttime LW TOA fluxes for FM1 Terra from the CERES Edition2 release. In Edition2, temporally varying gains are used, and there is also a correction in spectral response for the SW part of the TOT channel based upon a separate analysis using a 3-channel comparison with deep convective clouds. Unfortunately, because of the spectral difference between DCCs and all-sky radiances, adjustments to the SW part of the TOT based upon 3-channel consistency for DCCs does not translate to 3-channel consistency for all-sky. As a result, a trend of -2.7 Wm^{-2} per decade in day–night difference is observed for Edtion2, which also affects the day/night average LW trend.

In order to quantify the drift in the SW portion of the TOT channel, Fig. 14 compares a time series of FM1 day–night LW TOA flux differences between 30°S and 30°N over ocean

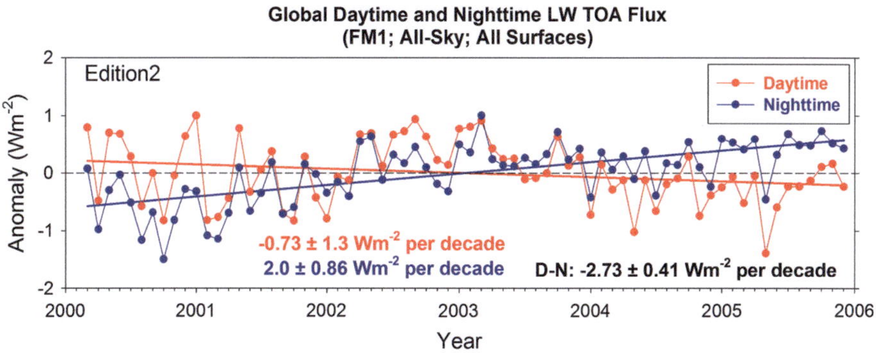

Fig. 13 All-sky global mean daytime and nighttime LW TOA flux anomalies for FM1 Terra from the CERES Edition2 release

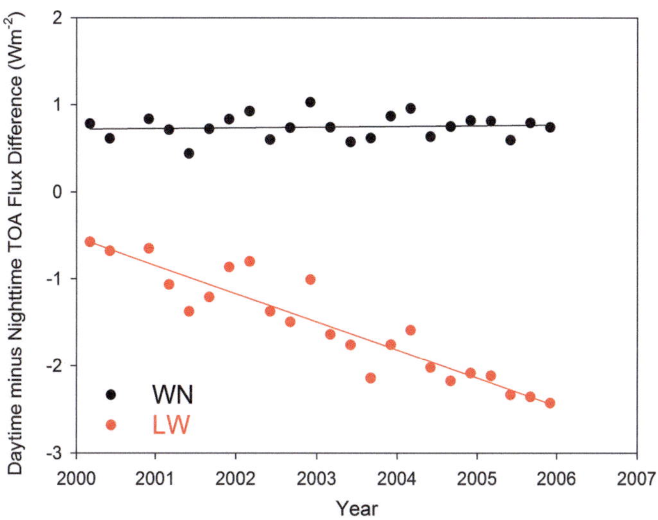

Fig. 14 FM1 day–night WN and LW TOA flux differences between 30°S and 30°N over ocean. For LW, the latest gains and SW spectral response function corrections are applied, but no change is made to the TOT spectral response function

after applying the latest gains and SW spectral response function correction with day–night TOA flux differences derived from WN channel radiances. The LW daytime–nighttime TOA flux differences show a decrease of 2 Wm^{-2} over 6 years, while WN day–night difference remain relatively constant. A 2 Wm^{-2} decrease in daytime–nighttime LW TOA flux difference exceeds expectation by an order-of-magnitude based upon observed temperature-humidity changes by AIRS level 2 data. Between 2002 and 2005, the change in the daytime–nighttime difference in temperature is $\approx$0.4 K for layers above 10 hPa, which corresponds to a negligible TOA LW flux change. The change in water vapor mixing ratio is 0.05 g kg^{-1} below 700 hPa, corresponding to a 0.2 Wm^{-2} LW TOA flux change.

An alternate representation of the discrepancy between LW and WN day–night differences is provided in Fig. 15, which shows the relationship between 1° zonal average

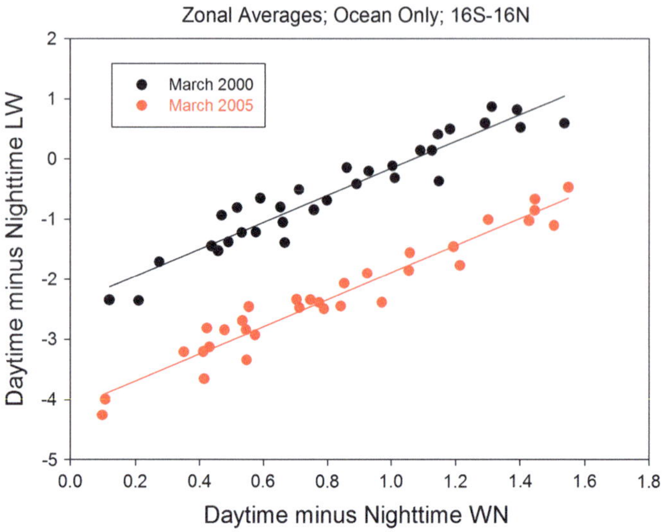

Fig. 15 Relationship between 1° zonal average LW and WN day–night unfiltered radiance differences from CERES FM1 for March 2000 and March 2005 over ocean between 16°S and 16°N

LW and WN day–night unfiltered radiance differences from CERES FM1 for March 2000 and March 2005 over ocean between 16°S and 16°N. While the slopes are the same for these 2 months, the March 2005 regression line is offset from that in March 2000. As an independent test of the temporal stability of the relationship between LW and WN daytime–nighttime TOA flux differences, we compared daytime–nighttime radiance differences for spectrally integrated AIRS radiances over the entire spectral range observed by AIRS and integrated AIRS radiances between 8.1 and 11.8 μm (CERES WN channel) for 6 September months between 2002 and 2008 for tropical ocean scenes. We find that the daytime–nighttime radiance differences for these two spectral ranges are highly linear and repeatable from year-to-year, and the linear regression coefficients show no statistically significant changes for this period.

To compensate for the day–night drift in CERES LW unfiltered radiance, we adjust the SW portion of the TOT channel spectral response function to remove the offset from mission start in the day–night LW and WN regression relation. The adjustments to the SW portion of the TOT channel are derived for each month following the same approach as for the SW channel based upon Eq. 1. Figure 16a, b show deseasonalized daytime and nighttime LW TOA flux anomalies for 30°S–30°N for the crosstrack CERES Aqua (Fig. 16a) and Terra (Fig. 16b) after adjusting the SW portion of the TOT spectral response function in each month. Daytime and nighttime LW TOA flux anomalies show consistent changes with no significant day–night drift.

Appendix 2: Reason for CERES Terra and Aqua LW Drift

Comparisons between CERES Terra and Aqua for July 2002–February 2010 (Table 1) show SW "trend" consistency of $0.04 \pm 0.27\ \mathrm{Wm^{-2}}$ in the tropics and $0.12 \pm 0.25\ \mathrm{Wm^{-2}}$ for the globe. In the LW, differences are much larger: the Terra minus Aqua difference is $-0.39 \pm 0.21\ \mathrm{Wm^{-2}}$ for 30°S–30°N, and for the globe, it is –

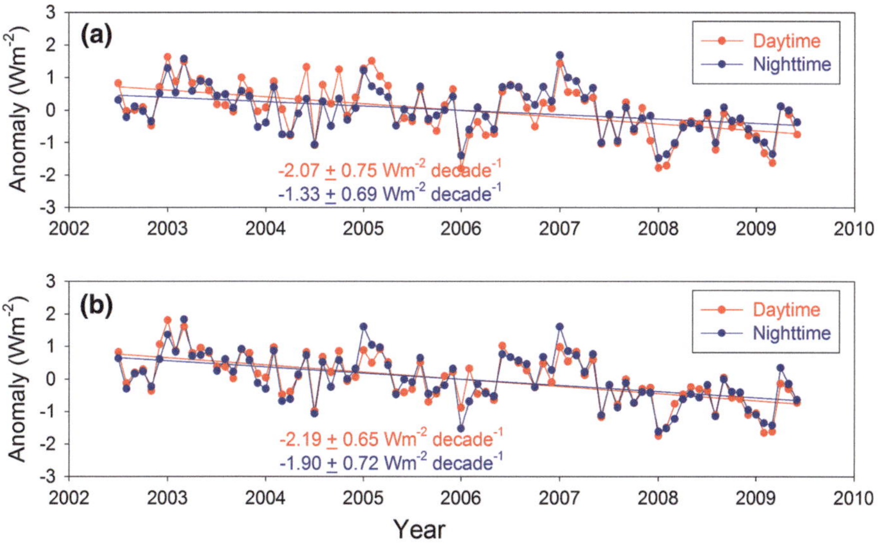

Fig. 16 Deseasonalized daytime and nighttime LW TOA flux anomalies for 30°S–30°N for crosstrack CERES **a** Aqua and **b** Terra after correcting for changes in SW and TOT channel spectral response function changes

0.63 ± 0.36 Wm^{-2}. This appendix provides an explanation why differences are so much larger in the LW.

To understand the cause of the LW discrepancy, it is helpful to consider some background information. Terra and Aqua carry two CERES instruments each. Early in both missions, one instrument on each spacecraft was placed in crosstrack mode to optimize spatial sampling, and the other instrument was either in a rotating azimuth plane (RAP) scan mode or alongtrack (AT) mode. Only crosstrack CERES data are used in producing Level-3 gridded data products because crosstrack provides global coverage daily. CERES RAP data are used to optimize angular sampling in order to construct CERES angular distribution models, and CERES AT data are used for validating the CERES angular models. To ensure the gimbals on each instrument were fully exercised, each instrument's scan mode was alternated from RAP to crosstrack and back on a 3-month cycle during the first 2 years of Terra and first year of Aqua. Eventually, FM1 on Terra and FM4 on Aqua were permanently placed in crosstrack mode, and FM2 and FM3 in RAP/AT mode. It was later realized that operating the instruments in RAP mode exposes the optics to increased UV exposure and molecular contamination. Preventative measures were taken by restricting the instruments from scanning in the RAM direction.

Both CERES instruments on Terra have been functioning nominally since the beginning of the mission in December 1999. With the exception of 10 months (04/00; 08/00–10/00; 02/01–04/01; 08/01–10/01; 01/06–02/06), FM1 has been in crosstrack mode throughout the Terra mission. Unfortunately, the SW channel of CERES FM4 on Aqua failed in March 2005. Consequently, with the exception of 6 months (11/02–01/03; 05/03–07/03), FM4 was the main crosstrack instrument from July 2002–March 2005, and FM3 has been in crosstrack mode from April 2005 onwards. The CERES team placed all 4 CERES instruments (FM1–FM4) on the same radiometric scale at their respective beginning of missions (Szewczyk et al. 2011). A one-time adjustment to FM2 was applied to match FM1 in March of 2000. Similarly, adjustments to FM3 and FM4 were made in July 2002 to match FM1. All-sky

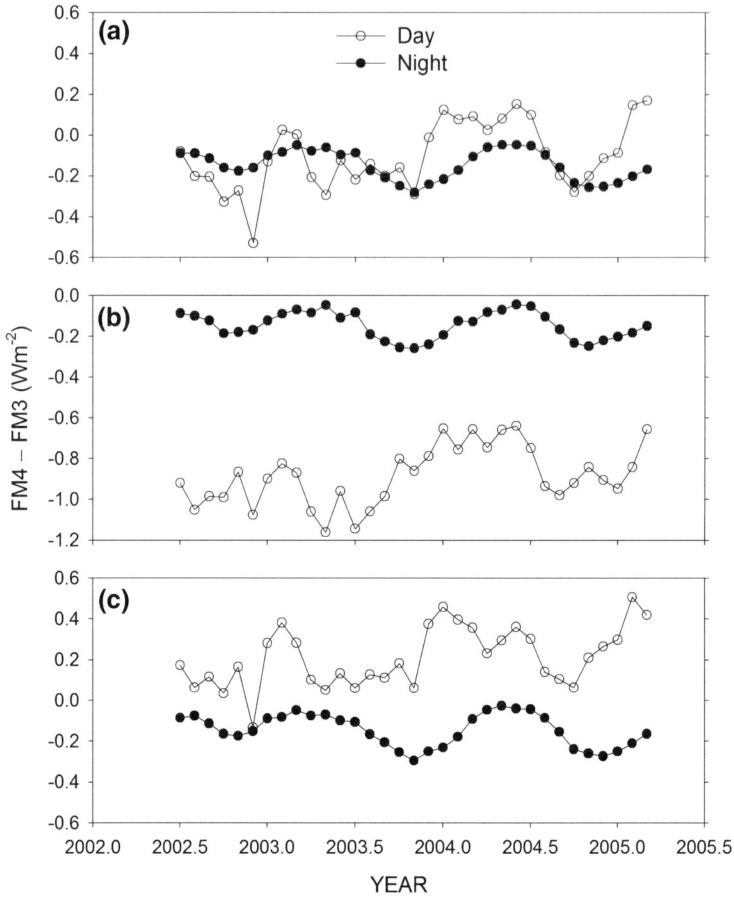

Fig. 17 Mean difference between FM4 and FM3 coincident nadir LW footprint pairs for all-sky conditions during daytime and nighttime for **a** all surfaces; **b** land only and **c** ocean only

data over all surface types were used to derive the radiometric scale adjustments. Each instrument is fully autonomous after that without any further corrections.

As a check for any discontinuities in the transition between FM4 and FM3, TOA fluxes from the two CERES instruments on Aqua are directly compared with one another at nadir over the exact same area and at the same time. Figure 17a shows FM3 and FM4 LW direct comparison results during nighttime and daytime for all-sky conditions over all surface types. At night, the FM4–FM3 difference at the beginning of the mission is -0.1 Wm^{-2} and grows to -0.2 Wm^{-2} by March 2005. Because of this difference, continuing the FM4 record with FM3 in April 2005 introduces a spurious 0.2 Wm^{-2} increase in the middle of the record, which contributes to the negative slope in the Terra minus Aqua anomaly difference. During daytime, FM4 and FM3 are closer to one another at the time of the FM4 to FM3 transition, but this is largely due to compensation between land and ocean, as shown in Fig. 17b, c, which provide FM4–FM3 direct comparison results for land and ocean, respectively. During daytime, the FM3 LW TOA flux exceeds FM4 by $\approx$0.8 Wm^{-2} over land/desert (Fig. 17b), while FM3 is lower than FM4 by $\approx$0.3 Wm^{-2} over ocean (Fig. 17c).

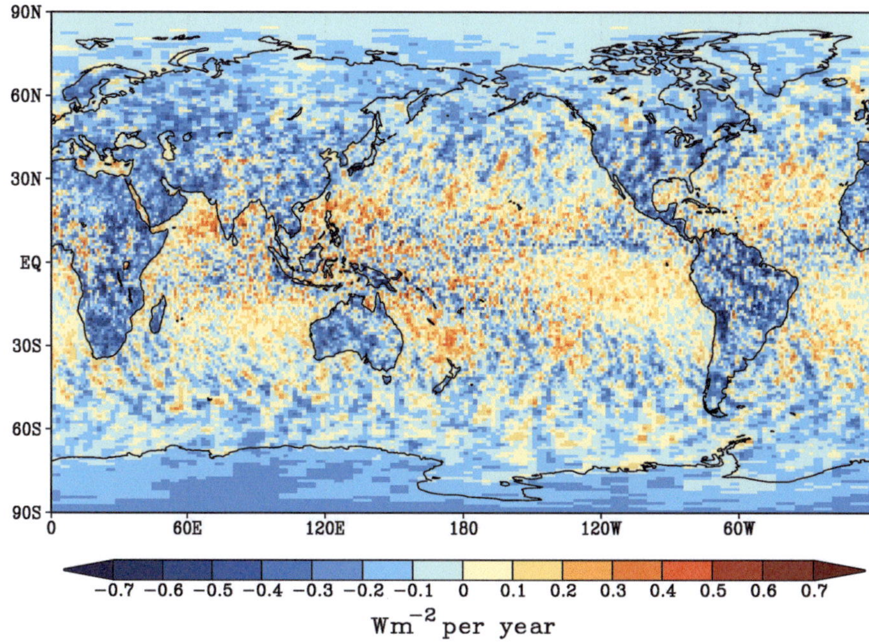

Fig. 18 Slope of Terra minus Aqua all-sky LW TOA flux anomalies for July 2002–June 2009 from SSF1deg-lite_Ed2.5 (units: Wm^{-2} per year)

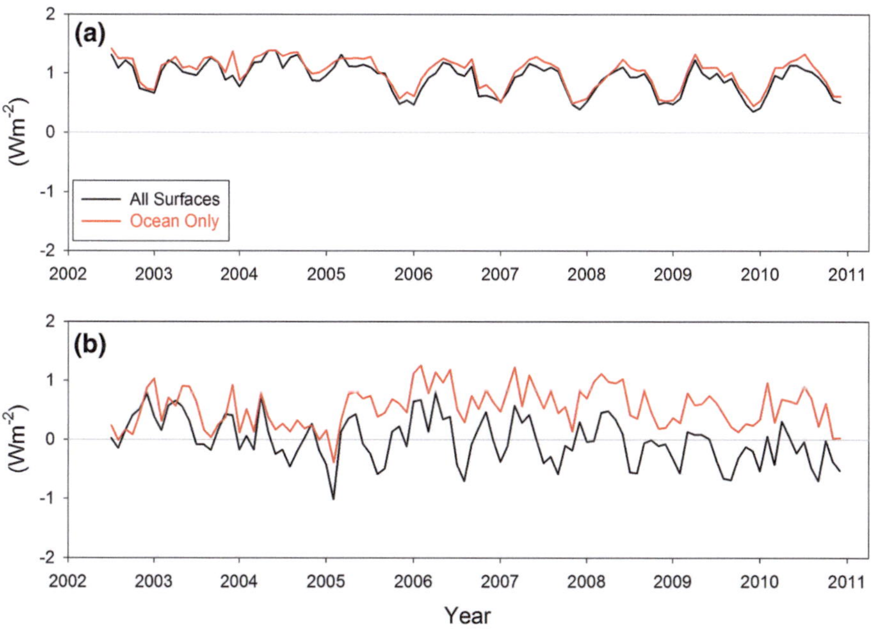

Fig. 19 Terra minus Aqua all-sky LW TOA flux difference over 60°S–60°N for **a** nighttime and **b** daytime based on the SSF1deg-lite_Ed2.5 product

These marked discontinuities between FM3 and FM4 cause large regional differences between Terra and Aqua anomalies. Figure 18 shows a regional map of the slope of Terra minus Aqua all-sky LW TOA flux anomaly difference from the CERES SSF1deg-lite_Ed2.5 data product. Consistent with results in Fig. 17a–c, the slope is negative over land and is mostly positive over ocean. An alternate representation is provided in Fig. 19a, b, which shows Terra minus Aqua all-sky LW TOA flux differences for 60°S–60°N at night (Fig. 19a) and during daytime (Fig. 19b). At night, Terra LW TOA fluxes exceed Aqua fluxes by ≈ 1 Wm^{-2} for both ocean and all surface types, whereas the daytime results depend on surface type. There is also a noticeable jump in the ocean-only LW TOA flux difference (red line in Fig. 19b) in spring 2005 at the time when the Aqua crosstrack instrument transitions from FM4 to FM3.

The CERES team is currently developing an improved approach for placing FM3 and FM4 on the same radiometric scale that removes the large scene-type inconsistencies found in Figs. 17, 18, and 19. The CERES team anticipates reprocessing the Aqua record with this improvement in a future release.

References

Allan RP, Slingo A (2002) Can current climate model forcings explain the spatial and temporal signatures of decadal OLR variations? Geophys Res Lett 29(7):1141–1144

Barnes WL, Pagano TS, Salomonson VV (1998) Prelaunch characteristics of the moderate resolution imaging spectroradiometer (MODIS) on EOS-AM1. IEEE Trans Geosci Remote Sens 36:1088–1100

Clark LG, Dibattista JD (1978) Space qualification of optical-instruments using NASA long duration exposure facility. Opt Eng 17:547–552

Clement AC, Burgman R, Norris JR (2009) Observational and model evidence for positive low-level cloud feedback. Science 325:460–464

Dessler AE (2010) A determination of the cloud feedback from climate variations over the past decade. Science 330(6010):1523–1527. doi:10.1126/science.1192546

Di Girolamo L, Menzies A, Zhao G, Mueller K, Moroney C, Diner DJ (2010) Multi-angle imaging spectroradiometer (MISR) level 3 cloud fraction by altitude algorithm theoretical basis. JPL D-62358, Jet Propulsion Laboratory, Pasadena, 23 pp. Available from: http://eospso.gsfc.nasa.gov/eos_homepage/for_scientists/atbd/docs/MISR/MISR_CFBA_ATBD.pdf

Doelling DR, Keyes DF, Rutan DA, Nordeen ML, Morstad D, Sun M, Loeb NG, Wielicki BA, Young DF (2012) Geostationary enhanced temporal interpolation for CERES flux products. J Appl Meteor Climatol (submitted)

Eitzen ZA, Xu K-M, Wong T (2011) An estimate of low-cloud feedbacks from variations of cloud radiative and physical properties with sea surface temperature on interannual time scales. J Clim 24:1106–1121

Fu Q, Liou K-N (1992) On the correlated-k distribution method for radiative transfer in nonhomogenous atmospheres. J Atmos Sci 49:2139–2156

Fu Q, Liou K-N (1993) Parameterization of the radiative properties of cirrus clouds. J Atmos Sci 50: 2008–2025

Gottschalk J, Bell GD (2009) The Madden–Julian oscillation. Bull Am Meteorol Soc 90:S78–S79

Hartmann DL, Larson K (2002) An important constraint on tropical cloud-climate feedback. Geophys Res Lett 29:1951. doi:10.1029/2002GL015835

Hooker SB, Esaias WE, Feldman GC, Gregg WW, McClain CR (1992) An overview of SeaWiFS and ocean color. SeaWiFS Tech. Rep. Series, vol 1, NASA Tech. Memo. 104566, 785 pp

Kato S, Sun-Mack S, Miller WF, Rose FG, Chen Y, Minnis P, Wielicki BA (2010) Relationships among cloud occurrence frequency, overlap, and effective thickness derived from CALIPSO and CloudSat merged cloud vertical profiles. J Geophys Res 115:D00H28. doi:10.1029/2009JD012277

L'Ecuyer TS, Jiang JH (2010) Touring the atmosphere aboard the A-Train. Phys Today 63:36–41

Lin B, Wielicki B, Chambers L, Hu Y, Xu K-M (2002) The Iris hypothesis: a negative or positive cloud feedback? J Clim 15:3–7

Loeb NG, Kato S, Loukachine K, Manalo-Smith N (2005) Angular distribution models for top-of-atmosphere radiative flux estimation from the clouds and the earth's radiant energy system instrument on the Terra satellite. Part I: methodology. J Atmos Oceanic Technol 22:338–351

Loeb NG, Priestley KJ, Kratz DP, Geier EB, Green RN, Wielicki BA, Hinton POR, Nolan SK (2001) Determination of unfiltered radiances from the clouds and the earth's radiant energy system (CERES) instrument. J Appl Meteorol 40:822–835
Loeb NG et al (2007) Multi-instrument comparison of top-of-atmosphere reflected solar radiation. J Clim 20:575–591
Loeb NG, Wielicki BA, Doelling DR, Smith GL, Keyes DF, Kato S, Manalo-Smith N, Wong T (2009) Toward optimal closure of the Earth's top-of-atmosphere radiation budget. J Clim 22:748–766
Madden RA, Julian PR (1971) Detection of a 40–50 day oscillation in the zonal wind in the tropical Pacific. J Atmos Sci 28:702–708
Madden RA, Julian PR (1994) Observations of the 40–50 day tropical oscillation: a review. Mon Weather Rev 122:814–837
Matthews (2009) In-flight spectral characterization and calibration stability estimates for the clouds and the earth's radiant energy system (CERES). J Atmos Ocean Tech 26:1685–1716
Matthews G, Priestley KJ, Spence P, Cooper D, Walikanen D (2005) Compensation for spectral darkening of short wave optics occurring on the cloud's and the earth's radiant energy system. In: Butler JJ (ed) Earth observing systems X, Proceedings of SPIE, vol 5882. SPIE, Bellingham, doi: 10.1117/12.618972
Minnis P, Sun-Mack S, Young DF, Heck PW, Garber DP, Chen Y, Spangenberg DA, Arduini RF, Trepte QZ, Smith WL Jr, Ayers JK, Gibson SC, Miller WF, Chakrapani V, Takano Y, Liou KN, Xie Y (2011) CERES edition-2 cloud property retrievals using TRMM VIRS and Terra and Aqua MODIS data, part I: algorithms. IEEE Trans Geosci Remote Sens 49:4374–4400
Philander SGH (1990) El Niño, La Niña and the southern oscillation. Academic Press, San Diego
Remer LA et al (2005) The MODIS aerosol algorithm, products and validation. J Atmos Sci 62:947–973. doi:10.1175/JAS3385.1
Salomonson VV, Barnes WL, Maymon PW, Montgomery HE, Ostrow H (1989) MODIS: advanced facility instrument for studies of the earth as a system. IEEE Trans Geosci Remote Sens 27:145–153
Shankar MS, Thomas S, Priestley KJ (2010, Aug) Pre-launch characterization of spectral response functions for the clouds and earth's radiant energy system (CERES) instrument sensors. In: Proceedings of SPIE, earth observing systems XV, vol 7807, 780702
Suarez MJ (2005) Documentation and validation of the Goddard earth observing system (GEOS) data assimilation system—version 4, vol 26. Tech. Rep. Series on Global Modeling and Data Assimilation, NASA/TM-2005-104606, 181 p
Susskind J, Molnar G, Iredell L, Loeb NG (2012) Interannual variability in outgoing longwave radiation as observed by AIRS and CERES. J Geophys Res (submitted)
Szewczyk ZP, Priestley KJ, Walikainen DR, Loeb NG, Smith GL (2011) Putting all CERES Scanners (Terra/Aqua) on the same Radiometric Scale. In: Remote sensing of clouds and atmosphere, vol XIII, Proceedings of SPIE, Paper ERS11-RS04-5
Thomas S, Priestley KJ, Manalo-Smith N, Loeb NG, Hess PC, Shankar M, Walikainen DR, Szewcyzk ZP, Wilson RS, Cooper DL (2010, Aug) Characterization of the clouds and the earth's radiant energy system (CERES) sensors on the Terra and Aqua spacecraft. In: Proceedings of SPIE, earth observing systems XV, vol 7807, 780702
Trenberth KE, Fasullo JT, O'Dell C, Wong T (2010) Relationships between tropical sea surface temperature and top-of-atmosphere radition. Geophys Res Lett 37:L03702
Wielicki BA et al (2002) Evidence for large decadal variability in the tropical mean radiative energy budget. Science 295:841–844
Wolter K, Timlin MS (1998) Measuring the strength of ENSO events: how does 1997/1998 rank? Weather 53:315–324
Wong T, Wielicki BA, Lee RB III, Smith GL, Bush KA, Willis JK (2006) Reexamination of the observed decadal variability of the earth radiation budget using altitude-corrected ERBE/ERBS nonscanner WFOV data. J Clim 19:4028–4040
Woods T, Rottman G, Harder J, Lawrence G, McClintock B, Kopp G, Pankratz C (2000) Overview of the EOS SORCE mission. SPIE Proc 4135:192–203
Xue Y, Reynolds RW (2009) Sea surface temperatures in 2008. Bull Am Meteorol Soc 90:S47–S48
Young DF, Minnis P, Doelling DR, Gibson GG, Wong T (1998) Temporal interpolation methods for the clouds and the earth's radiant energy system (CERES) experiment. J Appl Meteorol 37:572–590
Zelinka MD, Hartmann DL (2010) Why is longwave cloud feedback positive? J Geophys Res 115:D16117. doi:10.1029/2010JD013817
Zelinka MD, Hartmann DL (2011) The observed sensitivity of high clouds to mean surface temperature anomalies in the tropics. J Geophys Res 116:D23103. doi:10.1029/2011JD016459

Surv Geophys (2012) 33:387–393
DOI 10.1007/s10712-011-9167-6

Estimating Global Energy Flow from the Global Upper Ocean

John M. Lyman

Received: 25 May 2011 / Accepted: 1 December 2011 / Published online: 16 December 2011

Abstract The relative significance of short multi-year linear trends in the global integral of 0–700 m ocean heat content anomaly (OHCA) is investigated by examining the overlapping segments of the 16-year OHCA curve from Lyman et al. (Nature 465:334–337, 2010). Segments of 4 years and less are found not to be significantly different from each other or from 0 at the 90% confidence interval. Likewise, short 5- to 7-year segments are not statistically different from each other. Ten-year and longer trends are significant and provide a useful comparison for satellite observations of the radiation imbalance at the top of the atmosphere.

Keywords Ocean heat content · Top of the atmosphere (TOA) · Expendable bathythermograph (XBT) · Warming

1 Introduction

Approximately 90% of the radiation imbalance at the top of the atmosphere (TOA) from 1993 to 2003 was absorbed by the upper 0–700 m of the global oceans (Bindoff et al. 2007). Hence, the majority of the global energy budget can be estimated as a balance between the net energy flux at the TOA and the change in the global integral of ocean heat content anomaly (OHCA) in the upper 0–700 m. Ten- to 40-year trends in the upper ocean widely range from 0.3 to 0.6 Wm^{-2} depending on the method and period for which the trend was computed. (Here and throughout the text, the ocean uptake of heat is expressed as an equivalent radiative forcing over the total area of the Earth.) (Willis et al. 2004; Domingues et al. 2008; Levitus et al. 2009; Ishii and Kimoto 2009; Lyman et al. 2010). These estimates preclude deep OHCA changes, which have been shown in modeling

J. M. Lyman
Joint Institute for Marine and Atmospheric Research, University of Hawaii at Manoa, Honolulu, HI, USA

J. M. Lyman (✉)
NOAA/Pacific Marine Environmental Laboratory, Seattle, WA, USA
e-mail: John.Lyman@noaa.gov

studies to be important to the TOA energy imbalance (Katsman and van Oldenborgh 2011; Palmer et al. 2011; Meehl et al. 2011; Trenberth and Fasullo 2011).

Estimates of annual upper OHCA depend on the large-scale in situ monitoring of global ocean temperature. Of the many observational platforms that sample the ocean, there are only a few that sample the global ocean frequently enough and on large enough spatial scales to have a significant impact on the global integral of annual OHCA (Lyman and Johnson 2008). For the early part of the record (1971–2003), expendable bathythermographs (XBTs) dominate global in situ coverage. These instruments are largely constrained to ship tracks that under-sample the southern hemisphere (Fig. 1, middle panel). They measure temperature and infer depth from time, many of which extend to just below 700 m. Currently (2004–2010), the near global coverage of the Argo array of robotic autonomous floats provides the bulk of in situ measurements used in estimating the global integral of OHCA (Fig. 1, bottom panel), providing near global coverage of the ice-free 0–1,000 m oceans since 2004. As older floats that only extend to 1,000 m are replaced, the depth of the Argo array extends to its target depth of 2,000 m. Recently, biases have been detected in both XBTs and Argo profiling floats. While the known Argo biases have been largely removed, the exact way to correct the XBT bias remains an open topic of research (Gouretski and Koltermann 2007; Wijffels et al. 2008; Ishii and Kimoto 2009; Levitus et al. 2009; Gouretski and Reseghetti 2010) and is the largest source of uncertainty in the 0–700 m OHCA curve (Lyman et al. 2010).

While the warming signal in the upper ocean is clearly significant over decades, the significance of OHCA changes over shorter periods is not clear. von Schuckmann et al. (2009) estimated a significant 6-year trend for the 0–2,000 m ocean from 2003 to 2008 to be 0.5 ± 0.1 Wm^{-2}, based solely on Argo data. Yet, estimates of the 0–700 m OHCA over this same period do not appear to show significant trends (Levitus et al. 2009; Lyman et al. 2010; Johnson et al. 2011).

Here, we examine the significance of multi-year trends from different segments of the 1993–2008 global 0–700 m OHCA curve that appeared in Lyman et al. (2010) to see whether short multi-year trends are significantly different from each other. This period spans both the XBT and Argo eras of sampling. The uncertainty used in the trend estimates takes into account error due to sampling, mapping method, climatology choice, and choice of XBT bias correction. The data and methods used in the analysis are described in Sect. 2. The significance of the interannual trends is analyzed in Sect. 3.

2 Data and Methods

The OHCA estimate is composed of in situ temperature measurements from both the World Ocean Database (WOD, Boyer et al. 2009) and the US Argo Global Data Assembly Centers (GDAC, http://www.usgodae.org/argo/argo.html). Mooring, XBT, and conductivity, temperature and depth (CTD) observations are taken from the WOD, while profiling float data are only taken from the US Argo GDAC. The Argo floats with potential surface pressure drifts have been removed (Barker et al. 2011), and XBT corrections have been applied (Lyman et al. 2010).

As in Lyman et al. (2010), the estimates of the heat content are computed from the vertical integral of heat at a given location. Then, a mean and seasonal cycle are removed. The resulting heat content is then binned into one-year bins and objectively mapped. The resulting maps are globally integrated to produce the global OHCA curves shown in Fig. 2 of Lyman et al. (2010). This methodology is explained in detail in Lyman et al. (2010).

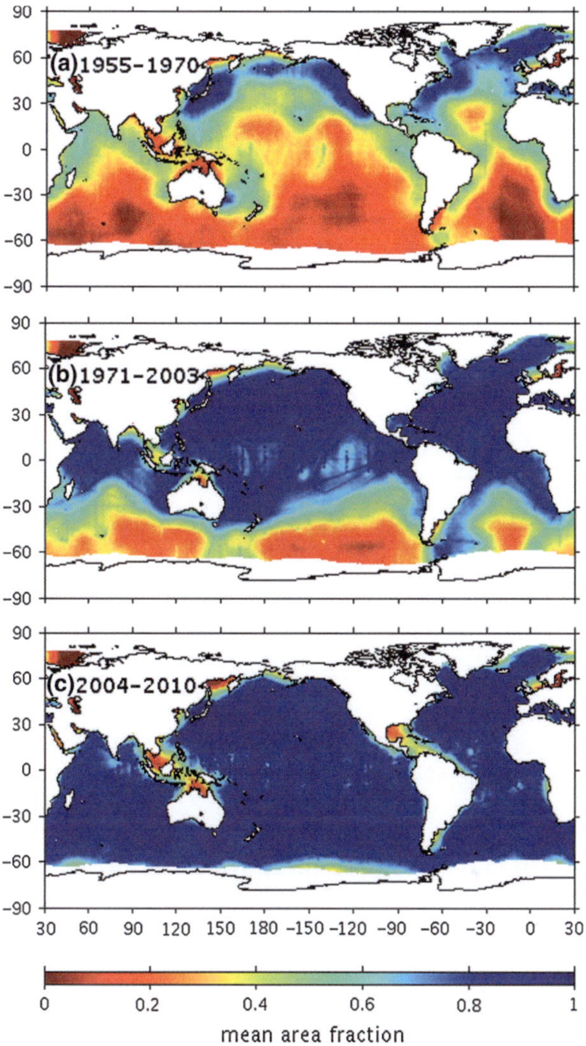

Fig. 1 Mean fraction of the 0–700 m ocean with in situ measurements of temperature, averaged from 1955 to 1970 (*top panel*), 1971–2003 (*middle panel*), and 2004–2010 (*bottom panel*). This figure is updated from Lyman and Johnson (2008)

Uncertainties are expressed at the 90% confidence interval unless stated otherwise. The largest source of uncertainty comes from the spread in the global integral in OHCA from the multiple XBT bias corrections that were available when the uncertainty estimates were made (Lyman et al. 2010). These bias corrections attempt to remove systematic errors hopefully leaving random error behind; however, there is a chance that some biases remain. Estimating XBT biases is a current area of research and is constantly evolving. Trends are computed as weighted least square fits and are expressed as an equivalent radiative forcing over the total area of the Earth. Degrees of freedom (DOF) are computed from the autocorrelation of the time series. For the shorter multi-year trends, DOF were

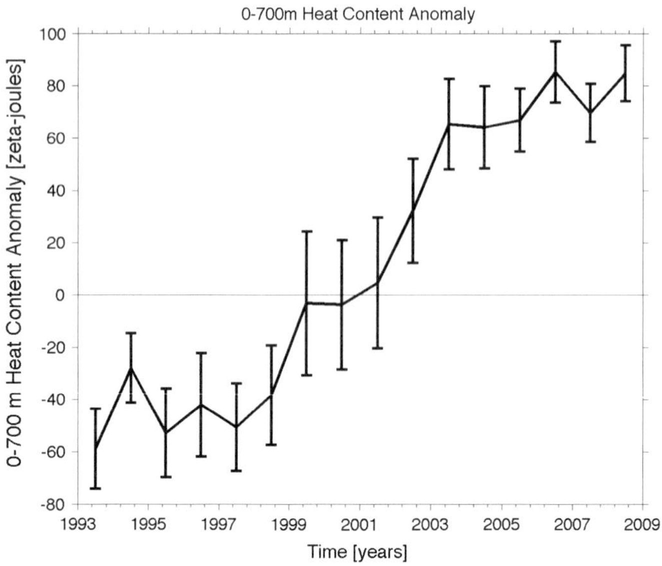

Fig. 2 0–700 OHCA variation, measured in zeta Joules, zJ, 10^{21} J, from Lyman et al. (2010) (Fig. 2 *black line*). The *error bars* contain uncertainty from mapping choice, XBT-bias correction + QC (quality control) choice, sampling error, and choice of climatology. These uncertainties are expressed at one standard error. The long term (16-year) trend is 0.64 ± 0.11 Wm^{-2}

estimated by the mean DOF over all possible segments of the OHCA curve. It is possible that this is an underestimate of the DOF, and the uncertainties could be larger.

3 Multi-Year Trends in Global Upper Ocean Heat Content

The 0–700 m OHCA curve in Fig. 2 has a significant 16-year trend of 0.64 ± 0.11 Wm^{-2} (expressed as an equivalent radiative forcing over the total area of the Earth.). Looking at the OHCA curve alone (ignoring the error bars), there are periods where the upper ocean heat uptake was low (1993–1999 and 2003–2008) and a period when the ocean uptake was larger (1999–2003). However, when the uncertainties on the OHCA curve are taken into account, it is not clear that these trends over shorter record lengths are significantly different from each other.

Short 4-year segments are never significantly different from each other and from zero (Fig. 3, green line). The longer, steep 5-year segments are only slightly different from zero (Fig. 3, light blue line), with large $\sim$1 Wm^{-2} uncertainties (Fig. 4, light blue line). The XBT bias correction is by far the largest contributor to the uncertainty (Fig. 4, dashed lines). Over time, it is possible that this uncertainty will reduce as a consensus is reached about the XBT bias correction, increasing the significance of trends with short record lengths. As the segments length increases the uncertainties decrease and the trend approaches the 16-year trend. Even so, the 6- and 7-year segments still have relatively large uncertainties near $\sim$0.5 Wm^{-2} (Fig. 4). This makes the flat segments of the curve in 1993–1999 and 2003–2008, with trends of 0.18 ± 0.51 and 0.24 ± 0.49 Wm^{-2} respectively, indistinguishable from the steep 1999–2003 segment, which warms at 1.19 ± 1.13 Wm^{-2}.

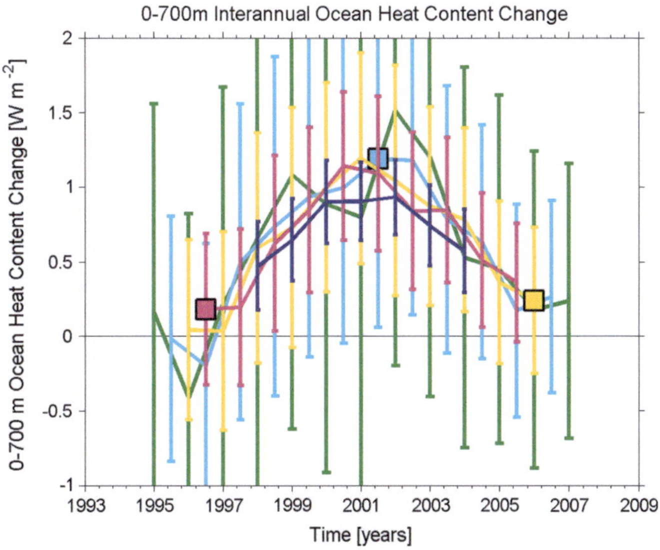

Fig. 3 Interannual tends in 0–700 m OHCA (Fig. 2) expressed as an equivalent radiative forcing over the total area of the Earth, for 4-year (*green*, 2 DOF), 5-year (*light blue*, 2.5 DOF), 6-year (*yellow*, 2.8 DOF), 7-year (*pink*, 3 DOF), and 10-year (*purple*, 3 DOF) segments. Time is centered on the middle of segment. *Squares* mark the 5-year 1999–2003 segment (*light blue*, 1.19 ± 1.13 Wm^{-2}), the 6-year 2003–2008 segment (*yellow*, 0.24 ± 0.49 Wm^{-2}), and the 7-year 1993–1999 segment (*pink*, 0.18 ± 0.51 Wm^{-2})

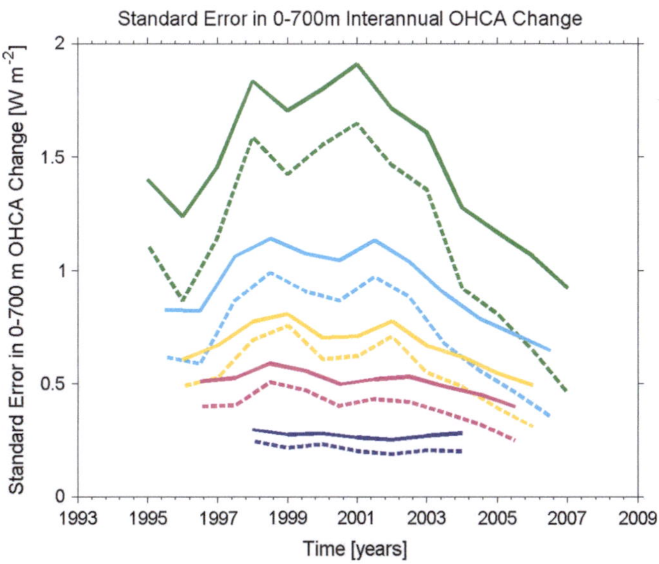

Fig. 4 Uncertainties shown in *error bars* of Fig. 3 (*solid lines*) and uncertainties due to XBT correction (*dashed lines*), colors follow Fig. 3

4 Discussion

The 0–700 m OHCA record from 1993 to 2008 has significant 10-year trends that give useful estimates of the net TOA radiation imbalance. Direct measurements of TOA radiation imbalance from the clouds and the Earth's radiant energy system (CERES) observations are not capable of constraining the mean flux (Loeb et al. 2009). Likewise, shorter, multi-year variability that is constrained by CERES is not measurable in OHCA. Hopefully, estimates of the upper OHCA curve will be able to provide a constraint on shorter timescale variability in TOA radiation imbalance as better and better estimates of the XBT bias correction become available and the uncertainty on upper OHCA decreases.

Currently, the steep sections of the upper OHCA curve (1999–2003) are not significantly different from the flat sections (1993–1999 and 2003–2008). Ideally, in the absence of mapping, quality control (QC) and XBT biases uncertainty the current Argo array could detect smaller year-to-year changes in upper OHCA. However, for the 16 years discussed here, XBTs remain crucial to computing OHCA as they link the Argo era of sampling with past decades.

This analysis examines the significance of the global integral of upper OHCA. On regional spatial scales, short multi-year and even seasonal to monthly variability in OHCA can be orders of magnitude larger than the global integral. The uncertainty of regional variability is different from the global integral, and its significance needs to be considered separately.

The Argo array now extends to depths of 2,000 m, providing excellent coverage of the upper ocean. Estimates in the deep abyssal ocean (>4,000 m) gleaned from sparsely spaced ship board measurements show measurable warming over decades but are woefully inadequate to look at short annual to 5-year variability (Purkey and Johnson 2010). Models have estimated that the deep ocean can uptake up to 35% of the imbalance at the TAO over 8-year periods (Katsman and van Oldenborgh 2011) and that deep (0–4,000 m) estimates of OHCA are necessary to capture the heat content signal within the oceans (Palmer et al. 2011; Meehl et al. 2011; Trenberth and Fasullo 2011).

In order to quantify both the magnitude and uncertainty of uptake of heat and truly close the Earth's energy budget, the observing system needs to be expanded into the deep (>2,000 m) and ice-covered oceans. A deep extension of Argo would supplement the current array, improving our ability to monitor short multi-year changes in global OHCA.

Acknowledgments The bulk of the in situ data used herein was provided through the *World Ocean Database 2009* (http://www.nodc.noaa.gov). Float data were collected and made freely available by Argo (a pilot program of the Global Ocean Observing System) and contributing national programs (http://www.argo.net/). Gregory C. Johnson provided helpful comments and insight. The NOAA Climate Program Office and the NOAA Office of Oceanic and Atmospheric Research provided support for this research. The findings and conclusions in this article are those of the authors and do not necessarily represent the views of the National Oceanic and Atmospheric Administration. JIMAR contribution number 11–375. PMEL contribution number 3722.

References

Bindoff NL et al (2007) Observations: oceanic climate change and sea level. In: Solomon S et al (eds) Climate change 2007: the physical science basis. Contribution of Working Group I to the Fourth Assessment Report of the Intergovernmental Panel on Climate Change. Cambridge University Press, Cambridge, United Kingdom and New York, NY, USA

Barker PM, Dunn JR, Domingues CM, Wijffels SE (2011) Pressure sensor drifts in Argo and their impacts. J Atmos Ocean Tech 28:1036–1049. doi:10.1175/2011JTECHO831.1
Boyer TP, Antonov JI, Baranova OK, Garcia HE, Johnson DR, Locarnini RA,Mishonov AV, Seidov D, Smolyar IV, Zweng MM (2009) World Ocean Database 2009. In: Levitus S (ed) NOAA Atlas NESDIS 66, U.S. Gov. Printing Office, Washington, DC, 216 pp, DVDs
Domingues CM, Church JA, White NJ, Gleckler PJ, Wijffels SE, Barker PM, Dunn JR (2008) Improved estimates of upper-ocean warming and multidecadal sea-level rise. Nature 453:1090–1093. doi:10.1038/nature07080
Gouretski VV, Koltermann KP (2007) How much is the ocean really warming? Geophys Res Lett 34:L01610. doi:10.1029/2006GL027834
Gouretski VV, Reseghetti F (2010) On depth and temperature biases in bathythermograph data: development of a new correction scheme based on analysis of global ocean database. Deep-Sea Res 57:812–833
Ishii M, Kimoto M (2009) Revaluation of historical ocean heat content variations with time-varying XBT and MBT depth bias corrections. J Oceanogr 65:287–299. doi:10.1007/s10872-009-0027-7
Johnson GC, Lyman JM, Willis JK, Levitus S, Boyer T, Antonov J, Palmer MD, Good SA (2011) Global oceans: ocean heat content. In: Blunden J, Arndt DS, Baringer MO (eds) State of the climate in 2010, Bull Am Meteorol Soc 92:6:S81–S84. doi:10.1175/1520-0477-92.6.S1
Katsman C, van Oldenborgh GJ (2011) Tracing the upper ocean's "missing heat". Geophys Res Lett 38:L14610. doi:10.1029/2011GL048417
Levitus S, Antonov J, Boyer T, Locarnini RA, Garcia HE, Mishonov AV (2009) Global ocean heat content 1955–2007 in light of recently revealed instrumentation problems. Geophys Res Lett 36:L07608. doi:10.1029/2008GL037155
Loeb NG, Wielicki BA, Doelling DR, Smith GL, Keyes DF, Kato S, Manalo-Smith N, Wong T (2009) Toward optimal closure of the earth's top-of-atmosphere radiation budget. J Clim 22:748–766. doi:10.1175/2008JCLI2637.1
Lyman J, Johnson G (2008) Estimating annual global upper ocean heat content anomalies despite irregular in situ ocean sampling. J Clim 21:5629–5641. doi:10.1175/2008JCLI2259.1
Lyman J, Good S, Gouretski V, Ishii M, Johnson G, Palmer M, Smith D, Willis J (2010) Robust warming of the global upper ocean. Nature 465:334–337. doi:10.1038/nature09043
Meehl G, Arblaster J, Fasullo J, Hu A, Trenberth K (2011) Model-based evidence of deep-ocean heat uptake during surface-temperature hiatus periods. Nature Clim Change 1:360–364. doi:10.1038/nclimate1229
Palmer MD, McNeall DJ, Dunstone NJ (2011) Importance of the deep ocean for estimating decadal changes in earth's radiation balance. Geophys Res Lett 38:L13707. doi:10.1029/2011GL047835
Purkey SG, Johnson GC (2010) Warming of global abyssal and deep southern ocean between the 1990s and the 2000s: contributions to global heat and sea level rise budgets. J Clim 23:6336–6351. doi:10.1175/2010JCLI3682.1
Trenberth KE, Fasullo JT (2011) Tracking earth's energy: from El Niño to global warming. Surv Geophys, Accepted. doi:10.1007/s10712-011-9150-2
von Schuckmann K, Gaillard F, Traon PYL (2009) Global hydrographic variability patterns during 2003–2008. J Geophys Res 114:C09007. doi:10.1029/2008JC005237
Wijffels SE, Willis J, Domingues CM, Barker P, White NJ, Gronell A, Ridgway K, Church JA (2008) Changing expendable bathythermograph fall rates and their impact on estimates of thermosteric sea level rise. J Clim 21:5657–5672. doi:10.1175/2008JCLI2290.1
Willis JK, Roemmich D, Cornuelle B (2004) Interannual variability in upper ocean heat content, temperature, and thermosteric expansion on global scales. J Geophys Res 109:C12036. doi:10.1029/2003JC002260

Surv Geophys (2012) 33:395–412
DOI 10.1007/s10712-012-9179-x

Uncertainty Estimate of Surface Irradiances Computed with MODIS-, CALIPSO-, and CloudSat-Derived Cloud and Aerosol Properties

Seiji Kato · Norman G. Loeb · David A. Rutan · Fred G. Rose · Sunny Sun-Mack · Walter F. Miller · Yan Chen

Received: 9 June 2011 / Accepted: 20 January 2012 / Published online: 23 February 2012

Abstract Differences of modeled surface upward and downward longwave and shortwave irradiances are calculated using modeled irradiance computed with active sensor-derived and passive sensor-derived cloud and aerosol properties. The irradiance differences are calculated for various temporal and spatial scales, monthly gridded, monthly zonal, monthly global, and annual global. Using the irradiance differences, the uncertainty of surface irradiances is estimated. The uncertainty (1σ) of the annual global surface downward longwave and shortwave is, respectively, 7 W m^{-2} (out of 345 W m^{-2}) and 4 W m^{-2} (out of 192 W m^{-2}), after known bias errors are removed. Similarly, the uncertainty of the annual global surface upward longwave and shortwave is, respectively, 3 W m^{-2} (out of 398 W m^{-2}) and 3 W m^{-2} (out of 23 W m^{-2}). The uncertainty is for modeled irradiances computed using cloud properties derived from imagers on a sun-synchronous orbit that covers the globe every day (e.g., moderate-resolution imaging spectrometer) or modeled irradiances computed for nadir view only active sensors on a sun-synchronous orbit such as Cloud-Aerosol Lidar and Infrared Pathfinder Satellite Observation and CloudSat. If we assume that longwave and shortwave uncertainties are independent of each other, but up- and downward components are correlated with each other, the uncertainty in global annual mean net surface irradiance is 12 W m^{-2}. One-sigma uncertainty bounds of the satellite-based net surface irradiance are 106 W m^{-2} and 130 W m^{-2}.

Keywords Surface net irradiance · Surface radiative energy budget

1 Introduction

Estimating the surface irradiance is important in understanding the energy cycle of the globe for several reasons. The sum of surface net irradiance and other surface enthalpy

S. Kato (✉) · N. G. Loeb
Climate Science Branch, NASA Langley Research Center, Hampton, VA 23681-2199, USA
e-mail: seiji.kato@nasa.gov

D. A. Rutan · F. G. Rose · S. Sun-Mack · W. F. Miller · Y. Chen
Science System & Applications Inc, Hampton, VA, USA

(sensible and latent heat) fluxes is the energy flux through the lower boundary of the atmospheric column and the upper boundary of an ocean column. Therefore, the global mean net surface irradiance balances with the sum of the surface latent and sensible heat fluxes and ocean heating rate (Wong et al. 2006). In addition, the radiative net energy deposition in the atmosphere and vertical and horizontal profiles of the energy deposition determine the dynamics in the atmosphere. Understanding the top-of-atmosphere (TOA) surface and atmospheric irradiances quantitatively is, therefore, necessary to quantitatively understand the dynamics, which in turn controls cloud feedback processes (Wielicki et al. 1995). The global mean surface irradiance estimate is only possible through modeling surface irradiances. In earlier studies, cloud properties derived from passive satellite instruments combined with radiative transfer models have been used to estimate surface irradiance (e.g., Pinker and Laszlo 1992; Zhang et al. 1995, and a summary is given by Kandel and Viollier 2010).

While surface irradiances computed with satellite-derived and modeled cloud properties have been compared in earlier studies (e.g., Hatzianastassiou et al. 2005; Su et al. 2008; Stephens 2011), the uncertainty of surface irradiances averaged over different temporal and spatial scales, such as monthly or annual, and regional, zonal, or global, is not well understood. The purpose of this paper is to extend the study by Kato et al. (2011) to estimate uncertainties of surface irradiance components in various spatial and temporal scales ($1^\circ \times 30^\circ$ or $1^\circ \times 1^\circ$ gridded, 1° zonal, and global spatial scales and monthly and annual temporal scales).

In this study, we define the uncertainty as a range of surface irradiances in which the true value resides at a 68% probability. Our goal is different from estimating the error of a specific surface irradiance estimate, although we need to have a specific surface irradiance estimate to attach the uncertainty. Taylor and Kuyatt (1994) describe the difference nicely; the result of a measurement (modeled irradiance) could have a negligible error because it can unknowably be very close to the truth even though it may have a large uncertainty. There are two possible ways of estimating the modeled surface irradiance uncertainty. One way is to estimate the uncertainty of the input variables, perturb inputs by the uncertainty amount, and compute the irradiance using a radiative transfer model. The irradiance change by the sensitivity study is considered as the uncertainty due to input variables. Some earlier studies (e.g., Zhang et al. 1995; Zhang et al. 2004; Kim and Ramanathan 2008) used this approach. A second approach is to use surface observations and compute the root mean square (RMS) difference of modeled and observed surface irradiances. We primarily take the former approach in this study, but briefly examine how uncertainties derived by the two approaches differ.

Sections 2 and 3 present a brief overview of surface longwave and shortwave irradiance estimates, respectively, to understand the range of the global annual mean values. Section 4 briefly discusses the computation method using CALIPSO (Winker et al. 2010), CloudSat (Stephens et al. 2008), moderate resolution imaging spectrometer (MODIS) and Clouds and the Earth's Radiant Energy System (CERES) data. Section 5 analyzes the uncertainty of surface irradiances for various temporal and spatial scales by comparing two modeled irradiances (sensitivity approach). Section 6 uses surface observations to estimate modeled surface irradiance uncertainties (surface observation approach). Section 7 combines uncertainties of all surface irradiance components and discusses the uncertainty in the global annual net surface irradiance.

2 Global Annual Mean Surface Downward Longwave Irradiance

Stephens et al. (2011) provide a summary of the global annual mean surface longwave upward and downward irradiance estimated from satellite observations, reanalysis, and ground-based observations. The global annual mean downward longwave irradiance estimated from satellite observations (GEWEX SRB, ISCCP-FD, CERES) ranges from 342 to 348 W m^{-2} (Stephens et al. 2011). The global annual mean downward longwave irradiance estimated by reanalyses ranges from 324 to 340 W m^{-2} (Stephens et al. 2011).

Wild et al. (2001) compared the global annual mean surface downward longwave irradiance computed in general circulation models (GCMs) with surface observations in Global Energy Balance Archive (GEBA) and Baseline Surface Radiation Network (BSRN, Ohmura et al. 1998) data sets. Their results show that the modeled global annual mean surface downward longwave irradiance by GCMs varies more than 40 W m^{-2}, ranging from 303 to 344 W m^{-2}. GCM-derived surface downward longwave irradiances are less than observed irradiances especially under dry and cold conditions. Their results suggest that the difference is caused by the underestimation of surface downward longwave irradiances under cloud-free conditions.

The global annual mean surface downward longwave irradiance under clear-sky conditions computed with satellite-based observations (passive sensor-derived clouds) and that from reanalyses shown in Stephens (2011) range from 309 to 326 W m^{-2}, which is a smaller variation than all-sky values (324–348 W m^{-2}). In addition, among values given by Stephens (2011), all but two clear-sky values agree to within a few W m^{-2}. All-sky global annual mean surface downward longwave irradiances from reanalyses tend to be smaller than satellite-based estimates (Stephens 2011), indicating that the difference is caused by clouds, especially low-level clouds. This suggests that cloud properties used in reanalyses are different from those derived from satellites and that the cloud difference is a source of the difference between satellite-based and reanalysis-based estimates for all-sky conditions.

When the global annual mean surface downward longwave irradiance is estimated using MODIS-derived cloud properties by the Ed2 CERES cloud algorithm (Minnis et al. 2011), the global annual mean value is smaller by 3.6 W m^{-2} compared with the irradiance estimated using CALIPSO and CloudSat-derived cloud properties (Kato et al. 2011). The exact value of the surface downward longwave irradiance difference caused by passive and active sensor–derived clouds depends on the cloud base height estimated from passive sensors. For example, the increase of the global mean surface downward longwave irradiance estimated by Zhang et al. (2004) using climatological cloud vertical profiles (Wang et al. 2000) from International Satellite Cloud Climatology Project C1 input data (ISCCP-FC) is 1.8 W m^{-2}.

Though these differences due to cloud base are significant, studies using satellite-derived cloud properties by Zhang et al. (2006) and by Kato et al. (2011) show that the largest uncertainty in computing the global annual mean surface downward longwave irradiance is caused by uncertainty in near-surface air temperature and precipitable water. The uncertainty in the global annual mean surface downward longwave irradiance caused by surface temperature and water vapor amount estimated by Kato et al. (2011) is ± 4.5 and ± 5.2 W m^{-2}, respectively. Both are larger than the uncertainty due to cloud base height found in Kato et al. (2011).

3 Global Annual Mean Surface Shortwave Irradiance

Comparisons of surface shortwave irradiances from previous studies are more problematic due to different assumptions made regarding TOA solar insolation. To avoid the effect of the insolation difference in various estimates, we use the surface absorptance, which is the net surface shortwave irradiance (downward minus upward) divided by the TOA insolation, to compare surface shortwave irradiances from earlier studies. The global annual mean surface absorptance ranges from 0.471 to 0.548 (Table 1). The absorptance given by L'Ecuyer et al. (2008) is the first estimate of the global annual mean surface downward shortwave irradiance using CloudSat-derived cloud properties. However, it is probably biased high because clouds that were undetected by CloudSat are missing in their computations. If we exclude the maximum and minimum values shown in Table 1, the absorptances range from 0.471 to 0.495. This corresponds to a range of 8.2 W m^{-2}, using an insolation of 341.3 W m^{-2}.

Uncertainty in the surface downward shortwave irradiance is due to uncertainties in column water vapor amount, cloud and aerosol properties and their vertical profiles, treatment of diurnal cycle of clouds, the surface albedo, and the insolation. Based on a sensitivity study done by Zhang et al. (1995), if all surface downward shortwave irradiance changes considered in their study (atmospheric, surface, and cloud properties) are independent random errors with unknown sign, the uncertainty in the global annual mean irradiance is 15.3 W m^{-2} [out of 193.4 W m^{-2} (Rossow et al. 1995), based on an average from April 1985 through January 1989 using every third month]. Rossow (1995) also compared modeled surface downward shortwave irradiances with surface observations. The bias and RMS differences are from 10 to 20 W m^{-2} and from 10 to 15 W m^{-2}, respectively. These values are consistent with modeled and observed surface downward shortwave irradiance difference given by Kato et al. (2008) at three Atmospheric Radiation Measurement (ARM) sites, 20.5 W m^{-2} for Manus, 12.6 W m^{-2} for Southern Great Plains (SGP), and 6.4 W m^{-2} for Barrow, Alaska.

There are also uncertainties in the observations of irradiance at the Earth's surface due to instrumentation. As pointed out by Gulbrandsen (1978), if the temperature gradient within the instrument is not corrected, surface downward shortwave irradiances measured by pyranometers can be biased low. Once the temperature gradient within a pyranometer is eliminated (e.g., Haeffelin et al. 2001), modeled clear-sky surface downward shortwave

Table 1 Global annual mean surface shortwave irradiance estimates

	SW down (W m^{-2})	SW up (W m^{-2})	NET (W m^{-2})	TOA insolation (W m^{-2})	Surface absorptance
Gupta et al. (1999)	185	24	161	341.3	0.472
Zhang et al. (2004)	189	24	165	341.8	0.483
Hatzianastassiou et al. (2005)	172	23	149	341.8	0.436
Kim and Ramanathan (2008)			164	348.1	0.471
L'Ecuyer (2008)[a]	218	25	193	352	0.548
Wang and Pinker (2009)			168	341.7	0.492
Trenberth et al. (2009)	184	23	161	341.3	0.472
Kato et al. (2011)	192	23	169	341.3	0.495

[a] Only area covered by the CloudSat ground track and 82°N to 82°S is considered

irradiances agree with observations to within 1% (Michalsky et al. 2006) when accurate temperature and humidity profiles, aerosol properties, and surface albedo are used.

4 Top-of-Atmosphere (TOA) and Surface Radiative Flux Estimate Using CALIPSO- and CloudSat-Derived Clouds

Active instruments can detect clouds better than passive sensors. Active instruments can also observe cloud base heights, while the cloud base height derived from passive sensors relies on climatological vertical cloud profiles (Zhang et al. 2004) or an empirical relationship based on the cloud physical thickness, cloud top temperature, and optical thickness (Minnis et al. 2010, 2011). Properly determining the cloud base height improves the accuracy of the surface downward longwave irradiance. In addition, active sensors can detect multi-layer clouds and can improve cloud detections, particularly at nighttime over the polar regions. Furthermore, better cloud top height derived from CALIPSO and CloudSat leads to a better estimate of the emission component in near-infrared channels, which can improve MODIS-derived cloud properties (Kato et al. 2011).

Improvements in modeled TOA and surface irradiances using Cloud-Aerosol Lidar with Orthogonal Polarization (CALIOP)- and CloudSat Cloud Profiling Radar (CPR)-derived cloud and aerosol properties compared with irradiances computed with MODIS only are investigated by Kato et al. (2011) using 1 year of CALIPSO, CloudSat, CERES, and MODIS combined data. When 1 year of TOA instantaneous irradiances computed with CALIOP- and CPR-derived properties are averaged globally, the reflected shortwave irradiance decreases by 12.5 W m^{-2} (5.0%, relative to MODIS only) and the longwave irradiance increases by 2.5 W m^{-2} (1.1%, relative to MODIS only). As a consequence, both the reflected shortwave and longwave irradiances computed with CALIOP- and CPR-derived properties agree better with CERES-derived TOA irradiances to within 0.5 W m^{-2} (out of 237.8 W m^{-2} CERES value) for reflected shortwave and 2.6 W m^{-2} (out of 240.1 W m^{-2} CERES value) for longwave irradiances. Differences in monthly zonal mean irradiance are, however, larger at some latitudes than the difference of the global mean irradiances. Therefore, a good agreement with the global annual mean CERES-derived irradiance is, in part, because of compensating errors. Nevertheless, the result of Kato et al. (2011) indicates that the accuracy of modeled global annual mean irradiances improves when CALIOP- and CPR-derived cloud and aerosol properties are used. A summary of surface irradiances estimated by Kato et al. (2011) is shown in Table 2. Note that a bias of 1.5 W m^{-2} in the surface downward longwave irradiance, which is explained in Sect. 5 and in Kato et al. (2011), is expressed explicitly in Table 2.

5 Uncertainty Estimates

In this section, we calculate the longwave and shortwave surface upward and downward irradiances using two different methods and assume the difference between these methods defines the irradiance uncertainty primarily due to cloud input differences. In computations with CALIPSO and CloudSat, the B1 enhanced algorithm (Minnis 2010) is used to derive cloud properties from MODIS radiances. In computations with MODIS only, the Ed2 CERES cloud algorithm (Minnis 2011) is used to derive cloud properties from MODIS radiances. Cloud properties derived from MODIS are used in both cases, but in the first (B1 enhanced), the passive retrieval is enhanced by the addition of the cloud information

Table 2 Global annual mean surface irradiance estimated by Kato et al. (2011)

	Irradiance (W m^{-2})
Longwave down	345.4 + 1.5
Longwave up	398.1
Shortwave down	191.9
Shortwave up	22.8

supplied by CALIPSO and CloudSat, whereas in the second (Ed2), the passive retrieval stands alone. Aerosol optical thicknesses are given by the MODIS level 2 aerosol product, MYD04 (Remer et al. 2005), in both computations. CALIPSO provides the height of aerosol layers for computations with CALIPSO and CloudSat data. Aerosol type determines optical properties of aerosol for shortwave and longwave computations. When CALIPSO detects dust layers, dust aerosols are used in irradiance computations, while the aerosol type in the MODIS only irradiance computations relies on an aerosol transport model. Ocean surface albedos are from Jin et al. (2004) with slightly different foam parameters in computations with MODIS only and with CALIPSO and CloudSat. Broadband land surface albedos are inferred from MODIS narrowband albedos (Moody et al. 2005) for computations with CALIPSO and CloudSat data. MODIS only computations use the clear-sky TOA albedo derived from CERES measurements (Rutan et al. 2009).

Figures 1 and 2 show, respectively, the surface downward longwave irradiance difference and the relative difference of the surface downward shortwave irradiance. We calculate the RMS difference of these two sets of irradiances averaged over four different temporal and spatial scales, gridded, 1° zonal, and global spatial scales and monthly and annual temporal scales. We use a 30° longitude $\times$1° latitude grid box because this assures at least one sample a day in a grid box by CALIPSO and CloudSat. CALIPSO and CloudSat provide better cloud properties, but do not cover the entire grid box every day as does MODIS. The assumption is that the RMS difference represents the irradiance bounds in which the true gridded monthly mean irradiance computed with perfect cloud and aerosol properties resides at a 68% probability. We expect a similar uncertainty in the modeled irradiance computed for a 1° $\times$ 1° degree grid box with full swath data because most grid boxes are sampled daily. The RMS differences of two sets of irradiances are used to estimate the uncertainties. For example, the RMS difference of gridded monthly mean irradiances is computed to estimate the gridded irradiance uncertainty. Similarly, the RMS of 1° zonal monthly mean irradiances is computed to estimate the monthly zonal irradiance uncertainty. Estimated uncertainties are summarized in Table 3. Uncertainty estimates are given for ocean and land separately because of the significant differences in clouds and atmospheric properties over each surface type. Their combined uncertainty is, at times, smaller than ocean or land alone due to a partial cancelation of errors when irradiances over both surface types are averaged over a 1° zone. The uncertainty of the monthly gridded value is the average of ocean and land values weighted by, respectively, 0.7 and 0.3, instead of using ocean and land-mixed grid box values. Uncertainty in the surface upward irradiance over polar regions can be very large because identifying the presence of snow and ice is sometimes difficult. For this reason, the uncertainty of regional surface upward shortwave irradiance does not apply over the polar regions.

Uncertainty in the surface upward longwave irradiances is due primarily to uncertainty in the skin temperature. Both modeled irradiances with and without active sensors share the

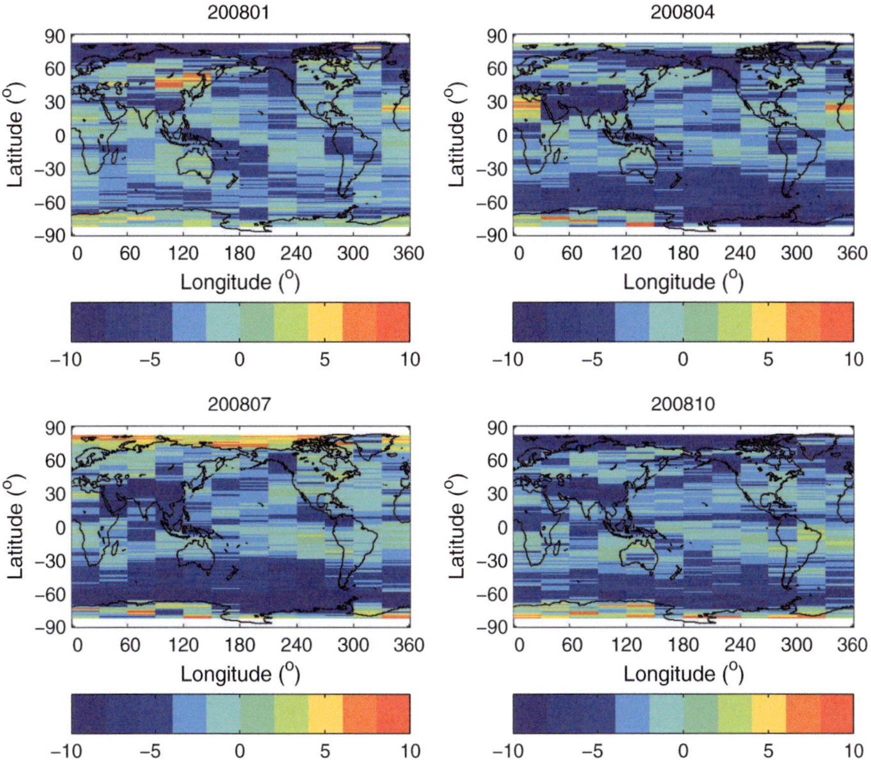

Fig. 1 Monthly mean surface longwave downward irradiance difference gridded in 1° latitude by 30° longitude grids. The difference is defined as the irradiance computed with MODIS only minus the irradiance computed with CALIPSO and CloudSat

same Goddard Earth Observing System (GEOS-5)-derived, (Rienecker et al. 2008; Bloom et al. 2005) surface skin temperature as an input. Instead of using the irradiance RMS difference, therefore, we compute the RMS skin temperature difference between that extracted from GEOS-5 and the skin temperature derived from MODIS when the MODIS, CALIPSO, and CloudSat data all indicate the scene is clear. This allows us to estimate the uncertainty in the upward longwave irradiance using the RMS surface skin temperature difference and the mean surface skin temperature.

Because the level 2 footprint products we used in this study do not account for the diurnal cycle of insolation, the RMS difference of shortwave irradiances is computed as the relative difference from the mean value. We then multiply the relative RMS difference by the mean irradiance estimated with diurnal cycle of insolation [values extracted from the CERES AVG product (Kato et al. 2011)]. For example, the surface downward shortwave uncertainty is computed by multiplying the mean of relative RMS differences shown in Fig. 2 by a mean irradiance accounting for a diurnal cycle shown in Table 3.

Uncertainties caused by aerosol optical properties and precipitable water in the surface downward shortwave irradiance are not included in the difference between active sensor- and passive sensor-derived surface downward shortwave irradiances described above. According to Kim and Ramanathan (2008), uncertainty in the global annual mean net surface shortwave irradiance under clear-sky conditions due to aerosol optical thickness,

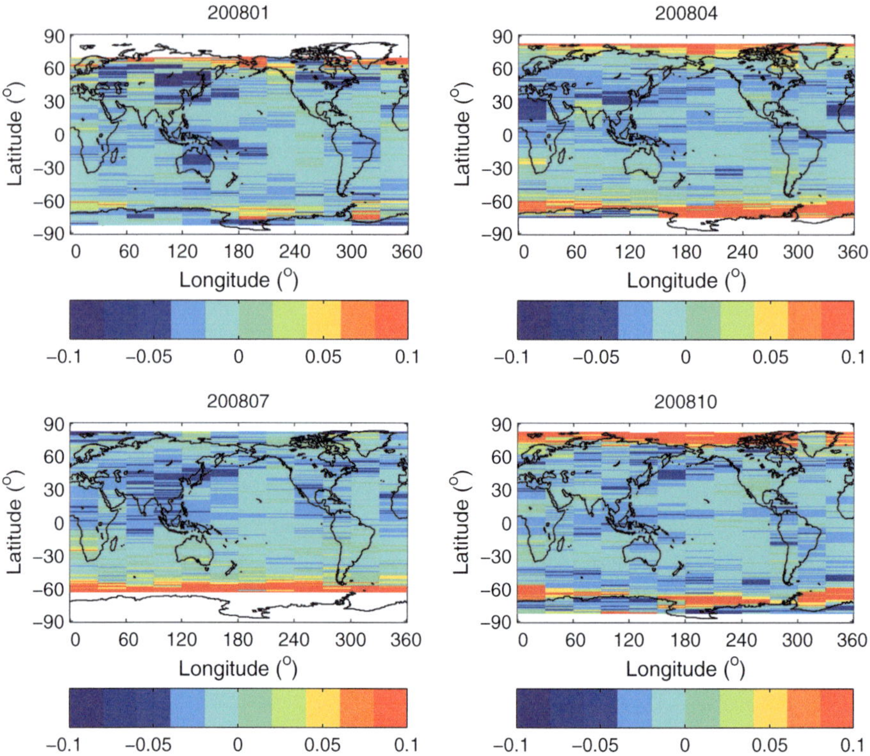

Fig. 2 Monthly mean surface shortwave downward irradiance relative difference gridded in 1° latitude by 30° longitude grids. The relative difference is defined as the irradiance computed with MODIS only minus the irradiance computed with CALIPSO and CloudSat divided by the irradiance computed with CALIPSO and CloudSat

single scattering albedo, precipitable water, and ozone amount is, respectively, 1.5 W m^{-2}, 1.5 W m^{-2}, 1.3 W m^{-2}, and 0.4 W m^{-2}. If these uncertainties are independent, the sum of these uncertainties leads to a 2.9 W m^{-2}, 2.7 W m^{-2}, 3.4 W m^{-2} uncertainty in the global annual mean net surface shortwave irradiance for land plus ocean, ocean, and land, respectively, after converting the net uncertainty to the downward irradiance uncertainty using corresponding global annual mean surface albedos. Considering that these uncertainties are in addition to that derived from irradiance differences primarily due to clouds, we take the square root of the sum of squares of the value and the RMS difference due to clouds for the global annual mean surface downward shortwave uncertainty. For all other temporal- and spatial-scale uncertainties listed in Table 3, we simply add the value to the RMS difference due to clouds. Table 4 summarizes input variables considered in estimating the uncertainties.

Table 3 shows that the uncertainty in different temporal and spatial scales differs. When the error in passive sensor-derived cloud properties depends on cloud type and the probability distribution of cloud type occurrence changes month to month, the irradiance uncertainty computed with them is relatively large. However, even though a particular cloud type occurs in a grid box, the probability distribution of cloud type tends to be constant when the probability is computed over a larger area and a longer time period. If

Table 3 Summary of uncertainties in the irradiance computed with satellite-derived cloud and aerosol properties, in Wm^{-2}

		Mean value	Estimated uncertainty			
			Monthly gridded	Monthly zonal	Monthly global	Annual global
Downward longwave	Ocean + land	345	14	11	7	7
	Ocean	354	12	10	7	7
	Land	329	17	15	8	7
Upward longwave	Ocean + Land	398	15	8	3	3
	Ocean	402	13	9	5	5
	Land	394	19	15	5	4
Downward shortwave	Ocean + Land	192	10	8	6	4
	Ocean	190	9	8	5	4
	Land	203	12	10	7	5
Upward shortwave	Ocean + Land	23	11	3	3	3
	Ocean	12	11	3	3	3
	Land	53	12	8	6	6

the magnitude of the error decreases by shifting a cloud type distribution with a specific cloud type preferentially occurring in the distribution to a uniform distribution among all cloud types, the bias error tends to cancel. Because the cloud distribution tends to be uniform and stable with time, if we use larger spatial and longer temporal scales, the uncertainty decreases at increasing temporal and spatial scales.

The discussion of how the bias error with known sign is separated from the uncertainty follows, using the uncertainty estimates given in Kato et al. (2011). The cloud base height detected by CloudSat can cause a 1.1 W m^{-2} bias error in the global annual mean downward longwave irradiance due to precipitation. If three-hourly skin temperature and one-hourly temperature and humidity profiles are used as opposed to one-hourly skin temperature and temperature and humidity profiles, then the global annual mean surface downward irradiance decreases by 2.6 W m^{-2} due to missing peak values during daytime (Kato et al. 2011). The uncertainty in irradiance caused by the uncertainty in the near-surface temperature and precipitable water is ±6.9 W m^{-2}. Based on this, the range of the global annual mean surface downward longwave irradiance computed with CALIPSO- and CloudSat-derived cloud and aerosol properties is from −5.4 to 8.4 W m^{-2} (+2.6−1.1 ± 6.9 W m^{-2}). If the global annual mean surface downward longwave irradiance is estimated using only MODIS-derived cloud and aerosol properties, bias errors of cloud fraction and cloud base height cause a −3.4 W m^{-2} irradiance bias (Kato et al. 2011). This leads to the range of the passive sensor-derived global annual mean surface downward longwave irradiance of −0.2–14.0 W m^{-2} (+2.6 + 3.4−1.1 ± 6.9 W m^{-2}). If we treat these bounds as one standard deviation (i.e., 2 out of 3 chance that the modeled irradiance lies in the interval), then the uncertainty expressed is 6.9 W m^{-2} (the mean of the upper and lower bound) for both irradiances (Taylor and Kuyatt 1994). Note that bias error does not contribute to the uncertainty expressed in this way. The range where the true value resides for a specific irradiance estimate is computed by adding bias errors of the specific estimate to the uncertainty. Therefore, uncertainties shown in Table 3 are considered as values after known biases have been removed. The overall uncertainty in Table 4,

Table 4 Uncertainty in the global annual mean surface irradiances (Ocean + Land), in Wm^{-2}

	Bias error with known sign	Uncertainty	References
Surface downward longwave irradiance			
Clouds base height	−1.1[a]		Kato et al. (2011)
Temporal interpolation	−2.6		Kato et al. (2011)
Surface temperature		4.5	Kato et al. (2011)
Precipitable water		5.2	Zhang et al. (2006), Kato et al. (2011)
Interannual variability		0.8	Kato et al. (2011)
Overall uncertainty		6.9	
Surface upward longwave irradiance			
Surface skin temperature		3.2	This study
Interannual variability		0.4	This study
Overall uncertainty		3.2	
Surface downward shortwave irradiance			
Clouds		2.8	This study
Aerosol optical thickness[b]		1.7	Kim and Ramanathan (2008)
Aerosol single scattering albedo[b]		1.7	Kim and Ramanathan (2008)
Precipitable water[b]		1.5	Kim and Ramanathan (2008)
Ozone[b]		0.5	Kim and Ramanathan (2008)
Interannual variability		0.3	This study
Overall uncertainty		4.0	
Surface upward shortwave irradiance			
Albedo		3.4	This study
Interannual variability		0.1	This study
Overall uncertainty		3.4	

[a] 1.1 $W m^{-2}$ is when active sensors are used but −3.6 Wm^{-2} when only passive sensor is used

[b] Surface albedo of 0.12 is used to estimate the downward shortwave uncertainty from the net shortwave uncertainty given by Kim and Ramanathan (2008)

therefore, does not include bias errors with known sign. One other factor that affects the uncertainty is the TOA irradiance derived from CERES observations. We assume that modeled surface irradiances are not independent of the TOA irradiance derived from CERES observations. In other words, modeled TOA irradiances need to be consistent with TOA CERES-derived irradiances to within their uncertainties.

6 Comparison with Surface Observations

Modeled and observed surface irradiances have been compared in many studies (e.g., Rossow and Zhang 1995; Rose et al. 2006; Charlock et al. 2006; Wang and Pinker 2009; Niu et al. 2010). The RMS difference between modeled and observed irradiances at surface sites can be used as a measure of uncertainty. The noise of temporal and spatial mismatch between observed surface irradiances, however, dominates especially shortwave irradiances in the comparisons. The uncertainty might be overestimated, if the surface site does not represent the grid box where the site is located when the RMS difference of modeled and observed surface irradiance is used as the uncertainty.

Figure 3 shows comparison of modeled monthly 1° × 1° gridded surface longwave and shortwave downward irradiances with surface observations. Modeled irradiances are extracted from the edition 2 CERES AVG product. The 26 grid boxes used in the sample have surface observation sites located within their boundaries. Surface sites and grid boxes are selected such that surface properties in the vicinity of the site adequately represent the entire grid box ("Appendix"). For example, surface sites on islands or at high elevations with variable terrain are excluded. Surface observations are averaged over a month. Five years of data from March 2000 through February 2005 are used. The RMS difference for the surface downward longwave and shortwave irradiances is, respectively, 9.4 and 8.5 W m^{-2}.

We can separate the RMS difference into three components,

$$r^2 = r_s^2 + r_t^2 + r_m^2, \tag{1}$$

where r is the RMS difference computed with the difference between modeled gridded monthly mean irradiance and observed monthly mean irradiance, r_s is the RMS difference due to variability of surface irradiance within the grid (spatial sampling), r_t is the uncertainty due to temporal resolution of modeled irradiance, and r_m is modeling error and the noise due to matching modeled irradiances with surface observations. The spatial RMS error r_s arises because observations at a surface site measure the irradiance at one location while satellite observations cover the entire grid box. The temporal sampling based on satellite observation is typically limited to several times a day with a sun-synchronous orbit or once every 3 h or 1 h with a geostationary satellite. Modeling error includes the error in the inputs and assumptions in the model. In addition, the difference caused by matching instantaneous modeled irradiance with surface observations at satellite overpass time, as

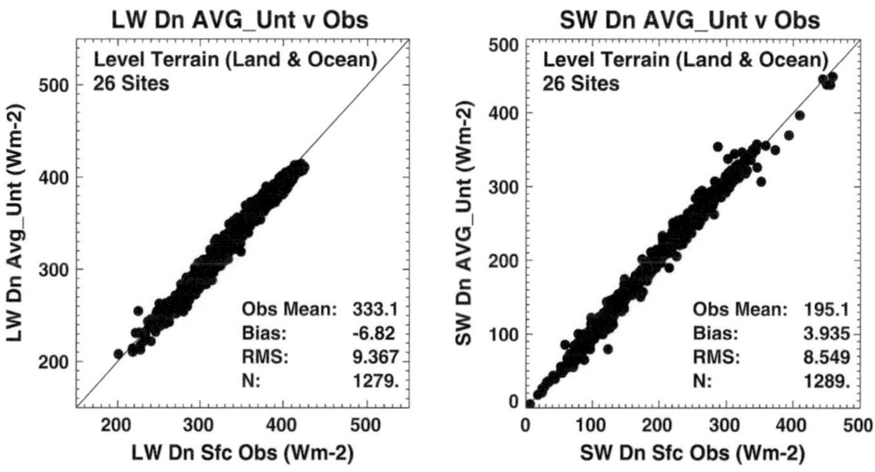

Fig. 3 Comparison of monthly mean observations at 26 surface sites and 1° × 1° monthly mean modeled surface downward longwave irradiance (*left*) and surface downward shortwave irradiance (*right*). Modeled irradiances are extracted from Ed2 CERES AVG product. Data from March 2000 through October 2005 are used

well as the uncertainty in surface observations are included in r_m because the noise is not easily separated from the modeling error.

The RMS difference shown in Fig. 3 is the sum of these three components. When the surface properties at a surface observation site do not represent the surface property of the grid where the surface site is located, r_s is large ("Appendix"). If surface sites indicated by green bars in Figs. 5 and 6 are used, however, r_s is almost negligible for the monthly mean surface downward longwave irradiance and 2.6 W m^{-2} for the monthly mean surface downward shortwave irradiance. Comparisons are made between instantaneous modeled irradiances and observed irradiances at the satellite overpass time, and scaling the instantaneous irradiance difference to the daily mean value gives the r_m term. The temporal sampling term r_t can be inferred from modeled and observed daily mean irradiance comparison, or if we assume the relationship given by Eq. 1, r_t can also be derived from r, provided r_s and r_m are known. We estimate the three components from monthly mean modeled and observed irradiances comparisons at 26 sites where the surface sites represent the grid box surface properties relatively well (Table 5). The values of r_t in Table 5 are derived from r, r_s, and r_m, which are the residual of the relationship given by Eq. 1 for the RMS difference and $r_t = r - r_s - r_m$ for the bias. The RMS differences might be considered as spatial sampling noise r_s, temporal sampling noise r_t, and instantaneous modeling error r_m for a typical grid box. Note that a large downward shortwave irradiance r_t bias is due to a larger r_m, which is a result of matching the TOA reflected shortwave irradiance with the CERES-derived reflected shortwave irradiance (Rose et al. 1997). The difference between modeled and CERES-derived TOA irradiances, therefore, affects the modeled surface irradiance.

The RMS difference shown in Fig. 3 is computed with monthly mean gridded and observed surface irradiances. In principle, if we have a surface site in a grid box, we can remove r_t and r_m by comparing gridded monthly mean modeled irradiance with surface observations. Then, the uncertainty of modeled surface irradiance is r_s because the modeled irradiance cannot be evaluated better than the spatial sample noise. If we have at least one surface site in each grid box and ignore measurement uncertainty, then the uncertainty of modeled irradiance, in principle, can be much smaller and equal to r_s. Because most grid boxes do not have surface sites, the uncertainty of monthly gridded mean modeled irradiance is somewhat equivalent to the RMS difference shown in Fig. 3. Figure 4 shows a comparison of the RMS difference from Fig. 3 and the uncertainty of the gridded monthly mean surface downward longwave and shortwave irradiances from Table 3. Uncertainties of gridded monthly mean downward longwave and shortwave irradiances listed in Table 3

Table 5 Monthly gridded mean irradiance uncertainties derived from modeled and observed surface irradiances[a], in Wm^{-2}

	Total[b], r	Spatial sampling, r_s	Temporal sampling[c], r_t	Modeling error, r_m
LW down RMS	9.7	1.3	6.7	6.9
SW down RMS	8.4	4.1	2.7	6.8
LW down mean bias	−7.0	−0.1	Negligible	−6.9
SW down mean bias	3.8	−2.3	−0.6	6.7

[a] Derived from 26 surface sites used in Fig. 3

[b] Values derived from Fig. 3 scaled by 345 (from Table 3)/333 (mean LW observation value) and 192 (from Table 3)/195 (mean SW observation value)

[c] Values derived from, r, r_s, and r_m

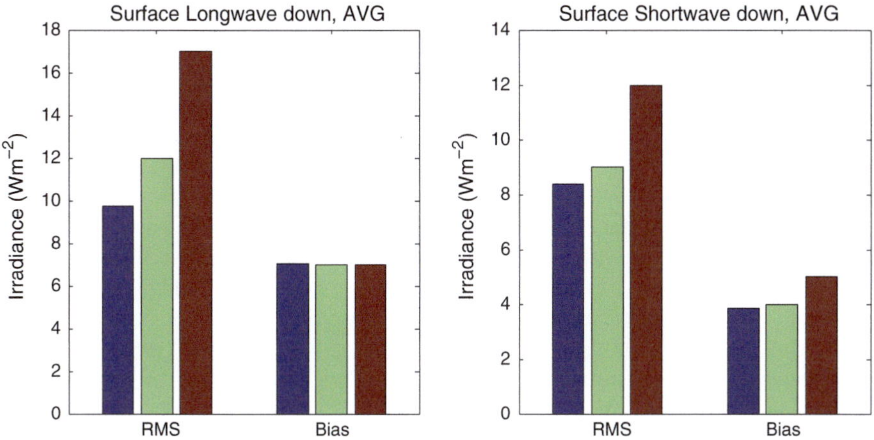

Fig. 4 (*Left*) Comparison of the root square mean (RMS) error (*left blue bar*) and bias error (*right blue bar*) derived from Fig. 3 with the uncertainty of the monthly gridded surface downward longwave irradiance over ocean (*left green bar*) and over land (*left brown bar*) and with the uncertainty annual global surface downward longwave irradiance over ocean (*right green bar*) and over land (*right brown bar*). The uncertainties are from Table 3. *Right*). Same as the left plot but for surface downward shortwave irradiance

are larger than the RMS difference. This indicates that our estimate is on the conservative side according to the comparison with surface observations.

The bias difference of monthly gridded mean irradiances and observations shown in Fig. 3 is not zero. This suggests that even when many modeled irradiances are averaged for large regions such as zonal or global or a longer time scale such as annual mean, the irradiance error does not completely cancel out. As a consequence, the global annual mean irradiance uncertainty might be comparable to the mean bias shown in Fig. 3. The comparison shown in Fig. 4 is consistent with this expectation.

7 Discussion and Conclusions

The uncertainty of the downward longwave irradiance is the largest uncertainty among the global annual mean surface irradiance component uncertainties (Table 3). The global annual mean surface downward longwave irradiance is 345.4 W m^{-2} with an estimated bias error of -1.5 W m^{-2} (Kato et al. 2011). Therefore, the 1σ range is 339.9–353.9 W m^{-2} ($345.4 + 1.5 \pm 7$ W m^{-2}). The spread of the global annual mean surface downward longwave irradiance using satellite observations discussed in Sect. 2 is 6 W m^{-2} (342–348 W m^{-2}), which is smaller than the 7 W m^{-2} uncertainty estimated in this study.

The global annual mean surface downward shortwave irradiance estimated in Kato et al. (2011) is 191.9 W m^{-2}. One-σ bounds of the global annual mean downward shortwave irradiance are 187.9–195.9 W m^{-2} (with the insolation of 341.3 W m^{-2}). The uncertainty of the global annual mean surface up and downward shortwave irradiances estimated in this study is, respectively, 3 and 4 W m^{-2}. If we assume upward and downward shortwave irradiance uncertainties are correlated, the uncertainty of global annual mean net surface shortwave irradiance is 7 W m^{-2}. One-σ bounds of the global annual mean surface absorptance are 0.475–0.516. The spread of the global annual mean surface net shortwave irradiance discussed in Sect. 3 is 8.2 W m^{-2} (the absorptance from 0.471 to 0.495), if we exclude the maximum and minimum values. The uncertainty estimated in this study is close to the spread of values estimated by earlier studies discussed in Sect. 3, but 50% (3 out of 6) of the values fall outside the one-σ bounds.

Given the uncertainty in Table 3, it is possible to estimate the uncertainty in the global annual net surface irradiance. However, the uncertainties in the four irradiance components are not independent. For example, the surface temperature uncertainty affects both upward and downward longwave irradiances. Similarly, if surface downward shortwave irradiance is overestimated, then surface upward shortwave irradiance is probably overestimated. We therefore assume that the shortwave and longwave components are independent, but that upward and downward irradiance uncertainties are correlated. Adding upward and downward components, squaring the sum of shortwave and longwave components, summing them, and taking the square root of the resulting value gives an uncertainty of the global annual mean net surface irradiance of 12 W m^{-2}. Therefore, 1σ bounds of the global annual mean net surface irradiance are 117.9 ± 12 W m^{-2} (from 105.9 to 129.9 W m^{-2}). Note that the -1.5 W m^{-2} bias error in the downward longwave irradiance (Kato et al. 2011) is subtracted from the mean value in this estimate.

Studies by Zhang et al. (2006) and Kato et al. (2011) indicate that when satellite-derived cloud and aerosol properties are used for computations, uncertainty in the surface downward longwave irradiance is dominated by the uncertainty in near surface temperature and precipitable water. Because the uncertainty in the surface downward longwave irradiance is the largest component of the uncertainty in the net surface irradiance, the improvement of near-surface temperature and precipitable water accuracy is necessary to improve the net surface irradiance estimate. The estimate of diurnal cycle of satellite-derived cloud properties relies on geostationary satellites, which provide less accurate cloud properties than MODIS. The process to incorporate geostationary satellite–derived cloud properties in the irradiance estimate minimizes the effect of the error in geostationary-derived cloud properties (Young et al. 1998). In addition, the diurnal cycle correction to the top-of-atmosphere global annual mean shortwave irradiance is small (Loeb et al. 2009). The uncertainty due to the error in the geostationary satellite–derived cloud properties is, however, not explicitly included in the current uncertainty estimate. Investigating the effect of geostationary–derived cloud properties on the surface downward shortwave irradiance is necessary to further understand the uncertainty in the surface net irradiance.

Acknowledgments We thank Drs. Graeme Stephens, Tristan L'Ecuyer, Martin Wild, Tom Charlock, and Tom Ackerman for useful discussions and two anonymous reviewers for providing constructive comments. We also thank Ms. Amber Richards for proof reading the manuscript. The work was supported by the NASA's CERES and NASA Energy and Water Cycle Study (NEWS) projects. CERES data are supplied by the NASA Langley Research Center Atmospheric Sciences Data Center. ARM data is made available through the U.S. Department of Energy as part of the Atmospheric Systems Research Program. NOAA SURFRAD (Augustine et al. 2000) and GMD data are made available through the NOAA's Earth System Research Laboratory/Global Monitoring Division - Radiation (G-RAD). The Woods Hole Oceanographic Institute, Upper Ocean Processes (UOP) group, supplies NTAS and STRATUS buoy data.

Appendix: Surface Observation Spatial Sampling Noise Estimate

To understand the spatial sampling noise within a 1 degree grid box, we use two sets of *modeled* surface irradiances extracted from CERES products, CRS and SYN. CRS is a level 2 product that contains instantaneous modeled surface irradiances at the local Terra overpass time. SYN is a level 3 product that contains temporally averaged estimates of surface irradiance, where hourly modeled irradiances represent the entire 1° × 1° grid box. We subset CRS irradiances within 25 km from surface sites and average them over a month. The resulting mean irradiance simulates surface observations at overpass time averaged over a month. We also subset SYN monthly mean irradiances from a subset of grid boxes that coincide with the Terra satellite overpass time. The 62 grid boxes were selected from a larger set used for CERES validation. Some contain surface observation sites and some simply represent ecosystems of interest. If the box contains a surface observation site, the 25-km-radius region from which CRS is averaged is centered at that site (though in this analysis no surface observations are used). If there is no surface site, then the CRS footprints fall within 25 km of the center of the 1° grid box. We then

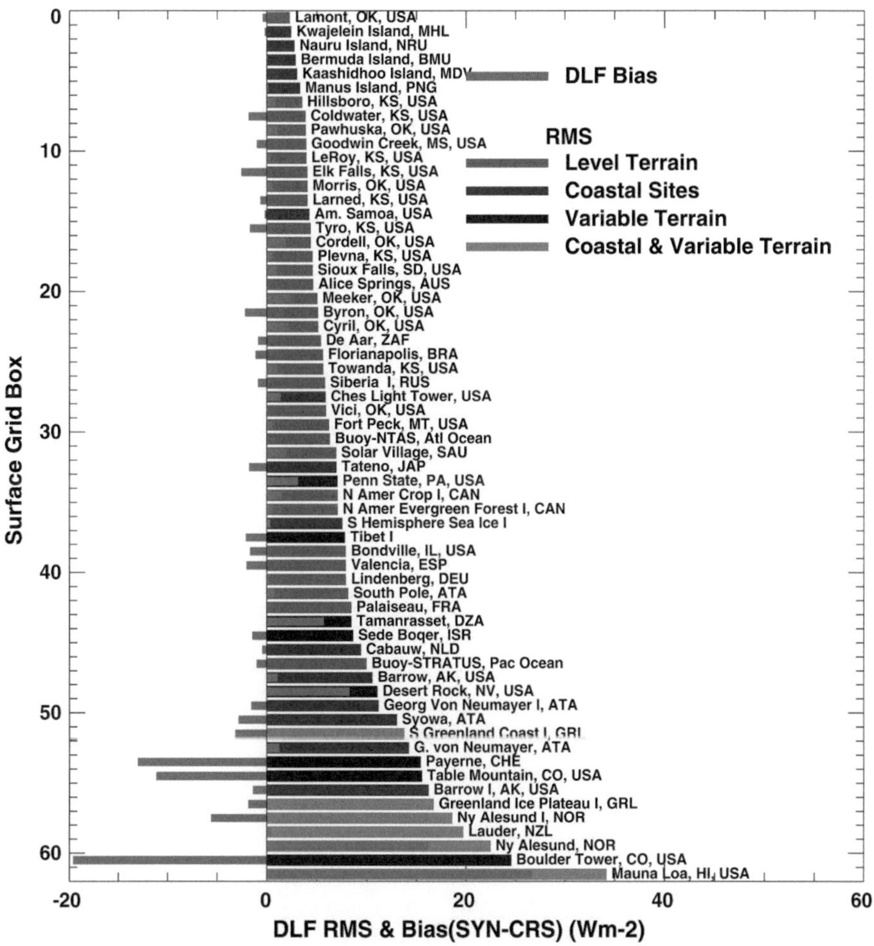

Fig. 5 Sampling noise estimate for 62 surface sites. The RMS difference and bias are estimated comparing modeled downward longwave irradiances at Terra local overpass time within 25 km from the surface location averaged over a month and 1° × 1° gridded monthly mean irradiances within 30 min from overpass time (i.e., an hour box contains the local overpass time). Six Julys from 2000 through 2005 are used

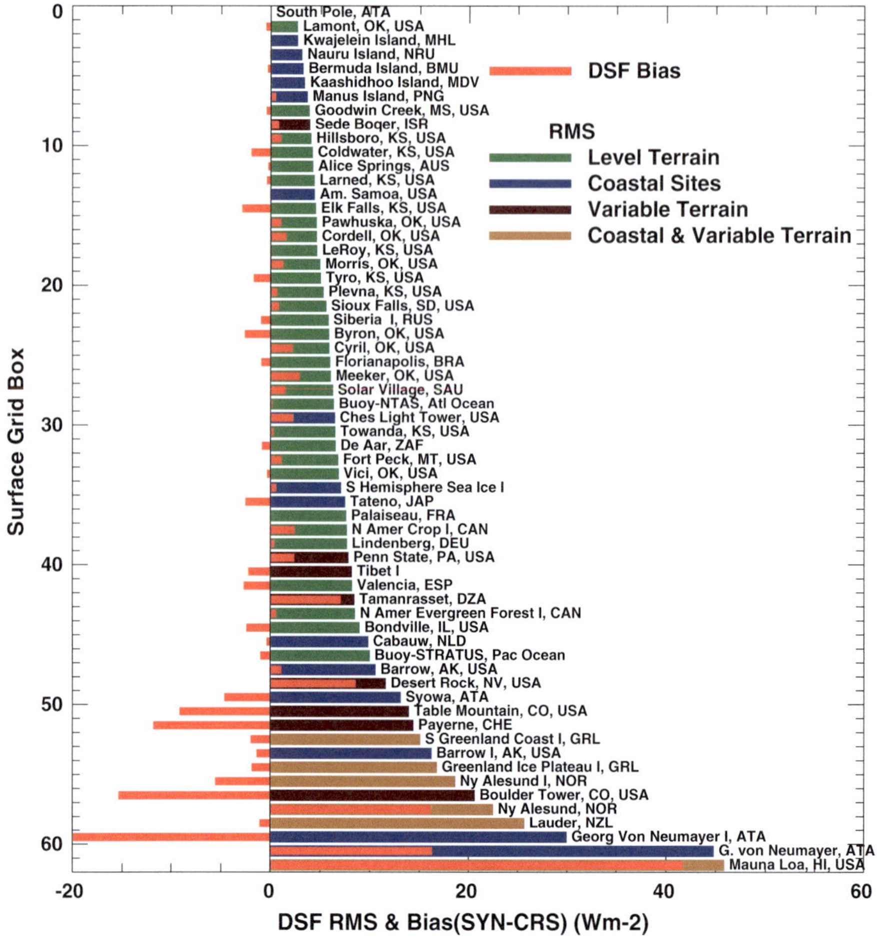

Fig. 6 Same as Fig. 5 but for the surface downward shortwave irradiance

compute the RMS difference between the CRS and SYN monthly means for each locations using 6 Julys (2000 through 2005). Figures 5 and 6 show the histogram of the RMS differences. Surface locations are separated into four different groups: uniform terrain, coastal, variable terrain, and coastal and variable terrain. The 26 sites used in Fig. 3 are a subset of those shown in Figs. 5 and 6 that contain surface observations and are designated as uniform terrain (Figs. 5, 6).

References

Augustine JA, DeLuisi JJ, Long CN (2000) SURFRAD-a national surface radiation budget network for atmospheric research. Bull Amer Met Soc 81(10):2341–2358

Bloom SA et al (2005) Documentation and validation of the Goddard Earth Observing System (GEOS) data assimilation system version-4, Technical Report Series on global modeling and data assimilation, 104606, 26

Charlock TP, Rose FG, Rutan DA, Jin Z, Kato S (2006) The global surface and atmosphere radiation budget: an assessment of accuracy with 5 years of calculations and observations. In: Proceedings of 12th conference on atmospheric radiation, 10–14, July, Madison Wisconsin. Available at http://snowdog.larc.nasa.gov/cave/

Gulbrandsen A (1978) On the use of pyranometers in the study of spectral solar radiation and atmospheric aerosols. J Appl Meteorol 17:899–904

Gupta SK, Ritchey NA, Wilber AC, Whitlock CH, Gibson GG, Stackhouse PW Jr (1999) A climatology of surface radiation budget derived from satellite data. J Clim 12:2691–2710

Haeffelin M, Kato S, Smith AM, Rutledge K, Charlock TP, Mahan JR (2001) Determination of the thermal offset of the Eppley precision spectral pyranometer. Appl Opt 40:472–484

Hatzianastassiou N, Matsoukas C, Fotiadi A, Pavlakis KG, Drakakis E, Hatzidmitriou D, Vardavas I (2005) Global distribution of Earth's surface shortwave radiation budget. Atmos Chem Phys 5:2847–2867

Jin Z, Charlock TP, Smith WL Jr, Rutledge K (2004) A look-up table for ocean surface albedo. Geophys Res Lett 31:L22301

Kandel R, Viollier M (2010) Observation of the Earth's radiation budget from space. C R Geosci 342:286–300. doi:10.1016/j.crte.2010.01.005

Kato SF, Rose G, Rutan DA, Charlock TP (2008) Cloud effects on the meridional atmospheric energy budget estimated from clouds and the Earth's radiant energy system (CERES) data. J Clim 4223–4241

Kato S, Rose FG, Sun-Mack S, Miller WF, Chen Y, Rutan DA, Stephens GL, Loeb NG, Minnis P, Wielicki BA, Winker DM, Charlock TP, Stackhouse PW, Xu K-M, Collins W (2011) Computation of top-of-atmosphere and surface irradiances with CALIPSO, CloudSat, and MODIS-derived cloud and aerosol properties. J Geophys Res 116:D19209. doi:10.1029/2011JD016050

Kim D, Ramanathan V (2008) Solar radiation budget and radiative forcing due to aerosols and clouds. J Geophys Res 113:D02203. doi:10.1029/2007JD008434

L'Ecuyer TS, Wood NB, Haladay T, Stephens GL, Stackhouse PW Jr (2008) Impact of clouds on atmospheric heating based on the R04 CloudSat fluxes and heating rate data set. J Geophys Res 113, D00A15, doi:10.1029/2008JD009951

Loeb GN, Wielicki BA, Doelling DR, Smith GL, Keyes DF, Kato S, Manalo-Smith N, Wong T (2009) Toward optimal closure of the Earth's top-of-atmosphere radiation budget. J Clim 22:748–766

Michalsky JJ, Anderson GP, Barnerd J, Delamere J, Gueymard C, Kato S, Kiedron P, McComiskey A, Ricchiazzi P (2006) Shortwave radiative closure studies for clear skies during the atmospheric radiation measurement 2003 aerosol intensive observation period. J Geophys Res 111, D14S90, doi:10.1029/2005JD006341

Minnis P, Sun-Mack S, Trepte QZ, Chang F-L, Heck PW et al (2010) CERES Edition 3 cloud retrievals. AMS 13th conference atmospheric radiation. Portland, OR, June 27–July 2, 5.4

Moody EG, King MD, Platnick S, Schaaf CB, Gao Feng (2005) Spatially complete global surface albedos: value-added datasets derived from Terra MODIS land products. IEEE Trans Geosci Remote Sens 43:144–158

Minnis P et al (2011) CERES Edition-2 cloud property retrievals using TRMM VIRS and terra and aqua MODIS data. Part I: Algorithms. IEEE Trans Geosci Remote Sens 49, 11, doi:10.1109/TGRS.2011.2144601 (in press)

Niu X, Pinker RT, Cronin MF (2010) Radiative fluxes at high latitude. Geophys Res Lett 37:L20811. doi:10.1029/2010GL044606

Ohmura A, Dutton EG, Forgan B, Frohlich C, Gilgen H, Hegne H, Heimo A, Konig-Langlo G, McArthur B, Muller G, Philipona R, Pinker R, Whitlock CH, Dehne K, Wild M (1998) Baseline surface radiation network (BSRN/WCRP): new precision radiometery for climate change research. Bull Amer Meteor Soc 79(10):2115–2136

Pinker RT, Laszlo I (1992) Modeling surface solar irradiance for satellite applications on global scale. J Appl Meteorol 31:194–211

Remer LA, Kaufman YJ, Tanré D, Mattoo S, Chu DM et al (2005) The MODIS aerosol algorithm, products, and validation. J Atmos Sci 62:947–973

Rienecker MM et al (2008) The GOES-5 data assimilation system-documentation of versions 5.0.1, 5.1.0, and 5.2.0. In: Suarez M (ed) NASA technical report series on global modeling and data assimilation, vol 27. NASA/TM-2008-105606

Rose F, Charlock T, Fu Q, Kato S, Rutan D, Jin Z (2006) CERES proto-edition 3 radiative transfer: tests and radiative closure over surface validation sites. In: Proceedings of 12th conference on atmospheric radiation (AMS), 10–14 July 2006, Madison, Wisconsin. Available at http://snowdog.larc.nasa.gov/cave/

Rose FG, Charlock TP, Rutan DA, Smith GL (1997) Tests of a constrainment algorithm for the surface and atmospheric radiation budget. Extended Abstract, 9th Conference of Atmospheric Radiation, Long Beach, CA, Feb 2–7

Rossow WB, Zhang Y-C (1995) Calculation of surface and top of atmosphere radiative fluxes from physical quantities based on ISCCP data set 2. Validation and first results. J Geophys Res 100(D1):1167–1197

Rutan D, Rose F, Roman M, Manalo-smith N, Schaaf C, Charlock T (2009) Development and assessment of broadband surface albedo from clouds and the earth's radiant energy system clouds and radiation swath data product. J Geophys Res 114:D08125. doi:10.1029/2008JD010669

Stephens GL, Wild M, Stackhouse P Jr, Ecuyer TL', Kato S (2011) The global character of the flux of downward longwave radiation. Submitted to J Clim

Su H, Wood EF, Wang H, Pinker RT (2008) Spatial and temporal scaling behaviour of surface shortwave downward radiation based on MODIS and in situ measurements. IEEE Geosci Remote Sens Let 5:542–546

Stephens GL et al (2008) CloudSat mission: performance and early science after the 1 year of operation. J Geophys Res 113, D00A18, doi:10.1029/2008JD009982

Taylor BN, Kuyatt CE (1994) Guidelines for evaluating and expressing the uncertainty of NIST measurement results, National Institute of Standard and Technology technical note 1297

Trenberth KE, Fasullo JT, Kiehl J (2009) Earth's global energy budget. Bull Amer Meteor Soc. doi: 10.1175/2008BAMS2634.1

Wang H, Pinker RT (2009) Shortwave radiative fluxes from MODIS: model development and implementation. J Geophys Res 114:D20201. doi:10.1029/2008JD010442

Wang J, Rossow WB, Zhang Y (2000) Cloud vertical structure and its variations from 20-year global rawinsonde dataset. J Clim 13:3041–3056

Wielicki BA, Cess RD, King MD, Randall DA, Harrison EF (1995) Mission to planet earth: role of clouds and radiation in climate. Bull Am Meteor Soc 76:2125–2153

Wild M, Ohmura A, Gilgen H, Morcrette J–J, Slingo A (2001) Evaluation of downward longwave radiation in general circulation models. J Clim 14:3227–3239

Winker DM, Pelon J, Coakley JA Jr, Ackerman SA, Charlson RJ, Colarco PR, Flamant P, Fu Q, Hoff R, Kittaka C, Kubar TL, LeTreut H, McCormick MP, Megie G, Poole L, Powell K, Trepte C, Vaughan MA, Wielicki BA (2010) The CALIPSO mission: a global 3D view of aerosols and clouds. Bull Am Meteorol Soc 91:1211–1229. doi:10.1175/2010BAMS3009.1

Wong T, Wielicki BA, Lee RB III, Smith GL, Bush KA, Willis JK (2006) Reexamination of the observed decadal variability of the earth radiation budget using altitude-corrected ERBE/ERBS nonscanner WFOV data. J Clim 19:4028–4040

Young DF, Minnis P, Doelling DR, Gibson GG, Wong T (1998) Temporal interpolation methods for the clouds and the earth's radiant energy system (CERES) experiment. J Appl Meteor 37:572–590

Zhang Y-C, Rossow WB, Lacis AA (1995) Calculation of surface and top of atmosphere radiative fluxes from physical quantities based on ISCCP data sets 1. Method and sensitivity to input data uncertainties. J Geophys Res 100:1149–1165

Zhang Y-C, Rossow WB, Lacis AA, Oinas V, Mishchenko MI (2004) Calculation of radiative fluxes from the surface to top of atmosphere based on ISCCP and other global data sets: refinements of the radiative transfer model and the input data. J Geophys Res 109:D19105. doi:10.1029/2003JD004457

Zhang Y-C, Rossow WB, Stackhouse PW Jr (2006) Comparison of different global information sources used in surface radiative flux calculation: radiative properties of the surface. J Geophys Res 112: D01102. doi:10.1029/2005JD007008

Surv Geophys (2012) 33:413–426
DOI 10.1007/s10712-011-9150-2

Tracking Earth's Energy: From El Niño to Global Warming

Kevin E. Trenberth · John T. Fasullo

Received: 17 May 2011 / Accepted: 26 September 2011 / Published online: 14 October 2011

Abstract The state of knowledge and outstanding issues with respect to the global mean energy budget of planet Earth are described, along with the ability to track changes over time. Best estimates of the main energy components involved in radiative transfer and energy flows through the climate system do not satisfy physical constraints for conservation of energy without adjustments. The main issues relate to the downwelling longwave (LW) radiation and the hydrological cycle, and thus the surface evaporative cooling. It is argued that the discrepancy is 18% of the surface latent energy flux, but only 4% of the downwelling LW flux and, for various reasons, it is most likely that the latter is astray in some calculations, including many models, although there is also scope for precipitation estimates to be revised. Beginning in 2000, the top-of-atmosphere radiation measurements provide stable estimates of the net global radiative imbalance changes over a decade, but after 2004 there is "missing energy" as the observing system of the changes in ocean heat content, melting of land ice, and so on is unable to account for where it has gone. Based upon a number of climate model experiments for the twenty-first century where there are stases in global surface temperature and upper ocean heat content in spite of an identifiable global energy imbalance, we infer that the main sink of the missing energy is likely the deep ocean below 275 m depth.

Keywords Earth's energy · Global warming · El Niño · Ocean heat content · Atmospheric radiation budget · Climate change

1 Introduction

Weather and climate on Earth are determined by the amount and distribution of incoming radiation from the sun. For a steady-state climate, global mean outgoing longwave

The National Center for Atmospheric Research is sponsored by the National Science Foundation.

K. E. Trenberth (✉) · J. T. Fasullo
National Center for Atmospheric Research, Boulder, CO 80307, USA
e-mail: trenbert@ucar.edu

radiation (OLR) necessarily balances the incoming absorbed solar radiation (ASR), but with redistributions of energy within the climate system to enable this to happen on a global basis. Incoming radiant energy may be scattered and reflected by clouds and aerosols or absorbed in the atmosphere. The transmitted radiation is then either absorbed or reflected at the Earth's surface. Radiant solar (shortwave) energy is transformed into sensible heat, latent energy (involving different water states), potential energy, and kinetic energy before being emitted as longwave infrared energy. Energy may be stored, transported in various forms, and converted among the different types, giving rise to a rich variety of weather or turbulent phenomena in the atmosphere and ocean. Moreover, the energy balance can be upset in various ways, changing the climate and associated weather.

Kiehl and Trenberth (1997) reviewed past estimates of the global mean flow of energy through the climate system and presented a best estimate of the budget based on various measurements and models, by taking advantage of various closure constraints. They also performed a number of radiative computations to examine the spectral features of the incoming and outgoing radiation and determined the role of clouds and various greenhouse gases in the overall radiative energy flows. At the top-of-atmosphere (TOA) values relied heavily on observations from the Earth Radiation Budget Experiment (ERBE) from 1985 to 1989, when the TOA values were approximately in balance.

Fasullo and Trenberth (2008a) provide an assessment of the global energy budgets at TOA and the surface, for the global atmosphere, and ocean and land domains based on a synthesis of satellite retrievals, reanalysis fields, a land surface simulation, and ocean temperature estimates. As well as ERBE data, they made use of the newly available Clouds and the Earth's Radiant Energy System (CERES) measurements. They constrained the TOA budget to match estimates of the global imbalance associated with changes in atmospheric composition and climate. They included an assessment of sampling errors and the differences between the ERBE and CERES measurements. There is an annual mean transport of energy by the atmosphere from ocean to land regions of 2.2 ± 0.1 PW (Petawatts $= 10^{15}$ W) primarily in the northern winter when the transport exceeds 5 PW. Fasullo and Trenberth (2008b) went on to evaluate the temporal and spatial characteristics of meridional atmospheric energy transports for ocean, land, and global domains, while Trenberth and Fasullo (2008) delved into the ocean heat budget in considerable detail and provided an observationally based estimate of the mean and annual cycle of ocean energy divergence and a comprehensive assessment of uncertainty.

2 The Global Energy Budget

Trenberth et al. (2009) updated the Kiehl and Trenberth (1997) global energy flow diagram (Fig. 1) based on CERES measurements from March 2000 to November 2005, which include improvements in retrieval methodology and hardware, including its exploitation of MODIS (Moderate Resolution Imaging Spectro-radiometer) retrievals for scene identification, and they discussed continuing sources of uncertainty. These results built on the results of Fasullo and Trenberth (2008a, b) to update other parts of the energy cycle and flows through the atmosphere. To help understand sources of error and the discrepancies among various estimates, a breakdown of the budgets into land and ocean domains and a consideration of the annual and diurnal cycles were included. The TOA energy imbalance can probably be most accurately determined from climate models, and Fasullo and Trenberth (2008a) deduced the imbalance to be 0.9 W m^{-2}, where the error bars are ± 0.5 W m^{-2}. Figure 1 is modified slightly from that originally published, as discussed

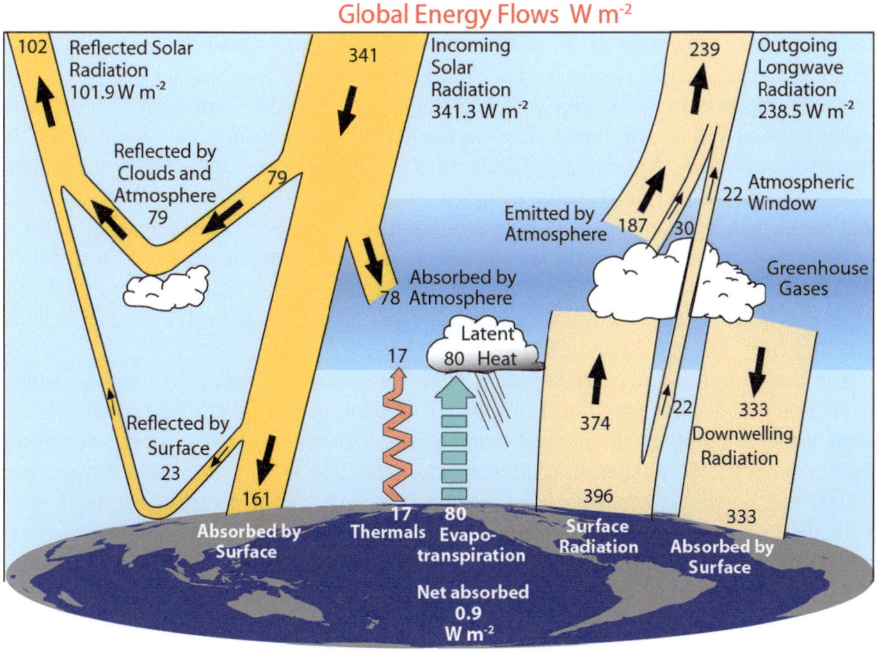

Fig. 1 The global annual mean earth's energy budget for 2000–2005 (W m^{-2}). The *broad arrows* indicate the schematic flow of energy in proportion to their importance. Adapted from Trenberth et al. (2009) with changes noted in the text

below. Uncertainties are discussed in Trenberth et al. (2009) and are mostly not dealt with here, except to highlight some sources of discrepancy with other studies.

In Fig. 1, use has been made of conservation of energy and the assumption that, on a time scale of years, the change in heat storage within the atmosphere is very small. Accordingly, the net radiation at TOA R_T is the sum of the ASR minus the OLR: $R_T = \text{ASR} - \text{OLR}$. In turn, the ASR is the difference between the incoming solar radiation and the reflected solar radiation. At the surface, the ASR has to be offset by the sensible heat and latent heat fluxes plus the net longwave radiation. The latter is made up of two large terms: the emitted radiation from the surface and the downwelling longwave radiation coming back from the atmosphere. Both at the surface and TOA the imbalance is the same and, as noted above, is estimated to be 0.9 W m^{-2}.

Updates in Trenberth et al. (2009) included revised absorption in the atmosphere by water vapor and aerosols, since Kim and Ramanathan (2008) found that updated spectroscopic parameters and continuum absorption for water vapor increased the absorption by 4–6 W m^{-2}. The sensible heat has values of 17, 27, and 12 W m^{-2} for the globe, land, and ocean (just over 70% of the Earth), and, even with uncertainties of 10%, the errors are only order 2 W m^{-2}. There is widespread agreement that the global mean surface upward longwave (LW) radiation is about 396 W m^{-2}, which is dependent on the skin temperature and surface emissivity (Zhang et al. 2006).

Global precipitation should equal global evaporation for a long-term average, and estimates are likely more reliable of the former. However, there is considerable uncertainty in precipitation over both the oceans and land (Trenberth et al. 2007b; Schlosser and Houser 2007). The latter is mainly due to wind effects, undercatch, and sampling, while the former is due to shortcomings in remote sensing. Global Precipitation Climatology Project (GPCP) values (Huffman et al. 2009) are considered most reliable for precipitation (Trenberth et al. 2007b), while results from CloudSat (e.g., Stephens and Haynes 2007) may help improve on these, with prospects mainly for increases in precipitation owing to undersampling low warm clouds. Consequently, the GPCP values are considered to be likely somewhat low. Accordingly, Trenberth et al. (2009) increased the GPCP values over oceans by 5% and the global value assigned was 80.0 W m^{-2} (2.76 mm/day). The latent heat flux values were apportioned between ocean and land as in Trenberth et al. (2007a) by assuming an observationally based runoff into the ocean of 40×10^3 km^3 $year^{-1}$ (Trenberth et al. 2007a). The land precipitation from GPCP is 2.06 mm/day, while the ocean precipitation was deduced to be 3.06 mm/day, reasonably close to estimates of latent heat flux from Yu and Weller (2007).

The downward and net LW radiation were computed as a residual. After the adjustments noted above for latent heat and better accounting for the effects of aerosols and water vapor in the ASR, the revised estimates are 333 and 63 W m^{-2} for the downward and net surface LW. Wild et al. (2001) proposed that 344 W m^{-2} is a best estimate from models but noted that considerable uncertainties exist and especially that there were problems in accurate simulation of thermal emission from a cold, dry, cloud-free atmosphere, and a dependence on water vapor content. The latter may relate to the formulation of the water vapor continuum. The correct simulation of low clouds is also a continuing challenge for models and is likely to also exist as a source of major model bias in downward LW flux. Costa and Shine (2011) have recomputed the amount of radiation from the surface that actually reaches space under clear sky and cloudy conditions, which is a preferred method than simply assigning a value based on the so-called atmospheric window, as Kiehl and Trenberth (1997) did, and the value globally is 22 W m^{-2}. This has been incorporated in the version of the global atmospheric flows in Fig. 1.

It has been argued that downward LW radiation is likely to be underestimated owing to the view from satellites which will miss underlying low clouds and overestimate cloud base height. Zhang et al. (2006) found that the surface LW flux was very sensitive to assumptions about tropospheric water vapor and temperatures but did not analyze the dependence on clouds. Yet, the characteristics of clouds on which the back radiation is most dependent, such as cloud base, are not well determined from conventional space-based measurements and hence the need for missions such as CloudSat (e.g., Stephens et al. 2002; Haynes and Stephens 2007). Preliminary downwelling LW radiation estimates based on Cloudsat variables indeed suggest values even higher (347 ± 7 W m^{-2}; ±1 σ) (Kato et al. 2011). Yet, considerable uncertainty remains, even in this estimate. For example, there are sources of error in how the 3-D heterogeneity of clouds is treated, and there is no unique way to treat the effects of overlap on the downward flux, although Cloudsat information can help. For clouds at all levels, emissivity assumptions will affect the estimated downward LW flux, and the amount of water vapor between the surface and the cloud base is a challenge to quantify. In the tropics, the effect of continuum absorption also strongly affects the impact of cloud emission on surface LW fluxes.

From surface closure constraints, however, it is very difficult to accommodate these higher estimates of the downwelling LW radiation, which if changed from 333 to 347 W m^{-2} would require an accommodation of an extra 14 W m^{-2} in other terms of the surface budget. The estimate of the incoming solar at the surface is, if anything, low, as the aerosol and water vapor absorption is now more likely to be slightly overestimated. Indeed, Kato et al. (2011) have a value 8 W m^{-2} higher, further exacerbating their surface energy imbalance. The sensible heat and outgoing surface LW radiation estimates are fairly well constrained. This leaves only the surface evaporative cooling, which is balanced by global precipitation, as the term to accommodate such an increase. The value used for the surface latent energy flux is already increased over available estimates, as noted above. Our current assessment is that further uncertainties might enable that value to increase to perhaps 85 W m^{-2}, but that seems to be the upper limit of current uncertainties in precipitation retrieval. The extra 14 W m^{-2} makes up 18% of the surface latent energy flux but is only 4% of the downwelling LW flux. Accordingly, our assessment is that it is most likely that the latter is seriously astray in some calculations, including many models.

3 Changes in Energy Balance Over the Past Decade

With the successes of CERES, variability in the net radiative incoming energy at the TOA can now be measured to within 0.1 W m^{-2} $year^{-1}$. Thus, a key objective is to track the flow of such anomalies through the system over time in order to address the question as to how variability in energy fluxes is linked to climate variability. The main energy reservoir is the ocean, and the exchange of energy between the atmosphere and ocean is ubiquitous, so that heat once sequestered can resurface at a later time to affect weather and climate on a global scale. Thus, a change in the energy balance has consequences, sooner or later, for the climate. Moreover, we have observing systems in place that nominally can measure the major storage and flux terms, but due to errors and uncertainty it remains a challenge to track anomalies with confidence.

A climate event, such as the drop in surface temperatures over North America in 2008 (Perlwitz et al. 2009), is often stated to be due to natural variability, as if this fully accounts for what has happened. Aside from weather events that primarily arise from instabilities in the atmosphere, natural climate variability has a cause. Its origins may be external to the climate system: a change in the sun, a volcanic eruption, or Earth's orbital changes that ring in the major glacial to interglacial swings. Or its origins may be internal to the climate system and arise from interactions between the atmosphere, oceans, cryosphere, and land surface, which depend on the very different thermal inertia of these components.

3.1 El Niño

As an example of natural variability, the biggest El Niño in the modern record by many measures occurred in 1997–1998. Successful warnings were issued a few months in advance regarding the unusual and disruptive weather across North America and around the world and were possible in part because the energy that sustains El Niño was tracked in the ocean by a new moored buoy observing system in the Tropical Pacific. Typically prior to an El Niño, in La Niña conditions, the cold sea waters in the central and eastern tropical Pacific create high atmospheric pressure and clear skies, with plentiful sunshine heating the ocean waters. The ocean currents redistribute the ocean heat which builds up in the tropical western Pacific Warm Pool until an El Niño provides relief (Trenberth et al. 2002). The spread of warm waters across the Pacific in collaboration with changing winds in turn promotes evaporative cooling of the ocean, moistening the atmosphere, and fueling tropical storms and convection over and around the anomalously warm waters. The changed atmospheric heating alters the jet streams and storm tracks and influences weather patterns for the duration of the event (Trenberth et al. 1998).

In 2007–2008, a strong La Niña event that spilled over to the 2008–2009 northern winter had direct repercussions for cooler weather across North America and elsewhere

(Perlwitz et al. 2009). But, by June 2009, the situation had reversed as the next El Niño emerged and grew to be a moderate event, with temperatures in the top 150 m of the ocean above normal by as much as 5°C across the equatorial Pacific in December 2009. Multiple storms barreled into Southern California in January 2010, consistent with expectations from the El Niño. The El Niño continued until May 2010 but abruptly reversed to become a strong La Niña by July 2010.

We can often recognize these changes once they have occurred, and they permit some level of climate forecast skill. But a major challenge is to be able to track the energy associated with such variations more thoroughly: Where did the heat for the 2009–2010 El Niño actually come from? Where did the heat suddenly disappear to during the La Niña? Past experience (Trenberth et al. 2002) suggests that global surface temperature rises at the end of and lagging El Niño, as heat comes out of the Pacific Ocean mainly in the form of moisture that is evaporated and which subsequently rains out, releasing the latent energy. Meanwhile, maximum warming of the Indian and Atlantic Oceans occurs about 5 months after the El Niño owing to sunny skies and lighter winds (less evaporative cooling), while the convective action is in the Pacific. This led to a vigorous hurricane season in the Atlantic in 2010 and extensive flooding in China and India in July, and Pakistan in August 2010 in association with the much above normal sea surface temperatures (SSTs), while the La Niña refocused action to occur in these regions and away from the Pacific domain. Very high SSTs in the Gulf of Mexico and tropical North Atlantic favored an active North Atlantic hurricane season and record rains in Colombia. Subsequently, the high SSTs around and north of Australia promoted the flooding in Queensland in December 2010 and January 2011, even as very cold conditions occurred in Europe and North America. Yet a holistic analysis of how such variations relate to the anomalous flow of energy through the climate system is generally lacking. Future work will benefit from recently released updates in the CERES dataset to span a full decade (e.g., EBAF ed2.5), but many key datasets continue to lack the duration and stability required when addressing these questions.

3.2 Anthropogenic Climate Change

The human influence on climate, arising mostly from the changing composition of the atmosphere (IPCC 2007), also affects energy flows. Increasing concentrations of carbon dioxide and other greenhouse gases have led to a post-2000 imbalance at the TOA of 0.9 ± 0.5 W m^{-2} (Trenberth et al. 2009) (Fig. 1) that produces “global warming” or, more correctly, an energy imbalance. Tracking how much extra energy has gone back to space (Murphy et al. 2009) and where this energy has accumulated is possible, with reasonable closure for 1993–2003 (IPCC 2007); see Fig. 2. Over the past 50 years, the oceans have absorbed about 90% of the total heat added to the climate system, while the rest goes to melting sea and land ice, and warming the land surface and atmosphere. Because carbon dioxide concentrations have further increased since 2003, the amount of heat subsequently being accumulated should be even greater.

While the planetary imbalance at TOA is too small to measure directly from satellite, instruments are far more stable than they are absolutely accurate with calibration stability <0.3 Wm^{-2} per decade (95% confidence) (Loeb et al. 2009). Tracking relative changes in Earth’s energy by measuring solar radiation in and infrared radiation out to space, and thus changes in the net radiation, seems to be at hand (Wong et al. 2009). This includes tracking the slight decrease in solar insolation from 2000 until 2009 with the ebbing 11-year sunspot cycle, enough to offset 10–15% of the estimated net human induced warming (Trenberth 2009).

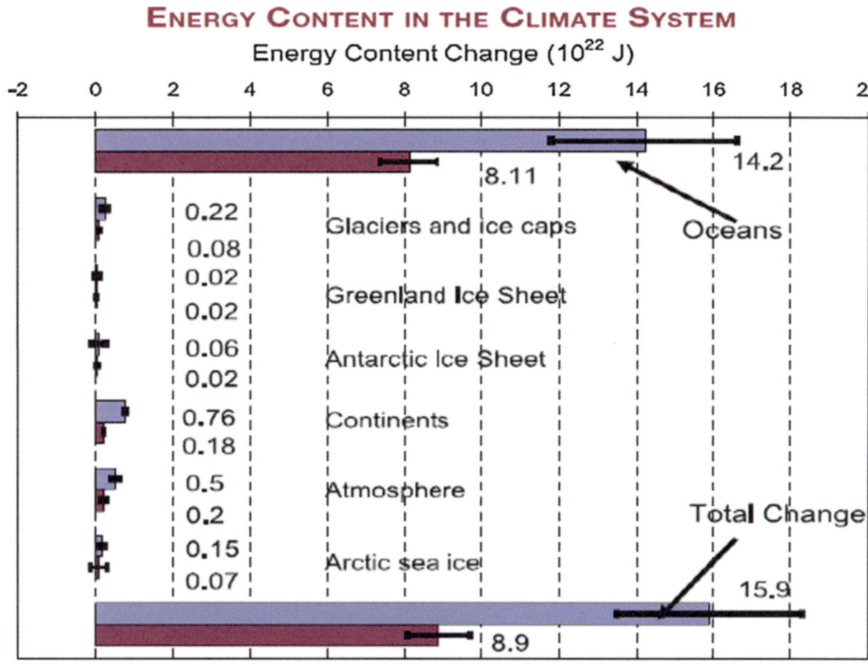

Fig. 2 Energy content changes in different components of the earth system for 2 periods (1961–2003 and 1993–2003). *Blue bars* are for 1961–2003; burgundy *bars* are for 1993–2003. Positive energy content change means an increase in stored energy (i.e., heat content in oceans, latent heat from reduced ice or sea ice volumes, heat content in the continents excluding latent heat from permafrost changes, and latent and sensible heat and potential and kinetic energy in the atmosphere). All *error* estimates are 90% CI. No estimate of confidence is available for the continental heat gain. Some of the results have been scaled from published results for the two respective periods. From (IPCC 2007, Figs. TS.15 and 5.4)

In 2008 for the tropical Pacific during La Niña conditions, extra TOA energy absorption was observed as expected (Wong et al. 2009); see Fig. 3. The Niño 3.4 SST index is also plotted on this figure, and the slightly delayed response of the OLR to cooler conditions in the record and especially in 2008 is clear. However, the decrease in OLR with cooler conditions is accompanied by an increase in ASR as clouds decrease in amount, leaving a pronounced net heating (>1.5 W m^{-2}, cf. Loeb et al. 2009) of the planet in the cooler conditions. And so this raises the question as to whether a coherent perspective that accounts for both TOA and ocean variability can be constructed from the available observations. By 2004, the ocean observing system had reached new capabilities, as some 3,000 Argo floats populated the ocean for the first time to provide regular temperature soundings of the upper 2,000 m, giving new confidence in the ocean heat content assessment. But ocean temperature measurements from 2004 to 2008 suggested a substantial slowing of the increase in global ocean heat content (Levitus et al. 2009; Lyman et al. 2010), precisely during the time when CERES estimates depict an increase in the planetary imbalance.

A complete assessment of all the energy in the climate system from multiple datasets (Trenberth and Fasullo 2010) is given in Fig. 4. In this figure, adapted from Trenberth and Fasullo (2010), an older preliminary version of CERES data was used and there are small differences and a slightly reduced overall trend compared with Fig. 3. The dashed black

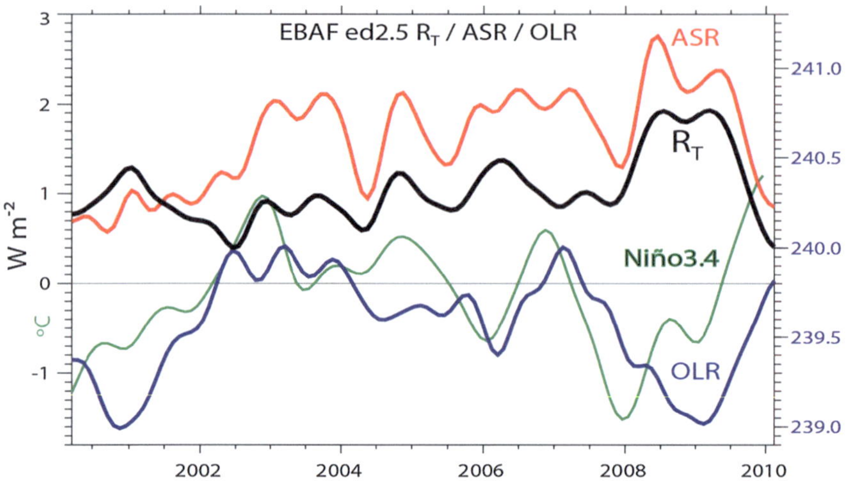

Fig. 3 Recently updated net radiation from the TOA from CERES as EBAF Ed2.5 http://www.ceres.larc.nasa.gov/products.php?product=EBAF. The ASR (*red*) and OLR (*blue*) are given on the *right axis*, and the net (R_T = ASR − OLR) is given on the *left axis* (note the change in *scale*) in W m^{-2}. Also shown is the Niño 3.4 SST index (*green*) (*left axis*, °C). The decadal low pass filter is a 13-term filter used in IPCC (2007), making it similar to a 12-month running mean

curve shows the revised CERES TOA radiation time series, updated to 2009 (from Fig. 3). Also the TOA radiation is not smoothed as much as in the published figure, thereby showing the interannual variability.

Since 1992, sea level observations from satellite altimeters at millimeter accuracy reveal a global increase of ~3.2 mm $year^{-1}$ as a fairly linear trend (Fig. 4), although with two main blips corresponding to an enhanced rate of rise during the 1997–1998 El Niño and a brief slowdown in the 2007–2008 La Niña. Since 2003, the detailed gravity measurements from Gravity Recovery and Climate Experiment (GRACE) of the change in glacial land ice and water show an increase in mass of the ocean. This so-called eustatic component of sea level rise may have compensated for the decrease in the thermosteric (heat related expansion) component (Cazenave et al. 2009; Leuliette and Miller 2009). However, for a given amount of heat, 1 mm of sea level rise can be achieved much more efficiently—by a factor of 40–70 typically—by melting land ice rather than expanding the ocean (Trenberth 2009). So although some heat has gone into the record breaking loss of Arctic sea ice, and some has undoubtedly contributed to unprecedented melting of Greenland (van den Broeke et al. 2009) and Antarctica (Chen et al. 2009), these anomalies are unable to account for much of the measured TOA energy imbalance (Fig. 4). This gives rise to the concept of "missing energy".

To emphasize the discrepancy, Fig. 5 presents an alternative version of Fig. 2 for 1992–2003, as a contrast to 2004–2008. The accounting for all terms and the net imbalance is compatible with physical expectations and climate model results. Here, the net imbalance is about 0.7 W m^{-2} at TOA for 1992–2003 (note this is simply the sum of the terms, and was not measured at TOA). However, for the 2004–2008 period, when we do have TOA radiation measurements, the decrease in solar radiation associated with the sunspot cycle and the quiet sun in 2008 contributes somewhat, but the Ocean Heat Content (OHC) change is a lot less than in the previous period (Lyman et al. 2010) and a residual imbalance term, the missing energy, is required.

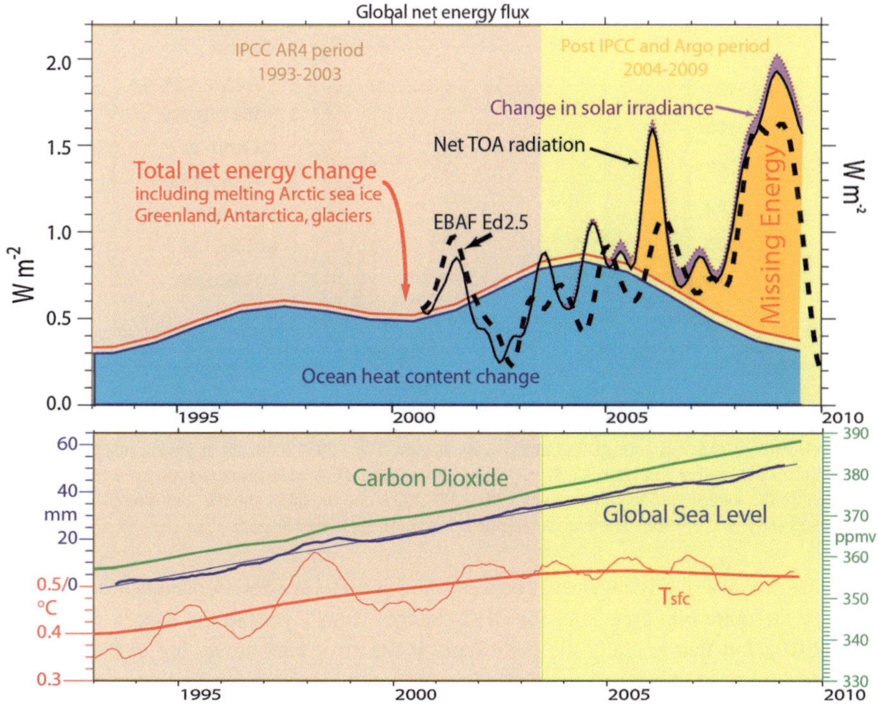

Fig. 4 The disposition of energy entering the climate system is estimated. The observed changes (*lower panel*; Trenberth and Fasullo 2010) show the 12-month running means of global mean surface temperature anomalies relative to 1901–2000 from NOAA [*red* (*thin*) and *decadal* (*thick*)] in °C (*scale lower left*), CO_2 concentrations (*green*) in ppmv from NOAA (*scale right*), and global sea level adjusted for isostatic rebound from AVISO (*blue*, along with linear trend of 3.2 mm/year) relative to 1993, scale at left in mm). Rates of change of global energy in W m^{-2} (*top panel*) are contrasted between the AR4-IPCC era 1993–2003 and the post-2003 Argo era. From 1992 to 2003, the decadal OHC changes (*blue*) along with the contributions from melting glaciers, ice caps, Greenland, Antarctica, and Arctic sea ice plus small contributions from land to atmosphere warming (*red*) suggest a total warming for the planet of 0.6 ± 0.2 W m^{-2} (95% *error bars*). After 2000, preliminary observations from TOA (*black*) referenced to the 2000 values, as used in Trenberth and Fasullo (2010), show an increasing discrepancy (*gold*) relative to the total warming observed (*red*). The quiet sun changes in total solar irradiance reduce the net heating slightly (*purple*), but a large energy component is missing (*gold*). The recently revised TOA radiation from CERES, known as EBAF Ed2.5, is shown by the *black dashed curve*. The decadal filter is from Trenberth et al. (2007b). Adapted from Trenberth and Fasullo (2010). The monthly global surface temperature data are from NCDC, NOAA: http://www.ncdc.noaa.gov/oa/climate/research/anomalies/index.html; the global mean sea level data are from AVISO satellite altimetry data: http://www.aviso.oceanobs.com/en/news/ocean-indicators/mean-sea-level/; and the CO_2 at Mauna Loa data are from NOAA http://www.esrl.noaa.gov/gmd/ccgg/trends/

One can question why, in Fig. 4, the CERES curve was attached to the total energy change curve at the beginning of the record and, indeed, this was arbitrary but based on the sense that the energy balance was reasonably explained at that time. Clearly, the variability in the CERES curve is at odds with that of the oceans, and in the original publication it was appropriately smoothed. But, by including the shorter-term variability, we can now see how much the changes in the CERES processing have altered things. It is difficult to assign error bars to all of the terms because structural errors are the main ones of importance; these arise from systematic sampling and instrumental errors for example. One interpretation of Fig. 4

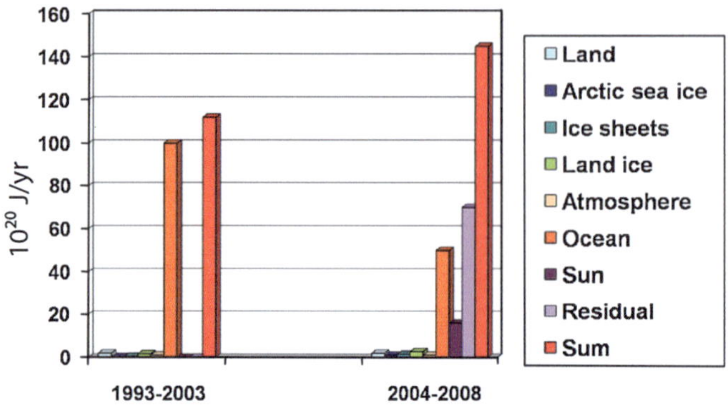

Fig. 5 The energy entering the climate system is estimated for the various components: warming of the atmosphere and land, OHC increase, melting of glaciers and ice caps (land ice), melting of the major ice sheets (Greenland and Antarctica), and changes in the sun. For 1993–2003, these are summed to give the total which is equivalent to about 0.7 W m^{-2}. For 2004–2008, TOA measurements are used to provide an increment to the total based on comparisons with 2000–2003, and the quiet sun has contributed, but the sum is achieved only if a spurious residual is included. Units are 10^{20} J/year

is that there is no real missing energy because the error bars are not adequately accounted for and they are quite big, especially for OHC changes from 1 year to the next (e. g., Lyman et al. 2010), but that is really the point here. If the error bars are so big that there is no mismatch, then the values are next to useless because we cannot say anything useful about interannual or longer-term variations in energy in the climate system. As discussed below, it seems likely that undersampling of the ocean, especially the deep ocean, may well account for the main discrepancy as it is manifested at only certain times (when La Niña is present).

3.3 Modeling Temperature Stasis for a Decade

Further inroads into this problem will no doubt become possible as datasets are brought up to date and refined. In the meantime, we have explored the extent to which this kind of behavior occurs in the latest version of the NCAR CCSM version 4. We have examined 5 runs for the twenty-first century under the Representative Concentration Pathways RCP 4.5, which is a forcing of the climate system under increasing greenhouse gases and changing aerosols into the twenty-first century under a stabilization of values that results in a radiative forcing of 4.5 W m^{-2} by the late twenty-first century, with constant forcing beyond. The details of this scenario do not matter for our current purposes. Figure 6 shows global mean surface temperature for each of the ensemble members. For one run (001), 2-decade-long periods are highlighted where there is a stasis in the surface warming. We have examined what happens with regard to energy flows during such intervals for all ensemble members and a consistent picture emerges.

At the TOA, the radiative imbalance is known exactly under these circumstances (Fig. 7). One example (Fig. 7, top left panel) shows that the net radiation at the TOA (R_T) is order 1 W m^{-2} into the climate system. Clearly the planet imbalance is considerable, despite the stasis at the surface, and so where is the energy going?

Examination of the changes in OHC shows clearly that this is the main sink. Indeed, the full-depth OHC continues relentlessly upwards (Fig. 7), with no hesitation at all. However, the upper OHC for the top 275 m shows the same stasis as for the surface temperature

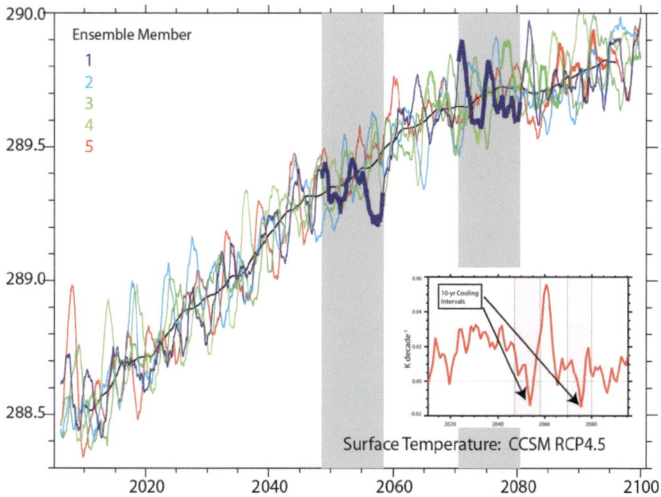

Fig. 6 Global mean surface temperatures (shown in degrees Kelvin) from CCSM4 for 5 runs into the twenty-first century, under the RCP4.5. The overall increasing temperatures are clear, and the average of the 5 runs is given by the *black line*. However, individual members of the ensemble show decades or longer with no increase: the *inset* at *lower right* shows the decadal rates of change for the *blue* run

(Fig. 7), in fact for this example the OHC (0–275 m) actually decreases. In this example (top right panel Fig. 7), roughly half of the heat is between 275 and 700 m depth, and the rest is below 700 m depth. Hence, the OHC integrated down to 700 m shows some slowdown, but the implication is that the missing heat is being deposited mainly in the region below 700 m depth. The time series also show that a swift recovery can occur at the end of the stasis period.

Further preliminary exploration of where the heat is going suggests that it is mainly in the Pacific between 40°S and 30°N and is associated with the negative phase of the Pacific Decadal Oscillation and/or La Niña events. However, this aspect is very preliminary and will be examined in much more detail elsewhere.

4 Conclusions

Closure of the observed energy budget over the past 5 years is elusive (Trenberth 2009; Trenberth and Fasullo 2010) although preliminary analysis of model results suggests that the ocean has absorbed considerably more heat than reported by observations, particularly below 700 m. Song and Colberg (2011) made similar conclusions using different reasoning based on satisfying the sea level change budget. Thus, state-of-the-art observations and basic analysis are unable to fully account for recent energy variability, since they either provide an incoherent narrative or imply error bars too large to make the products useful. Only by using conservation and physical principles can we infer the likely resolution. Was the May 2009 to May 2010 El Niño a manifestation of some of the missing energy reappearing? Certainly, the overall warmth of 2010 and its manifestations around the world, as the natural variability reinforced the global warming signal, would suggest this as a reasonable hypothesis. Or is the stasis continuing or restarting with a new La Niña taking over in the 2010–2011 northern winter? It may take a few years to gain perspective on this.

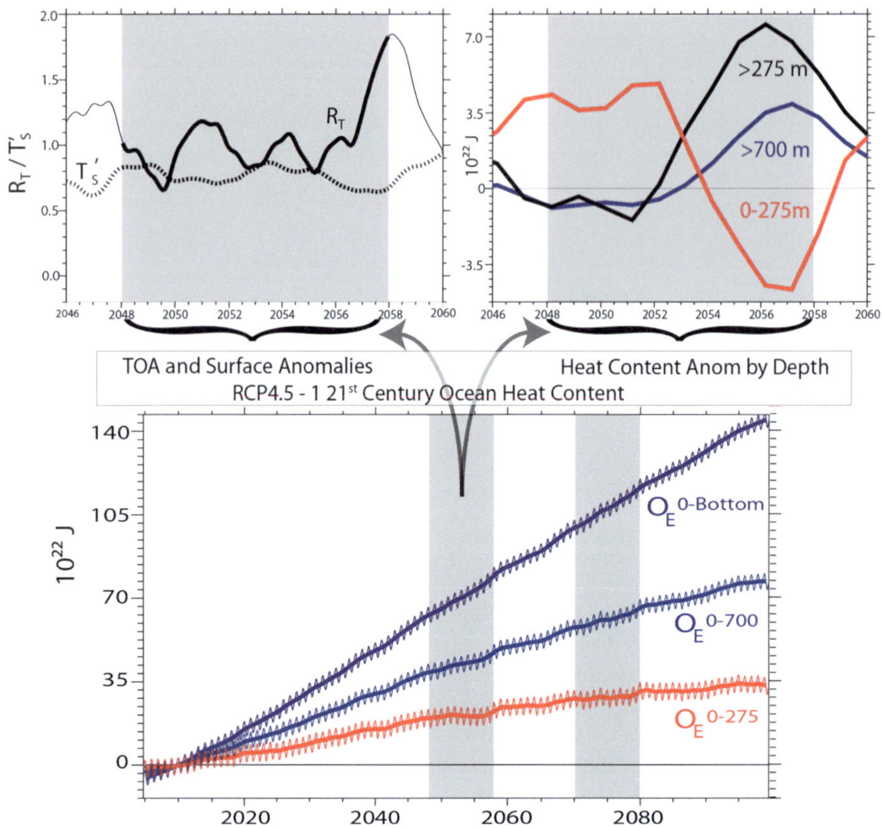

Fig. 7 The lower panel shows the perturbation relative to mean from 2005 to 2015 in OHC (O_E) for layers 0–275, 0–700 m depth, and for the entire *column* to the *bottom* of the ocean in 10^{22} J. The *thin curve* shows the full values and a 12-month running mean is also given. The *gray vertical bars* show times of stasis in the global mean surface temperature. The *top left panel* shows the TOA energy balance for the first stasis period 2048–2058 for the net radiation (R_T), along with the global mean surface temperature perturbation. The *top right panel* shows the OHC perturbation for 0–275, >275, and >700 m depth in 10^{22} J where the 20 years running mean has been removed

Proposals for addressing global warming now include geo-engineering whereby tiny particles are injected into the stratosphere to emulate the cooling effects of stratospheric aerosol of a volcanic eruption (Levitt and Dubner 2009). Implicitly, such proposals assume understanding and control of this process which requires detailed tracking of energy within the climate system. How can we understand whether the strong cold outbreaks of December 2009 and 2010 are simply a natural weather phenomenon, as they seem to be, or are part of some mysterious change in clouds or pollution, if we do not have adequate measurements? Tracking Earth's global energy and how it is partitioned is essential for understanding what is happening in the climate system and thus for attributing causes and predicting what comes next. It is vital information for planning adaptation to and coping with climate change.

Acknowledgments This research is partially sponsored by NASA under grant NNX09AH89G.

References

Cazenave A et al (2009) Sea level budget over 2003–2008: a reevaluation from GRACE space gravimetry satellite altimetry and Argo. Glob Planet Chang 65:83–88. doi:10:1-16/j.gloplacha.2008.10.004

Chen JL, Wilson CR, Blankenship D, Tapley BD (2009) Accelerated Antarctic ice loss from satellite gravity measurements. Nat Geosci 2:859–862

Costa SMS, Shine KP (2011) Outgoing longwave radiation due to directly-transmitted surface emission. J Atmos Sci (submitted).

Fasullo JT, Trenberth KE (2008a) The annual cycle of the energy budget: Pt I. Global mean and land-ocean exchanges. J Clim 21:2297–2313

Fasullo JT, Trenberth KE (2008b) The annual cycle of the energy budget: Pt II. Meridional structures and poleward transports. J Clim 21:2314–2326

Haynes JM, Stephens GL (2007) Tropical oceanic cloudiness and the incidence of precipitation: early results from CloudSat. Geophys Res Lett 34:L09811. doi:10.1029/2007GL029335

Huffman GJ, Adler RF, Bolvin DT, Gu G (2009) Improving the global precipitation record: GPCP version 2.1. Geophys Res Lett 36:L17808. doi:10.1029/2009GL040000

IPCC (2007) In: Solomon S et al (eds) Climate change 2007: the physical science basis. Cambridge University Press, New York, p 996

Kato S, et al. (2011) Improvements of top-of-atmosphere and surface irradiance computations with CALIPSO-, CloudSat-, and MODIS-derived cloud and aerosol properties. J Geophys Res 116. doi:10.1029/2011JD016050

Kiehl JT, Trenberth KE (1997) Earth's annual global mean energy budget. Bull Am Meteor Soc 78:197–208

Kim D, Ramanathan V (2008) Solar radiation and radiative forcing due to aerosols. J Geophys Res 113:D02203. doi:10.1029/2007JD008434

Leuliette EW, Miller L (2009) Closing the sea level rise budget with altimetry, Argo, and GRACE. Geophys Res Lett 36:L04608. doi:10.1029/2008GL036010

Levitt S, Dubner S (2009) Superfreakonomics: global cooling, patriotic prostitutes, and why suicide bombers should buy life insurance. Harper Collins, New York, p 288. ISBN: 10 0060889579

Levitus S et al (2009) Global ocean heat content 1955–2008 in light of recently revealed instrumentation problems. Geophys Res Lett 36:L07608. doi:10.1029/2008GL037155

Loeb NG et al (2009) Towards optimal closure of the earth's top-of-atmosphere radiation budget. J Clim 22:748–766

Lyman JM, Good SA, Gouretski VV, Ishii M, Johnson GC, Palmer MD, Smith DM, Willis JK (2010) Robust warming of the global upper ocean. Nature 465:334–337

Murphy DM et al (2009) An observationally based energy balance for the earth since 1950. J Geophys Res 114:D17107. doi:10.1029/2009JD012105

Perlwitz J, Hoerling M, Eischeid J, Xu T, Kumar A (2009) A strong bout of natural cooling in 2008. Geophys Res Lett 36:L23706. doi:10.1029/2009GL041188

Schlosser CA, Houser PR (2007) Assessing a satellite-era perspective of the global water cycle. J Clim 20:1316–1338

Song YT, Colberg F (2011) Deep ocean warming assessed from altimeters, gravity recovery and climate experiment, in situ measurements, and a non-Boussinesq ocean general circulation model. J Geophys Res 116:C02020. doi:10.1029/2010JC006601

Stephens GL, Haynes JM (2007) Near global observations of the warm rain coalescence process. Geophys Res Lett 34:L20805. doi:10.1029/2007GL030259

Stephens GL et al (2002) The CloudSat mission and the A-Train. Bull Am Meteor Soc 83:1771–1790

Trenberth KE (2009) An imperative for adapting to climate change: tracking earth's global energy. Curr Opin Env Sustain 1:19–27

Trenberth KE, Fasullo JT (2008) An observational estimate of ocean energy divergence. J Phys Oceanogr 38:984–999

Trenberth KE, Fasullo JT (2010) Tracking earth's energy. Science 328:316–317

Trenberth KE et al (1998) Progress during TOGA in understanding and modeling global tele connections associated with tropical sea surface temperatures. J Geophys Res 103:14291–14324

Trenberth KE, Caron JM, Stepaniak DP, Worley S (2002) Evolution of El Niño southern oscillation and global atmospheric surface temperatures. J Geophys Res 107(D8):4065. doi:10.1029/2000JD000298
Trenberth KE, Smith L, Qian T, Dai A, Fasullo J (2007a) Estimates of the global water budget and its annual cycle using observational and model data. J Hydrometeorol 8:758–769
Trenberth KE et al (2007b) Observations: surface and atmospheric climate change. In: Solomon S, Qin D, Manning M, Chen Z, Marquis MC, Averyt KB, Tignor M, Miller HL (eds) Climate change 2007. The physical science basis. Contribution of WG 1 to the fourth assessment report of the intergovernmental panel on climate change. Cambridge University Press, Cambridge, pp 235–336
Trenberth KE, Fasullo JT, Kiehl J (2009) Earth's global energy budget. Bull Am Meteor Soc 90:311–323
van den Broeke M et al (2009) Partitioning recent Greenland mass loss. Science 326:984–986
Wild M, Ohmura A, Gilgen H, Morcrette JJ, Slingo A (2001) Evaluation of downward longwave radiation in general circulation models. J Clim 14:3227–3239
Wong T, Stackhouse PW Jr, Kratz DP, Wilber AC (2009) Earth radiation budget at top-of-atmosphere [in "state of the climate in 2008"]. Bull Am Meteor Soc 90:S33–S34
Yu LS, Weller RA (2007) Objectively analyzed air-sea heat fluxes for the global ice-free oceans (1981–2005). Bull Am Meteor Soc 88:527–539
Zhang Y, Rossow WB, Stackhouse PW Jr (2006) Comparison of different global information sources used in surface radiative flux calculation: radiative properties of the near-surface atmosphere. J Geophys Res 111:D13106. doi:10.1029/2005JD006873

Surv Geophys (2012) 33:427–443
DOI 10.1007/s10712-012-9183-1

Changes in Earth's Energy Flows and Clouds in 228-Year Simulation with a High-Resolution AGCM

Masato Sugi

Received: 13 June 2011 / Accepted: 5 February 2012 / Published online: 14 March 2012

Abstract We have examined long-term changes in Earth's energy flows at top of the atmosphere (TOA) and at Earth's surface (land and ocean) by using 228-year simulation of a high-resolution global atmosphere model, MRI-AGCM3.2. It is found that the net downward short wave (SW) radiation (absorbed solar radiation, ASR) at TOA significantly increases during twenty-first century in agreement with a previous study. However, in the present study, the reason for the change is an increase in clear sky SW absorption by increased water vapor in the atmosphere, while it is a decrease in cloud amount in the previous study. It is also found that the long wave (LW) cloud radiative forcing for atmosphere is positive and increasing during twenty-first century in agreement with a previous study. The reason for the change in the present study is an increase in absorption by water vapor of the downward LW radiation emitted from clouds, while it is reductions of cloud amount in the middle troposphere in the previous study.

Keywords Earth's energy flow · Clouds · Climate change · High-resolution model

1 Introduction

The issue of how the radiative forcing induced by increased greenhouse gases (GHG) and aerosols changes the Earth's energy flows is one of the most important issues to understand the recent and future climate change. Trenberth and Fasullo (2009) (hereafter TF09) examined long-term changes in the Earth's energy flows at top of the atmosphere (TOA) by using Third Coupled Model Intercomparison Project (CMIP3) models. They found that the absorbed solar radiation (ASR) will significantly increase during twenty-first century under the A1B emission scenario mainly due to decreasing cloud amounts. They further pointed out that there is a large uncertainty in the projected changes in energy flow due to the large uncertainty in the clouds in the CMIP3 models.

M. Sugi (✉)
Japan Agency for Marine-Earth Science and Technology, Yokohama, Kanagawa-ken, Japan
e-mail: msugi@jamstec.go.jp

The change in energy flows at TOA is the change in energy input to the atmosphere and Earth's surface (land and ocean). Another important issue is how this energy input is partitioned into the atmosphere and the Earth's surface. Particularly, the global hydrological cycle is expected to be strongly affected by the change in energy input to the atmosphere. The global precipitation is constrained by the energy balance of the atmosphere rather than the availability of moisture (Mitchell et al. 1987; Stephens et al. 1994; Allen and Ingram 2002; Sugi and Yoshimura 2004). It has been shown that the global precipitation will increase but the tropical circulation will become weaker in the future due to global warming (Sugi et al. 2002; Held and Soden 2006; Vecchi and Soden 2007).

Stephens and Ellis (2008) (hereafter SE08) examined the long-term change in global precipitation in CMIP3 models. They found that global precipitation efficiency is reduced by a negative cloud radiative heating feedback through reduction of cloud amount in the middle troposphere. To examine the change in atmospheric energy budget, we need to know the changes in energy flows at the Earth's surface as well in addition to the changes at TOA. In the present study, we examine the long-term changes in Earth's energy flows not only at TOA and but also at the Earth's surface by using a long-term simulation of a high-resolution global atmosphere model to study the long-term changes in atmospheric energy budget.

High-resolution models (horizontal resolution of about 100 km or higher) have been extensively used in some climate change studies, particularly in the studies of the climate change impact on tropical cyclones. One interesting finding regarding the tropical cyclone and climate change is a reduction of global tropical cyclone (TC) frequency in a future GHG warmed climate (Bengtsson et al. 1996; Sugi et al. 2002). Later, this reduction of TC frequency has been confirmed by many high-resolution model studies (Knutson et al. 2010). An interesting but yet disputable issue is the reason for the global TC reduction. Sugi et al. (2002) argued that the reduction of global TC frequency is closely related to a weakening of tropical circulation.

If the changes in tropical circulations are closely related to the change in global atmospheric energy budget, and if the changes in clouds significantly affect the change in atmospheric energy budget, then the uncertainty in the clouds among the various models may be responsible for the uncertainty in the changes in tropical circulations. In this regard, we are concerned about how the uncertainty in clouds in a model may affect the energy budget of the atmosphere. In the present study, therefore, we examine the role of clouds in terms of cloud radiative forcing (CRF) at TOA and at the surface.

A brief description of the high-resolution model and the long-term simulation is given at Sect. 2. The evolution of global mean energy fluxes at TOA and the Earth's surface, and the atmospheric heating due to the convergence of the energy fluxes, as well as the evolution of global mean CRFs are presented in Sect. 3.

Global annual maps of the energy fluxes and CRFs are presented in Sect. 4, followed by discussion and conclusion in Sects. 5 and 6, respectively.

2 Model and Simulation

The model we used for the 228-year simulation is a high-resolution (horizontal grid size of 60 km) version of the new Meteorological Research Institute (MRI) Atmospheric General Circulation Model, MRI-AGCM3.2 (Mizuta et al. 2012). The model has 64 levels in the vertical with top at 0.01 hPa. The model was developed from the Japan Meterological Agency (JMA) Global Numerical Weather Prediction (NWP) Model (JMA, 2007) and shares many physical process subroutines, such as radiation scheme, PBL scheme, land surface scheme, and so on, with the NWP model, but some new physics schemes are incorporated. For MRI-AGCM3.2, we employed a new cumulus convection scheme developed by H. Yoshimura based on Tiedtke (1989) (Yukimoto et al. 2011). This convection scheme considerably improved the climatology of tropical convections. For the stratiform cloud scheme, Tiedtke (1993) cloud scheme is used. In this cloud scheme, cloud water and cloud amount are the prognostic variables. We do not use the subtropical marine stratocumulus scheme (Kawai and Inoue 2006) used in JMA NWP model. The direct effect of aerosols is calculated in the MRI-AGCM3.2, but indirect effect of aerosols is not considered in the present simulation.

The 228-year simulation was conducted using the 60-km resolution version of the MRI-AGCM3.2 with prescribed sea surface temperature (SST) and atmospheric concentration of greenhouse gas (GHG) including CO_2 and aerosols. For the years from 1873 to 2003, the observed SST and sea-ice (HadISST1, Rayner et al. 2003) and observed GHG concentration (NASA, 2004) are prescribed. For the years from 2004 to 2099, the ensemble average SST increase projected by CMIP3 models are added to the de-trended observed SST from 1979 to 2003. In this procedure, the interannual variation of SST during 1979-2003 is repeatedly used. Similarly, the reduction of sea-ice is considered based on the ensemble average sea-ice change projected by CMIP3 models. The GHG and aerosols for the years 2004–2099 are based on the IPCC Special Report on Emission Scenario (SRES) A1B scenario. Aerosols from volcanic eruptions are not included except for the Mt. Pinatubo eruption in 1991.

Table 1 shows the simulated global mean energy flux components at TOA and the Earth's surface, averaged over 20-year period at the end of nineteenth, twentieth, and twenty-first century. The simulated energy fluxes at the end of twentieth century are generally in good agreement with observations (Trenberth et al. 2009). One of the largest differences between the simulation and observation is seen in the latent heat flux at the surface. The simulated value of latent heat flux is 85.6 W m^{-2} (corresponding to global mean precipitation of 2.95 mm/day), which is considerably larger than the value 80 W m^{-2} (precipitation of 2.76 mm/day) estimated from observation. We have also noted some systematic biases in the simulation of radiation fluxes at TOA (Figures not shown). The net downward short wave radiation (ASR) shows a distinct negative bias over the tropical ocean and positive bias at high latitudes, particularly over the southern oceans. These systematic biases in ASR are very similar to those of CMIP3 models (Trenberth and Fasullo 2010).

It should be noted that the 228-year simulation was not originally designed for the present study, and the experimental design may not be most suitable to study the changes in Earth's energy flows and clouds. However, the simulation provides us with a good opportunity to examine the long-term changes in Earth's energy flows and role of clouds in detail by using a high-resolution state-of-the-art atmospheric model.

3 Evolution of the Global Annual Mean Energy Fluxes

Figure 1 shows the changes from the end of nineteenth century in global annual mean downward energy fluxes at TOA and at the Earth's surface (land and ocean) and net heating of the atmosphere. Figure 1a shows the changes in downward energy fluxes at TOA. The net downward short wave (SW) radiation (ASR) changes very little during twentieth century and then increases at a rate about 2 W m^{-2}/100 years toward the end of twenty-first century. During the period from 1970 to 2020, the net downward long wave

Table 1 Global mean energy fluxes averaged over 20 years at the end of nineteenth, twentieth and twenty-first century

	19C	20C	21C	20C–19C	21C–19C
DSWT	341.3	341.3	341.3	0.0	0.0
USWT	101.8	101.9	100.0	0.1	−1.8
ULWT	241.7	241.0	242.8	−0.7	1.1
DSWB	194.2	193.5	191.0	−0.7	−3.2
USWB	30.9	30.4	28.3	−0.5	−2.6
DLWB	332.7	336.5	355.0	3.8	22.3
ULWB	396.2	398.7	412.5	2.5	16.3
ULHB	84.9	85.6	90.6	0.7	5.7
USHB	17.3	17.0	16.2	−0.3	−1.1
NDSWT	239.6	239.4	241.4	−0.2	1.8
NDLWT	−241.7	−241.0	−242.8	0.7	−1.1
NDFLXT	−2.2	−1.5	−1.4	0.7	0.8
NDSWB	163.4	163.2	162.7	−0.2	−0.7
NDLWB	−63.5	−62.2	−57.5	1.3	6.0
NDLHB	−84.9	−85.6	−90.6	−0.7	−5.7
NDSHB	−17.3	−17.0	−16.2	0.3	1.1
NDFLXB	−2.3	−1.7	−1.6	0.6	0.7
AHSW	76.2	76.2	78.7	0.0	2.5
AHLW	−178.2	−178.8	−185.3	−0.6	−7.1
AHLH	84.9	85.6	90.6	0.7	5.7
AHSH	17.3	17.0	16.2	−0.3	−1.1

Unit is Wm^{-2}. The right two columns show the changes from nineteenth century to twentieth century and twenty-first century, respectively

D downward flux, *U* upward flux, *N* net downward flux (D-U), *SW* short wave radiation, *LW* long wave radiation, *SH* sensible heat, *LH* latent heat, *T* top of the atmosphere (TOA), *B* bottom of the atmosphere (surface), *FLX* total flux, *AH* atmospheric heating

(LW) radiation at (−OLR) is larger than nineteenth century by about 0.7 W m^{-2} due to the increased greenhouse gases (GHG). Then, it decreases (i.e., OLR increases) toward the end of twenty-first century, as the temperature of the Earth's surface and atmosphere increases. From 1970 and afterward toward the end of twenty-first century, the net downward radiation at TOA (R_T) is about 0.7 W m^{-2} larger than that of nineteenth century, indicating a continuous warming of the Earth. Generally, the changes in ASR, −OLR, and R_T shown in Fig. 1a agree well with the changes of them shown in TF09 for CMIP3 models.

Figure 1b shows the changes in downward energy fluxes at surface. From 1980 and afterward, the downward LW radiation and the upward LH flux significantly increase. These two flux changes basically balance each other. There is a small decrease in downward SW radiation, which is almost balanced by the decrease in upward sensible heat flux. Note that throughout the twenty-first century the total net downward energy flux at the surface continues to be about 0.7 W m^{-2} larger than that of nineteenth century, indicating a continuous warming of the Earth's surface.

Figure 1c shows the changes in net heating of the atmosphere. During the twenty-first century, there is a significant increase in LW cooling, which is mainly balanced by a

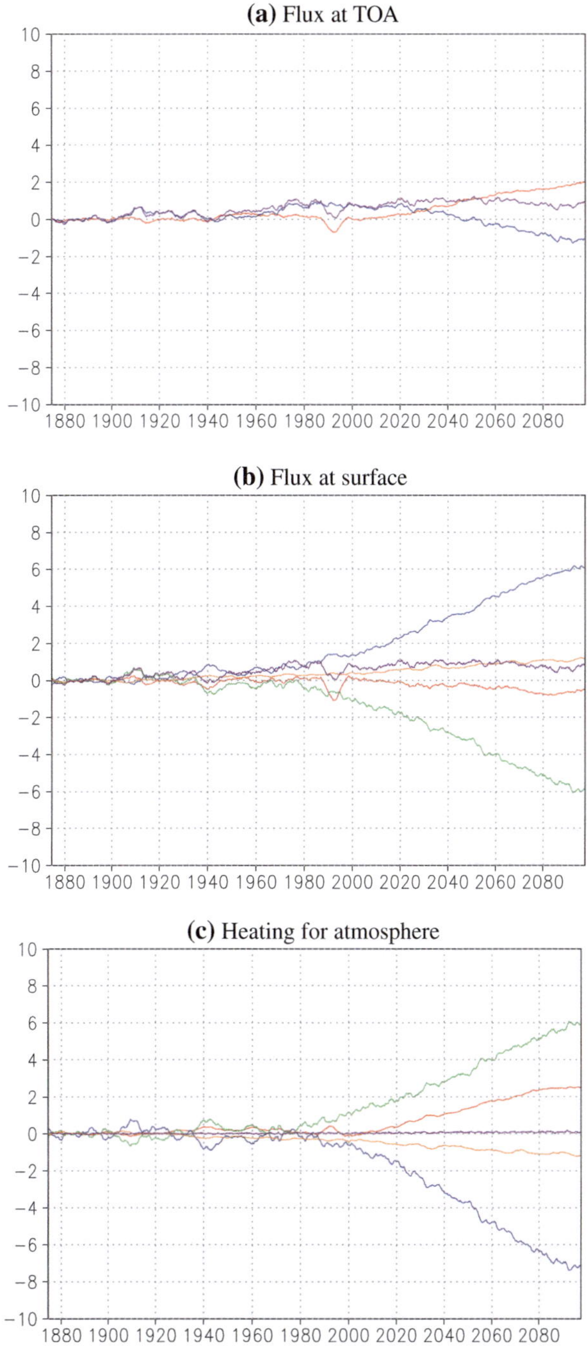

Fig. 1 Changes from the 20-year average at end of nineteenth century in global mean energy fluxes and heating of the atmosphere. Unit is W m^{-2}. **a** Net downward fluxes at top of the atmosphere. **b** Net downward fluxes at Earth's surface. **c** Net heating for the atmosphere. *Red* SW radiation, *blue* LW radiation, *orange* sensible heat, *green* latent heat, *purple* total energy flux or heating. Sixty-month running mean is applied for the curves to remove short time-scale variations

significant increase in latent heating. In the atmospheric energy balance, however, the increase in solar heating due to the increase in atmospheric moisture is significant and it is larger than the decrease in SH heating, and it is playing some important role in the atmospheric energy balance.

Table 2 shows the globally averaged CRF at TOA and Earth's surface and for the atmosphere, averaged over 20-year period at the end of nineteenth, twentieth, and twenty-first century. CRF represents the cloud effect on radiation and is defined as the difference between net downward radiation with cloud (R) and net downward clear sky radiation (CSR).

$$\mathrm{CRF} = R - \mathrm{CSR} \tag{1}$$

In the model, R and CSR are calculated at each grid point by a double call to the radiation routine with and without clouds. In this calculation, the same water vapor profile is used for R and CSR, and the water vapor effect of clouds is included as a part of cloud effect on radiation (Sohn and Bennartz 2008). Table 2 indicates that SW CRF at TOA is generally negative (cooling of atmosphere and surface), while LW CRF at TOA is generally positive (warming of atmosphere and surface). Table 2 also shows that SW CRF at surface is generally negative (cooling of surface), while LW CRF at surface is generally positive (warming of surface). On the other hand, both SW and LW CRF (cloud effect of radiative heating) for the atmosphere are generally positive (warming of atmosphere).

Figure 2 shows the changes from the end of nineteenth century in global annual mean cloud radiative forcing (ΔCRF) and clear sky radiation (ΔCSR). Note that the sum of ΔCRF and ΔCSR shown in Fig. 2 is equal to the changes in net downward radiation (or radiative heating of atmosphere) ΔR shown in Fig. 1.

$$\Delta R = \Delta\mathrm{CRF} + \Delta\mathrm{CSR} \tag{2}$$

Figure 2a shows the changes in CRF at TOA. During the twenty-first century, the negative SW CRF is increasing, indicating that the cooling effect of clouds is increasing toward the end of twenty-first century. On the other hand, LW CRF at TOA shows a slight decrease during twenty-first century, indicating that the warming effect of clouds is

Table 2 Global mean cloud radiative forcing (CRF) averaged over 20 years at the end of nineteenth, twentieth and twenty-first century

	19C	20C	21C	20C–19C	21C–19C
SWT	−43.3	−43.3	−44.8	0.0	−1.5
LWT	25.1	24.8	24.6	−0.2	−0.5
NETT	−18.2	−18.5	−20.2	−0.2	−2.0
SWB	−47.2	−47.1	−48.5	0.1	−1.3
LWB	21.9	21.4	20.0	−0.6	−1.9
NETB	−25.3	−25.8	−28.5	−0.5	−3.2
SWA	3.9	3.8	3.7	−0.1	−0.2
LWA	3.2	3.5	4.6	0.3	1.4
NETA	7.0	7.3	8.3	0.2	1.3

Unit is Wm^{-2}. The right two columns show the changes from nineteenth century to twentieth century and twenty-first century, respectively

SW short wave radiation, *LW* long wave radiation, *NET* net radiation, *T* CRF at top of the atmosphere (TOA), *B* CRF at bottom of the atmosphere (surface), *A* CRF for the atmosphere

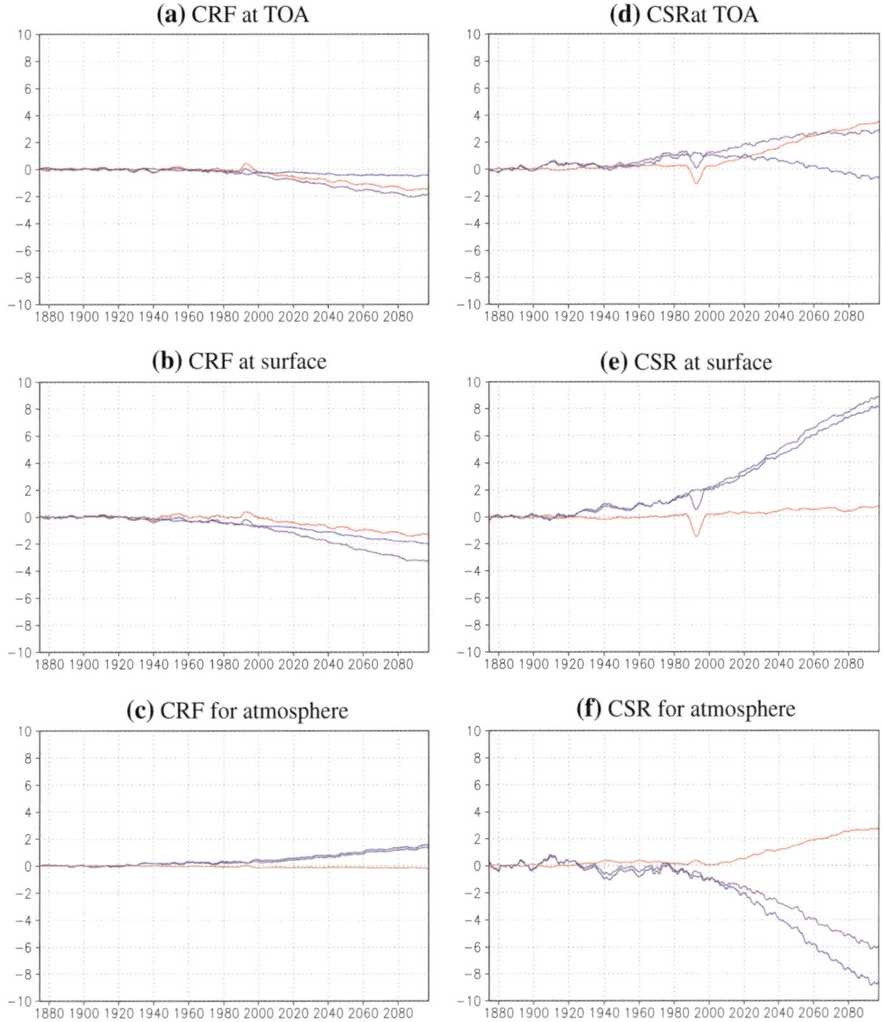

Fig. 2 Changes from the 20-year average at end of nineteenth century in global mean cloud radiative forcing (CRF) and clear sky radiation (CSR). Unit is W m^{-2}. **a** CRF at top of the atmosphere. **b** CRF at Earth's surface. **c** CRF for the atmosphere. **d** CSR at top of the atmosphere. **e** CSR at Earth's surface. **f** CSR for the atmosphere. *Red* SW, *blue* LW, *purple* net radiation CRF. Sixty-month running mean is applied for the curves to remove short time-scale variations

decreasing. Figure 2b shows the changes in CRF at the Earth's surface. SW CRF shows almost the same decrease as the decrease at TOA, while LW CRF shows a much larger decrease than that at TOA. Figure 2c shows the changes in CRF for the atmosphere (CRF at TOA minus CRF at surface), which represents the cloud effect of heating the atmosphere. For SW radiation, as the changes in CRF at TOA and CRF at surface are almost the same, the changes in SW CRF for the atmosphere are very small. On the other hand, for LW radiation, as there is much more decrease in CRF at surface than at TOA, LW CRF for the atmosphere (cloud effect of atmospheric radiative heating) is significantly increasing, and it is playing some important role in the global atmospheric energy balance. The small

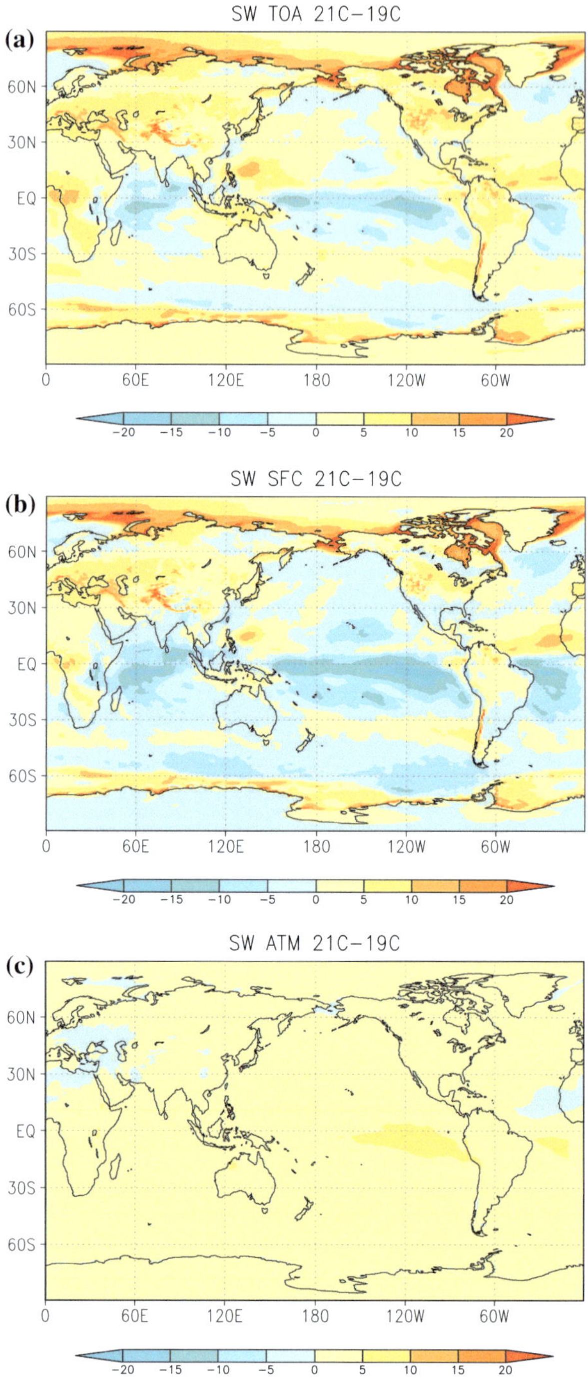

Fig. 3 Global maps of the changes from nineteenth to twenty-first century in annual mean SW radiation. Unit is W m^{-2}. **a** Net downward SW radiation at TOA. **b** Net downward SW radiation at surface. **c** Net SW heating of the atmosphere

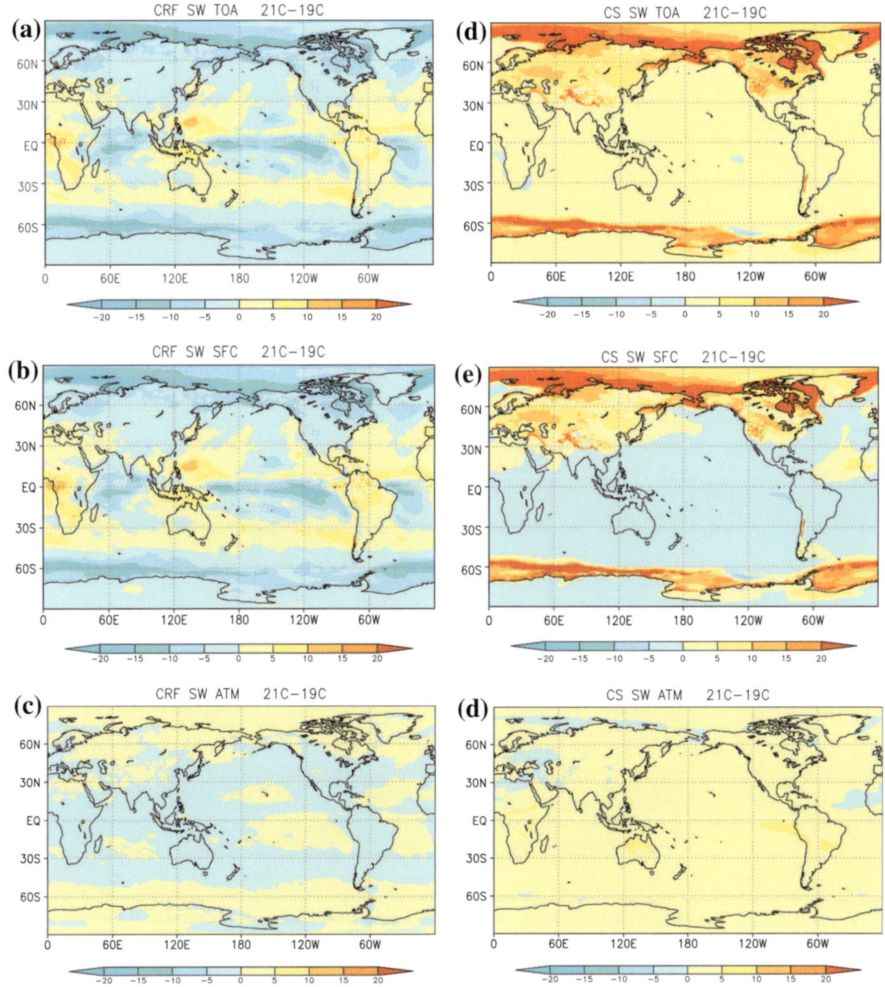

Fig. 4 Global maps of the changes from nineteenth to twenty-first century in annual mean SW CRF and SW CSR. Unit is W m^{-2}. **a** SW CRF at TOA. **b** SW CRF at surface. **c** SW CRF for the atmosphere. **d** SW CSR at TOA. **e** SW CSR at surface. **f** SW CSR for the atmosphere

change in SW CRF for the atmosphere and the significant increase in LW CRF for the atmosphere are consistent with the results of SE08.

Figure 2d shows the changes in CSR at TOA. During the twenty-first century, the SW CSR shows larger increase compared with the decrease in the SW CRF. This indicates that the change in SW radiation at TOA is dominated by the change in CSR (water vapor effect) rather than the change is CRF (cloud effect). Similarly, the change in LW radiation at TOA is also dominated by the change in CSR (water vapor effect) rather than the change in CRF (cloud effect). Figure 2e shows the changes in CSR at surface. During the twenty-first century, the SW CSR shows only a small increase, while the LW CSR shows a very large increase due to the increased GHG and water vapor. Figure 2f shows the changes in CSR for the atmosphere. Figures 1c and 2c, f indicate that both SW and LW heating of the atmosphere are dominated by CSR. The change in SW cloud effect on the atmosphere is

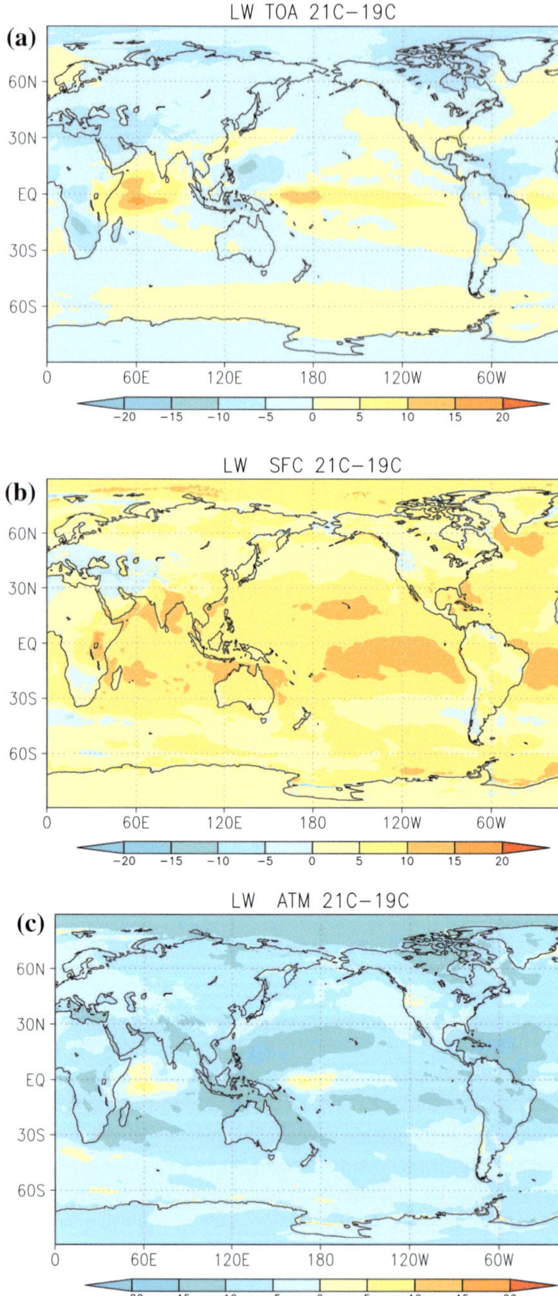

Fig. 5 Global maps of the changes from nineteenth to twenty-first century in annual mean LW radiation. Unit is W m^{-2}. **a** Net downward LW radiation at TOA. **b** Net downward LW radiation at surface. **c** Net LW heating of the atmosphere

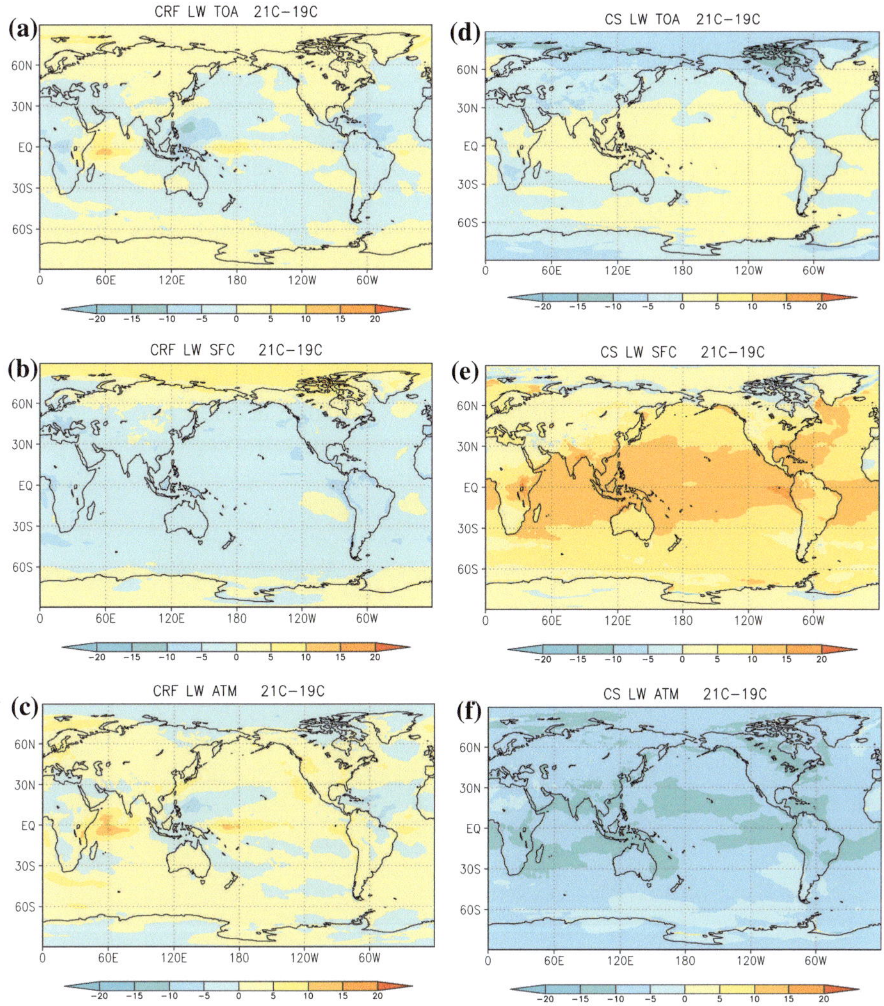

Fig. 6 Global maps of the changes from nineteenth to twenty-first century in annual mean LW CRF and LW CSR. Unit is W m^{-2}. **a** LW CRF at TOA. **b** LW CRF at surface. **c** LW CRF for the atmosphere. **d** LW CSR at TOA. **e** LW CSR at surface. **f** LW CSR for the atmosphere

very small, while the LW cloud effect of warming of atmosphere is increasing during the twenty-first century.

4 Global Maps of the Changes in Annual Mean Energy Fluxes

Figure 3a–c shows global maps of the changes from the end of nineteenth century in the annual mean downward radiation fluxes (a) at TOA, (b) at the Earth's surface (land and ocean), and (c) absorbed SW radiation by the atmosphere. The pattern of changes in SW radiation at TOA and surface are very similar to each other. The pattern also agrees with that shown by TF09. We can see a significant decrease of net downward SW radiation (increase of reflected SW radiation) at the tropical ocean region. This change is a result of the change in clouds (see Fig. 4). On the other hand, we can see a significant increase of net downward solar radiation (decrease of reflected solar radiation) at the polar sea-ice region, corresponding to a significant decrease of the sea-ice in twenty-first century. As shown in Fig. 3c, the change in the solar radiation absorbed by the atmosphere is rather uniform and weak positive. This indicates that a more SW radiation is absorbed by the overall increased atmospheric water vapor in twenty-first century.

Figure 4a–c shows the changes in SW CRF (a) at TOA, (b) at Earth's surface, and (c) for the atmosphere (cloud effect on SW atmospheric heating). The pattern of the changes in SW CRF at TOA and SW CRF at surface are very similar to each other. As a result, the SW CRF for the atmosphere, which is simply CRF at TOA minus CRF at surface, is generally small, in agreement with the result of SE08. This is because the major SW effect of clouds is reflection of solar radiation, and there is not so much contribution to the atmospheric heating, at least compared with the LW effect of clouds (see Fig. 2c). The negative SW CRF at TOA and surface is most significantly increasing over the equatorial

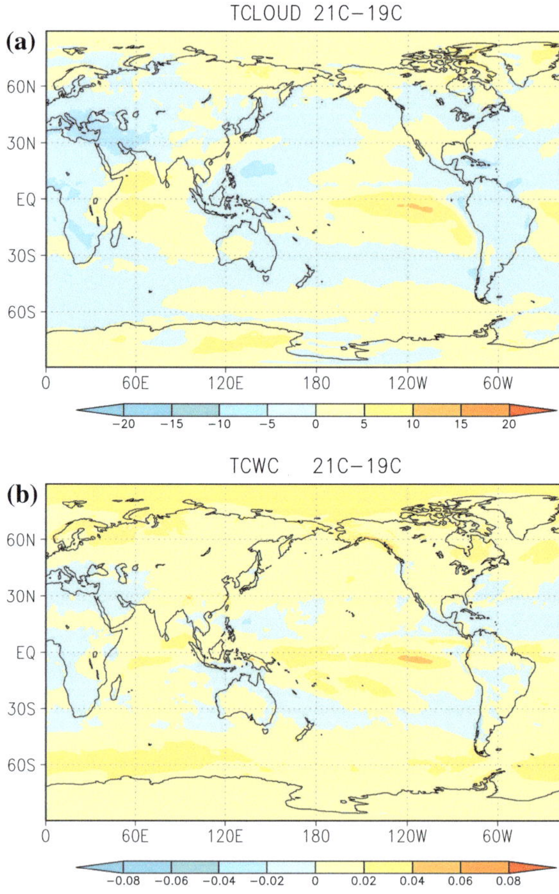

Fig. 7 Global maps of the changes from nineteenth to twenty-first century in annual mean column total cloud amount and column total cloud water content. **a** Cloud amount. Unit is %. **b** Cloud water content. Unit is kg m^{-2}

ocean regions and the polar sea-ice regions. On the other hand, we can see a region in the western North Pacific where the positive SW CRF is significantly increasing. This corresponds to a weakening of convections over this region (see next section for further discussion).

Figure 4d–f shows the changes in SW CSR (d) at TOA, (e) at Earth's surface, and (f) absorbed clear sky SW radiation by the atmosphere. In the polar sea-ice regions, as the effect of albedo decrease due to the reduction of sea-ice in summer (Fig. 4d, e) is larger than the effect of SW CRF change (Fig. 4a, b), the net downward SW radiation is increasing (Fig. 3a, b). Figure 4f shows a uniform warming of the atmosphere by clear sky SW radiation due to increased water vapor.

Figure 5a–c shows global maps of the changes from the end of nineteenth century in the annual mean net downward LW radiation (a) at TOA and (b) at Earth's surface, and (c) absorbed LW radiation by the atmosphere. The pattern of the change is generally in good agreement with that shown by TF09. We can see a decrease in net downward LW at TOA (increase of OLR) in most regions except for tropical oceans and Southern Ocean. The decrease of OLR over the tropical ocean corresponds well to the increase in clouds over the same region (see Fig. 6). As shown in Fig. 5b, c, the changes in net downward LW radiation at surface are mostly positive (warming of the surface), and the changes in absorbed LW radiation by the atmosphere are mostly negative (cooling of the atmosphere). These increases in surface warming and atmospheric cooling are simply due to the enhanced greenhouse effect by the increased GHG and water vapor. Note that there are small regions where the change in LW absorption is positive in equatorial Indian Ocean and equatorial western Pacific in Fig. 5c. These regions correspond to the regions where the change in LW cloud effect of warming the atmosphere is increasing (see Fig. 6c).

Figure 6a–c shows the changes in LW CRF (a) at TOA and (b) at surface, and (c) for the atmosphere (cloud effect on LW atmospheric heating). In the low and mid latitudes, LW CRF at TOA is generally decreasing except for equatorial ocean regions, where the LW cloud effect of warming is increasing. It should be noted that the pattern of the changes in LW CRF at TOA is similar (with opposite sign) to that of the changes in SW CRF at TOA (Fig. 4a), indicating that the changes in both SW and LW CRF at TOA are associated with the changes in clouds (see Fig. 7a). On the other hand, LW CRF at surface is rather uniformly decreasing (reduction of surface warming) except for the polar region, indicating that the changes in LW CRF at surface is not directly associated with the changes in clouds. This uniform decrease of the LW CRF at surface is probably due to the increased absorption by water vapor of the downward LW radiation emitted from high- and middle-level clouds. In the polar region, where there are not many high- and middle-level clouds, LW CRF at surface is increasing due to the increased low-level clouds. LW CRF for the atmosphere, which is equal to CRF at TOA minus CRF at surface, is generally increasing, with large increase over the equatorial ocean region and some decrease over the western North Pacific, tropical North Atlantic, and Arctic Ocean. This pattern of the changes in LW CRF for the atmosphere is generally in good agreement with SE08.

Figure 6d–f shows the changes in LW CSR (d) at TOA, (e) at Earth's surface, and (f) for the atmosphere (absorbed clear sky LW radiation by the atmosphere). Figure 6e, f shows significant increases in warming of the Earth's surface and cooling of the atmosphere due to the enhanced greenhouse effect by the increased GHG and water vapor. Note that the surface warming and atmospheric cooling are particularly large in the tropics where the water vapor increases most significantly.

5 Discussion

The global mean energy flux components at TOA and Earth's surface, averaged over the 20-year period at the end of nineteenth, twentieth and twenty-first century, are shown in Table 1. The right two columns show the changes from nineteenth to twentieth century and from nineteenth to twenty-first century, respectively. The most significant changes from nineteenth to twenty-first century are the 22.3 W m^{-2} increase in downward LW flux at surface and 16.3 W m^{-2} increase in upward LW flux at surface. As a result, the net downward LW flux increases by 6 W m^{-2} and mostly balanced by the increase in upward LH flux. In the change in energy balance at the surface, the contribution from changes in SW flux and SH flux is small. On the other hand, in the change in energy balance of the atmosphere, the contribution from the increase in SW absorption is significant.

To further examine the role of clouds in the changes in Earth's energy flows, the globally averaged CRF at TOA, Earth's surface and for the atmosphere, averaged over 20-year period at the end of nineteenth, twentieth and twenty-first century, are shown in Table 2. The magnitude of both the SW CRF and LW CRF reaches up to 10–30% of the respective radiation fluxes, indicating significant roles of clouds in the Earth's energy flows. Generally, SW CRF at TOA and surface is negative (cooling the atmosphere and surface), while LW CRF is positive (warming the atmosphere and surface). However, as shown in Table 2, both SW and LW CRF for the atmosphere (defined as CRF at TOA minus CRF at surface) are positive (warming the atmosphere).

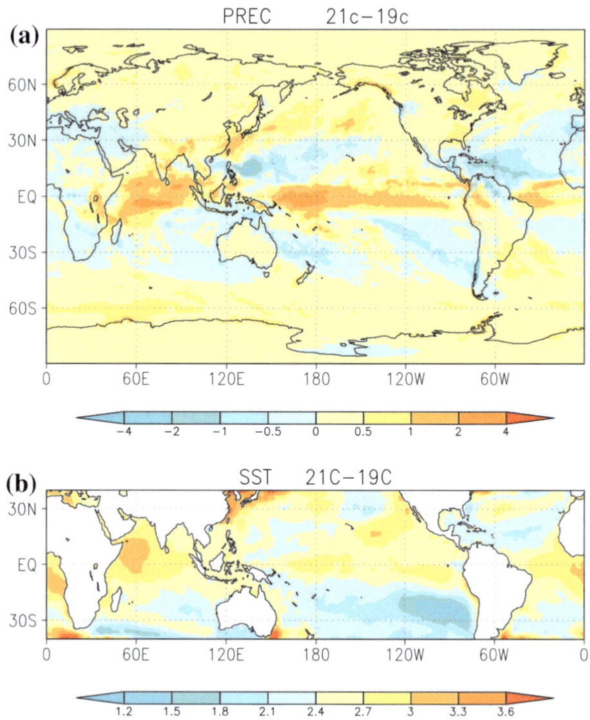

Fig. 8 Global maps of the changes from nineteenth to twenty-first century in annual mean precipitation and SST. **a** Precipitation. Unit is mm/day. **b** SST. Unit is °C

Regarding the changes from nineteenth to twenty-first century, we note that SW CRF decreases (negative CRF increases) both at TOA and surface, resulting in small changes in CRF for the atmosphere. On the other hand, LW CRF decreases only a little at TOA but it decreases significantly at surface, resulting in a significant increase (1.4 W m^{-2}) in CRF for the atmosphere (warming the atmosphere), which makes a significant contribution to the changes in the atmospheric energy balance.

Table 2 indicates that both the changes in globally averaged SW and LW CRF from nineteenth to twenty-first century are negative (cooling of atmosphere and surface is increasing). For SW CRF, this means the cloud effect (cooling) is increasing, but for LW CRF, this means the cloud effect (warming) is decreasing. Are clouds indeed increasing or decreasing in the model? In Fig. 7, we show the global maps of the changes from the end of nineteenth century in the annual mean column total cloud amount and column total cloud water content. We note that the pattern of the changes in cloud amount (Fig. 7a) generally agrees well with the pattern of the changes in LW CRF at TOA (Fig. 6a). The regions where the negative SW CRF is increasing generally correspond to the regions where cloud amount is increasing. However, there are wide areas in mid latitudes where cloud amount is decreasing but negative SW CRF is increasing. In such regions, we can see that cloud water content is increasing. In these areas, the increased cloud water content may increase the reflectivity of clouds and increase the negative SW CRF even though the cloud amount is decreased.

We have seen that both the LW and SW cloud effect increases from nineteenth to twenty-first century over the equatorial ocean region, while they decrease over the western Pacific and tropical North Atlantic. These positive and negative changes are closely related to the changes in activity of tropical convection. The map of changes in precipitation associated with the changes in convective activity is shown in Fig. 8. In Fig. 8, the map of changes in SST is also shown. We note that there are large SST increases over the Indian Ocean and central to eastern tropical Pacific (corresponding to a weakening of the cold tongue), while a relatively small SST increase over the western Pacific regions. These figures clearly demonstrate that the distribution of clouds and precipitation associated with convection is dominated by the distribution of changes in relative SST. As we use the SST increase projected by CMIP3 models, it is reasonable that the patterns of the precipitation change, and therefore, the pattern of changes in clouds associated with the tropical convections in our simulation are in good agreement with those projected by CMIP3 models. Indeed, the pattern of the precipitation change shown in Fig. 8 resembles the pattern of the change in precipitation by CMIP3 models (Meehl et al. 2007).

6 Conclusions

We have examined long-term changes in Earth's energy flows at TOA and surface by using the 228-year simulation of a high-resolution atmospheric model, MRI-AGCM3.2. It is found that the net downward SW radiation at TOA (ASR) significantly increases during twenty-first century, while the net downward LW radiation at TOA (−OLR) first increases toward the end of twentieth century then starts decreasing at around year 2020. These changes in ASR and −OLR are generally in good agreement with the changes in CIMIP3 models shown by TF09. The patterns of the changes in the present study in ASR and −OLR and in cloud amount are also similar to those of TF09. However, the reason for the changes, particularly for the change in global mean ASR, is very different. TF09 indicates that the significant increase in ASR at TOA during twenty-first century is mainly due to the reduction of clouds. On the other hand, in the present study, as the negative SW CRF at TOA is increasing, the effect of clouds on SW radiation (CRF) is decreasing ASR, and the main reason for the increase in ASR is the increase in clear sky SW absorption by increased water vapor in the atmosphere. This seems to suggest that the same energy balance could be achieved in different ways in different climate models. The large increase in the absorbed SW radiation by atmospheric water vapor can affect the atmospheric energy balance and in turn the hydrological cycle as well.

In the present study, the changes in both SW and LW CRF are negative and increase toward the end of twenty-first century. This is considered to result from a decrease in cloud amount and an increase in cloud water content. As the changes in global mean SW CRF at TOA and at surface are almost the same, the change in SW CRF for the atmosphere is very small. On the other hand, as the negative change in global mean LW CRF at surface is much larger than that at TOA, the change in LW CRF for the atmosphere is significantly positive. The positive change of LW CRF for the atmosphere is consistent with the result of SE08 for CMIP3 models. Moreover, the pattern of changes in LW CRF for the atmosphere in the present study also resembles that of SE08. However, the reason for the changes may be different. SE08 indicated that the decrease of mid- and low-level cloud amount is the reason for the warming of the atmosphere. On the other hand, we found that LW CRF at surface is uniformly decreasing due to an increased absorption by water vapor of the downward LW radiation emitted from clouds, and this is the reason for the increase in LW CRF for the atmosphere (warming of the atmosphere).

As shown in Meehl et al. (2007), the change in CRF varies from negative to positive among the models. The long-term changes in energy flows and the role of clouds may be different among the models with different CRF changes. In our model, the change in both SW and LW CRF is negative (cloud feedback is negative). The conclusions of the present study are based on one model with negative cloud feedback, and they may be different in a model with positive cloud feedback. It is important to understand how the models with different cloud representation produce different responses to the GHG forcing. For this purpose, it is recommended to examine the long-term changes in energy fluxes and CRF both at TOA and at surface in various models.

Acknowledgments This work was conducted under the framework of the "Projection of the Change in future Weather Extremes using Super-high-resolution Atmospheric Models" supported by the KAKUSHIN Program of the Ministry of Education, Culture, Sports, Science, and Technology (MEXT). The calculations were performed on the Earth Simulator.

References

Allen MA, Ingram WJ (2002) Constraints on future changes in climate and the hydrologic cycle. Nature 419:224–232. doi:10.1038/nature01092

Bengtsson L, Botzet M, Esch M (1996) Will greenhouse gas-induced warming over the next 50 years lead to higher frequency and greater intensity of hurricanes? Tellus 48A:57–73

Held IM, Soden BJ (2006) Robust responses of the hydrological cycle to global warming. J Clim 19:5686–5699

Kawai H, Inoue T (2006) A simple parameterization scheme for subtropical marine stratocumulus. SOLA 2:17–20. doi:10.2151/sola.2006-005

Knutson T, McBride J, Chan J, Emanuel K, Holland G, Landsea C, Held I, Kossin JP, Srivastava A, Sugi M (2010) Tropical cyclones and climate change. Nat Geosci. doi:10.1038/ngeo0779

Meehl GA et al. (2007) Global climate projections, in Climate Change 2007: the physical science basis, contribution of working group I to the fourth assessment report of the intergovernmental panel on climate change. In: Solomon S et al (ed) Cambridge University Press, New York, pp 748–849

Mitchell JFB, Wilson CA, Cunnington WM (1987) On CO_2 climate sensitivity and model dependence of results. Q J R Meteorol Soc 113:293–322

Mizuta R, Yoshimura H, Murakami H, Matsueda M, Endo H, Ose T, Kamiguchi K, Hosaka M, Sugi M, Yukimoto S, Kusunoki S, Kitoh A (2012) climate simulations using the improved MRI-AGCM with 20-km grid. J Meteorol Soc Jpn 90A:235–236

Rayner NA, Parker DE, Horton EB, Folland CK, Alexander LV, Rowell DP (2003) Global analysis of sea surface temperature, sea ice, and night marine air temperature since the late nineteenth century. J Geophys Res 108:4407

Sohn BJ, Bennartz R (2008) Contribution of water vapor to observational estimates of long wave cloud radiativa forcing. J Geophys Res 113:D20107

Stephens GL, Ellis TD (2008) Controls of global-mean precipitation increases in global warming GCM experiments. J Climate 21:6141–6155

Stephens G, Slingo A, Webb MJ, Minnett PJ, Daum PH, Kleiman L, Wittmeyer I, Randall DA (1994) Observations of the Earth's radiation budget in relation to atmospheric hydrology. 4. Atmospheric column radiative cooling over the worlds' oceans. J Geophys Res 99:18585–18604

Sugi M, Yoshimura J (2004) A mechanism of tropical precipitation change due to CO_2 increase. J Clim 17:238–243

Sugi M, Noda A, Sato N (2002) Influence of the global warming on tropical cyclone climatology: an experiment with the JMA Global Model. J Meteorol Soc Jpn 80:249–272

Tiedtke M (1989) A comprehensive mass flux scheme for cumulus parameterization in large scale models. Mon Wea Rev 117:1779–1800

Tiedtke M (1993) Representation of clouds in large scale models. Mon Wea Rev 121:3040–3061

Trenberth KE, Fasullo JT (2009) Global warming due to increasing absorbed solar radiation. Geophys Res Lett 36:L07706. doi:10.1029/2009GL037527

Trenberth KE, Fasulo T, Kiehl J (2009) Earth's global energy budget. Bull Amer Soc 90:311–323

Trenberth KE, Fasullo JT (2010) Simulation of present-day and twenty-first-century energy budgets of the southern oceans. J Clim 23:440–454

Vecchi GA, Soden BJ (2007) Global warming and the weakening of the tropical circulation. J Clim 20:4316–4340

Yukimoto et al. (2011) Meteorological Research Institute-Earth System Model v1 (MRI-ESM1)—model description. Tech Rep Meteorol Res Inst 64:1–83

Surv Geophys (2012) 33:445–451
DOI 10.1007/s10712-012-9193-z

Solar Forcing of Climate

C. de Jager

Received: 4 November 2011 / Accepted: 16 April 2012 / Published online: 9 May 2012

Abstract Solar activity is evident both in the equatorial activity centres and in the polar magnetic field variations. The total solar irradiance variation is due to the former component. During the extraordinarily long minimum of activity between sunspot cycles 23 and 24, the variations related to the equatorial field components reached their minimum values in the first half of 2008, while those related to the polar field variations had their extreme values rather at the end of 2009 and the first half of 2010. The explanation of this delay is another challenge for dynamo theories. The role of the open solar flux has so far been grossly underestimated in discussions of Sun-climate relations. The gradual increase in the average terrestrial ground temperature since 1610 is related both to the equatorial and polar field variations. The main component (0.077 K/century) is due to the variation of the total solar irradiance. The second component (0.040 K/century) waits for an explanation. The smoothed residual increase, presumably antropogenic, obtained after subtraction of the known components from the total increase was 0.31 K in 1999.

Keywords Solar forcing · Climate · Total and spectral solar irradiances

1 Introduction

The discussions in the workshop underlined the complexity of the Earth's atmosphere of which essential aspects are still far from being understood. This is in contrast to the simple and transparent structure of that of the Sun. But these thoughts made me remember a colloquium once held in an astronomical institute during which the lecturer claimed "A star is a very simple structure." Whereupon a voice from the audience interposed "You sir,

C. de Jager (✉)
Royal Netherlands Institute for Sea Research, P.O. Box 59, 1790 AB Den Burg, The Netherlands
e-mail: cdej@kpnplanet.nl

C. de Jager
Astronomical and Space Research Institutes at Utrecht, Den Burg, The Netherlands

would be a very simple structure too, at a distance of ten light-years." The Sun is at only eight light-minutes. That makes it the most complicated star that we know of, but along this line of reasoning, its atmosphere is still quite simple as compared to that of the Earth. In the following, I will discuss some of the essential elements and the resulting discussions of the presentations made at the workshop in session 3, which was devoted to solar forcing of climate.

2 Measuring Total Solar Irradiation and its Spectra

The energy received on Earth from the Sun relative to the non-solar energy (geothermal energy, tidal energy and waste heat from fossil fuel) stands as 1–0.0004. The primary components of the varying parts of the total solar radiation are the facular radiation and the sunspot darkening. These vary while the Sun rotates. Attempts to model that variation by Unruh show good agreement between model and observations. The contribution of the chromospheric network[1] is still to be examined.

So far, we possess data on the total solar irradiance (TSI) over three solar cycles. The observations, when performed simultaneously, show virtually the same time-dependent shapes, but they differ in the absolute values. This is a characteristic of absolute intensity measurements. Performing these is one of the most difficult experimental techniques. The resulting curves show a small degree of variability. Essentially, the shape of the Schwabe cycle of sunspot number variability is well reproduced in that of the TSI.

The variability is mainly due to the variation of the solar area covered with faculae, the brighter and hotter magnetic regions around sunspots that together constitute the activity centres. The size of the area covered with faculae and the number of such areas on the solar disc are, to a good approximation, proportional to the number of sunspots on the disc. While the spectrum of the non-magnetic part of the disk has its maximum intensity in the visual spectral range, the radiation from the activity centres has its maximum rather in the violet and ultraviolet parts. A substantial percentage of the enhanced radiation from the facular fields is emitted in the wavelength range below 400 nm. This is so because of the higher facular temperatures, about 10,000 K. This stands in contrast to the non-magnetic parts of the solar disk, of which the temperature is close to the solar effective temperature, $T_{eff} \approx 5{,}800$ K. All other magnetic parts of the Sun, such as the network, have higher temperatures. An important part of the far UV radiation is from the Lyman α line. It originates from higher parts of the chromosphere than all other contributions mentioned here. During Schwabe cycles, its amplitude varies by about 50 % (Lean and Woods 2010).

Dissipation of magnetic energy, transported towards the solar surface by the lower situated convective motions, is the main cause for the higher temperatures of the magnetic parts. Solar variability is mainly concentrated in the activity centres of which the main visual marking is the sunspots in their centres. It should be remarked, though, that a spot as such is a thoroughly inactive part of the Sun. The real variability, as presented in faculae, solar flares, flare surges, coronal mass ejections and the like, is concentrated in the surroundings of the spots. A still open problem is that of the emission by the network, whereby one should distinguish between the general solar network that covers the whole surface and the enhanced network that appears in the aftermath of the facular fields during their decline.

[1] A web-like pattern easily seen on the solar disk in the emissions of the red line of hydrogen (H-alpha) and in the ultraviolet due to the concentration of bundles of magnetic field lines under the effect of fluid motions.

3 The Recent Extreme Solar Minimum

Conspicuous is the very long-lasting recent minimum that we have witnessed in the past few years. The times, at which the various solar activity indicators reached their minimum values, ranged between early 2008 and late 2010. Interestingly, indices such as the UV flux and the MgII line core emission had their minima nearly at the same time (early 2008) as the sunspot number. This is understandable since the facular emission correlates well with sunspots. On the other hand, the galactic cosmic rays had their extreme values in the second part of 2009 and even as late as in 2010. The association with the cosmic ray flux suggests a physical cause related to the open solar flux (cf. "The Open Solar Flux" below) and the related emission of magnetic plasma.

A diagram presented by Lockwood, and referred to herewith as Fig. 1, best summarizes the above. It shows that indices related to the activity centres reached minimum values around the first half of 2008, while those related to coronal activity, the open solar flux and galactic cosmic rays, variables that are partly related to the polar activity and to solar particle fluxes had their extreme values around the end of 2009 and early 2010. Hence, contrary to the conventional picture that the polar magnetic fields reach their maximum simultaneously with minimum values of the equatorial fields, there was a time lag of nearly two years during the past sunspot minimum period. These observations must have their implication for our views on the physical mechanisms of the solar dynamo.

The extended minimum of 2008–2010 lasted longer than previous minima during the past century, and this associated with the very low polar magnetic field strengths during that period leads—on the basis of the Gnevyshev-Ohl rule—to the conclusion that the next solar activity maximum will be as late as 2013–2014 and fairly low ($\sim$60 in sunspot numbers; De Jager and Duhau 2009; Duhau and De Jager 2010). In this connection, we refer to Nandy et al. (2011) who found that a long-lasting minimum may be related to a situation in which the solar meridional sub-surface plasma streaming has an initial high velocity during the time before the preceding sunspot maximum and is followed by a strongly reduced velocity. I may remark, though, that the large variations in the meridional

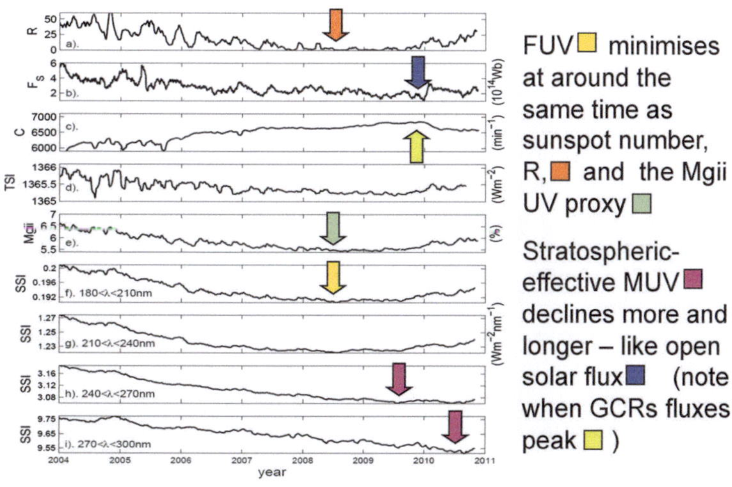

Fig. 1 Indices related to the activity centres (source and SOLSTICE data) shown by Lockwood during the workshop

velocities that are needed for explaining the observations were not observed in helioseismic observations.

Fröhlich drew attention to his observation that the TSI during the past solar minimum was lower than during the previous two minima. The difference with the earlier minima can be expressed in effective temperatures and is as small as 0.25 K. It is about equal to twice the mean error of the observed values and, understandably, this fact gives rise to dispute. Criticisms of the reality of the deep minimum are indeed implied in the contribution by Unruh. Also, Krivova et al. (2011) do not agree with this result. On the other hand, both Barnard et al. (2011) and Ball et al. (2012) confirm the low minimum between cycles 23 and 24.

The above observation may be compared with those of spectral irradiance. During the workshop, Unruh showed that the observed spectra, including their variability, could well be represented by numerical modelling during a full solar rotation. Although this would show that we well understand solar spectral variability, there is a remarkable recent development. Remarkable, because of its contrast with earlier observations, is the spectrum obtained by the SORCE-SIM team (Haigh et al. 2011). In the visual spectral range, the intensity appears to be less than in earlier minima while it is relatively higher in the ultraviolet part of the spectrum. These observations provoked discussion, since they were thought not to correspond with usual conjectures. Particularly, they are not confirmed in Unruh's analysis. An offhand explanation may be that these observations, if correct, are related to a yet to be explained very minor enhanced density of the upper photosphere and the chromosphere.

For the TSI, the situation is clearer, as the VIRGO observations agree well with the ACRIM-II/III data. Essentially, the differences between UV and visual spectral irradiance would be important for climate modelling (cf. Haigh et al. 2011).

4 The Open Solar Flux

The presentation by Rozanov at the workshop showed the results of the influence on climate of varying solar activity indices, especially the UV radiation but also the varying particles fluxes. Discussions during the workshop underlined the importance of the open solar flux (OSF) for the topic of Sun–Earth relations, as well as the need to include that notion in the relevant investigations. The OSF is usually defined as the magnetic flux emerging quasi-radially from a sphere around the Sun's centre situated at a distance of 2.5 solar radii, but this value does not receive unanimous agreement.

An important remark that emerged during the workshop was that instead of looking for correlations of the Earth's average surface temperature with the polar magnetic field variation, it may appear to be more rewarding to correlate average terrestrial surface temperatures with the open solar flux. Lockwood found that it correlates better with terrestrial surface temperatures than other indices do. An indication that the OSF is related to the polar magnetic fields may be extracted from a finding by Fröhlich that a good correlation is found when the comparison of TSI and the flux is restricted to the solar minima—at these times, the polar flux has maximal values.

In discussions during the workshop, Dudok de Wit showed a way to disentangle the open solar flux from the geomagnetic *aa* data. Over the past grand maximum of solar activity, the OSF shows a steady increase with time and, while the sunspot data reached their largest values during the extreme maximum of 1957–1958, the OSF had its maximum values only after 1980 according to Lockwood's presentation at the workshop. A recent attempt to link the open solar flux with total solar irradiance is based on the observations of cosmogenic radionuclides measured in ice cores (Steinhilber et al. 2009). The measurements cover a period of 9300 years.

If part of the Earth's temperature variation is related to the open solar flux then the conclusion seems unavoidable that part of the terrestrial warming must depend on particle emission, either from the Sun or else activated by the Sun. This aspect was discussed by Rozanov (cf. "Sun-Atmosphere Relationships"). The role of particles in Sun–climate relations has been discussed many times in the scientific literature, but in order to play a significant role, it should involve strong amplification mechanisms. An overlooked topic is the chemistry induced by precipitating particles and the role of aerosol formation by cosmic ray particles. An experiment called "CLOUD" is now under development at CERN which aims at simulating the processes of the interactions between cosmic ray particles and an experimental model simulating the Earth atmosphere with a blend of sulphuric acid and ammonia.

5 Sun-Atmosphere Relationships

Solar incoming radiation interacts with the atmosphere at various heights, depending on the wavelength and the character of the radiation (continuum versus lines). Particles of different kinds and energies are absorbed at different heights in the Earth's atmosphere. Rozanov discussed the energy deposition in the atmosphere, both electromagnetic and particle radiation. The heating rate in the atmosphere varies with height and with solar activity. The depth to where particles penetrate and the consequent atmospheric ionization are likely to vary. As to the charged particle inflow, it is only the galactic cosmic rays that are able to penetrate down to the troposphere. UV radiation is absorbed in stratospheric layers. The precise height and degree of absorption depend both on the wavelength and on the chemistry of the stratosphere. The resulting terrestrial surface temperature variation depends on the degree of coupling between the stratosphere and the lower layers. The forcing is not yet well defined, and the physical explanation of the solar effects is still beset with uncertainty.

Lockwood investigated periods of severe winters in England. A relative lack of cold winters appears to occur for periods of high solar activity. This correlation is associated with the occurrence of 'blocking events', which are large long-lived anticyclones that deflect the jet stream and thus may cause the occurrence of cold eastern winds over Europe.

Several authors have discussed the observed more or less gradual rise of the Earth's surface temperature since the last stages of the deep Maunder Minimum of solar activity, in around 1700 A,D. This was usually done by correlating the Earth's average surface temperature, or that of part of the Earth such as the Northern Hemisphere, with the total solar irradiance. Since the latter is only known for the last three Schwabe cycles, it is usually extended to earlier times by correlating it with the sunspot number. It is habitually found that these variations in TSI are too small to contribute effectively to terrestrial temperature variation.

By only considering the TSI and the sunspot number, the assumed solar variability that would influence climate is restricted to the sunspot belt, but the solar dynamo, which governs solar variability, is an intricate interplay between the toroidal sunspot belt's magnetic field and the poloidal (polar) field. The latter's principal manifestations are the solar wind emerging from coronal holes, the radiation from polar bright points and polar faculae and the like. Although the polar field does not show strong concentrations comparable with the sunspots, the total polar flux is comparable with that of the toroidal field. This is understandable because the one originates from the other. De Jager described an attempt to disentangle both correlations by studying seven recent average temperature data files and found that the average temperature increase over the period 1610–1970 (hence before the period of significant antropogenic warming) can be split in three components (cf. de Jager et al. 2010). Correlated with the equatorial field is an average increase of 0.077 K/century; the polar field component is 0.040 K/century, and a remaining part of 0.051 K/century is not of solar origin. The first is explained by the variation of the TSI provided one includes a positive water vapour feedback of a factor 2. The feedback effect was discussed by Rozanov, who showed that there is a fair amount of coupling between stratospheric warming and the resulting tropospheric effects. It was, though, not quantified.

The second component has not yet found a physical explanation, while the third must be due to atmospheric effects. After having subtracted the above-mentioned components from the recent average surface temperature values, a smoothed residual for 1999 was left of 0.31 K. A value derived relative to the average residuals for the period 1800—1950, and the total excess with regard to the period 1650–1750 would be 0.62 K (de Jager et al. 2010), while Lockwood found a value of 0.86 K for the excess of 2000–2009 with regard to 1650–1750. If both sets of data would be forming a homogeneous set, this would imply an increase for the period 1999–2009 of 0.24 K, but this increase is not supported by observations. The difference is most probably due to the use of different data sets.

Acknowledgments Thanks for the useful comments on an earlier version of this report are due to Roger Bonnet, Claus Fröhlich, Bas van Geel, Joanna Haigh and Thierry Dudok de Wit.

Appendix: Pilewskie's Comments on Solar Spectral Irradiance Forcing on Climate

Measurements of solar spectral irradiance (SSI) during the last solar cycle (cycle 23) from the NASA SORCE mission (Harder et al. 2009) have received considerable attention for suggesting that the Sun's output varies in ways not previously considered. SORCE data published to date shows that spectral irradiance in the ultraviolet region of the spectrum varied by a factor of 3–10 times more than it did in the previous cycles. However, this was compensated by bands in the visible and near-infrared where irradiance *increased* during the declining phase of the cycle. These new modes of spectral variability have been implemented in a number of climate and atmospheric model simulations (see, for example, Haigh et al. 2011). Although it has been suggested that modelled atmospheric response to the SORCE spectral variability agrees better with observations than simulations forced with Naval Research Laboratory (NRL) spectral irradiance reconstruction (Lean 2000), there are some inconsistencies among the models and with empirical evidence of solar cycle-induced temperature response. Perhaps, more importantly, however, it should be noted that validation of the SORCE spectral irradiance trends can only be provided by other irradiance measurements. Some papers (DeLand and Cebula (2012), Lean and DeLand (2012)) have called into question the consistency of the observed trends with other present and past measurements and with solar activity. Therefore, it is premature to speculate on the validity of the SORCE SSI trends until a thorough validation, now underway, is completed. Furthermore, the measurements published to date cover only part of a solar cycle and must be extrapolated over the full cycle for model analysis, adding additional uncertainty to the simulations (Garcia 2010). Until the SORCE SSI validation is complete, uncertainties in the measured irradiance and their trends are better quantified, and a full solar cycle has been observed, caution is urged in the interpretation of climate response to the recent observed trends in solar spectral irradiance.

References

Ball WT, Unruh YC, Krivova NA, Solanki S, Wenzler T, Mortlock DJ, Jaffe AH (2012) Reconstruction of total solar irradiance 1974-2009. Astron Astrophys. doi:10.1051/0004-6361/201118702

Barnard L, Lockwood M, Hapgood MA, Owens MJ, Davis CJ, Steinhilber F (2011) Predicting space climate change. Geophys Res Lett 38:L16103. doi:10.1029/2011GL048489

de Jager C, Duhau S (2009) Forecasting the parameters of sunspot cycle 24 and beyond. J Atmos Solar Terr Phys 71:239–245. doi:10.1016/j.jastp.2008.11.006

de Jager C, Duhau S, van Geel B (2010) Quantifying and specifying the solar influence on terrestrial surface temperature. J Atmos Solar-Terrestrial Phys 72:926–937. doi:10.1016/j.jastp.2010.04.011

DeLand MT, Cebula RP (2012) Solar UV variations during the decline of cycle 23. J Atmos Sol-Terr Phys 77:225–234. doi:10.1016/j.jastp.2012.01.007

Duhau S, de Jager C (2010) The forthcoming grand minimum of solar activity. J Cosmol 8:1983–1999

Garcia RR (2010) Atmospheric physics: solar surprise? Nature 467:668–669. doi:10.1038/467668a

Haigh JD, Winning AD, Toumi R (2011) An influence of solar spectral variations on radiative forcing of climate. Nature 467:696–699. doi:10.1038/nature09426

Harder JW, Fontena JM, Pilewskie P, Richard EC, Woods TN (2009) Trends in solar spectral irradiance variability in the visible and infrared. Geophys Res Lett 36. doi:10.1029/2008GL036797

Krivova NA, Solanki SK, Schmutz W (2011) Solar total irradiance in cycle 23. Astron Astrophys 529:A81. doi:10.1051/0004-6361/201016234

Lean J (2000) Evolution of the sun's spectral irradiance since the Maunder minimum. Geophys Res Lett 27:2425–2428. doi:10.1029/2000GL000043

Lean J, DeLand M (2012) How does the sun's spectrum vary? J Climate. doi:10.1175/JCLI-D-11-00571.1

Lean JL, Woods TN (2010) Solar spectral irradiance: measurements and models. In: Schrijver CJ, Siscoe GL (eds) Heliophysics III: evolving solar activity and the climates of space and earth. Cambridge University Press, Cambridge, pp 269–298

Nandy D, Muñoz-Jaramillo A, Martens PCH (2011) The unusual minimum of sunspot cycle 23 caused by meridional plasma flow variations. Nature 471:80–82. doi:10.1038/nature09786

Steinhilber F, Beer J, Fröhlich C (2009) Total solar irradiance during the Holocene. Geoph Res L 36:L19704. doi:10.1029/2009GL040142

Surv Geophys (2012) 33:453–473
DOI 10.1007/s10712-011-9168-5

Total Solar Irradiance Observations

Claus Fröhlich

Received: 7 August 2011 / Accepted: 5 December 2011 / Published online: 30 December 2011

Abstract The record of total solar irradiance (TSI) during the past 35 years has overlapping observations from space which can be merged to a composite, and three are available, namely the PMOD, the ACRIM and the IRMB composites. There are important differences between them, which are discussed in detail in order to find the best representation of solar variability during the last three cycles, for the following discussions of solar irradiance variability. Moreover, the absolute value of TSI from TIM on SORCE is 1,361 Wm^{-2}, substantially lower than the value 1,365 Wm^{-2}, which was observed by the classical radiometers. New results from specific experiments are now available, which are discussed in order to define the value to be used in, e.g., climate models. The most important issue regarding the recent TSI records is the low value observed during the minimum in 2009, which is 25% of a typical cycle amplitude lower than the value in 1996. The validity of this low value has been confirmed by comparing all existing TSI observations during cycle 23. On the other hand, activity indices, such as the sunspot number, the 10.7-cm radio flux (F10.7), the CaII and MgII indices and also the Ly-α irradiance or the frequency changes in low-order p modes, show a much smaller decreases relative to their respective typical cycle amplitude. It is most likely that an increasing contrast of the facular and network elements with decreasing magnetic field is responsible for this discrepancy. The value of TSI at minima is correlated with the open magnetic field of the Sun, B_R, at minima. Using B_R at minima, interpolated linearly in between as a fourth component of a proxy model based on the photometric sunspot index and on the MgII index improves the explanation of the variance of TSI over the full period of the last three solar cycles to 84.7%. Results from other models are also discussed.

Keywords Total solar irradiance · Solar variability · Solar activity

C. Fröhlich (✉)
Physikalisch-Meteorologisches Observatorium Davos, World Radiation Center,
7260 Davos Dorf, Switzerland
e-mail: cfrohlich@pmodwrc.ch

1 Introduction

Since late 1978, observations of total solar irradiance (TSI) with electrically calibrated radiometers (for details see, e.g., Fröhlich (2010)) from space have provided overlapping time series from different missions up to the present. This allows the construction of a continuous record over more than 30 years. The TSI time series used in this paper are from the following radiometers and missions:

- HF (Hickey-Frieden) on NIMBUS-7 of NOAA (National Oceanic and Atmospheric Administration), providing daily measurements from November 1978 until January 1993 (Hoyt et al. 1992);
- ACRIM I (Active Cavity Radiometer for Irradiance Monitoring) on SMM (Solar Maximum Mission) of NASA (National Aeronautics and Space Administration), providing daily measurements from February 1980 until June 1989 (Willson 1984);
- ACRIM type on ERBS (Earth Radiation Budget Satellite) of NASA, providing 2 weekly measurements from October 1984 until August 2003 (Lee et al. 1987);
- ACRIM II on UARS (Upper Atmosphere Research Satellite) of NASA, providing daily measurements from October 1991 until September 2001 (Willson 1994);
- VIRGO (Variability of solar Irradiance and Gravity Oscillations) with PMO6V (type 6V of radiometer developed by the Physico-Meteorological Observatory Davos, PMOD) and DIARAD (Dual Irradiance Absolute RADiometer, developed by Institut Royale Météorologique Belgique, IRMB) on SOHO (SOlar and Heliospheric Observatory) of ESA (European Space Agency) and NASA, providing daily measurements from February 1996 until present (Fröhlich et al. 1997);
- ACRIM III on ACRIMSAT of NASA, providing daily measurements from May 2000 until present (Willson 2001, 2011);
- TIM (Total Irradiance Monitor) on SORCE (SOlar Radiation and Climate Experiment) of NASA, providing daily measurements from February 2003 until present (Kopp et al. 2005).

Furthermore, there are TSI observations from radiometers within SOVIM (SOlar Variability and Irradiance Monitoring) within the SOLAR Experiment Package on the International Space Station available during its operation in 2008 (e.g., Mekaoui et al. 2010), and in early 2010, a French satellite, Picard, was launched by CNES (Centre National d'Etudes Spatiales). This spacecraft has radiometers within the PREMOS (PREcision MOnitor Sensor) and SOVAP (SOlar VAriability on Picard) packages on board, and results are expected soon (see, e.g., Thuillier et al. 2011, and references therein). Figure 1 gives a summary of the above-listed time series in their original forms. From the figure, it is evident that the absolute value of TSI is still controversial: the early difference between HF, ACRIM I and ERBS (Earth Radiation Satellite of NASA) is mainly due to the uncertainty of the aperture area, because, at that time, its determination was not really following metrology standards. Later, the observations were well within the stated uncertainty, of the order of $\pm 0.2\%$ until the observations with TIM on SORCE delivered a value lower by almost 5 Wm^{-2}. This difference is well outside the stated uncertainties and indicates that there is still a problem of our understanding of solar radiometry. Kopp et al. (2007) have developed a TSI radiometer facility (TRF) to investigate this problem, and some results have been published recently by Kopp and Lean (2011). From still unpublished results of the tests with TRF, the ACRIM III record has been updated to a new

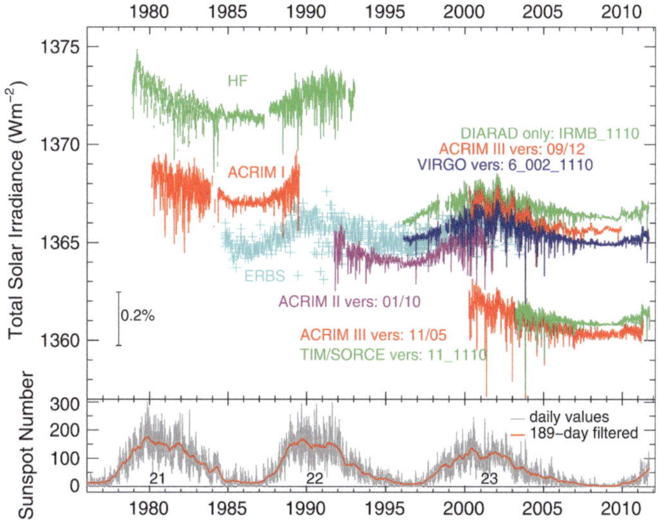

Fig. 1 *Top panel* Compared are daily averaged values of the Sun's total irradiance from radiometers on different space platforms as published by the instrument teams since November 1978. Note that the VIRGO TSI is determined from both VIRGO radiometers (PMO6V and DIARAD), whereas the DIARAD TSI is only based on this one alone. Note the new version of ACRIM III (Version 11/05) that reflect LASP/TRF determined scattering, diffraction and basic scale calibration adjustments. *Bottom panel* Sunspot number to illustrate the variability of solar activity for the three cycles

version 11/05[1] as shown in Fig. 1 (Willson 2011). The issue of the absolute value of TSI to be adopted will be discussed in Sect. 2.

From these data, three different composites have been constructed, and updated versions are shown in Fig. 2[2]. Similarities, but also important differences, are obvious from the plots; these are discussed in Sect. 2 and 3. Section 4 discusses the observed variability during the last three solar cycles, and Sect. 5 shows comparisons with different models of solar irradiance variability and discusses the implications for our understanding. I have described the observations and solar variability in several recent papers (e.g., Fröhlich 2009, 2011a, c) and will present here the most important aspects for climate research—both from the point of view of solar radiometry from space and for the interpretation of TSI variability. Moreover, I use updated versions of the data and present some new aspects of our understanding of solar irradiance variability.

2 Which Absolute Value of TSI Should Be Adopted?

The 'new' value of TSI by Kopp and Lean (2011) is around 1,360.8 ± 0.5 Wm^{-2}, whereas the canonical value amounts to 1,365.4 ± 1.3 Wm^{-2}. An obvious reason for this difference

[1] The versions of ACRIM follow the indications of http://www.acrim.com/Data{\%}20Products.htm as YY/MM, the date of release.

[2] Daily values available at: PMOD (01/1976–06/2011): ftp://ftp.pmodwrc.ch/pub/data/irradiance/composite/DataPlots/ext_composite_d41_62_1110.dat, ACRIM (11/1978–05/2011): http://www.acrim.com/RESULTS/data/composite/composite_acrim_1105.rtf and IRMB (1985–2008): http://remotesensing.oma.be/meteo/view/en/3385925-SARR.diarad.html

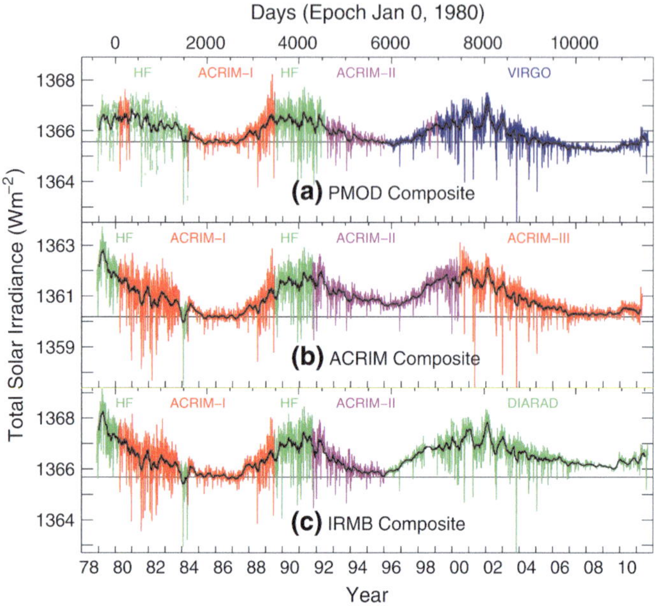

Fig. 2 Shown are the three composites: **a** PMOD (Fröhlich and Lean 1998, updated until June 2011; Fröhlich 2006), **b** ACRIM (Willson 1997, 2011; Willson and Mordvinov 2003) and **c** IRMB (Dewitte et al. 2004, updated with IRMB/DIARAD data)

could be the arrangement of the precision and view-limiting apertures: the 'classical' radiometers have the (smaller) precision aperture close to the cavity and the (larger) view-limiting one at the entrance of the instrument; in TIM this order is reversed. The classical arrangement may increase the sensitivity to stray-light from the view-limiting aperture, and the radiation reflected back into the cavity through the baffle, and also by heating the precision aperture and emitting IR radiation into the nearby cavity. Although these effects had been in principle accounted for in the characterization of the radiometers (see, e.g., Brusa and Fröhlich 1986), they seem to have been underestimated. The TSI Radiometer Facility (TRF) at LASP allows an overall assessment of these effects by comparing the measurements directly with a cryogenic reference radiometer (Kopp et al. 2007).

The work of the TRF is not complete, but already appears to require the changes discussed by Kopp and Lean (2011). The results for a spare PMO6V radiometer of VIRGO (in Kopp and Lean 2011, called VIRGO2) indicate a scattered light contribution of only 0.154%. Even by adding an estimated aperture heating of 0.03% (Fröhlich 2010), the effect is too small to explain the difference. Moreover, there are still no results available for the DIARAD radiometers and more investigations are needed to settle the PMO6V and DIARAD absolute scales and to enable a final correction of VIRGO scale. On the other hand, the recent TRF tests with an ACRIM III spare confirm the difference, and the new version 11/05 of the ACRIM data base yields values close to those of TIM. Although no final results for VIRGO are yet available, the PMOD composite may be adjusted as an intermediate solution to the TIM value during the 2008/9 minimum by multiplying it by 0.996894. The new ACRIM composite is somewhat lower than TIM ($\approx$0.05%), which is mainly due to a steeper decrease during the descending phase of cycle 23 (see also discussion in Sect. 3). The conclusion is that the TSI value during the recent minimum is indeed closer to 1,361 than to 1,365 Wm^{-2}.

3 Which TSI Composite Reproduces Best Solar Variability?

The ACRIM and IRMB composites use the data as published, whereas the PMOD composite introduces corrections of effects that could not be considered in the original data analysis because they were unknown at that time. The most obvious differences between the three composites of Fig. 2 are as follows: different amplitudes of cycles 21 and 22, and different trends (difference between the minima before and after the cycle) during cycle 22 and 23. In the following, these differences are discussed for each cycle separately by comparing the ACRIM and IRMB composite with the PMOD composite as shown in Fig. 3.

Cycle 21: During this cycle, we have data from two radiometers available, the HF and ACRIM I. As shown by Fröhlich (2006), both radiometers need corrections for the early increase first observed by the PMO6V radiometer of VIRGO, and for degradation by including exposure time and radiation dose in the analysis. For ACRIM I, the original degradation (Willson and Hudson 1991) has been revised by Fröhlich (2006). In contrast to all ACRIM instruments, the VIRGO and TIM experiments, the HF experiment has had no spare radiometer in space, less exposed to the Sun, to monitor the degradation. Thus, the early increase and degradation had to be determined by comparison with the revised ACRIM I and ERBE after the end of the operation of ACRIM I. For the time between November 1978 and February 1980, a proxy model (see, e.g., Fröhlich and Lean 2004; Fröhlich 2011a) is used; this is calibrated against the revised ACRIM I during 1980–1986. Together with the removal of the many slips in the HF record due to operational effects (for details, see Fröshlich 2006), an internally consistent correction of the HF results can be performed over all the years of its operation (1978–1993). Not only could the early increase and degradation be determined and corrected reliably, but also a long-term increase in its sensitivity similar to the one found for the DIARAD radiometer of VIRGO could be identified (for details, see Fröhlich 2006). For cycle 21, only the time series of the PMOD composite seems to show the solar variability adequately because the radiometers involved have been corrected for all possible long-term changes not included in the original evaluation. Recently, Scafetta and Willson (2009) and Scafetta (2010) insinuated again that the PMOD composite is incorrect, because the corrections applied to the original ACRIM I and HF time series have no basis at all and are only applied to get agreement with a proxy model originally proposed by Foukal and Lean (1988, 1990). As explained above, a proxy model is only used for comparison with HF before the start of ACRIM I, thus the statement of Scafetta and Willson (2009) and Scafetta (2010) has no firm ground and is simply wrong.

Cycle 22: During this cycle, a major problem arises from bridging the gap between ACRIM I and ACRIM II. For the PMOD composite, this is rather straightforward as an internally consistent HF time series is available, which automatically corrects for, e.g., the major slip in September 1989 (after a 2 day switch-off) by about 0.4 Wm^{-2}, first detected by Lee III et al. (1995). This paper identified a further slip in April 1990 which can no longer be verified and is most probably due to the overall increase in sensitivity. The ACRIM composite ignores these changes in the HF record and does not make the corresponding corrections, which increase the TSI value during the minimum 1996 relative to the one in 1986 by about 0.4 Wm^{-2}, the value found by Lee III et al. (1995) for the first slip in September 1989. The bridging of the ACRIM gap in the IRMB composite is done via ERBS, and the values during this period are averages of ERBS and HF. The minimum in 1996 is slightly higher than the one in 1986 and the one of PMOD composite, but the difference is much smaller than in the ACRIM composite. In Fig. 3, the slip of HF in September 1989 is obvious, as well as the sensitivity increases in HF over the mission.

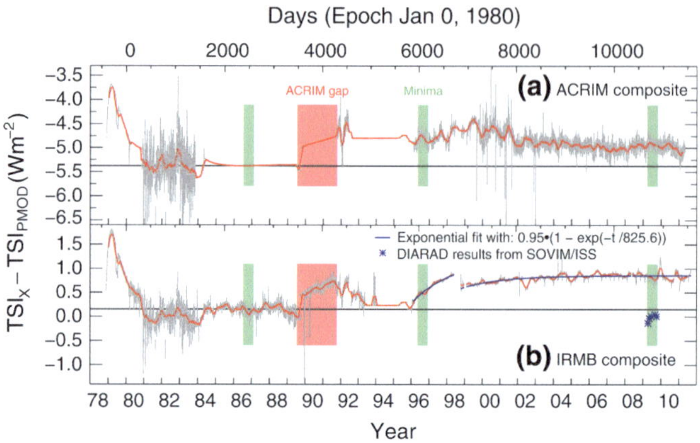

Fig. 3 Comparison of the ACRIM (**a**) and IRMB (**b**) with the PMOD composite (the full *red lines* are 81-day filtered values). The light *red shaded area* indicates the ACRIM gap period. The vertical light *green bars* indicate periods of 210 days centered on the solar minima. During the early period of the ACRIM gap, the slip in the HF data is obvious, as well as the overall sensitivity increases (for details see text and Fröhlich (2006)). The exponential fit to the IRMB data (1996–2010, *blue line*) demonstrates the non-exposure-dependent sensitivity increase in DIARAD. This function is essentially identical to the correction used in the VIRGO TSI analysis determined by comparison of the level-1.8[3] data of DIARAD and PMO6V. Also plotted are the independent results of a similar DIARAD radiometer within the experiment SOVIM on ISS (Mekaoui et al. 2010), which confirm the need for the exponential correction used in the VIRGO evaluation. The horizontal lines indicate the level of the minimum value in 1985/86

After the ACRIM gap, the PMOD composite uses the corrected HF data until the end of this time series, instead of the early ACRIM II data.

Cycle 23: During this cycle, the PMOD composite uses the VIRGO TSI, which is a combination of the results from the PMO6V and DIARAD radiometers (see, e.g., Fröhlich 2006). First, level-1.8 of PMO6V and DIARAD is determined, which corrects only for exposure-dependent changes, normally called degradation. The comparison of the two level-1.8 data shows that DIARAD has also a non-exposure-dependent increase in its sensitivity, which can only be determined by comparison with an independent radiometer. IRMB uses level-1.8 as final result and thus ignores the non-exposure-dependent sensitivity change. For the VIRGO evaluation, the difference of the level-1.8 time series of PMO6V and DIARAD for 1996–2003 was fitted to an exponential function with an amplitude of 0.852 Wm^{-2} and a time constant of 1005 days, and these coefficients are still used in the present-day VIRGO evaluation algorithm, which was introduced in version 6. The difference in the coefficients of the exponential function shown on Fig. 3 is partly due to the differences in calculating level-1.8 of DIARAD by IRMB and VIRGO, and partly due to the fact that coefficients still used have been determined during the first 7 years of operation only. The need for such a correction of DIARAD is also confirmed by the measurements with a similar DIARAD-type radiometer within the SOVIM experiment (Mekaoui et al. 2010) on the International Space Station (ISS), which lie very close to

[3] The VIRGO data levels are defined as follows: level-0 are the raw data as received; level-1 data are in physical units and corrected for all known effects as temperature, sun-spacecraft distance, etc.; level-1.8 are defined as corrected for degradation by comparison of the operational with a less exposed instrument of the same type; level-2 are the final results after all the possible changes during operation in space are included and, in the case of the radiometers, the PMO6V and DIARAD are harmonized.

VIRGO TSI. So, the origin of the difference between the IRMB and PMOD composite during cycle 23 as shown in Fig. 3 is explained by an instrumental effect of DIARAD, which is ignored by IRMB, although it has been known since the introduction of VIRGO version 6 in 2004 (Fröhlich 2005). As the IRMB composite during cycle 21 and 22 does not really contribute new insights in solar variability and with the problems during cycle 23, it should no longer be regarded as a reliable data set for solar cycle variability. Thus, it should certainly not be used as a counterargument to the decrease in VIRGO TSI during cycle 23 as put forward by, e.g., Krivova et al. (2011) or Shapiro et al. (2011).

Another way to check the validity of the low TSI value in 2008/9 and of the associated uncertainty analysis by Fröhlich (2009) is a comparison with ACRIM data. Composites of the two ACRIM III, versions 09/12 and 11/05, merged with ACRIM II, version 01/10, are used for the comparison as shown in the upper panels of Figs. 4 and 5. The main differences between the two ACRIM III versions can be summarized as follows:

- The removal of the annual modulation in version 09/12 increased the noise substantially;
- A strong modulation is observed during the overlap with ACRIM II and shortly after with a peak-to-peak amplitude of at least 300 ppm, which was not present in version 09/12;
- The degradation correction has also changed and there are now large changes in slope: stronger decrease during the maximum and first part of the descending phase of the cycle and a gradual increase over the minimum period. A similar increase is also present in the ACRIM III 09/12 but it is more gradual and the problem during the descending phase may be also influenced by the strong modulation mentioned before;
- The noise in version 11/05 seems to be at a constantly high level, whereas it increases from maximum to minimum for 09/12, another interesting difference.

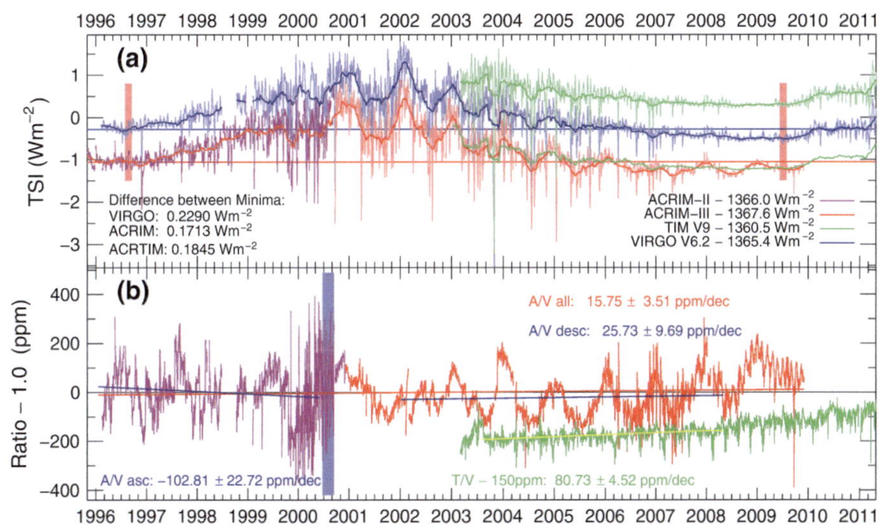

Fig. 4 Comparison of the ACRIM II (version 01/10) and ACRIM-III (version 09/12) composite with VIRGO and TIM TSI. The time series together with the difference between the minima in 1996 and 2008/2009 are shown in **a** and the ratio in **b**. ARITIM designates the composite of ACRIm-II, III and TIM. The *shaded areas* in **a** indicate the time of the minima (*light red*) and in **b** the period of overlap between ACRIM II and III within which a linearly tapered transition is performed (*light blue*)

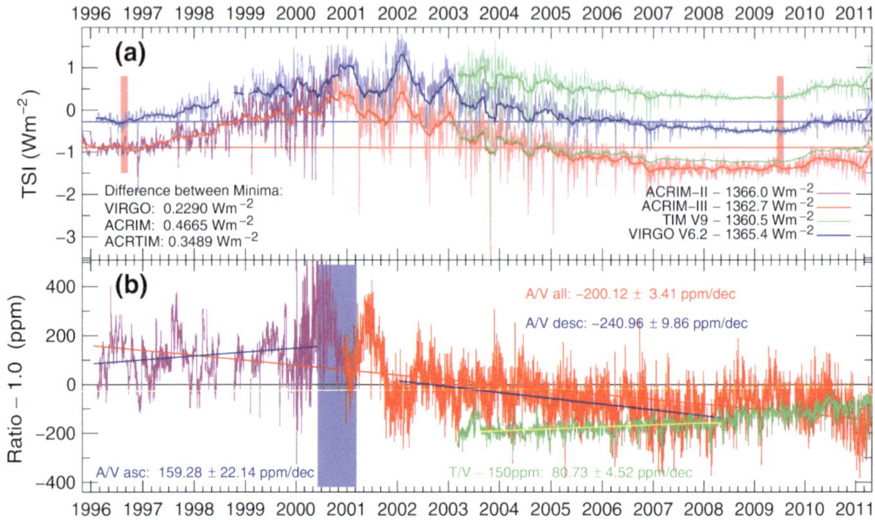

Fig. 5 Same as Fig. 4, but for the new ACRIM III (Version 11/05)

The scale change is not affected by these differences, which relate to the methods used to correct for degradation and other influences. Hence, there seems to be still some unresolved issues of the new evaluation of ACRIM III. Nevertheless, the comparison with VIRGO confirms within the uncertainties the low value of TSI as published by Fröhlich (2009), which is shown by the values on Figs. 4 and 5. Mainly for the version 09/12, it was important to replace the time series of ACRIM III after the launch of SORCE by the values from TIM, because the annual modulation was most probably not solar. The thus extended time series are called ACRTIM and the change between the minima is, for both versions of ACRIM III with 0.185 and 0.349 Wm^{-2}, respectively, within the VIRGO value of 0.228 ± 0.126 Wm^{-2}; this confirms the VIRGO result as an average. The new ACRIM III difference between minima is more than twice the VIRGO value, which is expected from the absolute difference to TIM, and this may indicate that the degradation correction is not yet final. A detailed description of the new evaluation will hopefully be published soon and answers the open questions.

As discussed above, the PMOD composite[4], shown in Fig. 6, is during all three cycles the most likely representation of solar variability; thus, I will use it for the further discussions. This choice is mainly due to our improved understanding of the changes of radiometers in space (and exposed to the Sun) experienced during the development of the procedures to evaluate VIRGO TSI. Effects such as the early increase, the role of the radiation dose for degradation or the possibility of non-exposure-dependent changes were still unrecognized by the HF and ACRIM teams developing their algorithms for the evaluation and thus were unknowingly ignored.

[4] The original and corrected HF and ACRIM data used for the construction of the PMOD composite as well as the time series used for comparison are available from ftp://ftp.pmodwrc.ch/pub/data/irradiance/composite/DataPlots/org2pmod-composite.dat and for the different time series used to determine VIRGO TSI are at ftp://ftp.pmodwrc.ch/pub/data/irradiance/virgo/TSI/virgo_tsi_d_v6_002_yymm.dat with the newest update indicated as yymm.

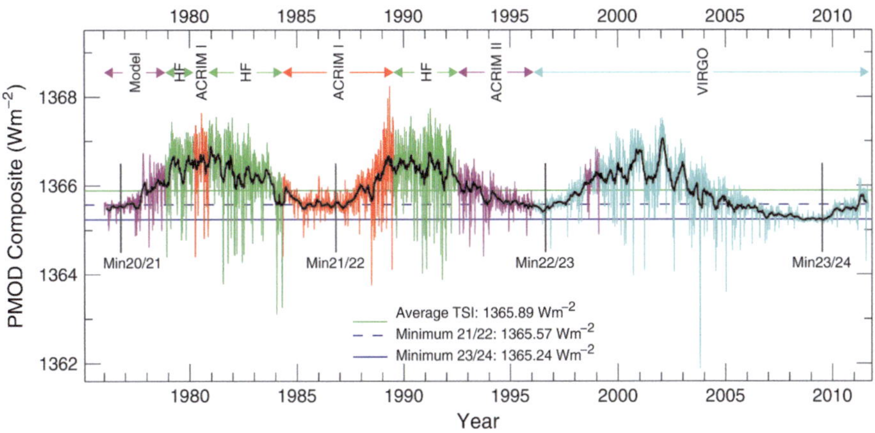

Fig. 6 Daily TSI of the PMOD composite (updated until early June 2011, Version 41_62_1110) and extrapolated with a proxy model back to 1976. The amplitudes of the three cycles decrease first and then increase again. The two horizontal lines indicate the value of the minima in 1986 and 2008, respectively. Note the low value of the 2008/9 minimum, which is 0.23Wm^{-2} lower than the previous one, or 25% in terms of the mean cycle amplitude

4 Discussion of TSI Variability During the Last Three Cycles

Figure 6 shows clearly that TSI varies on all time scales from days to decades. There are obvious changes on the solar rotation (27-day) timescale as, e.g., during the passage of active regions, with dips due to sunspots and rises due to faculae. There is also a modulation with the 11-year period with changing amplitude from cycle to cycle. The most important observation of the three cycles is the low value of TSI during the recent minimum relative to the ones before, almost 0.3 Wm^{-2} lower, which corresponds to one-fourth of the cycle amplitude. The amplitude here is defined as an equivalent of a sine-square wave, which can be easily calculated by multiplying the average of the time series for the period between the minima by two. This definition can be applied to any quantity considered, which makes it a comparable measure of the cycle amplitude. We need also an objective definition of the 'time of minimum' and I propose the first appearance of an important sunspot group of the new cycle (at high latitude and of opposite polarity), although the judgement for 'important' may be somewhat arbitrary. With this definition, the dates of the last four minima are at 16 October 1976, 25 October 1986, 21 August 1996 and 6 July 2009.

TSI variability may be compared to the variance of different indices like the sunspot number (SSN[5]), solar radio flux at 10.7-cm wavelength (F10.7[6]), the CaII K (Livingston et al. 2007, updated with recent observations by WL) or MgII (Viereck et al. 2010, updated with scaled SOLSTICE/SORCE data[7]) chromospheric indexes and the UV irradiance as observed in the Ly-α line at 121.5 nm[8]. The brightness of the chromosphere increases with the magnetic flux density beneath and is an indirect measure of the surface

[5] Available at ftp.ngdc.noaa.gov/STP/SOLAR_DATA/SUNSPOT_NUMBERS/INTERNATIONAL/daily.

[6] Available at ftp.ngdc.noaa.gov/STP/SOLAR_DATA/SOLAR_RADIO/FLUX/Penticton_Adjusted/daily.

[7] Available at http://lasp.colorado.edu/sorce/ssi_data/mgii/txt/sorce_mg_latest.txt.

[8] Available at http://lasp.colorado.edu/lisird/lya/.

magnetic flux density. Thus, the solar cycle modulation of the UV irradiance (for example Ly-α and the chromospheric indices) reflects the changing magnetic flux at the photosphere. The TSI amplitudes during the last three cycles are 0.9493, 0.9165 and 0.9124 Wm^{-2}, respectively. If the trend between minima is removed, these amplitudes become somewhat smaller: 0.9261, 0.8649 and 0.7954 Wm^{-2}. As the trend may have a different physical origin than that of the intra-cycle variation, the corrected amplitudes are more representative of the solar cycle variability, whereas the directly determined amplitude is a measure of the influence of TSI on the Earth's climate. The values are summarized in Table 1 and Fig. 7a and b, and the TSI amplitudes vary quite differently from those of, e.g., SSN and F10.7, each a kind of measure for the strength of the solar activity cycle. The amplitudes of these two parameters were about the same for cycle 21 and 22 and decreased by about 30% during cycle 23, whereas TSI shows a gradual decrease by about 6–7% per cycle; the CaII-K and MgII indices show a similar, but stronger gradual decrease (15–20% per cycle). It is interesting to note that for SSN and F10.7, the amplitude of cycle 20 was about the same or even somewhat smaller as the one observed for cycle 23, but the two cycles in between have almost the same amplitude. The main reason for this difference may be because the cycle modulation of SSN and F10.7 is mainly determined by the emergence of the strong bi-polar fields of active regions, mainly sunspots, and the modulation of irradiance and the chromospheric indices is mainly due to

Table 1 Cycle amplitudes and relative changes of minima in percent of the amplitude during cycle 22. The amplitudes are equivalent sine-squared with the trend during each cycle removed. The standard error of the average reflects the differences between the parameters, which is mainly due to the uncertainty in their long-term behaviour

Parameters	Cycle 21	Cycle 22	Cycle 23
Corrected cycle amplitude			
TSI	0.9261	0.8649	0.7954
B_R	1.368	1.532	1.211
SSN	139.4	135.3	92.9
F 10.7	124.3	122.5	88.3
p-mode freq		0.185	0.159
Ly-α	3.334	3.128	2.474
Ca II K	0.0128	0.0095	0.0073
Mg II	0.0174	0.0153	0.0128
Change of minima over each cycle in percent			
TSI	+5.37	−11.93	−27.06
B_R	+3.65	−18.32	−55.76
SSN	± 0.00	−4.10	−3.90
F 10.7	−1.50	± 0.00	−1.70
p-mode freq		−9.40	−4.90
Ly-α	+5.23	−8.39	−1.69
Ca II K	+0.10	+5.22	−4.31
Mg II	+0.01	+1.94	−4.56
Average SSN···Mg II	0.77 ± 2.58	−2.45 ± 5.84	−3.51 ± 1.45

The units are as follows: TSI (Wm^{-2}), the open field at 1 AU B_R (nT), sunspot number SSN (number), F 10.7 (sfu ≡ 10^{-22} Wm^{-2} Hz^{-1}), p-mode frequency changes (μHz), Ly-α (m Wm^{-2}), Ca II K and Mg II indices (core-to-wing ratio)

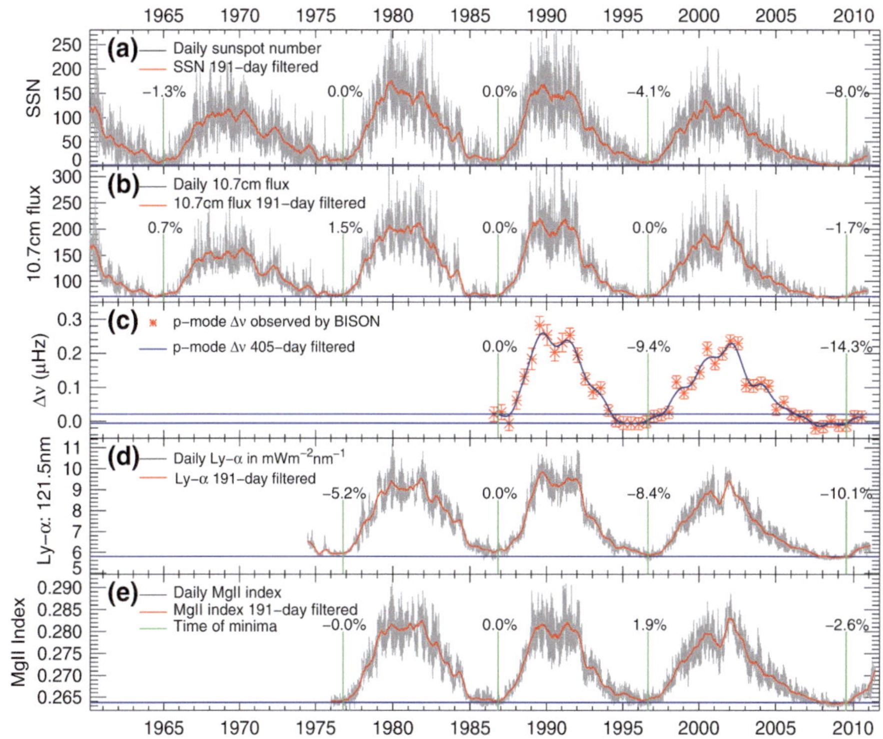

Fig. 7 Daily values of the time series of **a** SSN, **b** F 10.7, **c** frequency changes of the solar pressure-mode oscillations (p modes) as observed by BISON (BIrminham Solar Oscillations Network), (Broomhall et al. 2009), **d** Ly-α and **e** Mg II indices. The values indicated at solar minima (above the *vertical lines*) are the percent changes relative to the amplitude of cycle 22 and normalized to the minimum value between cycles 21 and 22 indicated by the *horizontal blue line*. In contrast to TSI of Fig. 6, these records show a factor of 4–5 smaller decrease from the 1996 to the 2009 minimum

the varying network in and outside active regions and to a lesser extent by the active region faculae. This is also reflected by the non-linear relation between F10.7 and the chromospheric indices or, e.g., Ly-α (Woods et al. 2000, found a square-root dependence). Also interesting is the hump in the p-mode frequency changes during the descending phase of cycle 22 and 23, which is also more pronounced in the SSN and F10-7 record than in Ly-α or MgII index (Fig. 7) and thus, most probably, due to sunspots and not faculae or network (Fröhlich et al. 2006; Broomhall et al. 2009).

The solar condition at minima is normally called the ‘quiet’ Sun, although it is not a well-defined state as shown by the observed differences between different minima. An example is the much ‘quieter’ Sun during the recent minimum compared to the two ones before. During this recent minimum, however, the magnetic field at the surface was by no means close to zero, but still quite important as shown by Schnerr and Spruit (2011). A major question is which part of these remaining magnetic fields depend on the dynamo deep inside, responsible for solar activity, and which part may be independently produced by a local dynamo at the surface, producing the emergence of bi-polar fields in ephemeral regions. For the discussion of the minimum values, the difference between the minimum value before and the one after a given cycle, relative to the cycle amplitude of the corresponding parameter, is used. The values in percent are summarized in Table 1 and indicated on Fig. 7

for the proxies, the chromospheric indices, Ly-α and the p-mode frequency changes and on Fig. 8 for TSI and B_R, the open magnetic field of the Sun at 1 AU. I have added to the 'proxies' the change in p-mode frequencies observed by the BISON programme (Broomhall et al. 2009), because these results represent an independent set of data showing also the influence of surface magnetic fields as a change in the low-order p-mode frequencies. The changes over the last two cycles are indeed within the standard deviation equal to the average changes of the last 6 parameters in Table 1, and they are most likely due to the surface magnetic fields (see, e.g., Broomhall et al. 2009). It is also interesting to note that the values at the dates of minima, in contrast to the other parameters, are already past the minimum value and on the ascending parts of the next cycle.

The value of TSI during the recent minimum is 27% lower than the previous one, compared to only 2–5% observed in the proxies, Ly-α and the chromospheric indices. The only solar parameter that shows also a quite large change is the open magnetic field of the Sun, B_R at 1AU, as shown in Fig. 8b. Interestingly, enough the values of TSI and B_R at the minima correlate very well as shown in Fig. 8c. The slope of the regression is 0.307 ± 0.130 $\mathrm{Wm}^{-2}/\mathrm{nT}$, and the standard deviation reflects the rather large uncertainties allocated to the minima values of TSI from Fröhlich (2009). The small deviations of the regression line from the points may indicate that these estimates are indeed quite conservative, so the low TSI during the last minimum may be more significant than just at the 1.5σ level. This correlation does not indicate a physical connection between TSI and B_R, but as B_R at minima is proportional to the solar polar field which is left over from the previous cycle and provides the starting field to the upcoming, it is an adequate measure of the strength of the prevailing solar activity. Hence, the correlation shows a relationship between

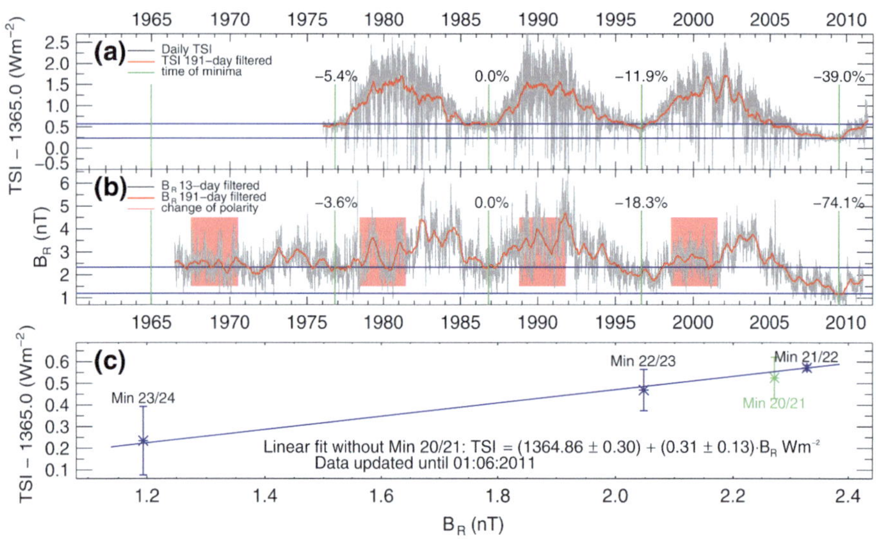

Fig. 8 Comparison of TSI and the open solar magnetic field. **a** daily values of TSI (as Fig. 1) and **b** the open magnetic field B_R at 1 AU. The *light red shaded* areas indicate the time of polarity change in the polar solar field. The correlation between the minimum values of TSI and those of B_R (*blue points* and the *blue line*) is shown on **c**. The *green point* is the extrapolated value of TSI not used in the calculation of the regression. The *error bars* are from a detailed analysis of the long-term uncertainties as determined by Fröhlich (2009) and the errors of the regression coefficients include these uncertainties, and hence are quite large (updated from Fig. 4 of Fröhlich, 2009)

the TSI minima and a measure of the strength of solar activity. A similar relationship can be determined for the chromospheric indices, Ly-α and the p-mode-frequency changes yielding an average of 7.1%/nT, which is indeed about 5 times smaller than for TSI (0.31 Wm^{-2}/nT = 35.8%/nT). These changes depend also on the strength of activity and the lower slope points to the few percent changes in the surface magnetic fields—here represented by the changes in the p-mode frequencies, but MDI (Michelson Doppler Imager on SOHO) magnetograms shows a very similar change between the minima 1996 and 2009.

An obvious conclusion is that there must be another effect on top of the direct influence of the surface magnetic fields, which changes TSI by a factor of about 4–5 times more than, e.g., the UV irradiance, although both are influenced by the same magnetic fields threading the solar surface. As proposed by Tapping et al. (2007) and Fröhlich (2009), an obvious and possibly naive explanation would be a decrease in the global temperature of the Sun by about 0.25 K, which would influence TSI much more than the spectral irradiance relative to their cycle amplitude. A similar argument was used by Kuhn et al. (1988) for the non-facular contribution to the irradiance increase from minimum to maximum. Later, Kuhn and Libbrecht (1991) identified this part as being due to the network which is in line with the present understanding of the cycle variation of TSI (see Sect. 5). Another explanation was recently put forward by Foukal et al. (2011) based on the effect of the increasing contrast of the network and faculae with decreasing magnetic flux density. This effect was observed in MDI data by Ortiz et al. (2006) by comparing the brightness near 676.8 nm with the corresponding magnetic field from the magnetograms, and the results have been used to, e.g., explain the difference between the influence of the long- and short-term Mg-II index in proxy models (e.g., Fröhlich and Lean 2004, see also Sect. 5). However, the limited resolution of MDI reduces the sensitivity of the magnetograms to small mixed-polarity flux due to cancellation, and hence, it underestimates also the contrast because the reference pixels with 'zero' flux are too bright. However, a test to replace the linearly increasing filling factor up to the saturation field of the facular and network contribution in the SATIRE model (Spectral And Total Irradiance REconstructions, e.g., Solanki et al. 2005, and Sect. 5) with the contrast function of Ortiz et al. (2006) failed: including the contrast increased the overall influence of faculae and network by only a few percent, instead of the factor of four.

Schnerr and Spruit (2011) evaluated high-resolution polarimetric observations with the Swedish 1-m Solar Telescope (SST) and Hinode and determined, using a model of the brightness to magnetic field dependence, its important contribution to TSI: during the recent minimum, this brightness effect increases TSI by as much as 0.15% over the value of a field-free Sun from the SST data of 23 May 2009. The difference between the trend over cycle 23 of the proxies and chromospheric indices is $\approx$3.5% and for TSI 27.0–3.5% or about 0.18 Wm^{-2}. Thus, out of the 0.15% only somewhat more than 0.01% are needed for the network and facular decrease as seen in TSI, which is varying with the strength of solar activity as indicated by the correlation with B_R. As suggested by Schnerr and Spruit (2011), the rest of 0.14%—or possibly somewhat less, because this minimum was not the lowest ever—is probably independent of solar activity and is due to a small-scale surface dynamo as proposed by MHD simulations (e.g., Vögler and Schüssler 2007; e.g., Pietarila Graham et al. 2009). This non-varying part includes also the magnetic flux from the ephemeral regions (e.g., Hagenaar et al. 2003, and references therein). That such a non-varying part exists was already noticed by Livingston et al. (2007) from the non-varying Ca II K observations in quiet Sun regions.

This interpretation of the difference of the trends of TSI and proxies or UV radiation is very important for reconstructions back to the Maunder Minimum or further: The minima of TSI change with B_R according to 0.307 Wm^{-2}/nT (27% of the cycle amplitude) and, e.g., the Ly-α irradiance by only 0.12 m Wm^{-2}/nT (3.5% of the cycle amplitude). The smaller effect on UV radiation is also confirmed by the CaII-K observations form Mt. Wilson since 1913, which show variability of the minima also of only about 4–5% (Foukal and Milano 2001; Bertello et al. 2010). This means that the removal of the magnetic field of the 'quietest' Sun is no longer a reasonable assumption for reconstructing TSI or UV radiation into the past. This idea had some notoriety as an explanation for a supposed bi-modal distribution of the activity of solar-like stars (Baliunas and Jastrow 1990), which turned out to be a flawed interpretation of the data. Still, the removal of all magnetic fields during the minima is erroneously propagated, e.g., by Shapiro et al. (2011) yielding extremely large variations, which can never be verified by, e.g., the Earth' temperature record during the last millennium (e.g., Foukal et al. 2006).

5 Comparison of TSI with Models

Proxy models based on sunspot darkening, represented by the photometric sunspot index (PSI), and brightening by faculae and network, represented by a chromospheric index, e.g., MgII, have been described for decades (e.g., Hudson 1988), with variable degrees of details. Here, I show updated results of an improved proxy model described in detail by Fröhlich (2011a, c). For this model, PSI has been improved and the MgII index is separated into short- and long-term parts, which reflect the solar rotation variance from the faculae and the solar cycle amplitude from mainly networks in and outside active regions, respectively. From the discussion above, it is clear that we need, in the proxy model for TSI over three cycles, a fourth component to account for the extra changes in TSI, assumed as trends between minima. So, the 4-component proxy model of TSI can be constructed as

$$S(t) = S_Q + \sum \Delta S = S_Q + a_F P_F(t) + a_N P_N(t) + a_S P_S(t) + a_T P_T(t), \qquad (1)$$

with the value of the quiet Sun, S_Q, and $a_{F,N,S,T}$ the four coefficients for the influence of faculae, network, sunspots and the trend. S_Q and $a_{F,N,S,T}$ are determined by a multi-linear regression between daily values of TSI and those of $P_{F,N,S,T}$. Because faculae and sunspots are both occurring together in active regions, the coefficients $a_{F,S}$ may not really represent the part of their contribution due to mutual compensation. A linear trend between minima may be assumed, similar to the change of the magnetic field in sunspot umbrae from 1996 until 2010 (Penn and Livingston 2006, extended to the present by Livingston, private communication). A closer look at these results, however, may indicate that the field was almost constant until about 2003, when the downward trend started. A similar behaviour is suggested by the trends observed in the VIRGO/SPM data (Fröhlich 2012). This would also explain the very good agreement with proxy or the SATIRE model during the period of the SORCE results, which started at the time of the slope change (see also below). The four components are shown in Fig. 9 and the results of the calibration with the PMOD composite in Fig. 10 and Table 2.

The explained variance, using this 4-component proxy model (84.7%) is quite high, especially for such a long period of time and considering the quite different behaviour of the three cycles. The linear fit to the difference between observed and model TSI over 36 years has a slope of only 10.7 ± 0.5 mWm^{-2}/decade or about 8 ppm/decade; compared with the downward trend during cycle 23 of more than 200 ppm/decade, this is very small. Together with the high correlation, this result demonstrates the importance of including the trend according to the TSI–B_R correlation and it is interesting that the result

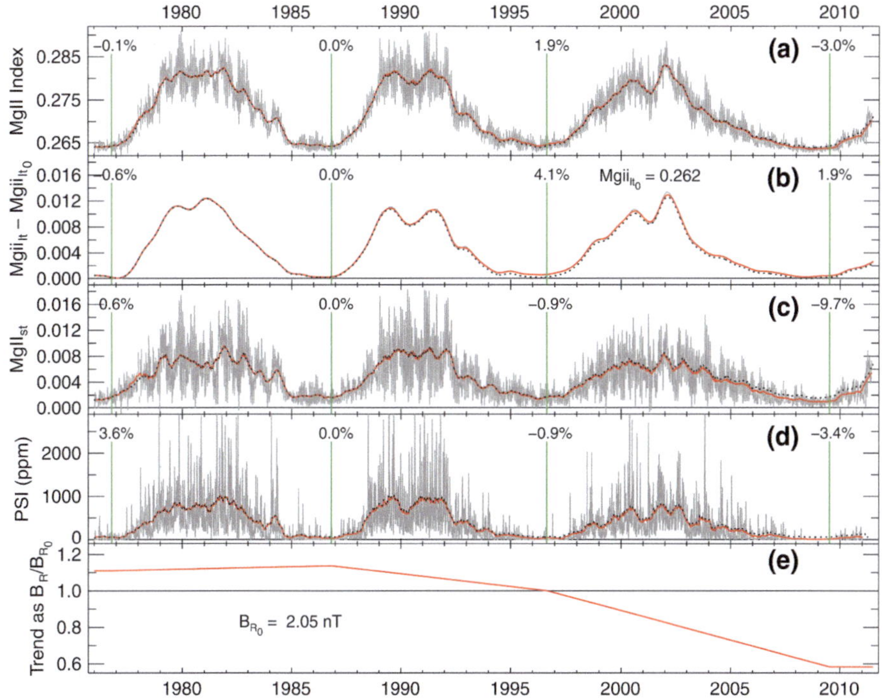

Fig. 9 Daily values of the Mg II index as observed are shown in **a** and decomposed into a long- (Mg II-lt) in **b** and a short-term (Mg II-st) part in **c**, and PSI in **d**. Times of the minima are given as *vertical green lines* and are labelled with the percentage deviation of that minimum from the value of the minimum around 1986 relative to the cycle 22 amplitude of the parameter. The *dotted lines* correspond to the parameter if its linear trend between minima is removed. **e** shows the trend of B_R as a ratio of daily values relative to the value during the 1996 minimum

(0.29 ± 0.01 Wm^{-2}/nT) is about 7% lower than the directly determined value of 0.31 ± 0.13 Wm^{-2}/nT from Fig. 8, which indicates that some of the directly determined slope is indeed explained by the trends in the Mg II index. Thus, 0.29 and not 0.31 Wm^{-2}/nT should be used for future long-term reconstructions of Steinhilber et al. (2009). The ratio of the coefficients for the long- and short-term Mg II index of the order of 1.5–1.7 has been known for some time (e.g., Fröhlich 2004, Fröhlich and Lean 2004), and a similar difference is also observed in the application of a MgII-based model to Ly-α by Woods et al. (2000). The fact that the same ratio is observed in chromospheric radiation means that the earlier explanation (Fröhlich and Lean 2004) by the difference in specific contrast for the network and faculae cannot be correct. It is simply due to the difference in area coverage of the two parts: the short-term part is from the active regions in a limited latitude band, whereas the long-term is distributed over the whole disk. The observed ratio of 1.61 corresponds to a latitude range of about $\pm 18°$, which seems a reasonable average of the activity range over the three cycles. The standard errors of the coefficients are all within 0.5 and 2.0% at the 1-σ level, which indicates that the multi-linear fit is quite reasonable. The results of the partial correlations show that the value for the long-term Mg II is highest with almost 60%, which is explained by dominance of the cycle amplitude. More interesting is the difference between the partial correlation of the short-term Mg II with nearly 20% for the facular

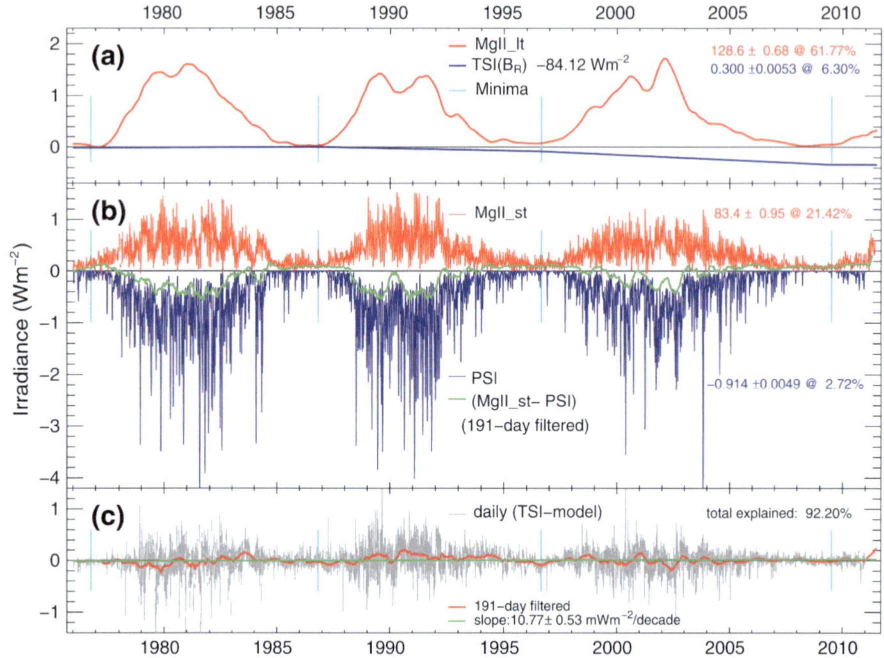

Fig. 10 The calibrated long-term Mg II-lt (*red*) together with the corresponding trend (*blue*) of the 'quiet Sun' is shown in **a**. The calibrated short-term Mg II-st (*red*) and PSI (*blue*) are shown in **b**. Note that all three panels have the same scale. The green curve of **b** shows the 191-day filtered difference between facular and sunspot influence. The net effect of active regions increases TSI during minima by $\approx 0.08\ \mathrm{Wm}^{-2}$, whereas during maxima TSI is dominated by sunspots and a decrease by $\approx 0.25\ \mathrm{Wm}^{-2}$ is observed. The daily difference between TSI and the model is shown in **c**

Table 2 Coefficients of the 4-component proxy model: S_Q is in Wm^{-2}, the others are in Wm^{-2} per Mg II units, in ppm for PSI and in nT for B_R

S_Q	a_F	a_N	$\frac{a_S}{S_Q}$	$\frac{a_T}{B_{R_0}}$	Ratio lt/st	r^2_{tot}
1364.80 ± 0.02	76.29	122.73	−0.8100	0.2902	1.6086	0.8467
Standard error 1-σ	1.0047	0.7237	0.004828	0.011626		
Standard error (%)	1.32	0.59	0.60	1.96		
r^2_a	0.1959	0.5894	0.0004	0.0611		

The standard errors for the coefficients are given as 1-σ values in the units of the coefficients and in percent

contribution and only 0.04% for PSI. This means that sunspots are relatively rare and so do not contribute much to the overall variance, which contradicts our intuition. Another interesting result is shown in Fig. 10b with the combined effect of sunspots and faculae being positive during the minima and negative during the rest of the cycle.

Kopp and Lean (2011) present results of a similar proxy model calibrated against TIM on SORCE, and get $r^2 = 0.92$ over the period 2003–2009. This indicates that a two- or three-component proxy model can be adjusted to any given period, but then may fail to explain the period back to 1980 of the PMOD, ACRIM and IRMB composite as shown in

Kopp and Lean (2011). Their conclusion is that the "current understanding of the sunspot and facular influences on total solar irradiance" is unable to explain the observed trend during cycle 23, which is fully supported by the above discussion, and hence does not question the observations. Also, the results of the SATIRE-S model from Ball et al. (2011), with a free parameter adjusted to fit the TIM values during 2003–2009, gets an even higher correlation ($r^2 = 0.97$) than that of Kopp and Lean (2011), indicating that the SATIRE-S is indeed more accurate in modelling the influence of sunspots by including pores and treating the umbra and penumbra separately and by using an appropriate center-to-limb variation for the faculae and network (Unruh et al. 1999). Krivova et al. (2011) present a similar application of SATIRE-S with similar conclusions, although they adjust the free parameter to the period of VIRGO TSI from 1996–2009 (same as used in the PMOD composite for this period). Due to the shape of the cycle variation, this fit is heavily weighted towards the decreasing part of the cycle and as Kopp and Lean (2011) they are not able to explain VIRGO TSI during the ascending phase of cycle 23. In contrast to Kopp and Lean (2011), however, they conclude that most probably the data analysis of VIRGO is flawed during the ascending part of cycle 23. This conclusion, however, contradicts earlier publications by the same group (e.g., Krivova et al. 2003; e.g., Wenzler et al. 2004), which showed that the PMOD composite, and hence VIRGO since 1996, is, according to their model, an excellent representation of TSI up to 2003, the end of their analysis at that time. This is further confirmed by Steinhilber (2010) who used synoptic maps from MDI and adjusted the free parameter during the period up to 2003 to enable a direct comparison with the results of Krivova et al. (2003), and the agreement during this period is very good, although the calibration of the MDI magnetograms data has changed from uncalibrated level-1.5 to the calibrated level-1.6. More importantly, Steinhilber (2010) also needs to add a trend to explain the large difference of TSI between the minima in 1996 and 2009. Thus, it is quite clear that the proxy models with PSI and Mg II index only as well as the SATIRE-S model are not able to explain the trend in TSI for reasons put forward by Foukal et al. (2011) and discussed in detail in Sect. 4. On the other hand, it also clear that models based on the observed surface magnetic fields from magnetograms or chromospheric indices are capable of explaining the UV irradiance like Ly-α, which show a 4-5 times smaller trend than TSI. But, as shown in Sect. 3, the trend is fully supported by the observations of ACRIM II and III and also of DIARAD (the basis for the IRMB composite during VIRGO), if its non-exposure-dependent sensitivity increase is included as shown in Fig. 3 and routinely used in the evaluation of VIRGO TSI.

6 Conclusions

The absolute value of TSI seems to have converged to the value presented by TIM on SORCE, which is now also supported by a new evaluation of ACRIM III from results of recent TRF comparison (Willson 2011). Although the corresponding comparison for the VIRGO radiometer, PMO6V and DIARAD, is still incomplete or not yet available, the already existing results point in the same direction, but the corrections may be smaller. Nevertheless, the recent minimum value of TSI is more likely closer to 1,361 Wm^{-2} than the canonical value of 1,365 Wm^{-2}. The PMOD composite may be adjusted to the new scale by multiplying the values by 0.996894.

Comparing the three available composites, it is clear that only the PMOD represents solar behaviour during cycle 21, because it critically reviewed the records of HF and ACRIM I and corrected them for effects not taken into account in the original analysis.

During cycle 22, the main difference occurs between ACRIM and PMOD, due to a slip in the HF record during the ACRIM gap which is corrected in the PMOD, but not in the ACRIM composite. Several independent observations support this slip and thus a corresponding correction is needed and the ACRIM record during this cycle is flawed by ignoring this effect. During cycle 23, the most important question is whether the difference between the minima of 1996 and 2009 of about a quarter of the cycle amplitude is real. The comparison of the VIRGO record with a ACRIM II-III composite and TIM supports the low value. During cycle 23, the IRMB composite suffers from a non-exposure-dependent increase in sensitivity of DIARAD on VIRGO, which is not accounted for in the IRMB analysis and hence it is not representing the solar cycle variability; the IRMB composite should therefore not be used for cycle 23 or beyond. Thus, the PMOD composite, which includes the corrected DIARAD within VIRGO TSI for cycle 23, can be used as reference TSI for the last three solar cycles. In summary, the trend-corrected amplitudes of the three cycles are 0.9261, 0.8649 and 0.7954 Wm^{-2} and the changes over the cycles, as differences between minima before and after the corresponding cycle, are 5.37, −11.93 and −27.06% relative to the TSI amplitude of cycle 22.

Proxies like the SSN, F10.7 or the chromospheric indices and the UV radiation, e.g. Ly-α, do show much smaller changes than TSI over each cycle with an average of the six parameters of 0.77 ± 2.58, -2.45 ± 5.84, $-3.51 \pm 1.45\%$ (Table 1). UV radiation, chromospheric proxies and surface magnetic fields show a 5 times smaller change than TSI, but both are related to the strength and the activity. The larger change in TSI is most likely related to the increased contrast of small magnetic fields, which is only observed for radiation emanating from the photosphere and not for those of the chromosphere, although the magnetic fields observed at the surface and their influence on the chromosphere are closely related. This effect is not taken into account in proxy models based on PSI and, e.g., Mg II index only, nor in the more sophisticated SATIRE-S model. Thus, these models need a further component taking the enhanced influence of the strength of the activity into account, by, e.g., a trend between the minima. More investigations are needed to specify the trend in more detail.

Quite obviously, such a trend is not needed to model, e.g., Ly-α or UV radiation at other wavelengths because their changes are only due to the magnetic field as observed at the surface and not amplified by, e.g., a changing contrast. This difference between the behaviour of total solar and UV irradiance has important implications for the reconstruction of irradiance back to, e.g., the Maunder Minimum. For TSI, the correlation with B_R, corrected for the changes already included by the magnetic field effect alone, can be used to determine its values at minima whenever B_R is known (see, e.g., Steinhilber et al. 2009). Based on the results of the 4-component proxy model and using the Ca II K observations from Mt. Wilson back to 1913 transformed into a MgII index, PSI from sunspot observations by the Royal Greenwich Observatory and B_R at minima from the aa-index, a reliable extension of the proxy model presented here back to 1914 is possible (Fröhlich 2011b).

Acknowledgments I thank the organizers of the ISSI Workshop "Observing and modelling Earth's Energy Flows" for an invitation to present a review of TSI observations and the two referees for their very helpful comments, improving the manuscript substantially. This work is supported by the Swiss National Science Foundation by providing support for travel, which is gratefully acknowledged. Moreover, this research would not have been possible without substantial contributions by the VIRGO and SOHO teams. SOHO is a cooperative mission between ESA and NASA, which was launched in late 1995 and is still in operation (as of July 2011).

References

Baliunas S, Jastrow R (1990) Evidence for long-term brightness changes of solar-type stars. Nature 348:520–523

Ball WT, Unruh YC, Krivova NA, Solanki S, Harder JW (2011) Solar irradiance variability: a six-year comparison between SORCE observations and the SATIRE model. Astron Astrophys 530:A71. doi: 10.1051/0004-6361/201016189

Bertello L, Ulrich RK, Boyden JE (2010) The mount Wilson Ca ii K plage index time series. Sol Phys 264:31–44. doi:10.1007/s11207-010-9570-z

Broomhall A, Chaplin WJ, Elsworth Y, Fletcher ST, New R (2009) Is the current lack of solar activity only skin deep? Astrophys J 700:162–165. doi:10.1088/0004-637X/700/2/L162

Brusa RW, Fröhlich C (1986) Absolute radiometers (PMO6) and their experimental characterization. Appl Opt 25:4173–4180. doi:10.1364/AO.25.004173

Dewitte S, Crommelynck D, Mekaoui S, Joukoff A (2004) Measurement and uncertainty of the long-term total solar irradiance trend. Sol Phys 224:209–216. doi:10.1007/s11207-005-5698-7

Foukal P, Milano L (2001) A measurement of the quiet network contribution to solar irradiance variation. Geophys Res Lett 28:883–886. doi:10.1029/2000GL012072

Foukal P, Ortiz A, Schnerr R (2011) Dimming of the 17th century sun. Astrophys J 733:38. doi: 10.1088/2041-8205/733/2/L38

Foukal PV, Lean J (1988) Magnetic modulation of solar luminosity by photospheric activity. Astrophys J 328:347–357. doi:10.1086/166297

Foukal PV, Lean J (1990) An empirical model of total solar irradiance variation between 1874 and 1988. Science 247:556–558. doi:10.1126/science.247.4942.556

Foukal P, Fröhlich C, Spruit H, Wigley TML (2006) Variations in solar luminosity and their effect on the earth's climate. Nature 443:161–166. doi:10.1038/nature05072

Fröhlich C (2004) Solar irradiance variability, in geophysical monograph 141: solar variability and its effect on climate. AGU, Washington, pp. 97–110. Chap. 2: Solar energy flux variations

Fröhlich C (2005) Solar irradiance variability since 1978. Memorie della Societa Astronomica Italiana 76:731–734

Fröhlich C (2006a) Solar irradiance variability since 1978: revision of the PMOD composite during solar cycle 21. Space Sci Rev 125:53–65. doi:10.1007/s11214-006-9046-5

Fröhlich C (2006b) Total solar irradiance variability: what have we learned from SOHO/VIRGO about solar cycle 23?. In: H. Lacoste (ed) Proceedings of SOHO 17: 10 Years of SOHO and Beyond, Giardini Naxos, Sicily, Italy, 7–12 May 2006 (ESA SP-217). ESA Publication Division, ESTEC, 2200 AG, The Netherlands, p36_froh. Also available from ftp://ftp.pmodwrc.ch/pub/Claus/Publications/SOHO17_cf_TSI_paper.pdf

Fröhlich C (2009) Evidence of a long-term trend in total solar irradiance. Astron Astrophys 501:27–30. doi: 10.1051/0004-6361/200912318

Fröhlich C (2010) Possible influence of aperture heating on VIRGO radiometry on SOHO. AGU Fall Meeting Abstracts, B874

Fröhlich C (2010) Solar radiometry, in observing photons in space. In: Huber MCE, Culhane JL, Pauluhn A, Timothy JG, Wilhelm K, Zehnder A (eds) ISSI Scientific Reports SR-009. ESA Communications, Noordwijk pp 525–540. Chap. 32

Fröhlich C (2011c) Total solar irradiance: what have we learned from the last three cycles and the recent minimum? Space Sci Rev 158. published online 11 May 2011. doi:10.1007/s11214-011-9780-1

Fröhlich C (2012) Spectral irradiance variation during cycle 23 from the three-channel VIRGO/SPM. Some preliminary results were presented at the SORCE meeting: ftp://ftp.pmodwrc.ch/pub/Claus/SORCE_2011/SPM_poster_cf.pdf. (in preparation)

Fröhlich C, Lean J (1998) Total solar irradiance variations: the construction of a composite and its comparison with models. In: Deubner FL, Christensen-Dalsgaard J, Kurtz D (eds) IAU symposium 185: new eyes to see inside the sun and stars. Kluwer, Dordrecht, pp 89–102

Fröhlich C, Lean J (2004) Solar radiative output and its variability: evidence and mechanisms. Astron Astrophys Rev 12:273–320. doi:10.1007/s00159-004-0024-1

Fröhlich C, Crommelynck D, Wehrli C, Anklin M, Dewitte S, Fichot A, Finsterle W, Jiménez A, Chevalier A, Roth HJ (1997) In-flight performances of VIRGO solar irradiance instruments on SOHO. Sol Phys 175:267–286. doi:10.1023/A:1004929108864

Fröhlich C, Appourchaux T, Chaplin WJ, Elsworth Y, Wachter R (2006) Solar cycle variability of total solar irradiance and P-mode frequencies, in SOHO-17. 10 years of SOHO and beyond. presented, but unpublished poster available from ftp://ftp.pmodwrc.ch/pub/Claus/SOHO-17/PMODES_A0poster.pdf

Fröhlich C (2011a) A four-component proxy model for total solar irradiance calibrated during solar cycles 21–23. Contrib Astron Obs Skalnate Pleso 41:113–132
Fröhlich C (2011b) Total solar irradiance during the last three cycles and what it means for the reconstruction back to 1915. EGU Abstracts,CL2.12. available from ftp://ftp.pmodwrc.ch/pub/Claus/EGU_2011/EGU2011_TSI_cf.ppt; a detailed publication is in preparation
Hagenaar HJ, Schrijver CJ, Title AM (2003) The properties of small magnetic regions on the solar surface and the implications for the solar dynamo(s). Astrophys J 584:1107–1119. doi:10.1086/345792
Hoyt DV, Kyle HL, Hickey JR, Maschhoff RH (1992) The NIMBUS-7 solar total irradiance: a new algorithm for its derivation. J Geophys Res 97:51–63
Hudson HS (1988) Observed variability of the solar luminosity. Ann Rev Astron Astrphys 26:473–508. doi:10.1146/annurev.aa.26.090188.002353
Kopp G, Lean JL (2011) A new, lower value of total solar irradiance: evidence and climate significance. Geophys Res Lett 38:L01706. doi:10.1029/2010GL045777
Kopp G, Lawrence G, Rottman G (2005) The total irradiance monitor (TIM): science results. Sol Phys 230:129–139. doi:10.1007/s11207-005-7433-9
Kopp G, Heuerman K, Harber D, Drake G (2007) The tsi radiometer facility: absolute calibrations for total solar irradiance instruments. In Society of photo-optical instrumentation engineers (SPIE) conference series 6677: 667709 doi:10.1117/12.734553
Krivova NA, Solanki SK, Schmutz W (2011) Solar total irradiance in cycle 23. Astron Astrophys 529:A81. doi:10.1051/0004-6361/201016234
Krivova NA, Solanki SK, Fligge M, Unruh YC (2003) Reconstruction of solar total and spectral irradiance variations in cycle 23: is solar surface magnetism the cause?. Astron Astrophys 399:1–4
Kuhn JR, Libbrecht KG (1991) Nonfacular solar luminosity variations. Astrophys J 381:35–37
Kuhn JR, Libbrecht KG, Dicke R (1988) The surface temperature of the sun and changes in the solar constant. Science 242:908–911
Lee RB III, Barkstrom BR, Cess RD (1987) Characteristics of the earth radiation budget experiment solar monitors. Appl Opt 26:3090–3096
Lee RB III, Gibson MA, Wilson RS, Thomas S (1995) Long-term total solar irradiance variability during sunspot cycle 22. J Geophys Res 100:1667–1675. doi:10.1029/94JA02897
Livingston W, Wallace L, White OR, Giampapa MS (2007) Sun-as-a-star spectrum variations 1974–2006. Astrophys J 657:1137–1149. doi:10.1086/511127
Mekaoui S, Dewitte S, Conscience C, Chevalier A (2010) Total solar irradiance absolute level from DIARAD/SOVIM on the international space station. Adv Space Res 45:1393–1406. doi:10.1016/j.asr.2010.02.014
Ortiz A, Domingo V, Sanahuja B (2006) The intensity contrast of solar photospheric faculae and network elements. II. Evolution over the rising phase of solar cycle 23. Astron Astrophys 452:311–319. doi:10.1051/0004-6361:20053869
Penn MJ, Livingston W (2006) Temporal changes in sunspot umbral magnetic fields and temperatures. Astrophys J 649:45–48. doi:10.1086/508345
Pietarila Graham J, Danilovic S, Schüssler M (2009) Turbulent magnetic fields in the quiet sun: implications of hinode observations and small-scale dynamo simulations. Astrophys J 693:1728–1735. doi:10.1088/0004-637X/693/2/1728
Scafetta N (2010) Climate change and its causes, a discussion about some key issues, in SPPI Original Paper (Science and Public Policy Institute, Haymarket, Va 20169, USA), pp 1–56. http://scienceandpublicpolicy.org/images/stories/papers/originals/climate_change_cause.pdf
Scafetta N, Willson RC (2009) ACRIM-gap and TSI trend issue resolved using a surface magnetic flux TSI proxy model. Geophys Res Lett 36:L05701. doi:10.1029/2008GL036307
Schnerr RS, Spruit HC (2011) The total solar irradiance and small scale magnetic fields. In: Kuhn JR, Harrington DM, Lin H, Berdyugina SV, Trujillo-Bueno J, Keil SL, Rimmele T (eds) Astronomical society of the Pacific conference series, vol 437, pp 167–175
Shapiro AI, Schmutz W, Rozanov E, Schoell M, Haberreiter M, Shapiro AV, Nyeki S (2011) A new approach to the long-term reconstruction of the solar irradiance leads to large historical solar forcing. Astron Astrophys 529:A67. doi:10.1051/0004-6361/201016173
Solanki SK, Krivova NA, Wenzler T (2005) Irradiance models. Adv Space Res 35:376–383. doi:10.1016/j.asr.2004.12.077
Steinhilber F (2010) Total solar irradiance since 1996: is there a long-term variation unrelated to solar surface magnetic phenomena? Astron Astrophys 523:A39. doi:10.1051/0004-6361/200811446
Steinhilber F, Beer J, Fröhlich C (2009) Total solar irradiance during the holocene. Geophys Res Lett 36:L19704. doi:10.1029/2009GL040142

Tapping KF, Boteler D, Charbonneau P, Crouch A, Manson A, Paquette H (2007) Solar magnetic activity and total irradiance since the maunder minimum. Sol Phys 246:309–326. doi:10.1007/s11207-007-9047-x
Thuillier G, Dewitte S, Schmutz W (2011) Picard team, the sun-climate connection through measurements and modeling: the Picard investigation. In: Miralles MP, Sánchez Almeida J (eds) The sun, the solar wind, and the heliosphere. IAGA Special Sopron Book Series, Springer, Berlin, pp 365
Unruh YC, Solanki SK, Fligge M (1999) The spectral dependence of facular contrast and solar irradiance variations. Astron Astrophys 345:635–642
Viereck RA, Snow M, Deland MT, Weber M, Puga L, Bouwer D (2010) Trends in solar UV and EUV irradiance: an update to the MgII Index and a comparison of proxies and data to evaluate trends of the last 11-year solar cycle. AGU Fall Meeting Abstracts: Poster GC21B-0877
Vögler A, Schüssler M (2007) A solar surface dynamo. Astron Astrophys 465:43–46. doi:10.1051/0004-6361:20077253
Wenzler T, Solanki SK, Krivova NA, Fluri DM (2004) Comparison between KPVT/SPM and SoHO/MDI magnetograms with an application to solar irradiance reconstructions. Astron Astrophys 427: 1031–1043. doi:10.1051/0004-6361:20041313
Willson RC (1984) Measurements of solar total irradiance and its variability. Space Sci Rev 38:203–242. doi:10.1007/BF00176830
Willson RC (1994) Irradiance Observations from SMM, UARS and ATLAS Experiments. In: Pap J, Fröhlich C, Hudson HS, Solanki S (eds) IAU Colloquium No. 143: the sun as a variable star: solar and stellar irradiance variations. Cambridge University Press, Cambridge, pp 54–62
Willson RC (1997) Total solar irradiance trend during solar cycles 21 and 22. Science 277: 1963–1965. see also comment by R. Kerr on page 1923 of the same issue of Science. doi:10.1126/science.277.5334.1963
Willson RC (2001) The ACRIMSAT/ACRIM III experiment: extending the precision, long-term total solar irradiance climate database. Earth Observer 13:14–17
Willson RC (2011) ACRIM3 and ACRIM composite results after 2010 Reflect LASP/TRF determined scattering, diffraction and basic scale calibration adjustments. Data available from http://www.acrim.com/RESULTS/data/acrim3/acrim3_sdc_ts_1104.rtf and http://www.acrim.com/RESULTS/data/composite/composite_acrim_1105.rtf
Willson RC, Hudson HS (1991) The Sun's luminosity over a complete solar cycle. Nature 351:42–44. doi:10.1038/351042a0
Willson RC, Mordvinov AV (2003) Secular total solar irradiance trend during solar cycles 21-23. Geophys Res Lett 30:1199. doi:10.1029/2002GL016038
Woods TN, Tobiska WK, Rottman GJ, Worden JR (2000) Improved solar Lyman α irradiance modeling from 1947 through 1999 based on UARS observations. J Geophys Res 105:27195–27216. doi:10.1029/2000JA000051

Surv Geophys (2012) 33:475–481
DOI 10.1007/s10712-011-9166-7

Solar Irradiance Models and Measurements: A Comparison in the 220–240 nm wavelength band

Yvonne C. Unruh · Will T. Ball · Natalie A. Krivova

Received: 25 July 2011 / Accepted: 1 December 2011 / Published online: 28 December 2011

Abstract Solar irradiance models that assume solar irradiance variations to be due to changes in the solar surface magnetic flux have been successfully used to reconstruct total solar irradiance on rotational as well as cyclical and secular time scales. Modelling spectral solar irradiance is not yet as advanced, and also suffers from a lack of comparison data, in particular on solar cycle time scales. Here, we compare solar irradiance in the 220–240 nm band as modelled with SATIRE-S and measured by different instruments on the UARS and SORCE satellites. We find good agreement between the model and measurements on rotational time scales. The long-term trends, however, show significant differences. Both SORCE instruments, in particular, show a much steeper gradient over the decaying part of cycle 23 than the modelled irradiance or that measured by UARS/SUSIM.

Keywords Sun: activity · Sun: irradiance · Spectral And Total Irradiance REconstructions (SATIRE) · SOlar Radiation and Climate Experiment (SORCE) · Upper Atmosphere Research Satellite/Solar Ultraviolet Spectral Irradiance Monitor (UARS/SUSIM)

1 Introduction

The solar irradiance, i.e. the solar energy flux received at the top of the Earth's atmosphere, changes on a wide range of time scales. Changes in total solar irradiance (TSI) have been monitored since 1978 and a number of composites combining data from different satellites

Y. C. Unruh (✉) · W. T. Ball
Astrophysics Group, Blackett Laboratory, Imperial College London, London SW7 2AZ, UK
e-mail: y.unruh@imperial.ac.uk

W. T. Ball
e-mail: william.ball08@imperial.ac.uk

N. A. Krivova
Max-Planck Institut für Sonnensystemforschung, 37191 Katlenburg-Lindau, Germany
e-mail: natalie@mps.mpg.de

are available (see, e.g., Domingo et al. 2009 for a discussion of TSI composites). Detailed modelling of the TSI has shown that solar irradiance changes on rotational to cyclical time scales can be well modelled as being related to changes in the solar surface magnetic flux alone (e.g., Preminger et al. 2002; Wenzler et al. 2006; Krivova et al. 2003; Ball et al. 2011).

Quantifying changes in solar spectral irradiance (SSI) is a more arduous task. Up until the launches of the Scanning Imaging Absorption spectrometer for Atmospheric Chemistry (SCIAMACHY) instrument on board ENVISAT and the SOlar Radiation and Climate Experiment (SORCE) satellite in 2002 and 2003, respectively, most of the spectral data were restricted to wavelengths below 400 nm, and more reliable data were only available for wavelengths below 300 nm (Woods et al. 1996). A compilation of UV irradiance measurements up to 2005, though not yet including SORCE data, was presented by DeLand and Cebula (2008). This compilation includes the authors' best estimate of which of the instruments yield the most reliable measurements at different times over the last three solar cycles, though no wavelength-by-wavelength cross-calibrations are applied at this stage.

The SOLar STellar Irradiance Comparison Experiment (SOLSTICE) (McClintock et al. 2005a, b) and Spectral Irradiance Monitor (SIM) (Harder et al. 2005a, b) onboard SORCE measure spectral solar irradiance between 115 and 320 nm and between 200 nm and 2.4 μm, respectively. Results from the decaying phase of cycle 23 have shown rather unexpected long-term behaviour, in particular a much steeper UV decline than that measured in cycle 22 by the Solar Ultraviolet Spectral Irradiance Monitor (SUSIM) and by SOLSTICE on board the Upper Atmosphere Research Satellite (UARS). In addition, SORCE/SIM also suggested an anticorrelation of SSI between wavelengths of $\sim$400 to $\sim$700 nm with TSI (Harder et al. 2009).

Model-based reconstructions of the solar UV irradiance have been presented by a number of authors, including Krivova et al. (2006, 2009) and Morrill et al. (2011). Here, we present reconstructions of the UV irradiance over the last solar cycle for wavelengths between 220 and 240 nm using the Spectral And Total Irradiance REconstructions—Satellite era (SATIRE-S) model. We compare the UV reconstructions with data from UARS/SUSIM and UARS/SOLSTICE, as well as to data from SIM and SOLSTICE onboard SORCE. A more in-depth analysis covering a wider spectral range as well as solar cycles 21 and 22 will be presented in Ball et al. (in preparation).

2 SATIRE-S

The reconstructions presented here are obtained with SATIRE-S (Krivova et al. 2011) and assume that all changes in solar irradiance are due to changes in the solar surface magnetic flux. The emergent magnetic flux forms dark and bright surface features that can be identified from magnetograms and continuum images. Having identified the magnetic features and knowing their contrast with respect to the quiet Sun (see Unruh et al. 1999), the irradiance change can be calculated by summing the different contributions over the solar surface.

SATIRE-S has one free parameter, B_{sat}; it effectively determines the field strength above which the facular contrast saturates and depends on the resolution and image characteristics of the magnetograms and continuum images (see Krivova et al. 2011 for more detail).

3 Solar UV Irradiance from 1999 to 2010

The reconstructions presented here are based on magnetograms and continuum images from the Michelson Doppler Imager (MDI) onboard the Solar and Heliospheric Observatory (SoHO). We consider the period between 1999 and 2009 after the SoHO 'vacation'. During this time, we have previously found excellent agreement between the Physikalisch-Metereologisches Observatorium Davos TSI composite (Fröhlich 2000, 2006) and the SATIRE-S reconstructions. The free parameter, B_{sat}, was fixed so that the regression slope of the TSI values calculated with SATIRE-S and those of the PMOD composite was unity for times between February 1999 and February 2010. The resulting value for B_{sat} is 424 G and the correlation coefficient over this period is $r = 0.983$ ($r^2 = 0.966$). Note that this value of B_{sat} can not be compared directly to earlier values (e.g., Krivova et al. 2003) on account of the changes in the magnetogram calibrations introduced for the MDI level 1.8 data.

The UV irradiances integrated between 220 and 240 nm are shown in Fig. 1 for times between 1999 and 2010. This spectral region has been selected as it had previously been found that the SATIRE-S calculations reproduce the UARS/SUSIM measurements well, despite SATIRE's reliance on Local Thermodynamic Equilibrium (LTE) which tends to be a poor approximation in this wavelength region. For other UV wavelengths, irradiance changes can be derived from the integrated 220 nm to 240 nm irradiance by linear regression (Krivova et al. 2006).

The SATIRE-S reconstructions are shown in blue, the UARS/SUSIM measurements are plotted in black. The error in the absolute measurements is significantly larger than the error in the relative changes of the irradiance (see, e.g., Woods et al. 1996). The SUSIM and SATIRE-S data have thus been normalised so that their mean values after the SoHO vacation agree. The SORCE/SIM and SORCE/SOLSTICE data are overplotted in red and purple, respectively. They have been normalised using the mean values during their overlap period (May 2005 to May 2010); they have also been shifted by 0.02 for ease of comparison with SUSIM and SATIRE. The inset shows a comparison between UARS/SOLSTICE (green), UARS/SUSIM (black) and SATIRE-S (blue, based on Kitt Peak

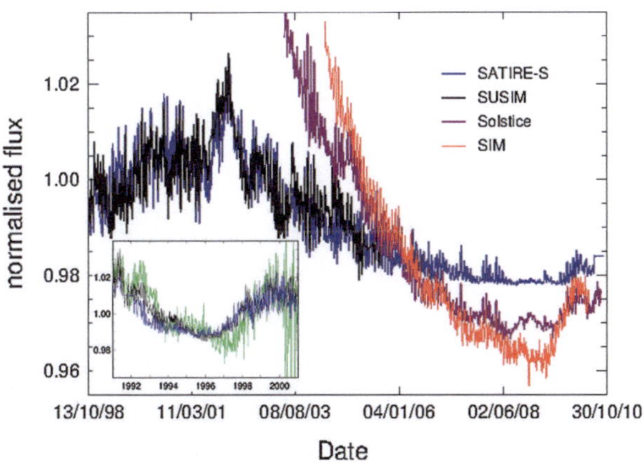

Fig. 1 Plot of the normalised solar UV irradiance integrated between 220 and 240 nm as modelled with SATIRE-S (*blue line*) and measured with UARS/SUSIM (*black*), SORCE/SOLSTICE (*purple*) and SORCE/SIM (*red*). The *inset* shows data going back to 1991: SATIRE-S (*blue*), UARS/SUSIM (*black*) and UARS/SOLSTICE (*green*)

images pre-1999) during the lifetime of the UARS/SOLSTICE instrument. As the data quality of UARS/SOLSTICE deteriorated from about 2001 onwards (as e.g., indicated by the increase in spurious downward spikes), the irradiances shown on the inset have been normalised to their pre-2000 mean.

The UV irradiance measured with both SORCE instruments shows a much steeper long-term trend than that found with SUSIM, either during cycle 23 or cycle 22 (see Krivova et al. 2009 for an in-depth discussion of UV irradiance measurements during cycle 22).

4 Discussion

4.1 Solar Rotational Time Scales

The modelled and the measured UV irradiances show excellent agreement on rotational time scales. This is illustrated in more detail in Fig. 2, where the irradiances are plotted over the shorter timespan of approximately 2 years. The irradiance changes by approximately 1% as the active regions rotate in and out of view and we find that the variability amplitude is well matched between the different instruments and the model. Indeed, on rotational time scales, the scatter between the irradiances measured with the different instruments is comparable with the scatter between the modelled and observed irradiance, as illustrated by the results of a regression analysis. The slopes and correlation coefficients are listed in Table 1. Rows 2–7 are for detrended data sets where the cyclical variability has been removed and only rotational variability retained. The number of data points for which the correlation was calculated is also listed in Table 1; note that the UARS/SUSIM measurements cover a much longer time span compared with SORCE/SIM. UARS/SOLSTICE data have not been included in the analysis here as they do not overlap with SORCE/SIM. We note, however, that the UARS/SUSIM and UARS/SOLSTICE rotational variability agree well with each other and also with SATIRE-S.

On rotational time scales, we find that the best agreement is between SATIRE-S and SORCE/SOLSTICE; the model is able to explain at least 83% of the rotational variability. This suggests that, on these time scales and for the instruments considered here, SORCE/SOLSTICE has the lowest instrumental noise. The higher instrumental noise of SIM is not too surprising as the 220–240 nm wavelength band is relatively near the blue edge of the SIM instrument (see also the discussion in Unruh et al. 2008). Considering the instrument

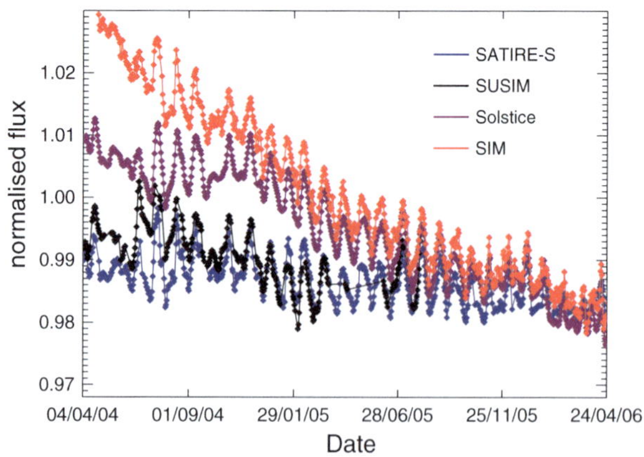

Fig. 2 A 2-year section of the normalised solar UV irradiance. The quantities plotted are as in Fig. 1, but showing the period between April 2004 and April 2006 when UARS and SORCE measurements overlap in more detail

Table 1 Correlation coefficients (column 4), regression slopes (column 5) and number of overlapping data points considered (column 6) for the SATIRE-S, UARS/SUSIM, SORCE/SIM and SORCE/SOLSTICE data

[1] Row	[2] Data set 1	[3] Data set 2	[4] r (r^2)	[5] Slope	[6] Points	[7] Time span
1	SATIRE-S	SUSIM	0.92 (0.85)	1.03	1,641	1999/02/19–2005/07/31
2	SATIRE-S	SUSIM	0.88 (0.77)	1.04	1,641	1999/02/19–2005/07/31
3	SATIRE-S	SIM	0.76 (0.58)	1.14	1,665	2004/04/22–2009/10/30
4	SATIRE-S	SOLSTICE	0.91 (0.83)	0.97	523	2003/05/14–2005/07/31
5	SUSIM	SOLSTICE	0.75 (0.56)	0.89	523	2003/05/14–2005/07/31
6	SUSIM	SIM	0.72 (0.52)	0.96	293	2004/04/22–2005/07/31
7	SIM	SOLSTICE	0.80 (0.64)	0.91	1,665	2004/04/22–2009/10/30

The results in the first row are for the normalised data and include the long-term cyclical trend; the remaining rows are all for detrended data

designs and capabilities, SORCE/SOLSTICE should yield better results for wavelengths below 240 nm, while SORCE/SIM is the instrument of choice above approximately 290 nm (J. Harder, private communication). Probably due to the loss of one of the reaction wheels in 2008, the noise level of SORCE/SIM increased slightly towards the end of the mission (see the period around cycle minimum in Fig. 1). This is borne out by the correlation analysis. Restricting the analysis to times between April 2004 and January 2008 results in higher correlation coefficients ($r = 0.81$ vs. 0.76 for the comparison with SIM and $r = 0.84$ vs. 0.80 for the comparisons with and SOLSTICE). At the same time, the correlation slopes improve to values nearer unity; in the case of SATIRE-S versus SIM (see row 3) the slope decreases from 1.14 to 1.05, while it increases from 0.91 to 0.97 for the SIM to SOLSTICE comparison (see row 7).

The rather low correlation coefficient for UARS/SUSIM with either SORCE/SIM or SORCE/SOLSTICE is mainly due to the short time span considered and due to what appears to be increased noise in SUSIM after mid 2003. The correlation cofficient between the detrended SATIRE-S and SUSIM time series for the time span after 2003 falls to $r = 0.79$ ($r^2 = 0.63$) compared with the more typical $r = 0.88$ ($r^2 = 0.77$) for the period starting in 1999.

4.2 Solar Cycle Variability

The most striking feature in Fig. 1 is the much more pronounced long-term trend observed by the two SORCE instruments compared with UARS/SUSIM and also with SATIRE-S. As already pointed out by Krivova et al. (2006, 2009) for the descending phase of cycle 22 and the rising phase of cycle 23, we find that the SATIRE-S reconstructions are in excellent agreement with the SUSIM measurements in the 220–240 nm range. But while SUSIM and SATIRE-S suggest a difference of the order of 4% between cycle minimum and maximum in cycle 22, SORCE/SIM and SOLSTICE show a significantly larger change during part of the declining phase of cycle 23.

The decline observed with SORCE/SOLSTICE is of the order of 6% between May 2003 and the cycle 23/24 minimum. A similar decline is seen for SORCE/SIM between May 2004 and the cycle minimum. We note that, in terms of the TSI change, the period between May 2003 (and indeed also May 2004) and the cycle minimum only corresponds to approximately half of the total cycle change in TSI. The $\sim$6% decline in the 220–240 nm integrated irradiance would, thus, be a lower limit and suggest an overall cycle change of the order of 12%. This is compared to the ~4% change in UARS/SUSIM and SATIRE-S (recall that the typical rotational variability measured by all instruments out of minimum is of the order of 1–2% in this wavelength range).

The relative accuracy of both UARS instruments over a time scale of a few years has conservatively been estimated to be of the order of 1% (Woods et al. 1996). In the 220–240 nm region, the SORCE/SOLSTICE accuracy and long-term stability is considered to be better than that of SIM (Harder et al. 2010). Snow et al. (2005) estimate the SORCE/SOLSTICE stability to be of the order of 1% per year currently, though it is expected that, pending outstanding corrections, SORCE/SOLSTICE will achieve a long-term stability of 0.5% per year by the end of the mission (M. Snow, private communication).

We, thus, find that the cycle variability observed with UARS/SUSIM, with UARS/SOLSTICE, and modelled with SATIRE-S agrees within the estimated uncertainty. Taking the 1% per year uncertainty at face value, the observed trends for SORCE lie just outside the long-term instrumental uncertainties and it is not clear how to reconcile the UARS and SORCE measurements in this wavelength region. Note that, while considerably steeper overall, the trends observed with the two SORCE instruments do not always agree well. The slope seen in SORCE/SIM during the first 9 months is substantially steeper than that recorded by SORCE/SOLSTICE. Both instruments show good long- and short-term agreement over a time span of approximately 2 years (February 2005 to January 2007) where they both record a decline of approximately 1% per year. Their subsequent trends, however, disagree at the estimated 1% stability level: SORCE/SIM shows a deeper minimum combined with a sharper decline and rise in and out of the minimum.

5 Summary

The typical rotational variability in the 220–240 nm band is of the order of 1%. On solar rotational time scales, we find excellent agreement between the irradiances recorded by UARS/SUSIM, SORCE/SOLSTICE and SORCE/SIM, and those reconstructed from SoHO/MDI images using the SATIRE-S model.

On yearly time scales and longer, the observed trends with the newer SORCE instruments, SORCE/SOLSTICE and SORCE/SIM, are much more pronounced than those measured with UARS/SUSIM or modelled with SATIRE-S. It is currently unclear how these trends can be reconciled and whether the differences are purely instrumental, or due to a change in the Sun's behaviour. One might argue that, after 14 years in space, UARS/SUSIM suffered from instrumental degradation during the declining phase of cycle 23. This seems unlikely, however, when considering the long-term trends measured during the second half of cycle 22 that are of a similar magnitude as those in cycle 23 and, furthermore, agree with those modelled by SATIRE-S. While it is simplistic to assume that UV irradiance trends will scale linearly with TSI, the approximately fourfold larger drop in the 220–240 nm irradiance during cycle 23 when the change in the TSI was comparable to that in cycle 22 remains intriguing.

Acknowledgments The authors would like to thank ISSI for the hospitality and lively meeting. The authors would also like to thank Jerry Harder for providing the SORCE/SIM data, as well as Linton Floyd for helpful discussions and information on SUSIM data and Marty Snow for information on the SORCE/SOLSTICE data. This work was supported by the *Deutsche Forschungsgemeinschaft, DFG* project number SO 711/1-3 and by the NERC SolCli consortium grant; we are indebted to Sami Solanki and Thomas Wenzler for many useful discussions.

References

Ball WT, Unruh YC, Krivova NA, Solanki S, Harder JW (2011) Solar irradiance variability: a six-year comparison between SORCE observations and the SATIRE model. Astron Astrophys 530:A71

DeLand MT, Cebula RP (2008) Creation of a composite solar ultraviolet irradiance data set. Journal of Geophysical Research (Space Physics) 113:A11103

Domingo V, Ermolli I, Fox P, Fröhlich C, Haberreiter M, Krivova N, Kopp G, Schmutz W, Solanki SK, Spruit HC, Unruh Y, Vögler A (2009) Solar surface magnetism and irradiance on time scales from days to the 11-year cycle. Space Sci Rev 145:337–380

Fröhlich C (2000) Observations of irradiance variations. Space Sci Rev 94:15–24

Fröhlich C (2006) Solar irradiance variability since 1978. Revision of the PMOD composite during solar cycle 21. Space Sci Rev 125:53–65. doi:10.1007/s11214-006-9046-5

Harder J, Lawrence G, Fontenla J, Rottman G, Woods T (2005a) The spectral irradiance monitor: scientific requirements, instrument design, and operation modes. Sol Phys 230:141–167

Harder JW, Fontenla J, Lawrence G, Woods T, Rottman G (2005b) The spectral irradiance monitor: measurement equations and calibration. Sol Phys 230:169–204

Harder JW, Fontenla JM, Pilewskie P, Richard EC, Woods TN (2009) Trends in solar spectral irradiance variability in the visible and infrared. Geo Phys Res Lett 36:L07801

Harder JW, Thuillier G, Richard EC, Brown SW, Lykke KR, Snow M, McClintock WE, Fontenla JM, Woods TN, Pilewskie P (2010) The SORCE SIM solar spectrum: comparison with recent observations. Sol Phys 263:3–24. doi:10.1007/s11207-010-9555-y

Krivova NA, Solanki SK, Fligge M, Unruh YC (2003) Reconstruction of solar irradiance variations in cycle 23: is solar surface magnetism the cause? Astron Astrophys 399:L1–L4

Krivova NA, Solanki SK, Floyd L (2006) Reconstruction of solar UV irradiance in cycle 23. Astron Astrophys 452:631–639

Krivova NA, Solanki SK, Wenzler T, Podlipnik B (2009) Reconstruction of solar UV irradiance since 1974. J Geophys Res (Atmos) 114:D00I04

Krivova NA, Solanki SK, Unruh YC (2011) Towards a long-term record of solar total and spectral irradiance. J Atmos Sol Terr Phys 73:223–234

McClintock WE, Rottman GJ, Woods TN (2005a) Solar-stellar irradiance comparison experiment II (SOLSTICE II): instrument concept and design. Sol Phys 230:225–258

McClintock WE, Snow M, Woods TN (2005b) Solar-stellar irradiance comparison experiment II (SOLSTICE II): pre-launch and on-orbit calibrations. Sol Phys 230:259–294

Morrill JS, Floyd L, McMullin D (2011) The solar ultraviolet spectrum estimated using the Mg II index and Ca II K disk activity. Sol Phys 269:253–267

Preminger DG, Walton SR, Chapman GA (2002) Photometric quantities for solar irradiance modeling. J Geophys Res (Space Phys) 107(A11):1354. doi:10.1029/2001JA009169

Snow M, McClintock WE, Rottman G, Woods TN (2005) Solar stellar irradiance comparison experiment II (SOLSTICE II): examination of the solar stellar comparison technique. Sol Phys 230:295–324. doi:10.1007/s11207-005-8763-3

Unruh YC, Solanki SK, Fligge M (1999) The spectral dependence of facular contrast and solar irradiance variations. Astron Astrophys 345:635–642

Unruh YC, Krivova NA, Solanki SK, Harder JW, Kopp G (2008) Spectral irradiance variations: comparison between observations and the SATIRE model on solar rotation time scales. Astron Astrophys 486:311–323. doi:10.1051/0004-6361:20078421

Wenzler T, Solanki SK, Krivova NA, Fröhlich C (2006) Reconstruction of solar irradiance variations in cycles 21–23 based on surface magnetic fields. Astron Astrophys 460:583–595

Woods TN, Prinz DK, Rottman GJ et al (1996) Validation of the UARS solar ultraviolet irradiances: Comparison with the ATLAS 1 and 2 measurements. J Geophys Res (Space Phys) 101(D6):9541–9570

Surv Geophys (2012) 33:483–501
DOI 10.1007/s10712-012-9192-0

Influence of the Precipitating Energetic Particles on Atmospheric Chemistry and Climate

E. Rozanov · M. Calisto · T. Egorova · T. Peter · W. Schmutz

Received: 2 September 2011 / Accepted: 14 April 2012 / Published online: 3 May 2012

Abstract We evaluate the influence of the galactic cosmic rays (GCR), solar proton events (SPE), and energetic electron precipitation (EEP) on chemical composition of the atmosphere, dynamics, and climate using the chemistry-climate model SOCOL. We have carried out two 46-year long runs. The reference run is driven by a widely employed forcing set and, for the experiment run, we have included additional sources of NO_x and HO_x caused by all considered energetic particles. The results show that the effects of the GCR, SPE, and EEP fluxes on the chemical composition are most pronounced in the polar mesosphere and upper stratosphere; however, they are also detectable and statistically significant in the lower atmosphere consisting of an ozone increase up to 3 % in the troposphere and ozone depletion up to 8 % in the middle stratosphere. The thermal effect of the ozone depletion in the stratosphere propagates down, leading to a warming by up to 1 K averaged over 46 years over Europe during the winter season. Our results suggest that the energetic particles are able to affect atmospheric chemical composition, dynamics, and climate.

Keywords Energetic particles · Atmospheric chemistry · Dynamics · Climate · Modeling

1 Introduction

Understanding of the past climate change and forecasting of the future climate requires a clear quantification of the contribution from all relevant anthropogenic and natural forcing. The contribution of energetic particles to climate change is a not well known and

E. Rozanov (✉) · T. Egorova · W. Schmutz
Physikalisch-Meteorologisches Observatorium Davos/World Radiation Center, Davos, Switzerland
e-mail: e.rozanov@pmodwrc.ch

E. Rozanov · T. Peter
Institute for Atmospheric and Climate Science ETH, Zurich, Switzerland

M. Calisto
International Space Science Institute (ISSI), Bern, Switzerland

controversial issue among many other natural factors. Therefore, the energetic particles were not taken into account neither for the extensive evaluation of the ozone trends (Eyring et al. 2007; SPARC CCMVal 2010) nor for the analysis of the climate changes (IPCC 2007). The aim of this paper is to analyze the impact of the energetic particles and convince the community that the energetic particles are able to change the mean state of the atmospheric chemistry and climate.

According to their sources and energy spectra, precipitating energetic particles can be divided into galactic cosmic rays (GCR), solar protons (SP), and low- and high-energy electrons (LEE and HEE). All these particles are able to ionize neutral molecules (e.g., N_2 and O_2) in the Earth's atmosphere, but the distribution of ionization rates in space and time depend on the particle type because of their interactions with the variable heliomagnetic and geomagnetic fields.

The GCR originate from outside of the solar system and consist mostly of protons with kinetic energy spanning the interval from 1 MeV up to at least 5×10^{13} MeV (e.g., Bazilevskaya et al. 2008). The GCR are able to ionize neutral molecules in the Earth's upper-to-middle atmosphere almost everywhere, but the penetration depth increases toward the polar atmosphere, where the effect of the geomagnetic field on charged particle motion means that more, and lower-energy, particles reach the atmosphere. The ionization rate by GCR reaches its maximum around 15 km (Usoskin et al. 2010). The ionization in the lower troposphere near the equator can only be produced by highly (>15 GeV) energetic particles. Modulation of the cosmic ray flux by the solar activity is more pronounced for less energetic (<1 GeV) particles, while a highly (above 100 GeV) energetic GCR flux remains almost intact (e.g., Bazilevskaya et al. 2008). This modulation leads to a pronounced variability of the GCR-induced ionization rate, which maximizes during the minimum of the solar activity cycles mostly over high geomagnetic latitudes (Usoskin and Kovaltsov 2006; Calisto et al. 2011).

Solar energetic protons are generated in the solar corona or in interplanetary space by coronal mass ejections (CME) driven shocks and/or during impulsive energy releases in solar flares reaching kinetic energy typically up to 500 MeV (Reames 1999; Cane et al. 2006).

After their generation and propagation, the protons drift along the magnetic lines ionizing the thermosphere (>80 km), mesosphere (50–80 km), and, in some extreme cases, the stratosphere (15–50 km) mostly over the polar regions (Jackman et al. 2008 and references therein). Solar proton events (SPE) are sporadic, but their frequency tends to increase during the maximum activity of the Sun when the probability of CME is higher, while their fluence has an opposite tendency (Barnard and Lockwood 2011).

Low-energy (1–30 keV) auroral electrons almost continuously precipitate from the Earth's magnetospheric plasmasheet (Brasseur and Solomon 2005). Their energy is mostly deposited in the lower thermosphere (above 90 km) inside the auroral ovals located over the 60°–75° geomagnetic latitude band (Baker et al. 2001; Barth et al. 2003). The strength of the auroral electron precipitation defined as the total hemispheric precipitating energy flux depends on the geomagnetic activity level and reaches its maximum during the declining phase of the solar activity cycle (Emery et al. 2008; Marsh et al. 2007).

High-energy (from 30 keV up to several MeV) electrons originate from the outer radiation belts (e.g., Bazilevskaya et al. 2008). Their precipitation to the atmosphere is caused by complex wave-particle interaction and mostly confined to the subauroral latitudes (Millan and Thorne 2007; Lam et al. 2010). The penetration depth of high-energy electrons varies from 90 km for 30 keV electrons and down to 50 km for 2 MeV electrons (Turunen et al. 2009). Meredith et al. (2011) pointed out that their fluxes can be substantially enhanced during the geomagnetic perturbations associated with enhanced dynamical pressure of the solar wind caused by CMEs and high-speed solar wind streams (HSSWS). The relation of high-energy precipitation events to solar activity is not well defined yet, but in general, the frequency of the CME-driven electron precipitation should be higher around solar maximum, while the probability of HSSWS-driven electron precipitation increases considerably when coronal holes extend to low latitudes, such that HSSWS impact on Earth, which occurs most often in the declining phase of the solar cycle (Richardson et al. 2000). The ionization induced by the energetic particle precipitation leads to the production of nitrogen and hydrogen oxides via the chain of chemical processes involving electrons, positive, and negative ions (Rusch et al. 1981; Aikin 1994; Solomon et al. 1981; Turunen et al. 2009; Egorova et al. 2011). The subsequent fate of the nitrogen and hydrogen oxides produced by particles depends on their lifetime. The response of the short-lived HO_x is highly localized in time and space and can be observed only during the particle precipitation event in the region where the ionization occurs. NO_x is more stable and can be transported by atmospheric winds; therefore, NO_x enhancement can be observed after the event has started and far from the production area. This mechanism is particularly important during the polar night when the lifetime of NO_x produced by solar protons and magnetospheric electrons is longer and they are transported by the downward air motion inside the polar vortex toward the middle stratosphere (Funke et al. 2005; Seppälä et al. 2007a). The NO_x and HO_x enhancement caused by the particles has important implications for the ozone concentration in the stratosphere. Below the tropopause, that is, below about 15 km, the ozone production can be enhanced by additional NO_x due to photosmog reactions (e.g., Wang et al. 1998), while in the middle atmosphere, additional amount of active nitrogen and hydrogen oxides intensifies ozone destruction (Brasseur and Solomon 2005). The ozone changes in the stratosphere lead to the perturbation of the net radiative heating and temperature regime, which can have implications for the dynamical state of the vortex and surface climate (Gray et al. 2010).

Many of the theoretically expected particle effects were confirmed by observation data analysis. The enhancement of NO in the auroral zone has been observed by SNOE measurements (Barth et al. 2001). NO_x and HO_x enhancement in the mesosphere and stratosphere due to in situ ionization was detected by satellite measurements mostly during powerful precipitation events (e.g., López-Puertas et al. 2005; Seppälä et al. 2007b; Damiani et al. 2008; Funke et al. 2011; Verronen et al. 2011). The enhancement of NO_x in the stratosphere after substantial geomagnetic perturbation was detected by many groups (e.g., Callis et al. 1998; Siskind et al. 2000; Randall et al. 2007; Clilverd et al. 2009). Its proper attribution, however, is still under investigations, because the separation of the direct production and transport of NO_x from above is not straightforward due to incomplete information of the energy spectrum of precipitating particles (López-Puertas et al. 2006, Sinnhuber et al. 2011). The production of NO_x and HO_x by GCR has not yet been confirmed observationally due to a rather small magnitude of the enhancement in the polar upper troposphere/lower stratosphere (UTLS) region, where the effects of GCR should be the most pronounced. The ozone depletion in the polar mesosphere and stratosphere after intensive SPE was observed by different satellite instruments (López-Puertas et al. 2005; Seppälä et al. 2007b; Funke et al. 2011). Significant correlation of the surface air temperature with the level of geomagnetic activity, which is represented by the Ap index, was discovered in the reanalysis data by Seppälä et al. (2009).

The modeling of particle effects has been mostly aimed at the study of short-term powerful particle precipitation events, which are characterized by a large signal to noise ratio and therefore can be readily identified in the model results and compared to observations. A large number of simulations of the atmospheric changes caused by SPE have been performed using models of different complexity (see Jackman et al. 2008 and references therein, Reddmann et al. 2010, Egorova et al. 2011) and a successful reproduction of the main observed features have been recently reported by Funke et al. (2011).

During the last years, attempts to estimate the importance of the particle precipitation for the long-term changes of ozone and climate using sophisticated chemistry-climate models (CCM) have been made. Langematz et al. (2005) found significant response of the stratospheric ozone caused by low-energy electrons. Rozanov et al. (2005) simulated the ozone and climate response to enhancement of nitrogen and hydrogen oxides caused by electrons with low and high energies and found significant response of the atmosphere and surface climate. Despite the application of an idealized forcing and absence of other particle types, these studies inspired further investigations in this direction. Jackman et al. (2009) estimated the long-term climate influence of the SPE and did not find pronounced response of the total column ozone and climate, which is expected due to low frequency of powerful SPE and a lack of continuous effects of GCR and energetic electrons as well as a relatively stable solar activity level during the considered period. The importance of NO_x production by auroral electrons was illustrated by Marsh et al. (2007) using a parameterization of the ionization rates based on geomagnetic Kp index, which classifies solar particle effects on the Earth's magnetic field. Baumgaertner et al. (2009, 2011) developed a parameterization for the NO_x production by low-energy electrons and studied the atmospheric response to the level of geomagnetic activity. They found substantial depletion of ozone in the polar winter stratosphere ($\sim$15–50 km) followed by a distinctive surface air temperature and sea-level pressure responses in agreement with previous investigations. Calisto et al. (2011) simulated the effects of the GCR using the CCM SOCOL v2.0 with continuously varying boundary conditions (transient mode) with the latest parameterization (Usoskin et al. 2010) of the GCR-induced ionization rates. They reported small ($\sim$3 %) but statistically significant ozone depletion in the Northern hemisphere lower polar stratosphere and a winter time warming in the eastern part of Europe and in Russia. A first attempt to include several types of precipitating particles (GCR, SPE, and low- to middle-energy electrons) has been made by Semeniuk et al. (2011). They reported a wintertime polar stratosphere ozone reduction up to 10 % and substantial dynamical changes over the Southern hemisphere caused by the energetic particles. Unfortunately, Semeniuk et al. (2011) did not analyze the surface air temperature response and the contribution of all energetic particles to climate change. The results presented in this paper and the fact that just Semeniuk et al. (2011) have investigated the influence of the GCRs, SPEs, and EEPs clearly show that more simulations driven by all known energetic particles are necessary to estimate their significance for the ozone trends and climate change analysis.

In this paper, we study the combined effects of the GCR, solar protons, and low-energetic electrons using the global chemistry-climate model SOCOL v2.0 (Schraner et al. 2008) focusing on the changes of atmospheric chemistry and dynamics caused by the inclusion of additional NO_x and HO_x sources. The description of the model and experimental setup including the applied parameterizations of the NO_x and HO_x production by different energetic particles are presented in Sect. 2. The results of the reference and experimental runs are analyzed in Sect. 3. Section 4 presents the summary and an outlook.

2 Model Description and Experimental Setup

The CCM SOCOL consists of the global circulation model MA-ECHAM4 and the chemistry-transport model MEZON. MA-ECHAM4 (Manzini et al. 1997) is a spectral model with T30 horizontal truncation, that is, the model has a horizontal resolution resulting in a grid spacing of about 3.75°; in the vertical direction, the model has 39 levels in a hybrid sigma-pressure coordinate system spanning the model atmosphere from the surface to 0.01 hPa.

The chemical-transport model MEZON (Rozanov et al. 1999; Egorova et al. 2003) exploits the same vertical and horizontal resolution and treats 41 chemical species of the oxygen, hydrogen, nitrogen, carbon, chlorine and bromine groups, which are coupled by 140 gas-phase reactions, 46 photolysis reactions and 16 heterogeneous reactions in/on aqueous sulfuric acid aerosols, water ice, and nitric acid trihydrate (NAT). The original version of the CCM SOCOL was described by Egorova et al. (2005). An extensive evaluation of the CCM SOCOL (Egorova et al. 2005; Eyring et al. 2007) revealed model deficiencies in the chemical-transport module and led to the development of the CCM SOCOL v2.0. The new features of the SOCOLv2.0 are the following: (1) all species are transported separately; (2) the mass conservation after each semi-Lagrangian transport step is maintained by the correction coefficients calculated for the chlorine, bromine, and nitrogen families instead for individual family members, but then applied to each individual species; (3) the mass fixer is applied to ozone only over the latitude band 40°S–40°N to avoid artificial mass loss in the polar lower stratosphere; (4) the water vapor removal by the highest ice clouds (between 100 hPa ($\sim$16 km) and the tropical cold point tropopause) is explicitly taken into account to prevent an overestimation of stratospheric water content; (5) the list of ozone-depleting substances is extended to 15 for the chemical treatment, while for the transport they are still clustered into three tracer groups; (6) the heterogeneous chemistry module was updated to include HNO_3 uptake by aqueous sulfuric acid aerosols, a parameterization of the liquid-phase reactive uptake coefficients and the NAT particle number densities are limited by an upper boundary of 5×10^{-4} cm^{-3} to take account of the fact that observed NAT clouds are often strongly supersaturated.

A comprehensive description of the CCM SOCOL v2.0 is presented by Schraner et al. (2008). The evaluation of CCM SOCOL in the framework of SPARC CCMVal-2 intercomparison campaign (SPARC CCMVal, 2010) showed substantial improvement of the representation of the transport and chemical processes in the model; however, some shortcomings in the simulation of gas transport still remains. It also showed that the internal variability of our model was found to be in good agreement with reanalyzed data.

The ionization rates due to the GCR have been parameterized using the recently developed CRAC:CRII (Cosmic Ray induced Cascade: Application for Cosmic Ray Induced Ionization) model extended toward the upper atmosphere (Usoskin et al. 2010). The model is based on a Monte Carlo simulation of the atmospheric cascade and reproduces the observed data within 10 % accuracy in the troposphere and lower stratosphere (Bazilevskaya et al. 2008; Usoskin et al. 2010). In the mesosphere ($\sim$50–80 km), the agreement between observed and simulated ionizations rates are not good because the ionization by other sources (solar radiation, precipitating soft particles of magnetospheric origin, etc.) becomes at least as important as by GCRs. The results of the CRAC:CRII model are parameterized (see Usoskin and Kovaltsov 2006) to give ion pair production rate as a function of the altitude (quantified via the barometric pressure), geomagnetic latitude (quantified via geomagnetic cutoff rigidity), and solar activity (quantified via the modulation potential θ).

For the solar protons, we applied daily averaged ionization rates from 1963 to 2008 compiled by Charles Jackman and available from the SOLARIS website (URL: http://www.geo.fu-berlin.de/en/met/ag/strat/forschung/SOLARIS/Input_data/ index.html) as a function of pressure between 888 hPa ($\sim$1 km above ground level) and 8×10^{-5} hPa

(~115 km). The ionization rates were introduced to the model over the polar region from 60° to 90° geomagnetic latitudes. The area of ionization by high-energy (>10 MeV) protons is well inside the model domain (Turunen et al. 2009) and their effects should be properly accounted for. Despite the fact that the model used in this study cannot directly include the production of NO_x and HO_x above 80 km (i.e. outside the model domain) by low-energy protons and concomitant electrons (Wissing et al. 2010) Funke et al. (2011) mention in their paper that the importance of electron contribution is still under discussion. Therefore, instead of the direct production, we have to parameterize the influx of NO_x produced by energetic electrons above the model top. We apply the Baumgaertner et al. (2009) parameterization of nitrogen oxides influx into the model domain based on the empirical relation with geomagnetic Ap index, which represents the geomagnetic activity using their "average excess NO_x" mode. The values of the Ap index were obtained from the National Geophysical Data Center (ftp://ftp.ngdc.noaa.gov). It has been shown by Funke et al. (2005) that the effect of the energetic electrons on the mesosphere and stratosphere is confined to the polar vortex. Therefore, the minimum absolute latitude of 55° has been used, that is, the products of the ionization by energetic electrons enter our model domain only at high latitudes. Further, downward propagation of ionization products depends on the presence of polar vortices and appropriate vertical transport, which guarantees that the ionization products properly affect the lower mesosphere and stratosphere. The model currently does not include high (>50 keV)-energy electrons that deposit their energy below 80 km because a proper parameterization for the high-energy electrons is not available. The lack of this process in the model can be justified by a small contribution of high-energy electron precipitation events to the total NO_x production by particles estimated by Sinnhuber et al. (2011) from observational data.

The GCR and SPE ionization rates cannot be directly used in the CCM SOCOL, which has no explicit treatment of ion chemistry; therefore, it is necessary to convert the ionization into the NO_x and HO_x production rates. Following Porter et al. (1976), we assumed that 1.25 nitrogen atoms are produced per ion pair, and 45 % of this is assumed to yield ground-state atomic nitrogen, while 55 % is assumed to go into $N(^2D)$ with instantaneous conversion to NO (see Introduction). The production of HO_x has been studied by Solomon et al. (1981) with a 1-D time-dependent model of neutral and ion chemistry. They parameterized the number of odd hydrogen particles produced per ion pair as a function of altitude and ionization for daytime, polar summer conditions of temperature, air density, and solar zenith angle. We implement these parameterizations in the CCM SOCOL to take into account the production of NO_x and HO_x induced by GCR and SPE from the ground up to the height of 0.01 hPa barometric pressure (altitude of ~ 80 km). The errors associated with this approach are within 10–20 % (Egorova et al. 2011), which is comparable with the accuracy of ionization rate calculations.

For this study, we have carried out two 46-year long runs with the CCM SOCOL v2.0 from 1960 to 2005. The reference run has been performed without the influence of the energetic particles, while the experimental run includes additional sources of NO_x and HO_x caused by the influence of GCR, SPE, and EEP. Both model experiments are driven by time-varying boundary conditions for the source gases, aerosol loading, solar irradiance, sea surface temperatures, and sea ice concentration identical to the CCMVal-2 REF-B1 experiment (SPARC CCMVal 2010; Morgenstern et al. 2010). The numerical experiments performed here are similar to the experiments described by Calisto et al. (2011), but the additional NO_x and HO_x sources are extended to take into account SPE and EEP in addition to GCR. The model integration time increases to 46 years, compared to Calisto et al. (2011), which had an integration time of 27 years. Our simulations are also close to the experiments analyzed by Semeniuk et al. (2011). These papers will be referred to as CA11 and SE11 hereafter.

3 Results

In this section, we analyze the difference between the climatological state of the model run with the influence of the energetic particles (experimental run) and the run without the influence of the energetic particles (reference run). All resulting changes in the atmospheric state are taken into account for the calculation of statistical significance, which is estimated using the Student's t test.

3.1 Nitrogen Oxides

Figure 1 illustrates the annual mean zonal mean NO_x response to the influence of the GCR, SPE, and EEP as a relative deviation of the experimental run from the reference run. The GCR, SPE, and EEP produce substantial amounts of NO_x. The simulated NO_x increase is strongest in the mesosphere. The pronounced NO_x enhancement of more than 1,000 % in the polar mesosphere (approximately 50–80 km) is an expected result of the direct NO_x production by the energetic particles. A significant enhancement of up to 400 % is depicted in the mesosphere over the middle and low latitudes, which can be attributed to the meridional transport. CA11 and SE11 demonstrated that there is no significant NO_x response to GCR in this area. EEP and SPE operate only over the polar regions. This effect is not really foreseen due to rather short NO_x lifetime outside the polar night area, but a similar feature has also been obtained by SE11. This NO_x enhancement can be explained by a rather fast transport in the mesosphere in comparison with the chemical lifetime of NO_x. The reliability of this mechanism is partially confirmed by the shape of the mesospheric NO_x enhancement which is not symmetrical in the latitudinal direction, showing more intensive increase in the Northern hemisphere, which is consistent with more dynamically active boreal cold seasons leading to more intensive meridional transport there. The southern polar hemisphere shows at about 10 hPa (~32 km), and at about

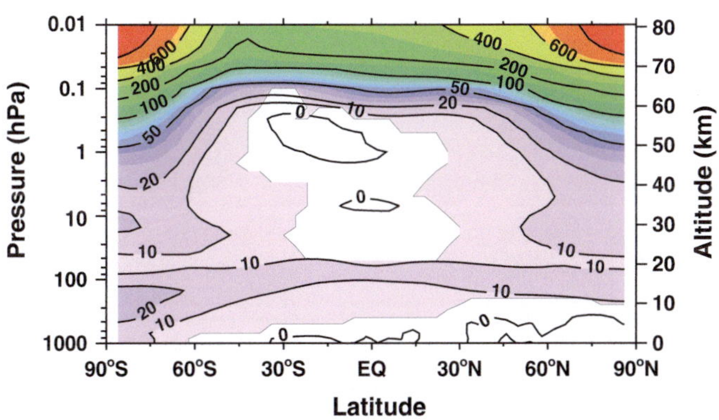

Fig. 1 Annual mean response of zonal mean NO_x, (%) to GCR, SPE, and EEP. Results are averaged from 1960 to 2005. *Solid contours* indicate positive changes. The *contours* are plotted for: 10, 20, 50, 100, 200, 400, 600, and 1,000 %. *Color* pattern indicates the regions where the changes are judged statistically significant at or better than 10 % level

200 hPa ($\sim$12 km), another statistical significant increase for NO_x. The enrichment at 10 hPa is a result of the combined SPE and EEP influence, whereas the NO_x at 200 hPa has been formed by GCR, which is also visible in the CA11 results. The enhancement of the tropospheric NO_x in the northern hemisphere is less than 10 % and statistically insignificant. This result does not agree with SE11, which reported a large (>30 %) and statistically significant increase in the tropospheric NO_x in both hemispheres. The reasons for this disagreement are not completely clear, but most probably it is the result of the simplified treatment of the tropospheric chemistry in the model applied by SE11 and the absence of other natural and anthropogenic NO_x sources in the troposphere.

The seasonal behavior of NO_x enhancement due to energetic particles averaged over the Northern and Southern polar region (60°–90°) is illustrated in Figs. 2 and 3, respectively.

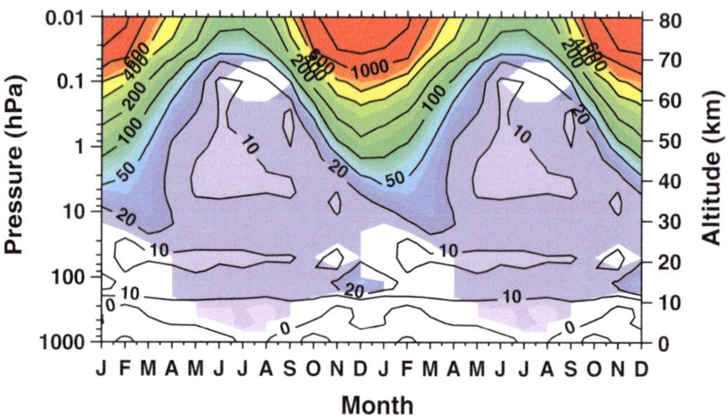

Fig. 2 Monthly mean response of zonal mean NO_x, (%) to GCR, SPE, and EEP averaged over 60°–90°N. Results are averaged from 1960 to 2005. *Solid contours* indicate positive changes. The *contours* are plotted for: 10, 20, 50, 100, 200, 400, 600, 1,000, and 2,000 %. *Color* pattern indicates the regions where the changes are judged statistically significant at or better than 10 % level. Twelve months of the monthly run climatology are repeated

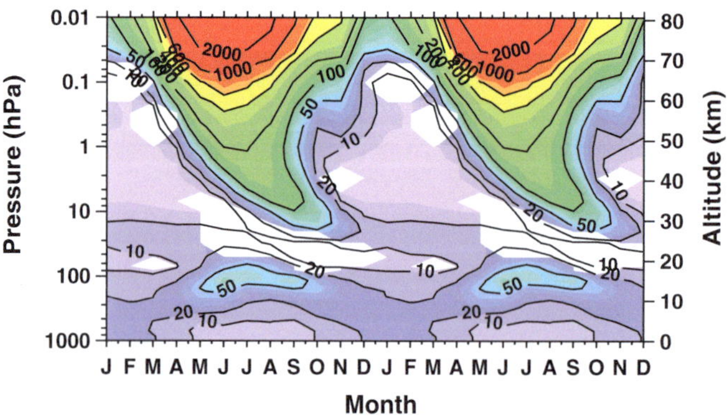

Fig. 3 Monthly mean response of zonal mean NO_x, (%) to GCR, SPE, and EEP averaged over 60°–90°S. Results are averaged from 1960 to 2005. *Solid contours* indicate positive changes. The *contours* are plotted for: 10, 20, 50, 100, 200, 400, 600, 1,000, and 2,000 %. *Color* pattern indicates the regions where the changes are judged statistically significant at or better than 10 % level. Twelve months of the monthly run climatology are repeated

The results in the mesosphere show an expected maximum response during the hemispheric winter when it is cold and dark. During this time, the downward propagation of thermospheric NO_x inside the vortex can operate and this NO_x, therefore, will not be depleted by solar irradiance. Stronger vortex and more intensive downward transport in the Southern hemisphere lead to a deeper penetration of the thermospheric NO_x. In this area, an increase of 50 % is visible down to 20 hPa (~26 km), while in the Northern hemisphere it reaches only 3 hPa (~40 km). The NO_x increase above 10 hPa (~32 km) during the hemispheric summer is smaller (only up to 100 %) due to the absence of the thermospheric NO_x influx and NO_x relaxation to the chemical equilibrium in the presence of the sunlight via NO photolysis followed by the cannibalistic reaction $N + NO \rightarrow N_2 + O$ (Brasseur and Solomon 2005). Similar distribution in the polar middle atmosphere has been obtained by SE11.

The seasonal cycle of NO_x enhancement is also visible in an altitude range from 12 km up to 18 km, where the maximum effects (20–50 % increase) take place during the winter. In the lower troposphere (up to 8 km), however, the maximum NO_x response occurs during summer time. These seasonal cycles are pronounced in the relative deviation and reflect seasonal behavior of the background abundances of NO_x, which reach minimum values during winter around 100 hPa (~16 km) and during summer in the lower troposphere. On the other hand, the contribution of the energetic particles, that is, GCR, SPE, and EEP in absolute values is seasonally more homogeneous.

Figure 4 illustrates the time evolution of the observed geomagnetic Ap index (red line) and the modeled NO_x produced by the energetic particles (black line) at ~45 km in the southern polar cap (60°–90°S). The shape of the NO_x enhancement closely follows the time series of Ap index showing distinctive spikes in 1994, 2000, 2003, and 2005. However, the magnitude of NO_x enhancement does not agree well with the magnitude of NO_x change obtained from the observation (Randall et al. 2007). In particular, the simulated amount of NO_x produced by the GCR, solar protons, and energetic electrons reach their maximum in 2000, while the observation shows the largest value of NO_x enhancement and Ap index in 2003. It means that the amount of NO_x reaching the stratosphere is modulated by the meteorological conditions and a slower downward propagation in 2003 than in 2000 (not shown) in the experimental run leads to a weaker NO_x enhancement during this year.

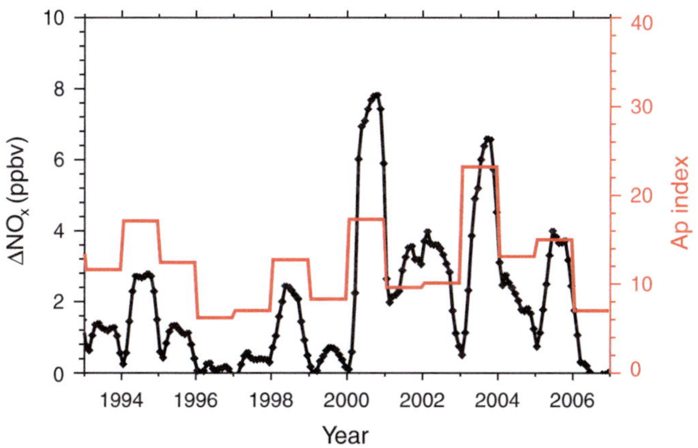

Fig. 4 Time evolution of the annual mean geomagnetic Ap index (*red line*) and the response of zonal mean NO_x (*black line*), (%) to GCR, SPE, and EEP at ~45 km averaged over 60°–90°S

Our results, compared with Baumgaertner et al. (2009), show that the parameterization of the low-energetic electrons is giving good results when comparing with satellite data. Less intensive effects in 2003 can be explained by the application of the "maximum NO_x excess" model by Baumgaertner et al. (2009) for the year 2003.

3.2 Hydrogen Oxides

Figures 5 and 6 illustrate the annual mean zonal mean OH and HO_x ($H + OH + HO_2$) response to the energetic particles as a relative deviation of the experimental run from the reference run. The influx of hydroxyl produced in the thermosphere (>80 km) by EEP was not included in the model due to the short lifetime of this species; therefore, the simulated OH and HO_x change is only due to direct production by SPE and GCR. The simulated OH increase is strongest in the middle stratosphere at about 16 km and in the troposphere going

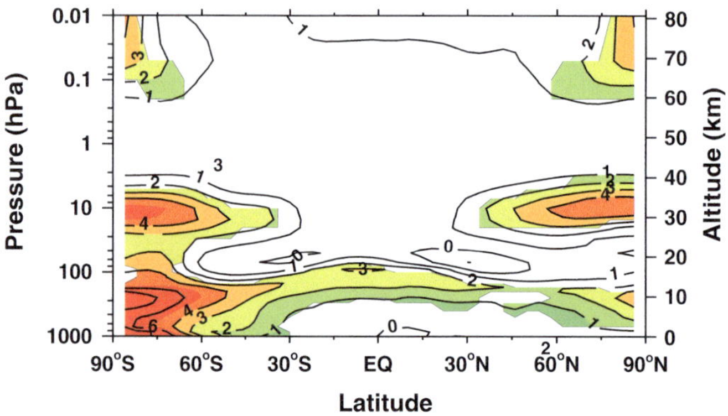

Fig. 5 Annual mean response of zonal mean OH, (%) to GCR, SPE, and EEP. Results are averaged from 1960 to 2005. The *contours* are plotted for: 1, 2, 3, 4, 6, and 8 %. *Color* pattern indicates the regions where the changes are judged statistically significant at or better than 10 % level

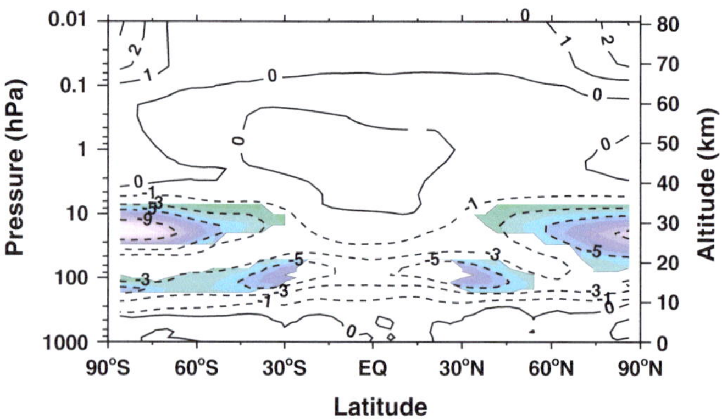

Fig. 6 Annual mean response of zonal mean HO_x, (%) to GCR, SPE, and EEP. Results are averaged from 1960 to 2005. *Solid contours* indicate positive, *dotted contours* negative changes. The *contours* are plotted for: −9, −5, −3, −1, 0, 1, and 2 %. *Color* pattern indicates the regions where the changes are judged statistically significant at or better than 10 % level

from surface up to about 12 km. It was shown by CA11 and SE11 that the production of OH by GCR is not important in the upper stratosphere (~50–60 km) and mesosphere; therefore, the small increase of OH there can be attributed to SPE effects. The OH increase around 10 hPa (~16 km) can be attributed to the combined effects of SPE and GCR. The simulated relative OH increase in the troposphere reaches its maximum in the unpolluted Southern hemisphere and in the uppermost troposphere, where its background concentration is small. The influence of the energetic particles on HO_x is negative in the upper troposphere/lower stratosphere domain because of the reaction chains: $NO_2 + HO_2 \rightarrow HNO_4$ and $NO + HO_2 \rightarrow NO_2 + OH$ followed by $NO_2 + OH + M \rightarrow HNO_3$ (Brasseur and Solomon, 2005). Our model indeed simulates a proper HNO_3 and HNO_4 increase (not shown) confirming these mechanisms. CA11 obtained a similar HO_x depletion due to GCR. A HO_x decrease in the upper troposphere/lower stratosphere was also found by SE11. Due to the absence of precipitating medium-energy-range electrons, which are capable of producing HO_x directly in the mesosphere (>80 km) and the upper stratosphere, our model does not produce the HO_x enhancement in the upper polar atmosphere obtained by SE11. However, the properties of these precipitating electrons are not well known and, at the moment, we do not have solid proof that such events are reasonably frequent and important.

3.3 Ozone and Temperature

The annual mean zonal mean ozone response to the energetic particles as a relative deviation of the experimental run from the reference run is presented in Fig. 7. The combined effect of GCR, SPE, and LEE consist of an ozone depletion going from about 26 km up to 80 km and an ozone increase in the troposphere, that is from the surface up to 12 km. The ozone depletion in the polar mesosphere (60–80 km) and middle stratosphere (~26–32 km) is explained by enhanced NO_x and OH concentrations, which accelerates known catalytical ozone destruction cycles (Brasseur and Solomon 2005; SE11). The ozone depletion in the polar mesosphere is the most pronounced and exceeds 10 % in the uppermost layer. The magnitude of the ozone depletion is substantially lower than the value reported by SE11. It can be explained by much larger OH increase simulated by

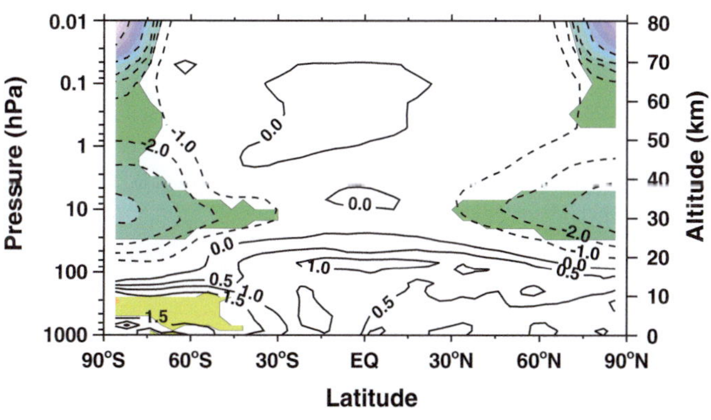

Fig. 7 Annual mean response of zonal mean O_3, (%) to GCR, SPE, and EEP. The *contours* are plotted for: −10.0, −6.0, −4.0, −3.0, −2.0, −1.0, 0.0, 0.5, 1.0, and 1.5 %. Results are averaged from 1960 to 2005. *Solid contours* indicate positive, *dotted contours* negative changes. *Color* pattern indicates the regions where the changes are judged statistically significant at or better than 10 % level

SE11 due to medium-energy-range electron precipitation. The ozone depletion in the middle stratosphere is a combined effect of all precipitating particles. Due to larger NO_x and OH enhancement in the Southern hemisphere, the magnitude of the ozone depletion at an altitude range of approximately 26–32 km is slightly greater, reaching 4 % over the pole. The area of depleted ozone extends to the middle latitudes due to the transport of polar air and local depletion caused by GCR. The simulated ozone depletion in the middle stratosphere is in agreement with SE11 data. The ozone depletion in the lower Northern stratosphere reported by CA11 is not well pronounced in our results probably because of the lower level of chlorine caused by a longer run covering the period from 1960 to 2005.

Additional NO_x in the troposphere (up to 16 km) can lead to an ozone production caused by so-called "photosmog" reactions (Wang et al. 1998; SE11). This "photosmog" reaction starts with the NO_2 photolysis ($NO_2 + hv \rightarrow O + NO$), which is then followed by ozone production through atomic and molecular oxygen recombination. Our model shows a statistical significant change only in the Southern hemisphere, where the natural level of NO_x is very small. The ozone increase there is about 1.5 %, which is smaller and less extended in space than SE11 estimations. As described earlier, the reason for this could be due to underestimated background concentrations of NO_x and O_3 in the model applied by SE11, resulting from both the use of an oversimplified tropospheric chemistry and missing sources of NO_x.

The seasonal march of the O_3 change caused by the energetic particles averaged over the Northern and Southern polar caps (60°–90°) is illustrated in Figs. 8 and 9. In general, the time evolution of the ozone depletion in the mesosphere and the stratosphere follows the time evolution of the NO_x enhancement shown in Figs. 2 and 3, starting with a maximum depletion during the winter and propagating down to the middle stratosphere up to early spring. Therefore, this behavior is related to the NO_x enhancement by the energetic electrons. The secondary maximum in the lower stratosphere in spring and further ozone depletion during spring and summer below 10 hPa (~32 km) is the result of a combined influence of the energetic electrons, solar protons, and GCR. Similar ozone depletion by up to 4 % during the summer season was also obtained by SE11. In the troposphere, the seasonal cycle differs between the hemispheres. In the Northern hemisphere, statistically

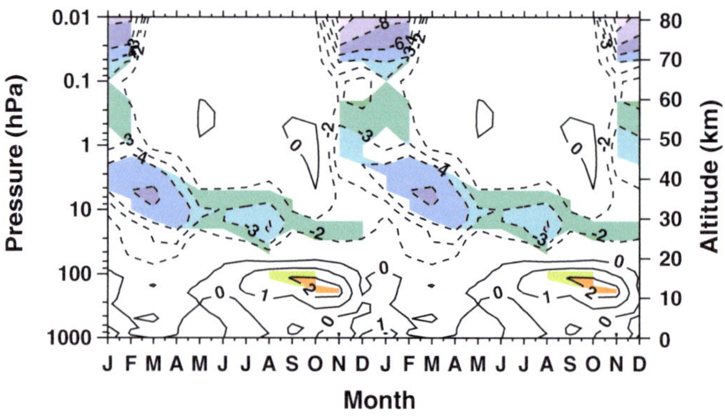

Fig. 8 Monthly mean response of zonal mean O_3, (%) to GCR, SPE, and EEP averaged over 60°–90°N. The *contours* are plotted for: −12.0, −10.0, −8.0, −6.0, −4.0, −3.0, −2.0, 0.0, 1.0, 2.0, and 3.0 %. Results are averaged from 1960 to 2005. *Solid contours* indicate positive, *dotted contours* negative changes. *Color* pattern indicates the regions where the changes are judged statistically significant at or better than 10 % level. Twelve months of the monthly run climatology are repeated

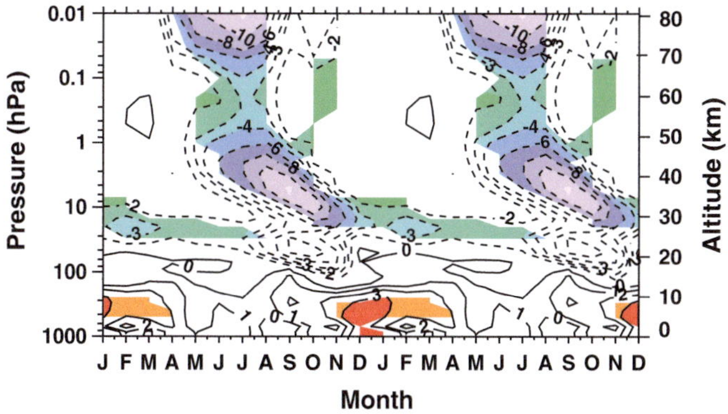

Fig. 9 Same as Fig. 8 but for 60°–90°S

significant ozone enhancement appears during late summer and fall seasons in the upper troposphere, while in the Southern hemisphere, the maximum ozone increase is confined to the middle troposphere and occurs during summer time. This behavior is explained by the seasonal cycle of background NO_x concentrations and availability of sunlight. In the Northern hemisphere, the NO_x-limited regime cannot be reached in the lower stratosphere due to continuous anthropogenic sources of NO_x. In the Southern hemisphere, the background NO_x concentration is always smaller and ozone production maximizes during the summer time when the photolysis of NO_2 is faster.

Ozone is capable of modifying the radiative balance in the atmosphere in short- and longwave spectral ranges. Tropospheric ozone is also an important greenhouse gas. Therefore, the described ozone changes can potentially alter radiative heating, thermal structure, and climate. The simulated monthly mean temperature changes due to the energetic particles over the northern high latitudes are illustrated in Fig. 10. As we can see, the ozone depletion shown in Fig. 8 can lead to some temperature changes varying from 0.5 K cooling in the stratosphere to 0.5 K warming in the mesosphere similar to the

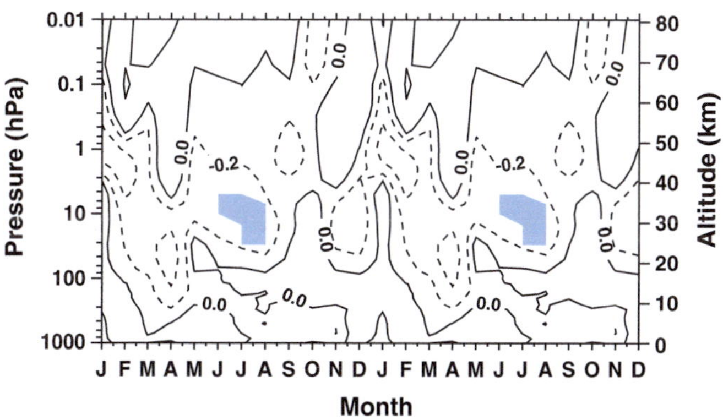

Fig. 10 Monthly mean response of zonal mean temperature, (K) to GCR, SPE, and EEP averaged over 60°–90°N. The *contours* are plotted for: −0.5, −0.2, 0, 0.5, and 0.8 K. Results are averaged from 1960 to 2005. *Solid contours* indicate positive, *dotted contours* negative changes. *Color* pattern indicates the regions where the changes are judged statistically significant at or better than 10 % level. Twelve months of the monthly run climatology are repeated

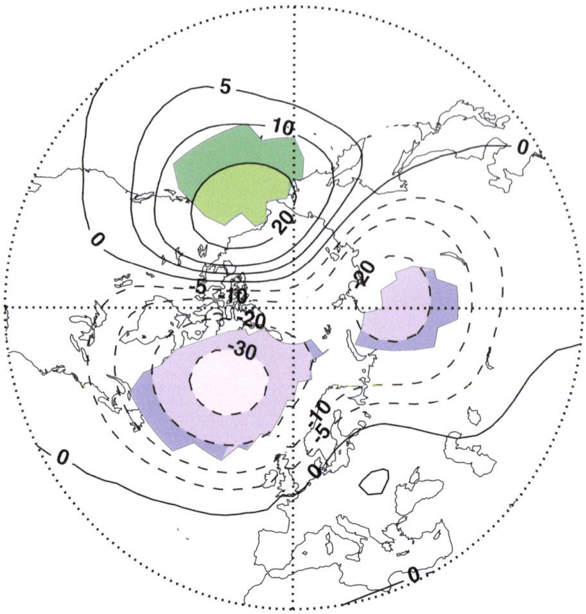

Fig. 11 The response of the geopotential height at 50 hPa, (m) to GCR, SPE, and EEP for boreal winter. Results are averaged from 1960 to 2005. *Solid contours* indicate positive, *dotted contours* negative changes. The *contours* are plotted for: −30, −20, −10, −5, 0, 5, 10, 20 m. *Color* pattern indicates the regions where the changes are judged statistically significant at or better than 10 % level

temperature response obtained by SE11. These changes, however, are not statistically significant in our simulations except for the 0.3 K cooling around 10 hPa ($\sim$32 km) during boreal summer. Nevertheless, such small temperature changes are able to slightly modify the dynamical state of the atmosphere and the strength of the polar vortex. The changes of the geopotential height at 50 hPa shown in Fig. 11 can reach 30 m and reflect an intensification of the polar stratospheric vortex. The downward propagation of this stratospheric perturbation can alter the pattern of the surface air temperature distribution. The details of the involved mechanisms have not been clearly identified (Shepherd, 2002), but the analyses of observational data (e.g., Baldwin and Dunkerton, 2001; Gray et al. 2010) suggest that the stratospheric vortex anomalies are able to propagate down to the troposphere affecting the surface temperature distribution. The response of the surface air temperature during boreal winter season over the Northern hemisphere to GCRs, SPEs, and EEPs is shown in Fig. 12. The pattern of the temperature changes consists of a warming averaged over 46 years of up to 1 K over Europe and a less pronounced warming over Central Russia. Some cooling spots appear over Greenland and the USA. This pattern can be attributed to a more positive phase of the Northern Annular mode. A similar pattern was also found and explained by a high level of geomagnetic activity by Seppälä et al. (2009) using observational data analysis and by Baumgaertner et al. (2011) from model experiments. The smaller magnitude of obtained surface air temperature response can be explained by the fact that we analyzed the response to the multi-year mean geomagnetic activity, while Seppälä et al. (2009) and Baumgaertner et al. (2011) used a composite analysis and analyzed the response during the years with high geomagnetic activity. The warming pattern obtained by Baumgaertner et al. (2011) is confined to northern Europe, while the warming in our simulation covers a wider area including central and southern Europe. The difference can be related to the absence of GCR in the model applied by Baumgaertner et al. (2011). It was pointed out by CA11 that GCR also are able to modify the surface air temperature distribution over Europe.

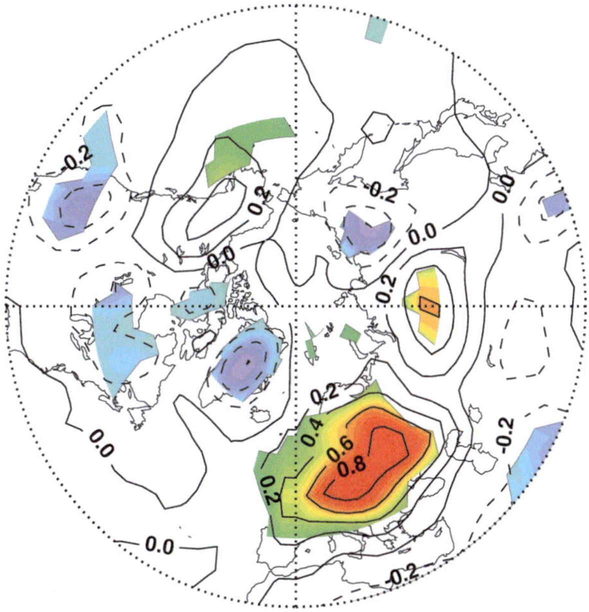

Fig. 12 The response of surface air temperature, (K) to GCR, SPE, and EEP for boreal winter. Results are averaged from 1960 to 2005. *Solid contours* indicate positive, *dotted contours* negative changes. The *contours* are plotted for: −0.6, −0.4, −0.2, 0.0, 0.2, 0.4, 0.6, and 0.8 K. *Color* pattern indicates the regions where the changes are judged statistically significant at or better than 10 % level. The basic reason that the surface temperature response is largest over Europe might be because they are rather sensitive to stratospheric perturbations (see Gray et al. 2010 and references therein)

4 Summary and Outlook

Using the CCM SOCOL v 2.0 in transient mode, we simulate the atmospheric state from 1960 up to 2005 with and without additional sources of NO_x and HO_x caused by three kinds of energetic particles: GCR, SPE, and EEP. We analyze the difference between the climatological states of these runs in order to establish the upper limit of their potential effects on climate. The results show a theoretically expected and consistent response of NO_x, HO_x, ozone, temperature, and dynamics in all considered atmospheric layers. This response consists of substantial increases of NO_x and OH mostly over the polar regions followed by ozone depletion and cooling leading to an intensification of the northern polar vortex and warming over Europe. An ozone increase resulting from GCR has been found in the Southern hemispheric troposphere. Our results are in general agreement with recently published papers (e.g., Calisto et al. 2011; Semeniuk et al. 2011; Baumgaertner et al. 2011) aimed at the understanding of particle precipitation effects on atmospheric chemistry and climate; however, some differences are also evident. Therefore, it is important to incorporate GCR, SPE, and EEP into CCM, which are designed to study atmospheric chemistry, dynamics and climate variability, and trends. Our results emphasize the importance of all types of precipitating particles and of a proper treatment of tropospheric chemistry.

Further efforts should be aimed at the improvement of the representation of energetic particles. For example, Rozanov et al. (2005) demonstrated that high-energy electron precipitation can have a large impact on the atmosphere using a rather simple approach. The comparison of our model results with those of Semeniuk et al. (2011) showed that the medium-energy electron precipitation is important for the simulation of the ozone response. There are some evidences that the frequency of these events is considerable (e.g., Meredith et al. 2011); however, the importance of high-energy electron precipitations still remains questionable (e.g., Sinnhuber et al. 2011). More detailed analysis of observations and modeling efforts are necessary to foster our understanding of these events and to develop a reliable parameterization of the ionization rates due to high-energy electron precipitation. It is worth noting that CA11 showed a comparable or even larger NH signal for surface temperature in the northern hemisphere due to GCR. Therefore, it is necessary to define which particle class is the most important for the surface climate. However, a proper answer to this question would require a larger set of simulations. The proper identification of the mechanism(s) responsible for the downward propagation of the stratospheric perturbation also remains a fundamental problem. Probably, additional simulations using strong artificial forcing would be necessary to clearly identify the chain of processes.

In the future, more attention should be paid to the prediction of possible changes of energetic particle precipitation. There are some indications that the Sun is entering a phase of low activity. There is a consensus that such a period (similar to Dalton or Maunder minimum) would be characterized by a lower total solar irradiance and higher GCR intensity; however, the changes of frequency and intensity of the electron precipitation events are not clear. Therefore, it is difficult to say whether the expected greenhouse warming will be enhanced or compensated by the decline of solar activity. The important direction of future studies should be the development of a proper model. At the moment, there is no model that is able to simulate all processes relevant to the effects of energetic charged particles on the atmosphere and climate. The existing CCM should be extended up in the vertical direction to treat particle precipitation on a physically solid basis to simulate relevant changes in the ocean dynamics and represent tropospheric chemistry in more detail.

Acknowledgments ER is partially supported by the Swiss National Science Foundation under grant CRSI122-130642 (FUPSOL). TE has received part of her funding from the European Commission's Seventh Framework Programme (FP7/20072013) under the grant agreement no.218816 (SOTERIA, project, http://www.soteria-space.eu). We acknowledge useful discussions within the ISSI Teams on Study of Cosmic Ray Influence upon Atmospheric Processes and Geospace Coupling to Polar Atmosphere.

References

Aikin A (1994) Energetic particle-induced enhancement of stratospheric nitric acid. Geophys Res Lett 21(10):859–862

Baker DN, Barth CA, Mankoff KE, Kanekal SG, Bailey SM, Mason GM, Mazur JE (2001) Relationships between precipitating auroral zone electrons and lower thermospheric nitric oxide densities: 1998–2000. J Geophys Res 106(A11):24465–24480. doi:10.1029/2001JA000078

Baldwin MP, Dunkerton TJ (2001) The solar cycle and stratosphere-troposphere dynamical coupling. Science 294:581–584. doi:10.1126/science.1063315

Barnard L, Lockwood M (2011) A survey of gradual solar energetic particle events. J Geophys Res 116:A05103. doi:10.1029/2010JA016133

Barth CA, Baker DN, Mankoff KD, Bailey SM (2001) The northern auroral region as observed in nitric oxide. Geophys Res Lett 28:1463–1466

Barth CA, Mankoff KD, Balley SM, Solomon SC (2003) Global observations of nitric oxide in the thermosphere. J Geophys Res 108:1027. doi:10.1029/2002JA009458

Baumgaertner AJG, Jöckel P, Brühl C (2009) Energetic particle precipitation in ECHAM5/MESSy1—Part 1: downward transport of upper atmospheric NOx produced by low energy electrons. Atmos Chem Phys 9:2729–2740. doi:10.5194/acp-9-2729-2009

Baumgaertner A, Seppälä A, Jöckel P, Clilverd MA (2011) Geomagnetic activity related NOx enhancements and polar surface air temperature variability in a chemistry climate model: modulation of the NAM index. Atmos Chem Phys 11:4521–4531

Bazilevskaya GA, Usoskin IG, Flückiger EO, Harrison RG, Desorgher L, Bütikofer R, Krainev MB, Makhmutov VS, Stozhkov YI, Svirzhevskaya AK, Svirzhevsky NS, Kovaltsov GA (2008) Cosmic ray induced ion production in the atmosphere. Space Sci Rev 137:149–173

Brasseur GP, Solomon S (2005) Aeronomy of the middle atmosphere. Springer, The Netherlands, p 644
Calisto M, Usoskin I, Rozanov E, Peter T (2011) Influence of galactic cosmic rays on atmospheric composition and dynamics. Atmos Chem Phys 11:4547–4556. doi:10.5194/acp-11-4547-2011
Callis LB, Natarajan M, Evans DS, Lambeth JD (1998) Solar atmospheric coupling by electrons (SOLACE): 1. Effects of the May 12, 1997 solar event on the middle atmosphere. J Geophys Res 103(D21): 28405–28419. doi:10.1029/98JD02408
Cane HV, Mewaldt RA, Cohen CMS, von Rosenvinge TT (2006) Role of flares and shocks in determining solar energetic particle abundances. J Geophys Res 111:A06S90. doi:10.1029/2005JA011071
IPCC (Intergovernmental Panel on Climate Change) (2007) The physical science basis, working group I contribution to the fourth assessment report of the IPCC, Intergovernmental Panel on Climate Change. Cambridge University Press, Cambridge, UK and New York, NY, USA
Clilverd MA, Seppälä A, Rodger CJ, Mlynczak MG, Kozyra JU (2009) Additional stratospheric NOx production by relativistic electron precipitation during the 2004 spring NOx descent event. J Geophys Res 114:A04305. doi:10.1029/2008JA013472
Damiani A, Storini M, Laurenza M, Rafanelli C (2008) Solar particle effects on minor components of the Polar atmosphere. Ann Geophys 26:361–370
Egorova T, Rozanov E, Zubov V, Karol IL (2003) Model for investigating ozone trends (MEZON). Izvestiya Atmos Ocean Phys 39:277–292
Egorova T, Rozanov E, Zubov V, Manzini E, Schmutz W, Peter T (2005) Chemistry-climate model SOCOL: a validation of the present-day climatology. Atmos Chem Phys 5:1557–1576
Egorova T, Rozanov E, Ozolin Y, Shapiro A, Calisto M, Peter T, Schmutz W (2011) The atmospheric effects of October 2003 solar proton event simulated with the chemistry–climate model SOCOL using complete and parameterized ion chemistry. J Atmos Solar Terr Phys 73:356–365. doi:10.1016/j.jastp.2010.01.009
Emery BA, Coumans V, Evans DS, Germany GA, Greer MS, Holeman E, Kadinsky-Cade K, Rich FJ, Xu W (2008) Seasonal, Kp, solar wind, and solar flux variations in long-term single-pass satellite estimates of electron and ion auroral hemispheric power. J Geophys Res 113:A06311. doi:10.1029/2007JA012866
Eyring V, Waugh DW, Bodeker GE, Cordero E, Akiyoshi H, Austin J, Beagley SR, Boville BA, Braesicke P, Brühl C, Butchart N, Chipperfield MP, Dameris M, Deckert R, Deushi M, Frith SM, Garcia RR, Gettelman A, Giorgetta MA, Kinnison DE, Mancini E, Manzini E, Marsh DR, Matthes S, Nagashima T, Newman PA, Nielsen JE, Pawson S, Pitari G, Plummer DA, Rozanov E, Schraner M, Scinocca JF, Semeniuk K, Shepherd TG, Shibata K, Steil B Stolarski RS, Tian W, Yoshiki M (2007) Multimodel projections of stratospheric ozone in the 21st century. J Geophys Res 112:D16303. doi:10.1029/2006JD008332
Funke B, López-Puertas M, Gil-Lopez S, von Clarmann T, Stiller GP, Fischer H, Kellman S (2005) Downward transport of upper atmospheric NOx into the polar stratosphere and lower mesosphere during the Antarctic winter 2003 and Arctic winter 2002/2003. J Geophys Res 112:D24308. doi: 10.1029/2005JD006463,2005
Funke B, Baumgaertner A, Calisto M, Egorova T, Jackman CH, Kieser J, Krivolutsky A, López-Puertas M, Marsh DR, Reddmann T, Rozanov E, Salmi S-M, Sinnhuber M, Stiller GP, Verronen PT, Versick S, von Clarmann T, Vyushkova TY, Wieters N, Wissing JM (2011) Composition changes after the "Halloween" solar proton event: the High-Energy Particle Precipitation in the Atmosphere (HEPPA) model versus MIPAS data intercomparison study. Atmos Chem Phys Discuss 11:9407–9514
Gray LJ, Beer J, Geller M, Haigh JD, Lockwood M, Matthes K, Cubasch U, Fleitman D, Harrison G, Hood L, Luterbacher J, Meehl GA, Shindell D, Shindell D, van Geel B, White W (2010) Solar influences on climate. Rev Geophys 48:RG4001. doi:10.1029/2009RG000282
Jackman CH, Marsh DR, Vitt FM, Garcia RR, Fleming EL, Labow GJ, Randall CE, López-Puertas M, Funke B, von Clarmann T, Stiller GP (2008) Short- and medium-term atmospheric constituent effects of very large solar proton events. Atmos Chem Phys 8:765–785. doi:10.5194/acp-8-765-2008
Jackman CH, Marsh DR, Vitt FM, Garcia RR, Randall CE, Fleming EL, Frith SM (2009) Long-term middle atmosphere influence of very large solar proton events. J Geophys Res D114:11304. doi:10.1029/2008JD011415
Lam MM, Horne RB, Meredith NP, Glauert SA, Moffat-Griffin T, Green JC (2010) Origin of energetic electron precipitation > 30 keV into the atmosphere. J Geophys Res 115:A00F08. doi:10.1029/2009JA014619
Langematz U, Grenfell JL, Matthes K, Mieth P, Kunze M, Steil B, Brühl C (2005) Chemical effects in 11-year solar cycle simulations with the Freie Universität Berlin Climate Middle Atmosphere Model with online chemistry (FUB-CMAM-CHEM). Geophys Res Lett 32:L13803. doi:10.1029/2005GL022686

López-Puertas M, Funke B, von Clarmann T, Fischer H, Stiller GP (2006) The stratospheric and mesospheric NOy in the 2002–2004 polar winters as measured by MIPAS/ENVISAT. Space Sci Rev 125:403–416. doi:10.1007/s11214-006-9073-2

López-Puertas M, Funke B, Gil-Lopez S, von Clarmann T, Stiller GP, Höpfner M, Kellmann S, Fischer J, Jackman CH (2005) Observation of NOx enhancement and ozone depletion in the Northern and Southern hemispheres after the October–November 2003 solar proton events. J Geophys Res 110: A09S43. doi:10.1029/2005JA011050

Manzini E, McFarlane NA, McLandress C (1997) Impact of the doppler spread parameterization on the simulation of the middle atmosphere circulation using the MA/ECHAM4 general circulation model. J Geophys Res 102:25751–25762

Marsh DR, Garcia RR, Kinnison DE, Boville BA, Sassi F, Solomon SC, Matthes K (2007) Modeling the whole atmosphere response to solar cycle changes in radiative and geomagnetic forcing. J Geophys Res 112:D23306. doi:10.1029/2006JD008306

Meredith NP, Horne RB, Lam MM, Denton MH, Borovsky JE, Green JC (2011) Energetic electron precipitation during high-speed solar wind stream driven storms. J Geophys Res 116:A05223. doi: 10.1029/2010JA016293

Millan RM, Thorne RM (2007) Review of radiation belt relativistic electron losses. J Atmos Solar Terr Phys 69:362–377

Morgenstern O, Giorgetta MA, Shibata K, Eyring V, Waugh DW, Shepherd TG, Akiyoshi H, Austin J, Baumgaertner AJG, Bekki S, Braesicke P, Brühl C, Chipperfield MP, Cugnet D, Dameris M, Dhomse S, Frith SM, Garny H, Gettelman A, Hardiman SC, Hegglin MI, Jöckel P, Kinnison DE, Lamarque JF, Mancini E, Manzini E, Marchand M, Michou M, Nakamura T, Nielsen JE, Olivié D, Pitari G, Plummer DA, Rozanov E, Scinocca JF, Smale D, Teyssedre H, Toohey M, Tian W, Yamashita Y (2010) Review of the formulation of present-generation stratospheric chemistry-climate models and associated external forcings. J Geophys Res 115:D00M02. doi:10.1029/2009jd013728

Porter HS, Jackman CH, Green AES (1976) Efficiencies for production of atomic nitrogen and oxygen by relativistic proton impact in air. J Chem Phys 65:1

Randall CE, Harvey VL, Singleton CS, Bailey SM, Bernath PF, Codrescu M, Nakajima H, Russell JM (2007) Energetic particle precipitation effects on the Southern Hemisphere stratosphere in 1991–2005. J Geophys Res 112:D08308. doi:10.1029/2006JD007696

Reames D (1999) Particle acceleration at the sun and in the heliosphere. Space Sci Rev 90:413–491

Reddmann T, Ruhnke R, Versick S, Kouker W (2010) Modeling disturbed stratospheric chemistry during solar-induced NOx enhancements observed with MIPAS/ENVISAT. J Geophys Res 115:D00I11. doi: 10.1029/2009JD012569

Richardson IG, Cliver EW, Cane HV (2000) Sources of geomagnetic activity over the solar cycle: relative importance of coronal mass ejections, high-speed streams, and slow solar wind. J Geophys Res 105(A8):18203–18213

Rozanov E, Schlesinger ME, Zubov V, Yang F, Andronova NG (1999) The UIUC three-dimensional stratospheric chemical transport model: Description and evaluation of the simulated source gases and ozone. J Geophys Res 104:11755–11781

Rozanov E, Callis L, Schlesinger M, Yang F, Andronova N, Zubov V (2005) Atmospheric response to NOy source due to energetic electron precipitation. Geophys Res Lett 32:L14811. doi:10.1029/2005 GL023041

Rusch D, Gerard G-C, Solomon S, Crutzen P, Reid G (1981) The effect of particle precipitation events on the neutral and ion chemistry of the middle atmosphere. 1. Odd nitrogen. Planet Space Sc 29:767–774

Schraner M, Rozanov E, Schnadt Poberaj C, Kenzelmann P, Fischer AM, Zubov V, Luo BP, Hoyle CR, Egorova T, Fueglistaler S, Brönnimann S, Schmutz W, Peter T (2008) Technical note: chemistry-climate model SOCOL: version2.0 with improved transport and chemistry/micro-physics schemes. Atmos Chem Phys 8:5957–5974

Semeniuk K, Fomichev VI, McConnell JC, Fu C, Melo SML, Usoskin IG (2011) Middle atmosphere response to the solar cycle in irradiance and ionizing particle precipitation. Atmos Chem Phys 11:5045–5077

Seppälä A, Verronen PT, Clilverd MA, Randall CE, Tamminen J, Sofieva V, Backman L, Kyrolä E (2007a) Arctic and Antarctic polar winter NOx and energetic particle precipitation in 2002–2006. Geophys Res Lett 34:L12810. doi:10.1029/2007GL029733

Seppälä A, Clilverd MA, Rodger CJ (2007b) NOx enhancements in the middle atmosphere during 2003–2004 polar winter: Relative significance of solar proton events and the aurora as a source. J Geophys Res 112:D23303. doi:10.1029/2006JD008326

Seppälä A, Randall CE, Clilverd MA, Rozanov E, Rodger CJ (2009) Geomagnetic activity and polar surface air temperature variability. J Geophys Res 114:A10312. doi:10.1029/2008JA014029

Shepherd TG (2002) Issues in stratosphere-troposphere coupling. J Meteorol Soc Jpn 80:769–792. doi:10.2151/jmsj.80.769

Sinnhuber M, Kazeminejad S, Wissing JM (2011) Interannual variation of NOx from the lower thermosphere to the upper stratosphere in the years 1991–2005. J Geophys Res 116:A02312. doi:10.1029/2010JA015825

Siskind DE, Nedoluha GE, Randall CE, Fromm M, Russell JM III (2000) An assessment of Southern Hemisphere stratospheric NOx enhancements due to transport from the upper atmosphere. Geophys Res Lett 27(3):329–332. doi:10.1029/1999GL010940

Solomon S, Rusch DW, Gerard J-C, Reid GC, Crutzen PJ (1981) The effect of particle precipitation events on the neutral and ion chemistry of the middle atmosphere, II, Odd hydrogen. Planet Space Sci 29(8):885–892

SPARC CCMVal (2010) SPARC report on the evaluation of chemistry-climate models In: Eyring V, Shepherd TG, Waugh DW (eds) SPARC, Toronto, Ontario, Canada, Tech. Rep. WCRP-132/WMO/TD-1526/SPARC Rep, p 5

Turunen E, Verronen PT, Seppälä A, Rodger CJ, Clilverd MA, Tamminen J, Enell CF, Ulich T (2009) Impact of different energies of precipitating particles on NOx generation in the middle and upper atmosphere during geomagnetic storms. J Atmos Sol Terr Phy 71:1176–1189. doi:10,1016/j.jastp.2008.07.05

Usoskin IG, Kovaltsov GA (2006) Cosmic ray induced ionization in the atmosphere: Full modeling and practical applications. J Geophys Res 111:D21206. doi:10.1029/2006JD007150

Usoskin IG, Kovaltsov GA, Mironova IA (2010) Cosmic ray induced ionization model CRAC:CRII: An extension to the upper atmosphere. J Geophys Res 115:D10302. doi:10.1029/2009JD013142

Verronen PT, Rodger CJ, Clilverd MA, Wang S (2011) First evidence of mesospheric hydroxyl response to electron precipitation from the radiation belts. J Geophys Res 116:D07307. doi:10.1029/2010JD014965

Wang Y, Jacob DJ, Logan JA (1998) Global simulation of tropospheric O_3-NO_x-hydrocarbon chemistry, 3. Origin of tropospheric ozone and effects of non-methane hydrocarbons. J Geophys Res 103:10757–10767

Wissing JM, Kallenrode M-B, Wieters N, Winkler H, Sinnhuber M (2010) Atmospheric ionization module Osnabrück (AIMOS): 2. Total particle inventory in the October–November 2003 event and ozone. J Geophys Res 115:A02308. doi:10.1029/2009JA014419

Surv Geophys (2012) 33:503–534
DOI 10.1007/s10712-012-9181-3

Solar Influence on Global and Regional Climates

Mike Lockwood

Received: 5 October 2011 / Accepted: 1 February 2012 / Published online: 16 May 2012

Abstract The literature relevant to how solar variability influences climate is vast—but much has been based on inadequate statistics and non-robust procedures. The common pitfalls are outlined in this review. The best estimates of the solar influence on the global mean air surface temperature show relatively small effects, compared with the response to anthropogenic changes (and broadly in line with their respective radiative forcings). However, the situation is more interesting when one looks at regional and season variations around the global means. In particular, recent research indicates that winters in Eurasia may have some dependence on the Sun, with more cold winters occurring when the solar activity is low. Advances in modelling "top-down" mechanisms, whereby stratospheric changes influence the underlying troposphere, offer promising explanations of the observed phenomena. In contrast, the suggested modulation of low-altitude clouds by galactic cosmic rays provides an increasingly inadequate explanation of observations.

Keywords Solar variability · Global climate change · Regional climate change · Blocking events · Maunder minimum · Total solar irradiance · UV spectral irradiance · Galactic cosmic rays · Open solar flux · Stratosphere–troposphere coupling

1 Introduction

Solar electromagnetic radiation powers Earth's climate system and, consequently, it is often naïvely thought that changes in this solar output must be responsible for changes in Earth's climate. However, the huge thermal time constant of the outer part of the Sun limits the variability in its surface temperature, and hence its total power output, which is

M. Lockwood (✉)
Department of Meteorology, University of Reading, P.O. Box 243, Earley Gate RG6 6BB, UK
e-mail: m.lockwood@reading.ac.uk

M. Lockwood
RAL Space, Rutherford Appleton Laboratory, Harwell Campus, Chilton, Didcot, Oxfordshire OX11 0QX, UK

dominated by visible and infrared emissions from the solar surface (the photosphere) (see review by Lockwood 2004). As a result, changes in solar power output on decadal, centennial and millennial timescales are limited to small changes in effective surface temperature (associated with magnetic fields) and potential, although as yet undetected, solar radius variations (see reviews by Solanki et al. 2005; Lockwood 2010). Larger percentage variations are seen in solar UV emissions (Lean et al. 1997) which arise from the lower solar atmosphere (the chromosphere) (Loukitcheva et al. 2009) and which influence the stratosphere in Earth's middle atmosphere between about 10 and 50 km (see review by Gray et al. 2010). Even more variable are solar X-rays and extreme ultraviolet (EUV) emissions that originate in the upper solar atmosphere (the corona) and dominate the behaviour of Earth's uppermost atmosphere (the thermosphere, above about 90-km altitude) (Le et al. 2011). In addition to these electromagnetic outputs, the Sun modulates energetic charged particle fluxes incident upon the Earth. Solar energetic particles (SEP) are emitted by solar flares and from the shock fronts that form ahead of super-sonic (and super-Alfvénic) ejections of material from the corona (Schwenn 2006). SEPs are incident upon Earth's atmosphere in polar regions where they enhance the destruction of stratospheric ozone (Jackman et al. 2006; 2008). Even more energetic are galactic cosmic rays (GCRs). These particles are not generated by the Sun; rather, they originate at the shock fronts emanating from violent galactic events such as supernovae explosions. However, the expansion of the shielding solar magnetic field into interplanetary space results in the Sun modulating the number of GCRs reaching Earth (see, for example, review by Potgieter 2008). Air ions generated by GCRs enable Earth's global electric (thunderstorm) circuit (Rycroft et al. 2008), and it has been proposed that they also modulate the formation of low-altitude clouds (Svensmark and Friis-Christensen 1997). The Sun also emits a continuous stream of low-energy charged particles called the solar wind (e.g., Marsch 2006). A small fraction of the solar wind energy incident on Earth is extracted by the geomagnetic field and deposited in the thermosphere at high latitudes (Cowley 1991; Thayer and Semeter 2004). This deposition changes the behaviour of the thermosphere globally (e.g., Fuller-Rowell et al. 2007), but this is an extremely low-density atmospheric layer, and there are no robust observations, nor any confirmed theory, that suggests these thermospheric variations are transmitted through the middle atmosphere to the troposphere below.

Both electromagnetic and charged particle emissions from the Sun are known to vary over the decadal-scale solar magnetic activity cycle, as do GCR fluxes (see review by Lockwood 2004). But any effects on climate are much more significant for any variations over longer timescales. This review discusses and evaluates potential effects on Earth's climate of variations in these solar emissions. "Top-down" mechanisms involve solar UV irradiance (or perhaps energetic particles) modulating stratospheric temperatures and winds which, in turn, may influence the underlying troposphere where Earth's climate and weather reside. These contrast with "bottom-up" effects in which the total solar irradiance (TSI, dominated by the visible and near-IR) variations cause surface temperature changes and upward coupling to the troposphere.

2 Historical Perspective

It is interesting to ponder what the author citation h-index of the astronomer Sir Frederick William Herschel (1738–1822) would be. Certainly, his 1801 paper (Herschel 1801), in which he speculated on a connection between sunspots and regional climate (for which he used the market price of wheat as a quantifiable proxy), has been cited a great many times and continues to be cited regularly today. Herschel appears to have been more aware of the limitations of his apparent correlation (writing "this prediction ought not to be relied on by anyone") than many who have subsequently cited his paper. Indeed, Herschel himself notes limitations to his speculation that many subsequent studies of solar influence on climate have failed to adequately consider. Common pitfalls include the following:

1. The potential for selection effects mean that one must ask, "Would a reported correlation coefficient be as high if an equivalent data set were substituted?" (For example, in Herschel's case, was the behaviour of the price of barley similar to that of wheat? If not, why not?) Selection effects can often arise (unintentionally or otherwise) from the use of restricted data intervals and/or the choice of which parameters to compare. Selection effects are the major reason why the specific issues (2)–(5) below are particular problems.
2. Sunspots are only indirectly related to the solar outputs that are relevant to climate. (Herschel attempts to build an argument about enhanced emission from sunspots which we now know to be incorrect).
3. The lack of suitable measurements of Earth's global climate on long timescales leads to the use of indirect proxies (Herschel used the price of wheat), and there are many other factors, unrelated to climate, which can influence such proxies.
4. The climate data used have often been for a limited season (in Herschel's case, this is effectively the growing season of wheat) which may not adequately reflect changes in the annual mean data.
5. The climate data used have often been regional, and not global (Herschel used the price of wheat in the UK and not on the global market); regional climates show large variations around the global mean.
6. The data sets are not homogeneous, and measurement series often contain instrumental drifts and discrete jumps. (Herschel notes this in respect of the sunspot data available to him).
7. Correlations can occur by chance, and the probability of this happening must be evaluated, allowing for the persistence in the data time series and the number of free parameters. Hence, the significance of any correlation must be determined (by comparison with a suitable noise model). In addition, tests are needed to prevent "data snooping", that is, exploiting a feature in the data that turns out to be a statistical or instrumental artefact. To ensure that wrong conclusions are not drawn from statistical coincidences, clear null hypotheses must be formulated and proper statistical tests performed (Yiou et al. 2010; Legras et al. 2010; Love et al. 2011).
8. Solar variability effects are convolved with other influences on climate (volcanoes, aerosols, land-use changes and greenhouse gases). For example, Herschel neglected the effects of volcanic eruptions that occurred during the interval that he studied.
9. The effects of internal climate variability have often been overlooked or underestimated. There are internal oscillations (particularly associated with the oceans on decadal timescales), and a common error is to "overfit" the data, thereby wrongly ascribing variations to an external solar origin. There are a huge number of publications looking at oscillations at periods close to that of the solar cycle and/or correlations with the solar cycle at a fixed lag. Tests of some of these reveal that they are unlikely to be driven by solar forcing and that they are more likely to reflect other natural cycles of the climate system or a harmonic combination of various multiyear cycles (Drijfhout et al. 1999; Moore et al. 2006).

10. Even a statistically significant correlation or detected periodic signal does not establish causality, neither do lags nor phase relationships for coupled systems involving stationary oscillations.
11. Lastly one must ask, "Is there a realistic mechanism or series of mechanisms that could have given rise to the correlation?" (Herschel's proposal invoked a chain of interactions, encompassing solar physics, climate science, agricultural science and the effect of supply-and-demand in economics!).

The reason why the citation rate for Herschel's speculation is so high is that a great many subsequent studies have explored the same ideas. The growth in modern-day interest can be traced to John A. "Jack" Eddy (1931–2009). Incidentally, Eddy's h-index is known and is modest, by modern standards, at 14; however, his most influential paper (Eddy 1976) has been cited 954 times (citation analysis using Web of Science, October 2011). This famous paper particularly focussed on European temperature records and the period of extremely low solar activity (when almost no sunspots were observed) known as the Maunder Minimum (circa 1645–1715). This topic is discussed further in Sect. 6 of the present review.

The academic reputation of the field of Sun-climate relations is poor because many studies do not address all, or even some of, the limitations listed above. It is also a field that in recent years has been corrupted by unwelcome political and financial influence as climate change sceptics have seized upon putative solar effects as an excuse for inaction on anthropogenic warming. In particular, figures and statistics with known errors, limitations and inaccuracies are repeatedly reproduced on the Internet and in the media (as discussed, for example, by Damon and Laut 2004), and publications are reported in a massively selective manner. None of this makes any difference to the scientific reality, or otherwise, of mechanisms connecting solar variability and Earth's climate; however, it does make evaluation of the evidence much more difficult. Recent reviews have been presented by Reid (2000), Rind (2002), Haigh (2003, 2007), Beer (2006), Foukal et al. (2006), de Jager (2008), Gray et al. (2010) and Lockwood (2004, 2010, 2012).

3 Solar Variability and Global Climate Response Timescales

Solar outputs reaching Earth undergo periodic variations over a range of timescales. These include the 27-day variation due to solar rotation, annual variations due to the ellipticity of Earth's orbit, the decadal-scale solar magnetic (sunspot) cycle and oscillations between grand solar maxima and minima on timescales of several centuries. How a key climate parameter, such as the Global Mean Air Surface Temperature (GMAST), varies in response to these solar variations will depend on its response timescale. Estimates of the GMAST response time have varied from less than 1 year to several decades and, in reality, the role of the oceans means that a spectrum of response times is to be expected. Knutti et al. (2008) discuss the time constants of different parts of the climate system: short time scales (1 year or less) apply to atmospheric adjustments and land surface processes; medium time scales (of order decades) to the melting of sea ice; and long time scales (many decades) to warming of the whole ocean surface layer. Ocean warming effects are complex because of mixing behaviour (Hansen et al. 1985), but an effective timescale of 5–20 years has been predicted by Dickinson and Schaudt (1998). The climate response modelling of Rind et al. (1999) and analysis of pre-industrial data by Waple et al. (2002) and Weber (2005) yield response times of up to 10 years. Held et al. (2010) use both an energy balance model (EBM) and a coupled atmosphere–ocean General Circulation Model (GCM) to investigate the implications of two time constants—a rapid response on a timescale of about 5 years and a slower "recalcitrant" response on a timescale of several hundred years. They show that the rapid response is currently dominant, but the longer timescale would become important if greenhouse gas emissions were to be stabilised. All these time constants mean that responses to 27-day solar variations will be damped to undetectable levels in GMAST, whereas the centennial solar variations should be reflected almost in full. The extent to which the solar-cycle variations are damped is still a matter of debate. Some studies find that the GMAST response to recent solar cycles may be 0.1 K or larger (White et al. 1997; Douglass and Clader 2002; Camp and Tung 2007; Lean and Rind 2009). However, some of the apparent responses found in regression studies of empirical data may not, in reality, be due to real associations and causality. This is because of one or more of the potential pitfalls discussed in Sect. 1 (such as overfitting so that internal variability is ascribed to external forcing, degeneracy between different forcing factors, understated uncertainties in regression coefficients and in the response to forcing). That this can be the case has been demonstrated clearly using model simulations (Ingram 2006; Benestad and Schmidt 2009; Stott and Jones 2009).

4 Global Climate Response

This section considers the multidecadal trends (e.g., Brönnimann et al. 2007) rather than the great many reports of putative solar-cycle variations observed in the troposphere (e.g., Camp and Tung 2007; Coughlin 2004; Gleisner et al. 2005). One reason why the global climate response timescales have recently become a consideration is the realisation that, since 1985, all the relevant solar outputs have been changing in the opposite direction to those needed to explain the rise in GMAST (Lockwood and Fröhlich 2007). Thus, for solar activity to contribute to recent warming, there must be a long GMAST response time constant. For example, Scafetta and West (2007) and Scafetta (2009) have employed multiregression fits of GMAST data that involve a short-timescale response (to allow the observed weak solar-cycle variations to be reproduced) plus a long-time-constant response. This is different from the Held et al. (2010) study discussed in the last section because the longer time-constant response dominates. Multiple regression analysis of this kind has been used several times previously (e.g., Santer et al. 2001; Lean 2006; Lockwood 2008), but can be misleading if inter-correlations exist between the data series that are input into the study or if genuine factors are omitted. In addition, naïve regression analyses may also be misleading due to convoluted nonlinear response associated with feedback processes, the use of non-optimal regression techniques or the presence of internal system variability (Benestad and Schmidt 2009). In general, all forcings need to be considered for the attribution of climate change and the effects of degeneracy between solar and volcanic forcings over recent centuries are a particular concern because major eruptions tend to occur at an average rate near one per solar cycle and even have a periodicity close to the 88 years of the solar Gliessberg cycle (Qu et al. 2011) and because bursts of volcanic eruptions and a lack of eruptions can cause long-term cooling and warming, respectively. Hence, such analyses need to be interpreted with a great deal of caution and judgement with regard to the choice of input variables and as to whether the conclusions are sensible. There are other potential pitfalls. For example, there is a strong possibility that the fits of Scafetta and West (2007) and Scafetta (2009) suffer from a "data snooping" problem as they are strongly dependent on a feature in the TSI data composite that they employed which has been argued to be a calibration error (Lockwood and Fröhlich 2008). Benestad and Schmidt (2009) present a full critique of the robustness of the fits; however, here, we simply look whether their implications are sensible or not. Scafetta and West (2007) and Scafetta (2009) find that solar change contributed over 60% of the temperature rise (i.e. more than 0.5°C) since pre-industrial times, and it is instructive to ask, "How does this agree with the changes in the various radiative forcings?"

Our best estimates of ΔS, the change in the average TSI between the Maunder Minimum and recent decades, have decreased over recent years (see review by Gray et al. 2010). Initially, ΔS was estimated from surveys of "Sun-like" stars (Baliunas and Jastrow 1990; Zhang et al. 1994), but results of more extensive surveys did not confirm the key result, nor did they support its application to our Sun (Hall and Lockwood 2004; Giampapa 2004). Subsequently, smaller ΔS estimates have been derived using sunspot numbers and the open magnetic flux of the Sun, which can be derived from historic geomagnetic data (see review by Lockwood 2004). These studies find ΔS to be comparable to the amplitude of recent solar cycles (i.e. 0.1%) (see Gray et al. 2010 and references therein) and such estimates are currently employed, for example, in the 2007 IPCC report. The more recent estimates of the change in TSI equate to a change in radiative forcing of $\Delta f_S = \Delta S(1 - A)/4$. Earth's albedo, A, is particularly poorly known (Charlson et al. 2005; Pallé et al. 2009) but using a best estimate of $A \approx 0.3$ (Loeb et al. 2009) gives $\Delta f_S \approx 0.24$ W m^{-2}. Two recent papers disagree with this consensus value, but do not agree on even the sign of the adjustment needed. Schrijver et al. (2011) argue that the consensus Δf_S is overestimated, whereas Shapiro et al. (2011) suggest it should be increased by a factor of at least 4, to a value near 1 W m^{-2}. Where these two papers differ is in their treatment of the "quiet" Sun between groups of sunspots ("active regions"), and Lockwood (2011b) argues that neither makes a compelling case for an adjustment to the consensus value at the present time.

From flask measurements and ice-core data, we can compute the radiative forcing caused by the observed rises in well-mixed trace (i.e. excluding water vapour) greenhouse gas (GHG) abundances over the same interval (Forster et al. 2007). The abundance of carbon dioxide has risen from 280 ppmv in pre-industrial times to 362.5 ppmv in 2000, which corresponds to +1.56 W m^{-2}. The corresponding radiative forcing contributions for other trace GHG are the following: +0.47 W m^{-2} for CH_4, +0.28 W m^{-2} for CFCs/HCFs/halons, and +0.14 W m^{-2} for NO_x. This gives a total radiative forcing by trace GHG of $\Delta f_G = 2.45$ W m^{-2}, which is ten times larger than our best estimate of the solar forcing Δf_S. Note that this estimate of Δf_G is known with considerable certainty from the analysis of absorption spectra. Persistent reports, mainly in the media and on the Internet, that absorption lines are saturated (such that adding more GHG does not cause any radiative forcing change) ignore a huge body of research that extends back to the 1950s and are demonstrably false when one considers the whole IR spectrum (Shine et al. 1995). Energy balance shows that, for steady state, the observed GMAST rise since the Maunder Minimum would correspond to a total forcing, including feedbacks (which include water vapour effects), of $\Delta f = 5.15$ W m^{-2} (Lockwood 2010). The uncertainty on this estimate is of order ±0.5 W m^{-2}, dominated by the uncertainty in the pre-industrial temperature, with a small contribution from the uncertainty in the power stored in the deep oceans.

Thus, to explain the warming with GHG forcing alone requires feedbacks to account for a forcing of $\Delta f - \Delta f_G \approx 2.67$ W m^{-2}, that is, a (positive) feedback factor of $a_G = \Delta f/\Delta f_G \approx 2.1$. To explain 60% of the warming with solar forcing requires a feedback factor of $a_S = 0.6 \times \Delta f/\Delta f_S \approx 12.8$ and a feedback factor of $a_G = 0.4 \times \Delta f/\Delta f_G \approx 0.8$ for greenhouse gases. In other words, the fit of Scafetta (2009) requires a huge positive feedback for the solar forcing but a negative feedback for the greenhouse forcing. Thus, this fit raises much bigger problems than it solves and makes no sense physically.

But not all multiple regression fits suffer from such problems. Figures 1 and 2 show the results of a fit to recent GMAST data (1954–2011) using the method of Lockwood (2008). Like that by Scafetta (2009), this fit allows for the climate response times (in this case by passing each input variation through a filter with a response time constant which is iterated in the fit); however, it differs in that it uses different input variations. In fact, the best-fit response time for each input was found to be near 1 year and constraining all inputs to the same response time generated an almost identical fit with fewer free parameters. The galactic cosmic ray flux is used to quantify the solar input as the data sequence is longer and more stably calibrated than the available TSI record, and there is a good (anti)correlation between the two (see Fig. 3). In addition, the El-Niño Southern oscillation (ENSO) can be regarded as an energy input into the surface layer that is quantified by the N3.4 index (based on the sea surface temperature in the equatorial Pacific). Figure 1 shows excellent agreement between observed and fitted variations for 1953–2011, and Fig. 2 shows that the dominant component of the fitted variation is anthropogenic and the solar contribution is small (and actually downward on average for this interval). The fit residuals shown in the bottom panel of Fig. 1 are attributed to internal variability. For this fit, we find that the contributions to the GMAST rise are in almost the same ratios as the radiative forcing changes over the interval. This means that the feedback factors required are approximately the same for all forcings (so $a_S \approx a_G \approx 2$). This multiple regression fit is subject to the same set of caveats as all others (i.e. selection effects of the choice of inputs, inter-correlations between inputs, nonlinear responses and overfitting of internal variability). However, it does demonstrate that a simple combination of known inputs can explain the observed GMAST variation (incidentally including any recent apparent

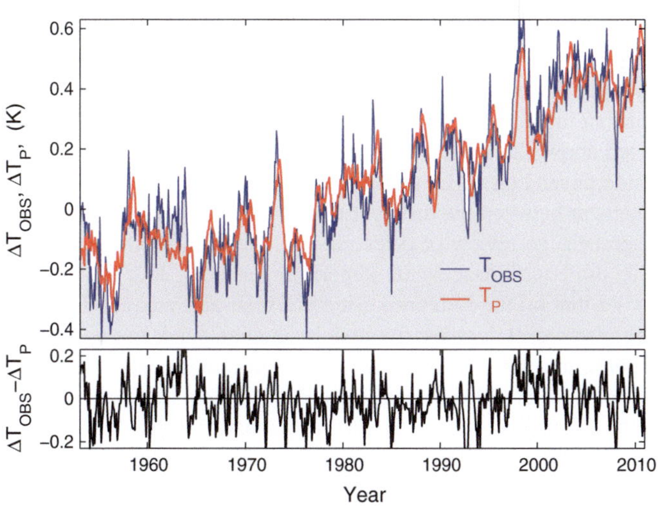

Fig. 1 *Top* observed (T_{OBS}, *blue*) and fitted (T_P, *red*) global mean air surface temperature (*GMAST*) variations for 1953–2011. The monthly observations are the HADCrut3v database (Brohan et al. 2006), and the fit is carried out using the procedure of Lockwood (2008) that allows for climate response times. *Bottom* the variation of the fit residuals ($T_{OBS} - T_P$)

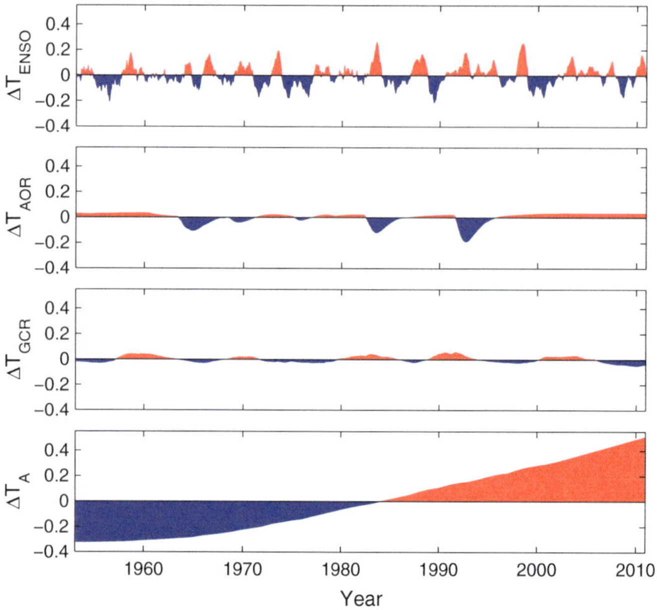

Fig. 2 The contributions to the fitted variation T_P in Fig. 1. Each *panel* uses the same *y*-axis scale to allow comparison (above/below average values are shown in *red/blue*), and each input parameter has been smoothed/lagged using its best-fit response time parameter. Fits are achieved with the Nelder-Mead search method. From *top* to *bottom* ENSO (input quantified by the N3.4 ENSO index), volcanoes (quantified by the global mean atmospheric optical depth, AOD); solar (quantified by GCR fluxes observed by neutron monitors), and anthropogenic (quantified by the GISS model E forcings for well-mixed trace greenhouse gases, land-use changes and man-made tropospheric aerosols). For comparison, the RMS fit residual $<(T_{OBS} - T_P)^2>^{1/2}$ is 0.10°C (see Fig. 1b)

plateau) very well indeed and in a way that is consistent with the known radiative forcing changes. Notice that this analysis has not made use of any numerical climate model results.

To be fully rigorous, one has to move to "detection–attribution" studies, in which the effect of varying an input to a general circulation ocean–atmosphere climate model (GCM) must be detected above the internal variability of the climate system, as modelled by the GCM. Results do depend on which GCM is used, although there is also now considerable degree of agreement between the different models and, indeed, with the results shown in Fig. 2. A comprehensive review of detection–attribution methods and results is given as part of the IPCC 4th Fourth Assessment Report (Hegerl et al. 2007). Some key conclusions in this context are that (a) the combined effects of solar and volcanic forcings has resulted in a net cooling over recent decades; (b) models simulate much less warming over the 20th century in response to solar forcing alone than for greenhouse gas forcing alone (Cubasch et al. 1997; Broccoli et al. 2003; Meehl et al. 2003, 2004; Tett et al. 1999, 2002); (c) in combination, the greenhouse effect on surface temperatures is dominant over the solar effect (Stone et al. 2007a, b), although there is some debate about the precise contributions (Stott et al. 2003); and (d) the changes in the altitude profiles also reveal a dominant greenhouse gas effect (Jones et al. 2003).

Coupled ocean–atmosphere climate models (GCMs) have developed rapidly over the last 2 decades, and this has altered the details of the attribution of solar effects. In general, the derived solar-induced changes in GMAST have been small, even for the larger drifts in solar

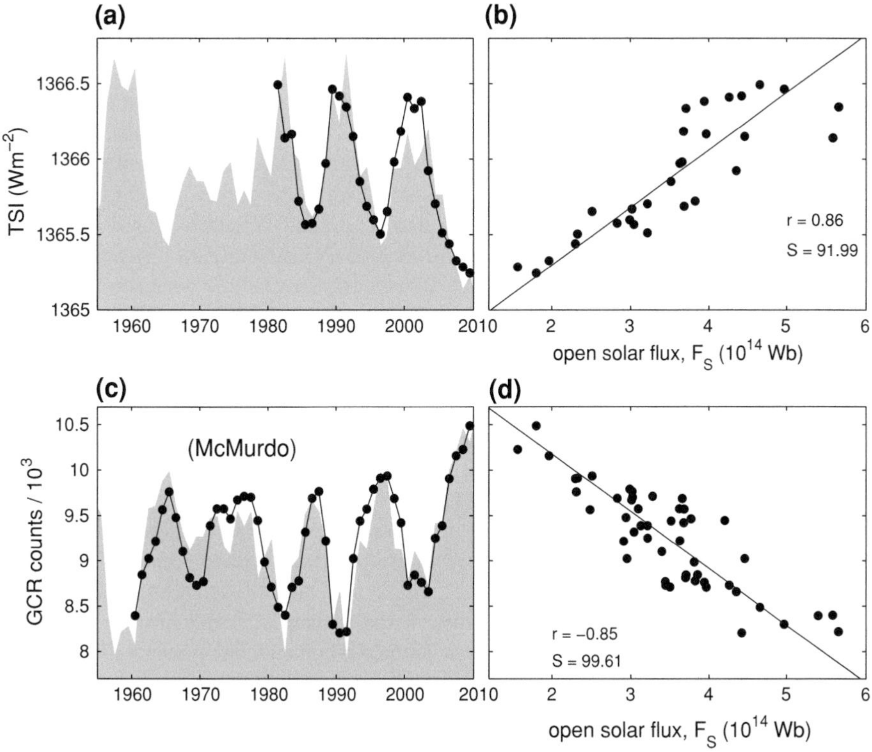

Fig. 3 The relationships between annual means of the open solar magnetic flux, F_S, and (*top*) the total solar irradiance, TSI (here quantified using the PMOD composite: Fröhlich 2009) and (*bottom*) the galactic cosmic ray (GCR) flux (here quantified using the counts detected by the McMurdo neutron monitor in Antarctica). The *left-hand panels* show the time series, and in both cases, the best-fit linear regression of F_S is shown by the *area shaded grey*. The *right-hand plots* show the scatter plots with the best-fit linear regression line. The correlation coefficients r and their significances S (compared against the AR (1) noise model) are given (from Lockwood et al. 2010a)

radiative forcing predicted by the early TSI reconstructions (Wigley and Raper 1990). Full comparisons are not possible because the TSI reconstructions have evolved at the same time as the complexity and resolution of the models has increased. As expected, weaker solar effects are found for the smaller ΔS inherent in the more recent TSI reconstructions (compare, for example, Wigley and Raper 1990 with Zorita et al. 2004 and/or Stendel et al. 2006). Examples of recent GCM experiments giving a relatively large "bottom-up" solar effect have been presented by Rind et al. (2008). The input solar forcing change between the Maunder Minimum and the present day used was 0.68 W m^{-2} (more than twice the consensus Δf_S discussed earlier) with an input of 1.9 W m^{-2} from total anthropogenic forcing, including Δf_G. This solar forcing is well outside the range of 0.06–0.30 W m^{-2} quoted by Hegerl et al. (2007), and the Rind et al. experiments generally gave a relatively large solar contribution (near 30%) to the total GMAST rise (the total GMAST rise also being larger than is observed). The tropical response, in particular, was much greater than observed, consistent with an excessive TSI change input to the experiments.

Rind et al. (2008) used a model with a coupled stratosphere at four different resolutions, to investigate various aspects of "top-down" solar cycle influence on the troposphere/

stratosphere system. The results show that the stratospheric response is highly repeatable and significant. Stratospheric westerly wind increases, driven by a greater equator to pole temperature gradient, were found to extend down into the troposphere and in some experiments the jet stream weakened and moved polewards. The predominant tropospheric response consisted of a warming of the troposphere, with coherent changes in regional precipitation. The tropospheric response was often not great, but was similar in all the different simulations. All runs revealed both top-down and bottom-up solar effects, but they account for only a small percentage of the total GMAST variance. A model described by Meehl et al. (2003, 2008, 2009) includes ocean–atmosphere coupling and with the stratosphere and suggests potential resonant connections between UV solar forcing and the ENSO oscillation, as also discussed by Emile-Geay et al. (2007) and White and Liu (2008a, b).

5 Paleoclimate Studies and Solar Proxies

For paleoclimate studies, it is important to bear in mind all the potential pitfalls listed in Sect. 2 of this review. However, there is an additional problem to consider with the data sequences, which is not a factor for direct observations, namely dating (van Geel and Mook 1989). Because record dates are uncertain, "wiggle matching" is often used (Kilian et al. 1995; Blaauw et al. 2003). Even if this is done with great care, such that the date of any one data point is not changed by more than the accurately computed dating uncertainty, there is inevitably a selection effect at work. This is because the adjustments are always made to improve correlations and not to degrade them. For this reason, there is a marked tendency for some correlations to degrade as the accuracy of the dating is improved.

An additional factor to bear in mind is that solar variability in paleoclimate studies is quantified using the abundance of cosmogenic isotopes, usually ^{14}C and ^{10}Be, found in terrestrial reservoirs. ^{10}Be is a spallation product of galactic cosmic rays hitting atmospheric O, N and Ar atoms; ^{14}C is produced by thermal neutrons, generated by cosmic rays, interacting with N. About two-thirds of the ^{10}Be are formed in the stratosphere and about one-third in the troposphere from where precipitation times into the reservoirs are typically 1 year and 1 week, respectively. The deposition into any one reservoir is influenced, to some extent, by climate conditions (e.g., Vonmoos et al. 2006; Field et al. 2006; Aldahan et al. 2008; Heikkila et al. 2008). On the other hand, the ^{14}C generated by GCRs takes part in the carbon cycle and is exchanged with the two major reservoirs, the oceans and the biomass. Hence, the transport and deposition into the reservoirs where they are detected (for example, ancient tree trunks for ^{14}C and ice sheets in the case of ^{10}Be) are vastly different for these two major cosmogenic isotopes. We can discount the possibility that matching variations in the isotope abundances in their respective reservoirs are similarly influenced by climate during their terrestrial life history, because the transport and deposition of each are so different. After allowance has been made to both series for the effect of geomagnetic field changes, the correlation between decadal means over the past millennium of the abundance of ^{10}Be and the production rate of ^{14}C is 0.81 (with the ^{10}Be leading, as expected, by about 10 years and a significance level exceeding 99%) (Bard et al. 1997). This correlation also applies to the more recent data. For example, the solar modulation potentials (see review by Usoskin 2008) for 1600–1960 derived from ^{10}Be and ^{14}C (as presented by Lockwood et al. 2011b) yield a correlation coefficient of 0.83. The ^{14}C data are the means from tree cores in many parts of the world, but the ^{10}Be data are from two just ice cores, one from Antarctica and another from Greenland. However, the influence of regional deposition variations on the ^{10}Be abundance is known to be small because the correlation with ^{14}C is high. Thus, it is concluded that the correlations that are found for both isotopes are associated with the one common denominator in their production, namely the incident cosmic ray flux.

The abundances of these isotopes are routinely used in paleoclimate studies as a proxy indicator of solar irradiance (Bard et al. 2000). Lockwood (2006) showed that the relationship between production of the ^{10}Be cosmogenic isotope and TSI is monotonic but may not be linear, depending on the effect of unknowns in the long-term quiet Sun irradiance behaviour. The flux of the GCRs that generate the cosmogenic isotopes is modulated by three following influences (Beer 2006): (1) the interstellar flux of GCRs incident on the heliosphere; (2) the shielding by the heliosphere; and (3) the shielding by the geomagnetic field. The spatial distribution of the interstellar GCR fluxes in our galaxy is on a sufficiently large scale compared to distances moved by our solar system through the galaxy, which means that we can neglect variations on timescales of Myr and smaller. The geomagnetic field shield has varied on timescales of 10 kyr. This variation has, in the main, been gradual during the Holocene, although there have been shorter-lived weakenings of the field (which may be geomagnetic reversal onsets that did not develop) such as the Laschamp event around 40 kyr ago. To allow studies of the remaining effect, solar variability, studies have corrected the cosmogenic isotope record to allow for the secular geomagnetic variations (e.g., Steinhilber et al. 2008, 2009).

Figure 3 investigates the relationship of GCRs and total solar radiance in modern data. The magnetic flux that leaves the top of solar atmosphere (the "open solar flux", F_S) is the source of magnetic field in the heliosphere, and it is irregularities in this heliospheric field which scatter GCRs and shield them away from Earth. As a result, there is a good inverse relationship between F_S and the flux of GCRs (Rouillard and Lockwood 2004), which is demonstrated by the lower panel in Fig. 3. The upper part of Fig. 3 looks at the relationship between F_S and the total solar irradiance, TSI. This relationship is more indirect than that with GCR flux in that F_S is a small, and potentially variable, fraction of the field that threads the photosphere and thereby modulates the visible/IR emission of Sun (via the darkening effects of sunspots and the brightening effect of smaller magnetic flux tubes called faculae: Solanki et al. 2005; Wenzler et al. 2006). F_S correlates well with TSI over the solar cycle (Lockwood and Stamper 1999; Lockwood 2002), and recent studies indicate that observed TSI does also reflect some of the longer-term drift in F_S (Fröhlich 2009). Modelling of the long-term variations (since the Maunder Minimum) in F_S and TSI (Vieira et al. 2011; Wang et al. 2005) provides support for the broad correlation between the two.

However, it should be noted that GCR fluxes and cosmogenic isotopes are not just indicators of TSI. Much of the magnetic field that modulates TSI as it passes through the photosphere also modulates solar UV emissions as it passes though the chromosphere. The UV part of the solar spectrum is much more variable than TSI and modulates ozone abundance, temperatures and winds in Earth's stratosphere (see review by Gray et al. 2010). Composites of observed UV spectral solar irradiance (SSI) had shown little variation in the spectral shape, such that at all wavelengths they exhibit similar temporal variations to TSI (Lean et al. 1997), and this has been reflected in centennial reconstructions of SSI (Lean 2000; Krivova et al. 2009). However, recent observations of the descent into the current "unusual" (for the space age) solar minimum (Lockwood 2010) cast doubt on this as they show more variability in the UV and considerable change in spectral shape (Harder et al. 2009). This implies that at the wavelengths responsible for stratospheric modulation, centennial variation may be greater than previously thought. These new data are from the SORCE satellite and are supported by observed changes in ozone abundance

(Haigh et al. 2010). If confirmed, this is a considerable change in our understanding in spectral irradiance variability as it means that TSI change is predominantly caused by UV changes and that the changes at IR/Visible wavelengths are actually in the opposite direction to those in the UV and TSI. This contrasts sharply with previous models in which the waveform of temporal variations was taken to be similar at all wavelengths (Lean 2000; DeLand and Cebula 2008; Krivova et al. 2009). At present, there is no consensus as to how these new results should be used and incorporated into models. Schmidt et al. (2011) argue that as the SORCE data come from one instrument (the Spectral Irradiance Monitor, SIM) and are for a limited period (the declining phase of cycle 23) it is premature to act upon them. However, it should be noted that the prior data were also limited in temporal and wavelength coverage, and there are doubts about their accuracy at some wavelengths (DeLand and Cebula 2008). Unfortunately, the wavelength band in which inter-calibrations can be carried out is very limited (Lockwood 2011a), and a homogenised data sequence of spectral irradiance will have to rely heavily on modelling (Krivova et al. 2011), at least until new space-based instrumentation can be commissioned. At the time of writing, the latest data from SIM (http://lasp.colorado.edu/sorce/data/) during the rising phase of solar cycle 24 show a reverse of the changes seen in the declining phase of cycle 23, despite the low activity levels of the new cycle (Owens et al. 2011). This implies that long-term calibration drift is not the cause of the radically different trends in the SIM data. SRPM (Solar Radiation Physical Modelling) analysis combines information of the features in the solar chromosphere with physics-based solar atmospheric spectral models to compute the emergent intensity spectrum, and initial work suggests that it is able to reproduce trends observed by SIM and that solar active regions are not the only cause of the observed variations (i.e. changes in the quiet-Sun internetwork/network radiance are involved).

The trends suggest that the open solar flux F_S may also be related to long-term variations in solar UV emissions (Lockwood et al. 2010b), and hence cosmogenic isotopes may be a proxy indicator of UV effects. In addition, cosmogenic isotopes would, of course, be a direct indicator of any cloud modulation by GCRs, where such a mechanism to be active, and F_S is highly anti-correlated with GCR fluxes (Rouillard and Lockwood 2004). Recent results relating to the GCR-cloud mechanism are discussed in Sect. 8.

This review focuses on recent work that uses the open solar flux F_S to quantify solar activity. From the discussion above, F_S can reveal effects associated with changes in TSI, UV spectral irradiance, GCRs or even SEP fluxes. In addition, F_S is highly correlated with mean geomagnetic activity levels (indeed it is reconstructed before the start of space observations using geomagnetic data, Lockwood et al. 2009; Lockwood and Owens 2011) and so is likely to be linked to reports of climate effects associated with geomagnetic activity (e.g., Seppälä et al. 2009; Bochníček and Hejda 2005; Thejll et al. 2003).

A great many studies have indicated that solar variations had an effect on pre-industrial climate throughout the Holocene (the warm interval since the last ice age). These studies have been carried out in many parts of the world and employ a huge variety of paleoclimate proxies. They will not be discussed here, and the reader is referred to the recent review by Lockwood (2012). The most interesting of these studies employed terrestrial data that are indicators of more than just local climate and/or where matching signals are seen at geographically separated sites. Some studies have used proxy data from very large numbers of sites to reconstruct temperature maps, cross-calibrating with early observational data (e.g., Mann et al. 2009). Nevertheless, a key point to note is that most paleoclimate evidence is local, or at best regional, and not global in scope.

Section 4 discussed how solar influence on climate on a global scale has been relatively minor and how studies that arrive at a different conclusion have been flawed and not robust. Section 6 reviews the growing evidence in modern data for solar influences on some regional and seasonal climates.

6 Regional Climate Effects

Using both a climate model and empirical reconstructions, Shindell et al. (2001a, b) found that even though solar-induced GMAST changes are small (below about 0.3°C), regional temperature changes associated with solar variability can be quite large. In their model, these occur primarily through a forced shift towards the low index state of the North Atlantic oscillation (NAO) as solar irradiance decreases. This leads to lower temperatures over the northern hemisphere continents, especially in winter (1–2°C), in agreement with historical records and proxy paleoclimate data for surface temperatures.

For post-industrial measurements, North and Stevens (1998) applied optimal signal detection theory to data from 36 regions around globe but found the solar signal was small and not highly significant. On the other hand, multivariate fits by Lean and Rind (2008) and Lean (2010) do find strong solar responses in certain regions. Figure 4 shows the multiple regression maps to modern AST (Air Surface Temperature) data by Lean (2010). These are derived in a similar way, and using the same inputs, as the fits to global means of AST (i.e. the GMAST) shown in Figs. 1 and 2. The maps in Fig. 4 are for annual mean data. The top panel shows that the response to a strong El-Niño event is, obviously, dominated by the equatorial Pacific, but there is also strong North Pacific cooling and continental North American warming. The second panel shows that a large equatorial volcanic eruption produces widespread cooling from 40°S to 70°N, especially in North America and the North Atlantic Ocean, but strengthened westerly winds, and a more positive phase of the NAO produces warming in the northern Eurasian continent. The bottom panel shows that the response to anthropogenic effects is widespread and particularly strong in the northern hemisphere at middle to high latitudes, as predicted by GCMs (e.g., Stott et al. 2000). The third panel shows that the response to solar variability is strongest at middle latitudes (near 40°) in both the northern and southern hemispheres, in the vicinity of the interface of the Hadley and Polar cells (Gleisner and Thejll 2003; Haigh 2003). We would expect climate to respond to solar variability with distinct seasonality because the lower latitudinal gradient of solar energy in summer means that dynamical processes are then not as strong. Remember that, as for all such multiple regression fits, the procedure may have overfitted decadal-scale internal oscillations with the input solar variation and so may have overestimated the response. The most noticeable feature is that the northern hemisphere response is much greater than the southern, with a strong longitudinal variation such that the biggest effect is over Europe and western Asia (referred to hereafter as Eurasia). Statistically significant signatures of multidecadal solar activity changes in atmospheric temperatures in Eurasia have recently been reported by Kossobokov et al. (2010) and Le Mouël et al. (2008, 2009, 2010) but Yiou et al. (2010) and Legras et al. (2010) argue that they do not survive proper null-hypothesis testing.

Climate fluctuations for different regions tend to cancel, giving global means that are much less variable than regional ones and which more clearly reveal trends. Figure 5 stresses how different the behaviour of a regional/seasonal climate in Europe can be from the global or hemispheric means. The plot uses winter (December–January–February, DJF) means, T_{DJF}, of the Central England Temperatures (CET) which is the world's longest instrumental record and extends back to 1659, at the start of the Maunder Minimum (Manley 1974; Parker et al. 1992). The CET measurements cover a spatial scale of order

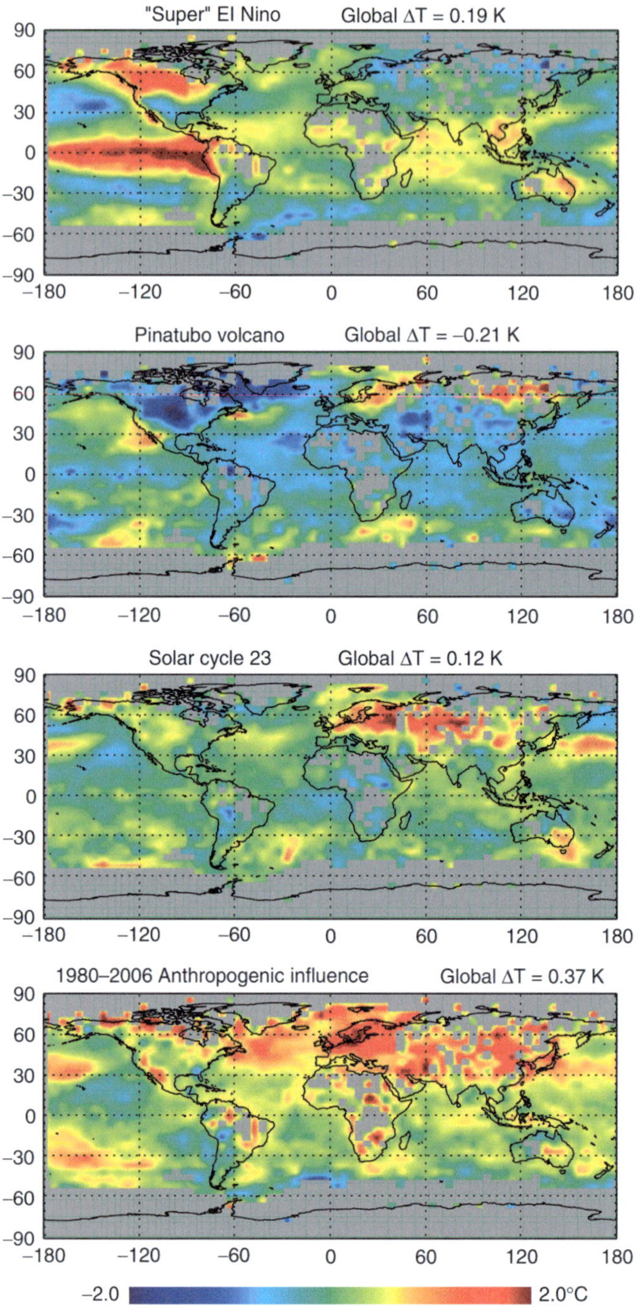

Fig. 4 Maps of the global responses to given features in the input data series in Fig. 2 from multiple regression analysis, which give a GMAST contribution of ΔT which is given at the *top* of *each panel*: (from *top* to *bottom*) **a** the "super-El-Niño" event of 1998, **b** the Pinatubo volcanic eruption, **c** solar cycle 23 and **d** the net anthropogenic forcing over the period 1980–2006. These are very similar in form to the maps of the regression coefficients of GMAST presented by Lean and Rind (2008) for the annual time series data shown in Fig. 2 (after Lean 2010)

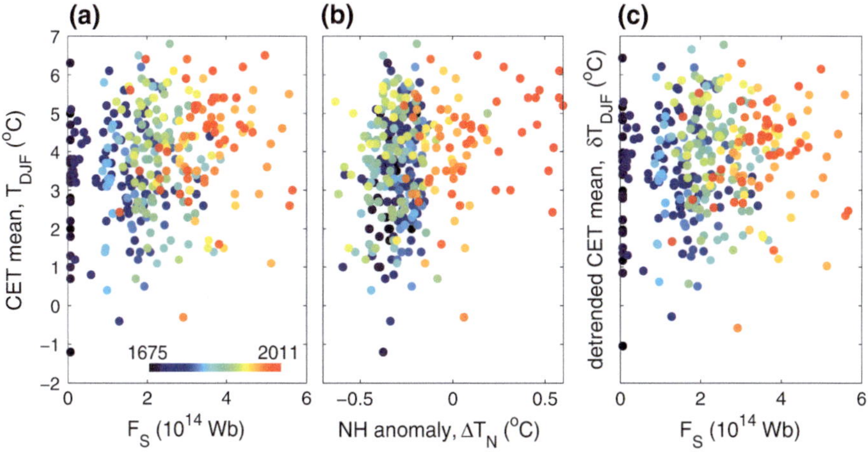

Fig. 5 Scatter plots of mean winter Central England temperature (CET), T_{DJF}, as a function of **a** open solar flux, F_S and **b** the hemispheric mean surface temperature anomaly ΔT_N (the HADCRUT3v reconstruction is here extended to before 1850 using the median of a basket of proxy paleoclimate reconstructions). Points are *coloured* according to the date using the scale shown in **a**. The *line* in **b** is the ordinary least squares linear regression fit, and the *grey area* bounds the uncertainty range set by the range of the proxy reconstructions for ΔT_N prior to 1850. The correlation coefficients for **a** and **b** are low at 0.23 and 0.25 (significant at the 99.1 and 99.6% levels using the AR (1) noise model). In the case of **b**, this demonstrates the great variability of a regional/seasonal temperature around a global-scale mean. **c** is the same as **a**, for the detrended winter CET data δT_{DJF}, where the detrending allows for the hemispheric change ΔT_N and uses the regression line shown in **b**. **c** shows that, even after allowing for the hemispheric trend, most of the coldest winters in central England occur at lower solar activity, as quantified by F_S. Lockwood et al. (2010a) test this against the null hypothesis that F_S has no effect on winter CET values and show that this tendency is statistically significant at around the 95% level. Several tests were carried out on the differences between distributions of F_S above and below various thresholds in δT_{DJF}. These were then repeated for the differences between distributions of δT_{DJF} above and below various thresholds in F_S

300 km which makes it a "small regional" climate indicator. In winter, the NAO, and associated changes in thermal advection, contribute to a large fraction of the observed variability of CET (Karoly and Stott 2006). Comparison of parts (a) with (b) of Fig. 5 shows that T_{DJF} correlates with both open solar flux F_S and the hemispheric mean surface temperature ΔT_N. However, the correlation is very weak in both cases. Lockwood et al. (2010a) show that there is a subtle solar influence on winter CET in that it is statistically significant that more cold winters, relative to the hemispheric trend, occur when the open solar flux is low (i.e. when solar irradiances are thought to be low and GCR fluxes are high). This can be seen by the lack of points in the bottom-right corner of Fig. 5c which was found to be significant at more than the 95% level using 6 different tests against the null hypothesis. Figure 6 shows that both surface temperature and surface pressure in winter (DJF means) in ERA-40+ re-analysis data for 1957–2001 show a clear influence of F_S (Woollings et al. 2010b). Figure 6a shows that difference in the means for low solar activity (lower tercile of the F_S distribution) minus that for high solar activity (the upper tercile of that distribution). Temperatures are lower in Eurasia by up to 4°K, but note that they are higher in Greenland by up to 3°K. Figure 6b shows the corresponding plot, sorted using the NAO index. In fact, the temperature drops in central Europe are not as large for the NAO sort, being up to about 2°K. The significances of the differences exceed 95% over a more extensive area in Eurasia for the solar than the NAO pattern. It is also interesting to

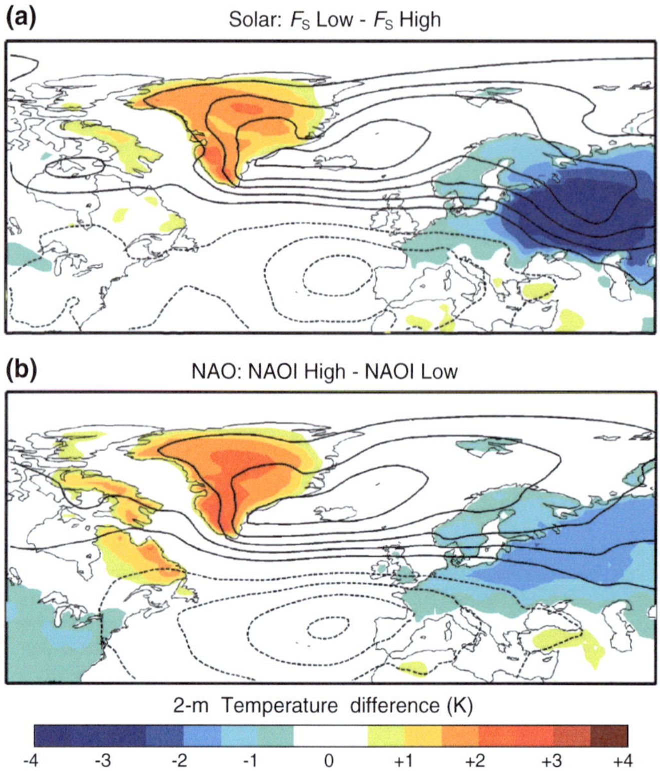

Fig. 6 **a** Composite difference maps of winter (DJF) means of mean surface level pressure (MSLP, contours 1 hPa apart) and 2-m temperature (*colour map*) between low- and high-solar conditions defined by the lower and upper terciles of the open solar flux, F_S. **b** Scaled anomalies associated with the NAO for comparison, in this case the two subsets are for the lower and upper terciles of the NAO index. Data are from the European Centre for Medium-range Weather Forecasting ERA-40 reanalysis data set (Uppala et al. 2005) extended to cover 44 complete winters (1957/1958 to 2000/2001) using operational data (from Woollings et al. 2010b). A bootstrap test was employed, resampling the sets of winters in order to estimate the sampling uncertainty, to test whether the solar pattern in **a** was different from that associated with the NAO in **b**: this found that the two patterns are significantly different at the 93% level

note that the perturbation circulation pattern for the solar effect is similar to, but still significantly (>93%) different from, that for the NAO, reaching further to the East.

Early instrumental records from the Maunder Minimum indicate an increased frequency of easterly winds influencing the UK winter temperatures during exceptionally cold winters (Slonosky et al. 2001). This is consistent with the circulation perturbation in Fig. 6 and has also been deduced from indirect proxies (Luterbacher et al. 2001), including the spatial patterns of changes in recorded harvest dates (Wanner et al. 1995). This strongly suggests a link with the incidence of long-lived winter "blocking" events in the eastern Atlantic at low solar activity. Blocking episodes comprise extensive and quasi-stationary anticyclones which can persist for several weeks, leading to extended cold periods in winter as the mild maritime westerly winds are replaced by continental north-easterlies and the land surface cools under cloudless skies. These changes are associated with a meander in the jet stream in the upper troposphere and lower stratosphere, such that its usual eastward (westerly) flow becomes south-westward around the longitude of the event centre (Woollings et al. 2010a). Long-lived Atlantic blocking events at more eastward locations have been found to be more prevalent at sunspot minimum than at higher solar activity, and this leads to colder winters in Europe (Barriopedro et al. 2008). Note, however, that to the west of the event centre warm air is drawn up from the south-west and hence if the blocking event centre moves exceptionally far to the east temperatures in western Europe may rise rather than fall: the region where CET is measured is quite close to the average longitude of blocking event centres and, although it shows a net cooling effect for lower solar activity, this is much smaller than the effect found further to the east. This evidence suggests that solar-induced perturbations modulate the North Atlantic jet stream and hence the occurrence of blocking. Other evidence supports this idea. For example, the effects of the changed position and frequency of blocking events may be seen as a manifestation of modes of low-frequency circulation variability which have been found to respond to solar activity giving increased/decreased frequencies of easterly/westerly circulation patterns over Europe under low-solar activity conditions (Huth et al. 2006, 2008). Figure 7 (from Woollings et al. 2010b) studies changes in the occurrence of DJF blocking events, using the means of the blocking index of Berrisford et al. (2007) from the same data set as used in Fig. 6. The anomaly, compared to the overall mean, for the high (upper tercile) F_S subset is shown on the left and for low (lower tercile) F_S subset the right. For the high F_S winters, the blocking occurrence is up to 6% lower than the average, in a restricted region in the eastern North Atlantic and over North-Western Europe. On the other hand, for the low F_S winters, the same region shows the blocking occurrence is up to 8% higher. Monte-Carlo tests show these changes are significant at the 99% level. This modulation of the occurrence of winter blocking events is highly significant and provides a good explanation of the modulation of the surface temperatures and pressures seen in Fig. 6.

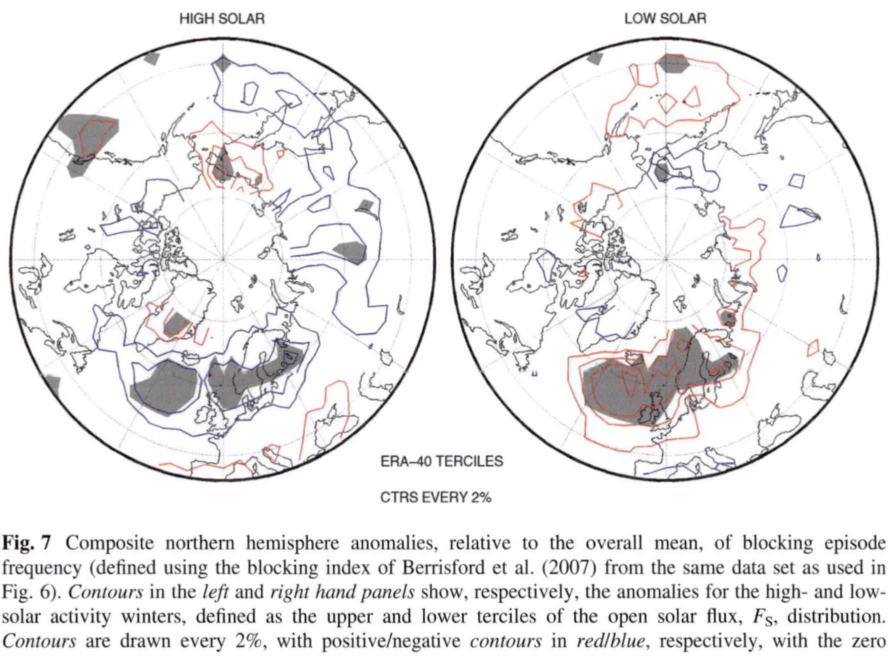

Fig. 7 Composite northern hemisphere anomalies, relative to the overall mean, of blocking episode frequency (defined using the blocking index of Berrisford et al. (2007) from the same data set as used in Fig. 6). *Contours* in the *left* and *right hand panels* show, respectively, the anomalies for the high- and low-solar activity winters, defined as the upper and lower terciles of the open solar flux, F_S, distribution. *Contours* are drawn every 2%, with positive/negative *contours* in *red/blue*, respectively, with the zero contour omitted. The *grey area* shows where the anomalies are significant at the 99% level, defined using a Monte-Carlo technique (from Woollings et al. 2010b)

Winter CET values are known to be strongly modulated by the NAO (Karoly and Stott 2006), and modelling has shown that stratospheric trends over recent decades, along with downward links to surface, are indeed strong enough to explain much of the prominent trend in the NAO and hence regional winter climate in Europe between the 1960s and the 1990s (Scaife et al. 2005). Furthermore, tropospheric blocking episodes are known to influence the stratosphere, and there have been strong suggestions that the coupling works both ways, such that they are also influenced by the stratosphere (Woollings et al. 2010a). In addition, various empirical studies have reported solar modulation of the NAO (Ruzmaikin and Feynman 2002; Thejll et al. 2003). The top-down mechanisms implied are discussed further in Sect. 7 and are likely to be connected to the centennial scale variations and phenomena seen in Europe during the Maunder Minimum. For example, the GCM runs experiments by Rind et al. (2008) produce a relative negative phase of the NAO during the Maunder Minimum which contributes to predicted lower European winter temperatures at that time.

Mann et al. (2009) studied regional climate on a global basis using paleoclimate data from tree trunks, ice sheets, corals and sediments from over 1,000 sites. They searched for solar influence on pre-industrial climate by comparing the difference between the "little ice age" pattern (which they define as 1400–1700 AD) and that for the "mediaeval maximum" (which they take to be 950–1250 AD): They find it shows a pattern that has some similarities to Fig. 4c. The reconstruction by Steinhilber et al. (2009) of TSI during the Holocene from cosmogenic isotopes gives a difference of 0.5 W m^{-2} between the means for these two intervals. However, the temporal variations deduced by Mann et al. do not agree particularly well with the Steinhilber et al. TSI reconstruction. For example, they do not show any feature at the time of the solar Maunder Minimum (circa 1645–1715 AD). Mann et al. argue that the spatial pattern is consistent with a relative positive/negative NAO-AO atmospheric circulation anomaly during the mediaeval maximum/little ice age, respectively, associated with annular bands of sea-level pressure anomalies that are positive/negative in the subtropics and mid-latitudes, and negative/positive at subpolar latitudes. This has been seen in model simulations as a dynamical response to imposed solar irradiance effects on ozone photochemistry (Shindell et al. 2001a, b).

Detection–attribution techniques have been applied on a regional and seasonal basis. Studies to date have generally found the solar contributions to be smaller than those of volcanoes and sulphate aerosols as well as smaller than the effects of greenhouse gas changes and land-use change. Hegerl et al. (2011) used a multifingerprint technique to detect a solar response in Europe, but in summer temperatures rather than the winter response discussed above. This was not robust to all tests and so remains speculative. It will be interesting to see whether these results change as models with better representation of the stratosphere are employed and when better characterisation of spectral solar irradiance becomes possible.

7 "Top-down" Solar Forcing

The dynamical coupling across the tropopause (Shepherd 2002) means that stratospheric forcing can influence the underlying troposphere (e.g., Hartley et al. 1998; Song and Robinson 2004; Matthes et al. 2006; Haigh and Blackburn 2006), and under some circumstances, robust tropospheric responses are indeed predicted by models. Observational studies show that stratospheric perturbations (e.g., stratospheric warmings) typically last at least a month, and on average, tropospheric effects are seen approximately 2 weeks later (e.g., Baldwin and Dunkerton 2001). Note, however, that the strong, two-way coupling across the tropopause means that the appearance of the change first in the stratosphere is not necessarily an indication that the stratospheric changes are driving those in the troposphere. The latitudes of tropospheric jet streams have been shown to be sensitive to the solar forcing of stratospheric temperatures (Haigh 1994, 2003; Poore et al. 2004). This could occur through disturbances to the stratospheric polar vortex (Gray et al. 2004), which can propagate downwards to affect the tropospheric jets (Plumb and Semeniuk 2003) or through the effects of stratospheric temperature changes on stratospheric winds and the refraction of tropospheric eddies (Kushner and Polvani 2004; Simpson et al. 2009). Models (e.g., Shindell et al. 2001a, b) predict that perturbations can descend from the stratosphere to the surface by altering the propagation of planetary waves coming up from the surface, an effect that has been observed in data (Perlwitz and Harnik 2003). However, models that do not yield realistic stratospheric dynamics fail to capture these wave flux changes. The tropospheric response is reduced when planetary waves are suppressed in the stratosphere by additional damping or when the strength of the stratospheric jet is increased. The models show that the stratosphere could play a crucial role in recent trends. For example, Scaife et al. (2005) have demonstrated that stratospheric trends over recent decades, along with downward links to surface, are indeed strong enough to explain much of the prominent trend in the NAO and hence regional winter climate in Europe between the 1960s and the 1990s.

A wide variety of tropospheric responses has been suggested. Recent numerical modelling by Ineson et al. (2011) has simulated the effect of the large decline in the stratosphere-effective UV band inferred from the recent SIM/SORCE measurements by Harder et al. (2009) and Haigh et al. (2010). As discussed in Sect. 5, there is debate as to whether these observations exaggerate the variability of the UV irradiance but, on the other hand, the modelling maintains constant ozone concentrations, and this will underestimate the effect. The modelled response to the decreasing solar UV begins in the upper stratosphere and lower mesosphere, consistent with ERA-40+ reanalysis data. This signal is largest in the tropics (Frame and Gray 2010; Lockwood et al. 2010b) and so gives a relative decrease in the pole-to-equator temperature gradient for lower solar activity. This response is reproduced in the model and require a weak easterly wind anomaly in the subtropical zonal mean circulation in the upper stratosphere. As predicted and inferred by Kodera (2002) and Kodera and Kuroda (2005), this anomaly amplifies as it propagates polewards and downwards during autumn and winter, giving a weaker polar vortex in December–January at solar minimum. Ineston et al. use the modelled Eliassen-Palm (EP) flux divergence to show that this downward propagation and amplification of the wind anomaly is associated with altered planetary wave activity.

The number and fluence of solar energetic particle events (SEPs) also vary with the level of solar activity quantified by F_S (Barnard et al. 2011). SEPs are directed towards Earth poles by the geomagnetic field where they generate nitrates on hitting Earth's atmosphere as demonstrated, for example, in observations by Mlynczak et al. (2003). These nitrates are seen to migrate downwards over time (Siskind et al. 2000; Funke et al. 2005), resulting in catalytic destruction of stratospheric ozone (Callis et al. 2000). Large SEP events generate considerable transient polar ozone depletions, which persist in winter until UV irradiance of the polar stratosphere resumes (Jackman et al. 2006, 2008; Randall et al. 2005, 2007). Recently, Seppälä et al. (2009) reported changes in polar surface temperatures which they associate with large SEP events. It should be noted that they used geomagnetic activity indices as indicators of the probability of SEP fluence, and these correlate highly with the open solar flux F_S. Thus, it is likely they are observing the same phenomenon as reported by Lockwood et al. (2010a), Woollings et al. (2010b) and Bochníček and Hejda (2005) but are using an interpretation based on the model predictions of SEP effects by Rozanov et al. (2005, 2012). Similarly, the effect reported by Lu et al. (2008) which they attributed to the solar wind could well be via a correlation with open solar flux. In this context, it is interesting to note that the effects of SEPs may also be caused by other precipitating energetic particles, including GCRs or energetic electrons energised by the solar wind (see review by Rozanov et al. 2012, and references therein). Chemical and dynamical modelling of the effects of energetic particles has yielded a surface temperature perturbation pattern similar to that shown in Fig. 6 (Rozanov et al. 2012).

8 Cosmic Ray Modulation of Clouds

A long-standing suggestion is that cosmic rays can aid the formation of clouds (see reviews by Ney 1959; Carslaw et al. 2002; Kirkby 2007). This would be a highly significant effect because clouds modulate Earth's energy balance by changing the albedo A and by enhancing the infrared greenhouse trapping effect. The balance between these two competing effects is dependent on the cloud altitude: For low clouds the albedo effect dominates (such that more cloud is a cooling effect), whereas for high-altitude clouds the greenhouse trapping effect dominates (so that more cloud gives warming). A great many papers have been written about low-altitude global cloud cover arguing that it shows a solar-cycle variation, consistent with GCR modulation of the growth of cloud condensation nuclei (CCN) (Svensmark and Friis-Christensen 1997; Svensmark 2007; Marsh and Svensmark 2000, 2003, 2004). These studies have been based on the available satellite data after 1983, which give global coverage, and so still cover something less than three solar cycles. Other studies find no significant effect (Sloan and Wolfendale 2008; Kernthaler et al. 1999; Kristjánsson and Kristiansen 2000; Sun and Bradley 2002, 2004; Kulmala et al. 2010). A third set of studies find significant correlations but only in restricted regions (Erlykin et al. 2010; Kristjánsson et al. 2008), whereas a fourth set report a solar-cycle variation but find it to be more consistent with solar UV irradiance variability (de Jager and Usoskin 2006; Erlykin et al. 2010; Udelhofen and Cess 2001). Voiculescu et al. (2006) argue that both UV and GCRs have an effect. Harrison (2008) detected a weak 1.68-year oscillation in the long (1947–1990) diffuse fraction data series from the Lerwick (maritime) station—an oscillation found in the open solar flux and GCR fluxes but not in the SSI or TSI variations (Rouillard and Lockwood 2004). Hence, this supports a direct effect of GCRs rather than irradiance changes. It is expected that any effect would be more significant in clean maritime air where there is a shortage of CCNs for water vapour to condense on, and air ions generated by GCRs may be more significant factor (Kazil et al. 2006; Laken et al. 2009).

The arguments against such direct cosmic ray-cloud connection have been: 1. Given the atmospheric supersaturations, there is no established mechanism that can cause the effect. 2. The inter-calibrations involved in the composite cloud-cover data set used render it unsuitable for this type of long-term variability analysis (e.g., Klein and Hartmann 1993). 3. The data sequences are too short and so the significance of the correlations is low. (For example, Sun and Bradley (2002, 2004) show that the uniquely high cosmic ray-cloud-cover correlation in the period 1983–1999 over the Atlantic Ocean is greatly weakened when an extended data set is used.) 4. Periods of low geomagnetic field, particularly the Laschamp event (Nöel and Tarling, 1975) which at 40 kyr ago is the most recent and well studied, gave enhanced GCR fluxes in the Earth's atmosphere but did not influence climate in the Greenland area (Beer 2006). Of these objections, only 4 argues that such a mechanism is not effective: 1–3 all argue that the evidence in its favour is inadequate. In relation to 4, there is some evidence that the failure to see an effect of the Laschamp event (Wagner et al. 2001) may have been a local characteristic of the climate in the Greenland area (Christla et al. 2004). Other studies give indications of an influence of the geomagnetic field on pre-industrial climate. Recent reports of associations between the main geomagnetic field and climatic changes over the last three millennia (Haug et al. 2001; Gallet and Genevey 2007 Gallet et al. 2005, 2006; Courtillot et al. 2007) would be significant, if they survived robust null-hypothesis testing, because they would be via GCR effects and unrelated to solar irradiance changes. These changes have also been linked to latitudinal motions of the inter-tropical convergence zone (Haug et al. 2001). Harrison and Stephenson (2006) found evidence for a small GCR effect on diffuse fraction measurements at a variety of maritime UK meteorological stations (for 1951–2000) and Harrison et al. (2011) have found the base of some type of clouds is lowered by higher cosmic ray fluxes. As these are homogeneous sequences of simple measurements, this does counter objection (2) to some extent; however, the effect was small.

In addition to the solar-cycle variation in GCRs, there are sudden decreases in GCR fluxes over periods of a few days called Forbush decreases. These have been reported as giving responses in cloud data, both ground-based (Veretenenko and Pudovkin 1997) and from satellites (Svensmark et al. 2009), in "compositing" analyses (also called "superposed epoch" and "Chree" analysis). The problem with these studies (particularly if a satellite data set is employed) is that there are very few large Forbush decreases, and the results appear to be dominated by one or two events in the intervals of available satellite data. This means that statistical significance is very low. The long delay of the putative effect is also a major puzzle. Brown (2008) reports that 3% of cloud variability on short timescales could be associated with GCRs, whereas Calogovic et al. (2010) find no response at all. The potential importance of this effect is that there are long-term changes in the open solar flux F_S (Lockwood et al. 2009; Lockwood and Owens 2011) that yields the long-term changes in GCRs (McCracken and Beer 2007; Muscheler et al. 2007; Steinhilber et al. 2008) which could introduce long-term changes in climate if the mechanism was to be sufficiently effective.

The best chance of answering objection (1) is the CLOUD experiment at CERN (Duplissy et al. 2010), which has just begun to study the "ion-aerosol clear-air" mechanism. The first results have just been published (Kirkby et al. 2011). The results have some fascinating new insights into effects in the boundary layer, but do not say much about the growth of water droplets in low-altitude cloud in the troposphere. Through the influence of sulphuric acid, ionisation was shown to enhance the rate of water droplet growth at the temperatures characteristic of middle troposphere. However, the ion-induced aerosols only grew to about 2 nm, and to grow into droplets large enough to influence incoming or outgoing radiation to Earth, would be far too slow because there is not enough sulphuric acid in the atmosphere (Pierce and Adams 2009). As yet, the experiment is not reproducing the true mix of constituents in the mid-troposphere, so the results are highly provisional. Other proposed mechanisms involving the atmospheric electric field have been proposed (Harrison 2000) but have not yet been investigated.

The cloud-cosmic ray suggestion increasingly fails to match observations. The correlation between the solar-cycle variation and global cloud-cover estimates has degraded rapidly in the more recent data (Gray et al. 2010). For each piece of evidence in its favour, an equal or greater number of studies fail to find the effect. Only studies finding rather subtle effects on cloud edges survive proper null-hypothesis testing. Lastly, we note that GCR-cloud effects would have shown up in the multivariate fits of GMAST shown in Figs. 1 and 2, which found only a weak solar response (indeed in these fits, the solar input was even quantified using GCR fluxes). In addition, even if the GCR-cloud mechanism was responsible for the effect that was seen, the global regression map (Fig. 4c) generates a major issue. Larger GCR fluxes would have been expected near the poles, where geomagnetic shielding of GCRs is weaker and with no longitudinal effects (with the possible exception of in the south Atlantic geomagnetic anomaly). Hence, additional effects, such as spatial variations in aerosols, would have to be invoked to explain maps such as shown in Fig. 4c.

9 The Future

We have no predictive models of the solar dynamo (see review by Weiss and Thompson 2009) and so can only make empirical analogue forecasts of future solar output changes. From solar-induced variations of cosmogenic isotopes over the past 10^4 years, Lockwood (2010) and Barnard et al. (2011) have deduced that there is an 8% chance that the Sun will return to Maunder Minimum conditions within 50 years. The recent evolution of solar cycle 24 indicates that the Sun may well be following such a trajectory (Owens et al. 2011). Feulner and Rahmstorf (2010) and Jones et al. (2012) have used GCMs and EBMs to predict that this will offset anthropogenically rising global temperatures by no more than about 0.2°C in the year 2100, relative to what would happen if the solar output remained constant. Similarly, Lean and Rind (2009) find that the solar decline would delay the arrival at a given temperature level by no more than about 5 years. Thus, these predictions show that continued solar decline will do little to alleviate anthropogenically driven global warming. However, the decline should do much to end the debate about the fraction of global warming that can be attributed to solar change. For the first time since about 1900, long-term solar and anthropogenic trends are now in opposite directions. Non-robust fits will fail sooner rather than later because of the change in solar behaviour. Thus, the next few years will give us much better estimates of the solar contribution to both global and regional climate change. For global temperature rise, there is every indication that these new estimates will, if anything, be smaller that previous estimates. On the other hand, there are indications that some regional climates will be more susceptible to solar changes (Lockwood et al. 2011a, b). Understanding spectral irradiance variability and differentiating "top-down" and "bottom-up" solar forcings will be needed as these will have very different effects on the spatial patterns of the responses and will behave differently in combination with other changes, such as sea ice loss. Key advances in this area will be made in modelling, with higher resolution in time and space, better understanding of numerical noise and of the relationship of natural climate variability to that in the model behaviour, fewer parameterisations, better definition of the inputs forcings and inclusion of more biological responses alongside the physical and chemical mechanisms.

References

Aldahan A, Hedfors J, Possnert G, Kulan A, Berggren A-M, Söderström C (2008) Atmospheric impact on beryllium isotopes as solar activity proxy. Geophys Res Lett 35:L21812. doi:10.1029/2008GL035189

Baldwin MP, Dunkerton TJ (2001) Stratospheric harbingers of anomalous weather regimes. Science 294:581–584. doi:10.1126/science.1063315

Baliunas S, Jastrow R (1990) Evidence for long-term brightness changes of solar-type stars. Nature 348:520–523. doi:10.1038/348520a0

Bard E, Raisbeck G, Yiou F, Jouzel J (1997) Solar modulation of cosmogenic nuclide production over the last millennium: comparison between ^{14}C and ^{10}Be records. Earth Planet Sci Lett 150(3/4):453–462. doi:10.1016/S0012-821X(97)00082-4

Bard E, Raisbeck G, Yiou F, Jouzel J (2000) Solar irradiance during the last 1200 years based on cosmogenic nuclides. Tellus 52B:985–992. doi:10.1034/j.1600-0889.2000.d01-7.x

Barnard L, Lockwood M, Hapgood MA, Owens MJ, Davis CJ, Steinhilber F (2011) Predicting space climate change. Geophys Res Lett 38:L16103. doi:10.1029/2011GL048489

Barriopedro D, Garcia-Herrera R, Huth R (2008) Solar modulation of Northern Hemisphere winter blocking. J Geophys Res 113:D14118. doi:10.1029/2008JD009789

Beer J (2006) Solar variability and climate change. Solar Variability Earth Clim 76:751–754

Benestad RE, Schmidt GA (2009) Solar trends and global warming. J Geophys Res 114:D14101. doi: 10.1029/2008JD011639

Berrisford P, Hoskins BJ, Tyrlis E (2007) Blocking and Rossby wave-breaking on the dynamical tropopause in the Southern Hemisphere. J Atmos Sci 64:2881–2898. doi:10.1175/JAS3984.1

Blaauw M, Heuvelink GBM, Mauquoy D, van der Plicht J, van Geel B (2003) A numerical approach to 14C wiggle-match dating of organic deposits: best fits and confidence intervals. Quat Sci Rev 22:1485–1500. doi:10.1016/S0277-3791(03)00086-6

Bochníček J, Hejda P (2005) The winter NAO pattern changes in association with solar and geomagnetic activity. J Atmos Sol-Terr Phys 67:17–32. doi: 10.1016/j.jastp.2004.07.014

Broccoli AJ et al (2003) Twentieth-century temperature and precipitation trends in ensemble climate simulations including natural and anthropogenic forcing. J Geophys Res 108(D24):4798. doi:10.1029/2003JD003812

Brohan P, Kennedy JJ, Haris I, Tett SFB, Jones PD (2006) Uncertainty estimates in regional and global observed temperature changes: a new dataset from 1850. J Geophys Res 111:D12106. doi:10.1029/2005JD006548

Brönnimann S, Ewen T, Griesser T, Jenne R (2007) Multidecadal signal of solar variability in the upper troposphere during the 20th century. Space Sci Rev 125:305–315. doi:10.1007/s11214-006-9065-2

Brown BH (2008) Short-term changes in global cloud cover and in cosmic radiation. J Atmos Sol-Terr Phys 70(7):1122–1131. doi:10.1016/j.jastp.2008.02.003

Callis LB, Natarajan M, Lambeth J (2000) Calculated upper stratospheric effects of solar UV flux and NO_y variations during the 11-year solar cycle. Geophys Res Lett 27:3869–3872. doi:10.1029/2000JD900117

Calogovic J, Albert C, Arnold F, Beer J, Desorgher L, Flueckiger EO (2010) Sudden cosmic ray decreases: no change of global cloud cover. Geophys Res Lett 37:L03802. doi:10.1029/2009GL041327

Camp CD, Tung KK (2007) Surface warming by the solar cycle as revealed by the composite mean difference projection. Geophys Res Lett 34:L14703. doi:10.1029/2007GL030207

Carslaw KS, Harrison RG, Kirkby J (2002) Cosmic rays, clouds and climate. Science 298:1732–1737. doi: 10.1126/science.1076964

Charlson RJ, Valero FPJ, Seinfeld JH (2005) In search of balance. Science 308(5723):806–807. doi: 10.1126/science.1108162

Christla M, Manginia A, Holzkäampera S, Spötlb C (2004) Evidence for a link between the flux of galactic cosmic rays and Earth's climate during the past 200 000 years. J Atmos Sol-Terr Phys 66:313–322. doi:10.1016/j.jastp.2003.12.004

Coughlin K (2004) Eleven-year solar cycle signal throughout the lower atmosphere. J Geophys Res 109:D21105. doi:10.1029/2004JD004873

Courtillot V, Gallet Y, Le Mouël J-L, Fluteau F, Genevey A (2007) Are there connections between the Earth's magnetic field and climate? Earth Planet Sci Lett 253:328–339. doi:10.1016/j.epsl.2006.10.032

Cowley SWH (1991) Acceleration and heating of space plasmas—basic concepts. Annales Geophys 9(3):176–187

Cubasch U et al (1997) Simulation of the influence of solar radiation variations on the global climate with an ocean-atmosphere general circulation model. Clim Dyn 13(11):757–767. doi:10.1007/s003820050196

Damon PE, Laut P (2004) Pattern of strange errors plagues solar activity and terrestrial climate data. EOS, Trans Am Geophys Union 85:370–374. doi:10.1029/2004EO390005

de Jager C (2008) Solar activity and its influence on climate. Netherlands J Geosci Geol Mijnbouw 87(3):207–213

de Jager C, Usoskin I (2006) On possible drivers of sun-induced climate changes. J Atmos Sol-Terr Phys 68:2053–2060. doi:10.1016/j.jastp.2006.05.019

DeLand MT, Cebula RP (2008) Creation of a composite solar ultraviolet irradiance data set. J Geophys Res 113:A11103. doi:10.1029/2008JA013401

Dickinson RE, Schaudt KJ (1998) Analysis of timescales of response of a simple climate model. J Clim 11:97–106. doi:10.1175/1520-0442(1998)011<0097:AOTORO>2.0.CO;2

Douglass DH, Clader BD (2002) Climate sensitivity of the Earth to solar irradiance. Geophys Res Lett 29(16). doi:10.1175/1520-0442(1998)011<0097:AOTORO>2.0.CO;2

Drijfhout SS, Haarsma RJ, Opsteegh JD, Selten FM (1999) Solar-induced versus internal variability in a coupled climate model. Geophys Res Lett 26:205–208. doi:10.1029/1998GL900277

Duplissy J, Enghoff MB, Aplin KL et al (2010) Results from the CERN pilot CLOUD experiment. Atmos Chem Phys 10(4):1635–1647. doi:10.5194/acp-10-1635-2010

Eddy JA (1976) The maunder minimum. Science 192(4245):1189–1202. doi:10.1126/science.192.4245.1189

Emile-Geay J, Cane M, Seager R, Kaplan A, Allmasi P (2007) El Niño as a mediator of the solar influence on climate. Paleoceanography 22:PA3210. doi:10.1029/2006PA001304

Erlykin AD, Sloan T, Wolfendale AW (2010) Correlations of clouds, cosmic rays and solar irradiation over the Earth. J Atmos Sol-Terr Phys 72(2/3):151–156. doi:10.1016/j.jastp.2009.11.002

Feulner G, Rahmstorf S (2010) On the effect of a new grand minimum of solar activity on the future climate on Earth. Geophys Res Lett 37:L05707. doi:10.1029/2010GL042710

Field CV, Schmidt GA, Koch D, Salyk C (2006) Modeling production and climate-related impacts on Be-10 concentrations in ice cores. J Geophys Res 111:D15107. doi:10.1029/2005JD006410

Forster P, Ramaswamy V, Artaxo P, Berntsen T, Betts R, Fahey DW, Haywood J, Lean J, Lowe DC, Myhre G, Nganga J, Prinn R, Raga G, Schulz M, Van Dorland R (2007) Changes in atmospheric constituents and in radiative forcing. In: Solomon S, Qin D, Manning M, Chen Z, Marquis M, Averyt KB, Tignor M, Miller HL (eds) Climate change 2007: the physical science basis. Contribution of working group I to the fourth assessment report of the intergovernmental panel on climate change. Cambridge University Press, Cambridge

Foukal P, Fröhlich C, Spruit H, Wigley TML (2006) Variations in solar luminosity and their effect on the Earth's climate. Nature 443:161–166. doi:10.1038/nature05072

Frame T, Gray LJ (2010) The 11-year solar cycle in ERA-40 data: an update to 2008. J Clim 23:2213–2222. doi:10.1175/2009JCLI3150.1

Fröhlich C (2009) Evidence of a long-term trend in total solar irradiance. Astron Astrophys 501(3):L27–L30. doi:10.1051/0004-6361/200912318

Fuller-Rowell T, Codrescu M, Maruyama N, Fredrizzi M, Araujo-Pradere E, Sazykin S, Bust G (2007) Observed and modeled thermosphere and ionosphere response to superstorms. Radio Sci 42(4): RS4S90. doi:10.1029/2005RS003392

Funke B, Lopez-Puertas M, Gil-Lopez S, von Clarmann T, Stiller G, Fischer H, Kellman S (2005) Downward transport of upper atmospheric NO_x into the polar stratosphere and lower mesosphere during the Antarctic 2003 and Arctic 2002/2003 winters. J Geophys Res 110:D24308. doi:10.1029/2005JD006463

Gallet Y, Genevey A (2007) The Mayans: climate determinism or geomagnetic determinism? EOS Trans Am Geophys Union 88:129–130. doi:10.1029/2007EO110001

Gallet Y, Genevey A, Fluteau F (2005) Does Earth's magnetic field secular variation control centennial climate change? Earth Planet Sci Lett 236:339–347. doi:10.1016/j.epsl.2005.04.045

Gallet Y, Genevey A, Le Goff M, Fluteau F, Eshraghi SA (2006) Possible impact of the Earth's magnetic field on the history of ancient civilizations. Earth Planet Sci Lett 246:17–26. doi:10.1016/j.epsl.2006.04.001

Giampapa M (2004), Stellar analogs of solar activity: the sun in a stellar context. In: Redi I, Güdel M, Schmutz W (eds) The sun, solar analogs and the climate. Springer, Berlin, pp 305–423, ISBN: 3-540-23856-5

Gleisner H, Thejll P (2003) Patterns of tropospheric response to solar variability. Geophys Res Lett 30:17129. doi:10.1029/2003GL017129

Gleisner H, Thejll P, Stendel M, Kaas E, Machenhauer B (2005) Solar signals in tropospheric re-analysis data: comparing NCEP/NCAR and ERA40. J Atmos Sol-Terr Phys 67(8–9):785–791. doi:10.1016/j.jastp.2005.02.001

Gray LJ, Crooks S, Pascoe C, Sparrow S, Palmer M (2004) Solar and QBO influences on the timing of stratospheric sudden warmings. J Atmos Sci 61:2777–2796. doi:10.1175/JAS-3297.1

Gray LJ, Beer J, Geller M, Haigh J, Lockwood M, Matthes K, Cubasch U, Fleitmann D, Harrison G, Hood L, Luterbacher J, Marsh N, Shindell D, van Geel B, White W (2010) Solar influences on climate. Rev Geophys 48:RG4001. doi:10.1029/2009RG000282

Haigh JD (1994) The role of stratospheric ozone in modulating the solar radiative forcing of climate. Nature 370:544–546. doi:10.1038/370544a0

Haigh JD (2003) The effects of solar variability on the earth's climate. Philos Trans R Soc Lond A 361(95):111. doi:10.1098/rsta.2002.1111

Haigh JD (2007) The sun and the earth's climate. Living Rev Solar Phys 4:2. URL (cited on 26 September 2009): http://www.livingreviews.org/lrsp-2007-2

Haigh JD, Blackburn M (2006) Solar influences on dynamical coupling between the stratosphere and troposphere. Space Sci Rev 125:331–344. doi:10.1007/s11214-006-9067-0

Haigh JD, Winning AR, Toumi R, Harder JW (2010) An influence of solar spectral variations on radiative forcing of climate. Nature 467(7316):696–699. doi:10.1038/nature09426

Hall JC, Lockwood GW (2004) The chromospheric activity and variability of cycling and flat activity solar-analog stars. Astrophys J 614(1):942–946. doi:10.1086/423926

Hansen J, Russell G, Lacis AA, Fung I, Rind D, Stone P (1985) Climate response times: dependence on climate sensitivity and ocean mixing. Science 229:857–859. doi:10.1126/science.229.4716.857

Harder JW, Fontenla JM, Pilewskie P, Richard EC, Woods TN (2009) Trends in solar spectral irradiance variability in the visible and infrared. Geophys Res Lett 36:L07801. doi:10.1029/2008GL036797

Harrison RG (2000) Cloud formation and the possible significance of charge for atmospheric condensation and ice nuclei. Space Sci Rev 94:381–396. doi:10.1023/A:1026708415235

Harrison RG (2008) Discrimination between cosmic ray and solar irradiance effects on clouds, and evidence for geophysical modulation of cloud thickness. Proc R Soc A 464(2098):2575–2590. doi:10.1098/rspa.2008.0081

Harrison RG, Stephenson DB (2006) Empirical evidence for a nonlinear effect of galactic cosmic rays on clouds. Proc R Soc A 462:1221–1233. doi:10.1098/rspa.2005.1628

Harrison RG, Ambaum MHP, Lockwood M (2011) Cloud base height and cosmic rays. Proc R Soc A 467:2777–2791. doi:10.1098/rspa.2011.0040

Hartley DE, Villarin JT, Black RX, Davis CA (1998) A new perspective on the dynamical link between the stratosphere and troposphere. Nature 391:471–474. doi:10.1038/35112

Haug G, Hughen K, Sigman D, Peterson L, Röhl U (2001) Southward migration of the intertropical convergence zone through the Holocene. Science 293:1304–1308. doi:10.1126/science.1059725

Hegerl GC, Zwiers FW, Braconnot P, Gillett NP, Luo Y, Marengo Orsini JA, Nicholls N, Penner JE, Stott PA (2007) Understanding and attributing climate change. In: Solomon S, Qin D, Manning M, Chen Z, Marquis M, Averyt KB, Tignor M, Miller HL (eds) Climate change 2007: the physical science basis. Contribution of working group I to the fourth assessment report of the intergovernmental panel on climate change. Cambridge University Press, Cambridge

Hegerl G, Luterbacher J, González-Rouco F, Tett SFB, Crowley T, Xoplaki E (2011) Influence of human and natural forcing on European seasonal temperatures. Nat Geosci 4:99–103. doi:10.1038/ngeo1057

Heikkila U, Beer J, Feichter J (2008) Modeling cosmogenic radionucleides Be-10 and Be-7 during the maunder minimum using the ECHAM5-HAM general circulation model. Atmos Chem Phys 8: 2797–2809. doi:10.5194/acp-8-2797-2008

Held IM, Winton M, Takahashi K, Delworth T, Zeng F, Vallis GK (2010) Probing the fast and slow components of global warming by returning abruptly to preindustrial forcing. J Clim 23(9):2418–2427. doi:10.1175/2009JCLI3466.1

Herschel W (1801) Observations tending to investigate the nature of the Sun, in order to find the causes or symptoms of its variable of light and heat; with remarks on the use that may possibly be drawn from solar observations. Philos Trans R Soc Lond 91:265–318

Huth R, Pokorná L, Bochníček J, Hejda P (2006) Solar cycle effects on modes of low-frequency circulation variability. J Geophys Res 111:D22107. doi:10.1029/2005JD006813

Huth R, Kyselý J, Bochníček J, Hejda P (2008) Solar activity affects the occurrence of synoptic types over Europe. Ann Geophys 26:1999–2004. doi:10.5194/angeo-26-1999-2008

Ineson S, Scaife AA, Knight JR, Manners JC, Dunstone NJ, Gray LJ, Haigh JD (2011) Solar forcing of winter climate variability in the northern hemisphere. Nature Geosci 4(11):753–757. doi:10.1038/NGEO1282

Ingram WJ (2006) Detection and attribution of climate change, and understanding solar influence on climate. Space Sci Rev 125(1–4):199–211. doi:10.1007/s11214-006-9057-2

Jackman CH, DeLand MT, Labow GJ, Fleming EI, Lopez-Puertas M (2006) Satellite measurements of middle atmospheric impacts by solar proton events in solar cycle 23. Space Sci Rev 125:381–391. doi:10.1007/s11214-006-9071-4

Jackman CH, Marsh DR, Vitt FM, Garcia RR, Fleming EL, Labow GJ, Randall CE, Lopez-Puertas M, Funke B, von Clarmann T, Stiller GP (2008) Short- and medium-term atmospheric constituent effects of very large solar proton events. Atmos Chem Phys 8:765–785. doi:10.5194/acp-8-765-2008

Jones GS, Tett SFB, Stott PA (2003) Causes of atmospheric temperature change 1960–2000: a combined attribution analysis. Geophys Res Lett 30:1228. doi:10.1029/2002GL016377

Jones GS, Lockwood M, Stott PA (2012) What influence will future solar activity changes over the 21st century have on projected global climate changes? J Geophys Res (in press). doi:10.1029/2011JD017013

Karoly DJ, Stott PA (2006) Anthropogenic warming of central England temperature. Atmos Sci Lett 7:81–85. doi:10.1002/asl.136

Kazil J, Lovejoy ER, Barth MC, O'Brien K (2006) Aerosol nucleation over oceans and the role of galactic cosmic rays. Atmos Chem Phys 6:4905–4924. doi:10.5194/acp-6-4905-2006

Kernthaler SC, Toumi R, Haigh JD (1999) Some doubts concerning a link between cosmic ray fluxes and global cloudiness. Geophys Res Lett 26(7):863–865. doi:10.1029/1999GL900121

Kilian MR, van der Plicht J, van Geel B (1995) Dating raised bogs: new aspects of AMS ^{14}C wiggle matching, a reservoir effect and climatic change. Quat Sci Rev 14:959–966. doi:10.1016/0277-3791(95)00081-X

Kirkby J (2007) Cosmic rays and climate. Surv Geophys 28(5/6):333–375. doi:10.1007/s10712-008-9030-6

Kirkby J, Curtius J, Almeida J et al (2011) Role of sulphuric acid, ammonia and galactic cosmic rays in atmospheric aerosol nucleation. Nature 476(7361):429–433. doi:10.1038/nature10343

Klein SA, Hartmann DL (1993) Spurious changes in the ISCCP dataset. Geophys Res Lett 20:455–458. doi:10.1029/93GL00211

Knutti R, Krähenmann S, Frame DJ, Allen MR (2008) Comment on ''Heat capacity, time constant, and sensitivity of Earth's climate system'' by S. E. Schwartz. J Geophys Res 113:D15103. doi:10.1029/2007JD009473

Kodera K (2002) Solar cycle modulation of the North Atlantic Oscillation: implications in the spatial structure of the NAO. Geophys Res Lett 29(8):1218. doi:10.1029/2001GL014557

Kodera K, Kuroda Y (2005) A possible mechanism of solar modulation of the spatial structure of the North Atlantic Oscillation. J Geophys Res 110:D02111. doi:10.1029/2004JD005258

Kossobokov V, Le Mouël J-L, Courtillot V (2010) A statistically significant signature of multidecadal solar activity changes in atmospheric temperatures at three European stations. J Atmos Sol-Terr Phys 72:595–606. doi:10.1016/j.jastp.2010.02.016

Kristjánsson JE, Kristiansen J (2000) Is there a cosmic ray signal in recent variations in global cloudiness and cloud radiative forcing? J Geophys Res 105(D9):11851–11863. doi:10.1029/2000JD900029

Kristjánsson JE, Stjern CW, Stordal F et al (2008) Cosmic rays, cloud condensation nuclei and clouds—a reassessment using MODIS data. Atmos Chem Phys 8(24):7373–7387. doi:10.5194/acp-8-7373-2008

Krivova NA, Solanki SK, Wenzler T, Podlipnik B (2009) Reconstruction of solar UV irradiance since 1974. J Geophys Res 114:D00I04. doi:10.1029/2009JD012375

Krivova NA, Solanki SK, Unruh YC (2011) Towards a long-term record of solar total and spectral irradiance. J Atmos Sol-Terr Phys 73:223–234. doi:10.1016/j.jastp.2009.11.013

Kulmala M, Riipinen I, Nieminen T, Hulkkonen M, Sogacheva L, Manninen HE, Paasonen P, Petäjä T, Dal Maso M, Aalto PP, Viljanen A, Usoskin I, Vainio R, Mirme S, Mirme A, Minikin A, Petzold A, Hõrrak U, Plaß-Dülmer C, Birmili W, Kerminen V-M (2010) Atmospheric data over a solar cycle: no connection between galactic cosmic rays and new particle formation. Atmos Chem Phys 10:1885–1898. doi:10.5194/acp-10-1885-2010

Kushner PJ, Polvani LM (2004) Stratosphere-troposphere coupling in a relatively simple AGCM: the role of eddies. J Clim 17:629–639. doi:10.1175/1520-0442(2004)017<0629:SCIARS>2.0.CO;2

Laken B, Wolfendale A, Kniveton D (2009) Cosmic ray decreases and changes in the liquid water cloud fraction over the oceans. Geophys Res Lett 36:L23803. doi:10.1029/2009GL040961

Le Mouël J-L, Courtillot V, Blanter E, Shnirman M (2008) Evidence for a solar signature in 20th-century temperature data from the USA and Europe. C R Geosci 340:421–430. doi:10.1016/j.crte.2008.06.001

Le Mouël J-L, Blanter E, Shnirman M, Courtillot V (2009) Evidence for solar forcing in variability of temperature and pressure in Europe. J Atmos Solar-Terr Phys 71:1309–1321. doi:10.1016/j.jastp.2009.05.006

Le Mouël J-L, Kossobokov V, Courtillot V (2010) A solar pattern in the longest temperature series from three stations in Europe. J Atmos Sol-Terr Phys 72:62–76. doi:10.1016/j.jastp.2009.10.009

Le H, Liu L, He H, Wan W (2011) Statistical analysis of solar EUV and X-ray flux enhancements induced by solar flares and its implication to upper atmosphere. J Geophys Res 116:A11301. doi:10.1029/2011JA016704
Lean J (2000) Evolution of the Sun's spectral irradiance since the Maunder Minimum. Geophys Res Lett 27:2425–2428. doi:10.1029/2000GL000043
Lean J (2006) Comment on "Estimated solar contribution to the global surface warming using the ACRIM TSI satellite composite" by N. Scafetta and B. J. West (2006). Geophys Res Lett 33:L15701. doi:10.1029/2005GL025342
Lean J (2010) Cycles and trends in solar irradiance and climate. Wiley Interdiscip Rev Clim Change 1(1):111–122. doi:10.1002/wcc.018
Lean JL, Rind DH (2008) How natural and anthropogenic influences alter global and regional surface temperatures: 1889 to 2006. Geophys Res Lett 35:L1870. doi:10.1029/2008GL034864
Lean JL, Rind DH (2009) How will Earth's surface temperature change in future decades? Geophys Res Lett 36:L15708. doi:10.1029/2009GL038932
Lean JL, Rottman GJ, Kyle HL, Woods TN, Hickey JR, Puga LC (1997) Detection and parameterization of variations in solar midand near-ultraviolet radiation (200–400 nm). J Geophys Res 102(29):939. doi:10.1029/97JD02092
Legras B, Mestre O, Bard E, Yiou P (2010) On misleading solar-climate relationship. Clim Past Discuss 6:767–800. doi:10.5194/cpd-6-767-2010
Lockwood M (2002) An evaluation of the correlation between open solar flux and total solar irradiance. Astron Astrophys 382:678–687. doi:10.1051/0004-6361:20011666
Lockwood M (2004) Solar Outputs, their variations and their effects of Earth. In: Haigh JD, Lockwood M, Giampapa MS, Rüedi I, Güdel M, Schmutz W (eds) The sun, solar analogs and the climate. Proceedings of Saas-Fee advanced course. Springer, Berlin, vol 34, pp 107–304, ISBN: 3-540-23856-5
Lockwood M (2006) What do cosmogenic isotopes tell us about past solar forcing of climate? Space Sci Rev 125:95–109. doi:10.1007/s11214-006-9049-2
Lockwood M (2008) Recent changes in solar outputs and the global mean surface temperature. III. Analysis of contributions to global mean air surface temperature rise. Proc R Soc A 464(2094):1387–1404. doi:10.1098/rspa.2007.0348
Lockwood M (2010) Solar change and climate: an update in the light of the current exceptional solar minimum. Proc R Soc A 466(2114):303–329. doi:10.1098/rspa.2009.0519
Lockwood M (2011a) Was UV spectral solar irradiance lower during the recent low sunspot minimum? J Geophys Res 116:D16103. doi:10.1029/2010JD014746
Lockwood M (2011b) Solar physics: shining a light on solar impacts. Nat Clim Change 1:98–99. doi:10.1038/nclimate1096
Lockwood M (2012) Solar induced climate effects. In: Robert M (ed) Encyclopaedia of sustainability science and technology. Springer, Berlin, ISBN 978-0-387-89469-0
Lockwood M, Fröhlich C (2007) Recent oppositely-directed trends in solar climate forcings and the global mean surface air temperature. Proc R Soc (Lond) A 463(2086):2447–2460. doi:10.1098/rspa2007.1880
Lockwood M, Fröhlich C (2008) Recent oppositely-directed trends in solar climate forcings and the global mean surface air temperature: II. Different reconstructions of the total solar irradiance variation and dependence on response timescale. Proc R Soc (Lond) 464(2094):1367–1385. doi:10.1098/rspa.2007.0347
Lockwood M, Owens MJ (2011) Centennial changes in the heliospheric magnetic field and open solar flux: the consensus view from geomagnetic data and cosmogenic isotopes and its implications. J Geophys Res 116:A04109. doi:10.1029/2010JA016220
Lockwood M, Stamper R (1999) Long-term drift of the coronal source magnetic flux and the total solar irradiance. Geophys Res Lett 26:2461–2464. doi:10.1029/1999GL900485
Lockwood M, Rouillard AP, Finch ID (2009) The rise and fall of open solar flux during the current grand solar maximum. Appl J 700:937–944. doi:10.1088/0004-637X/700/2/937
Lockwood M, Harrison RG, Woollings T, Solanki SK (2010a) Are cold winters in Europe associated with low solar activity? Environ Res Lett 5:024001. doi:10.1088/1748-9326/5/2/024001
Lockwood M, Bell C, Woollings T, Harrison RG, Gray LJ, Haigh JD (2010b) Top-down solar modulation of climate: evidence for centennial-scale change. Environ Res Lett 5:034008. doi:10.1088/1748-9326/5/3/034008
Lockwood M, Harrison RG, Owens MJ, Barnard L, Woollings T, Steinhilber F (2011a) The solar influence on the probability of relatively cold UK winters in the future. Environ Res Lett 6:034004. doi:10.1088/1748-9326/6/3/034004

Lockwood M, Owens MJ, Barnard L, Davis CJ, Steinhilber F (2011b) The persistence of solar activity indicators and the descent of the sun into maunder minimum conditions. Geophys Res Lett 38:L22105. doi:10.1029/2011GL049811

Loeb NG, Wielicki BA, Doelling DR, Kato S, Wong T, Smith GL, Keyes DF, Manalo-Smith N (2009) Toward optimal closure of the earth's top-of-atmosphere radiation budget. J Clim 22(3):748–766. doi: 10.1175/2008JCLI2637

Loukitcheva MA, Solanki SK, White SM (2009) The relationship between chromospheric emissions and magnetic field strength. Astron Astrophys 497(1):273–285. doi:10.1051/0004-6361/200811133

Love JJ, Mursula K, Tsai VC, Perkins DM (2011) Are secular correlations between sunspots, geomagnetic activity, and global temperature significant? Geophys Res Lett 38:L21703. doi: 10.1029/2011GL049380

Lu H, Jarvis MJ, Hibbins RE (2008) Possible solar wind effect on the northern annular mode and northern hemispheric circulation during winter and spring. J Geophys Res 113:D23104. doi: 10.1029/2008JD010848

Luterbacher J et al (2001) The late maunder minimum (1675–1715)—a key period for studying decadal scale climatic change in Europe. Clim Change 49:441–462. doi:10.1023/A:1010667524422

Manley G (1974) Central England temperatures: monthly means 1659 to 1973. Q J R Meteorol Soc 100:389–405. doi:10.1256/smsqj.42510

Mann ME, Zhang ZH, Rutherford S, Bradley RS, Hughes MK, Shindell D, Ammann C, Faluvegi G, Ni FB (2009) Global signatures and dynamical origins of the little ice age and medieval climate anomaly. Science 326(5957):1256–1260. doi:10.1126/science.1177303

Marsch E (2006) Solar wind responses to the solar activity cycle. Adv Space Res 38(5):921–930. doi: 10.1016/j.asr.2005.07.029

Marsh N, Svensmark H (2000) Cosmic rays, clouds, and climate. Space Sci Rev 94:215–230. doi: 10.1023/A:1026723423896

Marsh N, Svensmark H (2003), Galactic cosmic ray and El Nino Southern Oscillation trends in international satellite cloud climatology project D2 low-cloud properties. J Geophys Res 108(D6):4195. doi: 10.1029/2001JD001264

Marsh N, Svensmark H (2004) Comment on 'Solar influences on cosmic rays and cloud formation: a reassessment' by Bomin Sun and Raymond S. Bradley. J Geophys Res 109:D14205. doi:10.1029/2003JD004063

Matthes K, Kuroda Y, Kodera K, Langematz U (2006) Transfer of the solar signal from the stratosphere to the troposphere: Northern winter. J Geophys Res 111:D06108. doi:10.1029/2005JD006283

McCracken KG, Beer J (2007) Long-term changes in the cosmic ray intensity at Earth, 1428–2005. J Geophys Res 112:A10101. doi:10.1029/2006JA012117

Meehl GA, Washington WM, Wigley TML, Arblaster JM, Dai A (2003) Solar and greenhouse gas forcing and climate response in the twentieth century. J Clim 16:426–444. doi:10.1175/15200442

Meehl GA et al (2004) Combinations of natural and anthropogenic forcings in 20th century climate. J Clim 17:3721–3727. doi:10.1175/1520-0442(2004)017<3721:CONAAF>2.0.CO;2

Meehl GA, Arblaster JM, Branstator G, van Loon H (2008) A coupled air-sea response mechanism to solar forcing in the Pacific region. J Clim 21:2883–2897. doi:10.1175/2007JCLI1776.1

Meehl GA, Arblaster JM, Matthes K, Sassi F, van Loon H (2009) Amplifying the Pacific climate system response to a small 11-year solar cycle forcing. Science 325:1114–1118. doi:10.1126/science.1172872

Mlynczak M et al (2003) The natural thermostat of nitric oxide emission at 5.3 μm in the thermosphere observed during the solar storms of, April 2002. Geophys Res Lett 30:2100. doi:10.1029/2003GL017693

Moore J, Grinsted A, Jevrejeva S (2006) Is there evidence for sunspot forcing of climate at multi-year and decadal periods? Geophys Res Lett 33:L17705. doi:10.1029/2006GL026501

Muscheler R, Joos F, Beer J, Muller SA, Vonmoos M, Snowball I (2007) Solar activity during the last 1000 yr inferred from radionucleide records. Quat Sci Rev 26:82–97. doi:10.1016/j.quascirev.2006.07.012

Ney EP (1959) Cosmic radiation and the weather. Nature 183:451–452. doi:10.1038/183451a0

Nöel M, Tarling DH (1975) The Laschamp geomagnetic 'event'. Nature 253:705–707. doi:10.1038/253705a0

North GR, Stevens MJ (1998) Detecting climate signals in the surface temperature record. J Clim 11:563–577. doi:10.1175/1520-0442(1998)011<0563:DCSITS>2.0.CO;2

Owens MJ, Lockwood M, Barnard L, Davis CJ (2011) Solar cycle 24: implications for energetic particles and the probability of a new Maunder Minimum. Geophys Res Lett 38:L19106. doi:10.1029/2011GL049328

Pallé E, Goode PR, Montañés-Rodriguez P (2009) Interannual variations in Earth's reflectance 1999–2007. J Geophys Res 114:D00D03. doi:10.1029/2008JD010734
Parker DE, Legg TP, Folland CK (1992) A new daily Central England temperature series, 1772–1991. Int J Clim 12:317–342. doi:10.1002/joc.3370120402
Perlwitz J, Harnik N (2003) Observational evidence of a stratospheric influence on the troposphere by planetary wave reflection. J Clim 16:3011–3026. doi:10.1175/1520-0442(2003)016<3011:OEOASI>2.0.CO;2
Pierce JR, Adams PJ (2009) Can cosmic rays affect cloud condensation nuclei by altering new particle formation rates? Geophys Res Lett 36:L09820. doi:10.1029/2009GL037946
Plumb RA, Semeniuk K (2003) Downward propagation of extra-tropical zonal wind anomalies. J Geophys Res 108. doi:10.1029/2002JD002773
Poore RZ, Quinn TM, Verardo S (2004) Century-scale movement of the Atlantic intertropical convergence zone linked to solar variability. Geophys Res Lett 31:L12214. doi:10.1029/2004GL019940
Potgieter MS (2008) The modulation of galactic cosmic rays in the heliosphere: theory and models. Space Sci Rev 83(1–2):147–158. doi:10.1023/A:1005014722123
Qu WZ, Huang F, Du L, Zhao JP, Qin T, Cao Y (2011) The periodicity of volcano activity and its reflection in some climate factors. Chin J Geophys Chin Edn 54(3):643–655. doi:10.3969/j.issn.0001-5733.2011.03.003
Randall CE, Harvey VL, Manney GL, Orsolini Y, Codrescu M, Sioris C, Brohede S, Haley C, Gordley L, Zawodny J, Russell J III (2005) Stratospheric effects of energetic particle precipitation in 2003–2004. Geophys Res Lett 32:L05802. doi:10.1029/2004GL022003
Randall CE, Harvey V, Singleton CS, Bailey S, Bernath P, Codrescu M, Nakajima H, Russell JM III (2007) Energetic particle precipitation effects on the Southern Hemisphere stratosphere in 1992–2005. J Geophys Res 112:D08308. doi:10.1029/2006JD007696
Reid G (2000) Solar variability and the Earth's climate: introduction and overview. Space Sci Rev 94:1–11. doi:10.1023/A:1026797127105
Rind D (2002) The Sun's role in climate variations. Science 296:673–678. doi:10.1126/science.1069562
Rind D, Lean J, Healy R (1999) Simulated time-dependent climate response to solar radiative forcing since 1600. J Geophys Res 104:1973–1990. doi:10.1029/1998JD200020
Rind D, Lean J, Lerner J, Lonergan P, Leboissitier A (2008) Exploring the stratospheric/tropospheric response to solar forcing. J Geophys Res Atmos 113:D24103. doi:10.1029/2008JD010114
Rouillard AP, Lockwood M (2004) Oscillations in the open solar magnetic flux with period 1.68 years: imprint on galactic cosmic rays and implications for heliospheric shielding. Ann Geophys 22:4381–4396. doi:10.5194/angeo-22-4381-2004
Rozanov EV, Callis L, Schlesinger E, Yang F, Andronova N, Zubov VA (2005) Atmospheric response to NOy source due to energetic electron precipitation. J Geophys Res 32:L14811. doi:10.1029/2005GL023041
Rozanov E, Calisto M, Egorova T, Peter T, Schmutz W (2012) The influence of the precipitating energetic particles on atmospheric chemistry and climate. Surv Geophys. doi:10.1007/s10712-012-9192-0
Ruzmaikin A, Feynman J (2002) Solar influence on a major mode of atmospheric variability. J Geophys Res 107:4209. doi:10.1029/2001JD001239
Rycroft MJ, Harrison RG, Nicoll KA, Mareev EA (2008) An overview of earth's global electric circuit and atmospheric conductivity. Space Sci Rev 137(1–4):83–105. doi:10.1007/s11214-008-9368-6
Santer BD, Wigley TML, Doutriaux C, Boyle JS, Hansen JE, Jones PD, Meehl GA, Roeckner E, Sengupta S, Taylor KE (2001) Accounting for the effects of volcanoes and ENSO in comparisons of modelled and observed temperature trends. J Geophys Res 106:28033–28059. doi:10.1029/2000JD000189
Scafetta N (2009) Empirical analysis of the solar contribution to global mean air surface temperature change. J Atmos Sol Terr Phys 71(17–18):1916–1923. doi:10.1016/j.jastp.2009.07.007
Scafetta N, West BJ (2007) Phenomenological reconstructions of the solar signature in the Northern Hemisphere surface temperature records since 1600. J Geophys Res 112(D24):D24S03. doi:10.1029/2007JD008437
Scaife AA, Knight JR, Vallis GK, Folland CK (2005) A stratospheric influence on the winter NAO and North Atlantic surface climate. Geophys Res Lett 32:L18715. doi:10.1029/2005GL023226
Schmidt GA, Jungclaus JH, Ammann CM, Bard E, Braconnot P, Crowley TJ, Delaygue G, Joos F, Krivova NA, Muscheler R, Otto-Bliesner BL, Pongratz J, Shindell DT, Solanki SK, Steinhilber F, Vieira LEA (2011) Climate forcing reconstructions for use in PMIP simulations of the Last Millennium (v1.1). Geosci Model Dev Discuss 4:2451–2467, www.geosci-model-dev-discuss.net/4/2451/2011/. doi:10.5194/gmdd-4-2451-2011

Schrijver CJ, Livingston WC, Woods TN, Mewaldt RA (2011) The minimal solar activity in 2008–2009 and its implications for long-term climate modeling. Geophys Res Lett 38:L06701. doi:10.1029/2011GL046658

Schwenn R (2006) Space weather: the solar perspective. Living Rev Solar Phys 3(2). URL (cited on 17/1/2012): http://www.livingreviews.org/lrsp-2006-2

Seppälä A, Randall CE, Clilverd MA, Rozanov E, Rodger CJ (2009) Geomagnetic activity and polar surface air temperature variability. J Geophys Res 114:A10312. doi:10.1029/2008JA014029

Shapiro AI, Schmutz W, Rozanov E, Schoell M, Harberreiter M, Shapiro AV, Nyeki S (2011) A new approach to the long-term reconstruction of the solar irradiance leads to large historical solar forcing. Astron & Astrophys 529:A67. doi:10.1051/0004-6361/201016173

Shepherd TG (2002) Issues in stratosphere-troposphere coupling. J Met Soc Jpn 80:769–792. doi:10.2151/jmsj.80.769

Shindell DT, Schmidt GA, Miller RL, Rind D (2001a) Northern hemisphere winter climate response to greenhouse gas, ozone, solar, and volcanic forcing. J Geophys Res 106:7193–7210. doi:10.1029/2000JD900547

Shindell DT, Schmidt GA, Mann ME, Rind D, Waple A (2001b) Solar forcing of regional climate change during the maunder minimum. Science 294(5549):2149–2152. doi:10.1126/science.1064363

Shine KP, Fouquart Y, Ramaswamy V, Solomon S, Srinivasan J (1995) Radiative forcing. In: Houghton JT et al (eds) Climate change 1994. Cambridge University Press, Cambridge, Chapter 4, pp 163–203, ISBN 0 521 55055 6

Simpson IR, Blackburn M, Haigh JD (2009) The role of eddies in driving the tropospheric response to stratospheric heating perturbations. J Atmos Sci 66(5):1347–1365. doi:10.1175/2008JAS2758.1

Siskind DE, Nedoluha GE, Randall CE, Fromm M, Russell JM III (2000) An assessment of Southern Hemisphere stratospheric NOx enhancements due to transport from the upper atmosphere. Geophys Res Lett 27:329–332. doi:10.1029/1999GL010940

Sloan T, Wolfendale AW (2008) Testing the proposed causal link between cosmic rays and cloud cover. Environ Res Lett 3(2):024001. doi:10.1088/1748-9326/3/2/024001

Slonosky VC, Jones PD, Davies TD (2001) Instrumental pressure observations and atmospheric circulation from the 17th and 18th centuries. Lond Paris Int J Climatol 21:285–298. doi:10.1002/joc.611

Solanki SK, Krivova NA, Wenzler T (2005) Irradiance models. Adv Sp Res 35:376–383. doi:10.1016/j.asr.2004.12.077

Song YC, Robinson WA (2004) Dynamical mechanisms for stratospheric influences on the troposphere. J Atmos Sci 61:1711–1725. doi:10.1175/1520-0469(2004)061<1711:DMFSIO>2.0.CO;2

Steinhilber F, Abreu JA, Beer J (2008) Solar modulation during the Holocene. Astrophys Space Sci Trans 4(1–6):2008. doi:10.5194/astra-4-1-2008

Steinhilber F, Beer J, Fröhlich C (2009) Total solar irradiance during the Holocene. Geophys Res Lett 36:L19704. doi:10.1029/2009GL040142

Stendel M, Mogensen IA, Christensen JH (2006) Influence of various forcings on global climate in historical times using a coupled atmosphere-ocean general circulation model. Clim Dyn 26(1):1–15. doi:10.1007/s00382-005-0041-4

Stone DA, Allen MR, Stott PA (2007a) A multi-model update on the detection and attribution of global surface warming. J Clim 20:517–530. doi:10.1175/JCLI3964.1

Stone DA, Allen MR, Selten F, Kilphuis M (2007b) The detection and attribution of climate change using an ensemble of opportunity. J Clim 20:504–516. doi:10.1175/JCLI3966.1

Stott PA, Jones GS (2009) Variability of high latitude amplification of anthropogenic warming. Geophys Res Lett 36:L10701. doi:10.1029/2009GL037698

Stott PA, Tett SFB, Jones GS, Allen MR, Mitchell JFB, Jenkin GJ (2000) External control of 20th century temperature by natural and anthropogenic forcings. Science 290(5499):2133–2137. doi:10.1126/science.290.5499.2133

Stott PA, Jones GS, Mitchell JF (2003) Do models underestimate the solar contribution to recent climate change? J Clim 16:4079–4093. doi:10.1175/1520-0442(2003)016<4079:DMUTSC>2.0.CO;2

Sun B, Bradley RS (2002) Solar influences on cosmic rays and cloud formation: a reassessment. J Geophys Res 107(D14):4211. doi:10.1029/2001JD000560

Sun B, Bradley RS (2004) Reply to comment by N. D. Marsh and H. Svensmark on "Solar influences on cosmic rays and cloud formation: a reassessment". J Geophys Res 109:D14206. doi:10.1029/2003JD004479

Svensmark H (2007) Cosmoclimatology: a new theory emerges. Astron Geophys 48:118–124. doi:10.1111/j.1468-4004.2007.48118.x

Svensmark H, Friis-Christensen E (1997) Variations of cosmic ray flux and global cloud coverage—a missing link in solar-climate relationships. J Atmos Terr Phys 59:1225–1232. doi:10.1016/S1364-6826(97)00001-1
Svensmark H, Bondo T, Svensmark J (2009) Cosmic rays decreases affect atmospheric aerosols and cloud. Geophys Res Lett 36:L15101. doi:10.1029/2009GL038429
Tett SFB et al (2002) Estimation of natural and anthropogenic contributions to twentieth century temperature change. J Geophys Res 107(D16):4306. doi:10.1029/2000JD000028
Tett SFB et al (1999) Causes of twentieth-century temperature change near the Earth's surface. Nature 399:569–572. doi:10.1038/21164
Thayer JP, Semeter J (2004) The convergence of magnetospheric energy flux in the polar atmosphere. J Atmos Sol-Terr Phys 66(10):807–824. doi:10.1016/j.jastp.2004.01.035
Thejll P, Christiansen B, Gleisner H (2003) On correlations between the North Atlantic Oscillation, geopotential heights and geomagnetic activity. Geophys Res Lett 30(6):1347. doi:10.1029/2002GL016598
Udelhofen PM, Cess RD (2001) Cloud cover variations over the united states: an influence of cosmic rays or solar variability? Geophys Res Lett 28(13):2617–2620. doi:10.1029/2000GL012659
Uppala SM et al (2005) The ERA-40 re-analysis. Q J R Meteorol Soc 131(612):2961–3012. doi:10.1256/qj.04.176
van Geel B, Mook WG (1989) High resolution 14C dating of organic deposits using natural atmospheric 14C variations. Radiocarbon 31:151–155
Veretenenko VS, Pudovkin MI (1997) Effects of the galactic cosmic ray variations on the solar radiation input in the lower atmosphere. J Atmos Terr Phys 59(14):1739–1746. doi:10.1016/S1364-6826(96)00183-6
Vieira LEA, Solanki SK, Krivova NA, Usoskin I (2011) Evolution of the solar irradiance during the Holocene. Astron Astrophys 531:A6. doi:10.1051/0004-6361/201015843
Voiculescu M, Usoskin IG, Mursula K (2006) Different response of clouds to solar input. Geophys Res Lett 33:L21802. doi:10.1029/2006GL027820
Vonmoos M, Beer J, Muscheler R (2006) Large variations in Holocene solar activity: constraints from Be-10 in the Greenland Ice core project ice core. J Geophys Res 111:A10105. doi:10.1029/2005JA011500
Wagner G, Lvingstone DM, Masarik J, Muscheler R, Beer J (2001) Some results relevant to the discussion of a possible link between cosmic rays and the Earth's climate. J Geophys Res 106(D4):3381–3387. doi:10.1029/2000JD900589
Wang Y-M, Lean JL, Sheeley NR Jr (2005) Modeling the Sun's magnetic field and irradiance since 1713. Appl J 625:522–538. doi:10.1086/429689
Wanner H, Pfister C, Brázdil R, Frich P, Frydendahl K, Jónsson T, Kington J, Lamb HH, Rosenørn S, Wishman E (1995) Wintertime European circulation patterns during the late Maunder Minimum cooling period (1675–1704). Theor Appl Clim 51:167–175. doi:10.1007/BF00867443
Waple AM, Mann ME, Bradley RS (2002) Long-term patterns of solar irradiance forcing in model experiments and proxy-based surface temperature reconstructions. Clim Dyn 18:563–578. doi:10.1007/s00382-001-0199-3
Weber SL (2005) A timescale analysis of the Northern Hemisphere temperature response to volcanic and solar forcing. Clim Past 1:9–17. doi:10.5194/cp-1-9-2005
Weiss NO, Thompson MJ (2009) The solar dynamo. Space Sci Rev 144(1–4):53–66. doi:10.1007/s11214-008-9435-z
Wenzler T, Solanki SK, Kirova NA, Fröhlich C (2006) Reconstruction of solar irradiance variations in cycles 21–23 based on surface magnetic fields. A&A 460:583–595. doi:10.1051/0004-6361:20065752
White WB, Liu Z (2008a) Non-linear alignment of El Niño to the 11-yr solar cycle. Geophys Res Lett 35:L19607. doi:10.1029/2008GL034831
White WB, Liu Z (2008b) Resonant response of the quasi-decadal oscillation to the 11-yr period signal in the Sun's irradiance. J Geophys Res 113:C01002. doi:10.1029/2006JC004057
White WB, Lean J, Cayan DR, Dettinger MD (1997) Response of global upper ocean temperature to changing solar irradiance. J Geophys Res 102(C2):3255–3266. doi:10.1029/96JC03549
Wigley TML, Raper SCB (1990) Climatic change due to solar irradiance changes. Geophys Res Lett 17:2169–2172. doi:10.1029/GL017i012p02169
Woollings T, Charlton-Perez A, Ineson S, Marshall AG, Masato G (2010a) Associations between stratospheric variability and tropospheric blocking. J Geophys Res 115:D06108. doi:10.1029/2009JD012742
Woollings T, Lockwood M, Masato G, Bell C, Gray L (2010b) Enhanced signature of solar variability in Eurasian winter climate. Geophys Res Lett 37:L20805. doi:10.1029/2010GL044601
Yiou P, Bard E, Dandin P, Legras B, Naveau P, Rust HW, Terray L, Vrac M (2010) Statistical issues about solar-climate relations. Clim Past Discuss 6:461–487. http://www.clim-past-discuss.net/6/461/2010/

Zhang Q, Soon WH, Baliunas SL, Lockwood GW, Skiff BA, Raddick RR (1994) A method of determining possible brightness variations of the Sun in past centuries from observations of solar-type stars. Astrophys J 427:L111–L114. doi:10.1086/187377

Zorita E, von Storch H, Gonzalez-Rouco F, Cubasch U, Luterbacher J, Legutke S, Fischer-Bruns I, Schlese U (2004) Climate evolution in the last five centuries simulated by an atmosphere-ocean model: global temperatures, the North Atlantic Oscillation and the Late Maunder Minimum. Meteor Zeit 13:271–289. doi:10.1127/0941-2984/2004/0013-0271

Surv Geophys (2012) 33:557–564
DOI 10.1007/s10712-011-9157-8

The Role of Water Vapour in Earth's Energy Flows

Richard P. Allan

Received: 5 September 2011 / Accepted: 28 October 2011 / Published online: 24 November 2011

Abstract Water vapour modulates energy flows in Earth's climate system through transfer of latent heat by evaporation and condensation and by modifying the flows of radiative energy both in the longwave and shortwave portions of the electromagnetic spectrum. This article summarizes the role of water vapour in Earth's energy flows with particular emphasis on (1) the powerful thermodynamic constraint of the Clausius Clapeyron equation, (2) dynamical controls on humidity above the boundary layer (or free-troposphere), (3) uncertainty in continuum absorption in the relatively transparent "window" regions of the radiative spectrum and (4) implications for changes in the atmospheric hydrological cycle.

Keywords Water vapour · Hydrological cycle · Radiative processes · Climate

1 Introduction

Water vapour is of central importance to energy flows within the climate system, by modulating the transmission of radiative energy between the surface, atmosphere and space and also through transferring latent heat from the surface (evaporation) to the atmosphere (precipitation)[1] following transport of moisture within the atmosphere (Fig. 1). There are many aspects of these processes that are well understood, based upon robust physics, detailed process modelling and measurements ranging from laboratory experiments up to satellite observations of the entire globe (e.g., Held and Soden 2006; Sherwood et al. 2010a, b). One of the fundamental controls on the climate system is the Clausius Clapeyron equation which provides a powerful constraint on how saturated moisture content varies with air temperature:

[1] Latent heat released by condensation is retained in the atmosphere only after the water is removed by precipitation to the surface before re-evaporation in the atmosphere can occur.

R. P. Allan (✉)
Department of Meteorology/National Centre for Atmospheric Science,
University of Reading, Reading, UK
e-mail: r.p.allan@reading.ac.uk

$$\frac{1}{q_s}\frac{\mathrm{d}q_s}{\mathrm{d}T} \approx \frac{1}{e_s}\frac{\mathrm{d}e_s}{\mathrm{d}T} = \frac{L}{R_v T^2} = \begin{cases} 0.14\,\mathrm{K}^{-1} & \mathrm{T} = 200\,\mathrm{K} \\ 0.07\,\mathrm{K}^{-1} & \mathrm{T} = 273\,\mathrm{K}, \\ 0.06\,\mathrm{K}^{-1} & \mathrm{T} = 300\,\mathrm{K} \end{cases} \tag{1}$$

where q_s is the saturation specific humidity in kg water vapour per kg moist air, e_s is the saturation water vapour pressure (Pa), T is air temperature (K), L is the latent heat of vapourisation (2.5×10^6 J kg^{-1}), R_v is the gas constant for water vapour (461 J K^{-1} kg^{-1}) and the first term approximates to the second term[2]. Equation 1 therefore predicts increases in water vapour of order 6–7%/K for temperatures close to the surface assuming fixed relative humidity; further discussion is given by O'Gorman and Muller (2010).

Observations and modelling indicate that column integrated water vapour, averaged over sufficiently large scales, increases approximately exponentially with atmospheric temperature (Raval and Ramanathan 1989) as predicted by (1). The absorption of infra-red radiation by water vapour increases approximately in proportion to the logarithm of its concentration. These two powerful constraints generate an amplifying effect on changes in Earth's climate, enhancing the response of surface temperature to a radiative forcing or internal variability, and the resultant positive water vapour feedback is relatively "forgiving" in the sense that substantial excursions away from these basic physical constraints are required to alter its nature.

2 Radiative effects of water vapour

Absorption of radiation by water vapour in distinct bands across the electromagnetic spectrum, relating to vibrational and rotational modes of molecular excitation, is reasonably well characterised by detailed measurements and theory. Nevertheless, there remains a question as to the physical basis of radiative transfer away from the absorption line centres, important for the atmospheric "windows" in which radiative energy is more able to pass relatively unimpeded through the atmosphere compared to band centres. Although the continuum absorption in these windows, and its dependence on pressure and temperature, is now much better observed and characterised, there remains a lively debate on the underlying physical causes of this continuum. The historical perspective underpinning the development of a theory of water vapour continuum absorption has led to recent developments and experimental campaigns leading to new insights into this uncertain aspect of how water vapour influences the flows of radiative energy through the atmosphere (Shine et al. 2012). Although improved representation of the water vapour continuum is likely to reduce uncertainty in Earth's radiative energy budget only marginally, it may have implications for the retrieval of physical quantities from space and therefore how we observe Earth's climate system. There also remain important questions as to the consistency of water vapour absorption of solar radiation in climate models, with implications for responses of the hydrological cycle (Takahashi 2009).

3 Dynamical controls on the water vapour greenhouse effect

While the radiative effects of water vapour contribute uncertainty to the simulation of radiative fluxes, a greater source of uncertainty in simulating radiative fluxes regionally is

[2] The first term is larger than the second term by a factor $q_s p/\epsilon e_s = p/(p - e_s(1-\epsilon)) = 1 + q_s((1/\epsilon) - 1) \sim 1 + 0.6q_s$ assuming fixed pressure, where $\varepsilon = 0.622$ is the ratio of gas constants for dry air and water vapour. For a surface pressure of 1,000 hPa and using the August–Roche–Magnus empirical approximation for $e_s(T)$, the first term exceeds the second term by around 0.2% at 273 K and by 1% at 300 K.

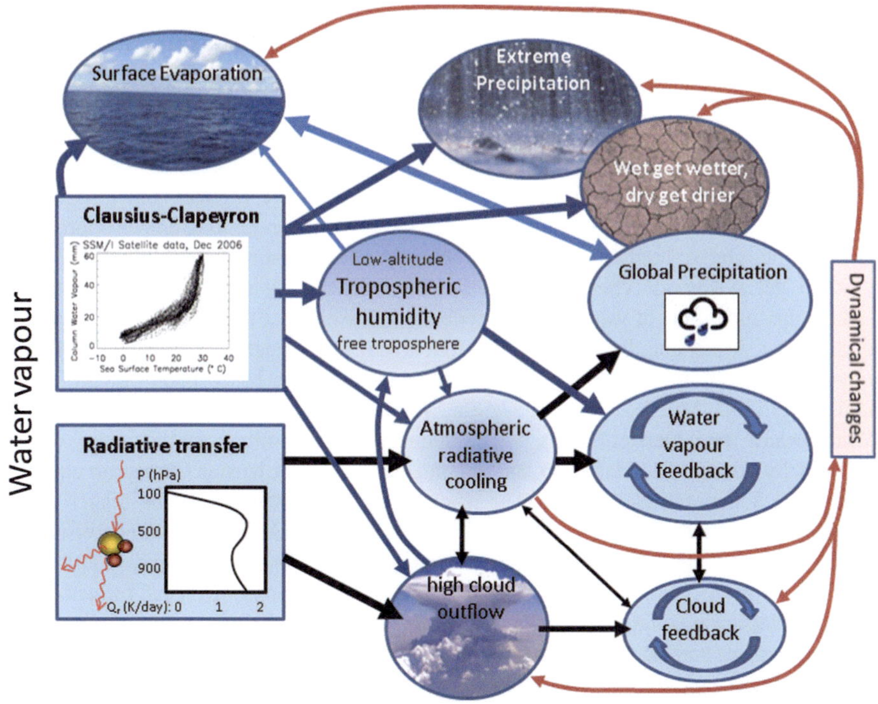

Fig. 1 The role of water vapour in Earth's energy flows: a simplistic schematic representation of some of the links between water vapour and energetic processes in the climate system. *Arrow* colours denote the primary driving physical process, including radiative transfer (*black*), thermodynamics (*blue*) and dynamics (*red*)

the distribution and concentration of water vapour throughout the atmosphere. The maintenance of the present-day distribution of relative humidity ($RH = e/e_s$), a diagnostic of how close the atmospheric water vapour pressure (e) is to the saturation vapour pressure (often expressed as a percentage), is also fundamental to positive water vapour feedback. Relative humidity within the subsiding branches of the tropical circulation is crucial in modulating the emission of infra-red radiation to space (Spencer and Braswell 1997; Allan et al. 1999; Brogniez et al. 2005).

The present day distribution of humidity is reasonably well approximated using a "point of last saturation" perspective, specifically that the amount of water vapour is determined by the temperature at which the considered parcel of air last experienced 100% relative humidity (Pierrehumbert 1998; Sherwood et al. 2010b). This is also well approximated by considering the circulation as being driven by subsidence induced by radiative cooling in the descending branches of the tropical circulation (e.g., Folkins et al. 2002), but may only apply to certain regions of the upper troposphere since mixing with lower-level humidity and transport from the mid-latitudes have also been shown to dominate in model simulations and reanalyses (Galewsky et al. 2005). Cloud microphysics may also modify the humidity distribution of the free-troposphere. However, modelling studies demonstrate that cloud ice plays a relatively minor role in moistening the upper troposphere compared to water vapour (John and Soden 2006). This suggests that it is inherently large-scale processes that maintain and control the relative humidity distribution at scales that are therefore resolvable by climate models simulations.

The precise dynamical controls on the "free" tropospheric humidity (above the influence of the boundary layer) require improved characterization since they may have important bearing on the overall responses of the humidity distribution (and indirectly on cloud), and hence the flows of longwave radiative energy, in a warming climate. Important in regulating these energy flows are the driest regions of the upper troposphere. Roca et al. (2012) find a dual control on one of the driest regions of the upper troposphere, located above the eastern Mediterranean: air is advected from both tropical convective regimes and mid-latitude disturbances, and the relative contribution of these two source regions exerts an appreciable affect on the relative humidity over these regions (see also Galewsky et al. 2005).

Small yet systematic changes in relative humidity in a warming climate contribute towards the magnitude of water vapour feedback and aspects of the hydrological cycle and energy flows. It has long been appreciated that a warming climate is associated with small reductions in subtropical free-tropospheric relative humidity, based upon climate model simulations (e.g., Mitchell et al. 1987; Sherwood et al. 2010a) and recent observations (Minschwaner and Dessler 2004), which reduces the magnitude of the water vapour feedback by around 5% relative to that anticipated from the Clausius Clapeyron equation alone (e.g., Soden and Held 2006). The mechanism for this change can be related to the outflow temperature of deep convection in the tropics and linked to the altitude of neutral buoyancy. This is determined by the decline in longwave radiative cooling with altitude in the upper troposphere due to reduced water vapour concentrations, again determined by the Clausius Clapeyron equation (Zelinka and Hartmann 2010). The robust physics relating to radiative transfer and Eq. 1 appears to explain positive feedback from high altitude clouds (Fig. 1) in the tropics simulated by climate models (Zelinka and Hartmann 2010; Soden and Vecchi 2011).

4 The atmospheric hydrological cycle

Near to the surface, small increases in relative humidity with warming, combined with diminishing wind stress (Gastineau and Soden 2011) and reduced near surface temperature lapse rate, appear to explain a reduced dependence of surface evaporation on temperature (2–3%/K) compared to that expected from the Clausius Clapeyron equation alone, sometimes referred to as a "muted response" (Richter and Xie 2008). Detecting the small changes in relative humidity over the oceans simulated by climate model projections is difficult. However, declines in relative humidity over moisture-limited land regions as global temperatures rise may be more apparent: projected and observed increases in land minus ocean temperature contrast (e.g., Lambert and Chiang 2007) coupled with the dominance of the oceans as a source of moisture for precipitation over land (e.g., Gimeno et al. 2010) may have contributed to recent reductions in relative humidity over land since 2000 (Simmons et al. 2010).

The present-day observed distribution of water vapour within the atmosphere is also an important diagnostic of the hydrological cycle. Substantial model biases in this distribution (John and Soden 2007) may therefore be symptomatic of inadequacies in the simulation of the global water cycle. The atmospheric hydrological cycle is intimately linked with Earth's energy flows through water vapour, from its direct impact upon the surface and atmospheric radiative budgets and also its influence on clouds and precipitation through convergence of moisture (Mitchell et al. 1987; Sherwood et al. 2010b). The radiative properties of water vapour, including continuum absorption and emission, are crucial in explaining an enhanced longwave radiative cooling of the atmosphere (Stephens and Ellis 2008; Allan

2009; Previdi 2010) and reduced radiative cooling of the surface as the climate warms. Recent advances in understanding of the links between precipitation and radiative energy budgets (e.g., Andrews et al. 2010) are discussed in detail by O'Gorman et al. (2012).

The dominance of radiative-convective processes in the tropics for maintaining temperature lapse rate (the decline in air temperature with altitude) helps to determine the increased rate of radiative cooling with increasing surface and atmospheric temperature (Lambert and Webb 2008; Allan 2009) which in turn explains why global precipitation increases with warming in the present climate at the rate of around 2–3%/K in climate models, a "muted" response relative to the increases in low-level water vapour of around 6–7%/K (Allan 2011). The differing rates of increase in water vapour and precipitation are reconciled in climate model simulations through changes in tropical circulation, in particular a reduction in the strength of the Walker circulation (Vecchi et al. 2006), although both external forcing and internally generated variability are thought to have contributed to an observed weakening over the twentieth century (Power and Kociuba 2011).

The physical basis for decreased tropical circulation can be understood in terms of the atmospheric mass flux (M) relating precipitation (P, here in units of kg/s) with low-level specific humidity (q) ignoring the small return flow of moisture at upper levels (Held and Soden 2006) (although see also Zahn and Allan 2011):

$$P \approx Mq. \tag{2}$$

Clearly, for Eq. 2 to be true, if P is constrained by the energy balance to rise at 2–3%/K and q is constrained to rise in accordance with Eq. 1 at around 6–7%/K, then M must diminish (Held and Soden 2006). Another perspective, relating to earlier discussion of radiatively driven subsidence, links vertical motion, w (km/day), to radiative cooling, Q (K/day), and the deviation from dry adiabatic lapse rate, or static stability, σ (K/km) (Knutson and Manabe 1995):

$$w = \frac{Q}{\sigma}. \tag{3}$$

In the tropics, rising temperatures are associated with a conservation of moist adiabatic lapse rate, leading to increased static stability. Therefore, the amount of subsidence required to balance Q through adiabatic heating is diminished in a warmer climate. Since the increase in stability averaged over the troposphere scales with boundary layer moisture, q, in accordance with Eq. 1 (6–7%/K), while the radiative cooling enhances at 2–3%/K, the stability changes dominate and the tropical circulation reduces (Held and Soden 2006), analogous to the situation in Eq. 2 (see also Chou and Chen 2010).

Of importance to climate impacts relating to the role of water vapour in Earth's energy flows is how precipitation and its extremes will respond to warming regionally over decadal time-scales. Locally, large-scale precipitation is dependent upon convergence of low-level moisture from surrounding regions. The Clausius–Clapeyron equation therefore provides a substantial constraint upon changes in intense precipitation events (Trenberth 2011; Sugiyama et al. 2010; Allan et al. 2010; O'Gorman and Schneider 2009; Bengtsson et al. 2009). O'Gorman and Schneider (2009) argue that the condensation rate (c) at a given level in the atmosphere associated with an extreme precipitation event scales as:

$$c = -\omega \frac{\mathrm{d}q_s}{\mathrm{d}p}\bigg|_{\theta^*}, \tag{4}$$

where θ^* is the equivalent potential temperature, denoting the change in q_s along a moist adiabat, and ω is the vertical pressure velocity in Pa/s. Assuming that almost all of the surface moisture is condensed in the storm, Eq. 4 scales with surface q_s at the rates indicated in Eq. 1. Changes in ω also strongly influence c and therefore precipitation and latent heating within storms.

The increased supply of low-level water vapour for intense rainfall is maintained in wet regions by the convergence of moisture transported from dry ocean basins of high evaporation (Gimeno et al. 2010) and may be understood in terms of the moisture balance equation (e.g., Held and Soden 2006):

$$\frac{\mathrm{d}(P-E)}{\mathrm{d}T} \approx \alpha(P-E), \tag{5}$$

where E is surface evaporation and α is equal to the terms in Eq. 1. Using this approach, Held and Soden (2006) demonstrate that Precipitation minus Evaporation ($P - E$) patterns are enhanced in a warming world. Since evaporation changes are relatively uniform, this explains why Eq. 5 is manifest most strongly as contrasting precipitation responses in the wet and dry regions of the tropics (e.g., Chou et al. 2007; Allan et al. 2010). An alternative perspective is to again consider the energy balance: latent heating through condensation of water droplets and eventual precipitation must be balanced by cooling through radiative processes and dry static energy transport, assuming other terms such as sensible heating are small (Muller and O'Gorman 2011); this may also have bearing upon the viability and stability of monsoon systems (Levermann et al. 2009).

5 Concluding remarks

While thermodynamics and radiative transfer theory provide a robust physical basis for water vapour feedback and changes in the hydrological cycle (Fig. 1), some of the details of the role of water vapour on Earth's energy flows remain to be clarified, in particular with regard to changes in the global water cycle. The Clausius Clapeyron constraint upon precipitation intensity appears to be less clear over tropical regions in models due to a varied response in updraft velocity (ω in (4)) within simulated extreme precipitation events (O'Gorman and Schneider 2009; Allan et al. 2010; Sugiyama et al. 2010) and there are substantial deviations from the anticipated 6–7%/K response in observational datasets (Lenderink and van Meijgaard 2010; Haerter et al. 2010) relating to spatial and temporal sampling scales as well as contrasting physical mechanisms such as atmospheric stabilisation or storm invigoration through intense latent heating or moisture limitation (Hardwick-Jones et al. 2010).

It is also plausible that temporary increases in the strength of the tropical circulation may yield larger precipitation responses than anticipated (e.g., Wentz et al. 2007) relative to secular trends (e.g., Semenov and Bengtsson 2002). There are also some limited observationally based analyses which suggest increases in tropical circulations (e.g., Sohn and Park 2010; Zahn and Allan 2011) although these changes may be sensitive to the assimilation of erroneous temperature lapse rate observations in the reanalyses employed (Held and Soden 2006). It is not currently clear what mechanisms drive decadal variability, nor whether the observing systems are of sufficient stability to capture such changes accurately (Trenberth 2002). A further related example relates to stratospheric water vapour which also appears to show substantial decadal variability, potentially influencing decadal temperature trends (Solomon et al. 2010), but a physical mechanism for the variation remains unclear. The characteristics of decadal fluctuations in global circulation, precipitation and upper tropospheric/lower stratospheric water vapour and their causative mechanisms currently remain an open question in relation to the role of water vapour in Earth's energy flows.

Acknowledgments Thanks to Paul O'Gorman, Remy Roca, Keith Shine, Brian Soden and two anonymous reviewers for comments and corrections on the original manuscript and for their input, along with B.-J. Sohn, Andrew Dessler and many others, to the International Space Sciences Institute Workshop Session on Observing and modelling Earth's energy flows. R Allan was supported by the UK Natural Environment Research Council PREPARE project (NE/G015708/1).

References

Allan RP (2009) Examination of relationships between clear-sky longwave radiation and aspects of the atmospheric hydrological cycle in climate models, reanalyses, and observations. J Climate 22:3127–4145

Allan RP (2011) Human influence on rainfall. Nature 470:344–345

Allan RP, Shine KP, Slingo A, Pamment JA (1999) The dependence of clear-sky outgoing longwave radiation on surface temperature and relative humidity. Q J R Meteorol Soc 125:2103–2126

Allan RP, Soden BJ, John VO, Ingram W, Good P (2010) Current changes in tropical precipitation. Environ Res Lett 5. doi:10.1088/1748-9326/5/2/025205

Andrews T, Forster PM, Boucher O, Bellouin N, Jones A (2010) Precipitation, radiative forcing and global temperature change. Geophys Res Lett 37:L14701. doi:10.1029/2010GL043991

Bengtsson L, Hodges KI, Keenlyside N (2009) Will extra-tropical storms intensify in a warmer climate? J Climate 22:2276–2301

Brogniez H, Roca R, Picon L (2005) Evaluation of the distribution of subtropical free tropospheric humidity in AMIP-2 simulations using METEOSAT water vapor channel data. Geophys Res Lett 32:L19708

Chou C, Chen C (2010) Depth of convection and the weakening of tropical circulation in global warming. J Climate 23:3019–3030

Chou C, Tu J, Tan P (2007) Asymmetry of tropical precipitation change under global warming. Geophys Res Lett 34:L17708. doi:10.1029/2007GL030327

Folkins I, Kelly KK, Weinstock EM (2002) A simple explanation of the increase in relative humidity between 11 and 14 km in the tropics. J Geophys Res 107:4736. doi:10.1029/2002JD002185

Galewsky J, Sobel A, Held I (2005) Diagnosis of subtropical humidity dynamics using tracers of last saturation. J Atmos Sci 62:3353–3367

Gastineau G, Soden BJ (2011) Evidence for a weakening of tropical surface wind extremes in response to atmospheric warming. Geophys Res Lett 38:L09706. doi:10.1029/2011GL047138

Gimeno L, Drumond A, Nieto R, Trigo RM, Stohl A (2010) On the origin of continental precipitation. Geophys Res Lett 37:L13804

Haerter JO, Berg P, Hagemann S (2010) Heavy rain intensity distributions on varying time scales and at different temperatures. J Geophys Res 115:D17102

Hardwick-Jones R, Westra S, Sharma A (2010) Observed relationships between extreme sub-daily precipitation, surface temperature, and relative humidity. Geophys Res Lett 37:L22805

Held IM, Soden BJ (2006) Robust responses of the hydrological cycle to global warming. J Climate 19: 5686–5699

John VO, Soden BJ (2006) Does convectively-detrained cloud ice enhance water vapor feedback? Geophys Res Lett 33:L20701

John VO, Soden BJ (2007) Temperature and humidity biases in global climate models and their impact on climate feedbacks. Geophys Res Lett 34:L18704. doi:10.1029/2007GL030429

Knutson TR, Manabe S (1995) Time-mean response over the tropical Pacific to increased CO_2 in a coupled ocean-atmosphere model. J Climate 8:2181–2199

Lambert FH, Chiang JCH (2007) Control of land-ocean temperature contrast by ocean heat uptake. Geophys Res Lett 34:L13704

Lambert FH, Webb MJ (2008) Dependency of global mean precipitation on surface temperature. Geophys Res Lett 35:L16706. doi:10.1029/2008GL034838

Lenderink G, van Meijgaard E (2010) Linking increases in hourly precipitation extremes to atmospheric temperature and moisture changes. Environ Res Lett 5:025208

Levermann A, Schewe J, Petoukhov V, Held H (2009) Basic mechanism for abrupt monsoon transitions. Proc Nat Acad Sci 106:20,572–20,577
Minschwaner K, Dessler AE (2004) Water vapor feedback in the tropical upper troposphere: model results and observations. J Climate 17:1272–1282
Mitchell J, Wilson CA, Cunnington WM (1987) On CO_2 climate sensitivity and model dependence of results. Q J R Meteorol Soc 113:293–322
Muller CJ, O'Gorman PA (2011) An energetic perspective on the regional response of precipitation to climate change. Nat Climate Change 1:266–271
O'Gorman PA, Muller CJ (2010) How closely do changes in surface and column water vapor follow Clausius-Clapeyron scaling in climate change simulations. Environ Res Lett 5:025207. doi: 10.1088/1748-9326/5/2/025207
O'Gorman PA, Schneider T (2009) The physical basis for increases in precipitation extremes in simulations of 21st-century climate change. Proc Nat Acad Sci 106:14,773–14,777
O'Gorman PA, Allan RP, Byrne MP, Previdi M (2012) Energetic constraints on precipitation under climate change. Surv Geophys. doi:10.1007/s10712-011-9159-6
Pierrehumbert RT (1998) Lateral mixing as a source of subtropical water vapor. Geophys Res Lett 25:151–154
Power SB, Kociuba G (2011) What caused the observed 20th century weakening of the Walker circulation? J Climate (in press). doi:10.1175/2011JCLI4101.1
Previdi M (2010) Radiative feedbacks on global precipitation. Environ Res Lett 5:025–211
Raval A, Ramanathan V (1989) Observational determination of the greenhouse effect. Nature 342:758–761
Richter I, Xie SP (2008) The muted precipitation increase in global warming simulations: a surface evaporation perspective. J Geophys Res 113:D24118. doi:10.1029/2008JD010561
Roca R, Guzman R, Lemond J, Meijer J, Picon L, Brogniez H (2012) Tropical and extra-tropical influences on the distribution of free tropospheric humidity over the intertropical belt. Surv Geophys (this volume)
Semenov V, Bengtsson L (2002) Secular trends in daily precipitation characteristics: greenhouse gas simulation with a coupled AOGCM. Climate Dyn 19:123–140
Sherwood SC, Ingram W, Tsushima Y, Satoh M, Roberts M, Vidale PL, O'Gorman PA (2010a) Relative humidity changes in a warmer climate. J Geophys Res 115:D09104. doi:10.1029/2009JD012585
Sherwood SC, Roca R, Weckwerth TM, Andronova NG (2010b) Tropospheric water vapor, convection, and climate. Rev Geophys 48:RG2001
Shine KP, Ptashnik IV, Rädel G (2012) The water vapour continuum: brief history and recent developments. Surv Geophys (this volume)
Simmons AJ, Willett KM, Jones PD, Thorne PW, Dee DP (2010) Low-frequency variations in surface atmospheric humidity, temperature, and precipitation: inferences from reanalyses and monthly gridded observational data sets. J Geophys Res 115:D01110. doi:10.1029/2009JD012442
Soden BJ, Held IM (2006) An assessment of climate feedbacks in coupled ocean-atmosphere models. J Climate 19:3354–3360
Soden BJ, Vecchi G (2011) The vertical distribution of cloud feedback in coupled ocean-atmosphere models. Geophys Res Lett 38:L12704. doi:10.1029/2011GL047632
Sohn BJ, Park SC (2010) Strengthened tropical circulations in past three decades inferred from water vapor transport. J Geophys Res 115:D15112. doi:10.1029/2009JD013713
Solomon S, Rosenlof KH, Portmann RW, Daniel JS, Davis SM, Sanford TJ, Plattner GK (2010) Contributions of stratospheric water vapor to decadal changes in the rate of global warming. Science 327:1219–1223
Spencer RW, Braswell WD (1997) How dry is the tropical free troposphere? Implications for global warming theory. Bull Am Meterol Soc 78:1097–1106
Stephens GL, Ellis TD (2008) Controls of global-mean precipitation increases in global warming GCM experiments. J Climate 21:6141–6155
Sugiyama M, Shiogama H, Emori S (2010) Precipitation extreme changes exceeding moisture content increases in MIROC and IPCC climate models. Proc Nat Acad Sci 107:571–575
Takahashi K (2009) The global hydrological cycle and atmospheric shortwave absorption in climate models under CO_2 forcing. J Climate 22:5667–5675
Trenberth KE (2002) Changes in tropical clouds and radiation. Science 296:2095a
Trenberth KE (2011) Changes in precipitation with climate change. Climate Res 47:123–138
Vecchi GA, Soden BJ, Wittenberg AT, Held IM, Leetmaa A, Harrison MJ (2006) Weakening of tropical pacific atmospheric circulation due to anthropogenic forcing. Nature 441:73–76
Wentz FJ, Ricciardulli L, Hilburn K, Mears C (2007) How much more rain will global warming bring? Science 317:233–235
Zahn M, Allan RP (2011) Changes in water vapor transports of the ascending branch of the tropical circulation. J Geophys Res 116:D18111. doi:10.1029/2011JD016206
Zelinka MD, Hartmann DL (2010) Why is longwave cloud feedback positive? J Geophys Res 115:D16117

Surv Geophys (2012) 33:535–555
DOI 10.1007/s10712-011-9170-y

The Water Vapour Continuum: Brief History and Recent Developments

Keith P. Shine · Igor V. Ptashnik · Gaby Rädel

Received: 22 June 2011 / Accepted: 21 December 2011 / Published online: 27 January 2012

Abstract The water vapour continuum is characterised by absorption that varies smoothly with wavelength, from the visible to the microwave. It is present within the rotational and vibrational–rotational bands of water vapour, which consist of large numbers of narrow spectral lines, and in the many 'windows' between these bands. The continuum absorption in the window regions is of particular importance for the Earth's radiation budget and for remote-sensing techniques that exploit these windows. Historically, most attention has focused on the 8–12 μm (mid-infrared) atmospheric window, where the continuum is relatively well-characterised, but there have been many fewer measurements within bands and in other window regions. In addition, the causes of the continuum remain a subject of controversy. This paper provides a brief historical overview of the development of understanding of the continuum and then reviews recent developments, with a focus on the near-infrared spectral region. Recent laboratory measurements in near-infrared windows, which reveal absorption typically an order of magnitude stronger than in widely used continuum models, are shown to have important consequences for remote-sensing techniques that use these windows for retrieving cloud properties.

Keywords Earth radiation budget · Water vapour spectroscopy · Water dimers · Remote sensing

1 Introduction

It is well established that water vapour is the most important gaseous absorber in the Earth's atmosphere—it is responsible for about 60% of the clear-sky greenhouse effect in the thermal infrared (at wavelengths greater than 4 μm) and for about 60% of the

K. P. Shine (✉) · I. V. Ptashnik · G. Rädel
Department of Meteorology, University of Reading, Earley Gate, Reading RG6 6BB, UK
e-mail: k.p.shine@reading.ac.uk

I. V. Ptashnik
V.E. Zuev Institute of Atmospheric Optics, SB RAS, 1, Academician Zuev Square, Tomsk 634021, Russia

atmospheric absorption of incoming solar radiation (wavelengths less than 4 μm) (e.g., Kiehl and Trenberth 1997). The absorption pervades the entire spectrum, quite weakly in the visible and ultra-violet, but more strongly throughout the infrared and microwave. Water vapour plays a key role in determining the present-day radiation budget and, because its abundance in the atmosphere is dependent on temperature, it can also play an important role in climate change. Model simulations indicate that the water vapour feedback (e.g., Manabe and Wetherald 1967) approximately doubles the response of surface temperature to the imposition of an external forcing (such as due to increases in carbon dioxide). Water vapour plays an important role in remote sensing—whether as a target species of that remote sensing, or as 'interference' when, for example, the target is remote sensing of surface or cloud properties.

The pervasive absorption by water vapour is largely due to the fact that, in its ground state, the water molecule is strongly polar (unlike, for example, carbon dioxide) for three axes of rotation. This means that it can absorb and emit radiation associated with (1) changes in its rotational energy levels in the microwave and far infrared, (2) changes in vibrational and rotational levels at visible, near and mid-infrared wavelengths and (3) due to electronic transitions in the ultraviolet (which will not be considered further here). The polar nature of the water molecule is also of importance spectroscopically, as it influences the nature of collisions between molecules and facilitates the formation of complexes of water with other molecules, of which the water dimer[1] $(H_2O)_2$ is of most importance here (we will often refer to individual water molecules as monomers).

The spectrum of water vapour (see e.g., Goody and Yung 1989) is characterised by strong absorption bands caused either by pure rotation of the molecule or by particular vibrational transitions associated with various stretching and bending modes of the molecule. These bands are made up of many thousands of individual relatively narrow (under typical atmospheric conditions) spectral lines, associated with particular transitions. Between these bands are so-called windows regions with much weaker absorption.

In addition to the absorption due to the individual spectral lines, water vapour possesses a slowly varying component of absorption, known as the *continuum*, which pervades both the bands and the windows from the visible through the infrared to the microwave. The absorption due to the water vapour spectral lines within the bands is normally so strong that the continuum in these regions is of secondary importance to the radiation budget (although, as will be discussed, it has turned out to be very important for understanding the *causes* of the continuum); in the windows, by contrast, the continuum is often the dominant cause of wavelength-averaged absorption (at least in clear skies) and so is of much greater importance both for the radiation budget and for remote-sensing techniques.

The purpose of this paper is to review the evolution in understanding of the water vapour continuum and discuss recent developments, including those arising from the UK research consortium CAVIAR (Continuum Absorption at Visible and Infrared wavelengths and its Atmospheric Relevance). Despite the fact that the continuum's existence has been

[1] The water dimer, i.e. two water molecules bound by relatively weak hydrogen bond, results in a spectrum which differs from the water monomer. In the mid and far-infrared, this absorption is predominantly due to intermolecular oscillations between the two water molecules while, in the microwave, it is due to rotation of the entire dimer (or of the monomers within the dimer). In the near-infrared, the distortion of the O–H bond (relative to its monomer state) that links the two water monomers contributes to the changes in the intramolecular vibrational spectrum. Due to the high spectral density of the dimer ro-vibrational lines, they are expected to be so strongly overlapping in atmospheric conditions that they appear as a continuum. Thus, the dimer spectrum consists of sub-bands in which heavily overlapped lines of individual stretching and bending modes have the form of a continuum.

recognised for many decades, until recently, there were many regions of the spectrum where the nature of its absorption was poorly characterised, because of lack of detailed measurements. Further there is ongoing debate and controversy as to the underlying causes of the continuum. Section 2 will provide a brief history of the development of understanding and the origins of the controversy and will review recent developments, focusing on the near-infrared region. Section 3 will provide an example of the consequences of these recent developments for the remote sensing of cloud properties, and Sect. 4 will conclude with a discussion of the outstanding issues.

2 Historical Developments in Understanding

This section discusses what we perceive to be some of the major developments in understanding of the water vapour continuum, from an atmospheric science perspective. We make no claim that this is a comprehensive review, as this would be impossible in the space (and indeed time) available—for almost every study we cite, the advice 'and see references therein' should be taken as implicit. The discussion is mostly restricted to the infrared, where the impact on the atmospheric radiation budget is strongest, although studies in the visible and microwave are important for applications (and in particular remote sensing) and for understanding the mechanisms which cause the continuum. We note that some of our perspective may be viewed as contentious by some, such as the nature of the debate on the continuum.

2.1 Pre-1966

It was Tyndall (1861) who first established experimental evidence that water vapour absorbs infrared radiation. Within about 30 years, wavelength-resolved spectra began to become available (as summarised by Fowle 1917). In some previous work, including our own, Hettner (1918) has been credited as first measuring absorption due to water vapour in the 8–12 μm (mid-infrared) window. However, we now believe that Rubens and Aschkinass (1898) were the first. The strength of the measured absorption in both of these studies is broadly consistent with modern measurements of the water vapour continuum in this window, insofar as we can establish the measurement conditions. Rubens and Aschkinass (1898) noted that, in contrast to other regions of the spectrum, they were unable to detect marked absorption bands between 9 and 11 μm (and hence hinting at the continuum nature of the absorption), but they attributed this to the lack of resolving power in their instruments. Hettner (1918) presented what seems to be the first detailed spectrum of water vapour across a broad wavelength range. Figure 1, from his paper, shows the band/window structure referred to above. In the near-infrared windows, Hettner could not detect any absorption, and only by measuring through long paths of pure water vapour could he detect any absorption in the mid-infrared window. In subsequent years, the work of Rubens and Aschkinass (1898) and Hettner (1918) was often referred to in the atmospheric science literature although the atmospheric relevance of the measured absorption in the mid-infrared window was questioned. For example, both Fowle (1917) and Simpson (1928) contended that water vapour did not absorb significantly in the mid-infrared window under atmospheric conditions in spite of the earlier measurements, although Brunt (1932) was more willing to accept its importance. In his comprehensive 'note', Brunt (1932) noted that the positions of absorption lines in the mid-infrared window existed 'on a background of

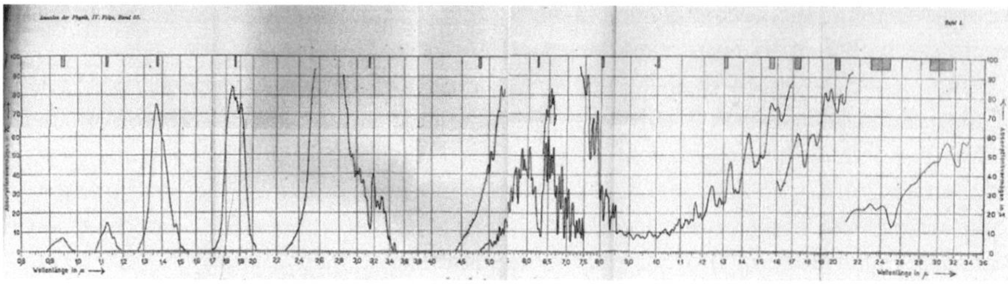

Fig. 1 Absorption spectrum of water vapour obtained by laboratory observations by Hettner (1918), which displays the characteristic 'band' and 'window' structure of that absorption. To obtain measureable absorption in the 8–12 μm region, Hettner had to use increased concentrations of water vapour, and hence two curves are plotted in the vicinity of this window

continuous absorption' and tentatively attributed it to ionisation, although this is now known to be impossible at such wavelengths.

It was Elsasser (1938a, b, c) who made the next important advance. Elsasser recognised firstly that the absorption in the mid-infrared window detected by Hettner's laboratory measurements must be relevant to the atmosphere in order to understand the amount of infrared radiation emitted from clear skies to the surface. He also recognised that such emission was not at wavelengths at which water vapour possessed rotation or ro-vibration bands. From being puzzled by the physical explanation for Hettner's measurements in Elsasser (1938a), he quickly (Elsasser 1938b, c) posited a theory that the continuous absorption could originate from the accumulated *far wings* of strong water monomer lines in the neighbouring absorption bands. Using the Lorentzian line shape for the collision-broadened lines and assuming these lines had half-widths of 0.25 cm^{-1} (now known to be too large—air-broadened half-widths are more typically 0.05–0.1 cm^{-1} at standard temperature and pressure), he computed the absorption in the mid-infrared window resulting from the far wings of rotational lines; indeed, he noted that his result was equivalent to assuming a single very strong line at 200 cm^{-1}. Atmospheric observations in the following few years (e.g., Adel 1939; Adel and Lampland 1940) confirmed the existence of emission in this window and demonstrated that Elsasser's parameterisation (although this word was not used) was effective at describing it. However, Cowling (1942) doubted whether the Lorentz formula was valid at wavenumbers far from the line centre, and both he and Strong (1941) felt that it remained an open question as to whether the absorption was continuous or whether it was due to (weak) bands of water vapour lines within the window itself. Later, Gebbie et al. (1951) presented evidence from atmospheric measurements, which leant more support to it being continuous. Anthony (1952) described atmospheric observations of an absorption being somewhat weaker than derived by Elsasser, but consistent with what would be expected had narrower, and more realistic, values of half-width been used by Elsasser.

Roach and Goody (1958) presented further atmospheric measurements of the absorption in the mid-infrared window, contrasting a (relatively) clean and a polluted environment, in an attempt to separate out the contributions of water vapour and aerosols to the continuum (Roach and Goody may have been the first to use the word 'continuum' to emphasise its spectrally smooth nature in contrast to the selective absorption by water vapour lines). They concluded that in the cleaner environment the observed continuum absorption in the mid-infrared window could not be due to aerosols or droplets and that Elsasser's far-wing hypothesis was 'plausible'; however, they too noted that agreement between observations and theory may be fortuitous, given the limitations in understanding of far-wing line shapes.

Bignell et al. (1963) made atmospheric measurements not only in the mid-infrared window, but at longer wavelengths, between 13 and 20 μm. In both spectral regions, after accounting for the effect of aerosols, Bignell et al. (1963) deduced that the absorption coefficient (in m^2 kg^{-1} or some related unit) was proportional to vapour pressure itself, an important result as it indicated that the optical depth [the product of the absorption coefficient and the abundance of the absorber along the path (in, for example, kg m^{-2})] is therefore proportional to the *square* of the vapour pressure—they tentatively ascribed this to the fact that the half-widths of self-broadened lines of water vapour are greater than those of air-broadened lines. However, this required a ratio of the self- to air-broadened widths of around 30, much higher than the modern value of around 5. Using updated spectral line data, they showed that agreement with Elsasser's far line-wing theory was 'not unsatisfactory' in the mid-infrared window, but the theory led to a factor of 3 underestimate at 20 μm. Bignell et al. (1963) also investigated the use of more sophisticated line shapes than the Lorentzian—one of these, the Van Vleck and Weisskopf (1945) 'full Lorentz' shape improved agreement at 20 μm, but led to worse agreement in the mid-infrared window. A similar result was obtained independently and discussed by Zhevakin and Naumov (1963) when applying the Van Vleck and Weisskopf (1945) profile and its modifications for comparison with experimental data in microwave and mid-infrared regions.

Bolle (1963) also reported measurements of continuum emission at 16–26 μm from the Jungfraujoch high-alpine station (where the dry conditions enabled easier observation of the emission between spectral lines) but did not relate these measurements to existing continuum theory.

2.2 1966–1980

Viktorova and Zhevakin (1966) and Penner and Varanasi (1967) began a debate that has not been fully settled to the present day.

Viktorova and Zhevakin (1966) were one of the first to suggest that 'anomalous' absorption in microwindows of the microwave and mm-wave spectral regions, detected in various experiments, could be successfully explained by water dimers rather than by uncertainty in the parameters or shape of water monomer lines. This suggestion was based on quantum-mechanical calculations of the water dimer rotational spectrum using the symmetric-top harmonic-oscillator approximation. Similar calculations were performed later by different authors (e.g., Braun and Leidecker (1974); Vigasin and Chlenova (1984)) and were recently shown to be in a reasonable agreement with the modern state-of-art quantum-chemical and molecular dynamic calculations (see Sect. 2.3).

A year later, Penner and Varanasi (1967) noted from laboratory measurements that the absorption coefficient in the mid-infrared window was not a linear function of vapour pressure as would be expected for the far wings of water monomer spectral lines; they proposed that the cause of the absorption was 'intermolecular coupling' rather than being a water monomer contribution. In Varanasi et al. (1968), they also detected a negative exponential temperature dependence of the mid-infrared continuum absorption coefficient in pure water vapour, with the exponential factor close to the dissociation energy of the water dimer (between 3 and 5 kcal/mol or ~2,000 K). This led the authors to extend their view and to note that the absorption was due to an 'association of water molecules due to hydrogen bonding'.

Until quite recently, the negative exponential temperature dependence of the pure water vapour cross-section (in cm^2 mol^{-1} atm^{-1}) has been one of the main arguments in favour of the water dimer hypothesis of the water vapour continuum. It is worth noting that the absorption cross-section in the Lorentzian wings of strong spectral lines of the monomer would usually be only very weakly dependent on temperature. More sophisticated line shapes (see later discussion) might, however, have the appropriate dependence.

Bignell's (1970) analysis of his laboratory measurements appears to have been the first to clearly distinguish between what would later be called the 'foreign continuum' (due to the interaction of water molecules with other molecules—notably nitrogen and oxygen in our atmosphere) and the 'self-continuum' (due to the interaction of two water molecules).[2] In this period, the latter was commonly referred to as 'e-type' absorption, to denote that the absorption coefficient was proportional to vapour pressure (conventional symbol e). Bignell's measurements demonstrated again that the self-continuum absorption coefficient was proportional to the vapour pressure and, following Varanasi et al. (1968), demonstrated that it possessed a negative temperature difference. His continuum measurements covered three spectral regions: 17–21 μm, 8–12 μm and, for what seems to be the first time, 4 μm. Bignell (1970) also performed a detailed analysis of previous laboratory and field observations in an attempt to verify the existence of e-type absorption. The foreign continuum in the mid-infrared window is (on a molecule-per-molecule basis) much weaker than the self-continuum but, in the atmosphere, this is offset by the fact that nitrogen is far more abundant than water vapour—only at vapour pressures, more than around 10 hPa is the self-continuum expected to dominate over the foreign continuum.

Bignell (1970) noted that the vapour pressure and temperature dependencies were consistent with the dimer theory of the continuum. He distanced himself from the far-wing theory he had earlier supported (Bignell et al. 1963) because, whilst it explained the e-type nature of the absorption, it failed to explain its negative temperature dependence. Bignell (1970) briefly commented that his newly derived continuum values would increase atmospheric radiative cooling rates in moist atmospheres by 1–2 K day^{-1}. He noted that the continuum might explain the large cooling rates measured by Cox (1969) in a tropical atmosphere. Soon after, Cox (1973) himself showed that this was indeed the case, presenting one of the first clear demonstrations of the effect of the continuum on cooling rates.

Between 1970 and 1980, a number of experimental papers confirmed both the quadratic pressure dependence and strong negative temperature dependence of the water vapour self-continuum. We note below only some of the most frequently cited.

Burch et al. (e.g., Burch 1970; Burch and Gryvnak 1978) began a series of measurements of the self- and foreign-continuum absorption that would extend into the 1980s, with an accuracy unprecedented for that time, in the whole mid-infrared spectral region (wavelengths around 4–25 μm) at different temperatures. They demonstrated with certainty that the Lorentzian wings of water vapour lines were far from adequate to model continuum absorption, both within 6-μm band and in the mid-infrared and in the 4-μm window. The continuum data from the experiments of Burch and colleagues are still used for comparison with modern experiments.

A series of laboratory and field experiments by Aref'ev et al. (1975) and Aref'ev and Dianov-Klokov (1977) on the temperature and pressure dependence of continuum absorption in the mid-infrared window using a CO_2 laser broadly supported and extended previous measurements. From both a review of earlier measurements (Adiks et al. 1975) and the analysis of their experimental data, the authors concluded that the continuum could not be caused by aerosol (including water aerosol) or by collisional broadening of

[2] In the literature, the continuum is frequently referred to as 'self-broadened' and 'foreign-broadened', but this is presumptuous as to the cause of the continuum (as was pointed out by Cormier et al. 2005)—'self continuum' and 'foreign continuum' are more neutral and are becoming more widely adopted now.

monomer lines and were strongly supportive of the dimer hypothesis. These data have also been often used in developing models that go beyond the Lorentzian far-wing assumption.

Montgomery (1978) measured the water self-continuum absorption near 8 μm in a wide temperature region (333–473 K). This work was often cited and used to validate various models as an example of the complicated temperature dependence of the continuum. Montgomery found that the negative temperature dependence of the continuum absorption coefficient turned to nearly temperature-independent or even a slightly positive temperature dependence above 400 K. He suggested that such behaviour could be explained by a combination of dimer contribution at lower temperatures and far-wing (collisional broadening) contribution at higher temperatures.

The failure of the Lorentzian line shape to predict the magnitude, temperature and pressure dependence of the water vapour continuum absorption contributed to the popularity of the dimer hypothesis. On the other hand, as noted above, it had long been recognised that the Lorentzian line shape is rather simplistic in its origin and that it might be expected to fail in modelling very far wings of absorption lines. New models of the line profile, especially far from the line centre, started to appear (see e.g., the discussion in Goody and Yung 1989). An integral representation of the absorption coefficient of spectral lines, developed earlier by adopting a quantum-mechanical formalism in the works of Anderson (1949) and Tsao and Curnutte (1962), was applied to specifically develop a far-wing approximation within a statistical approach in Fomin and Tvorogov (1973). Tvorogov and Nesmelova (1976) suggested an alternative far-wing model based on the semi-classical asymptotic approach, fully developed later by Nesmelova et al. (1986). In general, theoretical far-wing models of the continuum developed in that period were characterised by a number of adjustable parameters fit to experimental data, which, however, did not always guarantee that they would agree with experimental data at different temperatures or wavenumbers.

With regards to the atmospheric implications, the lack of a compelling physical framework to explain the continuum was the likely reason why the more empirical approach of Roberts et al. (1976) became widely used. This approach was based on a detailed analysis of several published and unpublished long path-length measurements of the continuum absorption in the mid-infrared window. They suggested that the temperature dependence of the continuum cross-section could be represented by a simple exponential form using just one parameter derived by fitting to the known experimental data. Their representation was incorporated into the LOWTRAN radiation code, which was frequently used in this era, and this subsequently led to its common adoption in other radiation codes; it was often referred to as the RSB continuum, after the initial letters of the surnames of Roberts et al. (1976). Roberts et al. were agnostic as to cause of the continuum, noting that the dimer or 'any reasonable line shape incorporating the strongly attractive interaction between water molecules during their collision' would give the observed temperature dependence, but they did not elaborate.

2.3 1980–2000

This period was characterised by two important developments. Firstly, work by Clough et al. (1980, 1989, 1992) led to a continuum model, called CKD (after the names of the authors Clough et al. 1989). Its successor MT_CKD (see Sect. 2.4) is currently the most widely adopted in atmospheric applications.[3] Secondly, a whole series of reports of further

[3] The CKD and MT-CKD models include the continua of several gases, in addition to water vapour, but the focus here will be on water vapour.

laboratory measurements by Burch and colleagues (e.g., Burch 1981, 1985; Burch and Gryvnak 1980; Burch and Alt 1984) presented what were then the most detailed measurements of the continuum, in terms of ranges of wavelengths and temperatures, many of which provided the backbone to the Clough et al. (1989) work. Other laboratory and field measurements across a wide range of conditions in this period contributed to further assessment and improvement of continuum models (e.g., Han et al. 1997; Tobin et al. 1996, 1999; Varanasi and Chudamani 1987). Grant (1990) presented a review of a wide range of measurements in the mid-infrared window.

The CKD model established a more formal approach to the definition of the continuum, which had validity throughout the microwave and infrared. To achieve this, it adopted as its basis the Van Vleck and Huber (1977) line-shape function, which is equivalent to the Lorentz shape in the infrared. They applied a so-called wavenumber-dependent 'chi function', which modified the line shape. The justification for this chi function was to account for the finite duration of collisions (the Lorentz line shape assumes them to be instantaneous); in practice, its form was empirical and chosen so as to fit the observed continuum. In the case of the self-continuum, this led to the lines being assumed super-Lorentzian (i.e. the chi function was greater than unity) from around 10–1,000 cm^{-1} from line centre and sub-Lorentzian at greater distances. The line-shape model was incapable of explaining or describing the observed negative temperature dependence of the continuum; instead, CKD adopted a purely empirical approach whereby a temperature dependence was imposed on the chi function by least-squares fitting to the observations of Burch and colleagues.

The CKD model also adopted a more formal (although essentially a quite arbitrary) definition in order to separate local line absorption from the continuum, which has been much used in later work (see especially Fig. 1 of Clough et al. 1989). This has allowed a more straightforward comparison of different derivations of the continuum. The line contribution is taken to be the Lorentz profile only within 25 cm^{-1} of the line centre,[4] but with the value of the line absorption at 25 cm^{-1} (sometimes called the 'pedestal' or 'plinth') subtracted from this line profile. Hence, part of the line absorption is included within the definition of the continuum. One rationale for this approach is that it leads to the resulting continuum being smoother. The extent to which this pedestal contributed to the continuum was not quantified in these early works, although Paynter et al. (2009) have shown that, within the near-IR bands of water vapour, around 5–7% of the self-continuum is due to this line absorption—it will be much less in window regions, where the line contribution is, by definition, much smaller. This definition causes the continuum to be dependent on the spectral line database of the monomer that is used in its derivation (particular near band centres) and necessitates ongoing efforts to improve knowledge of the monomer spectral line parameters.

Clough et al. (1980, 1989) demonstrated that the use of the Lorentz line approximation to represent the absorption in the far wings of the water vapour lines led to an *overestimate* of the strength of the continuum in the mid-infrared window and to a significant underestimate within water vapour bands; Varanasi and Chudamani (1987) noted the overestimate in the mid-infrared region, although the extent of the overestimate depends on the assumed vapour and air pressures and temperature.

The parameterisation used in the CKD model was very successful compared to all previous empirical and theoretical models and became the 'industry standard' in radiation codes within the atmospheric sciences. It allowed a reasonably comprehensive description

[4] This approximately corresponds to the applicability limit of the impact approximation, assuming an average duration of collision $\sim 10^{-12}$ s.

of continuum absorption in rather broad spectral regions using just one modified Lorentzian line shape with a few fitted parameters. The weakness of this approach, however, was the ad-hoc nature of these parameters, which therefore could not be verified from independent sources nor could the CKD model be directly used to support or refute any particular hypothesis for the cause of the continuum.

Clough et al. (1992) clearly demonstrated the impact of the CKD continuum on thermal infrared irradiances and cooling rates. Whilst the impact on lower tropospheric cooling rates via the self-continuum could be broadly anticipated from the work referred to in Sect. 2.2, Clough et al. (1992) also demonstrated that the foreign continuum, acting in the rotational bands of water vapour, led to a 25% enhancement of cooling rates in the upper troposphere, compared to a case with water monomer lines only.

During this period, the water vapour continuum played a 'bit part' in another controversy, concerning the amount of solar radiation absorbed by clouds—some analyses of observations indicated a cloud absorptance of typically 30%, whilst conventional theory and many observations indicated (and still indicates) a value closer to 10%. The possibility that a hypothetical enhanced water vapour continuum (coupled to the increased path length because of multiple scattering within the cloud) could impact on the absorption of solar radiation by clouds was raised (Arking 1999; Stephens and Tsay 1990); however, the lack of spectrally resolved measurements prevented any progress on this hypothesis. The more recently derived values of the continuum in the near-infrared windows (see Sect. 2.4) would not be able to explain very large cloud absorptances.

This period was also characterised by what was a quite bitter controversy between the far wing and dimer explanations for the continuum. More sophisticated far-wing theories were developed (e.g., Nordstrom and Thomas 1980; Thomas and Nordstrom 1985). The most widely known emerged from the quantum-mechanical approach of Fano (1963) and Rosenkranz (1985, 1987) and its later development by Ma and Tipping (e.g., Ma and Tipping 1991, 1992; Tipping and Ma 1995) based on the quasi-static approximation for the far wings. In later works (e.g., Ma and Tipping 1999, 2002), Ma and Tipping significantly updated and developed their formalism making it applicable for describing the continuum in different cases of intermolecular interaction (i.e. for different gases) and spectral regions. This theory seemed to be reasonably good in describing the experimentally derived water vapour continuum in the mid-infrared spectral region including the wing of the rotational band (see e.g., Cormier et al. 2002); however, the model significantly underestimates continuum absorption within bands. The authors of the model explain this by the fact that their far-wing formalism is inapplicable closer to band centres.

The dimer explanation of the continuum also underwent further development during this period. The dimer hypothesis was largely supported at the time by a single characteristic of the observations—the negative temperature dependence. Hinderling et al. (1987) presented a comprehensive laboratory study of the mid-infrared continuum based on laser photo-acoustic spectroscopy. Their measurements of the temperature dependence of the continuum between −20°C and +70°C were consistent, but not conclusively so, with a dimer interpretation for absorption. They also performed an analysis of absorption in supersaturated air in a diffusion chamber. They inferred that the absorption cross-section for the dimer was about 30 times smaller than was required to explain the observed continuum absorption and concluded that the mid-infrared continuum was predominantly due to local absorption lines and far wings of distant lines. However, their inference appears to be based on a somewhat arbitrary separation of the observed absorption between dimers and far wings at one position in their chamber, and a different separation could have led them to a quite different conclusion. Nevertheless, it would be interesting to see the novel methodology of Hinderling et al. (1987) repeated with more modern instrumentation, including direct monitoring of the vapour pressure in the chamber, and interpreted in the context of modern understanding of dimer abundance and ab initio estimates of dimer absorption strength.

One of the most common arguments against the possibility of a significant water dimer contribution in the mid-infrared window, however, was an analysis of positions and intensities of intermolecular transitions (e.g., Bohlander et al. 1980). Indeed, the wings of librational intermolecular oscillations are expected decrease rapidly at wavelengths shorter than about 15 μm. However, it was shown by Vigasin and Chlenova (1984), on the basis of the symmetric-top harmonic approximation, that the whole of the mid-infrared window 8–13 μm should be filled with overlapping overtones and combinational bands of intermolecular oscillations of the dimer. Moreover, the intensities of these bands, estimated using the harmonic (Vigasin and Chlenova 1984) and anharmonic approximations (Yukhnevich and Tarakanova 1988), were found to be comparable with the experimental continuum absorption (see Fig. 2). However, recent more sophisticated quantum-chemical (Scribano and Leforestier 2007) and molecular dynamic calculations (Lee et al. 2008) have shown that the expected contribution from these bands to the continuum in the mid-infrared window is markedly less (Fig. 2) although still not negligible—from 15 to 40%.

2.4 Post-2000

In 2003, a major new version of CKD continuum model was issued, which represented the first recomputation of the entire self and foreign continua since the original CKD model was developed in the 1980s. The new model, called MT_CKD (the MT standing for two of the originators of the new version—Mlawer and Tobin—see Mlawer et al. (2012) and Delamere et al. (2010)), is based on an updated formulation and has been widely adopted within the atmospheric science community. As noted by Mlawer et al. (2012), the original

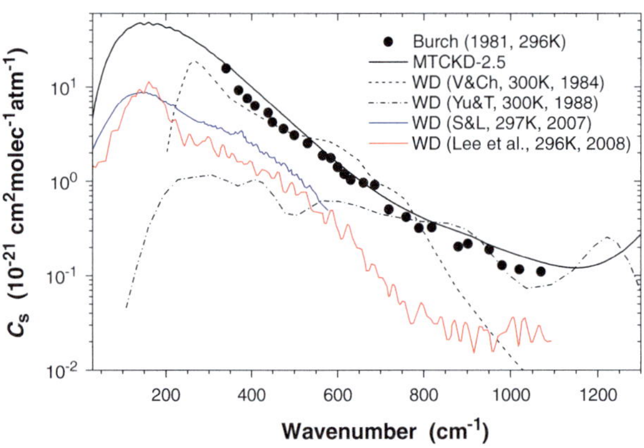

Fig. 2 Smoothed spectra of the calculated WD absorption cross-section from Vigasin and Chlenova (1984) (*V&Ch*), Yukhnevich and Tarakanova (1988) (*Yu&T*), Scribano and Leforestier (2007) (*S&L*) and Lee et al. (2008) in the far infrared and mid-infrared atmospheric window. The Burch (1981) experimental continuum and the MTCKD-2.5 continuum model are shown for comparison. The Lee et al. cross-section curve is shown a factor 1.86 lower than in the original work, whilst *V&Ch* and *Yu&T* are shown a factor 5/3 higher, to convert all data to the equilibrium constant used by *S&L* (0.05 atm^{-1} at 296 K). In the spectra calculated by *V&Ch* and *Yu&T*, the contribution from pure rotational water dimer absorption was entirely ignored, which explains a rapid decrease of the calculated absorption below 200–250 cm^{-1}

CKD model had undergone successive revisions, as new measurements had become available which, however, were not consistent with the original line shape used in CKD. In particular, it became increasingly difficult to justify the use of the super-Lorentzian component of the adopted line shape (see Sect. 2.3). Instead of the exclusively far-wing mechanism adopted in CKD, MT_CKD was based on the contributions from two components: a broad collision-induced absorption (CIA), which is considered to be dominant only within bands, and a (sub-Lorentzian) line-wing absorption, which should be dominant in windows. As discussed by Mlawer et al. (2012), in the time since the original development of MT_CKD, significant differences between earlier versions of MT_CKD and recent observations (by almost an order of magnitude in the 4-μm window) have been noted; these new measurements have not yet been used to reformulate the underlying basis of MT_CKD model but have instead been applied as simple scaling factors to the original MT_CKD values in certain spectral regions. As will be noted below, there is strong evidence that even with these revisions, MT_CKD underestimates continuum strength in the near-infrared windows.

Vigasin (2000) suggested a simple parameterisation of continuum absorption in the mid-infrared and microwave spectral regions using just a few physical parameters related to the water complexes. He has shown that this model is in a good agreement with known measurements.

Interest in the role of water dimers in the atmospheric science literature was reinvigorated by Vaida et al. (2001), and in a subsequent series of papers by co-authors (e.g., Schofield and Kjaergaard 2003; Kjaergaard et al. 2008; Garden et al. 2008) focusing on the near-infrared region. These papers applied quantum chemistry methods to perform ab initio calculations of both the potential energy surface and the dipole moment function for the water dimer, in order to then produce computations of the strengths and positions of the vibrational bands of the water dimer. These, together with knowledge of the atmospheric abundance of the dimer,[5] could then be used to calculate the radiative impact. However, a serious and remaining uncertainty in such analyses is that rotational structure of the infrared water dimer bands has not yet been directly computed; hence, assumptions have to be made as to the width and shape of the individual vibrational bands. Since the centres of the dimer bands are generally coincident with the strong water monomer bands, knowledge of the width and shape of the dimer bands is crucial in determining their ability to absorb in window regions and thus whether they can explain the continuum in those windows. By adopting various assumptions, Vaida et al. (2001) demonstrated the additional global-mean atmospheric absorption of solar radiation by dimers could be up to a few Wm^{-2}. Subsequently, Kjaergaard et al. (2003) also examined complexes of water vapour with other molecules (for example, nitrogen or oxygen) and indicated that these could be of similar, or possibly greater, importance than the dimer.

One important outcome of these calculations was that it indicated that, whilst the dimer spectrum had a continuum nature, it still possessed broad spectral structures, particularly near band centres. This opened up the prospect for attempts to detect these in-band features in laboratory spectra even at close to atmospheric conditions. This possibility was first

[5] Under atmospheric conditions, the partial pressure of water dimers is typically 3 orders of magnitude lower than the vapour pressure of the monomer. The abundance of the dimer decreases with increasing temperature but the exact value of the equilibrium constant that links the abundance of dimer to the monomer, and its temperature dependence, particularly in atmospheric conditions, has significant uncertainties (see e.g., Ptashnik 2008). Note that, while higher-order clusters of water vapour likely exist, their abundance is probably too low (for example the trimer is believed to be approximately three orders of magnitude less abundant than the dimer) to have a significant radiative impact.

pointed out and successfully realised by Ptashnik et al. (2004). This experimental approach has its own difficulties, as the separation of the dimer from the monomer signature within particular bands requires reasonably good knowledge of the monomer spectrum, and there is the possibility of continuum-like artefacts in the analysis due to, for example, large systematic errors in water monomer line parameters. Nevertheless, analysis of the temperature dependence of the residual absorption cross-section allowed a clear distinction between the contribution of water dimers (which has a strong negative T-dependence) and the cumulative local contribution from water monomer lines (which is nearly independent of temperature). Our own work (Ptashnik et al. 2004; Paynter et al. 2007) indicated well-pronounced absorption spectral features that are in a reasonable agreement, in both strength and position, with those expected from the calculated dimer spectra (Schofield and Kjaergaard 2003; Garden et al. 2008). Ptashnik (2008) reviews the evidence that similar spectral features could also be seen in earlier measurements of, for example, Burch (1985), Vetrov and Yukhnevich (1975), and Poberovsky (1976).

The experimental work also indicated some broad spectral features that could not be explained by the theoretical dimer spectra. Recently, Ptashnik et al. (2011a) suggested that this could be explained by the fact that the theoretical spectra accounted only for truly bound dimers. In addition to these bound dimers, it was proposed by Vigasin (1985, 1991, 2003, 2010) and Epifanov and Vigasin (1997), by a statistical partitioning of molecular pairs in phase space, that shorter-lived quasi-bound dimers should be of equal importance in atmospheric conditions. A simplified calculation (Ptashnik et al. 2011a) indicated that these quasi-bound dimers might explain the additional spectral features in the observed near-room temperature spectra of the continuum. This may also help to explain the wavelength variation of the temperature dependence of the continuum and non-exponential temperature dependence of the continuum at particular wavenumbers in the mid-infrared window detected by Baranov et al. (2008). Classical- and quantum-mechanical calculations by Schenter et al. (2002) for H_2O–H_2O pairs and 3D classical trajectory simulations by Lokshtanov et al. (2005) provide independent support for the statistical partitioning approach of Vigasin and hence support the view that quasi-bound water dimers play an important role at close to room temperature. Ptashnik et al. (2011a) also reviewed the role of dimers in explaining the continuum in the mid- and far-infrared—bound dimers may contribute 15–40%, with an expected additional contribution from the quasi-bound dimers.

Given the uncertainties in the width and shape of the vibrational bands of the dimer, the situation in the windows, which is more important from an atmospheric radiation point of view, is less clear. The continuum is much weaker than in the bands, and its accurate detection is an experimental challenge. Unlike in the bands, where the main source of error comes from the subtraction of the local line contribution, in the windows the baseline drifts in the spectroscopic measurements could lead to continuum-like artefacts if not properly taken into account. However, recent laboratory observations (Baranov and Lafferty 2011; Ptashnik et al. 2011b, 2012; Baranov 2011) demonstrate that the strength of both the self- and foreign-continuum is much stronger (by up to around an order of magnitude) in many near-infrared windows than given in the most recent version of the MT_CKD model. These measurements, which are summarised in Fig. 3, are not inconsistent with the dimer explanation, but they are far from conclusive given the poor understanding of the shape and width of the dimer bands.

These new analyses of laboratory observations also indicate that the observed continuum absorption in all near-infrared windows is much stronger than is predicted by the quasi-static far-wing theory (Tipping and Ma 1995) (see Fig. 3). Since the parameters of this theory are directly related to fundamental physical values, such as the monomer geometry and parameters associated with the intermolecular potential, it seems likely that

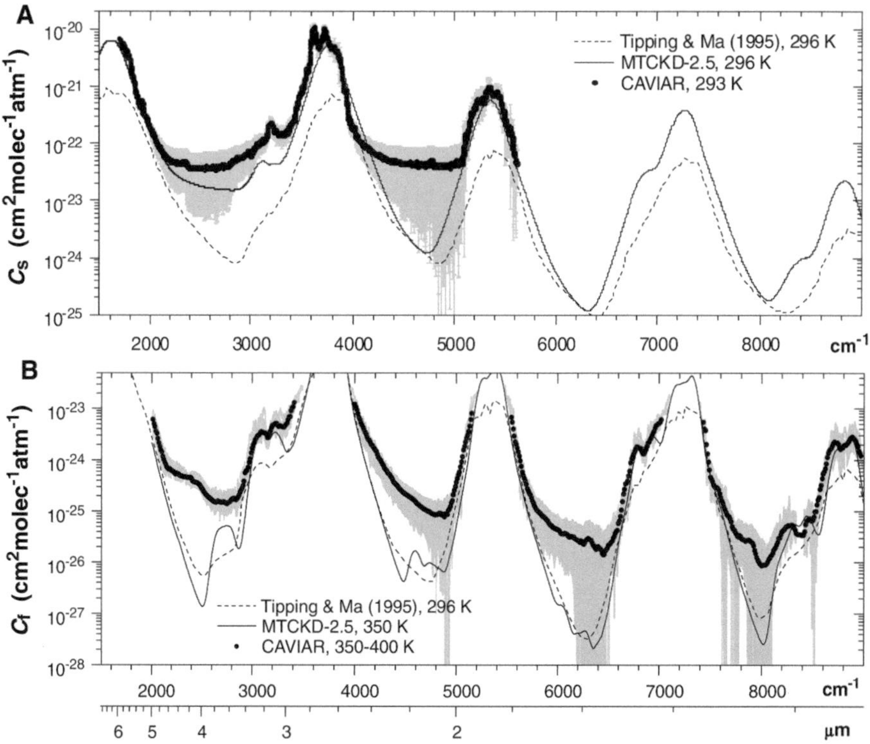

Fig. 3 Self- and foreign-continuum cross-section (*upper and lower panels* respectively) as retrieved from laboratory measurements in Ptashnik et al. (2011b, 2012) (CAVIAR), along with their associated uncertainties, compared to the far-wing model of Tipping and Ma (1995) and the MTCKD-2.5 continuum model

it will be difficult to fit it to the new experimental values. In the mid-infrared window, the recent calculation by Ma et al. (2008) is in a good agreement with experiment of Baranov et al. (2008) at wavenumbers greater than 10 μm but underestimates the continuum by up to 30–40% at lower wavelengths. The existing far-wing model also cannot explain the observed temperature dependence of the self-continuum in this window (Ma et al. 2008), or the non-exponential temperature dependence at particular wavenumbers.

Ma et al. (2008) and Leforestier et al. (2010) also presented theoretical estimates of the temperature dependence expected for far wings of water monomers, for dimer absorption and for collision-induced absorption. It was pointed out that the strongest temperature dependence is expected for water dimers. These temperature dependencies are likely to prove useful in further analysis of recent experimental results in the far- and mid-infrared spectral regions.

Another far-wing theory, based on asymptotic semi-classical approach (Nesmelova et al. 1986; Bogdanova Ju and Rodimova 2010; Klimeshina et al. 2011), shows a much better agreement with experimental data in the 4-μm window at high temperatures ($\sim$350 K) than Tipping and Ma (1995). However, at room temperature, this model also strongly underestimates the self-continuum in this window (Klimeshina et al. 2011).

In this period, cavity ring-down spectroscopy, which is particularly suited to detecting weak absorption, began to be actively applied to the continuum. Cormier et al. (2002)

presented evidence that the spectral variation in the mid-infrared window was consistent with the Ma and Tipping (1999) far-wing line theory, whilst Cormier et al. (2005) presented one of the most detailed analyses of the temperature dependence in this window at near-atmospheric conditions, which they found to be in good agreement with Vigasin's (2000) parameterisation for water complexes. They also presented evidence that the foreign continuum was around 250 times weaker than the self-continuum, a value smaller by an order of magnitude than that assumed in MT_CKD.

New Fourier transform spectroscopy measurements from aircraft over a particularly wide range of atmospheric conditions (Taylor et al. 2003) led to suggestions for important modifications to the CKD-type models in the mid-infrared window. Rowe and Walden (2009) reported refined measurements of the foreign continuum in the 6.3-μm band of water vapour from ground-based measurements in the Antarctic, which similarly indicated the need for modifications. Serio et al. (2008) reported atmospheric measurements of the foreign continuum in the far infrared, which indicated good agreement with MT_CKD at wavelengths less than 20 μm but found a weaker continuum, by typically around 20%, at longer wavelengths. More recently, airborne field measurements (coupled with satellite data) as part of the CAVIAR consortium have provided important new data with which to assess continuum models. Newman et al. (2012) show that changes in the strength of the self-continuum between MT_CKD versions 2.4 and 2.5 at around 3.6 μm lead to much better agreement with their observations; at 6.3 μm, there is broad agreement with both the foreign continuum in MT_CKD and the laboratory measurements of Ptashnik et al. (2012), but Newman et al. (2012) present evidence that there may be a temperature dependence in the strength of the foreign continuum at atmospheric temperatures. Green et al. (2012) assessed the strength of the foreign continuum in the far infrared, finding that whilst at wavelengths less than 30 μm, their retrievals agreed well with MT_CKD, discrepancies at 100 μm, where very few measurements have previously been made, reached 50%.

In summary, this period was characterised by an important and extensive revision to the CKD continuum model, strengthened evidence of the role of the dimer in explaining the continuum within bands and by laboratory measurements that improved the characterisation of the continuum both within bands and within windows.

3 Implications of New Laboratory Observations for Remote-Sensing Applications

In Ptashnik et al. (2011b, 2012), we estimated that our derived continuum adds about 1.25 W m^{-2} (or about 1.5%) to the global-mean clear-sky shortwave absorption, relative to that calculated using the MT-CKD model. These measurements also have significant implications for remote-sensing techniques that use these windows. Reflectance measurements in the near-infrared windows have long been used as a way of retrieving cloud particle size (Nakajima and King 1990); for example, the 2.1-μm channel on the MODIS (moderate resolution imaging spectroradiometer) satellite instrument is used as part of routine cloud droplet-size retrievals (e.g., Zhang and Platnick 2011). Reflectance at these wavelengths is strongly sensitive to particle size, with larger droplets giving a smaller reflectance.

We explore the impact of increases in continuum strength on such retrievals. Qualitatively, an increase in absorption due to a strengthened water continuum would lead to a decrease in albedo—if unaccounted for in a retrieval procedure, this lower albedo could be interpreted as indicating droplet sizes are larger than they are in reality. Any such effect would be expected to be largest for low clouds in the tropics, because of the dependence of the self-continuum on the square of the vapour pressure.

The results presented here do not constitute any replication of the MODIS retrieval process, but rather they show that the strength of the continuum modifies the dependence of the top-of-atmosphere albedo on cloud droplet effective radius in the 2.1-μm window. The calculations are performed for one band (2.083–2.128 μm) of the 220-band version of the Edwards and Slingo (1996) radiative transfer code (1) using the CKD 2.1 water vapour continuum (which is identical to MT_CKD in this spectral region) and then (2) using the self- and foreign-continuum derived by Ptashnik et al. (2011b, 2012) from laboratory measurements. The self-continuum was extrapolated to atmospheric temperatures following Ptashnik et al. (2011b) whilst the foreign continuum was assumed to be independent of temperature.

The calculations are performed for a 1-km thick overcast cloud layer with a liquid water content of 1×10^{-3} kg m^{-3}. The albedo is calculated as a function of effective droplet radius (over the range 6–9 μm) for both continuum representations (the CKD continuum is so weak in this spectral region that the albedo is almost the same when the CKD continuum is excluded completely). The results are presented as the decrease in droplet effective radius, when introducing the new continuum; hence, this is a kind of 'quasi-retrieval'—an albedo is derived for a given droplet radius using the CKD 2.1 model, and then the change in effective radius is computed, which gives the same albedo when the new continuum is used. The decrease in effective radius is almost independent of the effective radius itself, at least over the range we have tested it. Figure 4 shows the results for three standard atmospheres [tropical, mid-latitude summer (MLS) and mid-latitude winter (MLW)] for two solar zenith angles (30° and 60°) and for clouds, located between approximately 700–800 hPa and 800–900 hPa.

As expected, the effect is largest for a tropical atmosphere when the cloud top is lower (because of the dependence on the vapour pressure), and also largest for the higher solar zenith angle (as the increased path length means that there is more continuum absorption)—for the tropical, 800 hPa cloud top and 60° solar zenith angle, the new continuum requires droplet radii to be more than 1.5 μm smaller to generate a given albedo, when the new continuum is used. Even in the MLS case, the effect is around 1 μm, whilst it is less than 0.3 μm for all the MLW cases.

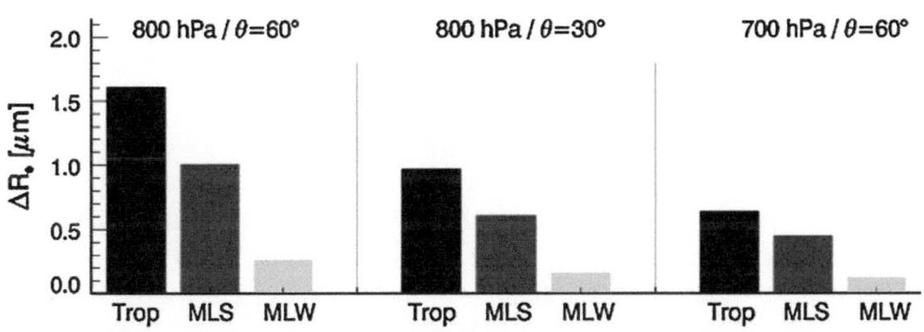

Fig. 4 Calculated decrease in effective cloud droplet radius (in μm) required to generate a given top-of-the-atmosphere albedo at 2.1 μm when the new values of the self- and foreign-continuum given in Ptashnik et al. (2011b, 2012) are used in place of the MT_CKD continuum model—this will approximately mimic the effect of the updated continuum on cloud effective radius retrievals from satellite in this spectral region. Results are shown for three atmospheres (tropical (Trop), mid-latitude summer (MLS) and mid-latitude winter (MLW)), two cloud-top pressures (800 and 700 hPa) and two solar zenith angles (θ) of 30° and 60°. Calculations are for a 1-km thick overcast cloud layer with a liquid water content of 1×10^{-3} kg m^{-3}

These results indicate that there may be a significant bias in MODIS, and similar, retrievals that use this 2.1-μm window region for cloud droplet-size retrieval and this bias will be dependent on location, time of day and cloud-top height. As is made clear by Zhang and Platnick (2011), there are multiple possible reasons for biases in cloud droplet radius retrievals from satellite measurements, relating to issues including cloud thickness and inhomogeneity. Nevertheless, it is interesting that Painemal and Zuidema (2011) find that in situ measurements of effective radius of marine stratocumulus are systematically smaller than those derived from the MODIS 2.1 μm channel, and Breon and Doutriaux-Boucher (2005) find similar differences between their satellite retrievals using a polarisation technique and those from MODIS. More detailed calculations of the impact in revisions in our knowledge of continuum absorption on satellite cloud droplet radius retrievals appear to be justified.

4 Discussion and Conclusions

In 1981, Burch commented that 'although the positions, intensities and widths of most of the significant absorption lines are well known, the H_2O absorption in the windows cannot be calculated from theoretical considerations alone … due to lack of knowledge about the shapes of the absorption lines and about possible absorption by dimers'. It is worth reflecting how much progress has been made in the intervening 30 years. In many ways, Burch's statement could be repeated, with some accuracy, word-for-word now. A particular difficulty is that whilst line-shape theory has advanced considerably, its experimental verification is exceedingly difficult because of the strong overlap between individual water monomer spectral lines.

It is certainly true that there remains heavy reliance on the empirically derived continuum within the window regions. However, within the absorption bands, the theoretically derived dimer strengths are showing considerable promise. They appear capable of explaining much of the observed continuum in the near-infrared bands and a significant fraction of that observed in the mid- and far-infrared. Within the near-infrared windows, dimers are a clear candidate mechanism, but further theoretical progress on the width and shape of the dimer bands is necessary for that hypothesis to progress. Existing far-wing theory does not appear able to explain the observed absorption. In the mid-infrared window, by contrast, both dimers (qualitatively) and far-wing theory seem able to explain many aspects of the observations—it seems likely that both mechanisms contribute to some extent and advances in theory and innovative experiments are required to make progress here.

One concern is that, even now, many of the laboratory experiments have to be performed in conditions not properly representative of atmospheric conditions—to generate large enough water paths in the laboratory, heated cells are required. Much progress has been made in this regard in recent years, with important implications for both the radiative budget and remote sensing, but extrapolation of these results to atmospheric conditions is not straightforward. Advances in instrumentation are needed and more 'redundancy' (i.e. repetition of measurements and analysis in different laboratories) would increase confidence that recently obtained results are robust.

Clearly, field observations of the continuum have played a considerable role in the evolution of understanding of the continuum, especially in the mid-infrared window where continuum absorption can be rather strong, and remain essential if the role of the continuum in real atmospheric conditions is to be understood. In the weaker near-infrared windows, separation of true water vapour continuum signal from instrumental artefacts and other contributors is a serious challenge.

Finally, the separation of the continuum absorption from that due to the line absorption by the water monomer is dependent on accurate knowledge of that line absorption, especially within bands—again, progress in recent years has been significant, but more remains to be done.

Acknowledgments This work was supported by the EPSRC/NERC CAVIAR consortium; IVP also acknowledges support from Russian National contract 02.740.11.5198. We thank the referees for their comments.

References

Adel A (1939) Atmospheric absorption of infrared solar radiation at the Lowell Observatory I. Astrophys J 89:1–9

Adel A, Lampland CO (1940) Atmospheric absorption of infrared solar radiation at the Lowell Observatory—III and IV. The spectral intervals 8.0–11.0 μ and 11.0–14.0 μ. Astrophys J 91:481–487

Adiks TG, Aref'ev VN, Dianov-Klokov VI (1975) Influence of molecular absorption on propagation of CO_2 laser radiation in terrestrial atmosphere (review). Sov J Quant Electron 5:481–488

Anderson PW (1949) Pressure broadening in the microwave and infra-red regions. Phys Rev 76:647–671

Anthony R (1952) Atmospheric absorption of solar infrared radiation. Phys Rev 85:674

Aref'ev VN, Dianov-Klokov VI (1977) Attenuation of 10.6-μm radiation by water vapor and the role of $(H_2O)_2$ dimers. Opt Spectrosc 42:488–492

Aref'ev VN, Dianov-Klokov VI, Radionov VF, Sizov NI (1975) Laboratory measurements of attenuation of CO_2 laser radiation by pure water vapor. Opt Spectrosc 39:560–561

Arking A (1999) Bringing climate models into agreement with observations of atmospheric absorption. J Clim 12:1589–1600

Baranov YI (2011) The continuum absorption in $H_2O + N_2$ mixtures in the 2000–3250 cm^{-1} spectral region at temperatures from 326 to 363 K. J Quant Spectrosc Radiat Transf. doi:10.1016/j.jqsrt.2011.06.005

Baranov YI, Lafferty WJ (2011) The water-vapor continuum and selective absorption in the 3–5 μm spectral region at temperatures from 311 to 363 K. J Quant Spectrosc Radiat Transf 112:1304–1313

Baranov YI, Lafferty WJ, Fraser GT, Ma Q, Tipping RH (2008) Water-vapor continuum absorption in the 800–1250 cm^{-1} spectral region at temperatures from 311 to 363 K. J Quant Spectrosc Radiat Transf 109:2291–2302

Bignell KJ (1970) The water-vapour infra-red continuum. Q J R Meteorol Soc 96:390–403

Bignell K, Saiedy F, Sheppard PA (1963) On the atmospheric infrared continuum. J Opt Soc Am 53:466–479

Bogdanova Ju V, Rodimova OB (2010) Line shape in far wings and water vapor absorption in a broad temperature interval. J Quant Spectrosc Radiat Transf 111:2298–2307

Bohlander RA, Emery RJ, Llewellyn-Jones DT, Gimmestad GG, Gebbie HA, Simpson OA, Gallagher JJ, Perkowitz S (1980) Excess absorption by water vapor and comparison with theoretical dimer absorption. In: Deepak A, Wilkerson TD, Ruhnke LH (eds) Atmospheric water vapor. Academic Press, New York, pp 241–254

Bolle HJ (1963) The 15–26 micron emission spectrum at Jungfraujoch (3570 m). Appl Optics 2:571–580

Braun C, Leidecker H (1974) Rotational and vibrational spectra for the H_2O dimer: theory and comparison with experimental data. J Chem Phys 61:3104–3113

Breon F-M, Doutriaux-Boucher M (2005) A comparison of cloud droplet radii measured from space. IEEE Trans Geosci Rem Sens 43:1796–1805

Brunt D (1932) Notes on radiation in the atmosphere. Q J R Meteorol Soc 58:389–420

Burch DE (1970) Investigation of the absorption of infrared radiation by atmospheric gases. Semi-annual technical report Philco-Ford Corporation Aeronutronic Division Newport Beach CA Rept. U-4784

Burch DE (1981) Continuum absorption by H_2O. Proc SPIE 277:28–39

Burch DE (1985) Absorption by H_2O in narrow windows between 3000 and 4200 cm^{-1}. US Air Force Geophysics Laboratory report AFGL-TR-85-0036 Hanscom Air Force Base, MA

Burch D, Alt R (1984) Continuum absorption by H_2O in the 700–1200 and 2400–2800 cm^{-1} windows. AFGL-TR-84-0128 Air Force Geophysics Laboratory Hanscom AFB, MA

Burch DE, Gryvnak DA (1978) Infrared absorption by CO_2 and H_2O. Report AFCRL-TR-78-0154 Air Force Cambridge Research Laboratory Hanscom AFB, MA

Burch DE, Gryvnak DA (1980) Continuum absorption by H_2O vapor in the infrared and millimetre wave regions. In: Deepak A, Wilkerson TD, Ruhnke LH (eds) Atmospheric water vapor. Academic Press, New York, pp 47–76

Clough SA, Kneizys FX, Davis R, Gamache R, Tipping R (1980) Theoretical line shape for H_2O vapor: application to the continuum. In: Deepak A, Wilkerson TD, Ruhnke LH (eds) Atmospheric water vapor. Academic Press, New York, pp 25–46

Clough SA, Kneizys FX, Davies RW (1989) Line shape and the water vapor continuum. Atmos Res 23:229–241

Clough SA, Iacono MJ, Moncet J-L (1992) Line-by-Line calculation of atmospheric fluxes and cooling rates: application to water vapor. J Geophys Res 97:15761–15785

Cormier JG, Ciurylo R, Drummond JR (2002) Cavity ringdown spectroscopy measurements of the infrared water vapor continuum. J Chem Phys 116:1030–1034

Cormier JG, Hodges JT, Drummond JR (2005) Infrared water vapor continuum absorption at atmospheric temperatures. J Chem Phys 122:114309

Cowling TG (1942) The absorption of water vapour in the far infra-red. Rep Prog Phys 9:29–41

Cox SK (1969) Observational evidence of anomalous infrared cooling in a clear tropical atmosphere. J Atmos Sci 26:1347–1349

Cox SK (1973) Infra-red heating calculations with a water vapour pressure broadened continuum. Q J R Meteorol Soc 99:669–679

Delamere JS, Clough SA, Payne VH, Mlawer EJ, Turner DD, Gamache RR (2010) A far-infrared radiative closure study in the Arctic: application to water vapor. J Geophys Res 115:D17106

Edwards JM, Slingo A (1996) Studies with a flexible new radiation code. 1. Choosing a configuration for a large-scale model. Q J R Meteorol Soc 122:689–719

Elsasser WM (1938a) Far infrared absorption of atmospheric water vapor. Astrophys J 87:497–507

Elsasser WM (1938b) Note on atmospheric absorption caused by the rotational water band. Phys Rev 53:768

Elsasser WM (1938c) New values for the infrared absorption coefficient of atmospheric water vapor. Mon Weather Rev 66:175–178

Epifanov SY, Vigasin AA (1997) Subdivision of phase space for anisotropically interacting water molecules. Mol Phys 90:101–106

Fano U (1963) Pressure broadening as a prototype of relaxation. Phys Rev 131:259–268

Fomin VV, Tvorogov SD (1973) Formation of the far wing contour of spectral lines broadened by a foreign gas; analysis of exponential decrease of continuous absorption beyond the band head of the 4.3 μm band of CO_2. Appl Opt 12:584–589

Fowle FE (1917) Water-vapor transparency to low-temperature radiation. Smithsonian Miscellaneous Collections 68:8 (68 p)

Garden A, Halonen L, Kjaergaard H (2008) Calculated band profiles of the OH-stretching transitions in water dimer. J Phys Chem 112:7439–7447

Gebbie HA, Harding WR, Hilsum C, Pryce AW, Roberts V (1951) Atmospheric transmission in the 1 to 14 μm region. Proc R Soc Lond A 206:87–107

Goody RM, Yung YL (1989) Atmospheric radiation: theoretical basis. Oxford University Press, Oxford

Grant WB (1990) Water vapor absorption coefficients in the 8–13-μm spectral region. Appl Opt 29:451–462

Green PD, Newman SM, Beeby RJ, Murray JE, Pickering JC, Harries JE (2012) Recent advances in measurement of the water vapour continuum in the far-IR spectral region. Phil Trans Roy Soc A (to appear). doi:10.1098/rsta.2011.0263

Han Y, Shaw JA, Churnside JH, Brown PD, Clough SA (1997) Infrared spectral radiance measurements in the tropical Pacific atmosphere. J Geophys Res 102:4353–4356

Hettner G (1918) Über das ultrarote Absorptionsspektrum des Wasserdampfes. Ann Phys 360:476–496

Hinderling J, Sigrist MW, Kneubühl FK (1987) Laser-photoacoustic spectroscopy of water-vapor continuum and line absorption in the 8 to 14 μm spectral window. Infrared Phys 27:63–120

Kiehl J, Trenberth KE (1997) Earth's annual global mean energy budget. Bull Am Meteor Soc 78:197–208

Kjaergaard HG, Robinson TW, Howard DL, Daniel JS, Headrick JE, Vaida V (2003) Complexes of importance to the absorption of solar radiation. J Phys Chem A 107:10680–10686

Kjaergaard H, Garden A, Chaban G, Gerber R, Matthews D, Stanton J (2008) Calculation of vibrational transition frequencies and intensities in water dimer: comparison of different vibrational approaches. J Phys Chem A 112:4324–4335

Klimeshina TE, Bogdanova YV, Rodimova OB (2011) Continuum absorption by water vapor in the atmospheric transparency windows 8-12 and 3-5 μm. Atmos Oceanic Opt (in press)
Lee M-S, Baletto F, Kanhere DG, Scandolo S (2008) Far-infrared absorption of water clusters by first-principles molecular dynamics. J Chem Phys 128:214506
Leforestier C, Tipping RH, Ma Q (2010) Temperature dependences of mechanisms responsible for the water-vapor continuum absorption. II. Dimers and collision-induced absorption. J Chem Phys 132:164302
Lokshtanov SE, Ivanov SV, Vigasin AA (2005) Statistical physics partitioning and classical trajectory analysis of the phase space in CO_2–Ar weakly interacting pairs. J Mol Struct 742:31–36
Ma Q, Tipping RH (1991) A far wing line shape theory and its application to the water continuum. J Chem Phys 95:6290–6301
Ma Q, Tipping RH (1992) A far wing line shape theory and its application to the foreign-broadened water continuum absorption III. J Chem Phys 97:818–828
Ma Q, Tipping RH (1999) The averaged density matrix in the coordinate representation: application to the calculation of the far-wing line shapes for H_2O. J Chem Phys 111:5909–5921
Ma Q, Tipping RH (2002) The frequency detuning correction and the asymmetry of line shapes: the far wings of H_2O–H_2O. J Chem Phys 116:4102–4115
Ma Q, Tipping RH, Leforestier C (2008) Temperature dependences of mechanisms responsible for the water-vapor continuum absorption: 1. Far wings of allowed lines. J Chem Phys 128:124313
Manabe S, Wetherald RT (1967) Thermal equilibrium of the atmosphere with a given distribution of relative humidity. J Atmos Sci 24:241–259
Mlawer EJ, Payne VH, Moncet J-L, Delamere JS, Alvarado MJ, Tobin DD (2012) Development and recent evaluation of the MT_CKD model of continuum absorption. Phil Trans Roy Soc A (to appear). doi: 10.1098/rsta.2011.0295
Montgomery GP (1978) Temperature dependence of infrared absorption by the water vapor continuum near 1200 cm^{-1}. Appl Opt 17:2299–2303
Nakajima T, King MD (1990) Determination of the optical thickness and effective particle radius of clouds from reflected solar radiation measurements. Part I: theory. J Atmos Sci 47:1878–1893
Nesmelova LI, Rodimova OB, Tvorogov SD (1986) Spectral line shape and intermolecular interaction. Nauka Novosibirsk, 216 pp
Newman SM, Green PD, Ptashnik IV, Gardiner TD, Coleman MD, McPheat RA, Smith KM (2012) Airborne and satellite remote sensing of the mid-infrared water vapour continuum. Phil Trans Roy Soc A (to appear). doi:10.1098/rsta.2011.0223
Nordstrom RJ, Thomas ME (1980) The water vapour continuum as wings of strong absorption lines. In: Deepak A, Wilkerson TD, Ruhnke LH (eds) Atmospheric water vapor. Academic Press, New York, pp 77–100
Painemal D, Zuidema P (2011) Assessment of MODIS cloud effective radius and optical thickness retrievals over the south-east Pacific with VOCALS-Rex in situ measurements. J Geophys Res (in press). doi: 10.1029/2011JD016155
Paynter DJ, Ptashnik IV, Shine KP, Smith KM (2007) Pure water vapor continuum measurements between 3100 and 4400 cm^{-1}: evidence for water dimer absorption in near atmospheric conditions. Geophys Res Lett 34:L12808
Paynter DJ, Ptashnik IV, Shine KP, Smith KM, McPheat R, Williams RG (2009) Laboratory measurements of the water vapor continuum in the 1200 cm^{-1}–8000 cm^{-1} region between 293 K and 351 K. J Geophys Res 114:D21301
Penner SS, Varanasi P (1967) Spectral absorption coefficients in the pure rotation spectrum of water vapor. J Quant Spectrosc Radiat Transf 7:687–690
Poberovsky AV (1976) Problemy fiziki atmosfery. Sbornik Trudov Univ Leningrad 13:81–87 (in Russian)
Ptashnik IV (2008) Evidence for the contribution of water dimers to the near-IR water vapour self-continuum. J Quant Spectrosc Radiat Transf 109:831–852
Ptashnik IV, Smith KM, Shine KP, Newnham DA (2004) Laboratory measurements of water vapour continuum absorption in spectral region 5000–5600 cm^{-1}: evidence for water dimers. Q J R Meteorol Soc 130:2391–2408
Ptashnik IV, Shine KP, Vigasin AA (2011a) Water vapour self-continuum and water dimers. 1. Review and analysis of recent work. J Quant Spectrosc Radiat Transf 112:1286–1303
Ptashnik IV, McPheat RA, Shine KP, Smith KM, Williams RG (2011b) Water vapor continuum absorption in near-infrared windows derived from laboratory measurements. J Geophys Res 116:D16305
Ptashnik IV, McPheat RA, Shine KP, Smith KM, Williams RG (2012) Water vapour foreign continuum absorption in near-infrared windows from laboratory measurements. Phil Trans Roy Soc A (to appear). doi:10.1098/rsta.2011.0218

Roach WT, Goody WM (1958) Absorption and emission in the atmospheric window from 770 to 1250 cm^{-1}. Q J R Meteorol Soc 84:319–331
Roberts RE, Selby JEA, Biberman LM (1976) Infrared continuum absorption by atmospheric water vapor in the 8–12-μm window. Appl Optics 15:2085–2090
Rosenkranz PW (1985) Pressure broadening of rotational bands, I. A statistical theory. J Chem Phys 83:6139–6144
Rosenkranz PW (1987) Pressure broadening of rotational bands, II. Water vapor from 300 to 1100 cm^{-1}. J Chem Phys 87:163–170
Rowe PM, Walden VP (2009) Improved measurements of the foreign-broadened continuum of water vapor in the 6.3 μm band at −30 C. Appl Opt 48:1358–1365
Rubens H, Aschkinass E (1898) Beobachtungen über Absorption und Emission von Wasserdampf und Kohlensäure im ultrarothen Spectrum. Ann Phys 300:584–601
Schenter GK, Kathmann SM, Garrett BC (2002) Equilibrium constant for water dimerization: analysis of the partition function for a weakly bound system. J Phys Chem A 106:1557–1566
Schofield DP, Kjaergaard HG (2003) Calculated OH-stretching and HOH-bending vibrational transitions in the water dimer. Phys Chem Chem Phys 5:3100–3105
Scribano Y, Leforestier C (2007) Contribution of water dimers absorption to the millimeter and far infrared atmospheric water continuum. J Chem Phys 126:234301
Serio C, Masiello G, Esposito F, Di Girolamo P, Di Iorio T, Palchetti L, Bianchini G, Muscari G, Pavese G, Rizzi R, Carli B, Cuomo V (2008) Retrieval of foreign-broadened water vapor continuum coefficients from emitted spectral radiance in the H_2O rotational band from 240 to 590 cm^{-1}. Opt Express 16:15816–15833
Simpson GC (1928) Further studies in terrestrial radiation. Mem R Meteorol Soc 3:1–26
Stephens GL, Tsay S-C (1990) On the cloud absorption anomaly. Q J R Meteorol Soc 116:671–704
Strong J (1941) Study of atmospheric absorption and emission in the infrared spectrum. J Franklin Inst 232:1–22
Taylor JP, Newman SM, Hewison TJ, McGrath A (2003) Water vapour line and continuum absorption in the thermal infrared-reconciling models and observations. Q J R Meteorol Soc 129:2949–2969
Thomas ME, Nordstrom RJ (1985) Line shape model for describing infrared absorption by water vapour. Appl Optics 24:3526–3530
Tipping RH, Ma Q (1995) Theory of the water vapor continuum and validations. Atmos Res 36:69–94
Tobin DC, Strow LL, Lafferty WJ, Olson WB (1996) Experimental investigation of the self- and N2-broadened continuum within the ν2 band of water vapor. Appl Opt 35:4724–4734
Tobin DC, Best FA, Brown PD, Clough SA, Dedecker RG, Ellingson RG, Garcia RK, Howell HB, Knuteson RO, Mlawer EJ, Revercomb HE, Short JF, van Delst PF, Walden VP (1999) Downwelling spectral radiance observations at the SHEBA ice station: water vapor continuum measurements from 17–26 micrometer. J Geophys Res 104:2081–2092
Tsao CJ, Curnutte B (1962) Line-widths of pressure broadened spectral lines. J Quant Spectrosc Radiat Transf 2:41–91
Tvorogov SD, Nesmelova LI (1976) Radiation processes in band wings of atmospheric gases. Izv Ros Akad Nauk Fiz Atmos Okeana 12:627–633
Tyndall J (1861) On the absorption and radiation of heat by gases and vapours and on the physical connexion of radiation, absorption and conduction. Phil Trans R Soc Lond 151:1–136
Vaida V, Daniel JS, Kjaergaard HG, Goss LM, Tuck AF (2001) Atmospheric absorption of near infrared and visible solar radiation by the hydrogen bonded water dimer. Q J R Meteorol Soc 127:1627–1643
Van Vleck JH, Huber DL (1977) Absorption, emission, and linebreadths: a semi-historical perspective. Rev Mod Phys 49:939–959
Van Vleck JH, Weisskopf VF (1945) On the shape of collision-broadened lines. Rev Mod Phys 17:227–236
Varanasi P, Chudamani S (1987) Self- and N_2-broadened spectra of water vapor between 7.5 and 15.5 μm. J Quant Spectrosc Radiat Transf 38:407–412
Varanasi P, Chou S, Penner SS (1968) Absorption coefficients for water vapor in the 600–1000 cm^{-1} region. J Quant Spectrosc Radiat Transf 8:1537–1541
Vetrov AA, Yukhnevich GV (1975) Some optical properties of high-density water vapors. Opt Spectrosc 39:273–275
Vigasin AA (1985) On the spectroscopic manifestations of weakly bound complexes in rarefied gases. Chem Phys Lett 117:85–88
Vigasin AA (1991) Bound, metastable and free states of bimolecular complexes. Infrared Phys 32:461–470
Vigasin AA (2000) Water vapor continuous absorption in various mixtures: possible role of weakly bound complexes. J Quant Spectrosc Radiat Transf 64:25–40

Vigasin AA (2003) Bimolecular absorption in atmospheric gases. In: Camy-Peyret C, Vigasin AA (eds) Weakly interacting molecular pairs: unconventional absorbers of radiation in the atmosphere. Kluwer, Netherlands, pp 23–47
Vigasin AA (2010) On the possibility to quantify contributions from true bound and metastable pairs to infrared absorption in pressurized water vapor. Mol Phys 108:2309–2313
Vigasin AA, Chlenova GV (1984) Water-dimer spectrum for wavelengths >8 μm, and extinction of radiation in the atmosphere. Izv Atmos Oceanic Phys 20:596–599
Viktorova AA, Zhevakin SA (1966) Absorption of micro-radiowaves in air by water vapor dimers. Rep Acad Sci USSR 171:1061–1064 (in Russian)
Yukhnevich GV, Tarakanova EG (1988) Some properties of the potential energy surface and vibrational spectrum of a strong hydrogen bond complex. J Mol Struct 117:495–512
Zhang Z, Platnick S (2011) An assessment of differences between cloud effective particle radius retrievals for marine water clouds from three MODIS spectral bands. J Geophys Res 116:D20215
Zhevakin SA, Naumov AP (1963) On the absorption coefficient of electromagnetic waves by water vapor in the region 10 μm–2 cm. Izv Vysshikh Uchebn Zavedenii Radiofiz 6:674–695 (in Russian)

Surv Geophys (2012) 33:565–583
DOI 10.1007/s10712-011-9169-4

Tropical and Extra-Tropical Influences on the Distribution of Free Tropospheric Humidity Over the Intertropical Belt

Rémy Roca · Rodrigo Guzman · Julien Lemond · Joke Meijer · Laurence Picon · Hélène Brogniez

Received: 18 June 2011 / Accepted: 7 December 2011 / Published online: 3 February 2012

Abstract Free tropospheric humidity (FTH) is a key parameter of the radiation budget of the Earth. In particular, its distribution over the intertropical belt has been identified as an important contributor to the water vapour feedback. Idealized radiative transfer computations are performed to underscore the need to consider the whole probability distribution function (PDF) rather than the arithmetical mean of the FTH. The analysis confirmed the overwhelming role of the dry end of the PDF in the radiative perturbation of the top of atmosphere longwave budget. The physical and dynamical processes responsible for the maintenance of this dry part of the FTH distribution are reviewed, and the lateral mixing between the tropics and the extra-tropics is revealed as a major element of the dry air dynamics. The evolution of this lateral mixing in the framework of the global warming is discussed, and perspectives of work are listed as a mean of a conclusion.

Keywords Free troposphere · Humidity · Satellite · OLR · Advection–condensation paradigm

Abbreviations

AC	Advection-condensation
AIRS	Atmospheric Infrared Sounder
AR4	Fourth Assessment Report
BLH	Boundary Layer relative Humidity
BT	Brightness temperature
DJF	December, January and February
ECMWF	European Centre for Medium-Range Weather Forecasts
ERA-Interim	ECMWF ReAnalyses-interim
ERA-40	ECMWF ReAnalyses-40

R. Roca (✉) · R. Guzman · J. Lemond · J. Meijer · L. Picon
Laboratoire de Météorologie Dynamique, Université Pierre et Marie Curie, Tour 45-55 3ème étage, BP 99, 4 Place Jussieu, 75005 Paris, France
e-mail: roca@lmd.jussieu.fr

H. Brogniez
Laboratoire Atmosphères, Milieux, Observations Spatiales, Paris, France

FIR	Far infrared
FTH	Free tropospheric humidity
IPSL	Institut Pierre-Simon Laplace
ITCZ	Inter Tropical Convergence Zone
JJA	June, July and August
NCEP	National Center for Environmental Prediction
OLR	Outgoing Long wave Radiation
PDF	probability distribution function
RH	Relative Humidity
TS	Temperature of the surface
TOA	Top of atmosphere

1 Introduction

Relative Humidity (RH) in the free troposphere (above the boundary layer and below the tropopause) plays an important role in the climate system (Sherwood et al. 2010a). In particular, over the intertropical regions, the free tropospheric humidity distribution is considered to contribute strongly to the water vapour feedback (e.g., Allan 2012). The climatological patterns of humidity distribution are recalled in Fig. 1 where satellite-derived multiyear estimates of seasonal mean relative humidity at 500 hPa are shown. The

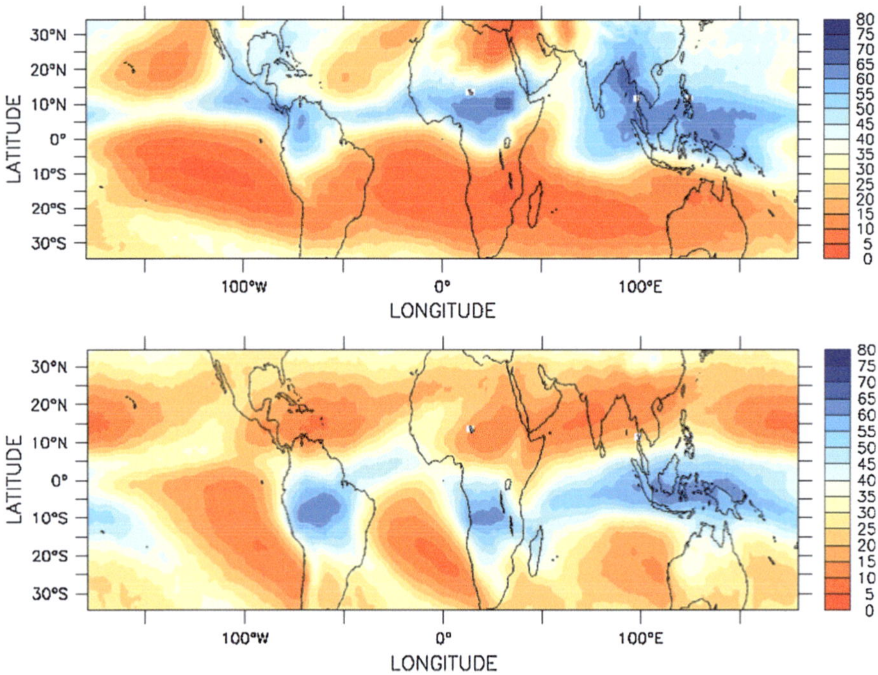

Fig. 1 Multiyear 2003–2007 average of seasonal mean relative humidity at 500 hPa from AIRS measurements for JJA (*top*) and DJF (*bottom*). Units are in percentage

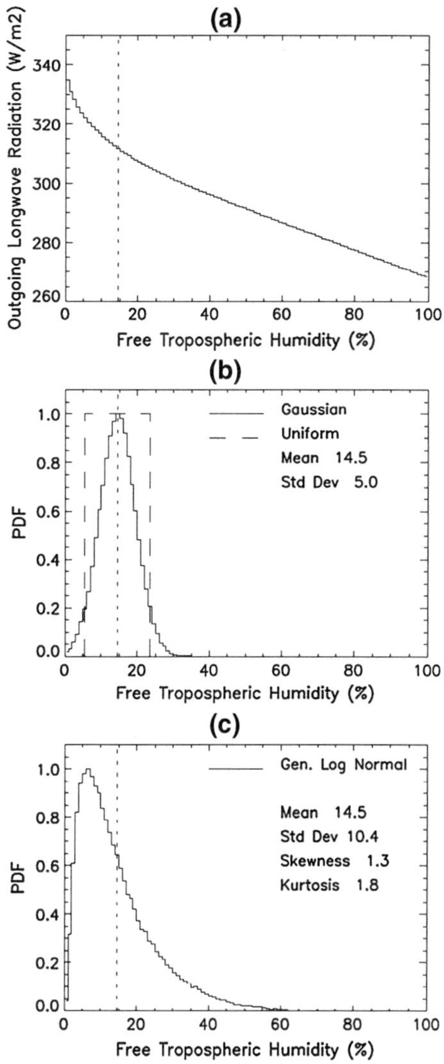

Fig. 2 **a** Outgoing longwave radiation as a function of the free tropospheric humidity simulated using a radiative transfer model and idealized profiles (see text for details). **b** two distributions (uniform and Gaussian) of FTH and **c** a log-normal distribution of FTH

intertropical region (roughly between 30°S and 30°N) is characterized by a strong meridional contrast: a very dry troposphere in the subsiding branches of the Hadley/Walker circulations (RH $< 10\%$) and a very moist region (RH $> 70\%$) associated with the ITCZ and the upward branch of the tropical circulation. The seasonal cycle is clearly depicted as well. The boreal summer migration of the ITCZ over the tropical land masses corresponding to the setting of the monsoon circulations over East Asia and West Africa is associated with an overall moistening of these areas. The subtropical winter cell appears more spread and dryer than the summer one, especially during the boreal summer. The extension and dryness of these regions have been highlighted as key elements of radiative-convective-dynamic equilibrium at play in the tropical climate (Pierrehumbert 1995).

The main reason for water vapour to be of importance to the energetics of the climate lies in the nonlinearity of the radiative transfer to the humidity. The outgoing longwave radiation (OLR) is indeed much more sensitive to a given perturbation in a dry rather than

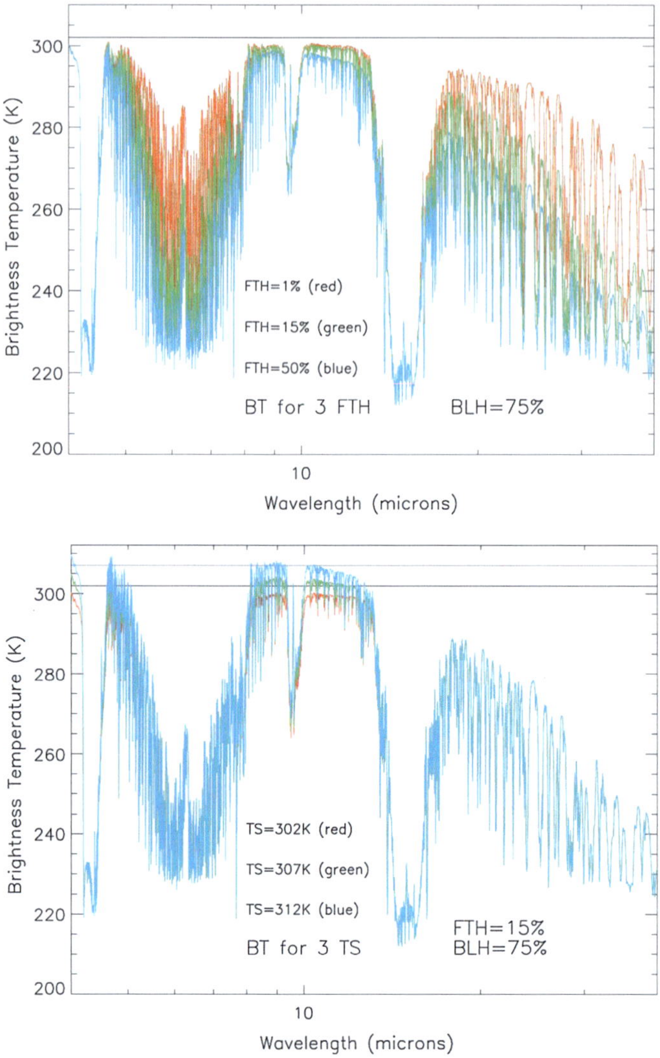

Fig. 3 *Top*: Brightness temperature (K) as a function of wavelength (micron) computed for a suite of idealized atmospheric profiles (see text for details). The colours correspond to the variation of the free tropospheric humidity parameters as indicated on the plot. *Bottom*: Same but for variation in the surface temperature

moist environment (Spencer and Braswell 1997; Held and Soden 2000; Sohn and Schmetz 2004), conferring a central role of the moisture distribution in these regions to the radiation budget of the planet and to the overall climate sensitivity. This nonlinearity is further illustrated in Fig. 2a that shows sensitivity of the OLR to idealized distribution of humidity in the free troposphere obtained from computations similar to that of Spencer and Braswell (1997). Radiative computations have been performed as discussed in Roca et al. (2005) but with the use of the MODTRAN code (Bernstein et al. 1996; Berk et al. 2000) instead of a broad-band model. The vertical profile of temperature is a multiyear summer average of Bay of Bengal conditions from ERA-40. Ozone and CO_2 are fixed at climatological values.

Relative humidity is constant in the boundary layer (at 75%), and a vertically constant RH profile is used over the troposphere and is varied from 1 to 99%. In the stratosphere, the RH profile is relaxed to climatological values (Roca et al. 2004). These computations confirm previous such calculations of the OLR sensitivity to FTH fluctuations and illustrate the nonlinearity of OLR to FTH at the dry end of the distribution (RH $< 15\%$). This non-linearity results from both the vibration–rotation bands of water vapour near 6.3 μm and the far infrared region ($\lambda > 18$ μm), the two regions that contribute the most to the whole OLR (e.g., Harries et al. 2008) and to its sensitivity to FTH (Fig. 3a). Overall, changing FTH from 1 to 15% decreases OLR by ~ 23 Wm^{-2}, while from 35 to 50%, it only decreases it by ~ 8 Wm^{-2}. Out of the 23 Wm^{-2} decrease, ~ 8 are accounted for by the 4- to 8-μm band, while ~ 10 Wm^{-2} are due to the 18- to 30-μm region. In the second case, both spectral regions contribute with ~ 1.5 Wm^{-2}. In the dry (moist) case, these two bands account for 80% (40%) of the radiative cooling. The spectral dependency of the OLR sensitivity to FTH is further illustrated in Fig. 3a. Changing FTH from 1 to 15% indeed reveals a cooling over both the vibration–rotation and FIR regions and hardly modifies the OLR over the window region. While moistening up to 50% does not strongly affect the OLR, the window region exhibits a few Kelvin cooling of similar magnitude to a 5-K change in the surface temperature (Fig. 3b). This suggests an enhanced role of the continuum absorption (Shine et al. 2012) in the moist case that would explain the more linear sensitivity in the moist case although it is difficult to decipher the continuum effect over the FIR region (Harries et al. 2008). Guzman et al. (2012) extended such analysis to various surface types (beyond oceanic conditions) and confirmed that, over the inter-tropical belt, the main determinants of the clear-sky OLR and its sensitivity can be resumed to FTH and the surface temperature. They underscore that as far as the variability of OLR is concerned, though, the FTH variability dominates the signal and the surface temperature variability can be ignored. Hence, because of its radiative importance, the understanding of the distribution of relative humidity in the free troposphere and, in particular at the dry end of the spectrum, requires a great deal of attention.

Of particular relevance to the study of the tropospheric humidity in the subsiding, dry regions is the advection–condensation (AC) paradigm (Pierrehumbert and Yang 1993; Sherwood 1996; Pierrehumbert et al. 2007). Such a perspective states that it is possible to account for the observed humidity distribution by only considering the large-scale transport and the saturation process. The only water vapour source is concentrated in the boundary layer, and the water sink is simply the saturation; the excess moisture is assumed to rain out without re-evaporating in the atmosphere. Similarly, any cloud materials' re-evaporation is neglected and is not accounted for in the water vapour source. While such strong assumptions might not be accurate in the vicinity of deep convection (Su et al. 2006; Sohn et al. 2008), it has been shown useful for the dry subsiding regimes (e.g., Schneider et al. 2010) where the humidity distribution is hence only driven by the large-scale dynamics and the atmospheric temperature field. Such a paradigm has been implemented in various ways over the last decade: from an idealized model of the PDF of relative humidity (Sherwood et al. 2006; Pierrehumbert et al. 2007; Ryoo et al. 2009; O'Gorman et al. 2011) to Lagrangian-based computations (Salathé and Hartmann 1997; Pierrehumbert 1998; Pierrehumbert and Roca 1998; Dessler and Sherwood 2000; Cau et al. 2007; Dessler and Minschwaner 2007; Brogniez et al. 2009) as well as the Eulerian version of this simplified view (Sherwood 1996; Galewsky et al. 2005; Couhert et al. 2010; Hurley and Galewsky 2010a; Wright et al. 2010). Despite its oversimplification of the problem, the advection–condensation paradigm has been successfully confronted to a large amount of observations over a variety of regions (Sherwood et al. 2010a) and has hence become a useful tool to investigate the dynamics of relative humidity in the troposphere.[1] The Lagrangian studies revealed the importance of the statistics of the last saturation of the air mass to discuss the origin and fate of humidity. Such statistics have been computed here for relative humidity at 500 hPa over a 20-year period using the large-scale winds and temperature fields from the NCEP reanalysis and back trajectories (Pierrehumbert 1998; Pierrehumbert and Roca 1998). The transport model is configured as in previous work (Roca et al. 2005). The reverse domain–filling integrations are performed using a 0.5°×0.5° grid over the 35°S–35°N area for all longitudes, every 6 h, over the whole period. The longitude, latitude and pressure of last saturations have been computed along with the relative humidity. These computations will here be used to investigate the tropical versus extra-tropical influences on the PDF of tropospheric humidity.

The paper is organized as follows. First, the importance considering the dry end of the PDF of relative humidity for radiative transfer considerations is explored. Tropical and extra-tropical influences on the distribution of relative humidity are further investigated in both current and warmer climates. A conclusion ends this survey.

2 The PDF of Relative Humidity in the Troposphere

2.1 PDF of RH and OLR

Energy constraints on planet Earth (i.e. applying the first law of thermodynamics) require that, at equilibrium, the Earth emits in the long wave as much radiation as its gets from the Sun. This budget approach is hence focused on the mean values of the OLR over the whole planet and over long time scales corresponding to the global radiative-convective equilibrium theory. While the mean OLR is the constrained parameter, owing to the nonlinearity of the clear-sky radiative transfer to water vapour (Figs. 2a, 3), the whole distribution of moisture has to be considered rather than its mean in order to link the distribution of humidity to that of radiation. To illustrate this, the OLR sensitivity to FTH curve (Fig. 2a) and four distributions of FTH for a dry case are considered (Fig. 2bc): a constant distribution with mean of 14.5%, an uniform distribution with mean of 14.5% bounded within plus or minus 5%, a Gaussian distribution with mean of 14.5% (and a 5% standard deviation) and a generalized log-normal distribution with a mean of 14.5% shown in Fig. 2c. The mean OLR corresponding to the constant distribution is 311 Wm^{-2}. The uniform and normal distribution yield to a mean OLR larger by 0.7 Wm^{-2} in both cases. The log-normal PDF, on the other hand, gives a 3-Wm^{-2} overestimation of the OLR with respect to the constant case. At the scale of the doubling of CO_2 problem, such a systematic bias could be significant depending on its geographical spread, which is explored next.

2.2 Time and Geographical Variations of the PDF of RH

At the intraseasonal scale, the PDF of RH reveals a variety of behaviours and strong departure from normal distribution. On the other hand, at the interannual scale, the moments of the intraseasonal PDF are normally distributed. The skewness of the intraseasonal PDF indeed is well characterized over our multiyear period by a mean and a standard deviation. Hence, in the following, we will make use of multiyear averages when mapping our variables. It is also important to note that, while we here focus on the relative

[1] And more recently for the stratospheric water vapour as well (Liu et al. 2010).

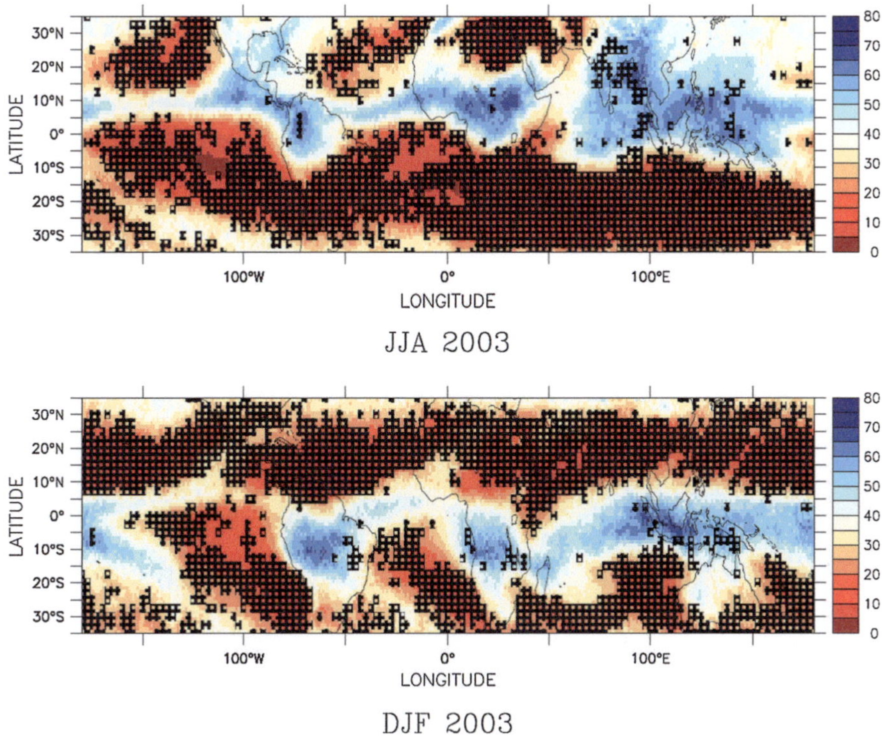

Fig. 4 Seasonal mean of relative humidity at 500 hPa for 2003 for JJA (*top*) and DJF (*bottom*) derived from AIRS measurements. Units are in percentage. The *black squares* represent the region where the intraseasonal PDF departs from a normal distribution according to a Kolmogorov–Smirnoff test at the 99% level

humidity distribution, the emphasis is set on the absolute humidity distribution underneath. The temperature distribution and its variability are neglected on the grounds that, within the tropical free troposphere, such temperature fluctuations are small and do not change significantly the relative humidity (Peixoto and Oort 1996). This is confirmed by recent investigation into the covariability of temperature and relative humidity in the free troposphere (Lemond 2009), but it should be kept in mind that this is a strong characteristic of the intertropical belt that is not valid elsewhere.

Departure from the normal distribution of the relative humidity is investigated through a Kolmogorov–Smirnoff (KS) test applied on the intraseasonal 500 hPa relative humidity distribution obtained from the daily average measurements of the AIRS instrument onboard the Aqua mission (Aumann et al. 2003; Fetzer et al. 2003). The regions corresponding to the dry areas (in the mean) depart significantly from normality, while the convective moist regions show a distribution of relative humidity within the season close to a Gaussian distribution (Fig. 4). For instance, over the moist Indonesian region (110–160°E and 0–15°N in summer; see Fig. 1) characterized by an multiyear average value of 58%, the KS test does not reject the normal assumption; indeed, the skewness and kurtosis are not departing much from zero (−0.53 and −0.33, respectively). Over the Mediterranean region, on the other hand (15:40°E and 25–35°N), the multiyear mean is low (11%) and the skewness and kurtosis strong (2.4 and 6, respectively) as also depicted in Fig. 7.

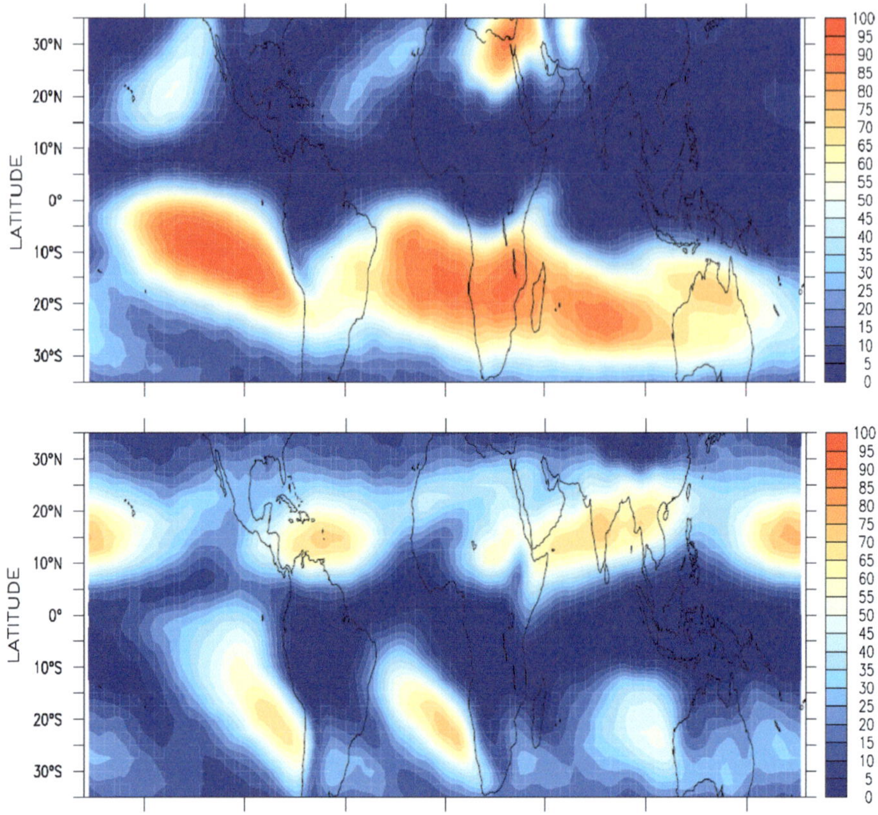

Fig. 5 Multi year 2003–2007 average of the frequency of occurrence of 500 hPa relative humidity below 10% within the season from AIRS measurements for JJA (*top*) and DJF (*bottom*). Units are percentage

The spatial variation of the PDF of the relative humidity in the free troposphere between the moist and dry regions is well explained by a simple theoretical implementation of the AC paradigm. Ryoo et al. (2009) indeed provided a generalized version of the model of Sherwood et al. (2006) for the humidity distribution that captures the difference in PDF between the dry and moist areas. It is assumed that relative humidity results from a uniform subsidence and random moistening events. Two parameters are used to quantify these processes: the ratio of drying time by subsidence to the interval time in between moistening events and a measure of the randomness of these events. In the convective moist region, the fitting of the model parameters to satellite estimates of relative humidity yields to large ratio and small randomness factor, indicating a rapid random moistening there, compatible with a local moistening by deep convection, and an almost Gaussian PDF. On the other hand, the nonconvective regions, which depart significantly from the normal distribution, are characterized by a small ratio and large randomness factor, suggesting a slow, less random moistening, likely linked to quasi-horizontal transport (Ryoo et al. 2009). Such analysis also indicates that a significant fraction of the intertropical belt area departs from the normal distribution, and hence, this departure, if not accounted for, could have a

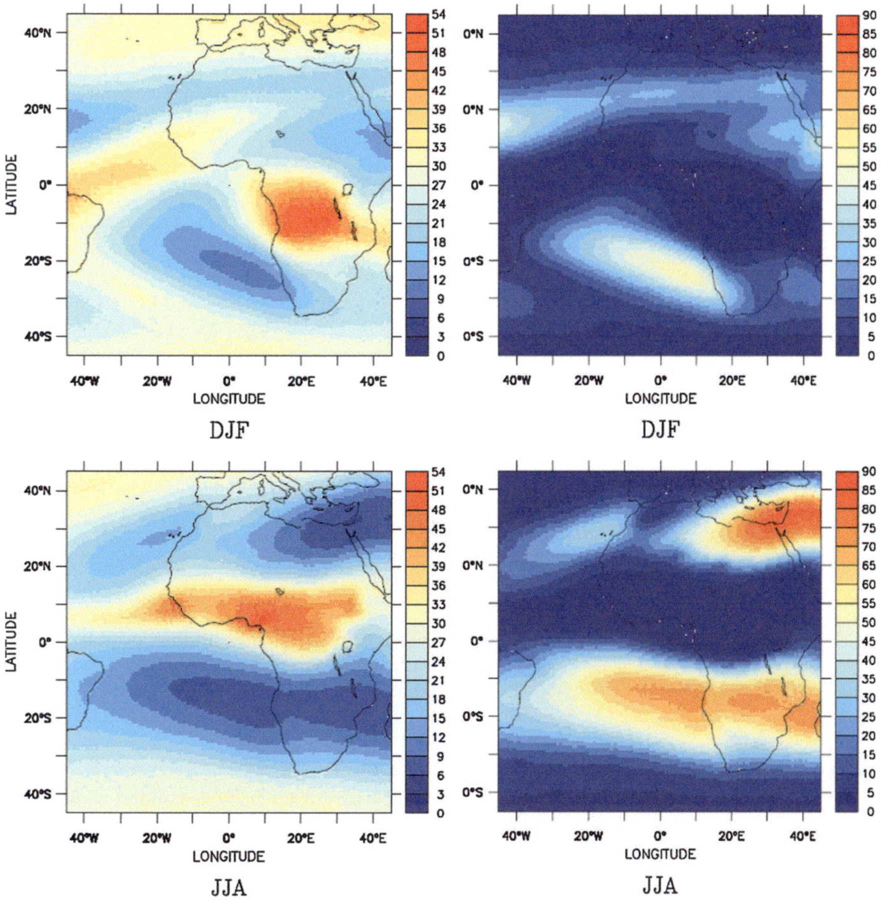

Fig. 6 Multiyear 1983–2004 average of the seasonal mean of FTH (*left panels*) and of the frequency of occurrence of FTH below 10% within the season (*right panels*). For DJF (*top*) and JJA (*bottom*), data are from METEOSAT measurements. Units are in percentage

significant impact on our understanding of OLR. In the next section, we explore a simple way to account for the PDF of the relative humidity.

2.3 A Simple Way to Account for the Non-normality of Relative Humidity Distribution

The need of handling the whole PDF of humidity instead of only the mean of the field implies the manipulation of the upper moments of the distribution (skewness and kurtosis). While the computations are straightforward, the comparison of two PDFs through the comparison of their 4 moments is not. Assuming a generalized log-normal distribution also requires 4 parameters to be fitted. It can be brought down to 2 parameters by imposing the lower and upper range limit of the distribution (0 and 100% for instance) at the cost of limiting the possible distributions. The simplified model (Ryoo et al. 2009) also comprises only two parameters, linked to the first two moments of the distribution. Still, the moments-to-moments comparison of PDFs remains difficult. Here, it is proposed to limit the analysis to a single parameter characterizing the PDF with

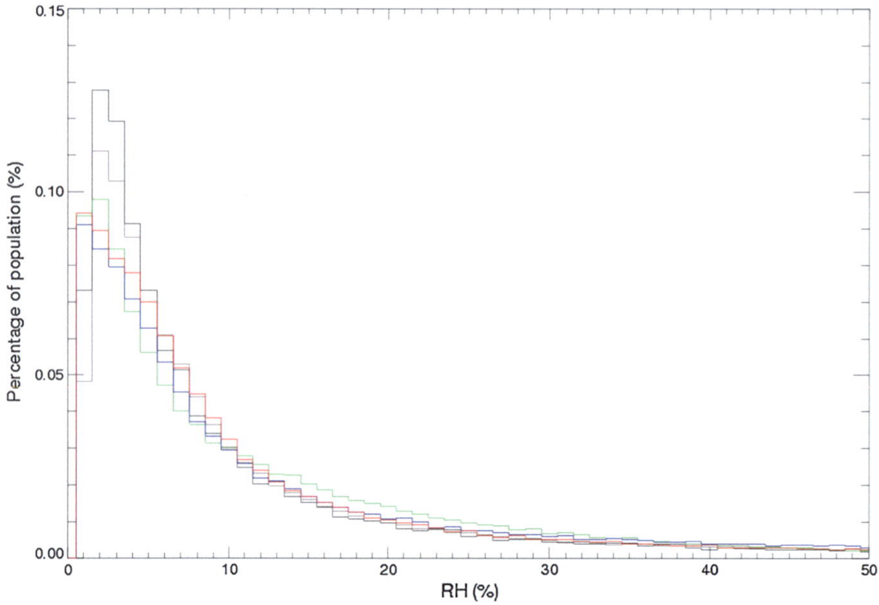

Fig. 7 Probability density functions (PDFs) of relative humidity at 500 hPa pressure level, for the JJA seasons over the 2003–2007 period, on the east Mediterranean area (15E:40E/25°N:35°N). PDFs are computed from AIRS (*black*), ERA-interim (*brown*), RH reconstructed (*green*), NCEP2 (*blue*), NCEP (*red*) estimations

emphasis on the dry foot of the distribution: the frequency of occurrence of RH below 10%, noted in the following as RHp10. Indeed, such a frequency parameter has been found to well discriminate the various PDF under considerations (Lemond 2009). Figure 5 shows the AIRS-based climatology of RHp10 for the 500 hPa level for boreal summer and winter periods. While the large-scale features seen in Fig. 1 are retained (contrast between boreal summer and winter, contrast between winter/summer hemisphere, moist and dry regions, etc), the patterns are here sharper. The gradients between the core regions of the winter cell subtropics and the poleward regions are stronger than when the mean is used. The delineation of the regions where dry air is prevalent and that are of importance to the radiation budget is also more effective. For instance, the boreal winter mean field (Fig. 1) suggests a dry extreme in the Eastern subtropical Pacific (90°W) between 10°S and 15°S, while the RHp10 computation reveals an homogeneous extreme region spreading from 17.5°S to 25°S. The frequency approach further helps to discriminate regions of identical mean. For instance, using the METEOSAT Free Tropospheric Humidity climatology (Roca et al. 2003; Brogniez et al. 2006; Brogniez et al. 2009; Roca 2011), the difference between the Eastern Mediterranean region and the southern Atlantic ocean, unseen in the classical climatology, is revealed in the frequency computations (Fig. 6). The frequency approach furthermore makes the comparisons between various humidity sources easier and is found to be a robust diagnostic not only for satellite-based relative humidity estimates but also for the models-based one (Fig. 7). In particular, the RHp10 map for the 500 hPa reconstructed relative humidity field using the Lagrangian modelling framework (See Introduction for details) reveals very realistic patterns (Fig. 8). Indeed, despite an overall low bias in the frequency of occurrence, the reconstructed field exhibits the patterns seen in the satellite measurements. The contrast between the winter and summer hemisphere is well captured as well as the regional details like the extended dry feature from the Sahel to the Bangladesh area during the boreal

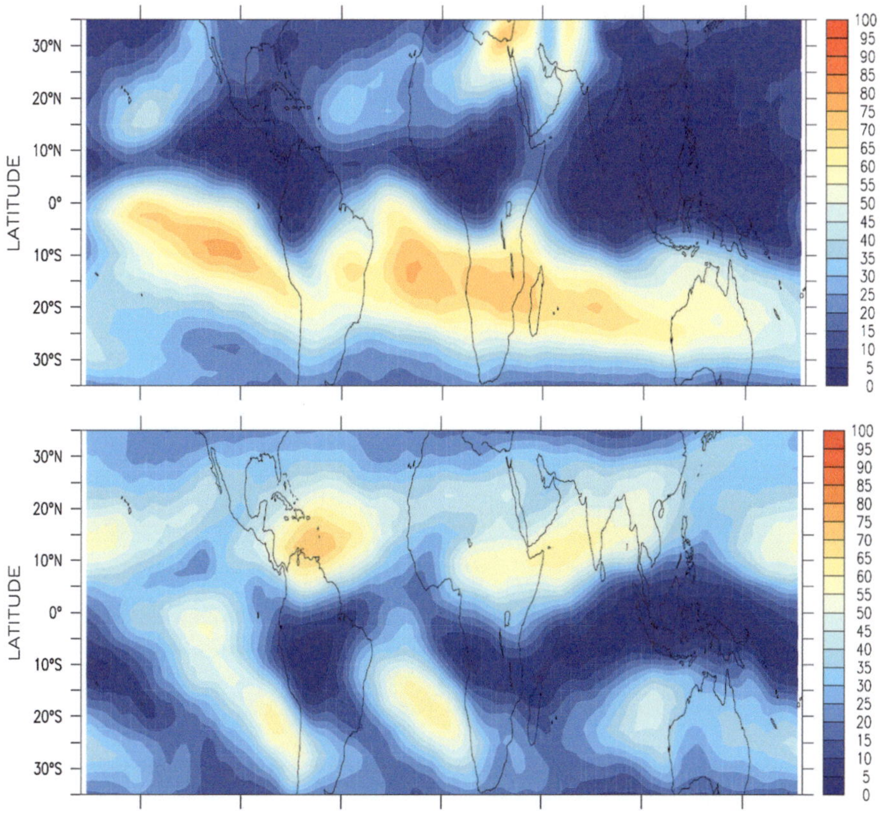

Fig. 8 Multi year 2003–2007 average of the frequency of occurrence of 500 hPa relative humidity below 10% within the season from the Lagrangian based reconstruction for JJA (*top*) and DJF (*bottom*). Units are percentage

winter. Similarly, the three dry patterns of the DJF summer hemisphere, east of the subtropical ocean basins, and their relative intensity are well reproduced. In boreal summer, the unique strong dry region over the Eastern Mediterranean region (Brogniez et al. 2009) is very well simulated so is the quasi-zonally extended subtropical dry region in the winter cell. These quick comparisons further confirm the usefulness of the advection–condensation paradigm and its present Lagrangian implementation for investigating the dry end of the relative humidity distribution through a frequency approach.

3 Extra-Tropical Influence on the Relative Humidity PDF

The tropical and extra-tropical influences of the *mean* humidity in the free troposphere have been investigated either from a water vapour budget (Schneider et al. 2006) or from a last saturation perspective (Galewsky et al. 2005) with the emphasis on the *zonal mean* view, offering contrasted conclusions (O'Gorman et al. 2011; see also discussion by Sherwood et al. 2010a). Here, the discussion is focused on the geographical patterns of the dry end of the relative humidity distribution.

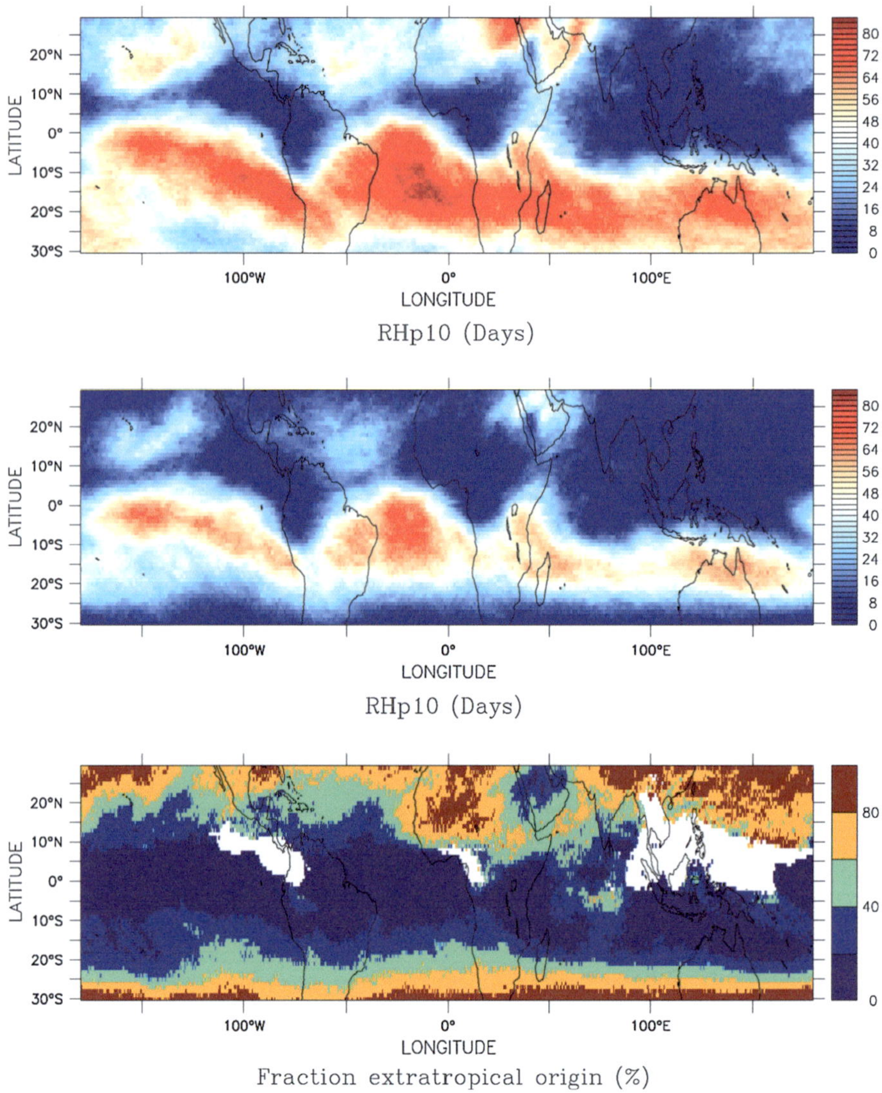

Fig. 9 Frequency of occurrence of 500 hPA relative humidity below 10% within the JJA 2003 season from the Lagrangian reconstruction. *Top*: full origin result. *Middle*: Tropical origin only results (latitude of last saturation equatorward of 35°). *Bottom*: the fraction of extra-tropical origins. Value with RHp10 lower than 10 have been blanked out. Units are in percentage. Lagrangian computations performed using NCEP temperature and winds information. See text for details

3.1 Current Climate

The RHp10 distribution of the reconstructed field for the boreal summer 2003 is compared to the RHp10 distribution obtained by keeping only the air masses that experienced last saturation within the intertropical belt (35°S–35°N) in Fig. 9. Excluding the extra-tropical last saturated air masses overall moistens the atmosphere. The domain averaged RHp10 decreases from 37 to 23% without the extra-tropical influence. While the patterns overall

Fig. 10 Multiyear average 1983–2004 of the zonal mean of the fraction of extra-tropical air masses for all (*plain line*) and for the one with relative humidity below 10% (*dashed line*) for winter (*top*) and summer (*bottom*). Lagrangian computations performed using NCEP temperature and wind information. See text for details

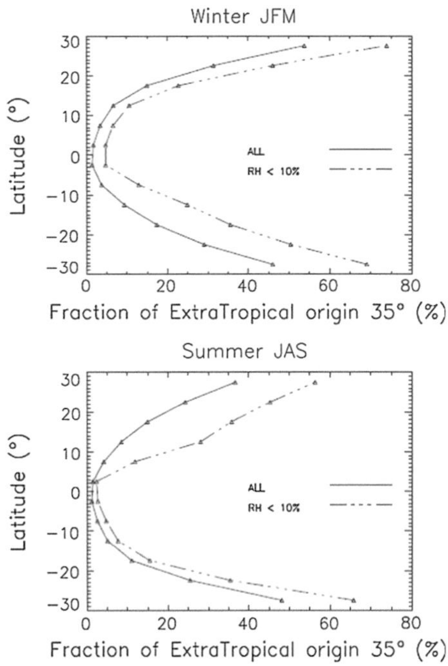

remain similar within the two computations, the driest areas nevertheless appear more impacted and less spread in the tropics only case (Fig. 9 middle). The very dry features in the subtropical south Atlantic is mainly built from tropical originating air with the fraction of extra-tropical influence less than 10% (Fig. 9c). Globally, the winter hemisphere seems to be characterized by a less pronounced influence of the extra-tropics than the summer one with fraction above 60% hardly reaching the 25°S parallel, while, over Sahara and over the Arabian Sea for instance, the 60% iso-line penetrates southward down to 10°N. This seasonal contrast appears as a robust characteristic of the 500 hPa relative humidity field and is confirmed when the whole 1983–2004 period is considered. Very dry air (RH < 10%) is relatively more frequently originating for the extra-tropics than the environmental air, at all latitudes (Fig. 10). Extra-tropical influence on dry air occurrence is stronger in the summer hemisphere and is significantly so closer to the tropics into the summer hemisphere as shown in Fig. 10.

Dry air is brought to these regions through subsidence that brings air from aloft, lateral mixing that brings air from the tropics and the extra-tropics (Pierrehumbert 1998). While the primary source of dry air is the extra-tropical cold processing by isentropic mixing (Yang and Pierrehumbert 1994; Galewsky et al. 2005), breaking Rossby waves have also been identified as a mechanism to bring dry extra-tropical air in the subtropics (Waugh 2005; Allen et al. 2009). During boreal winter, Cau et al. (2007) highlighted a suite of dynamic mechanisms related to transport of dry air into the subtropics (descending air masses associated with extra-tropical baroclinic systems, the subtropical anticyclone variability and the upper level jets dynamics). Such extra-tropical source is usually completed by a tropical upper tropospheric source region as far as dry air is concerned (e.g., Dessler and Minschwaner 2007) or local convective source from below. Focusing on the summer dry region of the Eastern Mediterranean region, Brogniez et al. (2009) showed that the local source could be ruled out in favour of the extra-tropical source as a better predictor of

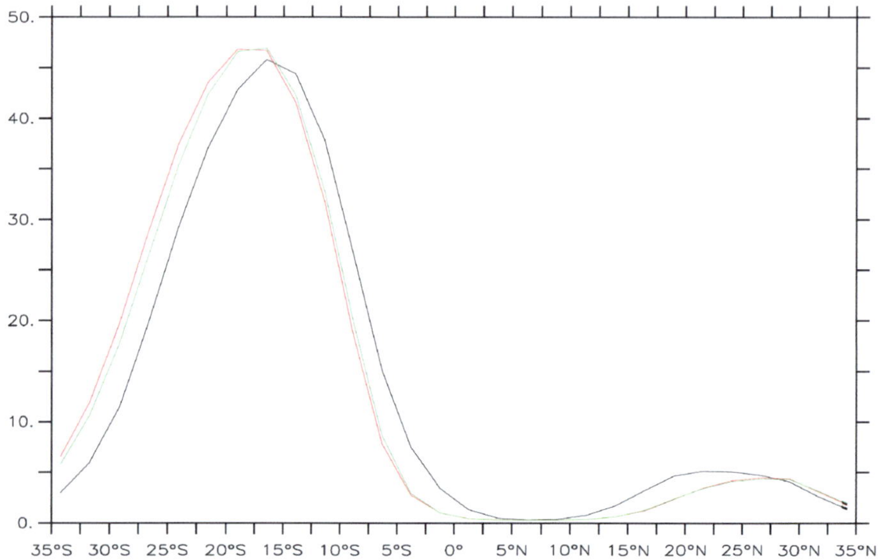

Fig. 11 Zonal mean of 500 hPa RHp10 in summer averaged over 20 years. In *black*, control run simulations (1970–2000), in *red* for A1B and *green* for A2B scenario (2070–2100)

the FTH interannual variability there. The difficulty still remains high in associating a given geographical source and process and their relative contributing role on the dry air production and calls for a dynamically based analysis of the climatology. Focusing on the dry foot of the PDF of humidity in the subtropics would help revealing the dominant mechanisms at play to explain the importance of the extra-tropical region as a source for it. The contribution of the extra-tropical source in the summer cell is indeed enhanced in boreal summer with respect to boreal winter (Fig. 10) and might suggest the importance of the more frequent breaking Rossby waves in the northern hemisphere during boreal summer.

3.2 Warmer Climate

The evolution of relative humidity in the free troposphere under climate change is important to the understanding of the role of the water vapour feedback in the climate system (Schneider et al. 2010). It is usually considered that climate model simulations act as if relative humidity remains unchanged, which is to be understood as the magnitude of the water vapour radiative feedback being similar in these climate models to the one in earlier radiative-convective climate change computations and even earlier energy balance calculations (Inamdar and Ramanathan 1998). Within the radiative-convective equilibrium framework, such constant RH behaviour explains the strong magnitude of the water vapour feedback. If RH were allowed to increase instead, the water vapour feedback would be diminished. Under the radiative-convective dynamic equilibrium assumption, it is the relative humidity of the dry subtropical regions that is key to the radiation budget. Analysis of relative humidity of climate model simulations of climate change scenario reveals an overall complex pattern of change with a notable feature relevant to our region of interest: the subtropical troposphere is the locus of a marked decrease in mean relative humidity

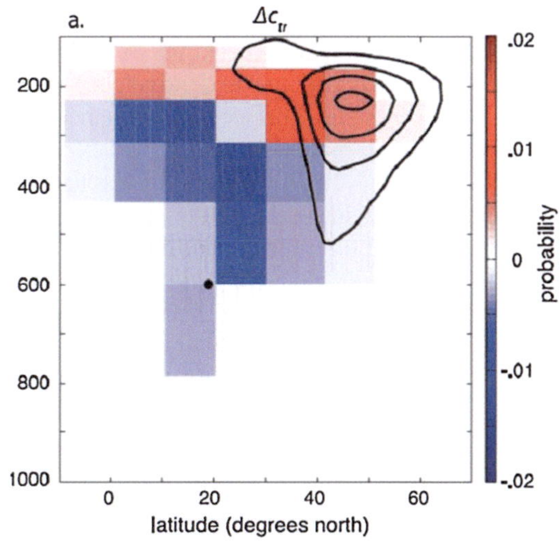

Fig. 12 DJF zonal means for global warming change in probabilities of last saturation for the RH minimum (19 N, 601 hPa), (contours are differences in the 2.5–6 day variance of the meridional wind, and contour interval is 5 m^2/s^2). The RH minimum is indicated by the *black circles*. From Hurley and Galewsky (2010a, b) copyright AGU

(Sherwood et al. 2010b). Two IPSL CM4 model (Marti et al. 2005) simulations of climate change within the AR4 effort have been analysed and compared to a control simulation. The statistics of RHp10 at 500 hPa further support the poleward shift of the relative humidity minima in both hemispheres together with a slight increase in the frequency of occurrence of dry air (Fig. 11). Various explanations have been presented to account for such poleward migration and extension of the Hadley cell (see Sherwood et al. 2010b for a summary).

As discussed all through this paper, such dry regions are under the control of the large-scale dynamics. The understanding of the relative humidity there requires the consideration of both the statistics of local temperature and the temperature of last saturation. Under an uniform warming of the atmosphere, it is straightforward to show that one does not expect any strong modifications of the RH distribution as far as the temperature change is small to moderate (Pierrehumbert et al. 2007). Trajectory computations under such uniform warming but with realistic circulations and last saturations statistics indeed confirmed the relative constant behaviour of the whole RH distribution. Nevertheless, one rather expects a nonuniform warming with increasing CO_2 concentration: stronger warming in the upper troposphere of the deep tropics (than in the mid troposphere) and small warming (if not a cooling) below the extra-tropical tropopause (Sherwood et al. 2010b). Such a differential lapse rate change is expected to perturb untrivially the last saturation statistics for the dry subtropical regions of the free troposphere. Experiments using climate change simulations and advection–condensation modelling indicate that, despite such complexity, the dynamics of air masses seem to change along with the differential lapse rate changes, in a way as to maintain near-constant distribution of relative humidity, be it at 500 hPa (Pierrehumbert et al. 2007) or all through the free troposphere (Brogniez and Pierrehumbert 2007). One reason for such a small response may lie in the relatively small-amplitude temperature changes that are occurring within these climate change simulations, and it would be interesting to extend such diagnostics to widely varying climates (O'Gorman and Schneider 2008). The investigation into the evolution of the last saturation statistics for the zonal mean relative humidity tropospheric minimum in the AR4 model (Huxley and Galewsky 2010b) reveals a poleward and upward shift (Fig. 12), which correspond to a contemporaneous shift in the mid-latitude baroclinic activity (Wright et al. 2010). These recent analyses indicate that the use of the AC paradigm along with the focus on some part of the whole PDF of the free tropospheric humidity is an encouraging step towards a better identification of the lead processes driving the water vapour feedback. This endeavour was once (even many times) considered as an unreachable perspective owing to the complexity of the sources (the so-called microphysics effect) and transport mechanisms (Renno et al. 1994; Couhert et al. 2010). It is nevertheless noted that the zonal averaging of the results is not well suited for investigating the dry end of the PDF due to the co-existence of moist and dry regimes at the same latitude (Figs. 1 and 4), which are fed by different physics and dynamics. The investigation of these subtropical features (like the summer time minimum of Brogniez et al. 2009) rather than the zonal mean in conjunction with the use of the adequate PDF characteristic instead of the first moment of these non-normal distributions is likely to help clarify the process at play in the maintenance of relative humidity under climate change.

4 Conclusions

This survey focused on our increased understanding of the drivers of the PDF of relative humidity in the free troposphere over the intertropical belt that the last decade has witnessed. The advection–condensation paradigm has shown its ability to clarify the dynamics of relative humidity in current climate, thanks to confrontation with a large number of mature (and maturing) observations. The dual origin of subtropical dry air, tropical deep convective regions and extra-tropical upper troposphere and the importance of this lateral mixing have also been recognized. The suite of available tools dedicated to the study of the relative humidity distribution is well furnished (idealized, Lagrangian, Eulerian frameworks). Incentive (through the use of a simple parameter) to go beyond the use of normal statistics for such non-normal moisture distribution in the subtropics has been provided. More generally, the need to avoid averaging (in time or zonally) the humidity fields to help sharpen the delineation of the physics underlying the distribution and its radiative impact onto OLR has been underscored. The interpretation of the water vapour feedback under the AC assumption is still linked to temperature through the Clausius–Clapeyron law although indirectly, through the statistics of last saturations. Differential temperature changes between the subtropical free troposphere and the main last saturation regions are identified as one of the major point to further clarify in the future. Already first attempts have shown encouraging results towards such an endeavour with the AR4 climate simulations and an appropriate diagnosis. Such efforts should help providing the important physical connection between the lapse rate and water vapour feedbacks.

The extra-tropical influence on the PDF of relative humidity is not confined to the subtropical regions nor is the importance of dry free tropospheric layers. In the deep tropics, dry air also has some connections to mid-latitudes (Yoneyama and Parsons 1999; Roca et al. 2005; Casey et al. 2009). There, dry air is important to the development of deep convection as surveyed in details in Del Genio (2011 this issue). The moisture distribution in the moist regions is also radiatively important through the modulation of the longwave cloud radiative forcing of high clouds (Sohn and Bennartz 2008) that can impact these deep convective systems' cloud feedback (Del Genio and Kovari 2002; Roca et al. 2005). The investigation into the importance of the lateral mixing and the extra-tropical source there is nevertheless made more difficult than that for the subtropical regions. One can nevertheless be optimistic with the current advent of a long-term record of space-based microwave measurements (e.g., John et al. 2011) that can probe the moisture vertical distribution in the near environment of convection and will definitely help these investigations, so will the upcoming Megha-Tropiques mission that will fly a dedicated payload to measure humidity profiles and TOA radiation in clear and cloudy regions over the intertropical belt (Roca 2011).

References

Allan RP (2012) The role of water vapour in earth's energy flows. Surv Geophys. doi:10.1007/s10712-011-9157-8 (this volume)

Allen G, Vaughan G, Brunner D, May TP, Heyes W, Minnis P, Ayers KJ (2009) Modulation of tropical convection by breaking Rossby waves. Quart J R Meteorol Soc 135:125–137. doi:10.1002/qj.349

Aumann HH, Chahine MT, Gautier C, Goldberg M, Kalnay E, McMillin L, Revercomb H, Rosenkranz PW, Smith WL, Staelin DH, Strow L, Susskind J (2003) AIRS/AMSU/HSB on the aqua mission: design, science objectives, data products and processing systems. IEEE Trans Geosci Remote Sensing 41(2):253–264

Berk A, Acharya PK, Bernstein LS, Anderson GP, Chetwynd Jr JH, Hoke ML (2000) Reformulation of the MODTRAN band model for finer spectral resolution. Proceedings of SPIE, vol 4049, Orlando, Florida, April, 2000

Bernstein LS, Berk A, Acharya PK, Robertson DC, Anderson GP, Chetwynd JH, Kimball LM (1996) Very narrow band model calculations of atmospheric fluxes and cooling rates. J Atmos Sci 53(19): 2887–2904

Brogniez H, Pierrehumbert R (2007) Intercomparison of tropical tropospheric humidity in GCMs with AMSU-B water vapor data. Geophys Res Lett 34:L17706. doi:10.1029/2007GL030967

Brogniez H, Roca R, Picon L (2006) A clear sky radiances archive from METEOSAT "water vapor" observations. J Geophys Res 111:D21109. doi:10.1029/2006JD007238

Brogniez H, Roca R, Picon L (2009) A study of the free tropospheric humidity interannual variability using meteosat data and an advection-condensation transport model. J Clim 22:6773–6787. doi:10.1175/2009JCLI2963.1

Casey Sean PF, Andrew ED, Courtney S (2009) Five-year climatology of mid troposphere dry air layers in warm tropical ocean regions as viewed by AIRS/Aqua. J Appl Meteor Climatol 48:1831–1842. doi:10.1175/2009JAMC2099.1

Cau P, Methven J, Hoskins B (2007) Origins of dry air in the tropics and subtropics. J Clim 20:2745–2759

Couhert A, Schneider T, Li J, Waliser DE, Tomkins AM (2010) The maintenance of the relative humidity of the subtropical free troposphere. J Clim 23:390–403. doi:10.1175/2009JCLI2952.1

Del Genio AD (2011) Representing the sensitivity of convective cloud systems to tropospheric humidity in general circulation models. Surv Geophys. doi:10.1007/s10712-011-9148-9

Del Genio AD, Kovari W (2002) Climatic properties of tropical precipitating convection under varying environmental conditions. J Clim 15:2597–2615. doi:10.1175/1520-0442(2002)015<2597:CPOTPC>2.0.CO;2

Dessler AE, Minschwaner K (2007) An analysis of the regulation of tropical tropospheric water vapor. J Geophys Res 112:D10120. doi:10.1029/2006JD007683

Dessler AE, Sherwood SC (2000) Simulations of tropical upper tropospheric humidity. J Geophys Res 105:20155–20163

Fetzer EJ, McMillin L, Tobin D, Aumann HH, Gunson MR, McMillan WW, Hagan DE, Hofstadter MD, Yoe J, Whiteman D, Barnes J, Bennartz R, Vemel H, Walden V, Newchurch M, Minnett P, Atlas R, Schmidlin F, Olsen ET, Goldberg M, Zhou S, Ding H, Smith W, Revercomb SRH (2003) AIRS/AMSU/HSB validation. IEEE Trans Geosci Remote Sens 41:418–431

Galewsky J, Sobel A, Held I (2005) Diagnosis of subtropical humidity dynamics using tracers of last saturation. J Atmos Sci 62:3353–3367

Guzman R, Picon L, Roca R (2012) A simple model of outgoing longwave radiation for the tropical ocean, land and desert. J Geophys Res (in preparation)

Harries J, Carli B, Rizzi R, Serio C, Mlynczak M, Palchetti L, Maestri T, Brindley H, Masiello G (2008) The far infrared earth. Rev Geophys 46:RG4004. doi:10.1029/2007RG000233

Held IM, Soden BJ (2000) Water vapor feedback and global warming. Annu Rev Energy Environ 25:441–475

Hurley JV, Galewsky J (2010a) A last saturation analysis of ENSO humidity variability in the subtropical Pacific. J Clim 23:918–931

Hurley JV, Galewsky J (2010b) A last-saturation diagnosis of subtropical water vapor response to global warming. Geophys Res Lett 37:L06702

Inamdar AK, Ramanathan V (1998) Tropical and global scale interactions among water vapor, atmospheric greenhouse effect, and surface temperature. J Geophys Res 103:32177–32194

John VO, Holl G, Allan RP, Buehler SA, Parker DE, Soden BJ (2011) Clear-sky biases in satellite infrared estimates of upper tropospheric humidity and its trends. J Geophys Res 116:D14108. doi:10.1029/2010JD015355

Lemond J (2009) Climatologie et variabilité de l'air sec de la troposphère libre intertropicale: analyse du climat actuel et de son évolution. PhD thesis, University Pierre and Marie Curie, Paris, France

Liu YS, Fueglistaler S, Haynes PH (2010) Advection-condensation paradigm for stratospheric water vapor. J Geophys Res 115:D24307. doi:10.1029/2010JD014352

Marti GE (2005) The new IPSL climate system model: IPSL-CM4. Report no 26, Institute Pierre Simon Laplace, Paris

O'Gorman PA, Schneider T (2008) The hydrological cycle over a wide range of climates simulated with an idealized GCM. J Clim 21:3815–3832

O'Gorman PA, Lamquin N, Schneider T, Singh MS (2011) The relative humidity in an isentropic advection–condensation model: limited poleward influence and properties of subtropical minima. J Atmos Sci 68:3079–3093. doi:10.1175/JAS-D-11-067.1

Peixoto JP, Oort AH (1996) The climatology of relative humidity in the atmosphere. J Clim 9(12):3443–3463

Pierrehumbert RT (1995) Thermostats, radiator fins, and the local runaway greenhouse. J Atmos Sci 52:1784–1806

Pierrehumbert RT (1998) Lateral mixing as a source of subtropical water vapor. Geophys Res Lett 25:151–154

Pierrehumbert RT, Roca R (1998) Evidence for control of Atlantic subtropical humidity by large scale advection. Geophys Res Lett 25:4537–4540

Pierrehumbert R, Yang H (1993) Global chaotic mixing on isentropic surfaces. J Atmos Sci 50:2462–2480

Pierrehumbert RT, Brogniez H, Roca R (2007) On the relative humidity of the atmosphere. In: Schneider T, Sobel AH (eds) The global circulation of the atmosphere. Princeton, Princeton University Press, pp 143–185

Renno N, Stoneet PH, Emanuel KA (1994) Radiative-convective model with an explicit hydrologic cycle 2. Sensitivity to large changes in solar forcing. J Geophys Res 99:17001–17020

Roca R (2011) The Megha-tropiques mission. MT special issue in QJRMS (in preparation)

Roca R, Brogniez H, Picon L, Desbois M (2003) Free tropospheric humidity observations from METEOSAT water vapor data, 83 AMS annual meeting, Long Beach, California, USA, in the CDROM, 9–13 February 2003

Roca R, Louvet S, Picon L, Desbois M (2004) A study of convective systems, water vapor and top of the atmosphere cloud radiative forcing over the Indian Ocean using INSAT-1B and ERBE data. Meteorol Atmos Phys. doi:10.1007/s00703-004-0098-3

Roca R, Lafore JP, Piriou C, Redelsperger JL (2005) Extratropical dry air intrusions into the West African monsoon mid troposphere: an important factor for the convective activity over the Sahel. J Atmos Sci 62:390–407

Ryoo JM, Ugasa T, Waugh DW (2009) PDFs of tropospheric humidity. J Clim 22:3357–3373

Salathé EP, Hartmann DL (1997) A trajectory analysis of tropical upper tropospheric moisture and convection. J Clim 10:2533–2547

Schneider T, Smith KL, O'Gorman PA, Walker CC (2006) A climatology of tropospheric zonal-mean water vapor fields and fluxes in isentropic coordinates. J Clim 19:5918–5933

Schneider T, O'Gorman PA, Levine XJ (2010) Water vapor and the dynamics of climate changes. Rev Geophys 48:RG3001. doi:10.1029/2009RG000302

Sherwood SC (1996) Maintenance of the free tropospheric tropical water vapor distribution, part II: simulation by large scale advection. J Clim 9:2919–2934

Sherwood SC, Kursinski ER, Read WG (2006) A distribution law for free tropospheric relative humidity. J Clim 19:6267–6277

Sherwood SC, Roca R, Weckwerth TM, Andronova NG (2010a) Tropospheric water vapor, convection, and climate. Rev Geophys 48:RG2001. doi:10.1029/2009RG00030

Sherwood SC, Ingram W, Tsushima Y, Satoh M, Roberts M, Vidale PL, O'Gorman PA (2010b) Relative humidity changes in a warmer climate. J Geophys Res 115:D09104

Shine KP, Ptashnik IV, Radel G (2012) The water vapour continuum: brief history and recent developments. Surv Geophys (this volume)

Sohn BJ, Bennartz R (2008) Contribution of water vapor to observational estimates of longwave cloud radiative forcing. J Geophys Res 113:D20107. doi:10.1029/2008JD01005
Sohn BJ, Schmetz J (2004) Water vapor-induced OLR variations associated with high cloud changes over the tropics: a study from meteosat-5 observations. J Clim 17:1987–1996
Sohn B-J, Schmetz J, Chung E-S (2008) Moistening processes in the tropical upper troposphere observed from meteosat measurements. J Geophys Res 113:D13109. doi:10.1029/2007JD009527
Spencer R, Braswell WD (1997) How dry is the tropical free troposphere? Implications for global warming theory. Bull Am Meteorol Soc 78:1097–1106
Su H, Read WG, Jiang JH, Waters JW, Wu DL, Fetzer EJ (2006) Enhanced positive water vapor feedback associated with tropical deep convection: new evidence from Aura MLS. Geophys Res Lett 33:L05709. doi:10.1029/2005GL025505
Waugh DW (2005) Impact of potential vorticity intrusions on subtropical upper tropospheric humidity. J Geophys Res 110:D11305. doi:10.1029/2004JD005664
Wright JS, Sobel A, Galewsky J (2010) Diagnosis of zonal mean relative humidity changes in a warmer climate. J Clim 23(17):4556–4569
Yang H, Pierrehumbert RT (1994) Production of dry air by isentropic mixing. J Atmos Sci 51:3437–3454
Yoneyama K, Parsons DB (1999) A proposed mechanism for the intrusion of dry air into the tropical western Pacific region. J Atmos Sci 56:1524–1546

Surv Geophys (2012) 33:585–608
DOI 10.1007/s10712-011-9159-6

Energetic Constraints on Precipitation Under Climate Change

Paul A. O'Gorman · Richard P. Allan · Michael P. Byrne · Michael Previdi

Received: 17 August 2011 / Accepted: 2 November 2011 / Published online: 23 November 2011

Abstract Energetic constraints on precipitation are useful for understanding the response of the hydrological cycle to ongoing climate change, its response to possible geoengineering schemes, and the limits on precipitation in very warm climates of the past. Much recent progress has been made in quantifying the different forcings and feedbacks on precipitation and in understanding how the transient responses of precipitation and temperature might differ qualitatively. Here, we introduce the basic ideas and review recent progress. We also examine the extent to which energetic constraints on precipitation may be viewed as radiative constraints and the extent to which they are confirmed by available observations. Challenges remain, including the need to better demonstrate the link between energetics and precipitation in observations and to better understand energetic constraints on precipitation at sub-global length scales.

Keywords Precipitation · Global change · Energetics

1 Introduction

It is not surprising that energetics play an important role in determining global-mean precipitation given that surface evaporation is ultimately driven by solar radiation.[1]

[1] See Peixoto and Oort (1992), Trenberth et al. (2009), and Stevens and Schwartz (2011) for reviews of the energy budget of the Earth.

P. A. O'Gorman (✉) · M. P. Byrne
Department of Earth, Atmospheric, and Planetary Sciences, Massachusetts Institute of Technology, 77 Massachusetts Avenue, Cambridge, MA 02139, USA
e-mail: pog@mit.edu

R. P. Allan
Department of Meteorology, National Centre for Atmospheric Science, University of Reading, Reading, UK

M. Previdi
Lamont-Doherty Earth Observatory, Columbia University, 61 Route 9W, Palisades, NY 10964, USA

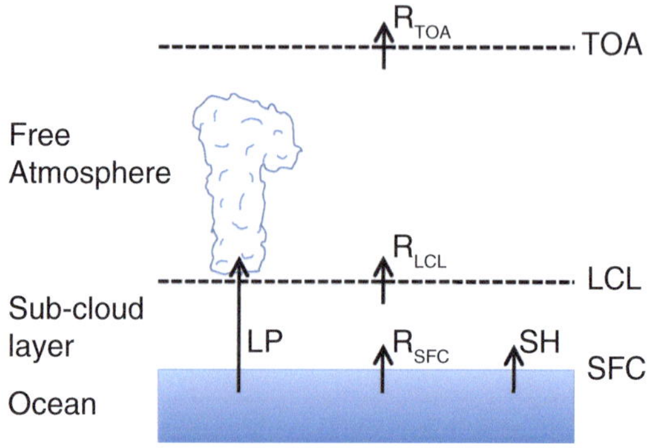

Fig. 1 Schematic of the energy budgets of the atmosphere and of the atmosphere above cloud base, as described in the text; LCL is the lifted condensation level. [Following Fig. 1 of Takahashi (2009a)]

In the context of climate change, we speak of energetics placing a constraint on precipitation in the sense that precipitation must change in such a way that the energy budget of the atmosphere (or the surface) continues to balance. We will work primarily with the energy budget of the atmosphere because its relatively small heat capacity facilitates analysis of transient changes in precipitation; the surface energy budget may also be usefully analyzed in a climate change context (Boer 1993; Richter and Xie 2008; Andrews et al. 2009; Lorenz et al. 2010). A schematic view of the global atmospheric energy budget is shown in Fig. 1. We begin by considering the energy budget of the entire atmosphere from the surface to the top of the atmosphere. The perturbation budget between two climates may be written as

$$L\delta P = \delta R_{\mathrm{TOA}} - \delta R_{\mathrm{SFC}} - \delta SH \tag{1}$$

where L is the latent heat of condensation (assumed constant for simplicity), P is the precipitation rate, R_{TOA} and R_{SFC} are the net radiation fluxes at the top of the atmosphere (TOA) and at the surface (SFC), respectively, SH is the surface sensible heat flux, and δ denotes the difference between the two climates.[2] Thus, changes in precipitation are directly related to changes in the radiative cooling of the atmosphere and the surface sensible heat flux. For a change in surface temperature, the resulting change in radiative cooling may be calculated to first order by assuming invariant relative humidity and a moist adiabatic lapse rate in the free troposphere. More generally, the changes in all radiative forcing agents, surface air temperature difference, and surface sensible heat flux must also be taken into account. This energetic approach is often more straightforward than understanding the precipitation response in terms of changes in large-scale circulations or convective mass fluxes that are themselves mediated by relatively subtle changes in atmospheric stability.

The energetic constraint described above was already discussed in some detail in the early work of Mitchell et al. (1987). Allen and Ingram (2002) showed how it could be used to understand the response of precipitation to climate change in different climate model simulations, and Pierrehumbert (2002) showed how the surface energy budget constrained precipitation in very warm climates. More recently, there has been considerable progress in using the energetic perspective to better quantify and understand several aspects of the response of precipitation to climate change: the feedbacks influencing the rate of change of precipitation under warming (e.g., Stephens and Ellis 2008; Previdi 2010), how the

[2] Positive fluxes of energy are upwards and all fluxes are averaged globally and over sufficiently long times that we may neglect changes in energy and water storage in the energy budget of a given climate state.

response of precipitation varies depending on the climate forcing agent (e.g., Andrews et al. 2010), and how precipitation responds on "fast" and "slow" timescales (e.g., Lambert and Faull 2007; Bala et al. 2010). In all of these cases, it is useful and natural to think of precipitation and evaporation as energy fluxes. Even for regional changes in precipitation, it is possible to usefully adopt an energetic approach (Muller and O'Gorman 2011).

The energetic constraint does not suggest that global-mean precipitation scales with the amount of water vapor in the atmosphere, although water vapor does play an important radiative role and can, therefore, be related to changes in precipitation through energetic arguments (Stephens and Ellis 2008; Allan 2009; Stephens and Hu 2010). Precipitation intensity or precipitation extremes, on the other hand, could be expected to be more directly affected by changes in water vapor concentrations and increase with warming at a faster rate than global-mean precipitation (Allen and Ingram 2002; Trenberth et al. 2003; Pall et al. 2007); recent results suggest that vertical gradients of specific humidity or surface specific humidities are relevant for precipitation extremes (O'Gorman and Schneider 2009a, b; Schneider et al. 2010; Muller et al. 2011; Romps 2011). Several authors have argued that this implies a reduction in the overall frequency of precipitation (e.g., Trenberth 1999, 2011; Allen and Ingram 2002; Trenberth et al. 2003; Stephens and Hu 2010). Decreases in precipitation frequency are seen in the subtropics in model simulations of warmed climates (e.g., Sun et al. 2007), and there is observational evidence for decreases in mean precipitation in tropical regions with mean descent (e.g., Allan et al. 2010), but it is important to note that the precipitation distribution could adjust in ways other than a simple decrease in frequency (Muller et al. 2011). Similar arguments have been made for a decrease in the strength of the circulation (e.g., Held and Soden 2006), although, again, the required magnitude of the circulation change is difficult to estimate from first principles (Schneider et al. 2010). We do not pursue the implications for precipitation frequency or circulation changes further, but rather focus on the radiative or energetic constraints themselves.

The purpose of this paper is to provide an overview of the modern understanding of energetic constraints on precipitation under climate change, to describe some of the major recent developments, and to examine in detail some of the key open questions. We begin by discussing the extent to which the atmospheric energetic constraint on global-mean precipitation may be viewed as a radiative constraint (Sect. 2) and to what extent energetics place an upper bound on global-mean precipitation (Sect. 3). We discuss the factors contributing to the rate of change of precipitation under climate change (Sect. 4), how these combine to determine the transient precipitation response (Sect. 5), and how the energetic perspective on precipitation changes may be extended to regional precipitation (Sect. 6). We then address the important issue of the extent to which observed changes in precipitation may be related to observed changes in net radiative cooling (Sect. 7). We conclude with a summary and a discussion of remaining challenges (Sect. 8).

2 Radiative or Energetic Constraint?

The energetic constraint on global-mean precipitation is often thought of as a radiative constraint in which latent heat release balances radiative cooling, but it clearly also involves the surface sensible heat flux (Eq. 1). Changes in surface sensible heat flux under climate change are not negligible; the upward sensible heat flux decreases in climate model simulations of greenhouse-gas driven warming (O'Gorman and Schneider 2008; Stephens and Ellis 2008; Liepert and Previdi 2009; Bala et al. 2010) and in response to increases in insolation (Bala et al. 2010), but may increase with warming in response to aerosol forcing (Liepert and Previdi 2009). One might, therefore, question if the radiative cooling of the atmosphere is really a constraint on precipitation, since the sensible heat flux could in principle adjust to accommodate a given radiative cooling (e.g., by changes in the air-sea temperature difference), so that a minimal model of the atmospheric or surface energy balance must take into account boundary layer transfer in order to constrain precipitation (Pierrehumbert 2002; Le Hir et al. 2009). Takahashi (2009a) proposed a simple model that avoids some of these difficulties by working with the atmospheric energy balance above cloud base (Fig. 1). Assuming that condensation only occurs above the lifted condensation level (LCL), it may be argued that the primary relevant balance is between latent heating and the radiative cooling above the LCL

$$L\delta P \simeq \delta R_{\mathrm{TOA}} - \delta R_{\mathrm{LCL}}. \tag{2}$$

This approximate balance relies on the lower level used (nominally the LCL) being below the level of substantial latent heating and above the level of substantial boundary layer dry sensible heat fluxes. It also neglects vertical dry static energy fluxes across the LCL associated with the large-scale circulation. We will refer to it as the free-atmospheric radiative constraint on precipitation (even though the LCL and the top of the boundary layer need not coincide).

Here, we test the accuracy of the free-atmospheric radiative constraint on precipitation in statistical-equilibrium simulations with an idealized general circulation model (GCM) over a wide range of climates (Fig. 2). The GCM is based on a version of the Geophysical Fluid Dynamics Laboratory (GFDL) dynamical core, but with simplified moist parameterizations,

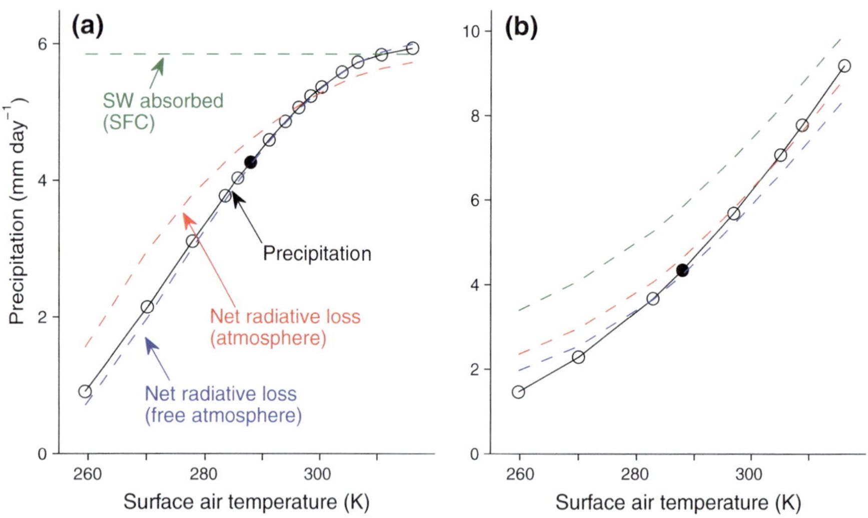

Fig. 2 Global-mean precipitation (*solid line with circles*) versus global-mean surface air temperature in two series of statistical-equilibrium simulations with an idealized GCM in which **a** the optical depth of the longwave absorber is varied and **b** the solar constant is varied. The *filled circles* indicate the reference simulation (common to both series) which has the climate most similar to present-day Earth's. The *red dashed lines* show the net radiative loss of the atmosphere, the *blue dashed lines* show the net radiative loss of the free atmosphere (above $\sigma = 0.86$), and the *green dashed lines* show the net absorbed solar radiation at the surface (all in equivalent precipitation units of mm day^{-1})

a two-stream semi-gray radiation scheme without cloud radiative effects, and a slab-ocean lower boundary condition (Frierson et al. 2006; Frierson 2007; O'Gorman and Schneider 2008). The climate is varied in one set of simulations by changing the longwave optical depth to mimic changes in greenhouse-gas concentrations with shortwave (SW) radiative heating held constant (Fig. 2a), as described in O'Gorman and Schneider (2008). We also present a new set of simulations in which the climate is varied by changing the solar constant over a range from 0.58 to 1.70 of the reference value (Fig. 2b). Although the radiation scheme is highly simplified, these simulations serve to illustrate the basic role of energetic constraints on precipitation over a wide range of climates—more realistic radiative transfer will be discussed in later sections. In addition, the GCM includes processes (boundary-layer turbulence and large-scale circulations) that could cause the free-atmospheric radiative constraint on global-mean precipitation (Eq. 2) to be inaccurate, and as such the idealized simulations provide a test of this constraint.

Global-mean precipitation increases with warming for both sets of simulations (Fig. 2a, b). The behavior in the limits of very low and high temperatures is broadly consistent with what might be expected from a simple analysis of the surface energy balance using bulk transfer formulae (Pierrehumbert 2002): precipitation tends to zero for sufficiently low temperatures because of low-specific humidities, and precipitation is limited at high temperatures by the availability of solar radiation at the surface (as discussed in the next section). In the "greenhouse-gas" simulations, the approximation of global-mean precipitation by the net radiative loss of the atmosphere (the dashed red line in Fig. 2a) is not accurate because of substantial changes in the surface sensible heat flux over the range of climates. The net radiative loss of the free atmosphere gives much better agreement (the dashed blue line in Fig. 2a). The free atmosphere is defined here as the atmosphere above the level $\sigma = 0.86$, where σ is pressure normalized by surface pressure. This level is chosen to give the best overall fit of radiative loss to precipitation and is taken to be representative of the nominal LCL in the approximate radiative constraint given by Eq. 2. But, the global-mean LCL is actually lower in the atmosphere ($\sigma \simeq 0.93$), a discrepancy which likely relates to the occurrence of substantial boundary layer fluxes at levels somewhat higher than the LCL. The rate of change of global-mean precipitation is 2.5% K^{-1} near the reference climate (the climate most similar to present-day Earth), compared to 1.5% K^{-1} for the net radiative loss of the atmosphere and 2.6% K^{-1} for the net radiative loss of the free atmosphere. The improvement gained from considering the free atmosphere is not as great in the solar series of simulations, but the overall accuracy of the free-atmospheric radiative constraint is not much worse; the rates of change at the reference climate are 3.2% K^{-1} for precipitation, 2.6% K^{-1} for the net radiative loss of the atmosphere, and 2.9% K^{-1} for the net radiative loss of the free atmosphere.

Our results provide some support for the approximation of global-mean precipitation under climate change using the net radiative loss of the free atmosphere. We will also find this to be a useful approximation in Sect. 4.2 when considering the effect on precipitation of black carbon aerosols at different levels in the atmosphere.

3 Upper Bound on Precipitation

Figure 2a shows that the rate of growth of global-mean precipitation with temperature declines sharply in the very warm climates of the "greenhouse-gas" series of simulations, despite the quasi-exponential increase with temperature of atmospheric water vapor content (O'Gorman and Schneider 2008), and suggesting the possibility of an upper bound on precipitation for a given solar constant (Pierrehumbert 1999, 2002; O'Gorman and Schneider 2008). In very warm climates, the atmosphere is optically thick in the longwave and the net longwave radiative flux at the surface becomes small. As discussed by Pierrehumbert (2002), the boundary layer becomes stable (with the near-surface air temperature greater than the surface temperature), and boundary layer turbulence is suppressed. To the extent that the surface sensible heat flux is small, the surface energy budget implies that the surface latent heat flux should then be equal to the absorbed shortwave radiation at the surface, corresponding to the horizontal line in Fig. 2a (the shortwave fluxes are held constant in this idealized set of simulations). The absorbed shortwave radiation is not exactly an upper bound because the surface sensible heat flux becomes directed downward providing additional energy for the evaporation of water (Pierrehumbert 2002). Similarly, in the simulations in which the solar constant is increased, precipitation approaches (but does not reach) the limit given by the absorbed shortwave radiation at the surface (Fig. 2b).

Thus, the energetic constraint on precipitation becomes relatively simple in the limit of very warm climates, although the exact amount of precipitation achievable may depend on poorly understood details of boundary-layer turbulence and radiative transfer. It is useful to remember the limiting case of very warm climates when considering the differences in scaling of water vapor concentrations and precipitation and, as discussed in the next section, the different responses of precipitation to solar and CO_2 forcings. The upper bound on global-mean precipitation given by the absorbed surface shortwave radiation is also relevant to very warm climates of the past, in which continental runoff plays an important role in the weathering thermostat (Pierrehumbert 2002; Le Hir et al. 2009). Le Hir et al. (2009) found that global-mean precipitation and continental runoff behaved similarly in the global mean as the climate changed, but it is not clear that this is a general result, and a better understanding of the constraints on precipitation over land is desirable.

4 Contributions to Changes in Precipitation

We next discuss the different contributions to changes in precipitation and how they lead to different responses of precipitation compared with temperature depending on the forcing agent. The response of global-mean precipitation to temperature change is clearly quite different for greenhouse-gas versus solar forcing, as evidenced by the leveling-off of precipitation at high temperatures in Fig. 2a compared with the continuous growth in Fig. 2b. Figure 3 [reproduced from Andrews et al. (2009)] shows a similar difference in behavior, but now for subsequent years following an instantaneous doubling of CO_2 or an instantaneous increase in solar irradiance in an atmospheric GCM coupled to a slab ocean. The response to CO_2 forcing is characterized by an initial decrease in precipitation, followed by a quasi-linear increase with temperature (e.g., Yang et al. 2003; Andrews et al. 2009). The response to solar forcing has a similar form but with a smaller initial decrease (extrapolating the precipitation curve to zero temperature change). As a result, the hydrological sensitivity, defined as the ratio of precipitation change to temperature change, is quite different at equilibrium for the CO_2 and solar forcings (compare the dotted lines in Fig. 3).[3] But, note that the slopes of the precipitation curves are similar for both forcings

[3] Interestingly, if the hydrological sensitivity is instead defined in terms of TOA radiative forcing rather than temperature change, it is not very different between solar and CO_2 forcing (Lambert and Faull 2007).

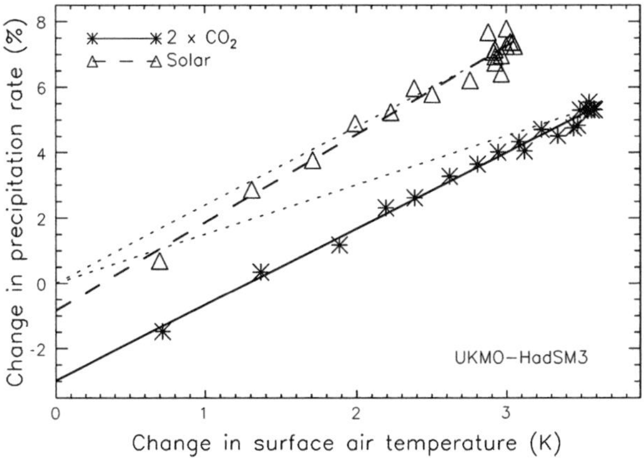

Fig. 3 Change in global-mean precipitation versus change in global-mean surface air temperature averaged over individual years subsequent to an instantaneous increase in CO_2 (*asterisks*) or solar irradiance (*triangles*) in simulations with the Hadley Centre Slab-Ocean Model version 3. There is a large initial decrease in precipitation in response to CO_2 forcing, but the slopes of the subsequent linear responses are similar for the solar and CO_2 simulations. *Dotted lines* show the slopes corresponding to the hydrological sensitivities calculated from the initial and final states [Fig. 7 from Andrews et al. (2009). © American Meteorological Society. Reprinted with permission]

(the solid and dashed lines in Fig. 3), so that the differential dependence of precipitation on temperature is similar.

The linear dependence of precipitation on temperature shown in Fig. 3 suggests that it may be useful to rewrite the atmospheric energy budget (Eq. 1), decomposing the right hand side (and the precipitation change) into a temperature-dependent part ($k\delta T$) and a temperature-independent part G,

$$L\delta P = k\delta T + G, \tag{3}$$

where k is a constant, and δT is the change in global-mean surface temperature (e.g., Allen and Ingram 2002). There are several ways to make this decomposition in practice. Previdi (2010) considers G to include the direct radiative forcing, while $k\delta T$ represents all feedbacks, including the change in surface sensible heat flux. Alternatively, one may decompose the response based on timescale into a "fast" component that occurs before sea surface temperatures (SSTs) respond substantially, and a "slow" component that increases in magnitude as the SSTs change on a multiyear timescale (Yang et al. 2003; Lambert and Faull 2007; Lambert and Webb 2008; Lambert and Allen 2009; Andrews et al. 2009, 2010; Takahashi 2009b; Bala et al. 2010; Andrews and Forster 2010; Frieler et al. 2011). The "slow" and "fast" responses may be calculated using fixed SST simulations or by regressing transient changes in precipitation and temperature (e.g., by calculating the slope and offset of the solid and dashed lines in Fig. 3).

4.1 Feedbacks

We first focus on the temperature-dependent part of the precipitation response and ask what physical processes contribute to its magnitude and the range of roughly 1.4–3.4% K^{-1} found in modern climate model simulations (Lambert and Webb 2008).

We make use of the radiative kernel technique (Soden et al. 2008) as applied to feedbacks on global-mean precipitation within the framework of the atmospheric energy budget by Previdi (2010). The simulations used are drawn from the World Climate Research Programme's (WCRP's) Coupled Model Intercomparison Project phase 3 (CMIP3) archive. We present results based on the climate change from 2001–2010 to 2101–2110 under an emissions scenario for greenhouse gases and aerosols (SRES A1B). The feedback analysis quantifies the contributions to $k\delta T$ in Eq. 3 from radiative feedbacks due to changes in tropospheric temperature, water vapor, clouds, and albedo, and from changes in surface sensible heat flux. We further decompose the temperature feedback into a Planck feedback (associated with vertically uniform warming) and a lapse rate feedback (associated with vertically nonuniform warming). We adopt the convention of Previdi (2010) that a positive feedback corresponds to a gain of energy for the atmospheric column and a negative feedback on precipitation. The feedbacks are shown in Fig. 4 for the nine climate models for which the necessary data were available [see Previdi (2010) for details]. This figure may be compared with Fig. 1 of Bony et al. (2006), which shows TOA radiative feedbacks rather than the atmospheric energy budget feedbacks shown here.

The water vapor feedback tends to suppress precipitation as a result of both increased shortwave heating and reduced longwave cooling [Previdi (2010); although Mitchell et al. (1987) and Hall and Manabe (2000) conclude that longwave radiative feedback enhances precipitation under warming based on different simulations and analysis]. The reduced longwave cooling is a residual of opposing effects of increases in specific humidity in the lower troposphere which tend to cool the atmospheric column and increases in specific humidity in the upper troposphere which tend to warm the atmospheric column (Previdi 2010). The effect of changes in water vapor concentrations is, therefore, strongly altitude dependent. Takahashi (2009b) argued that inter-model scatter in the change in clear-sky absorption of shortwave radiation by water vapor is important for inter-model scatter in the precipitation response and that the source of the inter-model scatter was not from different radiative transfer schemes or different changes in the amount of column water vapor. The

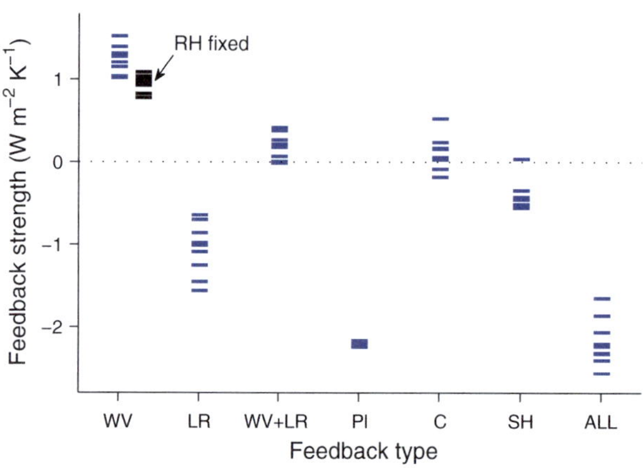

Fig. 4 Feedbacks (*blue dashes*) on the atmospheric energy budget in coupled simulations with nine climate models. *Positive values* indicate a gain in energy for the atmospheric column and a negative feedback on precipitation. Feedbacks shown are water vapor (WV), lapse rate (LR), the sum of water vapor and lapse rate (WV + LR), Planck (Pl), cloud (C), surface sensible heat flux (SH), and the sum ALL = WV + LR + Pl + C + SH. Albedo feedback is negligible and is not shown. *Black dashes* show the water vapor feedback for invariant relative humidity (RH)

water vapor feedback is slightly stronger and has greater inter-model scatter than the feedback that would result from an invariant relative humidity distribution (black dashes in Fig. 4). A consistent pattern of relative humidity changes is found in response to warming in climate model simulations (e.g., Mitchell et al. 1987; Sherwood et al. 2010), and although the fractional changes in relative humidity are not as large as the fractional changes in saturation vapor pressure, they may still be expected to affect radiative fluxes.

The lapse rate feedback is of similar strength to the water vapor feedback, but it tends to enhance precipitation. The magnitudes of the lapse rate and water vapor feedbacks are correlated between models, but they have opposite sign, so that the sum of the two feedbacks (WV + LR) has less inter-model scatter than might otherwise be expected. A similar relationship between lapse rate and water vapor feedbacks has been found for TOA radiative feedbacks (e.g., Bony et al. 2006; Soden et al. 2008).

The cloud feedback is calculated by adjusting the change in cloud radiative forcing to account for cloud-masking effects (Soden et al. 2008; Previdi 2010). It can be positive or negative depending on the model, but tends to suppress precipitation in the multi-model mean. Stephens and Ellis (2008) also found that cloud effects tended to mute the increase in precipitation in the multi-model mean. Although shortwave cloud radiative feedbacks contribute strongly to inter-model scatter in climate sensitivity, they might be expected to have less impact on the atmospheric cooling that controls changes in precipitation (Lambert and Webb 2008). Nonetheless, cloud radiative feedback is the single biggest contributor to inter-model scatter if the water vapor and lapse rate feedbacks are considered as one contribution.

With the exception of one model, the surface sensible heat flux feedback tends to increase precipitation and has relatively little inter-model scatter. The Planck feedback is large in magnitude and has almost the same value in all models considered (-2.1 to -2.2 W m^{-2} K^{-1}). The albedo feedback is very small in magnitude and is not shown. The sum of the feedbacks has a value of ~ -2W m^{-2} K^{-1} and must be combined with the forcing to give the change in precipitation found in these simulations. Compared with previous analyses of TOA radiative feedbacks (e.g., Bony et al. 2006), some of the primary differences are the negligible albedo feedback, the addition of the surface sensible heat flux, a possibly smaller contribution to scatter from the cloud feedback, and a possibly greater contribution from changes in relative humidity.

4.2 Dependence on Forcing Agent

We next consider the dependence of the precipitation response on the nature of the forcing agent. The direct radiative effect of increased concentrations of CO_2 is to decrease the net upwelling longwave radiation at the top of the atmosphere (TOA) and increase the net downwelling longwave radiation at the surface, with a net decrease in the radiative loss of the atmosphere [contributing to $G < 0$ in Eq. 3] (Ramanathan 1981; Mitchell et al. 1987). Consequently, if surface temperature is held fixed and CO_2 concentrations are increased, then global-mean precipitation decreases (e.g., Yang et al. 2003), consistent with Fig. 3. The direct radiative effect of increased insolation is much smaller in the atmospheric energy budget than in the TOA budget since much of the increased shortwave radiation passes through the atmosphere or is scattered back to space. Nonetheless, an increase in insolation does lead to increased shortwave absorption in the atmosphere, which tends to decrease precipitation. Because of the difference in the fast response of precipitation to solar and CO_2 forcing (Fig. 3), we expect global-mean precipitation to decrease in an idealized geoengineering experiment in which the effect on global-mean surface temperature of higher concentrations of greenhouse gases is offset by the effect of a decrease in solar constant. Bala et al. (2008) find that global-mean precipitation is reduced by $\sim 2\%$ for a doubling of CO_2 in such a geoengineering experiment, which is roughly consistent with what might be inferred from Fig. 3.

Figure 5a [based on Table 2 of Andrews et al. (2010)] shows that the slow temperature-dependent part of the response is quite similar for solar, CO_2, and aerosol, forcing, so that the different responses to different forcing agents arises primarily because of the fast component of the response.[4] Aerosols affect precipitation in a number of different ways (e.g., Ramanathan et al. 2001), but here we will discuss their effect on precipitation through their radiative role in the energetic constraint on precipitation. Figure 5a shows that sulfate aerosols yield a similar precipitation response to solar forcing, as might be expected given that they scatter shortwave radiation. However, increases in the burden of black carbon aerosols may decrease global-mean precipitation even as they increase global-mean surface temperature because of a strongly negative fast response (Fig. 5a).

The negative fast response of precipitation to black carbon aerosols results from absorption of shortwave radiation in the troposphere (induced changes in cloud radiative effects may also play a role). Whether or not the total precipitation response is negative depends on the level in the atmosphere at which the black carbon aerosols occur (Ming et al. 2010; Ban-Weiss et al. 2011). This was demonstrated by Ming et al. (2010) in a set of simulations in which the burden of black carbon aerosols was increased at different levels in the troposphere [Fig. 5b, which is based on Table 1 of Ming et al. (2010)]. Roughly speaking, precipitation decreases when black carbon aerosols are added in the free troposphere, but it increases when they are added near the surface. In all cases shown in Fig. 5b, the surface temperature increases, changes in shortwave absorption provide a negative contribution to the precipitation change, and changes in surface sensible heat flux and temperature provide a positive contribution.[5] If black carbon aerosols are added near the surface, the increased shortwave absorption is partially canceled by a decrease in the upward sensible heat flux, and precipitation increases because of the increase in radiative cooling related to the increase in temperature. The sensible heat flux response is considerably smaller if the aerosol is added in the free troposphere, and the increase in shortwave heating is then partially balanced by a decrease in latent heating (and precipitation).

The precipitation response to black carbon aerosols is more easily understood using the approximate balance of precipitation and free-atmospheric radiative cooling (Eq. 2) in which radiative heating near the surface and surface sensible heat fluxes do not enter. In the case of the addition of black carbon aerosols near the surface, the only effect on free atmosphere radiative cooling is through the temperature increase, and precipitation increases accordingly. When black carbon aerosols are added higher up in the atmosphere, they directly affect the free-atmospheric radiative heating and precipitation decreases.

[4] Andrews et al. (2010) show that the temperature dependence of the slow precipitation response is similar for nine different forcing scenarios. The precipitation sensitivities are normalized by a temperature change that is different for the slow and total responses because the fast response includes a change in land surface temperature and the slow response is calculated as the difference between total and fast responses.

[5] Ming et al. (2010) consider the change in surface sensible heat flux to be part of the fast or temperature-independent response.

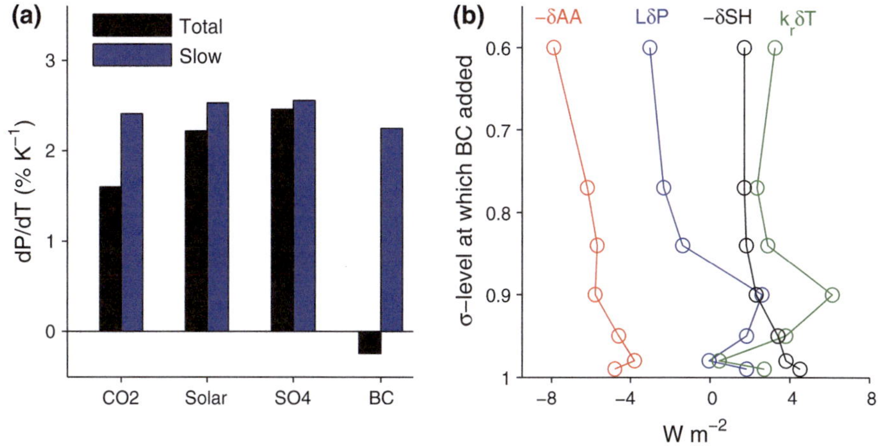

Fig. 5 **a** Sensitivities of the total and slow precipitation responses for different forcings (CO_2, solar, sulfate aerosols, and black carbon aerosols) in simulations with an atmospheric model coupled to a mixed-layer ocean. The total response is the sum of the slow and fast responses [based on Table 2 of Andrews et al. (2010)]. **b** Terms in the perturbation atmospheric energy budget in response to additional black carbon aerosols at different σ-levels in simulations with an atmospheric GCM coupled to a mixed-layer ocean. The terms shown are the changes in atmospheric absorption induced directly by the absorbing aerosols (δAA), radiative feedbacks ($k_r\delta T$) with surface temperature T and $k_r = 1.8$ W m^{-2} K^{-1}, the change in latent heating associated with precipitation $L\delta P$, and the change in upward sensible heat flux (δSH). We show $-\delta AA$ and $-\delta SH$ so that all terms are positive when contributing to a positive change in precipitation [based on Table 1 of Ming et al. (2010)]

5 Transient Changes in Precipitation

The discussion of the preceding section makes clear that changes in surface temperature are not sufficient to determine the equilibrium response of precipitation, but rather that changes in atmospheric radiative forcing must also be specified. Temperature changes may occur on an entirely different timescale to the radiative forcing, and the transient evolution of precipitation depends on the evolution of both radiative forcing and temperature. For example, we expect a slower rate of increase in precipitation with respect to temperature in a period in which both greenhouse-gas concentrations and temperature are increasing than in a period in which greenhouse-gas concentrations have stabilized but temperature continues to increase (Andrews and Forster 2010).

Some further nonintuitive properties of the transient precipitation response have recently been illustrated in coupled climate model simulations (Wu et al. 2010; Cao et al. 2011). Wu et al. (2010) showed that, in response to a change in the trend of atmospheric CO_2 concentrations from upward to downward, temperature begins to fall almost immediately, but the rate of increase in precipitation actually accelerates, before eventually decreasing after several decades. Cao et al. (2011) found a similar behavior in response to a rampdown of CO_2 concentrations [Fig. 6c; reproduced from Cao et al. (2011)] and also demonstrated an extreme limit of the same behavior in which CO_2 concentrations are changed in a step-like manner, first upward and then downward, resulting in downward and then upward spikes in precipitation (red lines in Fig. 6a, b; see also Fig. 3). As discussed by Wu et al. (2010) and Cao et al. (2011), these transient responses may be understood using the energetic constraint on precipitation. For example, a sudden decrease in CO_2 concentrations leads to an increase in radiative cooling and an increase in precipitation. Only on the slow timescale of the

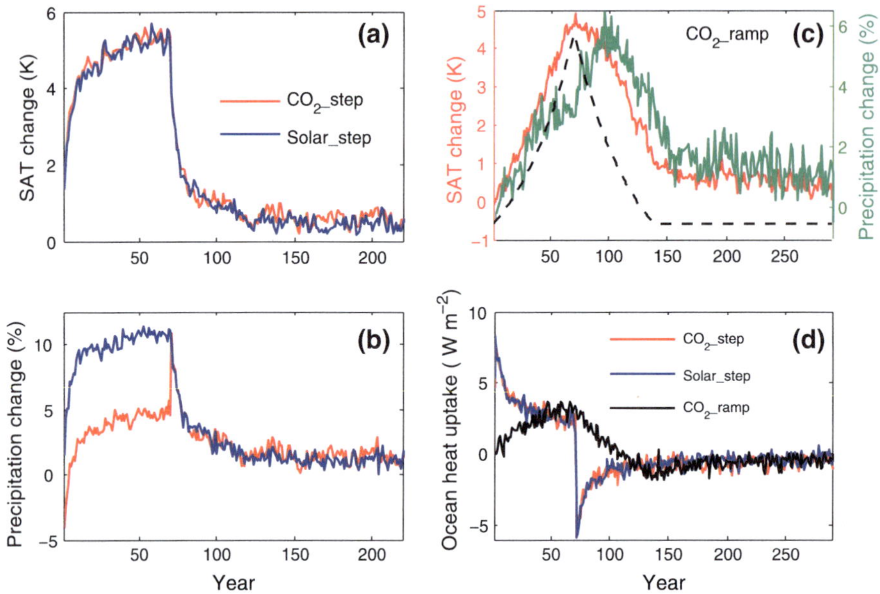

Fig. 6 Temporal variations in **a**, **c** surface air temperature, **b**, **c** precipitation, and **d** ocean heat uptake in three simulations with a coupled climate model (HadCM3L). In the "step" simulations (**a**, **b**, **d**), atmospheric CO_2 or the solar constant are instantaneously increased and then instantaneously returned to their initial value after 70 years (CO_2 is quadrupled or the solar constant is increased by 4.54%). In the CO_2 ramp simulation (**c**, **d**), atmospheric CO_2 is increased by 2% per year until quadrupling after 70 years, and then decreased by 2% per year until the original value is reached [*dashed black line* in (**c**)]. Note the downward and upward spikes in precipitation in the CO_2 step simulation at years 0 and 70, respectively [reproduction of Fig. 1 of Cao et al. (2011)]

ensuing temperature decrease does the radiative cooling and precipitation decrease. The precipitation and temperature responses are more qualitatively similar to one another in a simulation in which the step-like changes are in the solar constant (blue lines in Fig. 6a, b), consistent with the much smaller fast response of precipitation for solar forcing. (The magnitude of the changes in solar constant are chosen to give similar surface temperature variations as for the CO_2 simulation). From the surface energy budget perspective, the rate of ocean heat uptake should be expected to influence the transient response of evaporation and precipitation. However, the time history of ocean heat uptake is almost identical in the simulations with solar and CO_2 step-like changes (Fig. 6d), which shows that the sudden reduction in ocean heat uptake at year 70 is not necessarily sufficient to give a temporary increase in precipitation, depending on the nature of the radiative forcing.

The energetic perspective also makes clear that radiative feedbacks will be important for low-frequency variability of precipitation. For example, Hall and Manabe (2000) found a reduction in the interannual variability of global-mean precipitation in a climate model simulation in which the longwave radiative feedback of water vapor was suppressed.

6 Regional Changes in Precipitation

We have focused on the energetic constraint on global-mean precipitation, but regional changes in precipitation are of greater importance for impacts of climate change. Regional

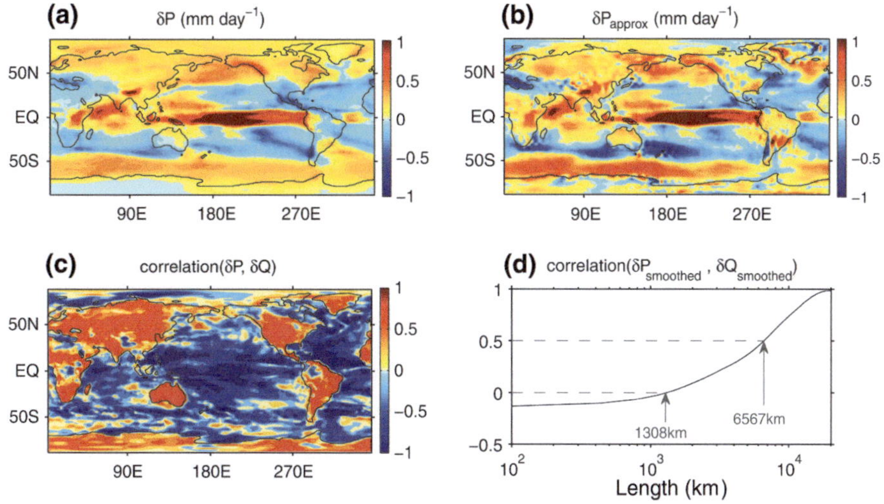

Fig. 7 Energetic perspective on local precipitation changes in the multi-model mean of CMIP3 simulations: **a** change in precipitation (mm day^{-1}), **b** an approximation $\delta P_{\mathrm{approx}} = \delta\langle P\rangle + \delta H_m/L$ that neglects changes in eddy dry static energy (DSE) fluxes and spatial variations in the change in diabatic cooling, **c** the inter-model correlation coefficient of the change in precipitation and diabatic cooling, and **d** the global-mean of this correlation coefficient as a function of the length scale of a spatial filter that is first applied to the changes in precipitation and diabatic cooling [see Muller and O'Gorman (2011) for details]

precipitation changes have been conventionally analyzed in terms of column water vapor balance, relating the difference between precipitation and evaporation to the water vapor flux convergence (e.g., Held and Soden 2006; Seager et al. 2010). The energy budget has sometimes been used as an additional constraint to account for changes in the circulation which are an important regional contributor to the changes in precipitation (Chou and Neelin 2004; Chou et al. 2009; Levermann et al. 2009; Chou and Chen 2010). Alternatively, the energetic approach to global-mean precipitation changes may be extended to local precipitation changes by including changes in horizontal transports of dry static energy (DSE),

$$L\delta P = \delta Q + \delta H, \tag{4}$$

where δH is the change in the vertical integral of the DSE flux divergence, and we will refer to $Q = R_{\mathrm{TOA}} - R_{\mathrm{SFC}} - SH$ as the diabatic cooling (excluding latent heating).[6] The vertical integral in δH is taken over the atmospheric column, although it could be approximated by a vertical integral over the troposphere. Increases in DSE flux divergence cool the atmospheric column and tend to increase precipitation for fixed diabatic cooling. Muller and O'Gorman (2011) used this local energy budget to analyze changes in precipitation in simulations drawn from the CMIP3 archive under the A1B emissions scenario (Fig. 7a). They showed that a simple approximation $\delta P_{\mathrm{approx}}$ for the change in local

[6] In the case of tropical precipitation extremes, the primary balance is between latent heating and DSE flux divergence (Muller et al. 2011).

precipitation δP is given in terms of the change in the vertical integral of the DSE flux divergence by the mean circulation (δH_m) and the change in global-mean precipitation ($\delta\langle P\rangle$), such that $L\delta P_{\text{approx}} = L\delta\langle P\rangle + \delta H_m$ (Fig. 7b). Further simple approximations may then be made to relate components of δH_m to changes in temperature and mid-tropospheric vertical velocity (Muller and O'Gorman 2011), resulting in approximate relations comparable to those used in the water vapor budget approach (e.g., Held and Soden 2006).

Locally, changes in diabatic cooling may be balanced by changes in precipitation or changes in DSE flux divergence. A fundamental question then arises as to what extent local changes in diabatic cooling δQ are in fact reflected in local changes in precipitation δP. For example, one might ask to what extent localized radiative forcing from aerosols will be related to a corresponding local change in precipitation. Muller and O'Gorman (2011) addressed this question by examining the inter-model correlation coefficient between changes in precipitation δP and changes in diabatic cooling δQ and found that, while δP and δQ were positively correlated over land, they were negatively correlated over ocean because of cloud and water vapor feedbacks (Fig. 7c). The scale dependency of the relationship between δP and δQ was addressed by smoothing over a range of length scales prior to calculation of the correlation coefficient. The global-mean correlation coefficient reaches a value of 0.5 for a smoothing length scale of $\sim$7,000 km (Fig. 7d), implying that only on relatively large scales or over land are δP and δQ strongly positively correlated. Further work is needed to understand the physical processes that contribute to the correlation between changes in precipitation and diabatic cooling at different length scales and to better understand the contrast between their relationship over land and ocean (cf. Trenberth and Shea 2005; Lambert and Allen 2009).

7 Observations of the Atmospheric Energetic Constraint on Precipitation

The energetic constraint on global precipitation is difficult to confirm observationally. Long-term (50 or more years) observations of precipitation (surface rain gauges) are primarily confined to northern hemisphere land regions (e.g., Min et al. 2011), while spatially complete ocean estimates are only available through satellite retrievals of infrared radiances since 1979 and microwave radiances since 1987 (Huffman et al. 2009) and contain substantial uncertainties relating to sampling and calibration (Adler et al. 2001). Reconstructions of past precipitation may be made using reanalyses or surface pressure and temperature measurements (e.g., Arkin et al. 2010), but they are limited by the homogeneity of input data and the physical relationships employed.

Estimates of Earth's radiative energy balance are limited to an even greater extent: satellite data have provided near-global estimates of variability for much of the period since 1985 (Wielicki et al. 2002; Loeb et al. 2009) but with substantial calibration and sampling issues (Trenberth 2002; Wong et al. 2006). Surface measurements are limited to solar radiometers available for a handful of locations over land since the 1950s (Wild 1999), increasing to around 50 well-calibrated longwave and shortwave radiation measurements for the recent decade (Ohmura et al. 1998). Therefore, estimating recent changes in the surface and atmospheric radiative balance is currently possible only through the additional use of reanalyses data sets combined with additional modeling (Zhang et al. 2004).

Current global variability in P and its driving variables are compared in Fig. 8. During this period, global-mean surface air temperature (T) varied by almost 1 K, primarily relating to El Niño Southern Oscillation (ENSO), but also a cooling in 1992 following the eruption of Mt. Pinatubo in June 1991 and a warming trend during the 1990s.

7.1 Observed Changes in Surface Temperature and Water Vapor

Atmospheric moisture is tightly coupled to global temperatures (Fig. 8a, b) as evidenced by column integrated water vapor (*W*) from the Special Sensor Microwave Imager [SSM/I; Wentz and Schabel (2000)] and the European Centre for Medium-range Weather Forecasts (ECMWF) Interim ReAnalysis [ERA Interim; Dee et al. (2011)] and also by surface-specific humidity (*q*) from HadCRUH observations (Willett et al. 2008). Global SSM/I estimates are constructed using ocean-only SSM/I data between 50°S–50°N and applying ERA Interim values elsewhere. ERA Interim *W* reduces with time compared with the SSM/I estimates, attributable to differences over the tropical oceans (John et al. 2009), and

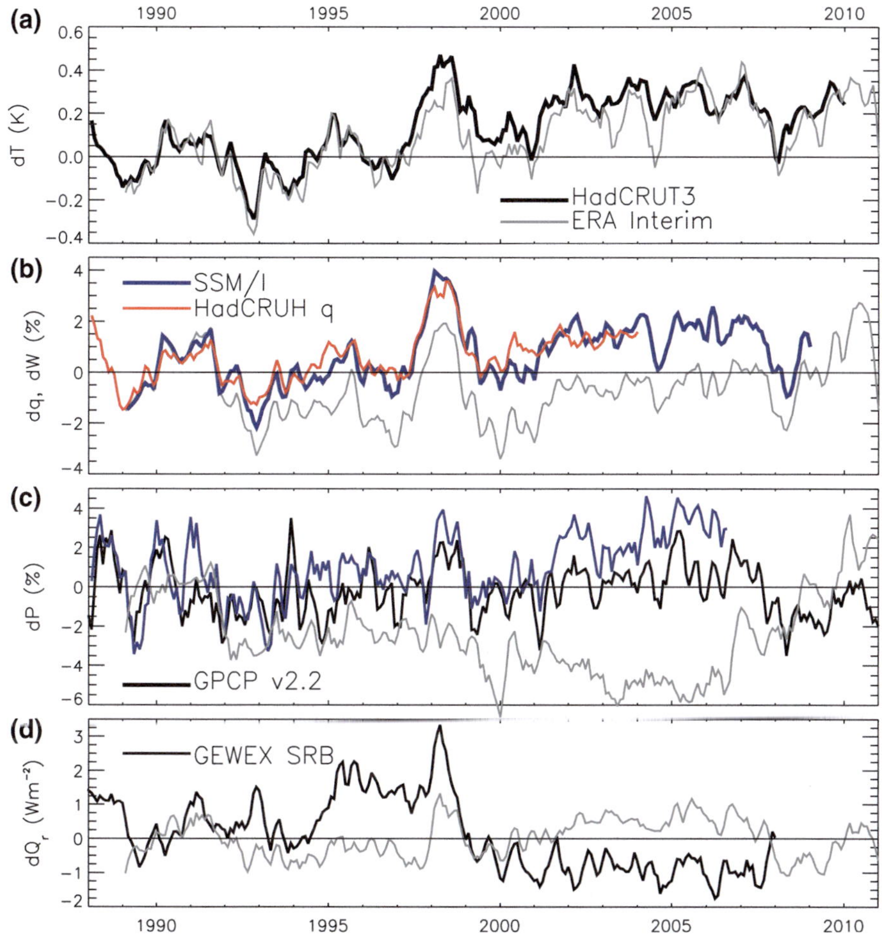

Fig. 8 Global-mean de-seasonalized monthly anomalies in **a** surface air temperature, **b** column integrated water vapor and surface-specific humidity, **c** precipitation and **d** total net radiative cooling of the atmosphere. A 3-month smoothing is applied for clarity. The mean anomaly for January 1989 to December 1990 is subtracted from each time series

Table 1 Linear least squares fits for selected deseasonalized global-mean monthly variables and data sets over the period 1989–2008

Variables (dy/dx)	Data set (y)	Data set (x)	dy/dx ± SE	r
dW/dT	SSM/I, ERA Interim	HadCRUT	$6.6 \pm 0.4\%\ K^{-1}$	0.87
dP/dT	GPCP	HadCRUT	$3.4 \pm 0.9\%\ K^{-1}$	0.40
dQ_r/dT	ERA Interim	ERA Interim	$2.5 \pm 0.3\ W\ m^{-2}K^{-1}$	0.71
LdP/dQ_r	GPCP	ERA Interim	1.0 ± 0.2	0.54

may be an artifact of the observing system as documented previously (Trenberth et al. 2001; Allan et al. 2002; Bengtsson et al. 2004).

Similarity between HadCRUH and SSM/I-ERA interim estimates of global moisture variability ($r = 0.86$) is striking, both demonstrating the robust relationship between low-level moisture and temperature ($dW/dT \sim 7\%\ K^{-1}$; Table 1), close to that expected from the Clausius–Clapeyron equation (O'Gorman and Muller 2010). This has been documented previously using surface observations (Willett et al. 2008), satellite data (Wentz and Schabel 2000; Santer et al. 2007; Allan 2009), and radiosonde soundings (Durre et al. 2009). The dependence of atmospheric water vapor amounts on temperature is important in accounting for observed contrasting regional precipitation responses (Zhang et al. 2007), in particular for the ascending and descending portions of the tropical circulation (Allan et al. 2010) and enhanced tropical seasonality (Chou et al. 2007), and for the observed increase in the intensity of precipitation (Rajeevan et al. 2008; O'Gorman and Schneider 2009a; Zolina et al. 2010; Allan et al. 2010; Min et al. 2011). Importantly for the energetic constraint on precipitation, increases in atmospheric water vapor content are also physically consistent with enhanced longwave radiative cooling of the atmosphere to the surface (Allan 2006; Stephens and Ellis 2008; Philipona et al. 2009) and enhanced shortwave radiative heating of the atmosphere (Allan 2009; Takahashi 2009b).

7.2 Observed Changes in Precipitation

Observed global precipitation changes (Fig. 8c) are from the Global Precipitation Climatology Project [GPCP v2.2; Huffman et al. (2009)], which combines infrared and SSM/I microwave radiances over ice-free oceans since 1988 and rain gauges over land, and an estimate from Wentz et al. (2007), who combined SSM/I data over ice-free oceans with GPCP data over remaining regions. Simulated precipitation from ERA Interim is also shown and displays a negative trend up until 2005, as noted by John et al. (2009), and a rapid increase over the period 2006–2010, variations which are known to be erroneous (Dee et al. 2011). Since the hydrological cycle in reanalyses is not well-constrained globally, their representation of decadal changes in water vapor and precipitation is unlikely to be realistic.

Global precipitation from GPCP shows a weak positive relationship with surface temperature from HadCRUT ($dP/dT = 3.4\%\ K^{-1}$) over the period 1989–2008 (Table 1), similar to values estimated by Adler et al. (2008) for the period 1979–2006. Allowing for autocorrelation using the method of Yang and Tung (1998), which estimates 36 degrees of freedom in the time series, this is only just significant at the 95% confidence level. The autocorrelation timescale for precipitation (6 months) is notably shorter than for water vapor and net radiation (both around 20 months); the timescale dependence of the controlling processes may provide further insight into the role of forcing, feedback, and response (Harries and Futyan 2006).

There is considerable sensitivity of precipitation response to the data set and time period chosen (Quartly et al. 2007; Wang et al. 2008; John et al. 2009). The relationship with temperature changes found above ($dP/dT = 3.4\%\ K^{-1}$) is weaker than calculated by Wentz et al. (2007), who estimated a 7% K^{-1} response for a shorter time period (Adler et al. 2008), and is also weaker than estimated changes in ocean evaporation (Yu 2007; Li et al. 2011). These estimates were based upon linear trends in precipitation and temperature which introduce substantial uncertainty given the short length of the climate record and the influence of ENSO and volcanic eruptions (Gu et al. 2007; Lambert et al. 2008); the estimated observed responses are, therefore, not inconsistent with the large spread in responses calculated from climate models using similar methods (Previdi and Liepert 2008; Liepert and Previdi 2009; Allan 2009). The larger response over the period 1987–2006 (Wentz et al. 2007) may be consistent with an apparent intensification of the Walker and Hadley circulation since 1979 (Sohn and Park 2010; Zahn and Allan 2011; Li et al. 2011) although declines in circulation have also been identified by other authors (Vecchi et al. 2006; Power and Smith 2007; Gastineau and Soden 2011), based upon reanalyses and satellite wind data. A blended reconstruction of twentieth century precipitation by Arkin et al. (2010), based upon gauge observations over land and surface temperature and pressure patterns over ocean, suggests a smaller hydrological sensitivity of 2.5% K^{-1}.

7.3 Estimating Net Radiative Cooling

The energetic constraint on precipitation discussed in previous sections implies that changes in global-mean precipitation depend both on changes in the radiative cooling of the atmosphere and changes in surface sensible heat flux. As discussed in Sect. 2, the net atmospheric radiative cooling above the sub-cloud layer may provide a more direct link to global-mean precipitation (Takahashi 2009a), but given the difficulties of generating such a data set, we instead use the approach of Allan (2006) and show estimates of the net radiative cooling of the atmosphere (Q_r, longwave radiative cooling minus shortwave radiative heating) in Fig. 8d.

There is poor agreement between estimates of Q_r from the NASA Global Energy and Water Cycle Experiment (GEWEX) Surface Radiation Budget (SRB) project, which uses cloud and radiation information retrieved from satellite and atmospheric profiles from reanalyses as input to radiative transfer models (Stackhouse et al. 2011) and the ERA Interim reanalysis project. Variability from the International Satellite Cloud Climatology Project (ISCCP) D2 radiative flux products (Zhang et al. 2004) is substantially larger (not shown), showing poor agreement with SRB and ERA Interim. John et al. (2009) also found larger differences in atmospheric longwave and shortwave radiative divergence between data sets for the tropical ocean. One of the problems with the approach adopted by SRB and ISCCP is that surface radiative fluxes are not well constrained by the satellite measurements, in particular for the longwave fluxes, and homogeneity of the ISCCP data set is questionable (Evan et al. 2007).

Although ERA Interim clouds are generated by model parameterizations, the simulation of changes in net radiation at the top of the atmosphere is reasonable based upon comparison with satellite data (Allan 2011) and, notwithstanding issues with drifts in W shown in Fig. 8b and the lack of account for changes in aerosol optical depth, appear physically reasonable based upon the relationship with surface temperature ($dQ_r/dT = 2.5\ W\ m^{-2}\ K^{-1}$, Table 1) and previous analysis (Allan 2009; John et al. 2009). Although clouds contribute to heating of the moist tropical troposphere and cooling of the atmospheric column for stratocumulus regions and higher latitudes (Sohn 1999), their impact upon decadal changes in the atmospheric radiative budget may be weak, as discussed by John et al. (2009). (For climate models, Fig. 4 shows that the cloud radiative feedback on precipitation is not different from zero to within the inter-model scatter.)

7.4 Observed Link Between Precipitation and Atmospheric Net Radiative Cooling

The relationship between GPCP P (converted into units of W m^{-2} by multiplying by the latent heat of condensation L) and ERA Interim Q_r, $LdP/dQ_r = 1.0 \pm 0.2$, is weak (Table 1) although statistically significant at the 95% confidence level allowing for autocorrelation and it is also physically reasonable. Accounting more carefully for changes in aerosol, cloud and water vapor may improve the observational constraint on precipitation changes. For example, decadal trends in aerosol optical depth associated with global "dimming" and "brightening" (Wild et al. 2005; Mishchenko et al. 2007) are thought to explain increases in rainfall over land during the 1990s (Wild et al. 2008; Wild and Liepert 2010).

Within the framework of the atmospheric energy budget constraint on precipitation, scattering aerosols influence atmospheric radiative cooling and precipitation through the surface temperature response, but absorbing aerosols also directly lead to radiative heating of the atmosphere and thus may affect precipitation independent of surface temperature changes (Lambert et al. 2008; Andrews et al. 2009; Ming et al. 2010; Ban-Weiss et al. 2011), as discussed in Sect. 4.2. A further influence on hydrological sensitivity relates to greenhouse gases. Decadal trends in P are influenced by the secular rises in CO_2 concentrations which act in a similar way to absorbing aerosols by radiatively heating the atmosphere. Based upon these arguments, rising CO_2 concentrations in the 2000s, combined with stable decadal temperature, should result in a negative precipitation trend; this is not immediately obvious from Fig. 8c.

8 Conclusions

We have given a review of many of the insights to be gained from the energetic perspective on the response of precipitation to climate change. A number of open questions remain, several of which we now discuss briefly.

While it is clear that the atmospheric energy budget constrains the possible changes in global-mean precipitation, especially in warm climates, there is still some uncertainty as to the nature of the constraint. For example, we have discussed whether it may be approximated as a purely radiative constraint by balancing latent heating with the radiative cooling of the free atmosphere. Results presented here based on simulations with an idealized GCM over a wide range of climates provide some support for this free-atmospheric radiative constraint. Further work is needed to evaluate its accuracy, although the necessary radiative fluxes near the top of the boundary layer are not readily available in global observational or climate model data sets. It would be straightforward to further test the accuracy of the free-atmospheric radiative constraint using simulations with a comprehensive climate model in which the necessary radiative fluxes were stored. Even if the free-atmospheric radiative constraint is not very accurate, it may still be useful conceptually. For example, we have shown that it gives a particularly simple explanation for the simulated response of precipitation to radiative forcing from black carbon aerosols (Sect. 4.2).

Considerable progress has been made in quantifying the different feedbacks that contribute to changes in global-mean precipitation. The results presented here [extending the analysis of Previdi (2010)] suggest that cloud radiative feedbacks are a primary contributor to the inter-model scatter in the response of precipitation. Further characterization of the sources of uncertainty in the response of precipitation is desirable, with the aim of clarifying the extent to which these are similar to the sources of uncertainty for climate sensitivity and of identifying the processes whose parameterization is most problematic in this context. The need to include the changes in surface sensible heat fluxes (unless the free-atmospheric radiative constraint is proven adequate) distinguishes the problem from that of TOA radiative feedbacks, and it is important to develop a better understanding of the response of surface sensible heat fluxes on different timescales and for different forcings (cf., Liepert and Previdi 2009).

The different responses to different forcing agents have also been described, including recent progress in quantifying the fast response of precipitation to different types of aerosol forcing. The vertical structure of changes in water vapor, black carbon aerosols, and clouds is expected to be important in determining the magnitude and even the sign of the precipitation response. The combination of slow and fast responses means that it is not necessarily straightforward to relate observed or simulated transient changes in precipitation to changes in temperature. An intriguing possibility is that it is more appropriate to relate changes in precipitation to changes in radiative forcing rather than changes in temperature; in addition to the closer agreement then found between hydrological sensitivities for different forcing agents (Lambert and Faull 2007), it may be more appropriate to relate changes in energy fluxes to one another than to changes in temperature.

We have discussed how the energetic perspective on global-mean precipitation changes may be extended to regional precipitation changes by including horizontal energy fluxes (DSE fluxes) in the analysis. One important question that could be addressed in such a framework is the extent to which energetics can be used to give a simple constraint on changes in precipitation over land (for example, in terms of radiative forcings and changes in surface temperature). Such a constraint would be particularly useful because many of the available historical observations of precipitation are over land rather than ocean. Lambert and Allen (2009) found that a simple regression model for precipitation changes that was adequate in the global-mean was not adequate over land alone, even when land-ocean energy transports were accounted for. Nonetheless, starting from the full local energy budget, it should be possible to systematically make approximations to derive the minimal energetic model needed to account for changes in precipitation over land.

Currently available observations do not allow us to definitively link changes in global-mean precipitation with changes in the radiative energy budget of the atmosphere. Uncertainties arise for both the observed changes in radiative fluxes and precipitation, and more extensive and longer-term observations are clearly desirable. The observations we do have raise a number of important questions. For example, what sets the different autocorrelation timescales of global-mean precipitation, column water vapor, and net radiative cooling? And what are the contributions of different forcing agents to the observed hydrological sensitivity over different time periods? These basic questions are important for understanding observations of ongoing changes in the global hydrological cycle.

Progress could be made on several of the open questions identified here using simulations which have recently become available from the Coupled Model Intercomparison Project phase 5. In particular, the experiments designed to probe fast and slow responses and the impacts of changes in clouds and aerosols could be used to better understand precipitation responses in different models and emissions scenarios.

Acknowledgments M. Byrne is supported through the MIT Joint Program on the Science and Policy of Global Change. R. Allan is funded through the National Environment Research Council PREPARE project (NE/G015708/1) and National Centre for Atmospheric Sciences. We thank the American Meteorological Society (AMS) and T. Andrews for reproduction of Fig. 3, the American Geophysical Union (AGU) and L. Cao for reproduction of Fig. 6, and C. J. Muller for Fig. 7. The SSM/I data were extracted from Remote Sensing Systems, the GPCP data from the NASA Goddard Space Flight Center, the HadCRUH and HadCRUT data from http://www.metoffice.gov.uk/hadobs/ and ERA Interim data from http://www.ecmwf.int. We acknowledge the modeling groups, the Program for Climate Model Diagnosis and Intercomparison (PCMDI) and the WCRP's Working Group on Coupled Modelling (WGCM) for their roles in making available the WCRP CMIP3 multi-model data set. Support of this data set is provided by the Office of Science, US Department of Energy.

References

Adler RF, Kidd C, Petty G, Morissey M, Goodman HM (2001) Intercomparison of global precipitation products: the third precipitation intercomparison project (PIP-3). Bull Am Meteorol Soc 82:1377–1396

Adler RF, Gu G, Wang JJ, Huffman GJ, Curtis S, Bolvin D (2008) Relationships between global precipitation and surface temperature on interannual and longer timescales (1979–2006). J Geophys Res 113:D22,104

Allan RP (2006) Variability in clear-sky longwave radiative cooling of the atmosphere. J Geophys Res 111:D22, 105

Allan RP (2009) Examination of relationships between clear-sky longwave radiation and aspects of the atmospheric hydrological cycle in climate models, reanalyses, and observations. J Clim 22:3127–3145

Allan RP (2011) Combining satellite data and models to estimate cloud radiative effect at the surface and in the atmosphere. Meteorol Appl 18:324–333

Allan RP, Slingo A, Ramaswamy V (2002) Analysis of moisture variability in the European Centre for medium-range weather forecasts 15-year reanalysis over the tropical oceans. J Geophys Res 107:4230

Allan RP, Soden BJ, John VO, Ingram WI, Good P (2010) Current changes in tropical precipitation. Environ Res Lett 5:025205

Allen MR, Ingram WJ (2002) Constraints on future changes in climate and the hydrologic cycle. Nature 419:224–232

Andrews T, Forster PM (2010) The transient response of global-mean precipitation to increasing carbon dioxide levels. Environ Res Lett 5:025212

Andrews T, Forster PM, Gregory JM (2009) A surface energy perspective on climate change. J Clim 22:2557–2570

Andrews T, Forster PM, Boucher O, Bellouin N, Jones A (2010) Precipitation, radiative forcing and global temperature change. Geophys Res Lett 37:L14701

Arkin PA, Smith TM, Sapiano MRP, Janowiak J (2010) The observed sensitivity of the global hydrological cycle to changes in surface temperature. Environ Res Lett 5:035201

Bala G, Duffy PB, Taylor KE (2008) Impact of geoengineering schemes on the global hydrological cycle. Proc Nat Acad Sci 105:7664–7669

Bala G, Caldeira K, Nemani R (2010) Fast versus slow response in climate change: implications for the global hydrological cycle. Clim Dyn 35:423–434

Ban-Weiss GA, Cao L, Bala G, Caldeira K (2011) Dependence of climate forcing and response on the altitude of black carbon aerosols. Clim Dyn doi:10.1007/s00382-011-1052-y

Bengtsson L, Hagemann S, Hodges KI (2004) Can climate trends be calculated from reanalysis data?. J Geophys Res 109:D11111

Boer GJ (1993) Climate change and the regulation of the surface moisture and energy budgets. Clim Dyn 8:225–239

Bony S, Colman R, Kattsov VM, Allan RP, Bretherton CS, Dufresne JL, Hall A, Hallegatte S, Holland MM, Ingram W, Randall DA, Soden BJ, Tselioudis G, Webb MJ (2006) How well do we understand and evaluate climate change feedback processes?. J Clim 19:3445–3482

Cao L, Bala G, Caldeira K (2011) Why is there a short-term increase in global precipitation in response to diminished CO_2 forcing?. Geophys Res Lett 38:L06703

Chou C, Chen CA (2010) Depth of convection and the weakening of tropical circulation in global warming. J Clim 23:3019–3030

Chou C, Neelin JD (2004) Mechanisms of global warming impacts on regional tropical precipitation. J Clim 17:2688–2701

Chou C, Tu JY, Tan PH (2007) Asymmetry of tropical precipitation change under global warming. Geophys Res Lett 34:L17708
Chou C, Neelin JD, Chen CA, Tu JY (2009) Evaluating the "Rich-Get-Richer" mechanism in tropical precipitation change under global warming. J Clim 22:1982–2005
Dee DP, Uppala SM, Simmons AJ, Berrisford P, Poli P, Kobayashi S, Andrae U, Balmaseda MA, Balsamo G, Bauer P, Bechtold P, Beljaars ACM, van de Berg L, Bidlot J, Bormann N, Delsol C, Dragani R, Fuentes M, Geer AJ, Haimberger L, Healy SB, Hersbach H, Hólm EV, Isaksen L, Kållberg P, Köhler M, Matricardi M, McNally AP, Monge-Sanz BM, Morcrette JJ, Park BK, Peubey C, de Rosnay P, Tavolato C, Thépaut JN, Vitart F (2011) The ERA-Interim reanalysis: configuration and performance of the data assimilation system. Q J R Meteorol Soc 137:553–597
Durre I, Williams CN Jr, Yin X, Vose RS (2009) Radiosonde-based trends in precipitable water over the Northern Hemisphere: an update. J Geophys Res 114:D05112
Evan AT, Heidinger AK, Vimont DJ (2007) Arguments against a physical long-term trend in global ISCCP cloud amounts. Geophys Res Lett 34:L04701
Frieler K, Meinshausen M, von Deimling TS, Andrews T, Forster P (2011) Changes in global-mean precipitation in response to warming, greenhouse gas forcing and black carbon. Geophys Res Lett 38:L04702
Frierson DMW (2007) The dynamics of idealized convection schemes and their effect on the zonally averaged tropical circulation. J Atmos Sci 64:1959–1976
Frierson DMW, Held IM, Zurita-Gotor P (2006) A gray-radiation aquaplanet moist GCM. Part I: static stability and eddy scale. J Atmos Sci 63:2548–2566
Gastineau G, Soden BJ (2011) Evidence for a weakening of tropical surface wind extremes in response to atmospheric warming. Geophys Res Lett 38:L09706
Gu G, Adler RF, Huffman GJ, Curtis S (2007) Tropical rainfall variability on interannual-to-interdecadal and longer time scales derived from the GPCP monthly product. J Clim 20:4033–4046
Hall A, Manabe S (2000) Effect of water vapor feedback on internal and anthropogenic variations of the global hydrologic cycle. J Geophys Res 105:6935–6944
Harries JE, Futyan JM (2006) On the stability of the Earth's radiative energy balance: response to the Mt. Pinatubo eruption. Geophys Res Lett 33:L23814
Held IM, Soden BJ (2006) Robust responses of the hydrological cycle to global warming. J Clim 19:5686–5699
Huffman GJ, Adler RF, Bolvin DT, Gu G (2009) Improving the global precipitation record: GPCP version 2.1. Geophys Res Lett 36:L17808
John VO, Allan RP, Soden BJ (2009) How robust are observed and simulated precipitation responses to tropical ocean warming. Geophys Res Lett 36:L14702
Lambert FH, Allen MR (2009) Are changes in global precipitation constrained by the tropospheric energy budget?. J Clim 22:499–517
Lambert FH, Faull NE (2007) Tropospheric adjustment: the response of two general circulation models to a change in insolation. Geophys Res Lett 34:L03701
Lambert FH, Webb MJ (2008) Dependency of global mean precipitation on surface temperature. Geophys Res Lett 35:L16706
Lambert FH, Stine AR, Krakauer NY, Chiang JCH (2008) How much will precipitation increase with global warming?. Eos Trans AGU 89:193–194
Le Hir G, Donnadieu Y, Goddéris Y, Pierrehumbert RT, Halverson GP, Macouin M, Nédélec A, Ramstein G (2009) The snowball Earth aftermath: exploring the limits of continental weathering processes. Earth Planet Sci Lett 277:453–463
Levermann A, Schewe J, Petoukhov V, Held H (2009) Basic mechanism for abrupt monsoon transitions. Proc Nat Acad Sci 106:20,572–20,577
Li G, Ren B, Yang C, Zheng J (2011) Revisiting the trend of the tropical and subtropical Pacific surface latent heat flux during 1977–2006. J Geophys Res 116:D10115
Liepert BG, Previdi M (2009) Do models and observations disagree on the rainfall response to global warming?. J Clim 22:3156–3166
Loeb NG, Wielicki BA, Doelling DR, Smith GL, Keyes DF, Kato S, Manalo-Smith N, Wong T (2009) Toward optimal closure of the Earth's top-of-atmosphere radiation budget. J Clim 22:748–766
Lorenz DJ, DeWeaver ET, Vimont DJ (2010) Evaporation change and global warming: the role of net radiation and relative humidity. J Geophys Res 115:D20118
Min SK, Zhang X, Zwiers FW, Hegerl GC (2011) Human contribution to more-intense precipitation extremes. Nature 470:378–381
Ming Y, Ramaswamy V, Persad G (2010) Two opposing effects of absorbing aerosols on global-mean precipitation. Geophys Res Lett 37:L13701

Mishchenko MI, Geogdzhayev IV, Rossow WB, Cairns B, Carlson BE, Lacis AA, Liu L, Travis LD (2007) Long-term satellite record reveals likely recent aerosol trend. Science 315:1543
Mitchell JFB, Wilson CA, Cunnington WM (1987) On CO_2 climate sensitivity and model dependence of results. Q J R Meteorol Soc 113:293–322
Muller CJ, O'Gorman PA (2011) An energetic perspective on the regional response of precipitation to climate change. Nat Clim Change 1:266–271
Muller CJ, O'Gorman PA, Back LE (2011) Intensification of precipitation extremes with warming in a cloud resolving model. J Clim 24:2784–2800
O'Gorman PA, Muller CJ (2010) How closely do changes in surface and column water vapor follow Clausius–Clapeyron scaling in climate-change simulations?. Environ Res Lett 5:025207
O'Gorman PA, Schneider T (2008) The hydrological cycle over a wide range of climates simulated with an idealized GCM. J Clim 21:3815–3832
O'Gorman PA, Schneider T (2009a) The physical basis for increases in precipitation extremes in simulations of 21st-century climate change. Proc Natl Acad Sci 106:14,773–14,777
O'Gorman PA, Schneider T (2009b) Scaling of precipitation extremes over a wide range of climates simulated with an idealized GCM. J Clim 22:5676–5685
Ohmura A, Dutton EG, Forgan B, Fröhlich C, Gilgen H, Hegner H, Heimo A, König-Langlo G, McArthur B, Müller G, Philipona R, Pinker R, Whitlock CH, Dehne K, Wild M (1998) Baseline surface radiation network (BSRN/WCRP): new precision radiometry for climate research. Bull Am Meteorol Soc 79:2115–2136
Pall P, Allen MR, Stone DA (2007) Testing the Clausius–Clapeyron constraint on changes in extreme precipitation under CO_2 warming. Clim Dyn 28:351–363
Peixoto JP, Oort AH (1992) Physics of climate. American Institute of Physics, New York
Philipona R, Behrens K, Ruckstuhl C (2009) How declining aerosols and rising greenhouse gases forced rapid warming in Europe since the 1980s. Geophys Res Lett 36:L02806
Pierrehumbert RT (1999) Subtropical water vapor as a mediator of rapid global climate change. In: Clarks PU, Webb RS, Keigwin LD (eds) Mechanisms of global climate change at millennial time scales. Geophys. Monogr. Ser., vol 112, American Geophysical Union, Washington, p 339
Pierrehumbert RT (2002) The hydrologic cycle in deep-time climate problems. Nature 419:191–198
Power SB, Smith IN (2007) Weakening of the Walker circulation and apparent dominance of El Niño both reach record levels, but has ENSO really changed?. Geophys Res Lett 34:L18702
Previdi M (2010) Radiative feedbacks on global precipitation. Environ Res Lett 5:025211
Previdi M, Liepert BG (2008) Interdecadal variability of rainfall on a warming planet. Eos Trans AGU 89:193–195
Quartly GD, Kyte EA, Srokosz MA, Tsimplis MN (2007) An intercomparison of global oceanic precipitation climatologies. J Geophys Res 112:D10121
Rajeevan M, Bhate J, Jaswal AK (2008) Analysis of variability and trends of extreme rainfall events over India using 104 years of gridded daily rainfall data. Geophys Res Lett 35:L18707
Ramanathan V (1981) The role of ocean-atmosphere interactions in the CO_2 climate problem. J Atmos Sci 38:918–930
Ramanathan V, Crutzen PJ, Kiehl JT, Rosenfeld D (2001) Aerosols, climate, and the hydrological cycle. Science 294:2119–2124
Richter I, Xie SP (2008) Muted precipitation increase in global warming simulations: A surface evaporation perspective. J Geophys Res 113:D24118
Romps DM (2011) Response of tropical precipitation to global warming. J Atmos Sci 68:123–138
Santer BD, Mears C, Wentz FJ, Taylor KE, Gleckler PJ, Wigley TML, Barnett TP, Boyle JS, Brüggemann W, Gillett NP, Klein SA, Meehl GA, Nozawa T, Pierce DW, Stott PA, Washington WM, Wehner MF (2007) Identification of human-induced changes in atmospheric moisture content. Proc Natl Acad Sci 104:15,248–15,253
Schneider T, O'Gorman PA, Levine XJ (2010) Water vapor and the dynamics of climate changes. Rev Geophys 48:RG3001
Seager R, Naik N, Vecchi GA (2010) Thermodynamic and dynamic mechanisms for large-scale changes in the hydrological cycle in response to global warming. J Clim 23:4651–4668
Sherwood SC, Ingram W, Tsushima Y, Satoh M, Roberts M, Vidale PL, O'Gorman PA (2010) Relative humidity changes in a warmer climate. J Geophys Res 115:D09104
Soden BJ, Held IM, Colman R, Shell KM, Kiehl JT, Shields CA (2008) Quantifying climate feedbacks using radiative kernels. J Clim 21:3504–3520
Sohn BJ (1999) Cloud-induced infrared radiative heating and its implications for large-scale tropical circulations. J Atmos Sci 56:2657–2672

Sohn BJ, Park SC (2010) Strengthened tropical circulations in past three decades inferred from water vapor transport. J Geophys Res 115:D15112
Stackhouse PW Jr, Gupta SK, Cox SJ, Zhang T, Mikovitz JC, Hinkelman LM (2011) 24.5-year SRB data set released. GEWEX News 21:10–12
Stephens GL, Ellis TD (2008) Controls of global-mean precipitation increases in global warming GCM experiments. J Clim 21:6141–6155
Stephens GL, Hu Y (2010) Are climate-related changes to the character of global-mean precipitation predictable?. Environ Res Lett 5:025209
Stevens B, Schwartz SE (2011) Observing and modeling Earth's energy flows. Surv Geophys (submitted)
Sun Y, Solomon S, Dai A, Portmann RW (2007) How often will it rain?. J Clim 20:4801–4818
Takahashi K (2009a) Radiative constraints on the hydrological cycle in an idealized radiative-convective equilibrium model. J Atmos Sci 66:77–91
Takahashi K (2009b) The global hydrological cycle and atmospheric shortwave absorption in climate models under CO_2 forcing. J Clim 22:5667–5675
Trenberth KE (1999) Conceptual framework for changes of extremes of the hydrological cycle with climate change. Clim Change 42:327–339
Trenberth KE (2002) Changes in tropical clouds and radiation. Science 296:2095a
Trenberth KE (2011) Changes in precipitation with climate change. Clim Res 47:123–138
Trenberth KE, Shea DJ (2005) Relationships between precipitation and surface temperature. Geophys Res Lett 32:L14703
Trenberth KE, Stepaniak DP, Hurrell JW, Fiorino M (2001) Quality of reanalyses in the tropics. J Clim 14:1499–1510
Trenberth KE, Dai A, Rasmussen RM, Parsons DB (2003) The changing character of precipitation. Bull Am Meteorol Soc 84:1205–1217
Trenberth KE, Fasullo JT, Kiehl J (2009) Earth's global energy budget. Bull Am Meteorol Soc 90:311–323
Vecchi GA, Soden BJ, Wittenberg AT, Held IM, Leetmaa A, Harrison MJ (2006) Weakening of tropical Pacific atmospheric circulation due to anthropogenic forcing. Nature 441:73–76
Wang JJ, Adler RF, Gu G (2008) Tropical rainfall-surface temperature relations using Tropical Rainfall Measuring Mission precipitation data. J Geophys Res 113:D18115
Wentz FJ, Schabel M (2000) Precise climate monitoring using complementary satellite data sets. Nature 403:414–416
Wentz FJ, Ricciardulli L, Hilburn K, Mears C (2007) How much more rain will global warming bring?. Science 317:233–235
Wielicki BA, Wong T, Allan RP, Slingo A, Kiehl JT, Soden BJ, Gordon CT, Miller AJ, Yang S, Randall DA, Robertson F, Susskind J, Jacobowitz H (2002) Evidence for large decadal variability in the tropical mean radiative energy budget. Science 295:841–844
Wild M (1999) Discrepancies between model-calculated and observed shortwave atmospheric absorption in areas with high aerosol loadings. J Geophys Res 104:27,361–27,371
Wild M, Liepert B (2010) The Earth radiation balance as driver of the global hydrological cycle. Environ Res Lett 5:025003
Wild M, Gilgen H, Roesch A, Ohmura A, Long CN, Dutton EG, Forgan B, Kallis A, Russak V, Tsvetkov A (2005) From dimming to brightening: decadal changes in solar radiation at Earth's surface. Science 308:847–850
Wild M, Grieser J, Schär C (2008) Combined surface solar brightening and increasing greenhouse effect support recent intensification of the global land-based hydrological cycle. Geophys Res Lett 35:L17706
Willett KM, Jones PD, Gillett NP, Thorne PW (2008) Recent changes in surface humidity: development of the HadCRUH dataset. J Clim 21:5364–5383
Wong T, Wielicki BA, Lee RB, Smith GL, Bush KA, Willis JK (2006) Reexamination of the observed decadal variability of the Earth radiation budget using altitude-corrected ERBE/ERBS nonscanner WFOV data. J Clim 19:4028–4040
Wu P, Wood R, Ridley J, Lowe J (2010) Temporary acceleration of the hydrological cycle in response to a CO_2 rampdown. Geophys Res Lett 37:L12705
Yang F, Kumar A, Schlesinger ME, Wang W (2003) Intensity of hydrological cycles in warmer climates. J Clim 16:2419–2423
Yang H, Tung KK (1998) Water vapor, surface temperature, and the greenhouse effect—a statistical analysis of tropical-mean data. J Clim 11:2686–2697
Yu L (2007) Global variations in oceanic evaporation (1958–2005): the role of the changing wind speed. J Clim 20:5376–5390
Zahn M, Allan RP (2011) Changes in water vapor transports of the ascending branch of the tropical circulation. J Geophys Res 116:D18111

Zhang X, Zwiers FW, Hegerl GC, Lambert FH, Gillett NP, Solomon S, Stott PA, Nozawa T (2007) Detection of human influence on twentieth-century precipitation trends. Nature 448:461–465
Zhang Y, Rossow WB, Lacis AA, Oinas V, Mishchenko MI (2004) Calculation of radiative fluxes from the surface to top of atmosphere based on ISCCP and other global data sets: refinements of the radiative transfer model and the input data. J Geophys Res 109:D19105
Zolina O, Simmer C, Gulev SK, Kollet S (2010) Changing structure of European precipitation: longer wet periods leading to more abundant rainfalls. Geophys Res Lett 37:L06704

Surv Geophys (2012) 33:609–617
DOI 10.1007/s10712-012-9182-2

The Role of Clouds: An Introduction and Rapporteur Report

Patrick C. Taylor

Received: 17 June 2011 / Accepted: 5 February 2012 / Published online: 7 March 2012

Abstract This paper presents an overview of discussions during the *Cloud's Role* session at the *Observing and Modelling Earth's Energy Flows Workshop*. N. Loeb and B. Soden convened this session including 10 presentations by B. Stevens, B. Wielicki, G. Stephens, A. Clement, K. Sassen, D. Hartmann, T. Andrews, A. Del Genio, H. Barker, and M. Sugi addressing critical aspects of the role of clouds in modulating Earth energy flows. Presentation topics covered a diverse range of areas from cloud microphysics and dynamics, cloud radiative transfer, and the role of clouds in large-scale atmospheric circulations patterns in both observations and atmospheric models. The presentations and discussions, summarized below, are organized around several key questions raised during the session. (1) What is the best way to evaluate clouds in climate models? (2) How well do models need to represent clouds to be acceptable for making climate predictions? (3) What are the largest uncertainties in clouds? (4) How can these uncertainties be reduced? (5) What new observations are needed to address these problems? Answers to these critical questions are the topics of ongoing research and will guide the future direction of this area of research.

Keywords Clouds · Cloud feedback · Cloud radiative forcing · Cloud uncertainties · Model cloud evaluation

1 Introduction

Radiation emanating from the Sun provides energy for all Earth processes. Earth system energy is found in the fast moving winds of the jet streams, in the enormous latent heat release in hurricanes, and in the vibrations of molecules located in the deep ocean. The manner in which energy is distributed throughout the Earth system affects global temperature, precipitation, and wind patterns. Tracking and modeling Earth's energy flows are thus necessary for understanding climate.

P. C. Taylor (✉)
NASA Langley Research Center, Hampton, VA 23681, USA
e-mail: patrick.c.taylor@nasa.gov

Clouds are a critical component of the climate system because they are a major modulator of energy flows through the Earth system. Clouds interact with solar and terrestrial radiative energy flows through reflection, absorption, and emission processes. The influence of clouds on solar and terrestrial radiative fluxes is traditionally defined as the difference between total- and clear-sky broadband fluxes, called the shortwave (SW) and longwave (LW) cloud radiative forcing (CRF), respectively (e.g., Ramanathan et al. 1989; Chen et al. 2000; Stephens 2005). The SW CRF quantifies the enhanced reflection of solar radiation, and removal of energy from the system, due to clouds with respect to clear sky. The LW CRF quantifies the reduction of energy leaving the system through LW emission to space by clouds with respect to clear sky. The Net top-of-atmosphere (TOA) cloud influence (Net CRF) is measured to be -20 W m^{-2} (Loeb et al. 2009) from Clouds and Earth's Radiant Energy System (CERES) observations, indicating an integrated cooling influence by clouds. Additionally, clouds influence Earth energy flows by acting as sources of latent heat and precipitation. Under equilibrium conditions, latent heating of the atmosphere by condensation within clouds and the sensible heating by surface fluxes must balance the net radiative cooling of the atmosphere by solar energy absorption and infrared energy emission and absorption. This condition provides an energetic constraint on the hydrologic cycle response to climate change, which couples the latent heating and atmospheric radiative cooling responses (Stephens and Ellis 2008). Further, vertical motion within clouds, especially deep convection, transports heat and momentum upward contributing to planetary-scale circulations, e.g., Hadley Circulation and the Madden Julian Oscillation (Hendon and Woodberry 1993). Clouds influence many facets of Earth energy flows and must be accurately represented in climate models to obtain reliable climate change projections.

General Circulation Model (GCM) projections of the Earth system response to an external climate forcing are fraught with uncertainty due to differences in the predicted cloud response. Forced with a doubling of CO_2, state-of-the-art GCMs exhibit a wide range of equilibrium climate sensitivities: ranging from +2.0 to larger than +4.5 K (Solomon et al. 2007). Most of this spread in climate sensitivity is due to climate feedback differences: temperature, water vapor, clouds, and surface albedo. Cloud feedbacks are considered the largest contributor to inter-model spread in climate sensitivity (Cess et al. 1990; Stephens 2005, Soden and Held 2006; Bony et al. 2006; Randall et al. 2007; Dufrense and Bony 2008). While all cloud types contribute to cloud feedback and significant uncertainties exist in all cloud processes (Stephens 2005), subtropical, marine boundary layer clouds contribute most significantly to the inter-model spread in cloud feedback (Bony and Dufresne 2005; Stephens 2005). As a result, all cloud-related processes, and specifically those important in subtropical marine boundary layer cloud regimes, must be better understood to reduce uncertainty in climate sensitivity.

Recent work suggests an additional cause of cloud-related climate sensitivity uncertainty, namely cloud forcing, cloud adjustment, or "fast feedback." Gregory et al. (2004) suggest that the significant inter-model climate sensitivity differences may result from differences in the radiative forcing. These differences are postulated to stem from a direct system response to the CO_2 forcing that occurs on short timescales, less than 1 year, before any significant surface temperature response. The cloud forcing can therefore be separated from feedback based on timescale. Clouds are suggested to represent a significant portion of the direct response to the CO_2 forcing (Andrews and Forster 2008; Gregory and Webb 2008). Therefore, the majority of the cloud impact on climate change may be a forcing rather than a feedback (Gregory et al. 2004; Andrews and Forster 2008; Gregory and Webb 2008). Whether this interpretation is realistic or strictly a model artifact is yet to be determined.

Clouds pose a unique challenge due to the large range of time and space scales at which cloud processes occur, and the complexities associated with observing cloud processes. Clouds occur in a multitude of different forms with each type influencing Earth energy flows differently. Low clouds, stratocumulus, and cumulus contribute strongly to the negative Net CRF because SW CRF $\gg$ LW CRF (Hartmann and Doelling 1991; Stephens 2005). High clouds, cirrus, and cirrostratus possess a positive Net CRF because LW CRF $\gg$ SW CRF (Hartmann and Doelling 1991; Stephens 2005). The Net CRF of deep convective clouds, however, is near zero due to a cancelation between large SW reflection and reduction in LW emission to space, SW CRF = LW CRF (Kiehl and Ramanathan 1990; Kiehl 1994). These different cloud forms also contribute differently to non-radiative energy flows: global latent heating and momentum transports. The vast range of time and space scales encompassed by cloud-related processes also contributes to the unique challenges of understanding clouds. Cloud processes occur on time scales from seconds to days and spatial scales from microns to thousands of kilometers. To account for sub-grid scale processes in GCMs, simplifying parameterizations are required, which adds considerable uncertainty in climate model simulations. An implicit assumption in parameterization is that a scale separation exists between resolved and unresolved processes such that the behavior of unresolved processes can be represented by resolved variables. In many cases, this assumption does not hold and the need to parameterize contributes to the unique challenge.

Lastly, complexities associated with cloud process observations are another component contributing to the challenge of understanding clouds. In situ microphysical cloud observations require expensive airborne instrumentation generally only operated in short duration field campaigns. This limits the ability to obtain statistical distributions of cloud process observations under a range of atmospheric conditions. However, one attempt has been made to organize an airborne instrumentation campaign focused on obtaining a representative statistical sample over an extended period of time (Vogelmann et al. 2008). Due to the limited nature of in situ cloud observations, most research studies use satellite cloud observations. Satellite platforms allow global sampling of clouds; however, the interpretation of measured radiances is required to analyze physical cloud properties (e.g., cloud cover, emission temperature, particle phase/size, and water path). To transform observed radiances into physical cloud properties requires assumptions about the atmospheric state and cloud microphysics that can lead to retrieval biases and influence scientific conclusions. As a result, the nature and complexity of cloud processes lead to difficulties in the quantitative characterization and observation of clouds making clouds a unique scientific challenge.

The purpose of the chapter is to assess the current state of cloud process understanding, identify modeling uncertainties in Earth energy flows associated with clouds, and to identify future observations that are necessary to reduce cloud-related uncertainties. Workshop presentations encompassed a variety of cloud observational and modeling challenges that will be discussed in the papers that follow. The remainder of this paper will summarize the various topics discussed at the workshop.

2 Discussion Summary

Participants raised several critical questions around which this discussion is organized. The topics discussed below address five questions that received the most discussion: (1) What is the best way to evaluate clouds in climate models? (2) How well do models need to represent clouds to be acceptable for making climate predictions? (3) What are the largest uncertainties in clouds? (4) How can these uncertainties be reduced? (5) What new observations are needed to address these problems?

2.1 What is the Best Way to Evaluate Clouds in Climate Models?

The question, how should GCMs be evaluated, received significant discussion. D. Hartmann and A. Clement presented model evaluation tools asserting that models must be tested against understood physics. The issue was raised, however, that climate models are routinely evaluated against observed mean state variables, which does not directly represent a test of model physics. The use of CRF for model evaluation provides insight into the model representation of cloud radiative characteristics and individual cloud regimes biases (e.g., Kiehl and Ramanathan 1990; Bony et al. 2004; Zhang et al. 2005). However, LW, SW, and Net CRFs are integrated quantities determined by convolving cloud optical properties and frequency of occurrence, which enables models to correctly simulate CRF due to compensating errors between cloud optical properties and frequency (e.g., Zhang et al. 2005). Such compensations are frequently found within GCMs, namely negative cloud frequency and positive cloud optical depth biases (e.g., Zhang et al. 2005). Therefore, direct evaluation of model cloud frequency and optical properties was recommended in discussion.

Satellite observations have proven useful for the identification of model cloud frequency of occurrence and optical property biases; however, due to different sampling and cloud definitions, this comparison is not straightforward. To address these inconsistencies, satellite instrument simulators have been employed in GCM simulations (e.g., Klein and Jakob 1999; Haynes et al. 2007). The International Satellite Cloud Climatology Project (ISCCP) satellite simulator (Klein and Jakob 1999) has been successfully used to evaluate climate models (Zhang et al. 2005; Williams and Webb 2009). Recent research has also extended the satellite simulator methodology to A-Train active lidar and radar remote sensors (e.g., Chepfer et al. 2010; Bodas-Salcedo et al. 2011). Discussion outlined the utility of the satellite simulator methodology for model cloud evaluation; however, further refinement is required (Mace et al. 2011).

Categorizing cloud properties by atmospheric dynamic and thermodynamic conditions is an effective manner to evaluate model cloud properties. This strategy is useful because cloud formation under different atmospheric conditions results in different cloud characteristics. Bony et al. (2004) and Bony and Dufresne (2005) illustrate this evaluation technique in the tropics using annual mean 500-hPa vertical velocity. Eitzen et al. (2008) and Eitzen et al. (2011) extended this approach to evaluate model cloud optical properties and radiative characteristics by applying a joint probability distribution methodology and dynamic and thermodynamic state diagnostics. Meeting participants acknowledged the usefulness of such evaluation techniques. A. Del Genio and H. Barker stressed the need to go beyond monthly or annual mean evaluation techniques and rigorously test model parameterizations on the timescale they operate.

A critical component of parameterization development and evaluation is accurate atmospheric state knowledge. R. Wood stated that large-scale weather reanalysis is inadequate for climate purposes. One identified example is the poor treatment for boundary layer clouds in weather reanalysis. Marine stratocumulus clouds are largely neglected in meteorological reanalysis because no "significant" weather occurs in these regions. These cloud regimes, however, are critical to climate. Participants agreed that improved atmospheric state information is required to improve cloud parameterizations, and that further work in this area is necessary.

To address question (1), the best evaluation tool is likely to depend on the goal of the model simulation. A metric that indicates skill at the global scale may not have the same significance at regional scales. Murphy et al. (2004) illustrated that cloud and radiation variables are the best discriminators between high and low sensitivity models. For climate change prediction purposes, the answer to question (1) must be tied to evaluation techniques that possess skill in discriminating between high and low climate sensitivity. The current generation Cloud Feedback Model Intercomparison Project 2 (CFMIP2) GCMs have implemented a full suite of satellite simulators, enabling the comprehensive GCM-observational comparisons of model cloud and radiative characteristics. This new wealth of information on GCM cloud and radiation quantities will undoubtedly shed light on relationships between clouds, radiation, and climate sensitivity. The search for GCM performance metrics relevant to climate sensitivity is an active and critical area of research.

2.2 How Well Do Models Need to Represent Clouds to Be Acceptable for Making Climate Predictions?

Directly related to methods of GCM cloud evaluation, B. Wielicki raised another question regarding cloud representation in climate models: when is a model's cloud representation good enough for climate change prediction? To answer this question, one could consider the use of root mean square error metrics for CRF, cloud occurrence, or cloud optical properties applied spatially. However, it is unknown whether these metrics are tied to climate sensitivity. B. Wielicki stated that a model's ability to reproduce the current mean climate state is not necessarily a good indicator of climate change prediction skill, but that decadal change is a skillful metric (Soden et al. 2008). A potential metric discussed in a presentation by A. Clement tests the ability of a model to reproduce the covariance of physically related variables, for example sea surface temperature and low cloud amount (Clement et al. 2009). An argument linking a variable covariance metric to climate sensitivity can be made considering that fundamental covariances between critical climate variables at monthly, seasonal, or annual time scales will likely shape longer time scale covariances the describe climate feedbacks. A seasonal covariance analysis has proved useful for the surface albedo feedback (Hall and Qu 2006); however, an observational link between cloud covariance and cloud feedback strength has not been shown. In pursuit of an observational quantity linked to cloud feedback strength, Dessler (2010) found that cloud feedbacks inferred from short-term TOA radiative flux variability, a single decade or El Nino, have almost no bearing on the long-term, climate scale cloud feedbacks. The fidelity with which GCMs must represent cloud processes to provide accurate climate change prediction is unknown, and more research is needed to address uncertainty in climate models.

2.3 What Uncertainties Exist in the Role of Clouds in Modulating Earth Energy Flows? How Can These Uncertainties Be Reduced?

Recent modeling studies have highlighted the need to better understand "fast feedback" in the climate system. "Fast feedback," also referred to as cloud adjustment or cloud forcing, is the direct response of clouds to a radiative forcing at time scales of less than a year. T. Andrews reviewed cloud adjustment in a presentation where he discussed a possible link with the carbon cycle through a physiological response by plants. Gregory et al. (2004) and Gregory and Webb (2008) illustrated this effect in a GCM using an instantaneous 4xCO_2

step forcing. Andrews and Forster (2008) have suggested that the inter-model cloud feedback spread may be misinterpreted as a spread in cloud adjustment. Expressed in workshop discussion by S. Schwartz, uncertainty exists in this concept stemming from the unrealistic step-function forcing applied to analyze the model cloud adjustment. However, Gregory and Webb (2008) show for one model that the cloud response is approximately linear in the forcing magnitude within methodological uncertainties. The result suggests that this model phenomenon behaves similarly under different forcing magnitudes and is likely to operate in smaller, more realistic CO_2 forcing scenarios. However, the question moving forward, discussed in the chapter by T. Andrews, is does this phenomenon occur in the Earth system or is this only a GCM phenomenon?

Low cloud feedback under anthropogenic climate change remains a significant source of inter-model spread in climate sensitivity stemming from uncertainties in model parameterizations that influence marine boundary layer clouds (Bony et al. 2004; Bony and Dufrense 2005; Stephens 2005; Bony et al. 2006; Randall et al. 2007). Additional uncertainty in these cloud regimes is due to the relationship between cloud albedo and mean cloud liquid water path (LWP). Stephens (2010) in a GEWEX newsletter summarized a model inconsistency with observations where subtropical boundary layer clouds are simulated too thick, which suggests a misrepresented cloud albedo feedback. The relationship between cloud albedo and LWP is nonlinear. Stephens (2010) illustrated that current generation climate models underestimate the sensitivity of cloud albedo to a small perturbation in LWP. This illustrates the need for cloud evaluation metrics that link directly to climate sensitivity.

Another significant uncertainty discussed in this session is how cloud–aerosol interactions change the influence of clouds on Earth energy flows. Cloud–aerosol interactions can significantly influence Earth energy flows by modulating the radiative energy budget through changes in cloud albedo and the hydrologic cycle by influencing precipitation and drizzle frequency (e.g., Rosenfeld 2006; Stevens and Feingold 2009). The major cloud–aerosol interactions are termed the first and second aerosol indirect effects, namely cloud albedo and cloud lifetime. Both of these effects can influence the CRF by altering the column albedo (more information is available in the chapter on *Aerosol Forcing*). The main point in the discussion, expressed by B. Stevens, G. Stephens, and R. Wood specifically, was that the aerosol–cloud indirect effects are a first-order cloud formation and cloud precipitation problem. One motivation for this statement is that aerosol indirect effects in models cannot be correct if cloud frequency is poorly simulated. G. Stephens noted in his presentation that recent A-train observations indicate a smaller global impact of aerosol indirect effects than in GCMs that include cloud–aerosol interactions, due to misrepresentations of large but compensating effects; this point is also made by Stevens and Feingold (2009). Further, G. Stephens demonstrated compelling evidence indicating that open cellular marine stratocumulus clouds produce a column brightening in the presence of shiptracks; however, closed cells do not. Future work in cloud–aerosol interactions from space will likely require satellite sampling that resolves cloud lifetime scale processes.

2.4 What New Observations are Needed to Address These Problems?

Meeting discussion expressed that current satellite observations are difficult to use for process studies due to temporal sampling limitations. R. Kahn identified aircraft campaigns as the best way to perform cloud process studies. However, A. Del Genio noted that a weakness of field campaigns is the short duration and limited domain. Meeting participants discussed that a potential way forward could be to design satellite or high altitude balloon missions capable of observing cloud processes. One possible future satellite observation strategy could be to launch a constellation of small satellites with the same measurement capabilities and observe the same location a small time interval apart. This observational strategy would enable the measurement of time tendency terms for cloud process studies from space and greatly enhance cloud statistics toward improving model parameterization.

The general consensus was that current satellite cloud observation technology is generally sufficient, while acknowledging that adding the capability of scanning active sensors would greatly enhance the ability to observe clouds from space. Because scanning active sensor technology is expensive, there must be a continued emphasis on the fusion of passive and active remote sensors (e.g., Kato et al. 2010). The current generation of active sensors provides accurate retrievals of cloud location, including top and base height, and optical properties but is unable to scan. Passive sensors scan and retrieve cloud properties over a larger volume; however, passive retrievals can suffer from biases in cloud location and cloud properties. All meeting participants agreed that given the current state of technology, the synergy between passive and active sensors is the best future path (e.g., A-Train, EarthCARE, and ACE).

Participants expressed a need for improved cloud data products accurate enough for use in long-term climate trend analysis. Current methodologies can result in time-dependent biases in cloud properties because of limitations in instrument accuracy and ancillary input sources needed to characterize atmospheric state and surface properties. Secondly, current satellite cloud data products are limited by the accuracy of the instruments. To better understand Earth's energy flows and the role of clouds in modulating these flows, the group discussion expressed the need for a climate observing and monitoring system to detect decadal time scale climate trends, particularly clouds. G. Stephens stated in his presentation that a climate observing strategy is likely required for future progress on cloud feedback. The capabilities of the current global observing system are limited by accuracy. Meeting participants generally agreed that a step forward would be to fly a high-accuracy, well-calibrated instrumentation package that provides a stable calibration standard for all satellite sensors, similar to the CLimate Absolute Radiance and REfractivity Observatory (CLARREO) concept proposed in the United States National Research Council Decadal Survey (NRC 2007). Research and planning are required to obtain cloud information capable of detecting trends in cloud properties that are critical to unraveling cloud feedback.

3 Summary

Understanding the role of clouds in modulating Earth's energy flows is a complicated and an extremely important endeavor. Outlined above are key discussion topics motivated by the central role clouds play in future climate change. The *Cloud's Role* session at the *Observing and Modelling Earth's Energy Flows Workshop* served to identify the key cloud-related uncertainties and to stimulate discussion on how to address these issues. Many aspects of the questions discussed above are unanswered and require further research. Resolving these problems will lead to improvements in our understanding of climate change, Earth energy flows, and the role of clouds. The subsections of this chapter will elucidate the critical role of clouds in modulating Earth energy flows from observational (K. Sassen and H. Barker) and modeling (T. Andrews, A. Del Genio, and M. Sugi) perspectives, assessing the current state of knowledge and providing perspectives on future directions.

References

Andrews T, Forster PM (2008) CO_2 forcing induces semi-direct effects with consequences for climate feedback interpretations. Geophys Res Lett 35:L04802. doi:10.1029/2007GL032273

Bodas-Salcedo A et al (2011) COSP: satellite simulation software for model assessment. Bull Amer Meteor Soc 92:1023–1043. doi:10.1175/2011BAMS2856.1

Bony S, Dufresne J-L (2005) Marine boundary layer clouds at the heart of tropical cloud feedback uncertainties in climate models. Geophys Res Lett 32:L20806. doi:10.1029/2005GL023851

Bony S, Dufresne J-L, Le Treut H, Morcrette J-J, Senior C (2004) On dynamic and thermodynamic components of cloud changes. Climate Dyn 22:71–86. doi:10.1007/s00382-003-0369-6

Bony S et al (2006) How well do we understand and evaluate climate change feedback processes? J Climate 19:3445–3482

Cess RD et al (1990) Intercomparison and interpretation of climate feedback processes in 19 atmospheric general circulation models. J Geophys Res 95:16601–16615

Chen T, Rossow WB, Zhang Y (2000) Radiative effects of cloud-type variations. J Climate 13:264–286

Chepfer H, Bony S, Winker D, Cesana G, Dufresne JL, Minnis P, Stubenrauch CJ, Zeng S (2010) The GCMOriented CALIPSO Cloud Product (CALIPSO-GOCCP). J Geophys Res 115:D00H16. doi:10.1029/2009JD012251

Clement AC, Burgman R, Norris JR (2009) Observational and model evidence for positive low-level cloud feedback. Science 325:460–464

Dessler AE (2010) A determination of the cloud feedback from climate variations over the past decade. Science 330:1523–1527. doi:10.1126/science.1192546

Dufrense J-L, Bony S (2008) An assessment of the primary sources of spread of global warming estimates from coupled atmosphere-ocean models. J Climate 21:5135–5144

Eitzen ZA, Xu K-M, Wong T (2008) Statistical analyses of satellite cloud object data from CERES. Part V: relationships between physical properties of marine boundary layer clouds. J Climate 21:6668–6688

Eitzen ZA, Xu K-M, Wong T (2011) An estimate of low-cloud feedbacks from variations of cloud radiative and physical properties with sea surface temperature on interannual time scales. J Climate 24:1106–1121

Gregory J, Webb M (2008) Tropspheric adjustment induces a cloud component in CO_2 forcing. J Climate 21:58–71

Gregory JM et al (2004) A new method for diagnosing radiative forcing and climate sensitivity. Geophys Res Lett 31. doi:10.1029/2003GL018747

Hall A, Qu X (2006) Using the current seasonal cycle to constrain snow albedo feedback in future climate change. Geophy Res Lett 33:L03502. doi:10.1029/2005GL025127

Hartmann DL, Doelling D (1991) On the net radiative effectiveness of clouds. J Geophys Res 96:869–891

Haynes JM, Marchand RT, Luo Z, Bodas-Salcedo A, Stephens GL (2007) A multi-purpose radar simulation package: quickbeam. Bull Amer Meteor Soc 88:1723–1727. doi:10.1175/BAMS-88-11-1723

Hendon HH, Woodberry K (1993) The diurnal cycle of tropical convection. J Geophys Res 98:16623–16637

Kato S, Sun-Mack S, Miller WF, Rose FG, Chen Y, Minnis P, Wielicki BA (2010) Relationships among cloud occurrence frequency, overlap, and effective thickness derived from CALIPSO and CloudSat merged cloud vertical profiles. J Geophys Res 115:D00H28. doi:10.1029/2009JD012277

Kiehl JT (1994) On the observed near cancellation between longwave and shortwave cloud forcing in tropical regions. J Climate 7:559–656

Kiehl JT, Ramanathan V (1990) Comparison of cloud forcing derived form the earth radiation budget experiment with that simulated by the NCAR community climate model. J Geophys Res 95: 11679–11698

Klein SA, Jakob C (1999) Validation and sensitivities of frontal clouds simulated by the ECMWF model. Mon Weather Rev 127(10):2514–2531

Loeb NG et al (2009) Toward optimal closure of the Earth's top-of-atmosphere radiation budget. J Climate 22:748–766

Mace GG et al (2011) Evaluation of the ISSCP simulator using ground-based remote sensing data. J Climate 24:1598–1612

Murphy JM et al (2004) Quantification of modelling uncertainties in a large ensemble of climate change simulations. Nature 430:768–772

National Research Council (2007) Earth science and applications from space: national imperatives for the next decade and beyond. National Academies Press, Washington, DC
Ramanathan V et al (1989) Cloud-radiative foring and climate: results from the Earth radiation budget experiment. Science 243:57–63
Randall DA et al (2007) Climate models and their evaluation. In: Solomon S, Qin D, Manning M, Chen Z, Marquis M, Averyt KB, Tignor M, Miller HL (eds) Climate change 2007: the physical science basis. Contribution of Working Group I to the fourth assessment report of the Intergovernmental Panel on Climate Change. Cambridge University Press, Cambridge, New York, NY
Rosenfeld D (2006) Aerosol-cloud interactions control of earth radiation and latent heat release budgets. Space Sci Rev 125:149–157. doi:10.1007/s11214-006-9053-6
Soden BJ, Held IM (2006) An assessment of climate feedback in coupled ocean-atmosphere models. J Climate 19:3354–3360
Soden BJ, Held IM, Colman R, Shell KM, Keihl JT, Shields CA (2008) Quantifying climate feedbacks using radiative kernels. J Climate 21:3504–3520
Solomon S et al (2007) Technical summary. In: Solomon S, Qin D, Manning M, Chen Z, Marquis M, Averyt KB, Tignor M, Miller HL (eds) Climate change 2007: the physical science basis. Contribution of Working Group I to the fourth assessment report of the Intergovernmental Panel on Climate Change. Cambridge University Press, Cambridge, New York, NY
Stephens GL (2005) Cloud feedbacks in the climate system: a critical review. J Climate 18:237–273
Stephens GL (2010) Is there a missing low cloud feedback in current climate models? GEWEX News 20:5–7
Stephens GL, Ellis TD (2008) Controls of global-mean precipitation increases in global warming GCM experiments. J Climate 21:6141–6155
Stevens B, Feingold G (2009) Untangling aerosol effects on clouds and precipitation in a buffered system. Science 461:607–612
Vogelmann AM et al (2008) RACORO science and operations plan. DOE/SC-ARM-0806, US Department of Energy, Office of Science, Office of Biological and Environmental Research, 39 pp. [Available online at http://www.arm.gov/publications/programdocs/doe-sc-arm-0806.pdf?id=29]
Williams KD, Webb MJ (2009) A quantitative performance assessment of cloud regimes in climate models. Climate Dyn 33:141–157. doi:10.1007/s00382-008-0443-1
Zhang MH et al. (2005) Comparing clouds and their seasonal variations in 10 atmospheric general circulation models with satellite observations. J Geophys Res 110:D15S02. doi:10.1029/2004JD005021

Surv Geophys (2012) 33:619–635
DOI 10.1007/s10712-011-9152-0

Cloud Adjustment and its Role in CO_2 Radiative Forcing and Climate Sensitivity: A Review

Timothy Andrews · Jonathan M. Gregory · Piers M. Forster · Mark J. Webb

Received: 21 April 2011 / Accepted: 10 October 2011 / Published online: 17 November 2011

Abstract Understanding the role of clouds in climate change remains a considerable challenge. Traditionally, this challenge has been framed in terms of understanding cloud feedback. However, recent work suggests that under increasing levels of atmospheric carbon dioxide, clouds not only amplify or dampen climate change through global feedback processes, but also through rapid (days to weeks) tropospheric temperature and land surface adjustments. In this article, we use the Met Office Hadley Centre climate model HadGSM1 to review these recent developments and assess their impact on radiative forcing and equilibrium climate sensitivity. We estimate that cloud adjustment contributes ~0.8 K to the 4.4 K equilibrium climate sensitivity of this particular model. We discuss the methods used to evaluate cloud adjustments, highlight the mechanisms and processes involved and identify low level cloudiness as a key cloud type. Looking forward, we discuss the outstanding issues, such as the application to transient forcing scenarios. We suggest that the upcoming CMIP5 multi-model database will allow a comprehensive assessment of the significance of cloud adjustments in fully coupled atmosphere–ocean-general-circulation models for the first time, and that future research should exploit this opportunity to understand cloud adjustments/feedbacks in non-idealised transient climate change scenarios.

Keywords Climate sensitivity · Radiative forcing · Cloud feedback · Tropospheric adjustment

T. Andrews (✉) · J. M. Gregory · M. J. Webb
Met Office Hadley Centre, FitzRoy Road, Exeter EX1 3PB, UK
e-mail: timothy.andrews@metoffice.gov.uk

J. M. Gregory
NCAS-Climate, University of Reading, Reading, UK

P. M. Forster
School of Earth and Environment, University of Leeds, Leeds, UK

1 Introduction

A useful tool for measuring the response of the climate system to an external forcing is its equilibrium climate sensitivity, defined as the equilibrium global mean surface temperature change in response to a step change CO_2 doubling (e.g., Knutti and Hegerl 2008). It also provides a useful metric for comparing the responses of different climate models to external forcings. The Coupled Model Intercomparison Project phase 3 (CMIP3) (Meehl et al. 2007) generation of climate models simulate a range of equilibrium climate sensitivity of 2.1–4.4 K (Randall et al. 2007). Cloud feedback has long been identified a major source of this uncertainty (e.g., Bony et al. 2006; Webb et al. 2006; Randall et al. 2007), and remains the least well understood feedback process (Stephens 2005).

A recent development on the role of clouds in climate change has been the possibility that some of the model uncertainty normally associated with cloud feedback may have been misdirected. Here, feedback is defined as the part of the change that evolves with global mean surface temperature. There is now a body of modelling work (Gregory and Webb 2008; Andrews and Forster 2008; Williams et al. 2008; Andrews et al. 2009; Doutriaux-Boucher et al. 2009; Dong et al. 2009; Colman and McAvaney 2011) showing that changes in clouds also influence climate through short timescale (days to weeks) non-feedback tropospheric temperature and land surface adjustments. These adjustments come about rapidly, before global mean surface temperature change, due to the change in atmospheric radiative heating caused by the introduction of the forcing agent and/or through land surface processes. Cloud feedbacks, as defined here, operate on a multi-annual to decadal timescale associated with changes in global mean surface temperature (e.g., Bony et al. 2006).

The idea of rapid non-feedback cloud adjustments to radiative forcing is not entirely new; it is analogous to how changes in atmospheric aerosols lead to rapid changes in cloudiness through microphysical processes (aerosol indirect effects) and/or local atmospheric heating (semi-direct effects). These aerosol-induced cloud adjustments are routinely included in estimates of aerosol radiative forcing as they improve estimates of long-term changes in surface temperature (e.g., Shine et al. 2003; Hansen et al. 2005; Lohmann et al. 2010).

Cloud adjustments are important for a number of reasons (see also, Colman and McAvaney 2011):

(1) *Predicting transient climate change* If a significant component of the cloud response under climate change can be considered a cloud adjustment, then a component of cloud changes will keep up with, ‘track’, changes in radiative forcing. On the other hand, cloud changes dominated by feedback processes will lag changes in radiative forcing on a multi-annual to decadal timescale due to the thermal inertia of the climate system. Correctly separating these processes is needed to correctly predict transient climate change (Gregory and Webb 2008).
(2) *Mechanisms of climate change* Feedbacks operate through large-scale changes in surface temperature (e.g., Bony et al. 2006) and so should be independent of the cause of the initial perturbation (assuming that the pattern of surface warming is the same). Cloud adjustments, on the other hand, will strongly depend on the nature of the forcing agent, such as its radiative properties, distribution in the atmosphere, microphysical properties and ability to interact with the carbon cycle (see, for example, Sect. 6.3).

(3) *Definition of radiative forcing* If adjustments are included in estimates of climate feedbacks then feedbacks may appear forcing dependent, limiting the utility of the radiative forcing concept in predicting surface temperature change. It is partly for this reason that there are now several radiative forcing definitions (see Shine et al. 2003; Hansen et al. 2005; Forster et al. 2007) that include various climate 'responses', tropospheric and cloud adjustment included.
(4) *Uncertainties in cloud feedback* Multi-model differences in cloud feedback may be smaller than previously thought if estimates are contaminated by uncertainties in cloud adjustments (Andrews and Forster 2008).

In the rest of this article, we will survey some of the above ideas in more detail, using the Met Office Hadley Centre climate model HadGSM1 to illustrate the concepts. Section 2 describes the model. Section 3 describes the forcing/feedback framework that is used to separate cloud adjustments from feedback. Section 4 quantifies the influence of cloud adjustment on the top-of-atmosphere (TOA) radiation balance. Section 5 estimates the impact of this on the model's equilibrium climate sensitivity. Section 6 discusses the key mechanisms and processes involved. Section 7 discusses the outstanding issues, applications to the real world, and future research directions. Finally, Sect. 8 presents a short summary.

2 Experimental Design

We use the atmospheric component of the Met Office Hadley Centre Global Environmental Model version 1 (HadGEM1) (Martin et al. 2006) coupled to a simple 50 m thermodynamic mixed-layer ('slab') ocean and sea-ice model (Johns et al. 2006). This setup is referred to as HadGSM1. Parallel to a control run with constant CO_2, a standard climate sensitivity experiment is performed in which CO_2 is instantaneously doubled and subsequently held constant. The model is run for 70 years, of which we regard the first 20 years as the transient part and the remaining 50 years as the new perturbed equilibrium state (Fig. 1). Note that this experiment is the same as that archived in CMIP3. All changes are determined by subtracting a linear fit through corresponding portions of the annual mean control data in order to remove any model drift and control noise from the climate change signal. All radiative fluxes are defined as positive downwards.

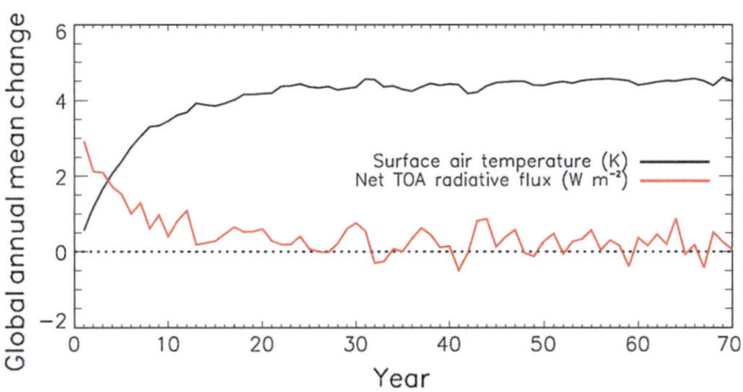

Fig. 1 Time series showing the change in global annual mean surface air temperature (in units of K—*black line*) and net top of atmosphere (TOA) radiative flux (in units of Wm^{-2}—*red line*) in response to CO_2 doubling at year zero

We also make use of a CO_2 doubling scenario in atmosphere-only mode, i.e. with climatological sea-surface-temperatures (SSTs) and sea-ice (see Sect. 3.2). The model is run for 10 years parallel to a control run. Note that this is the same experiment as used by Andrews et al. (2010). This experiment uses a slightly different version of the atmospheric component of HadGSM1, in that it has an improved representation of aerosols (see Bellouin et al. 2007), but we regard the two models as the same for practical purposes.

3 Separating Forcing from Feedback

The energy balance of the climate system can be expressed by a simple linear relationship between radiative forcing and climate response (e.g., see Boer and Yu 2003; Gregory et al. 2004; Gregory and Webb 2008). In response to an external forcing, the change in net TOA radiative flux, N (units Wm^{-2}), can be described by the radiative forcing, F (units Wm^{-2}), minus the response of the climate system, described by $\alpha\Delta T$, where ΔT is the change in global mean surface air temperature (units K) and α is the climate feedback parameter (units $Wm^{-2}\,K^{-1}$). Using this notation, the climate response opposes the imposed radiative forcing, and the heat balance can be described by,

$$N = F - \alpha\Delta T. \tag{1}$$

In the following section, we discuss this relationship and describe two independent methods that exploit it to evaluate the contribution of cloud adjustment to radiative forcing.

3.1 Forcing and Feedback Separated by Timescale

Figure 1 shows the change in global mean surface air temperature (ΔT, black line) and net TOA radiative flux (N, red line) during the 70 year $2 \times CO_2$ HadGSM1 simulation. Gregory et al. (2004) showed that the simple relationship (Eq. 1) between these two variables allows a separation of forcing from feedback according to timescale (see also, Gregory and Webb 2008). In the following, we discuss this relationship and use it to separate cloud adjustment from cloud feedback.

Figure 2a shows the global annual mean change in net TOA radiative flux as a function of ΔT during the first 20 years after CO_2 was doubled. The time evolution begins on the left of the plot where the y-axis intercept represents the increase in radiative flux into the climate system caused by the introduction of the forcing agent at $\Delta T = 0$, which we interpret as forcing according to Eq. 1 (Gregory et al. 2004). In this case, the forcing is $\sim$3 Wm^{-2} (the regression coefficients are given in Table 1). Gregory and Webb (2008) performed the same analysis across a range of CMIP3 models and found a range of values of $\sim$3–4 Wm^{-2}, HadGSM1 therefore lies at the lower end of this range. It is important to note that this diagnosed forcing is different from the traditional radiative forcing that allows stratospheric temperatures to adjust but keeps the surface and tropospheric temperatures and states held fixed (e.g., Forster et al. 2007). The approach used in Fig. 2 sets no constraints on tropospheric temperatures or state, only that globally averaged *surface* temperature is unchanged. Therefore, changes in the tropospheric lapse-rate, clouds, and any other variable that responds quicker than global mean ΔT, and has an impact on the TOA radiation balance, will be included in our forcing estimate.

The climate evolves in response to the forcing rightwards and downwards along the curve, i.e. ΔT increases and the rate of heat uptake by the system reduces (see Fig. 1, also). The new steady state will be achieved towards the bottom right of this curve when the

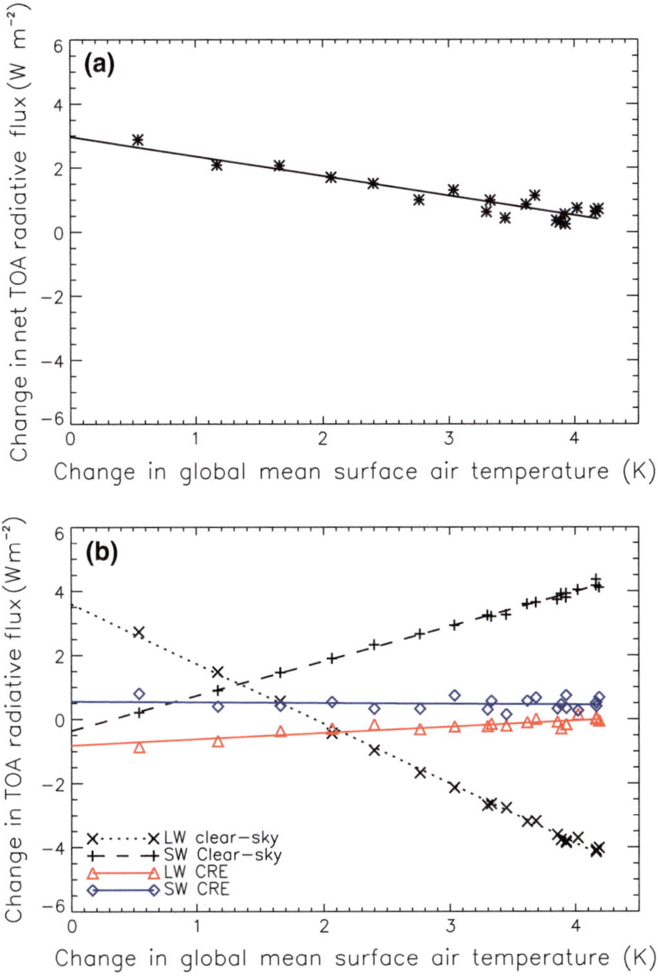

Fig. 2 **a** Change in global annual mean net TOA radiative flux as a function of global annual mean surface air temperature change during the first 20-years after CO_2 was instantaneously doubled. (**b**) as (**a**) but showing its breakdown into radiative flux components: *LW* clear-sky, *SW* clear-sky, *LW* cloud radiative effect and *SW* cloud radiative effect. All fluxes are defined as positive downwards. The slope and intercept of these curves provide a simple way of separating feedback (in units of $Wm^{-2}\ K^{-1}$) and forcing (in units of Wm^{-2}), respectively

perturbation in the rate of heat uptake reaches zero. The gradient of the curve measures the signed feedback parameter, $-\alpha$ in Eq. 1. Using this sign convention means a positive value represents a positive feedback, and a negative value represents a negative feedback. Overall, this quantity must be negative (i.e. oppose the forcing) for a stable system, as seen in Fig. 2a where it is equal to $\sim -0.6\ Wm^{-2}\ K^{-1}$ (Table 1).

The net radiative flux term (Fig. 2a) can be broken down into components (Fig. 2b and Table 1): longwave (LW) and shortwave (SW), as well as clear-sky and cloud radiative effect (CRE). The CRE term is defined as the difference between net downward radiative fluxes in all-sky (i.e. the observed meteorological conditions, including clouds if present) and clear-sky (i.e. assuming no cloud) conditions. For the purpose of this section, we use

Table 1 $2 \times CO_2$ forcing (in units of Wm^{-2}) and feedback (in units of $Wm^{-2}\,K^{-1}$) as diagnosed from the regression coefficients of Fig. 2 (intercept represents forcing, gradient represents feedback)

	Net	LW clear-sky	SW clear-sky	LW CRE	SW CRE	Net CRE
Forcing (Wm^{-2})	2.97 ± 0.36	3.60 ± 0.16	−0.36 ± 0.10	−0.82 ± 0.16	0.55 ± 0.26	−0.27 ± 0.28
Feedback ($Wm^{-2}\,K^{-1}$)	−0.61 ± 0.11	−1.87 ± 0.05	1.07 ± 0.03	0.19 ± 0.05	−0.02 ± 0.08	0.17 ± 0.08

Errors are 95% uncertainties from the regressions

changes in CRE as a measure of cloud adjustments and/or cloud feedback, noting that 'cloud masking' (see below) will affect these results and is addressed in Sect. 4. The dominant term in the forcing (intercept) is in the LW clear-sky, consistent with the radiative effect of a greenhouse gas (Fig. 2b; Table 1). LW clear-sky feedback is strong and negative (stabilising) and results from water vapour, lapse-rate and black body feedbacks, the last being dominant (e.g., Soden and Held 2006). SW clear-sky provides a positive feedback, consistent with reduced clear-sky albedo due to a loss of sea-ice and snow cover with increased global temperature, as well as enhanced absorption through the water vapour feedback. Gregory and Webb (2008) discuss these clear-sky terms in more detail across the CMIP3 models, so we will focus on the cloud terms in detail from now on.

LW CRE weakly increases with positive ΔT, as seen by the positive slope of the red line in Fig. 2b, indicating a positive LW cloud feedback of $0.19 \pm 0.05\ Wm^{-2}\,K^{-1}$. Across the CMIP3 generation of models, Gregory and Webb (2008) found a range of LW CRE feedback of ~ -0.1 to $+0.3\ Wm^{-2}\,K^{-1}$. In contrast, there is no statistically significant relationship between changes in SW CRE and global mean surface air temperature in this particular model (Table 1), indicating little or no SW cloud feedback. This is a robust feature across most of the models analysed by Gregory and Webb (2008) except for a few exceptions, such as the 'high climate sensitivity' version of the Model for Interdisciplinary Research on Climate 3.2 (MIROC3.2), which has a large SW CRE feedback, $0.62 \pm 0.05\ Wm^{-2}\,K^{-1}$. Net cloud feedback (sum of LW and SW CRE) is small in HadGSM1 (in comparison to other feedback terms), amounting to $0.17 \pm 0.08\ Wm^{-2}\,K^{-1}$ (Table 1).

Lack of SW CRE feedback is shown by the blue line in Fig. 2b being flat. The change in SW CRE is constant and positive, i.e. the blue line. This indicates a change in cloudiness that came about *before* changes in ΔT, as seen by the non-zero positive intercept of $0.55 \pm 0.26\ Wm^{-2}$. It is this rapid change in cloudiness that does not scale with ΔT that we interpret as a non-feedback cloud adjustment (Gregory and Webb 2008). The intercept, when combined with the traditional forcing estimate, allows us to quantify the effect of the rapid adjustment of clouds on the TOA radiation budget. This allows cloud adjustments to be easily incorporated into our forcing estimate.

A word of caution is that changes in CRE do not necessarily imply a change in cloud properties due to the way they are calculated using clear-sky fluxes (e.g., Soden et al. 2004; 2008). In Sect. 4, we will address this limitation on the CRE approach by quantifying the consequent 'cloud masking' effect and removing it from our CRE adjustment term.

3.2 Forcing Diagnosed by Preventing Feedback

The regression approach used in Sect. 3.1 to evaluate forcing and feedback is particularly useful as it does not require a special experiment beyond a standard instantaneous step change experiment. However, it does have its limitations. For example, if the signal is small then the regression results can be statistically uncertain and sensitive to the number of years included in the regression and the way in which the control is subtracted. It is also limited in providing regional information, although regressing local radiative fluxes against global mean surface temperature is possible (e.g., Gregory and Webb 2008), the results can include considerable noise if the variability is large.

Rather than regressing ΔT back to zero to evaluate forcing, another approach is to fix ΔT at zero from the outset. As $\Delta T = 0$ in such an experiment, Eq. 1 would imply that $N = F$, and hence forcing could be evaluated from the change in net TOA radiation balance. In practice, this condition is implemented by running an experiment in atmosphere-only mode, i.e. by replacing the ocean model with climatological SSTs (ideally derived from the coupled control simulation) and including the forcing agent (e.g., Shine et al. 2003; Hansen et al. 2005). The rationale for such an experiment is that it prohibits climate feedbacks as surface temperatures (or SSTs at least) are prevented from evolving, but allows for any change in stratospheric temperatures as well as the tropospheric lapse-rate and any other variable (e.g., cloud cover, precipitation, water vapour) to respond to the forcing agent (Shine et al. 2003; Hansen et al. 2005). We refer to this experiment as a 'fixed-SST experiment'. This approach is particularly useful as it provides a straight forward way of evaluating forcing/adjustments regionally, as well as reducing noise if the experiment is run for a sufficiently long time. However, it gives no information on feedbacks. Note that the aerosol community have routinely been using this method to incorporate cloud adjustments (e.g., indirect and/or semi-direct effects) into their forcing estimates, often referred to as 'quasi forcing' or 'radiative flux perturbation' (e.g., Rotstayn and Penner 2001; Lohmann et al. 2010).

A comparison between the regression intercepts and forcing diagnosed from the fixed-SST experiment is presented in Fig. 3. In agreement with Gregory and Webb (2008); Bala et al. (2010) and Colman and McAvaney (2011), who all provide similar comparisons but with different models, we conclude that fixed-SST experiments provide an independent method for evaluating forcing/adjustments and are largely in agreement with the regression approach. In practice, it means the regression approach is capturing the adjustment of the atmosphere and land surface to radiative forcing. The agreement is not perfect; however, for example the two methods do not agree in the SW clear-sky forcing term in this case

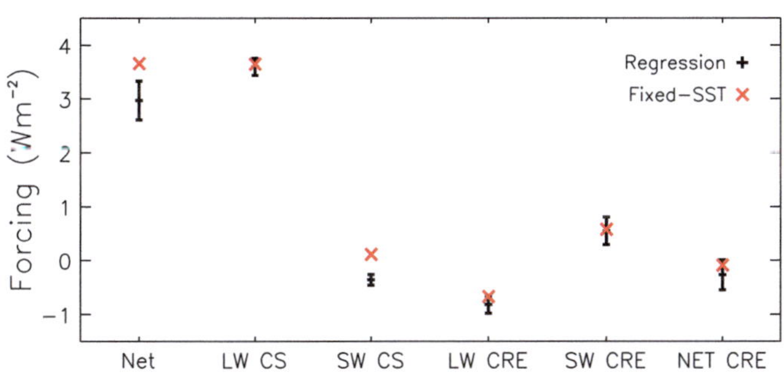

Fig. 3 Comparison of forcing terms evaluated using two independent methodologies. The regression approach separates forcing from feedback by timescale (see Sect. 3.1; Fig. 1), whereas the fixed-SST approach uses climatological SSTs and so prevents feedback by inhibiting surface temperature change (see Sect. 3.2). *Error bars* on the regression results represent 95% uncertainties in the regression intercepts. 'CS' stands for clear-sky

(and in other terms in the above studies). Differences are to be expected given that the methods impose $\Delta T = 0$ in different ways. The regression approach only requires *global* mean surface air temperature change to be zero, it sets no constraints on *local* surface temperature change. The fixed-SST method on the other hand sets $\Delta T = 0$ by imposing all *local* SST changes to be zero, but in practice there is a small amount of surface warming from the land surface and over sea-ice (see Fig. 4d). This could give rise to regional differences between the methods (Gregory and Webb 2008). Further discussion on these differences is given in Shine et al. (2003); Hansen et al. (2005); Gregory and Webb (2008) and Andrews et al. (2009).

4 Quantifying the Radiative Effect of Cloud Adjustment

The net CRE adjustment diagnosed in Sect. 3 is small and negative for HadGSM1, which is a common feature across most of the multi-model comparison in Gregory and Webb (2008). The interpretation therefore, could be that clouds adjust to the increased CO_2 in such a way as to reduce the TOA energy imbalance and so reduce the CO_2 radiative forcing and subsequent global temperature change. However, this neglects the instantaneous effects clouds have on the CO_2 radiative forcing. For example, instantaneous forcing calculations show that CO_2 radiative forcing is larger in clear-skies than in overcast conditions (Soden et al. 2008; Andrews and Forster 2008; Colman and McAvaney 2011). Andrews and Forster (2008) determined that the presence of clouds reduced global mean $2 \times CO_2$ instantaneous radiative forcing by $\sim$0.5 Wm^{-2} across a selection of CMIP3 models, which is in agreement with similar values calculated by Soden et al. (2008) using the geophysical fluid dynamics laboratory (GFDL) model and Colman and McAvaney (2011) using the Australian Bureau of Meteorology Research Centre (BMRC) model. This 'cloud masking' effect arises due to increased CO_2 instantaneously reducing TOA upward LW radiation, but emission from optically thick clouds remaining the same. Hence, CO_2 radiative forcing is larger in clear-skies than in all-skies, giving rise to an instantaneous CRE component. This instantaneous CRE masking effect must be removed to quantify the radiative effect of cloud adjustments.

We quantify the instantaneous CO_2 CRE masking term in HadGSM1 by using instantaneous forcing calculations (the difference between two calls of the radiation code: one based on the control configuration and one with the forcing agent introduced). Globally, we find the all-sky radiative forcing to be 0.56 Wm^{-2} smaller than the clear-sky radiative forcing, therefore doubling CO_2 instantaneously generates a -0.56 Wm^{-2} CRE term due to clouds shielding the impact of underlying CO_2 changes. Subtracting this masking term from the CRE terms in Sect. 3 leaves a change in TOA radiative flux that comes about due to a change in cloudiness, that is, the cloud adjustment that we are trying to quantify (Table 2). In HadSM1, cloud adjustment acts to enhance the initial radiative forcing by $\sim$0.5 Wm^{-2} globally. This is a common feature across most of the CMIP3 models analysed by Andrews and Forster (2008). The spatial patterns of these terms are presented in Fig. 4a, b, c (and are discussed further in Sect. 6). Locally, cloud adjustment can exceed 10 Wm^{-2} in places (Fig. 4c).

A final caution on cloud masking is that the separation of clear-sky and CRE will also influence our estimates of feedbacks through analogous masking effects (Soden et al. 2004, 2008). For example, clouds may shield much of the impact of changes in underlying surface albedo, leading to an overestimate in surface albedo feedback if computed in clear-skies, and hence a CRE feedback term despite no change in cloud properties (Soden et al.

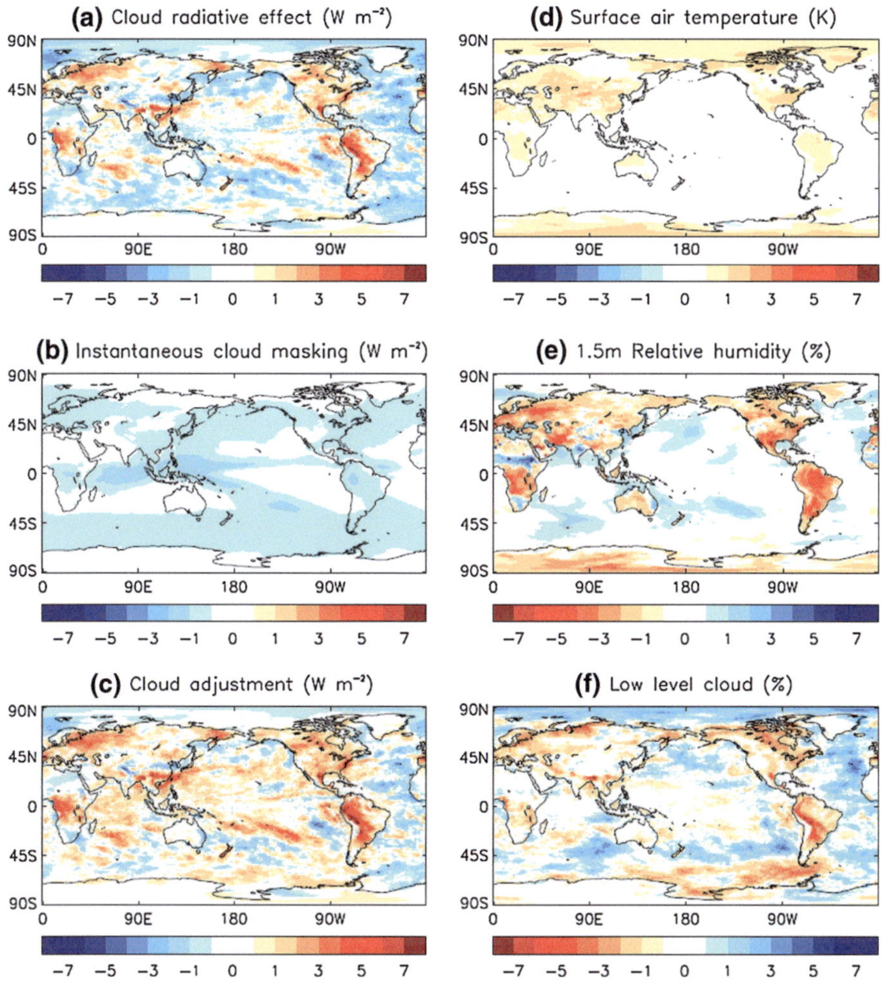

Fig. 4 Spatial pattern of the 2 × CO_2 net TOA **a** change in cloud radiative effect as diagnosed from the fixed-SST forcing experiment, **b** instantaneous cloud masking, and **c** cloud adjustment (all in units Wm^{-2}). Global mean values of these patterns are presented in Table 2. **d**, **e** and **f** show spatial patterns of the adjustments in surface air temperature (in unit of K), near-surface relative humidity (in units of %) and low level cloudiness (in units of %), as diagnosed from the change in the 2 × CO_2 fixed-SST experiment. **b** is calculated from instantaneous radiative forcing calculations, with and without clouds. **c** is calculated as the difference between **a** and **b**

2008). Moreover, if there are also rapid adjustments in the clear-sky terms, possibly arising from rapid adjustments in land-snow cover or sea-ice, then further cloud masking effects could contaminate our estimates of cloud adjustments. One method to avoid such issues altogether is to use a partial radiative perturbation (PRP) technique (e.g., Wetherald and Manabe 1988). This method makes no distinction between clear-sky and all-sky, it simply takes the change in any variable, such as cloud fraction, and imposes it on an offline radiation code to determine the radiative effect on the TOA radiation budget. Colman and McAvaney (2011) performed such an analysis with step change CO_2 experiments and combined the PRP approach with the regression and fixed-SST methods. They found a SW

Table 2 Changes in LW, SW and Net CRE as diagnosed from the 2 × CO_2 fixed-SST 'forcing' experiment

	LW CRE	SW CRE	Net CRE
Forcing (Wm^{-2})	−0.67	0.58	−0.09
Instantaneous masking (Wm^{-2})	−0.62	0.06	−0.56
Cloud adjustment (Wm^{-2})	−0.06	0.53	0.47

These forcing terms include an instantaneous cloud masking effect (determined from double radiation calls with and without clouds, see Sect. 4) and a cloud adjustment. All units are Wm^{-2}

cloud adjustment of 0.94 ± 0.43 Wm^{-2} and little/no LW cloud adjustment, qualitatively confirming the CRE adjustment results described here. Furthermore, their approach gives additional insight as the cloud terms can be broken down into cloud fraction and optical properties. They found virtually all of the SW cloud adjustment being the result of changes in cloud fraction.

5 Impact on Equilibrium Climate Sensitivity

The equilibrium climate sensitivity of this model is 4.4 K. Assuming linearity, we combine the net cloud adjustment term (0.47 Wm^{-2}) with the feedback parameter for this model (−0.61 $Wm^{-2}\ K^{-1}$) to estimate the contribution of cloud adjustment to the equilibrium climate sensitivity to be ~0.8 K, that is, ~18% of the total value. Similarly, Gregory and Webb (2008) showed that differences in cloud adjustment explained the 1.2 K difference in equilibrium climate sensitivity between two versions of the Met Office model HadSM3. Given multi-model differences in the magnitude of cloud adjustment (Gregory and Webb 2008; Andrews and Forster 2008), it is likely that this effect could explain some of the spread in equilibrium climate sensitivity across models. For example, the range of net cloud adjustment across models reported in Andrews and Forster (2008) is −0.52 to + 0.91 Wm^{-2}. Combining this range with a multi-model mean feedback parameter of −0.96 $Wm^{-2}\ K^{-1}$ (Gregory and Webb 2008) suggests the uncertainty in cloud adjustment alone could give rise to an uncertainty in multi-model equilibrium climate sensitivity of the order 1.5 K.

6 Mechanisms and Cloud Adjustment Processes

6.1 Timescale of Adjustments

A pertinent question is: what is the timescale of cloud adjustments, i.e. how fast *is* fast? Dong et al. (2009) used an ensemble of 4 × CO_2 fixed-SST experiments with daily diagnostics to investigate the transient behaviour of adjustments. They concluded that land surface and tropospheric adjustment (including cloud) are statistically indistinguishable from their equilibrium value after only 5 days. This is consistent with Doutriaux-Boucher et al. (2009) who used an ensemble regression approach (as in Fig. 2) with monthly data and found no deviation from linearity, implying that the adjustment timescale is considerably less than one month. Such a short timescale relative to the surface response is not surprising given the much smaller heat capacity, and therefore relaxation time, of the troposphere and land surface compared to the ocean.

6.2 Atmospheric Heating and Links to Precipitation

The immediate impact of increasing CO_2 is mostly felt in the troposphere. For example, Andrews et al. (2010) showed that the 3.5 Wm^{-2} instantaneous radiative forcing on doubling CO_2 is partitioned between the troposphere and surface by 2.8 and 0.7 Wm^{-2}, respectively. This sudden change in tropospheric radiative heating may alter the local atmospheric lapse-rate and stability, influencing convection and cloudiness and hence the TOA radiation budget (Dong et al. 2009). Over the global ocean, near-surface temperatures change little on the adjustment timescale, but the CO_2 induced tropospheric heating leads to significant warming above the surface, with maximum increases occurring around 800 hPa (Dong et al. 2009). Using fixed-SST experiments, both Dong et al. (2009) and Colman and McAvaney (2011) showed that this corresponds to reductions in relative humidity, and consequent reductions in cloud fraction. Colman and McAvaney (2011) further showed that this is largely confined to the tropics and subtropics, where reductions in cloud fraction occur at $\sim$800 hPa at the equator, but follow an arc towards the subtropical near-surface. These spatial changes also correlate well with the instantaneously induced change in tropospheric radiative heating (Colman and McAvaney 2011).

Note, however, that while CO_2 tropospheric heating leads to a tropospheric temperature adjustment that influences cloudiness (amongst other things), it is neither the temperature nor cloud changes that are the dominant terms in restoring the energy balance of the troposphere. It is changes in condensational heating and associated precipitation/surface evaporation that play this role (e.g., Mitchell et al. 1987; Allen and Ingram 2002; Lambert and Allen 2009). Global mean precipitation initially decreases following an increase in CO_2 through rapid tropospheric adjustment processes, then increases on a multi-annual timescale associated with changes in global surface temperature. This is now well documented and understood in modelling studies (e.g., Yang et al. 2003; Lambert and Webb 2008; Andrews et al. 2010), including the HadGSM1 model used here (Andrews et al. 2009).

6.3 Physiological Response of Plants to Increased CO_2

A plant's stomata control the exchange of carbon dioxide, oxygen and water vapour between the plant's leaf and the atmosphere. Under increased CO_2, the stomata do not open as wide, reducing the transfer of moisture and energy to the atmosphere through evapotranspiration, thus altering the surface energy balance and near-surface climate (e.g., Field et al. 1995; Sellers et al. 1996; Betts et al. 1997; Cox et al. 1999). Over land, this 'CO_2 physiological forcing' plays a considerable role in the near-surface and tropospheric response to increased CO_2 (e.g., Joshi et al. 2008; Joshi and Gregory 2008; Dong et al. 2009; Doutriaux-Boucher et al. 2009; Boucher et al. 2009; Andrews et al. 2011). The reduced surface latent heat flux to the atmosphere leads to a significant drying and warming of the boundary layer over land, with correspondingly large reductions in relative humidity and cloud fraction (Doutriaux-Boucher et al. 2009; Dong et al. 2009; Andrews et al. 2011).

A recent development has been the realisation that these physiological processes act on a similar timescale to tropospheric adjustment (Doutriaux-Boucher et al. 2009; Dong et al. 2009), and so any rapid cloud changes can be considered as part of cloud/tropospheric adjustment and so incorporated into the radiative forcing. For example, Doutriaux-Boucher et al. (2009) used a 12-member ensemble of 5-year 4 $\times$ CO_2 and 2 $\times$ CO_2 step experiments, with and without the stomatal conductance change, and followed a similar regression approach as used in Fig. 2 here. They found that the cloud fraction changes due to CO_2 physiological forcing influenced the intercepts (i.e. can be considered as rapid adjustments). Comparisons with estimates derived by fixing the stomatal conductance revealed that the physiological effect more than doubled the net TOA CRE cloud adjustment term that came about from the CO_2 radiative effects acting alone.

In HadGSM1, we expect a component of cloud adjustment to similarly arise from such a process. Figure 4e shows the change in the near-surface relative humidity in the $2 \times CO_2$ fixed-SST experiment. Consistent with the above processes there are large reductions in near-surface relative humidity over land, particularly over the Amazon and central African forest, as an immediate response to increased CO_2. Note that a parallel experiment with fixed stomatal conductance would be required to definitively attribute such changes to CO_2 physiological effects.

We investigate the impact of these changes on low level cloudiness using the International Satellite Cloud Climatology Project (ISCCP) simulator diagnostics (Klein and Jakob 1999; Webb et al. 2001). Figure 4f shows the adjustment in low level cloud amount (defined as cloud top pressure greater than 680 mb), as diagnosed from the fixed-SST experiment. We observe large reductions in low level cloud amount over land, consistent with Doutriaux-Boucher et al. (2009) and Andrews et al. (2011). Changes in low level cloudiness often influence the net CRE at the TOA more than equivalent changes in mid to high level clouds. This is because changes in low level clouds generate large changes in SW CRE as they are usually highly reflective, but have no/small compensating LW CRE changes as they are close (and therefore of similar temperature) to the surface. Changes in SW CRE due to changes in high clouds, however, are often accompanied by compensating changes in LW CRE. Consequently therefore, the spatial pattern of the net CRE cloud adjustment (Fig. 4c) closely follows the adjustments in low level cloud amount (Fig. 4f) over much of the globe, but particularly so over land.

7 Outstanding Issues and Future Research

Despite considerable progress in the last few years on the role of cloud adjustment in CO_2 radiative forcing (as surveyed above), there are still many outstanding issues to be addressed. Therefore, interpretations are likely to change as more detailed research is done. In the following section, we discuss some of these outstanding issues and suggest ways forward.

7.1 Atmosphere–ocean general circulation models (AOGCMs)

The results presented here have generally considered mixed-layer ocean models only. Mixed-layer ocean models are useful for comparing and evaluating the response to external forcing as it is computationally inexpensive to run many idealised experiments and they can often be run to equilibrium; something which is too computationally expensive for fully coupled atmosphere–ocean general circulation models (AOGCMs). AOGCMs on the other hand, are crucial for more realistic and transient scenarios. A key question, therefore, is how important are cloud adjustments in transient AOGCM simulations?

Firstly, we expect certain methods used to evaluate cloud adjustments (e.g., the regression approach) to depend on the ocean component if the transient behaviour described by Eq. 1 is different in mixed-layer and fully dynamic ocean models. For example, both Gregory et al. (2004) and Williams et al. (2008) observed non-linearity between radiative fluxes and changes in global mean surface temperature in the first few decades of idealised step AOGCM forcing experiments (equivalent to a non-linearity in our Fig. 2). Williams et al. (2008) interpreted the non-linearity as evidence of adjustments to CO_2 radiative forcing operating on a decadal timescale, identifying a non-linear stratocumulus cloud response as the local patterns of SSTs changed in the subtropical ocean basins, as its dominant cause. The advantage of including this decadal response into estimates of forcing is that it removes the apparent time variation of effective climate sensitivity (Williams et al. 2008) that was observed in previous studies (e.g., Senior and Mitchell, 2000). However, decadal adjustments imply an ocean component in its mechanism because the atmosphere does not have sufficient 'memory' for decadal responses. Therefore, rather than decadal adjustments to forcing, this might be better interpreted as a change in feedback (e.g., Winton et al. 2010).

Clearly, identifying a cloud feedback that depends on climate state, or a decadal cloud adjustment in response to forcing, is important for the same reasons outlined in the introduction of this article. The upcoming CMIP5 multi-model intercomparison project is requesting step change $4 \times CO_2$ experiments in AOGCMs for the first time (Taylor et al. 2009). This will present researchers with a unique opportunity to address this issue across models.

7.2 Transient Scenarios and Observations

One obvious difference between the real world and the idealised experiments highlighted here is the large step change CO_2 forcing applied to the model. The question, therefore, is do adjustments occur at smaller magnitude forcings and how are they realised in transient forcing scenarios, such as historical twentieth century simulations and future emission scenarios?

The problem in evaluating forcing/adjustments in transient forcing scenarios is that there is no clear way of separating forcing from feedback as they will both evolve together. This will present a challenge to the observing community as cloud feedback estimates may be contaminated by cloud adjustments. Forster and Taylor (2006) proposed a method for diagnosing the radiative forcing/adjustments from transient forcing scenarios, but it assumes that the feedback parameter (as diagnosed from idealised runs) is time invariant. As seen in Gregory et al. (2004) and Williams et al. (2008), it is not obvious that this should hold. Perhaps a better method for diagnosing forcing/adjustments in transient forcing scenarios would be to run the scenarios in fixed-SST mode, i.e. with climatological SSTs as in Sect. 3.2. Held et al. (2010) used a similar method to diagnose a radiative forcing time series in a historical simulation. This would provide a simple method of diagnosing cloud adjustments in transient scenarios and, if combined with a parallel fully coupled run, could be used to better isolate cloud feedback. Unfortunately, this was not part of the CMIP5 protocol, but it is hoped modelling groups may perform their own transient fixed-SST experiments for this purpose.

Another avenue to explore is possible linkages between the energy and water cycles. For example, the condensational heating associated with precipitation provides a link between the hydrological cycle and radiative processes such as cloud feedback (Stephens 2005; Stephens and Ellis 2008). An advantage of this approach is that precipitation adjustments, particularly in response to greenhouse gases or black carbon aerosol, are subject to energetic constraints and provide a more robust signal across climate models than cloud adjustments do, and therefore are potentially easier to observe in transient forcing scenarios (e.g., Lambert and Allen 2009; Andrews and Forster 2010; Frieler et al. 2011). However, there is still substantial disagreement on recent precipitation trends across datasets (e.g., Arkin et al. 2010), presenting a considerable challenge if attempting to tease out adjustment processes.

7.3 Consistencies Between Methods of Diagnosing Feedback

Another outstanding issue is the consistency of methods used to evaluate cloud feedbacks. There is now a large set of methods to consider, all giving different results: regression of changes in CRE against global surface temperature change (see Sect. 3.1), PRP techniques (discussed in Sect. 4), radiative kernels (e.g., Soden et al. 2008), uniform and patterned +4 K experiments that part of the CMIP5 experimental design (Taylor et al. 2009), and combinations thereof. Each methodology has its merits, but with cloud masking effects and cloud adjustments to consider, it is no longer straightforward to interpret what each methodology physically represents. For example, Andrews and Forster (2008) showed that accounting for cloud adjustment in CRE approaches reduces the range of model predicted cloud feedback. However, their values were still inconsistent with PRP or kernel approaches as they did not remove cloud masking errors. On the other hand, PRP and kernel approaches will be contaminated by cloud adjustment errors if they consider all cloud changes to be feedback.

One way forward would be to compute cloud feedback using all the above methodologies for a single model. By considering the role of cloud masking and/or adjustments, one might be able to unify the feedback methodologies and explain the relative importance of masking or adjustments in them.

7.4 Processes and Mechanisms

In Sect. 6, we discussed possible mechanisms of cloud adjustment to CO_2 radiative forcing, namely atmospheric heating and interactions with the biosphere. We identified low level cloudiness as a key cloud type in HadGSM1 cloud adjustment processes. However, there is still a considerable amount we do not know and possible mechanisms to explore, such as changes in dynamics and atmospheric stability. For example, one possible mechanism might be through circulation changes induced through the rapid warming of the land surface (see Fig. 4d). CMIP5 is requesting 3 hourly data for $4 \times CO_2$ fixed-SST experiments; this may allow us to probe the adjustment processes in detail. Furthermore, CMIP5 will contain a series of atmospheric model intercomparison project (AMIP) experiments conducted in connection with the cloud feedback model intercomparison project (CFMIP) (e.g., Bony et al. 2011) that will allow us to diagnose and investigate in detail the physical drivers of CO_2 induced cloud adjustments. It is also possible that basic physical processes could be investigated in simple radiative-convective models.

8 Summary

In response to increased CO_2, climate models suggest that not all cloud changes should be considered as feedback—a significant component of the change in cloud radiative effect is due to rapid cloud adjustments associated with the adjustment of the land surface and troposphere to radiative forcing. The effect of cloud adjustment on the TOA radiation budget can be routinely incorporated into our estimates of CO_2 radiative forcing by using either, (1) a regression approach that separates forcing from feedback according to time-scale, and/or, (2) fixed-SST experiments that use climatological SSTs to prevent climate feedback by inhibiting surface temperature change (see Sect. 3). Including tropospheric adjustments in our definition of radiative forcing is analogous to including stratospheric adjustment, and is routinely done for aerosol indirect/semi-direct effects.

In HadGSM1, we estimate that cloud adjustment contributes $\sim$0.5 Wm^{-2} to the CO_2 radiative forcing, which translates into $\sim$0.8 K of the 4.4 K equilibrium climate sensitivity of this model. As cloud adjustment varies across models, it is therefore likely to contribute to the range of equilibrium climate sensitivities across models. We stress, however, that cloud feedback is still important and remains a major source of uncertainty in climate change projections.

Looking forward, there are many outstanding issues to be addressed, such as the role of cloud adjustment and feedback in AOGCMs, how it is realised in realistic transient forcing scenarios, what the detailed mechanisms and processes are, and how the observing platform might be utilised to observe such processes. The upcoming CMIP5 multi-model database will provide a unique opportunity to test some of these ideas and provide further insights into cloud adjustment processes and their role in climate change.

Acknowledgments We thank the International Space Science Institute in Bern for hosting the workshop on 'Observing and modelling Earth's energy flows', and its participants for fruitful discussions. We acknowledge two anonymous reviews for helping to improve the clarity of the manuscript. Funds for this survey were provided by a NERC open CASE award with the University of Leeds and the Met Office. This work was supported by the Joint DECC/Defra Met Office Hadley Centre Climate Programme (GA01101).

References

Allen MR, Ingram WJ (2002) Constraints on future changes in climate and the hydrological cycle. Nature 419:224–232. doi:10.1038/nature01092

Andrews T, Forster PM (2008) CO_2 forcing induces semi-direct effects with consequences for climate feedback interpretations. Geophys Res Lett 35:L04802. doi:10.1029/2007GL032273

Andrews T, Forster PM (2010) The transient response of global-mean precipitation to increasing carbon dioxide levels. Environ Res Lett 5:025212. doi:10.1088/1748-9326/5/2/025212

Andrews T, Forster PM, Gregory JM (2009) A surface energy perspective on climate change. J Clim 22:2557–2570. doi:10.1175/2008JCLI2759.1

Andrews T, Forster PM, Boucher O, Bellouin N, Jones A (2010) Precipitation, radiative forcing and global temperature change. Geophys Res Lett 27:L14701. doi:10.1029/2010GL043991

Andrews T, Doutriaux-Boucher M, Boucher O, Forster PM (2011) A regional and global analysis of carbon dioxide physiological forcing and its impact on climate. Clim Dyn 36:783–792. doi:10.1007/s00382-010-0742-1

Arkin PA, Smith T, Sapiano M, Janowiak J (2010) The observed sensitivity of the global hydrological cycle to changes in surface temperature. Environ Res Lett 5:035201. doi:10.1088/1748-9326/5/3/035201

Bala G, Caldeira K, Nemani R (2010) Fast versus slow response in climate change: implications for the global hydrological cycle. Clim Dyn 35:423–434. doi:10.1007/s00382-009-0583-y

Bellouin N, O Boucher, JM Haywood, CE Johnson, A Jones, JG L Rae, S Woodward (2007) Improved representation of aerosols for HadGEM2. Hadley Centre Tech Note, 73, 43 pp

Betts RA, Cox PM, Lee SE, Woodward FI (1997) Contrasting physiological and structural vegetation feedbacks in climate change simulations. Nature 387:796–799

Boer GJ, Yu B (2003) Climate sensitivity and response. Clim Dyn 20:415–429. doi:10.1007/s00382-002-0283-3

Bony S et al (2006) How well do we understand and evaluate climate change feedback processes? J Clim 19:3445–3482

Bony S, Webb M, Bretherton C, Klein S, Siebesma P, Tselioudis G, Zhang M (2011) CFMIP: towards a better evaluation and understanding of clouds and cloud feedbacks in CMIP5 models. Clivar Exch 56(16):2
Boucher O, Jones A, Betts RA (2009) Climate response to the physiological impact of carbon dioxide on plants in the Met Office Unified Model HadCM3. Clim Dyn 32:237–249. doi:10.1007/s00382-008-0459-6
Colman R, McAvaney BJ (2011) On tropospheric adjustment to forcing and climate feedbacks. Clim Dyn 36:1649–1658. doi:10.1007/s00384-011-1067-4
Cox PM, Betts RA, Bunton CB, Essery RLH, Rowntree PR, Smith J (1999) The impact of new land surface physics on the GCM simulation of climate and climate sensitivity. Clim Dyn 15:183–203
Dong B, Gregory JM, Sutton RT (2009) Understanding land-sea warming contrast in response to increasing greenhouse gases. Part I: transient adjustment. J Clim 22:3079–3097
Doutriaux-Boucher M, Webb MJ, Gregory JM, Boucher O (2009) Carbon dioxide induced stomatal closure increases radiative forcing via a rapid reduction in low cloud. Geophys Res Lett 36:L02703. doi: 10.1029/2008GL036273
Field CB, Jackson RB, Mooney HA (1995) Stomatal responses to increased CO_2: implications from the plant to the global scale. Plant Cell Environ 18:1214–1225
Forster PM, Taylor KE (2006) Climate forcings and climate sensitivities diagnosed from coupled climate model integrations. J Clim 19:6181–6194
Forster P et al (2007) Changes in atmospheric constituents and in radiative forcing. In: Solomon S et al (eds) Climate change 2007: the physical science basis. Contribution of working group 1 to the fourth assessment report of the intergovernmental panel on climate change. Cambridge Univ. Press, Cambridge, pp 131–234
Frieler K, Meinshausen M, von Deimling TS, Andrews T, Forster P (2011) Changes in global-mean precipitation in response to warming, greenhouse gas forcing and black carbon. Geophys Res Lett 38:L04702. doi:10.1029/2010GL045953
Gregory JM, Webb M (2008) Tropospheric adjustment induces a cloud component in CO_2 forcing. J Clim 21:58–71
Gregory JM et al (2004) A new method for diagnosing radiative forcing and climate sensitivity. Geophys Res Lett 31:L03205. doi:10.1029/2003GL018747
Hansen J et al (2005) Efficacy of climate forcings. J Geophys Res 110:D18104. doi:10.1029/2005JD005776
Held IM, Winton M, Takahashi K, Delworth T, Zeng F, Vallis GK (2010) Probing the fast and slow components of global warming by returning abruptly to preindustrial forcing. J Clim 23:2418–2427. doi:10.1175/2009JCLI3466.1
Johns TC et al (2006) The new hadley centre climate model HadGEM1: evaluation of coupled simulations. J Clim 19:1327–1353
Joshi M, Gregory JM (2008) Dependence of the land-sea contrast in surface climate response on the nature of the forcing. Geophys Res Lett 35:L24802. doi:10.1029/2008GL036234
Joshi MM, Gregory JM, Webb MJ, Sexton DMH, Johns TC (2008) Mechanisms for the land/sea warming contrast exhibited by simulations of climate change. Clim Dyn 30:455–465. doi:10.1007/s00382-007-0306-01
Klein SA, Jakob C (1999) Validation and sensitivities of frontal clouds simulated by the ECMWF model. Mon Weather Rev 127:2514–2531
Knutti R, Hegerl GC (2008) The equilibrium sensitivity of the earth's temperature to radiation changes. Nature 1:735–743. doi:10.1038/ngeo337
Lambert FH, Allen MR (2009) Are changes in global precipitation constrained by the tropospheric energy budget? J Clim 22:499–517
Lambert FH, Webb MJ (2008) Dependence of global mean precipitation on surface temperature. Geophys Res Lett 35:L16706. doi:10.1029/2008GL034838
Lohmann U et al (2010) Total aerosol effect: radiative forcing or radiative flux perturbation? Atmos Chem Phys 10:3235–3246
Martin GM, Ringer MA, Pope VD, Jones A, Dearden C, Hinton TJ (2006) The physical properties of the atmosphere in the new Hadley centre global environmental model (HadGEM1). Part I: model description and global climatology. J Clim 19:1274–1301
Meehl GA, Covey C, Delworth T, Latif M, McAvaney B, Mitchell JFB, Stouffer RJ, Taylor KE (2007) The WCRP CMIP3 multi-model dataset: a new era in climate change research. Bull Amer Meteor Soc 88:1383–1394
Mitchell JFB, Wilson CA, Cunnington WM (1987) On CO_2 climate sensitivity and model dependence of results. Q J R Meteorol Soc 113:293–322

Randall DA et al (2007) Climate models and their evaluation. In: Solomon S et al (eds) Climate Change 2007: the physical science basis. Contribution of working group 1 to the fourth assessment report of the intergovernmental panel on climate change. Cambridge Univ. Press, Cambridge, pp 591–662
Rotstayn LD, Penner JE (2001) Indirect aerosol forcing, quasi forcing, and climate response. J Clim 14:2960–2975
Sellers PJ, Bounoua L, Collatz GJ, Randall DA, Dazlich DA, Los SO, Berry JA, Fung I, Tucker CJ, Field CB, Jensen TG (1996) Comparison of radiative and physiological effect of doubled atmospheric CO_2 on climate. Science 271:1402–1406. doi:10.1126/science.271.5254.1402
Senior CA, Mitchell JFB (2000) The time-dependence of climate sensitivity. Geophys Res Lett 27(17):2685–2688
Shine KP, Cook J, Highwood EJ, Joshi MM (2003) An alternative to radiative forcing for estimating the relative importance of climate change mechanisms. Geophys Res Lett 30(20):2047. doi: 10.1029/2003GL018141
Soden BJ, Held IM (2006) An assessment of climate feedbacks in coupled ocean-atmosphere models. J Clim 19:3354–3360
Soden BJ, Broccoli AJ, Hemler RS (2004) On the use of cloud forcing to estimate cloud feedback. J Clim 17(19):3661–3665
Soden BJ, Held IM, Colman R, Shell KM, Kiehl JT, Shields CA (2008) Quantifying climate feedbacks using radiative kernels. J Clim 21:3504–3520. doi:10.1175/2007JCLI2110.1
Stephens GL (2005) Cloud feedbacks in the climate system: a critical review. J Clim 18:237–273
Stephens GL, Ellis TD (2008) Controls of global-mean precipitation increases in global warming GCM experiments. J Clim 21:6141–6155
Taylor KE, Stouffer RJ, Meehl GA (2009) A summary of the CMIP5 experimental design. Available at http://www-pcmdi.llnl.gov/
Webb M, Senior CA, Bony S, Morcrette J–J (2001) Combining ERBE and ISCCP data to assess cloud in the Hadley Centre, ECMWF and LMD atmospheric climate models. Clim Dyn 17:905–922
Webb MJ et al (2006) On the contribution of local feedback mechanisms to the range of climate sensitivity in two GCM ensembles. Clim Dyn 27:17–38. doi:10.1007/s00382-006-0111-2
Wetherald RT, Manabe S (1988) Cloud feedback processes in a GCM. J Atmos Sci 45:1397–1415
Williams KD, Ingram WJ, Gregory JM (2008) Time variation of effective climate sensitivity in GCMs. J Clim 21:5076–5090. doi:10.1175/2008JCLI2371.1
Winton M, Takahashi K, Held IM (2010) Importance of ocean heat uptake efficacy to transient climate change. J Clim 23:2333–2344. doi:10.1175/2009JCLI3139.1
Yang F, Kumar A, Schlesinger ME, Wang W (2003) Intensity of hydrological cycles in warmer climates. J Clim 16:2419–2423

Surv Geophys (2012) 33:637–656
DOI 10.1007/s10712-011-9148-9

Representing the Sensitivity of Convective Cloud Systems to Tropospheric Humidity in General Circulation Models

Anthony D. Del Genio

Received: 7 May 2011 / Accepted: 8 September 2011 / Published online: 30 September 2011

Abstract Convective cloud variability on many times scales can be viewed as having three major components: a suppressed phase of shallow and congestus clouds, a disturbed phase of deep convective clouds, and a mature phase of transition to stratiform upper-level clouds. Cumulus parameterization development has focused primarily on the second phase until recently. Consequently, many parameterizations are not sufficiently sensitive to variations in tropospheric humidity. This shortcoming may affect global climate model simulations of climate sensitivity to external forcings, the continental diurnal cycle of clouds and precipitation, and intraseasonal precipitation variability. The lack of sensitivity can be traced in part to underestimated entrainment of environmental air into rising convective clouds and insufficient evaporation of rain into the environment. As a result, the parameterizations produce deep convection too easily while stabilizing the environment too quickly to allow the effects of convective mesoscale organization to occur. Recent versions of some models have increased their sensitivity to tropospheric humidity and improved some aspects of their variability, but a parameterization of mesoscale organization is still absent from most models. Evidence about the effect of these uncertainties on climate change projections suggests that climate modelers should make improved simulation of high and convective clouds as high a priority as better representations of low clouds.

Keywords Convection · Climate · Humidity · Diurnal cycle · Intraseasonal variability

1 Introduction

Spatial variations of insolation and sea surface temperature, and differences in the thermal inertia and albedo of land and ocean surfaces, create energy and water flows between the tropics and subtropics, between warm and cold oceanic regions, and between oceans and

A. D. Del Genio (✉)
NASA Goddard Institute for Space Studies, 2880 Broadway, New York, NY 10025, USA
e-mail: anthony.d.delgenio@nasa.gov
URL: http://www.giss.nasa.gov/staff/adelgenio.html

continents that have major effects on the planetary energy balance and feedbacks that determine the response of each to climate forcings (e.g., Trenberth et al. 2009; Fasullo 2010). Our understanding of these feedbacks comes primarily from simulations of different climates by general circulation models (GCMs). The energy and water transfers are accomplished by the large-scale atmospheric circulations that GCMs explicitly resolve, but the strength of these circulations and the energy and water that is transported depend crucially on parameterized small-scale processes whose treatment in GCMs is regarded to be highly uncertain.

Perhaps the most important of these parameterization uncertainties is that due to moist convection. Dynamical upward transport by convection removes excess heat from the surface more efficiently than longwave radiation is able to accomplish in the presence of a humid, optically thick boundary layer, and deposits it in the upper troposphere where it is more easily radiated to space, thereby affecting the planetary energy balance. Drying and moistening of the atmosphere by convection regulates the vertical profile of atmospheric water vapor and thus determines how much is transported horizontally. Furthermore, convection influences where clouds form and dissipate, thus affecting the planetary albedo and potentially giving rise to cloud and water vapor feedbacks that determine the global climate sensitivity to anthropogenic forcing.

Mapes et al. (2006) argue that moist convective variability can be understood in terms of three basic convective structures, or "building blocks" (Fig. 1). During suppressed conditions when the boundary layer is capped by a significant inversion and/or the free troposphere above is dry, shallow and midlevel-top ("congestus") convective clouds that heat and moisten the lower troposphere are most common. As the atmosphere humidifies and destabilizes, deep convection is eventually triggered, heating the entire column, while shallow clouds continue to be present. Finally, under the right environmental conditions, individual deep convective cells organize into mesoscale clusters with extensive stratiform rain regions and anvils that primarily heat the upper troposphere through mesoscale updrafts and latent heating but cool the lower troposphere as falling rain evaporates.

This sequence describes the $\sim$12- to 24-h lifecycle of individual convective clusters (Houze 2004; Futyan and Del Genio 2007), but Mapes et al. (2006) suggest that longer-term convective variability on time scales of days to several months can be understood by "stretching" the same set of three building blocks. In this view, longer-term periods of suppressed, disturbed, and mature convective conditions have different relative frequencies of occurrence of the three building blocks, thus shifting the convective heating profile downward and upward over time. Mapping of International Satellite Cloud Climatology Project (ISCCP) cloud classification and CloudSat/CALIPSO cloud profile occurrences onto many individual episodes of the Madden–Julian Oscillation (MJO) (Fig. 2) provides strong observational support for this idea (Chen and Del Genio 2009; Tromeur and Rossow 2010; Del Genio et al. 2011a). Morita et al. (2006) present evidence of similar behavior in Tropical Rainfall Measuring Mission (TRMM) microwave and radar rain top heights. Kiladis et al. (2005) show that diabatic heating anomalies regressed against eastward-propagating intraseasonal outgoing longwave radiation variability show exactly the type of upward-westward tilting heating pattern that is consistent with the stretched building block picture (Fig. 3). They also show that most of the heating can be decomposed into a first baroclinic mode with heating throughout the troposphere that captures the middle deep convective phase and a second baroclinic mode with opposite-signed heating in the upper and lower troposphere that in its opposite phases describes the shallow/congestus and mature stratiform components. A number of simple theories of tropical convectively

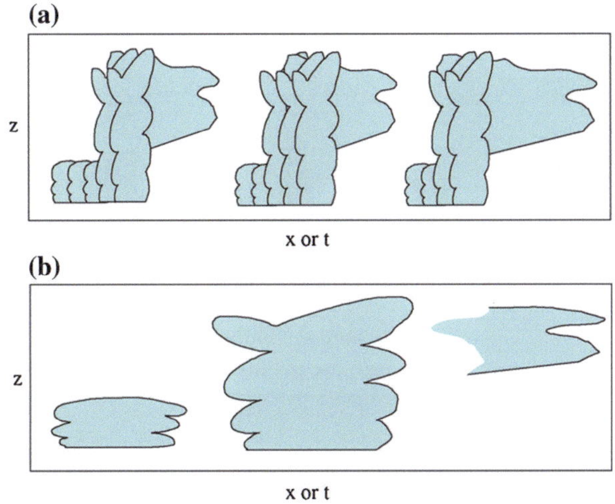

Fig. 1 Schematic illustration of the stretched building block hypothesis. Individual mesoscale convective systems (*upper panel*) consist of contributions from shallow/congestus clouds, deep convective clouds, and stratiform rain and anvil areas. On longer time scales, large-scale waves modulate environmental conditions so as to change the relative frequency of occurrence of the three building blocks, so that a smoothed, low-pass filtered view of the variability (*lower panel*) has a structure similar to that of an individual mesoscale convective system. Reprinted from Mapes et al. (2006) with permission from Elsevier

coupled waves invoke this dual baroclinic mode structure (e.g., Mapes 2000; Khouider and Majda 2006; Kuang 2008).

2 Implications for Cumulus Parameterization and Energy Flows

Despite decades of model development, cumulus parameterizations still have great difficulty simulating the observed variability of moist convection and its interaction with the general circulation. Viewed from the building block perspective, it is easy to understand why. Much of the history of cumulus parameterization has been devoted to diagnosing individual deep convective events and the heating profiles they produce. Although some cumulus parameterizations represent a spectrum of convective types of different depths (e.g., Arakawa and Schubert 1974; Moorthi and Suarez 1992; Donner 1993), these schemes tend to underestimate shallow and congestus convection, most likely because of design elements and validation approaches more appropriate to deep convection. Other schemes only simulate deep convection (e.g., Emanuel 1991; Zhang and McFarlane 1995), sometimes with an additional scheme for removing instability at midlevels (Hack 1994), thus necessitating a separate shallow convection parameterization. Only in the past decade or so have the latter received significant attention, largely as a result of insights gained from large-eddy simulation (LES) models (e.g., Siebesma et al. 2003). Organization of convection into clusters with mesoscale updrafts and downdrafts has only been attempted in one parameterization to date (Donner 1993).

The unsatisfactory state of modern cumulus parameterization was brought into the harsh light of day by Derbyshire et al. (2004). A set of idealized case studies with a thermodynamic structure unstable to deep convection was simulated by several cloud-resolving models (CRMs) with free troposphere humidities ranging from 25 to 90%. The CRMs

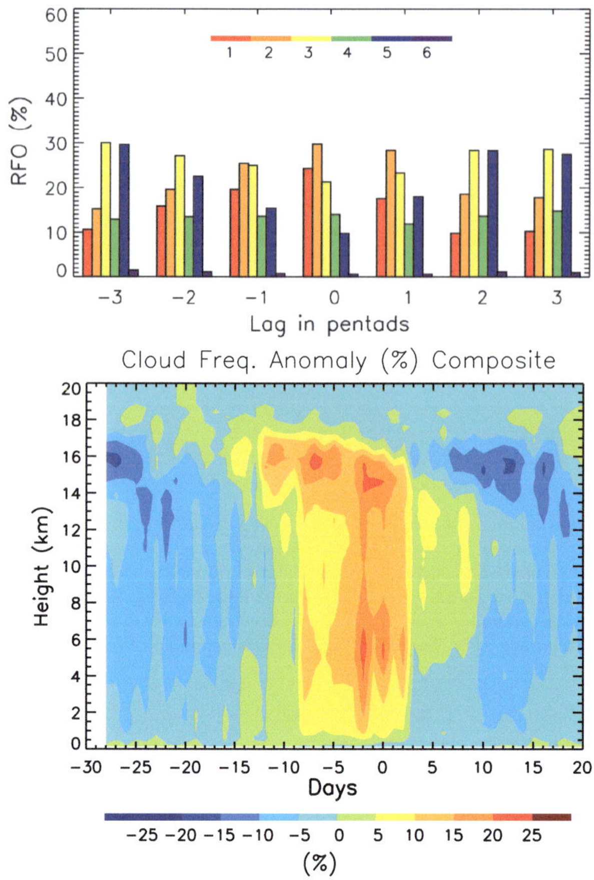

Fig. 2 Composite cloud frequency of occurrence as a function of MJO phase (expressed as lag in *pentads* relative to the peak). *Upper panel* ISCCP cloud regimes from k-means clustering analysis. *Red* deep convective, *orange* stratiform anvil, *yellow* congestus/disorganized convection, *green* isolated cirrus, *blue* shallow cumulus, *violet* stratocumulus. From Chen and Del Genio (2009). *Lower panel* CloudSat/CALIPSO GEOPROF-LIDAR cloud mask anomalies. From Del Genio et al. (2011a)

simulated shallow convection for the driest case and made a gradual transition to deep convection as tropospheric humidity increased. A number of single column models (SCMs) constructed from the parameterizations of GCMs were found to be much less sensitive to tropospheric humidity, some still producing deep convection in even the driest environment. This sensitivity of moist convection to environmental humidity has been indirectly observed as a sharp increase in precipitation with column water vapor at values of 40–55 mm (Bretherton et al. 2004). Holloway and Neelin (2009) show a similar sensitivity in soundings taken at the Department of Energy Atmospheric Radiation Measurement (ARM) Program site at Nauru Island and also show that the variation in column water is dominated by changes in the mid-troposphere (Fig. 4). Cloud radar profiles at Nauru and CloudSat/CALIPSO radar/lidar profiles demonstrate that convective penetration depth is limited in drier environments (Jensen and Del Genio 2006; Del Genio et al. 2011a). These results have been interpreted to indicate that the turbulent entrainment of drier air into cloudy convective updrafts is greatly underestimated by GCMs.

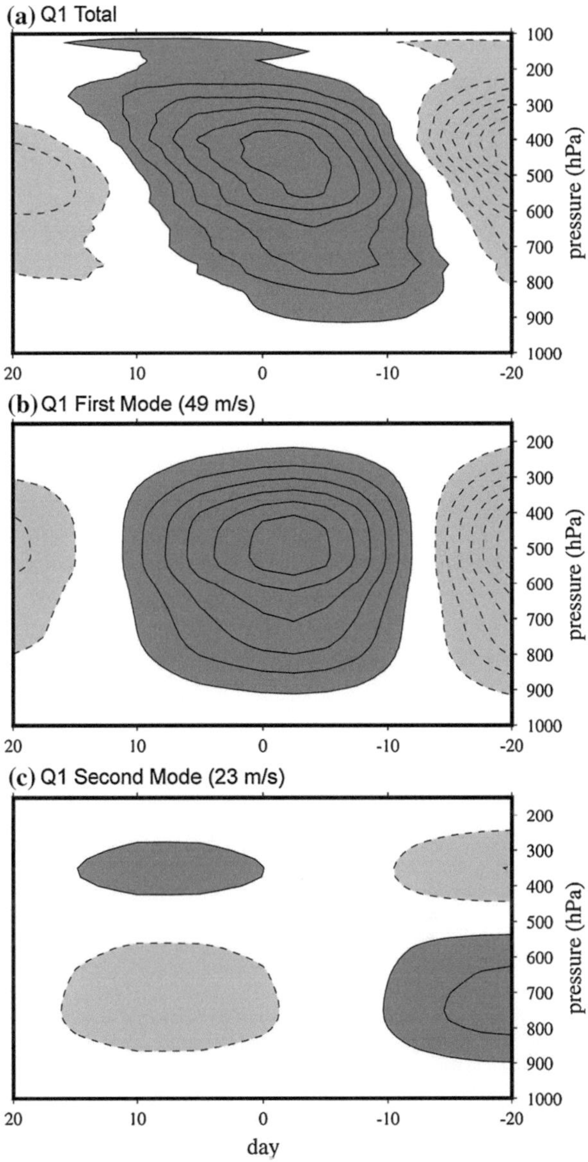

Fig. 3 **a** Observed diabatic heating anomaly profiles regressed against MJO-filtered outgoing longwave radiation during TOGA-COARE as a function of time relative to the MJO peak. **b** First baroclinic mode of the heating. **c** Second baroclinic mode of the heating. *Dark* (*light*) *shading* represent positive (negative) anomalies; the contour interval is 0.5°C day^{-1}. From Kiladis et al. (2005) (© Copyright 2005 AMS)

The relative insensitivity of GCM cumulus parameterizations to tropospheric humidity has potentially important consequences for modeling of energy transfers within the climate system. Sanderson et al. (2010), for example, analyzed a perturbed parameter ensemble of thousands of simulations of one GCM with different values for various free parameters in each ensemble member for current and doubled CO_2 forcing. They showed that the cumulus parameterization's entrainment coefficient had the greatest effect on model feedbacks and climate sensitivity (Fig. 5), including a large effect on the water vapor feedback via changes in the vertical profile of water vapor—a result not previously demonstrated in any GCM. This illustrates the mutual relationship between convection and

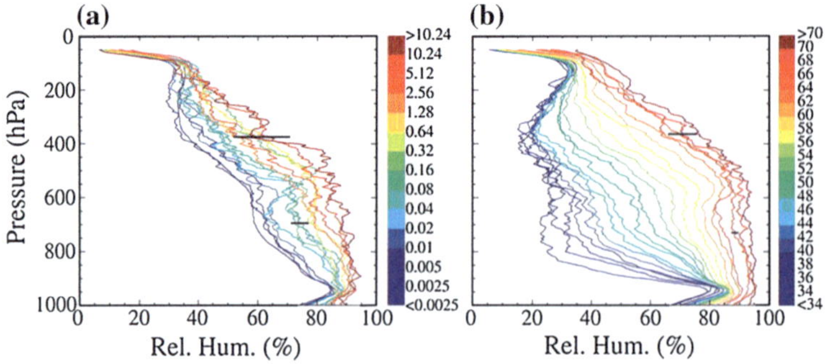

Fig. 4 Composite relative humidity profiles from soundings at the DOE ARM Nauru Island site binned by 1-h mean precipitation rate in mm h^{-1} (*left*) and by column water vapor in mm (*right*). The *horizontal bars* indicate the limits of the maximum and a typical standard error. From Holloway and Neelin (2009) (© Copyright 2009 AMS)

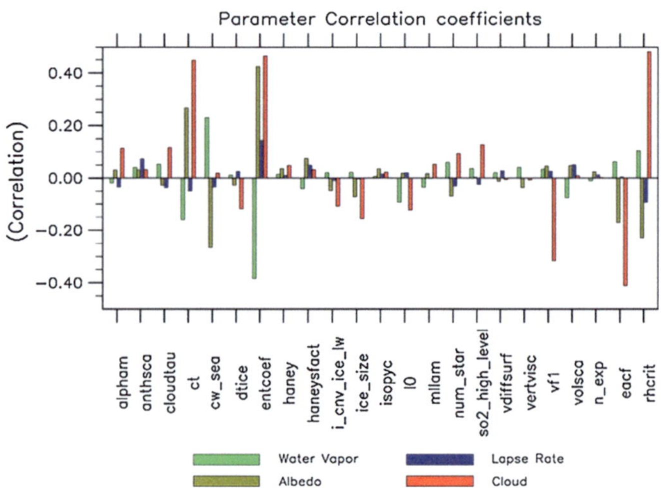

Fig. 5 Correlation coefficients between values of free parameters in the climateprediction.net perturbed parameter GCM ensemble and various global mean feedbacks derived from doubled CO_2 simulations with each ensemble member. The parameter "entcoef" is the convective entrainment coefficient. From Sanderson et al. (2010). Reprinted with permission of Springer

humidity: the depth of convection depends on how humid the surrounding environment is, but the detrainment of saturated air from convective clouds and the evaporation of falling convective rain in turn moisten the environment and thus control the humidity profile.

The effect of entrainment on the transition from shallow to deep convection is important for several aspects of current climate variability as well. For example, the diurnal cycle of continental precipitation exhibits peak rainfall in late afternoon or early evening in TRMM Precipitation Radar data (Hirose et al. 2008; Fig. 6), despite the fact that surface temperature and turbulent fluxes peak near noon. In some cases, this reflects geographic effects or propagation of systems to remote locations. In many situations, though, the atmosphere is simply unstable to deep convection by late morning but does not develop precipitating convection until hours later because entrainment of dry air limits convective penetration

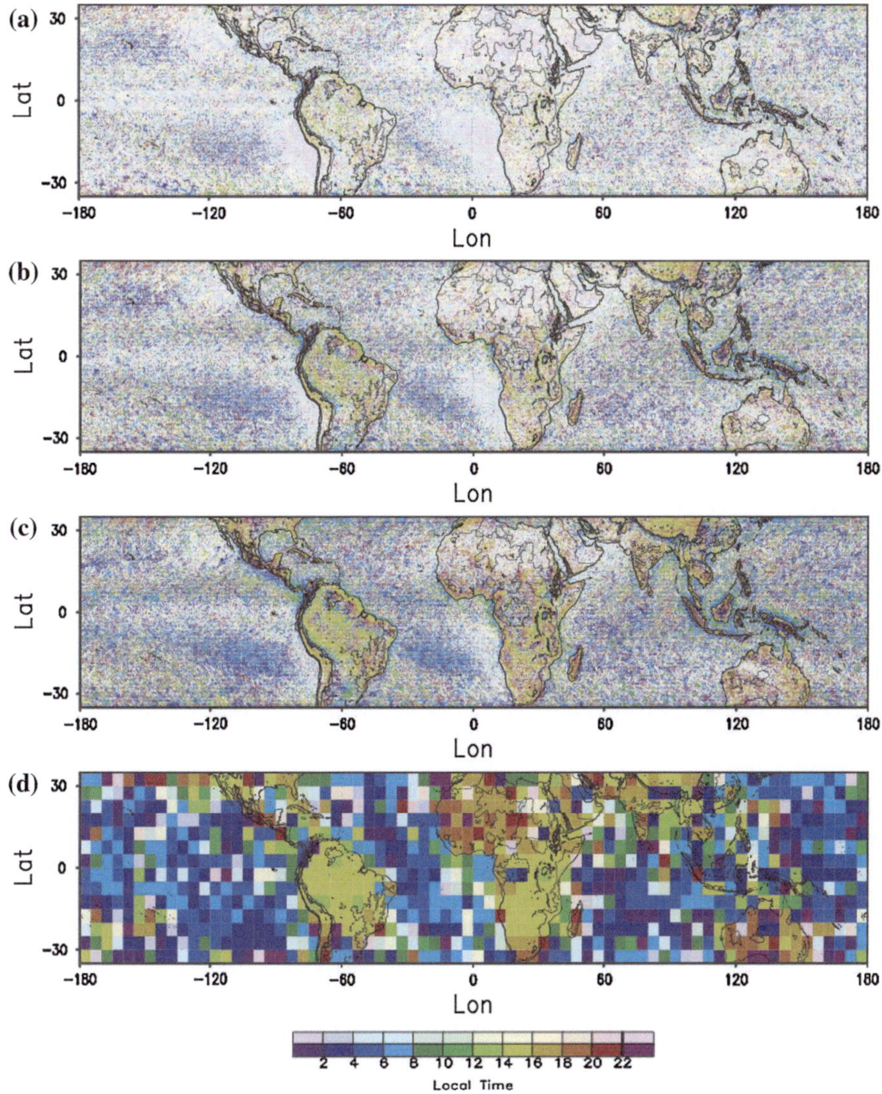

Fig. 6 Local time of maximum rainfall in TRMM Precipitation Radar data at 0.2° resolution for **a** 1998, **b** 1998–2000, and **c** 1998–2005, and **d** at 5° resolution for 1998–2005. From Hirose et al. (2008) (© Copyright 2008 AMS)

(Del Genio and Wu 2010; Zhang and Klein 2010). Guichard et al. (2004) and Dai (2006) showed that cumulus parameterizations uniformly simulate continental precipitation that peaks near noon, hours earlier than observed.

This error has potentially serious implications for the ability of GCMs to simulate the energy and water cycles. Convective clusters have large negative shortwave cloud forcing because of their extensive optically thick anvils. The shortwave forcing will be much stronger if it occurs at noon than if it occurs in late afternoon, because of the decrease in insolation over the course of the afternoon. Thus, we might expect the IPCC AR4 GCMs to have negative biases in absorbed sunlight over the continents. In fact, if anything, the opposite is true (Trenberth and Fasullo 2010), indicating that the GCMs' diurnal cycle errors are being overcompensated by other serious errors in model clouds. This example illustrates why monthly mean geographical distributions of climate parameters are poor indicators of model performance. The diurnal phase error also has ramifications for the surface water balance—rainfall near noon is more likely to evaporate rather than infiltrate into the soil compared with similar rainfall occurring in late afternoon. Do current GCMs dry out their surfaces too easily because of cumulus parameterization deficiencies, and might this affect projections of increasing drought in a warmer climate?

The relationship between convection and tropospheric humidity also appears to affect the MJO, the most important source of intraseasonal variability in weather in the Indian Ocean, Maritime Continent, and West Pacific regions of the tropics (Madden and Julian 1971). The MJO, a slowly eastward-propagating ($\sim$5 m s^{-1}) envelope of zonal wavenumber $\sim$1–3 that modulates the occurrence of precipitating deep convection, is distinct from other types of large-scale equatorial wave modes whose existence is predicted by simple shallow water theory (Wheeler and Kiladis 1999).

General circulation models have historically had great difficulty simulating the amplitude, phase speed, and propagation direction of the MJO (Lin et al. 2006). In part, this is due to the lack of a general consensus on the processes that drive the MJO. Some of the most promising ideas focus on the interaction between convection and tropospheric moisture (Bladé and Hartmann 1993; Hu and Randall 1994; Kemball-Cook and Weare 2001; Stephens et al. 2004; Benedict and Randall 2007). During the dry suppressed phase of the MJO, entrainment limits convective penetration depth, but the resulting shallow cumulus clouds moisten the atmosphere at the level at which they detrain, and also below via rain evaporation. This allows later convective events to rise through a more humid atmosphere and thus penetrate somewhat higher, further "recharging" tropospheric moisture, until the column is sufficiently humid to trigger the disturbed MJO phase. Eventually, deep convection organizes and "discharges" the built-up moisture through precipitation and compensating subsidence drying, returning the atmosphere to the suppressed MJO phase. The ISCCP and CloudSat/CALIPSO composites (Fig. 2) demonstrate that this gradual shift from more frequent shallow clouds to predominantly deep clouds does indeed occur in advance of the peak of the MJO, although by itself this does not prove causality. Nonetheless, it seems more than a coincidence that GCMs whose cumulus parameterizations are insensitive to tropospheric humidity poorly represent this phenomenon.

The MJO produces extended rainy and dry periods of several weeks duration and is thus of great practical importance for countries adjacent to the tropical Indian and West Pacific Oceans. It can also have remote impacts on midlatitude weather (Weickmann et al. 1985). This does not explain the unusual current level of research interest in the MJO, however. Instead, the prevailing view is that the MJO encapsulates much of our ignorance about cumulus parameterization in general. It leads us to question whether aspects of a changing climate and energy balance that depend on how convection interacts with the general circulation, or on how convection redistributes moisture vertically and regulates cloud altitude, are being predicted accurately by the current generation of climate models.

3 Parameterizing Entrainment in General Circulation Models

There are clear historical reasons for the apparent underprediction of entrainment by GCMs. Perhaps the most obvious requirement for a cumulus parameterization is that it produces deep convective clouds that often reach the tropopause, as commonly observed. Moist static energy in the tropics typically decreases to a mid-troposphere minimum and then increases upward to a tropopause value close to its surface value. Thus, air rising from the boundary layer undiluted by environmental air would lose buoyancy near the tropopause (Riehl and Malkus 1958). Cumulus parameterizations have thus usually assumed that a fraction of the upward mass flux represents protected cloud cores that do not interact with the environment (e.g., Arakawa and Schubert 1974). Warner (1970) showed that it was not possible to simultaneously predict the cloud top height of shallow cumuli (which implied nearly undilute ascent) and their subadiabatic liquid water content and liquid water variability (which implied substantial entrainment), but the "Warner paradox" did not affect parameterization development for decades.

The first acknowledgement by the GCM community that something was amiss in its entrainment assumptions was the study of Tokioka et al. (1988), who showed that, by implementing a non-zero minimum entrainment rate in the Arakawa-Schubert scheme, it was possible to produce an MJO-like disturbance in a GCM. In recent years, CRM studies that diagnosed entrainment rate from the rate of decrease with height of an otherwise conserved quantity (such as moist static energy or a passive tracer) have found that, in typical deep convective environments, undilute plumes do not exist (Khairoutdinov and Randall 2006; Kuang and Bretherton 2006; Romps and Kuang 2010a). Instead, the ability of deep convective parcels to reach the tropopause while entraining is due to overshoot beyond the level of neutral buoyancy combined with the enhancement of the parcel's kinetic energy by the release of the latent heat of fusion (Romps and Kuang 2010a). Even shallow cumuli appear to have a negligible probability of parcels rising to cloud top with no entrainment (Romps and Kuang 2010b).

Early cumulus parameterizations that calculated only the cumulus mass flux assumed that ascent terminated at the level of neutral buoyancy. More recent GCMs now include a diagnostic equation for convective updraft speed (Donner 1993; Jakob and Siebesma 2003; Del Genio et al. 2007) and thus account for overshoot. The issue then becomes how to parameterize the entrainment rate (which also affects the updraft speed). Most cumulus parameterizations in use in GCMs today assume either a single laterally entraining bulk plume (e.g., Gregory and Rowntree 1990) or a spectrum of laterally entraining plumes (e.g., Arakawa and Schubert 1974) with constant entrainment rate or a fixed inverse-height dependence (e.g., Jakob and Siebesma 2003; Neale et al. 2008).

The lateral entrainment concept has been questioned due to observations that suggest that mixing in clouds is primarily due to air that enters through cloud top and forms subcloud-scale penetrative downdrafts (e.g., Paluch 1979). This led to the development of episodic "buoyancy-sorting" parameterizations that assumed a variety of mixtures of cloud and environment air that would ascend or descend to their levels of vanishing buoyancy (e.g., Emanuel 1991). Lagrangian particle tracking of shallow cumulus in LES models has raised doubts about the interpretation of the Paluch data and suggested that most entrainment is in fact lateral (Heus et al. 2008). Several LES studies suggest that the observed properties of shallow cumulus can be explained, and the Warner paradox resolved, by an "intermittent entraining thermal" approach. In this view, many individual subcloud-scale elements rising from the boundary layer experience different entrainment histories; this naturally explains the observed cloud top, the subadiabatic liquid water, and its variability. There is disagreement, however, about whether the entrainment that different parcels experience is best described by applying a deterministic entrainment rate to parcels with different initial properties due to boundary-layer heterogeneity (Neggers et al.

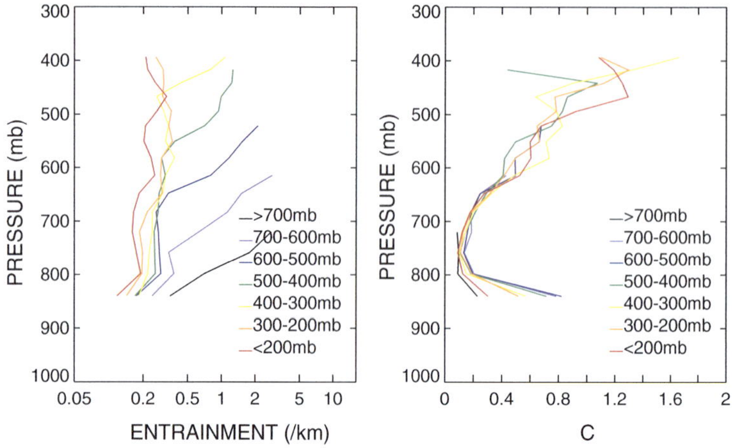

Fig. 7 *Left* CRM-inferred entrainment rate versus height for convective events penetrating to different pressure levels. *Right* inferred values of free parameter *C* in the Gregory (2001) entrainment parameterization for the convective events in the *left panel*. From Del Genio and Wu (2010)

2002) or by assuming stochastic entrainment with no boundary-layer variability (Romps and Kuang 2010b).

The stochastic approach may be a promising path for future parameterization development, especially given the variety of individual convective cloud depths observed by CloudSat/CALIPSO for a given amount of water vapor (Del Genio et al. 2011a). Until such parameterizations are available, however, the simple laterally entraining bulk plume provides a useful (albeit impressionist) way to capture the sensitivity of convection depth to tropospheric humidity on GCM space and time scales. The simplest response to the study of Derbyshire et al. (2004) is to just increase the specified entrainment rate assumed in a given GCM. This is unsatisfying, however, as parameterization development must ultimately move toward physically based approaches that can respond to variations in atmospheric state and have predictive capability for climate change.

Several interactive approaches to entrainment parameterization have in fact been proposed, and CRMs offer a platform for testing them. Del Genio and Wu (2010) diagnosed entrainment rates from the upward decrease in frozen moist static energy in a CRM simulation of daytime convection development over land, finding entrainment rates much larger than those typically assumed in GCMs and a decrease in the entrainment rate as convection deepened (Fig. 7, left), a behavior also inferred in previous studies (Grabowski et al. 2006; Khairoutdinov and Randall 2006; Kuang and Bretherton 2006). Del Genio and Wu compared the predictions of three proposed entrainment parameterizations to these results. A scheme proposed by Gregory (2001) based on a convective turbulence scaling view calculates the entrainment rate as $\varepsilon(z) = CB/w^2$, where B is the parcel buoyancy, w the convective updraft speed, and C a free parameter representing the fraction of the buoyant turbulent kinetic energy generation used for entrainment. ε, B, and w can all be diagnosed from the CRM updraft columns and the resulting C calculated. Figure 7 (right) shows that a single profile of C is valid for all types of convection from shallow to deep. Thus, the Gregory scheme may capture something fundamental about the relationship between cloud-scale buoyancy and smaller-scale turbulence via the inertial cascade (Grant and Brown 1999), although the question of what determines the vertical profile of C remains open.

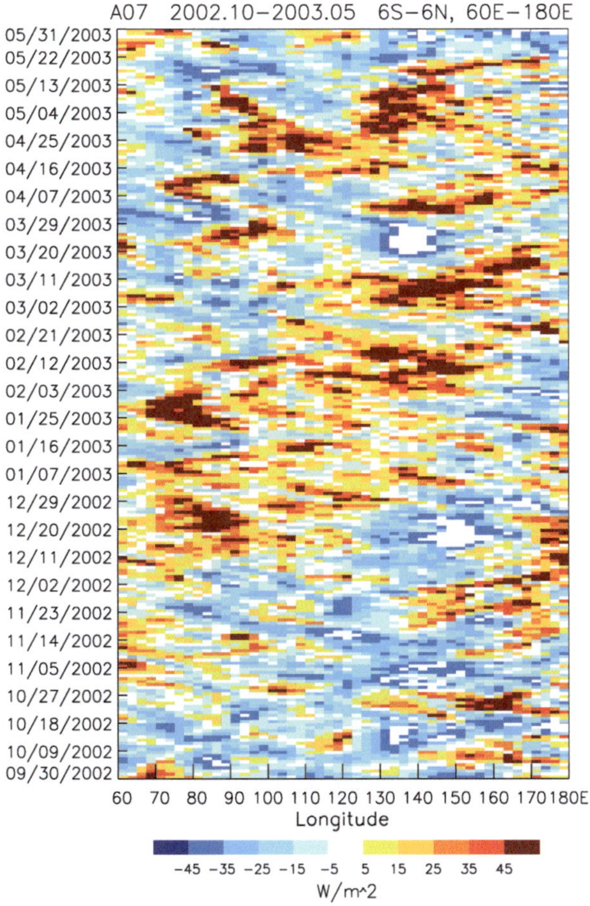

Fig. 8 Hovmöller diagram of outgoing longwave radiation (OLR) anomalies for the equatorial Indian Ocean–Maritime Continent–West Pacific region for an experimental version of the GISS Model E2 GCM that produces MJO-like variability. An example can be seen starting near 60°E in late January and reaching the dateline in late February, implying an eastward propagation speed of ~ 5 m s^{-1}. Positive anomalies indicate more high, thick cloud. From Del Genio et al. (2011a)

The GISS Model E2 GCM, which implemented the Gregory scheme with $C = 0.3$ and 0.6 for its two convective plumes, does not produce an MJO. However, with the value $C = 0.6$ for both plumes, MJO-like variability emerges (Kim et al. 2011a). An even more well-defined MJO occurs when both entrainment and rain evaporation into the environment are increased. An example from a radiatively balanced version of this latter model is shown in Fig. 8 (Del Genio et al. 2011a). Analysis of this model version indicates that unlike the control Model E2, which has a relatively dry mid-troposphere even when column water vapor is high and strong deep convection is occurring, the experimental version has a humid troposphere at all altitudes under heavily raining conditions and a drier upper troposphere when column water vapor is small (Fig. 9). This is consistent with the conclusion of Thayer-Calder and Randall (2009) about why the super-parameterization version of the Community Atmospheric Model simulates the MJO better than the version with a conventional cumulus parameterization.

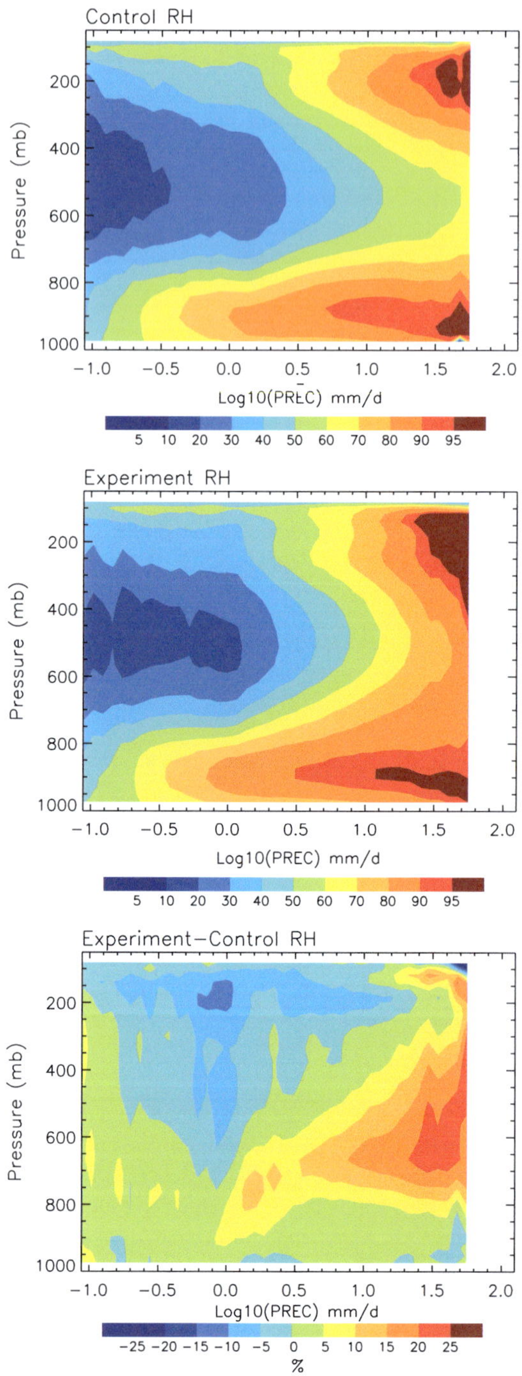

Fig. 9 Composite vertical profiles of relative humidity versus precipitation for the tropical Indian Ocean–Maritime Continent–West Pacific region for the GISS Model E2 GCM version that does not produce an MJO (*upper*), the experimental model version with stronger entrainment and rain evaporation seen in Fig. 8 that does produce MJO-like variability (*middle*), and the difference between the two (*lower*). From Del Genio et al. (2011a)

Another possibly relevant factor for GISS Model E2 is that the experimental version shifts the cumulus mass flux and heating peak downward, since stronger entrainment produces lower cloud top heights. This helps produce second baroclinic mode vertical heating structure (Fig. 3) that may be missing from the control model version. More shallow convection allows moist static energy imported into the column by large-scale moisture convergence to build up rather than immediately being exported by deep convection. Thus, the experimental model version may have smaller gross moist stability during the shallow-deep transition phase, destabilizing the MJO (Raymond et al. 2009).

The experience with the MJO in GISS Model E2 is consistent with effects that can be produced in many other GCMs by increased entrainment, a stronger trigger for deep convection initiation, or increased rain evaporation (Kim et al. 2009; Hannah and Maloney 2011). Why, then, do so many operational GCMs have dry middle tropospheres and poor MJOs? Kim et al. (2011b) show that physics changes that strengthen the MJO degrade other aspects of the model mean climate, e.g., excessive mean rainfall and intraseasonal rainfall variance. In other words, although a stronger link between convection and moisture increases large-scale variability, by itself it does not focus enough of it into the observed frequencies and wavenumbers. The reason for the MJO-mean state tradeoff appears to be that a parameterization that adequately suppresses convection in dry environments overly limits it in more favorable environments in which larger, more vigorous clouds entrain less. Furthermore, although stronger entrainment produces MJO-like variability in GISS Model E2, it does not by itself shift its noon peak in continental precipitation to later times. What is still missing from the models that might explain this?

4 The Next Parameterization Frontier: Mesoscale Organization

Figures 1, 2, and 3 offer a clue to the most glaring remaining deficiency of cumulus parameterizations. We have focused on the first two building blocks of convective systems, deep convective cells that produce first baroclinic mode heating throughout the troposphere and net drying, and shallow/congestus clouds in advance of deep convection that produce second baroclinic mode heating of low levels and moisten the troposphere. The third building block is the upper troposphere stratiform rain region and non-precipitating anvil that occurs as convective cells organize into mesoscale clusters during the mature phase of their lifecycle. Mesoscale updrafts produce condensation heating in the upper troposphere, while melting of falling snow and subsequent evaporation of stratiform rain cool and moisten the lower troposphere (Houze 2004). This produces a second baroclinic mode structure of opposite phase to that which occurs in the shallow/congestus phase.

Many GCMs include detrainment of condensate from their convective updrafts; how much depends on the particle size distribution, fall speeds, and the convective updraft speed (Del Genio et al. 2005). In most GCMs, though, the detrained condensate is treated simply as a stratiform cloud with no dynamics; only one GCM to date includes an estimate of mesoscale updrafts and downdrafts, but coincident with the parent convection (Donner 1993). Thus, we anticipate that most models have at best a very weak and brief second baroclinic mode heating period after deep convection is initiated (e.g., Del Genio et al. 2011b). Several simple models invoke a stratiform heating profile and its interaction with tropospheric humidity as central to the instability driving the MJO (e.g., Khouider and Majda 2006; Kuang 2008).

To move forward in this area, a parameterization must first determine whether convection organizes. Yuter and Houze (1998) discuss the "sustainability" of the environment that allows for continual regeneration of convection, leading to a gradual accumulation of air parcels from weakening convection that form the mesoscale stratiform rain and updraft region. Schumacher and Houze (2006) find that in addition to a warm, humid boundary layer that can maintain its thermodynamic state over long time periods, sustainability is enhanced by (1) high tropospheric humidity, which provides a more favorable environment for convection to deepen, and (2) moderate upper-level wind shear, which promotes spreading of condensate from convective updrafts into the stratiform rain region.

The boundary layer is thought to maintain its moist static energy through a quasi-equilibrium between the source due to surface turbulent fluxes of latent and sensible heat, and the sink due primarily to convective downdrafts that carry low moist static energy air from the mid-troposphere to the boundary layer (Emanuel et al. 1994). Downdrafts have gradually been introduced into cumulus parameterizations over the past two decades, but in a way that in retrospect may have caused as much harm as good. Since GCMs do not usually represent subgrid variability, cold dry downdraft air instantaneously mixes with the undisturbed moist humid boundary-layer air that generated the initial convection. This stabilizes the boundary later and suppresses further convection—might this be why GCMs have difficulty raining over land in late afternoon?

Qian et al. (1998) noted that downdraft air reaching the boundary layer actually remains distinct for hours and spreads, lifting undisturbed high moist static energy air at its edge and thus promoting further convection for hours, a point also emphasized by Tompkins (2001). Qian et al. developed a wake parameterization that accounted separately for the evolving properties of the downdraft cold pool. Grandpeix and Lafore (2010) use cold pool convergence to generate sufficient lifting energy to overcome convective inhibition. In an SCM, this delays triggering of the Emanuel deep convection scheme (which is normally initiated by positive convective available potential energy), producing a more realistic diurnal cycle (Rio et al. 2009). Mapes (2000) explored the competition between the stabilizing and destabilizing effects of downdrafts in a simple model of convectively coupled waves, asking, “Has a whole generation of cumulus parameterizations focused on the smaller of the two effects of downdrafts?”

The challenges of implementing a cold pool parameterization are the added prognostic variables (not to mention others associated with the resulting mesoscale updraft/downdraft) and one’s conceptions about how cold pools organize convection. Mapes and Neale (2011) suggest a novel approach to the former challenge. They define a single abstract dimensionless prognostic variable *org*, which captures whatever physics of organization one wishes to include as a forcing term, opposed by a relaxation with specified decay time that eliminates organization when forcing ceases. At each time step, the current value of *org* is used to diagnose whatever subgrid variability one wishes to affect convection.

One such effect proposed by Mapes and Neale (2011) is that since downdrafts from prior convection spawn new convection in the vicinity of the old, the new events rise through more humid air than the gridbox mean due to detrainment, rain evaporation, and cloud edge mixing from the older event—a possible explanation for the Schumacher and Houze (2006) tropospheric humidity effect on sustainability. Figure 10 shows a snapshot of weakly organizing convection in a CRM (Del Genio et al. 2011b). The left panel shows a classification of the convective and stratiform rain areas and a transition region in which detrainment and downdrafts occur. The right panel shows the corresponding 600-mb relative humidity. Very humid air occupies a much larger area than the convection itself, reducing the efficacy of entrainment and promoting further deep development. The domain mean humidity is 28% drier than convective updraft air, but the transition region air that surrounds the convective updrafts is only 12% drier.

Another proposed effect of the cold pool convergence is on the entrainment rate itself. Khairoutdinov and Randall (2006) show that after the onset of cold pools, the convergence organizes convecting air at larger scales than that of boundary-layer turbulence (Fig. 11). If

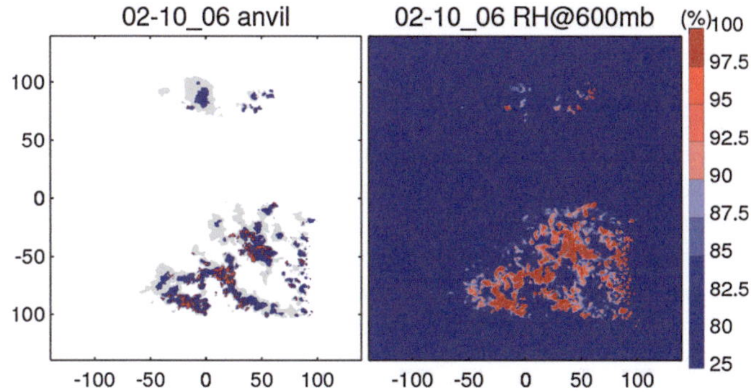

Fig. 10 *Left* instantaneous convective cluster cloud classification mask in the latitude–longitude plane in a CRM simulation centered on Darwin, Australia. *Red* deep convective updrafts. *Blue* transition region. *Gray* stratiform rain region. *Right* corresponding 600-mb relative humidity field. From Del Genio et al. (2011b)

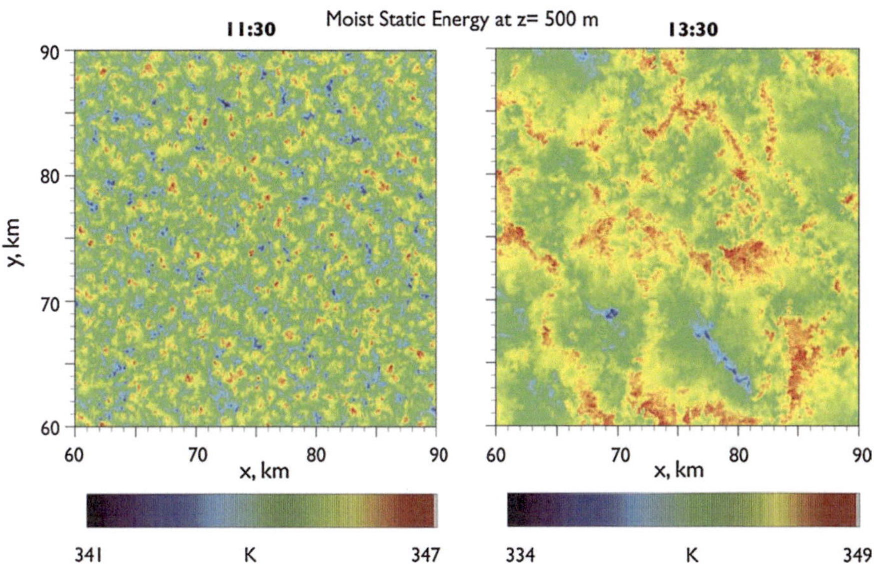

Fig. 11 Latitude–longitude moist static energy field at 500 m altitude at (*left*) 11:30 LST and (*right*) 13:30 LST, before and after the transition to deep convection respectively, in a CRM simulation of diurnal convective development over the Amazon. From Khairoutdinov and Randall (2006) (© Copyright 2006 AMS)

entrainment varies inversely with parcel size, entrainment rates should thus decrease as convection deepens and cold pools form, perhaps explaining the CRM-inferred decrease in entrainment with convection depth noted earlier. Del Genio and Wu (2010) have a different take on this process, suggesting that within the conceptual framework of the Gregory (2001) entrainment parameterization the enhancement of cloud base vertical velocity by cold pool convergence is what weakens entrainment.

These effects might perpetuate convection long enough to improve the diurnal cycle of continental convection. To simulate the second baroclinic mode heating profile required by several theories of the MJO, however, will require a separate parameterization of the dynamics of mesoscale updrafts and downdrafts. The lifetime of the stratiform rain region depends on the time-varying balance between the convective detrainment condensate source and the sink due to sedimentation of ice within the stratiform area. To get the source right, one must parameterize the persistence of convection after onset, which depends on how long the cold pool remains a distinct entity. This operates differently over ocean and land (Del Genio et al. 2011b): ocean cold pools recover at a rate dictated by surface latent and sensible heat fluxes, while cold pools over land quickly cool the land surface and suppress surface fluxes, so the cold pool remains close to its initial temperature while the undisturbed boundary layer cools diurnally to a similar temperature. The sink depends on properly characterizing the dynamics and microphysics of the stratiform rain region itself. In this region, updraft speeds are smaller than ice fall speeds—just the opposite of the convective region—leading to slow deposition growth as particles sediment. The ice eventually melts and evaporates, forming the mesoscale downdraft (Biggerstaff and Houze 1991). Once convection ceases, these processes determine how long the upper-level heating continues. The stratiform/anvil cloud directly affects the planetary energy balance (Clement and Soden 2005; Del Genio et al. 2005) and potentially cloud feedback (Yao and Del Genio 1999; Zelinka and Hartmann 2010) because of its large area, though it has little effect on water vapor feedback since the condensate is a small fraction of the total detrained water (Del Genio et al. 1991).

One difficulty with parameterizing mesoscale organization is that as the resolution of GCMs increases the size of convective clusters begins to exceed that of the gridbox. Numerical weather prediction models now routinely run at resolutions of 10–20 km, and even some climate models now have resolution finer than 100 km. By itself, increasing resolution does not appear to improve the simulation of the rainfall diurnal phase until a model gets down to almost convection-permitting (<10 km) resolution (Dirmeyer et al. 2011), suggesting that uncertainties in parameterizing cloud-scale processes are still important. However, as resolution increases, we might anticipate that some of the missing mesoscale updraft and downdraft circulation and associated second baroclinic mode heating will be resolved by the GCM dynamics. The question then becomes how to seamlessly navigate this transition from parameterized to resolved heating and precipitation without double-counting. Perhaps the transition region of detrainment, downdrafts, and cold pools between the convective and stratiform rain regions of mesoscale clusters (blue area in Fig. 10) will need to be parameterized at intermediate (mesoscale) resolution, but with communication between neighboring gridboxes that allows the resolved circulation to capture the vertical motions of the stratiform rain region. Even for current conventional cumulus parameterizations that do not treat organization, the assumption that convective updrafts occupy a very small area of the gridbox begins to break down as resolution increases. A general framework for adapting conventional parameterizations to operate at intermediate resolutions has been suggested by Arakawa et al. (2011).

5 Conclusions

Our ability to simulate energy flows in the Earth system is unfortunately limited by transports by small-scale processes such as moist convection that are still parameterized in global climate models. Historically, cumulus parameterization has received attention primarily for its effect on the diabatic heating profile and the general circulation. Only in the past decade has it become clear that tropospheric humidity acts as the convection throttle that regulates (a) where clouds form, thus controlling the planetary energy balance, (b) how energy is removed from the surface to higher altitude, and (c) how temporal variability in weather and planetary-scale lateral energy transports occurs. Cumulus parameterizations are beginning to change to reflect this new view, but adequately portraying convection in all its realizations remains a difficult problem. Progress is slower than might be expected because of the small number of scientists actively involved in model development and the financial and career roadblocks associated with pursuing such research (Jakob 2010).

To date, metrics for model evaluation have focused almost exclusively on time mean two-dimensional spatial distributions of easily observed parameters. It has become clear that such metrics have no predictive value for climate feedbacks and climate sensitivity (e.g., Collins et al. 2011). They are also probably not helpful for assessing most other important features of future climate projections, because temporal variability gives greater insight into the physical processes at work. In addition, most of what matters about climate change for users of climate model projections concerns specific events or anomalous regional weather patterns rather than mean fields. For the IPCC AR4 models, there has been some success in using interannual (Bony and Dufresne 2005) and decadal (Clement et al. 2009) variability to differentiate the fidelity of models with positive and negative low cloud feedbacks. The fact that the AR4 model sensitivity spread can be attributed largely to low cloud feedbacks does not mean that other cloud types do not contribute to uncertainty in climate sensitivity, however. Rather, the similar high and convective cloud response to climate change among the AR4 models in the Bony and Dufresne (2005) study may be due to shortcomings all the models have in common and thus may underestimate the inherent feedback uncertainty. Given the insensitivity of these models to tropospheric humidity and their failure to simulate the MJO and diurnal cycle, the results of perturbed parameter ensembles (Sanderson et al. 2010) should be a caution to the climate community to devote as much attention to the simulation of high and convective clouds as is currently being devoted to understanding low clouds.

It seems unlikely that it will ever be possible to establish a general set of metrics that can be used to anoint one subset of models as our most reliable indicators of all aspects of climate change. A more fruitful strategy might identify a subset of the most important climate change issues (climate sensitivity, Arctic sea ice loss, extreme precipitation, subtropical continental drought, etc.) and define separate metrics for each one based on appropriate aspects of current temporal variability that are diagnostic of the processes that cause the specific change. Diagnostics that reveal the interaction between convection and tropospheric humidity should be a high priority in any such effort.

Acknowledgments This work was originally presented at the Workshop "Observing and Modeling Earth's Energy Flows" held at the International Space Science Institute, Bern, Switzerland, 10–14 January 2011. The research described here was supported by the DOE Atmospheric System Research Program, the NASA Modeling and Analysis Program, the NASA Precipitation Measurement Missions, and the NASA CloudSat/CALIPSO Mission. I thank the anonymous reviewers for constructive comments that strengthened the paper.

References

Arakawa A, Schubert WH (1974) Interaction of a cumulus cloud ensemble with the large-scale environment. Part I. Theoretical formulation and sensitivity tests. J Atmos Sci 31:674–701

Arakawa A, Jung J-H, Wu C-M (2011) Toward unification of the multiscale modeling of the atmosphere. Atmos Chem Phys 11:3731–3742
Benedict JJ, Randall DA (2007) Observed characteristics of the MJO relative to maximum rainfall. J Atmos Sci 64:2332–2354
Biggerstaff MI, Houze RA Jr (1991) Kinematic and precipitation structure of the 10–11 June 1985 squall line. Mon Weather Rev 119:3034–3065
Bladé I, Hartmann DL (1993) Tropical intraseasonal oscillations in a simple nonlinear model. J Atmos Sci 50:2922–2939
Bony S, Dufresne J-L (2005) Marine boundary layer clouds at the heart of tropical cloud feedback uncertainties in climate models. Geophys Res Lett 32:L20806. doi:10.1029/2005GL023851
Bretherton CS, Peters ME, Back LE (2004) Relationships between water vapor path and precipitation over the tropical oceans. J Clim 17:1517–1528
Chen Y, Del Genio AD (2009) Evaluation of tropical cloud regimes in observations and a general circulation model. Clim Dyn 32:355–369
Clement AC, Soden B (2005) The sensitivity of the tropical-mean radiation budget. J Clim 18:3189–3203
Clement A, Burgman R, Norris J (2009) Observational and model evidence for positive low-level cloud feedback. Science 325:460–464
Collins M, Booth BBB, Bhaskaran B, Harris GR, Murphy JM, Sexton DMH, Webb MJ (2011) Climate model errors, feedbacks and forcings: a comparison of perturbed physics and multi-model ensembles. Clim Dyn 36:1737–1766
Dai A (2006) Precipitation characteristics in eighteen coupled climate models. J Clim 19:4605–4630
Del Genio AD, Wu J (2010) Entrainment and the diurnal cycle of continental convection. J Clim 23:2722–2738
Del Genio AD, Lacis AA, Ruedy RA (1991) Simulations of the effect of a warmer climate on atmospheric humidity. Nature 351:382–385
Del Genio AD, Kovari W, Yao M-S, Jonas J (2005) Cumulus microphysics and climate sensitivity. J Clim 18:2376–2387
Del Genio AD, Yao M-S, Jonas J (2007) Will moist convection be stronger in a warmer climate? Geophys Res Lett 34:L16703. doi:10.1029/2007GL030525
Del Genio AD, Chen Y, Kim D, Yao M-S (2011a) The MJO transition from shallow to deep convection in CloudSat/CALIPSO data and GISS GCM simulations. J Clim (in revision)
Del Genio AD, Wu J, Chen Y (2011b) Characteristics of mesoscale organization in WRF simulations of convection during TWP-ICE. J Clim (submitted)
Derbyshire SH, Beau I, Bechtold P, Grandpeix J-Y, Piriou J-M, Redelsperger J-L, Soares PMM (2004) Sensitivity of moist convection to environmental humidity. Q J R Meteorol Soc 130:3055–3079
Dirmeyer PA et al (2011) Simulating the diurnal cycle of rainfall in global climate models: resolution versus parameterization. Clim Dyn (in press). doi:10.1007/s00382-011-1127-9
Donner LJ (1993) A cumulus parameterization including mass fluxes, vertical momentum dynamics, and mesoscale effects. J Atmos Sci 50:889–906
Emanuel KA (1991) A scheme for representing cumulus convection in large-scale models. J Atmos Sci 48:2313–2335
Emanuel KA, Neelin JD, Bretherton CS (1994) On large-scale circulations in convecting atmospheres. Q J R Meteorol Soc 120:1111–1143
Fasullo JT (2010) Robust land–ocean contrasts in energy and water cycle feedbacks. J Clim 23:4677–4693
Futyan JM, Del Genio AD (2007) Deep convective system evolution over Africa and the tropical Atlantic. J Clim 20:5041–5060
Grabowski WW et al (2006) Daytime convective development over land: a model intercomparison based on LBA observations. Q J R Meteorol Soc 132:317–334
Grandpeix J-Y, Lafore J-P (2010) A density current parameterization coupled with Emanuel's convection scheme. Part I: the models. J Atmos Sci 67:881–897
Grant ALM, Brown AR (1999) A similarity hypothesis for shallow-cumulus transports. Q J R Meteorol Soc 125:1913–1936
Gregory D (2001) Estimation of entrainment rate in simple models of convective clouds. Q J R Meteorol Soc 127:53–72
Gregory D, Rowntree PR (1990) A mass flux convection scheme with representation of cloud ensemble characteristics and stability-dependent closure. Mon Weather Rev 118:1483–1506
Guichard F et al (2004) Modeling the diurnal cycle of deep precipitating convection over land with cloud-resolving models and single-column models. Q J R Meteorol Soc 130:3139–3172
Hack JJ (1994) Parameterization of moist convection in the National Center for Atmospheric Research community climate model (CCM2). J Geophys Res 99:5551–5568

Hannah WM, Maloney ED (2011) The role of moisture-convection feedbacks in simulating the Madden–Julian Oscillation. J Clim 24:2754–2770
Heus T, van Dijk G, Jonker HJJ, Van den Akker HEA (2008) Mixing in shallow cumulus clouds studied by Lagrangian tracking. J Atmos Sci 65:2581–2597
Hirose M, Oki R, Shimizu S, Kacji M, Higashiuwatoko T (2008) Finescale diurnal rainfall statistics refined from eight years of TRMM PR data. J Appl Meteorol Clim 47:544–561
Holloway CE, Neelin JD (2009) Moisture vertical structure, column water vapor, and tropical deep convection. J Atmos Sci 66:1665–1683
Houze RA Jr (2004) Mesoscale convective systems. Rev Geophys 42:RG4003. doi:10.1029/2004RG000150
Hu Q, Randall DA (1994) Low-frequency oscillations in radiative-convective systems. J Atmos Sci 51:1089–1099
Jakob C (2010) Accelerating progress in global atmospheric model development through improved parameterizations—challenges, opportunities and strategies. Bull Am Meteorol Soc 91:869–875
Jakob C, Siebesma AP (2003) A new subcloud model for mass-flux convection schemes: influence on triggering, updraft properties, and model climate. Mon Weather Rev 131:2765–2778
Jensen MP, Del Genio AD (2006) Factors limiting convective cloud-top height at the ARM Nauru Island climate research facility. J Clim 19:2105–2117
Kemball-Cook SR, Weare BC (2001) The onset of convection in the Madden–Julian oscillation. J Clim 14:780–793
Khairoutdinov M, Randall D (2006) High-resolution simulation of shallow-to-deep convection transition over land. J Atmos Sci 63:3421–3436
Khouider B, Majda AJ (2006) A simple multicloud parameterization for convectively coupled tropical waves. Part I: linear analysis. J Atmos Sci 63:1308–1323
Kiladis GN, Straub KH, Haertel PT (2005) Zonal and vertical structure of the Madden–Julian Oscillation. J Atmos Sci 62:2790–2809
Kim D et al (2009) Application of MJO simulation diagnostics to climate models. J Clim 22:6413–6436
Kim D, Sobel AH, Del Genio AD, Chen Y, Camargo SJ, Yao M-S, Kelley M, Nazarenko L (2011a) The tropical subseasonal variability simulated in the NASA GISS general circulation model. J Clim (submitted)
Kim D, Sobel AH, Maloney ED, Frierson DMW, Kang I-S (2011b) A systematic relationship between intraseasonal variability and mean state bias in AGCM simulations. J Clim. doi:10.1175/2011JCLI4177.1
Kuang Z (2008) A moisture-stratiform instability for convectively coupled waves. J Atmos Sci 65:834–854
Kuang Z, Bretherton CS (2006) A mass-flux scheme view of a high-resolution simulation of a transition from shallow to deep cumulus convection. J Atmos Sci 63:1895–1909
Lin J-L et al (2006) Tropical intraseasonal variability in 14 IPCC AR4 climate models. Part I: convective signals. J Clim 19:2665–2690
Madden R, Julian P (1971) Detection of a 40–50 day oscillation in the zonal wind in the tropical Pacific. J Atmos Sci 28:702–708
Mapes BE (2000) Convective inhibition, subgrid-scale triggering energy, and stratiform instability in a toy tropical wave model. J Atmos Sci 57:1515–1535
Mapes BE, Neale RB (2011) Parameterizing convective organization to avoid the entrainment dilemma. J Adv Model Earth Syst 3:M06004. doi:10.1029/2011MS000042
Mapes BE, Tulich S, Lin J, Zuidema P (2006) The mesoscale convection life cycle: building block or prototype for large-scale tropical waves? Dyn Atmos Ocean 42:3–29
Moorthi S, Suarez MJ (1992) Relaxed Arakawa-Schubert: a parameterization of moist convection for general circulation models. Mon Weather Rev 120:978–1002
Morita J, Takayabu YN, Shige S, Kodama Y (2006) Analysis of rainfall characteristics of the Madden–Julian Oscillation using TRMM satellite data. Dyn Atmos Ocean 42:107–126
Neale RB, Richter JH, Jochum M (2008) The impact of convection on ENSO: from a delayed oscillator to a series of events. J Clim 21:5904–5924
Neggers RAJ, Siebesma AP, Jonker HJJ (2002) A multiparcel model for shallow cumulus convection. J Atmos Sci 59:1655–1668
Paluch IR (1979) The entrainment mechanism in Colorado cumuli. J Atmos Sci 36:2467–2478
Qian L, Young GS, Frank WM (1998) A convective wake parameterization scheme for use in general circulation models. Mon Weather Rev 126:456–469
Raymond DJ, Sessions SL, Sobel AH, Fuchs Z (2009) The mechanics of gross moist stability. J Adv Model Earth Syst 1. doi:10.3894/JAMES.2009.1.9
Riehl H, Malkus JS (1958) On the heat balance in the equatorial trough zone. Geophysica 6:503–538

Rio C, Hourdin F, Grandpeix J-Y, Lafore J-P (2009) Shifting the diurnal cycle of parameterized deep convection over land. Geophys Res Lett 36:L07809. doi:10.1029/2008GL036779
Romps DM, Kuang Z (2010a) Do undiluted convective plumes exist in the upper tropical troposphere? J Atmos Sci 67:468–484
Romps DM, Kuang Z (2010b) Nature vs. nurture in shallow convection. J Atmos Sci 67:1655–1666
Sanderson BM, Shell KM, Ingram W (2010) Climate feedbacks determined using radiative kernels in a multi-thousand member ensemble of AOGCMs. Clim Dyn 30:175–190
Schumacher C, Houze RA Jr (2006) Stratiform precipitation production over sub-Saharan Africa and the tropical East Atlantic as observed by TRMM. Q J R Meteorol Soc 132:2235–2255
Siebesma AP et al (2003) A large eddy simulation intercomparison study of shallow cumulus convection. J Atmos Sci 60:1201–1219
Stephens GL, Webster PJ, Johnson RH, Engelen R, L'Ecuyer T (2004) Observational evidence for the mutual regulation of the tropical hydrological cycle and tropical sea surface temperatures. J Clim 17:2213–2224
Thayer-Calder K, Randall DA (2009) The role of convective moistening in the Madden–Julian Oscillation. J Atmos Sci 66:3297–3312
Tokioka T, Yamazaki K, Kitoh A, Ose T (1988) The equatorial 30–60 day oscillation and the Arakawa-Schubert penetrative cumulus parameterization. J Meteorol Soc Jpn 66:883–901
Tompkins AM (2001) Organization of tropical convection in low vertical wind shears: the role of cold pools. J Atmos Sci 58:1650–1672
Trenberth KE, Fasullo JT (2010) Simulation of present-day and twenty-first-century energy budgets of the southern oceans. J Clim 23:440–454
Trenberth KE, Fasullo JT, Kiehl J (2009) Earth's global energy budget. Bull Am Meteorol Soc 90:311–323
Tromeur E, Rossow WB (2010) Interaction of tropical deep convection with the large-scale circulation in the MJO. J Clim 23:1837–1853
Warner J (1970) On steady-state one-dimensional models of cumulus convection. J Atmos Sci 27:1035–1040
Weickmann KM, Lussky GR, Kutzbach JE (1985) Intraseasonal (30–60 day) fluctuations of outgoing longwave radiation and 250 mb streamfunction during northern winter. Mon Weather Rev 113:941–961
Wheeler M, Kiladis GN (1999) Convectively coupled equatorial waves: analysis of clouds and temperature in the wavenumber–frequency domain. J Atmos Sci 56:374–399
Yao M-S, Del Genio AD (1999) Effects of cloud parameterization on climate changes in the GISS GCM. J Clim 12:761–779
Yuter SE, Houze RA Jr (1998) The natural variability of precipitating clouds over the western Pacific warm pool. Q J R Meteorol Soc 124:53–99
Zelinka MD, Hartmann DL (2010) Why is longwave cloud feedback positive? J Geophys Res 115:D16117. doi:10.1029/2010JD013817
Zhang Y, Klein SA (2010) Mechanisms affecting the transition from shallow to deep convection over land: inferences from observations of the diurnal cycle collected at the ARM Southern Great Plains site. J Atmos Sci 67:2943–2959
Zhang GJ, McFarlane NA (1995) Sensitivity of climate simulations to the parameterization of cumulus convection in the Canadian Climate Centre general circulation model. Atmos Ocean 33:407–446

Surv Geophys (2012) 33:657–676
DOI 10.1007/s10712-011-9164-9

Computation of Solar Radiative Fluxes by 1D and 3D Methods Using Cloudy Atmospheres Inferred from A-train Satellite Data

H. W. Barker · S. Kato · T. Wehr

Received: 20 June 2011 / Accepted: 15 November 2011 / Published online: 29 December 2011

Abstract This study used realistic representations of cloudy atmospheres to assess errors in solar flux estimates associated with 1D radiative transfer models. A scene construction algorithm, developed for the EarthCARE mission, was applied to CloudSat, CALIPSO and MODIS satellite data thus producing 3D cloudy atmospheres measuring 61 km wide by 14,000 km long at 1 km grid-spacing. Broadband solar fluxes and radiances were then computed by a Monte Carlo photon transfer model run in both full 3D and 1D independent column approximation modes. Results were averaged into 1,303 (50 km)^2 domains. For domains with total cloud fractions $A_c < 0.7$ top-of-atmosphere (TOA) albedos tend to be largest for 3D transfer with differences increasing with solar zenith angle. Differences are largest for $A_c > 0.7$ and characterized by small bias yet large random errors. Regardless of A_c, differences between 3D and 1D transfer rarely exceed ± 30 W m^{-2} for net TOA and surface fluxes and ± 10 W m^{-2} for atmospheric absorption. Horizontal fluxes through domain sides depend on A_c with $\sim$20% of cases exceeding ± 30 W m^{-2}; the largest values occur for $A_c > 0.7$. Conversely, heating rate differences rarely exceed $\pm 20\%$. As a cursory test of TOA radiative closure, fluxes produced by the 3D model were averaged up to (20 km)^2 and compared to values measured by CERES. While relatively little attention was paid to optical properties of ice crystals and surfaces, and aerosols were neglected entirely, $\sim$30% of the differences between 3D model estimates and measurements fall within ± 10 W m^{-2}; this is the target agreement set for EarthCARE. This, coupled with the aforementioned comparison between 3D and 1D transfer, leads to the recommendation that EarthCARE employ a 3D transport model when attempting TOA radiative closure.

H. W. Barker (✉)
Cloud Physics and Severe Weather Research Section (ARMP), Environment Canada, 4905 Dufferin St., Toronto, ON M3H 5T4, Canada
e-mail: howard.barker@ec.gc.ca

S. Kato
NASA—Langley Research Center, Hampton, VA, USA

T. Wehr
European Space Agency, Noordwijk, The Netherlands

Keywords Cloud · Radiation · Climate · Satellite · EarthCARE · CloudSat · Cloud-Aerosol Lidar and Infrared Pathfinder Satellite Observations · The Moderate Resolution Imaging Spectroradiometer · Clouds and Earth's Radiant Energy System

1 Introduction

The ultimate boundary conditions of Earth's climate system are defined by the flow of solar radiation in and longwave emission out of the top-of-atmosphere (TOA). All else constituting climate and life falls between these boundaries. Understandably then, radiative transfer through, and optical properties of, the Earth-atmosphere system lie at the core of our concepts of climatic forcing and feedbacks. Moreover, many remote sensing techniques, both passive and active, involve measurement and subsequent model interpretation of (largely atmospheric) radiative transfer. Mathematical treatments of radiative transfer were developed initially by astrophysicists (e.g., Chandrasekhar 1950), but the latter half of the 20th century saw substantial advancements from Earth scientists in their response to the myriad conditions and scales of Earth's climate system. Arguably, these advancements are exemplified best by work centred on radiative transfer for cloudy atmospheres.

Over the past three decades there has been a string of diagnostic studies aimed at demonstrating errors associated with neglect of 3D radiative transfer within cloud, weather and climate models (e.g., McKee and Cox 1974; Davies 1978; Ellingson 1982; Welch and Wielicki 1984; Stephens 1988; Davis et al. 1990; Barker and Davies 1992; Cahalan et al. 1994; Marshak et al. 1995; O'Hirok and Gautier 1998; Barker et al. 1999, 2003; Cole et al. 2005a). Generally, these studies used small, and often selective, samples of 3D cloud fields derived either from idealized models, such as arrays of simple forms (Welch and Wielicki 1985; Kobayashi 1988) and scaling algorithms (Marshak and Davis 2005a), passive satellite data (Cahalan et al. 2005), aircraft data (Barker 1992; Räisänen et al. 2004) or limited-area domains simulated by cloud system-resolving models (Barker et al. 1999; Hinkelman et al. 2005). Ideally, a large number of 3D cloud fields based on observational data should be used to demonstrate errors due to neglect of 3D solar transfer. This would enable proper hypothesis tests to be carried out involving clouds whose geometric properties resemble most faithfully those that actually occur.

Cloud properties derived from synergistic retrievals using data from lidars, cloud-profiling radars and passive radiances provide, thus far, the most complete pictures of true cloud structure (ESA 2001; Stephens et al. 2002; Hogan et al. 2006; Winker et al. 2007; Dupont et al. 2011). However, because current active-passive systems point fixedly at either zenith (ground-based) or nadir (satellite-based), they provide 2D vertical cross-sections, not 3D domains. Those derived from satellite data have the advantage of providing a global perspective on a regular grid whose horizontal spacings are not subject to height-dependent advection rates. Their disadvantage is that they are less resolved than, and not as sensitive as, their surface counterparts.

To get around the issue of satellite retrievals being essentially 2D cross-sections, 3D fields can be constructed using the algorithm of Barker et al. (2011) that was developed for use by the EarthCARE satellite mission (ESA 2001). EarthCARE is slated for launch in 2015 and will carry a high-spectral resolution lidar, a 94-GHz cloud Doppler radar, a 7-channel imager and a broadband radiometer. Their algorithm combines conventional passive satellite imagery with nearby coeval information from active-passive profiles; *recipient* columns associated with off-nadir passive pixels get assigned, by proxy, an active-passive *donor* column associated with a nadir pixel thereby leading to production of a 3D domain of cloud.

This study represents a first step towards comparing 1D and 3D solutions for large samples of cloud fields derived from satellite data. A-train data and retrieval products (Kato et al. 2010) construction algorithm developed by Barker et al. (2011) to construct 3D cloudy atmospheres. A broadband solar transport solver (capable of affecting 1D and 3D solutions) was then applied to them. Reflected fluxes from the 3D model were compared to corresponding (20 km)^2 CERES values (Wielicki et al. 1996). For (50 km)^2 domains that resemble the size of weather and climate model grid-cells, flux differences arising from 1D and 3D solutions were compared.

In the following section the A-train dataset is described briefly. This is followed by descriptions of the 3D construction algorithm, the radiative transfer model and the nature of the simulations. Results are presented in Sect. 6, and conclusions and recommendations are in the final section.

2 Merged A-train Data

Data gathered by several A-train satellites were used to assess the importance of 3D solar radiative transfer. A merged A-train dataset has been compiled at NASA-Langley in which CloudSat (Stephens et al. 2002) and CALIPSO (Winker et al. 2007) profiles were mapped to 1-km resolution thereby associating each column with a 1-km MODIS pixel (Kato et al. 2010). It also includes the line of (20 km)^2 CERES broadband radiances (and fluxes) along the CloudSat-CALIPSO track as well as MODIS pixels either side of it out to $\sim$40 km.

A merged cloud mask was created by interpolating CloudSat's and CALIPSO's cloud masks onto CERES's arbitrary vertical grid which includes temperature, humidity and ozone (from the Goddard Earth Observing System—Data Assimilation System—GEOS-4; see Bloom et al. 2005; Kato et al. 2005). If a CloudSat or CALIPSO cloudy cell overlaps a CERES layer, the cell is designated to be filled with cloud.

Because a large fraction of columns have just cloud masks from CloudSat and CALIPSO, vertical profiles of cloud properties were set using MODIS quantities and the following algorithm. For each MODIS pixel along the CloudSat-CALIPSO track, effective visible optical depth τ_{eff}, effective particle size r_{eff} and particle phase were retrieved based on inversion of a 1D radiative transfer model (Minnis et al. 2008, 2010a, b). If there are n_{ice} and n_{liq} cells in a column, the former having temperatures <273 K and the latter >273 K, it is assumed that

$$\frac{3}{2}\left(\sum_{i=1}^{n_{\mathrm{ice}}}\frac{\mathcal{I}}{\rho r_{\mathrm{ice}}}\Delta z_i+\sum_{i=1}^{n_{\mathrm{liq}}}\frac{\mathcal{L}}{r_{\mathrm{liq}}}\Delta z_i\right)=\tau_{\mathrm{eff}} \quad (1)$$

where $\mathcal{I}$ and $\mathcal{L}$ are ice and liquid water contents, respectively, $\rho \approx 0.917$ g cm^{-3}, and Δz_i is layer thickness. If CloudSat's columnar classification (Sassen and Wang 2008) is cirrus, altostratus, altocumulus or deep convection, all ice cells in the column use $r_{\mathrm{ice}} = r_{\mathrm{eff}}$, and any liquid cells use $r_{\mathrm{liq}} = 10$ μm. When the columnar classification is stratus, stratocumulus, cumulus or nimbostratus, all liquid cells use $r_{\mathrm{liq}} = r_{\mathrm{eff}}$, and any ice cells use $r_{\mathrm{ice}} = 50$ μm (e.g., Yang et al. 2007). Equation 1 is then solved for water contents assuming that $\mathcal{I} = \mathcal{L}$ but restricting $\mathcal{I}$ to be no larger than 0.3 g m^{-3} (e.g., Protat et al. 2007).

This simple algorithm produces clouds that are, for the most part, locally homogeneous in the vertical. Obviously this can influence both 1D and 3D radiative transfer solutions, as well as their differences which are of concern here. Nevertheless, when clouds produced this way are operated on by 1D shortwave and longwave radiative transfer algorithms, local TOA flux estimates agree well with CERES data, particularly longwave values (Barker et al. 2011).

Without defending this vertical allocation scheme too much, it is worth noting that for geometrically thin clouds that get resolved into a small number of layers, such as many boundary layer clouds, the impact of vertical structure on 1D–3D transfer differences is a minor issue. The same goes for very cloudy scenes, which as will be seen constitute the majority of the cloudy (50 km^2) domains. Finally, preliminary studies, not reported here, using data from cloud system-resolving models suggest that vertical homogenization of water for towering convective clouds generally exaggerates 1D–3D differences, implying that results presented below are approximately an upper-bound for 1D–3D differences.

3 Constructing 3D Cloud Domains from A-train Data

For the convenience of readers, Barker et al. (2011) construction algorithm is reiterated in this section. A-train data make-up the 2D retrieved cross-sections (RXS). Let all pixels/columns along the RXS at positions $(i, 0)$, where for this study $i = 1$ was in the southern hemisphere of an ascending portion of an orbit, form the set of potential *donor* columns that can act as proxies for off-nadir *recipient* columns associated with pixels located at $\{(i,j) \mid j \in [-J,-1] \cup [1,J]\}$. For each recipient pixel, begin by computing

$$F(i, j;\, m) = \sum_{k=1}^{K} \left[\frac{r_k(i, j) - r_k(m, 0)}{r_k(i, j)} \right]^2 : m \in [i - m_1, i + m_2] \tag{2}$$

where r_k are MODIS radiances, and k denotes spectral interval. For this study $m_1 = m_2 = 200$ pixels which corresponds to searching for $\sim\pm 200$ km along the RXS beginning at the pixel on the RXS that is in the recipient's cross-track line. Next, $F(i, j; m)$ are ordered from smallest to largest to form the set

$$\begin{aligned} \{\min[F(i, j; m)], \ldots, \max[F(i, j; m)]\} = \{\mathcal{F}(i, j; 1), \ldots, \mathcal{F}(i, j; m_1 + m_2 + 1)\}: \\ \mathcal{F}(i, j; n) \leqslant \mathcal{F}(i, j; n + 1). \end{aligned} \tag{3}$$

Defining Euclidean distance between a potential donor at $(m,0)$ and recipient at (i, j) as

$$D(i, j; m) = \Delta L \sqrt{(i - m)^2 + j^2}, \tag{4}$$

where ΔL is imager resolution, let $D(i, j;m)$ and m go passively along with the ordering of $F(i, j; m)$ so that $\{\mathcal{F}(i, j; n)\}$ has an associated co-ordered set of distances $\{\mathcal{D}(i,j; n)\}$. One then solves for the index m^* as

$$\underset{m^* \in [1,(m_1+m_2+1)f]}{\arg\min} \quad \{\mathcal{D}(i, j; m^*)\} : f \in (0, 1), \tag{5}$$

which means: find the index m^* that corresponds to the smallest distance between recipient and those pixels that constitute the smallest $100f$% of $F(i, j; m)$. For this study $f = 0.03$ so that the smallest 3% of the $(m_1 + m_2 + 1)$ values of $F(i, j; m)$ are considered (provided they have the same surface type as the recipient). Knowing m^* leads directly to m and thus the donor column. Finally, all the properties associated with the column at $(m, 0)$ get replicated at (i, j). In addition to cloud properties, this includes profiles of temperature, moisture and aerosol, as well as surface conditions. The vast majority of cross-sections used here were completely over ocean and so there were 133 layers from the surface up to 65 km.

Figure 1 shows a schematic of this process which gets applied until the desired 3D scene is constructed. Clearly, RXS columns ($j = 0$) identify themselves so for them there is no need to apply the algorithm; the RXS forms the centre of the constructed domain.

Verification of the construction algorithm using A-train data is not straightforward. One can, however, go a certain distance by attempting to reconstruct the RXS itself. In so doing, one attempts to fill an RXS column at $(i, 0)$ by searching the RXS and applying (5) to potential donor pixels in $[i - m_1, i - n] \cup [i + n, i + m_2]$ which bars the first $\pm n$ pixels next to i; hence defining a *dead-zone* in the search process. This test is meant to mimic filling off-RXS columns that are $\pm n$ pixels away from the RXS. For example, when $n = 5$, searching for a proxy column begins five pixels away, just as for off-RXS pixels at $(i, \pm 5)$.

Figure 2 shows attempts to reconstruct a 400-km-long stretch of RXS. The upper image is the actual RXS merged cloud mask. This is an especially demanding case as it involves fairly dense multi-layer clouds. Over much of this domain passive-only retrievals would yield very little, if any, useful information about cloud vertical structure. Lower images are reconstructions for discrete values of $n = 1$, 5, 10 and 20. These results stem from using four spectral channels (0.62–0.67; 2.105–2.155; 8.4–8.7; and 11.77–12.27 μm) in (2). By $n = 5$, which corresponds to the outer edge of an 11-km-wide domain the likes of those to be computed for EarthCARE, it is clear that some error was creeping in; nevertheless, a significant amount of detail was captured. Even out at $n = 20$, multi-layers of clouds were replicated well. The region where the greatest difficulty was encountered between 100 and 200 km along the horizontal where clouds were transitioning or dying out entirely.

Figure 3 shows mean profiles of several variables accumulated out to n as functions of n for the field shown in Fig. 2. Results for accumulated fields, of widths $2n + 1$, are shown because the algorithm is intended to produce full 3D domains not single rows. For these accumulations averaging included the original RXS as it is included in constructed fields, yet gave double weight to the reconstructed lines so as to represent scene construction on both sides of the RXS. For clouds higher than 10 km, layer cloud fraction A_c, mean cloud

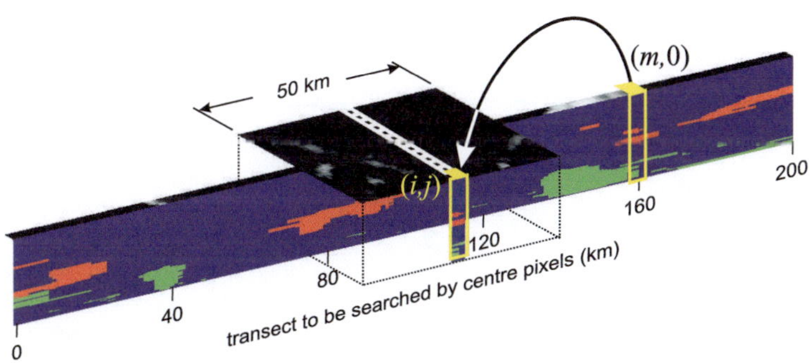

Fig. 1 The thin RXS is shown along with the sequence of MODIS visible pixels associated with it. The objective is to fill the volume marked by the wider MODIS swath with cloud properties drawn from the RXS. For example, the column associated with the pixel at $(m, 0)$ has been designated as the proxy for the pixel at (i, j) so the cloud-radiation attributes associated with $(m, 0)$ get donated to (i, j). The algorithm is applied until all desired off-RXS pixels are filled by donor RXS columns

Fig. 2 Topmost image is merged cloud mask (1-km horizontal resolution) for a stretch of tropical RXS along an orbit on 19 April 2007. *Black* and *grey* indicate ice and liquid, respectively. Sequences of *lower images* represent corresponding masks produced by the construction algorithm for various dead-zone lengths as listed

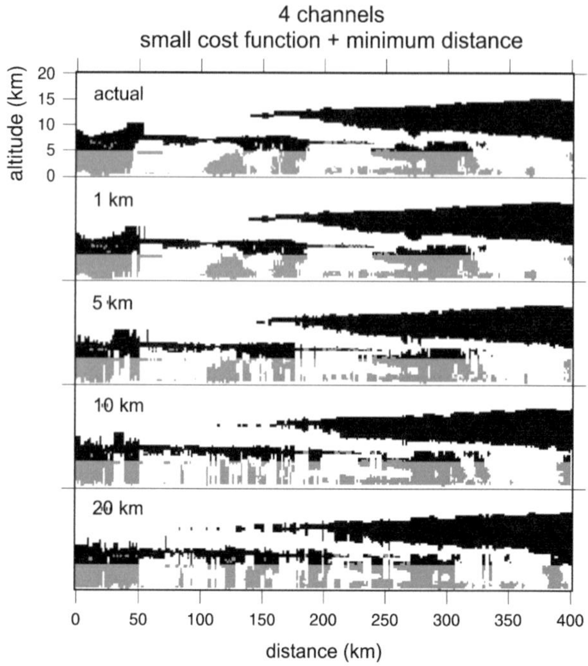

water contents $\langle w \rangle$, and $v = (\langle w \rangle / \sigma_w)^2$, where σ_w is standard deviation of w, were reproduced extremely well for all n. The largest errors on A_c occurred for $n = 20$ at altitudes near 5 km. $\langle w \rangle$ contents were reconstructed very well except for $n = 5$ and 10 for clouds between 2 and 5 km high where associated values of v were overly small, indicating that these errors were the result of having selected a small number of columns with anomalously large w. On the other hand, for $n = 20$, almost all reconstructed clouds below 10 km lacked sufficient horizontal variability as indicated by values of v being twice as large as they should be, despite corresponding $\langle w \rangle$ being fine. This was due to too many occurrences of a small number of RXS columns being used multiple times.

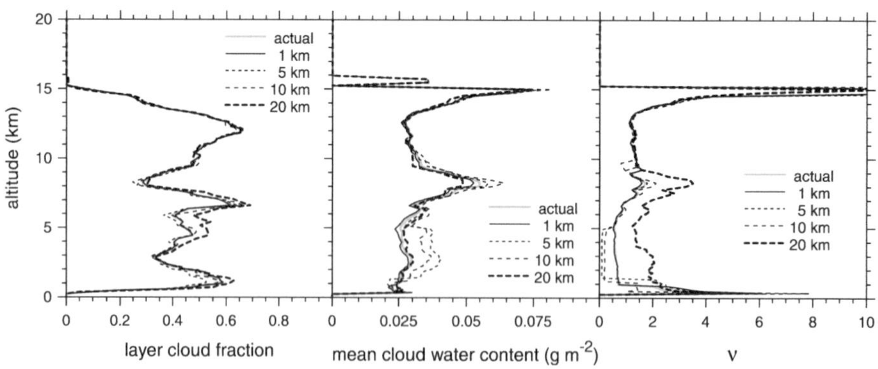

Fig. 3 Domain-average profiles of layer cloud fraction, cloud water content $\langle w \rangle$, and $v = (\langle w \rangle / \sigma_w)^2$, where σ_w is standard deviation of w, for the RXS (actual) shown in Fig. 2 as well as for reconstructed cross-sections corresponding to various dead-zone lengths as listed

As mentioned, the exemplary field used here was a difficult case of multi-layer tropical cloud. Most 400-km sections of cloud do not exhibit as much intricacy as this one. Numerous other examples were examined and almost all reconstructions performed equally well or better than those shown here. In general, the more extensive and planar the clouds and the fewer the number of definite layers, the better the reconstruction.

4 Radiative Transfer Model

Broadband shortwave (SW) fluxes and radiances were computed by a 3D Monte Carlo algorithm (Barker et al. 1998, 2003). For this study, full 3D results pertain to horizontal grid-spacings $\Delta L \simeq 1$ km. Cyclic horizontal boundary conditions were employed. Independent Column Approximation (ICA) results were produced by the same model using $\Delta L = 10^8$ km. Gaseous transmittances (H_2O, CO_2, O_3) were computed using the correlated *k*-distribution method with 31 quadrature points in cumulative probability space (Scinocca et al. 2008). Optical properties for liquid droplets (Dobbie et al. 1999; Lindner and Li 2000) and ice crystals (Fu 1996; Fu et al. 1998) were resolved into four bands. Aerosol effects were not included but Rayleigh scattering was. For each grid-cell a cumulative distribution of extinction was computed including Rayleigh scattering, gases and cloud particle sizes segregated into 23 radius bins progressing in width following a power law from 0–1 μm to 128.0–161.3 μm. Size-integrated Mie scattering functions were used for each bin, and phase functions for ice crystals were treated as though they were spherical droplets. To reduce variance in radiance estimates, Mie functions were used for a photon packet's first 5 scattering events with the Henyey-Greenstein function, with proper asymmetry parameter, thereafter if needed (Barker et al. 2003).

The majority of data used here were collected over the Pacific and Indian Oceans. Surface reflectance for oceans was represented by the Cox and Munk (1956) ergodic wave-facet model. Probability of reflection and direction of reflected photons were derived from zenith angle of incident photon and surface wind speed as reported in the A-train dataset (Bloom et al. 2005).

Land was treated as Lambertian with spectral albedos set to values inferred from MODIS data as reported in the A-train dataset (Rutan et al. 2009). For the model's 0.2–0.7 μm band, values obtained from 0.469 μm radiances were used. For the three near-IR bands, values for 1.24 μm were used.

5 Radiative Transfer Simulations

Radiative transfer simulations were performed on constructed domains that measured $\sim$61 km wide by $\sim$14,000 km long and consisted of 85.4×10^6 columns and $\sim 113.6 \times 10^6$ cells. Going much wider with the current version of the construction algorithm is not recommended (see Figs. 2, 3 as well as Barker et al. 2011). Less memory was required than one might think as optical properties were unique only for RXS profiles and each off-nadir profile was indexed to its RXS donor. While photons showered entire domains uniformly, the outer 5 km in the across-track direction and 500 km at the along-track ends were omitted from the analysis. The omitted portions did, however, serve in the 3D simulations as *buffer-zones* that at once shielded the inner-domain from potentially adverse effects set-up by cyclic horizontal boundary conditions and facilitated horizontal transport of photons through the sides of the inner-domain.

Simulations were performed using solar geometry at satellite time-of-crossing. As such, solar azimuth angle relative to A-train ground-track $\varphi_0(i, j)$ and solar zenith angle $\theta_0(i, j)$ were set to values reported in the merged A-train dataset. Photons were injected uniformly across constructed domains and given an initial weight of $\mu_0(i, j) = \cos\,\theta_0(i, j)$.

6 Results

Two lines of investigation were explored and these are discussed here. First, TOA reflected fluxes inferred from CERES radiances were compared to their simulated counterparts which were obtained by applying the 3D radiative transfer model to constructed A-train domains. Second, TOA reflected solar fluxes obtained by running the radiative transfer model in 3D mode were compared to results obtained by running it as a 1D model.

6.1 Reflected TOA Fluxes: CERES Versus Modelled

Comparison of CERES quantities to modelled values constitutes an attempt to perform a radiative closure experiment as CERES data were not used in the retrieval process. Figure 4 shows the CloudSat quick-look image for granule 999. While the constructed 3D domain was 14,000 km long and 61 km wide, the dots on the track run from ~44°S to ~72°N and demarcate the portion of the orbit considered for comparison.

Figure 5 shows TOA fluxes inferred from CERES radiances (Loeb et al. 2005, 2006) as well as model estimates for the constructed domain. Each value corresponds to a (20 km^2) CERES pixel. 256×10^6 photons were injected over the full domain making for $\sim 120 \times 10^3$ per CERES footprint with maximum Monte Carlo noise errors for TOA reflectance of ~0.002. Visually, the agreement is strikingly good for the most part, and encouraging too given the lack of synergistic retrieval and relatively little attention paid to ice scattering properties and surface albedo and no attention given to aerosols.

Figure 6 shows the cumulative histogram of differences between modelled–CERES fluxes (neglecting values between 50°N and 60°N where some cloud-related data were missing). Only about 30% of the differences fall within the ± 10 W m^{-2} goal of EarthCARE (ESA 2001; Domenech et al. 2011) and the bias for this particular orbit is -15 W m^{-2}. This large bias, however, arises primarily from the near-cloudless, high-Sun cases between 5°S and 25°N. Underestimation of albedo in this region could result from undetected small clouds (that were not included in the model), systematically too small ocean albedo (stemming from underestimated surface wind speeds) or neglect of aerosol. The plot also shows the cumulative histogram of differences between modelled–CERES nadir radiances scaled by π in order to be of similar magnitude to fluxes. The fact that histograms for fluxes and radiances are almost identical suggests that, for this orbit, there is no clear advantage to using either nadir radiances or fluxes to assess the retrievals. It is worth pointing out, however, that while radiances are a direct measurement CERES fluxes carry the additional error of radiance-to-flux conversion. It is expected that inclusion of off-nadir radiances, as in EarthCARE, will make for a more challenging assessment of retrievals.

Note also that measured radiances, and hence 1D MODIS retrievals, are subject to radiative smoothing and roughening (e.g., Marshak and Davis 2005b). Thus, when they are used at the 1 km scale to initialize a radiative transfer model, even a 3D model as here,

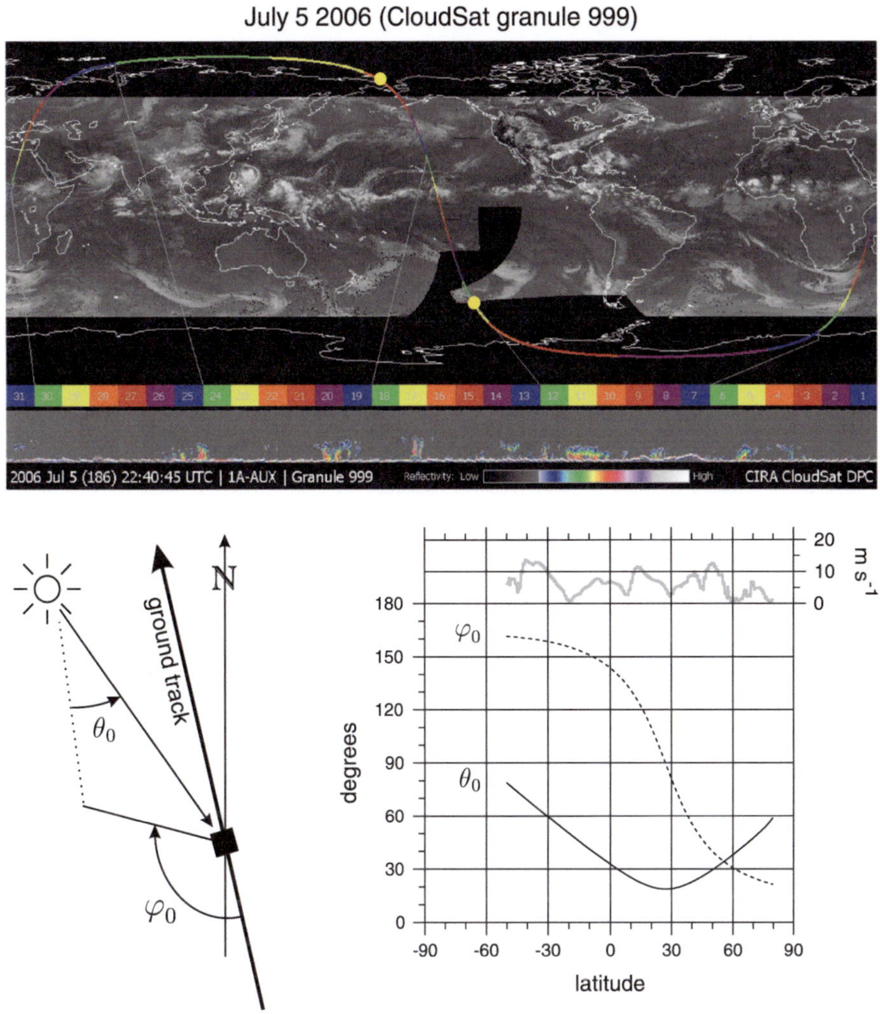

Fig. 4 Upper panel shows a CloudSat quick-look image for one of the orbits used here. The *line* indicates the satellite's ground-track and the *dots* on it mark the start (~44°S) to the end (~72°N) of the portion used for radiation calculations. Solar geometry is shown in the *lower left* while solar zenith θ_0 and azimuth φ_0 angles along with surface wind speed for the marked area on the quick-look image are shown on the graph

secondary smoothing and roughening takes place. This undoubtedly introduces errors into the comparison that are far from straightforward to estimate. The hope is that column properties retrieved by a robust synergistic algorithm will rely less on passive retrievals and thereby reduce smoothing/roughening errors. There is also the possibility that an expanded rendition of the construction algorithm might partially address the effects of horizontal transport (cf., Barker et al. 2011).

Before leaving this section it is interesting to give an example that goes well beyond the initial steps of radiative closure using domain averages and consider the much more demanding case of imagery reconstruction. Figure 7 shows a 50 × 150 km domain of MODIS cloud optical depth τ and its construction algorithm counterpart. Next to this is the

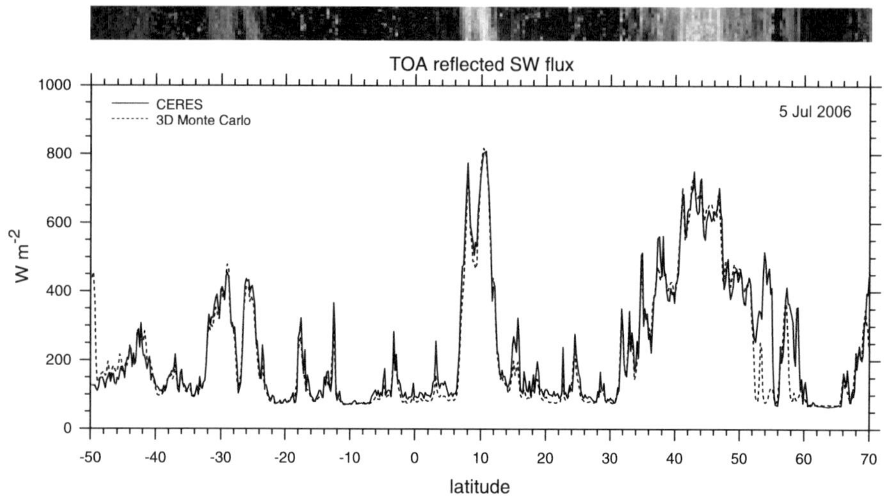

Fig. 5 CERES and 3D Monte Carlo estimates of TOA reflected solar fluxes for the marked portion of the orbit shown in Fig. 4. The 60-km-wide MODIS 0.645 μm image is shown along the *top*

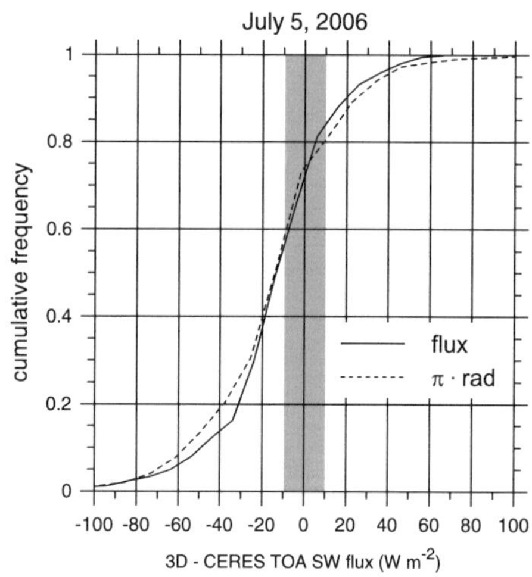

Fig. 6 Cumulative frequency distributions of (3D model–CERES) TOA reflected solar fluxes for the data shown in Fig. 5. Also shown is the distribution of differences between CERES nadir radiance and 3D Monte Carlo nadir radiances (multiplied, for plotting purposes, by π so as to put them on roughly the same scale as fluxes)

corresponding 0.645 μm MODIS image and nadir reflectances produced by the 3D model using constructed τ. While the reconstructed radiances are not perfect, the main features were captured well and will only be that much better once full synergistic retrieval comes online.

6.2 Reflected TOA Flux Differences: 1D Versus 3D

So, while a serious attempt at radiative closure using the current A-train dataset would be premature, on account of known difficiencies in retrieved properties it is likely that cloud structural features are defined well enough to conduct a comparison between 1D and 3D mean solar fluxes for mesoscale-size domains. In so far as the RXSs and their 3D constructed counterparts capture realistic cloud structure down to the 1 km scale with

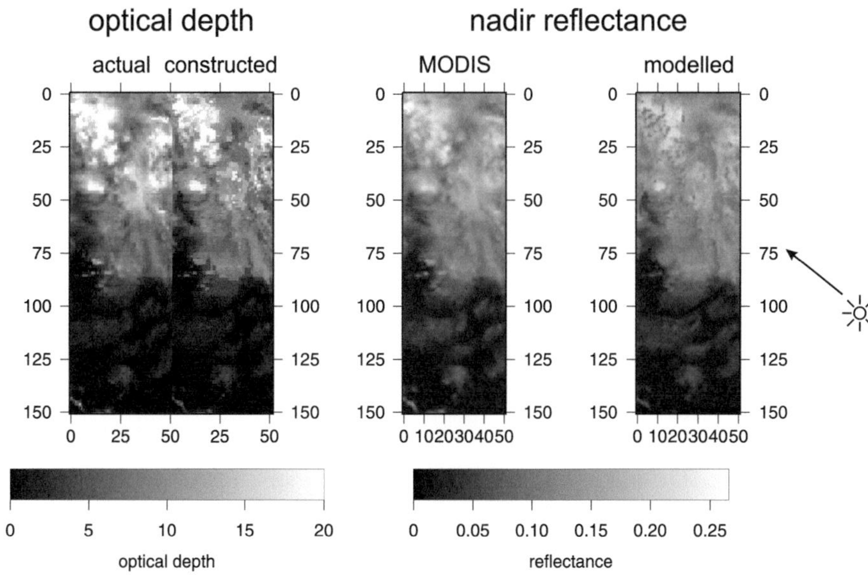

Fig. 7 Image on the *extreme left* (actual) shows cloud visible optical depths inferred by applying an inverse 1D RT model to MODIS radiances for a 50 × 150 km scene. Next to it is its 3D constructed version. To their *right* are MODIS radiances at 0.645 μm, and next to it corresponding radiances from the 3D RT model using the constructed field. Solar zenith angle was 40° and the Sun was coming in from the direction shown

sufficient accuracy, results presented here constitute a fair assessment of radiometric errors due to neglect of 3D solar transport.

When comparing domain-average characteristics of 1D and 3D radiative transfer it is common to employ standalone domains that are assuredly isolated in space via cyclic horizontal boundary conditions. This is because one knows from the start that all photons either get absorbed by the domain's atmosphere or underlying surface, or exit through the top. An arbitrarily delineated domain in the real atmosphere, however, is not isolated, cyclic boundary conditions do not apply and a clean comparison between 1D and 3D transfer becomes complicated—the more so as the domain's horizontal extent diminishes. For this reason, particular attention was paid in these experiments to fluxes through the vertical sides of domains.

Results for each simulation were averaged into 50 × 50 km sub-domains, a nominal area representing weather and climate model grid-cells. As each simulation used 256×10^6 photons, each sub-domain received $\sim 750 \times 10^3$ photons. This translates into very small Monte Carlo errors for domain-average fluxes and heating rates, ~2.5 times smaller than those for results presented in the previous sub-section.

Results were partitioned into sub-domains with total cloud fractions A_c: (i) $0.05 < A_c < 0.3$; (ii) $0.3 \le A_c \le 0.7$; and (iii) $A_c > 0.7$. This is because previous aforementioned studies have suggested that cloud structural effects should maximize at intermediate cloud amounts and taper off for larger and smaller A_c. Data from five orbits were used and the number of 50 × 50 km sub-domains for the three ranges of A_c were: (i) 205; (ii) 246; and (iii) 852, for a total of 1,303 sub-domains.

Figure 8 shows flux differences (3D–1D) as functions for μ_0 for the three ranges of A_c. For small cloud fractions the only clear signal is greater reflected flux at TOA when 3D transfer was considered. The same trend is visible for $A_c \in [0.3,0.7]$. This is the well-documented effect of cloud-side illumination for $\mu_0 \lesssim 0.8$ (Welch and Wielicki 1985). Conversely, at large μ_0 there is a slight tendency for 3D transfer to produce smaller reflectances due to photons emerging from cloud-sides in predominantly downwelling directions. Additionally for $A_c \in [0.3,0.7]$ there are trends for larger atmospheric absorption and weaker surface absorption (irradiance) by 3D transfer as μ_0 decreases. Again, this is due to increased cross-sectional area of cloud presented to direct-beam as μ_0 decreases. The vast majority of differences for TOA reflected and surface absorption are smaller than ± 20 W m^{-2}; for atmospheric absorption, differences are largely confined to ± 5 W m^{-2}.

When cloud fractions become greater than 0.7, the dominant feature is a large scatter of positive and negative differences that cluster around zero with a very slight tendency for 3D transfer to yield smaller reflectances by about 3 W m^{-2} and larger atmospheric absorption by ~ 2 W m^{-2}. These values are not too dissimilar to those obtained by Cole

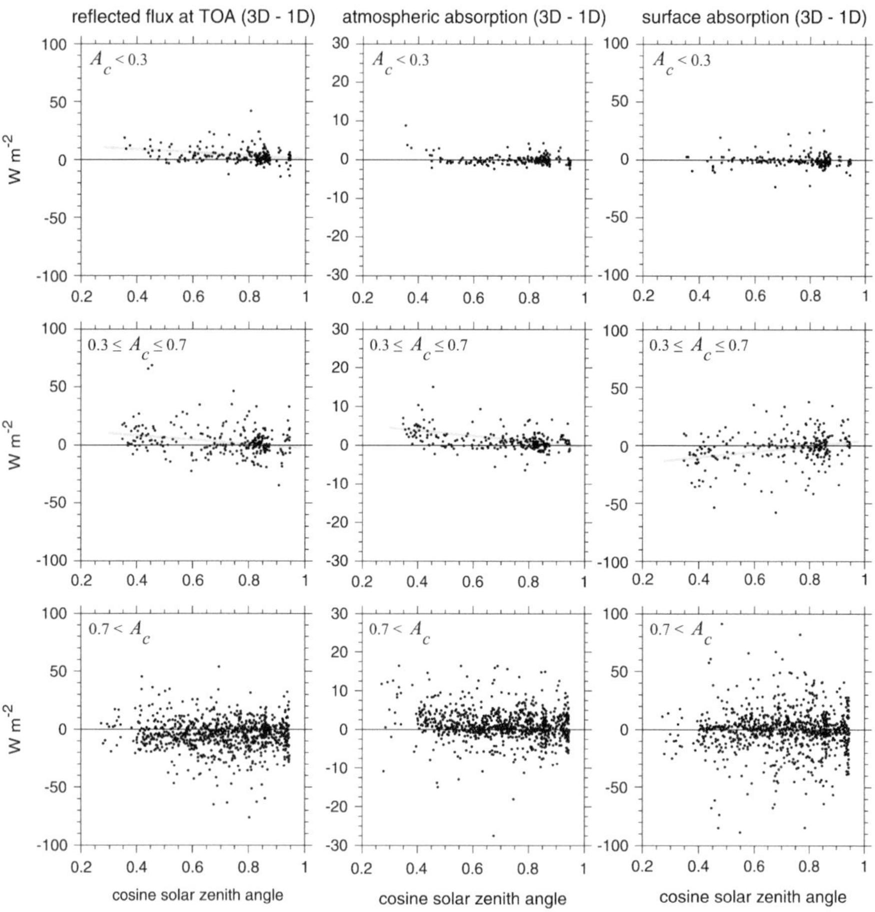

Fig. 8 Differences between fluxes predicted by 3D and 1D transfer for (50 km^2) scenes constructed from A-train data. *Left column* is for reflected flux at TOA, *centre* is for atmospheric absorption, while the *rightmost* is for surface absorption. *Each row* corresponds to scenes with A_c as listed on the plots. *Grey lines* are least squares fits, with the correlation coefficient r > 0.1

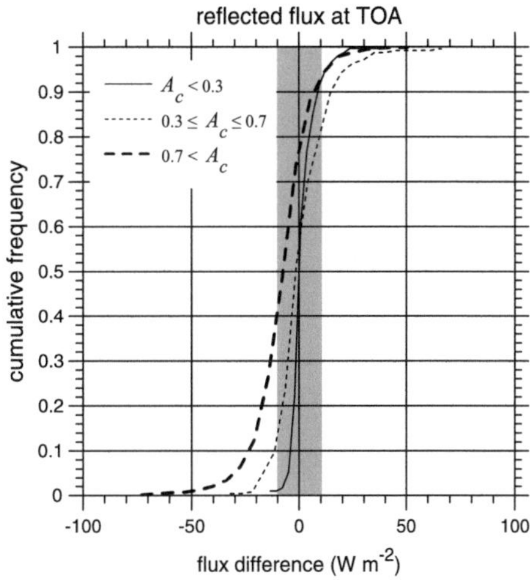

Fig. 9 Cumulative frequency distributions for differences in reflected solar flux at TOA as predicted by a 3D transport model and its 1D counterpart for 50 × 50 km constructed scenes (i.e. the values shown on the *left column* of Fig. 8)

et al. (2005b) using 2D model atmospheres, and possibly stem from photons undergoing internal reflections between multiple cloud layers thereby increasing photon pathlengths, number of scattering events, and hence absorption. As μ_0 decreases, the direct beam can undercut layers of clouds when 3D transfer is admitted; for 1D transfer, however, photons must diffuse through upper clouds before undergoing internal reflections.

Figure 9 shows cumulative frequency distributions of differences in reflected flux at TOA predicted by 3D and 1D radiative transfer for all sub-domains partitioned into the three cloud fraction ranges. Median values for $A_c < 0.7$ are very close to zero but for cloudier scenes it is -8 W m^{-2}. It can be expected that errors in TOA reflectance will exceed ±10 W m^{-2} for 10–30% of scenes having $A_c < 0.7$ if 1D rather than 3D transfer is used. This rate jumps to ~50% of scenes with $A_c > 0.7$. Preliminary results suggest that these percentages increase by 15–20% when smaller domains the size of CERES footprints are considered. The implication therefore for EarthCARE is that 30–65% of its constructed scenes can expect errors in computed reflected short wave (SW) flux at TOA to exceed its ±10 W m^{-2} target if 1D rather than 3D transfer is adhered to. More dramatic errors in excess of ±50 W m^{-2} appear as though they will be very rare regardless of cloud conditions.

Figure 10 shows surface and atmospheric absorption histograms that correspond to those in Fig. 9. Regarding surface absorption, the most interesting point is that errors due to use of 1D transfer are often large (i.e. greater than 20 W m^{-2}) for $A_c > 0.3$ and are dominated by overestimation relative to 3D values. This stems from no illumination of cloud-sides in 1D calculations, especially apparent for $A_c \in [0.3, 0.7]$. Biases in total atmospheric absorption due to use of 1D models appear to be of secondary importance for (50 km^2) domains as they rarely exceed ±10 W m^{-2}. Naturally more extreme localized differences exist at smaller scales but they are largely averaged out at 50 km.

6.3 Horizontal Transport Through Domain Sides

For a column of atmosphere associated with a 50 × 50 km sub-domain, let R, T and A be net fluxes out the top, at the surface and absorbed in the column. Since the

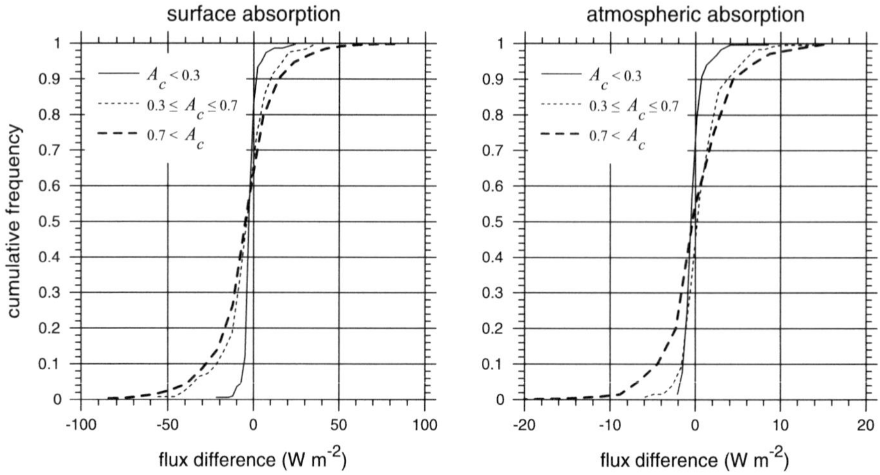

Fig. 10 As in Fig. 9 except these distributions are for atmospheric and surface absorption (i.e. data shown in the *centre* and *rightmost columns* of Fig. 8)

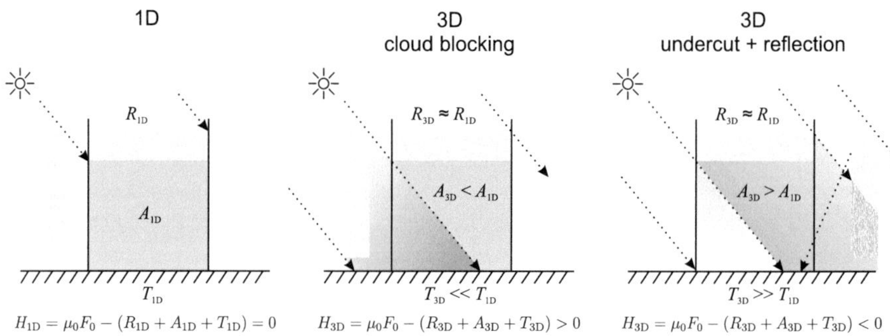

Fig. 11 Schematic diagram illustrating horizontal transport H of solar radiation. The *column* of concern is in the *centre* of each figure and consists of an overcast cloud. R, A and T denote reflectance, atmospheric and surface absorptance, respectively. *Left figure* shows 1D results where there is no transfer through the column's vertical sides and so $H_{1D} = 0$. The other figures show the impact on H_{3D} due to clouds outside the column of concern

sub-domains considered here were effectively standalone and not subject to cyclic boundary conditions,

$$R + T + A \neq \mu_0 F_0 \tag{6}$$

holds in general where F_0 is incident flux on a normal surface at the TOA. Horizontal transport of radiation through the sides of a sub-domain can be defined as

$$H = \mu_0 F_0 - (R + T + A). \tag{7}$$

Marshak and Davis (2005b) provided an in-depth discussion of H for solar transfer through cloudy atmospheres with (7) corresponding to their (12.13). Figure 11 illustrates how to interpret (7). For 1D transfer (6) is an equality and so $H = 0$. For 3D transfer, however, if

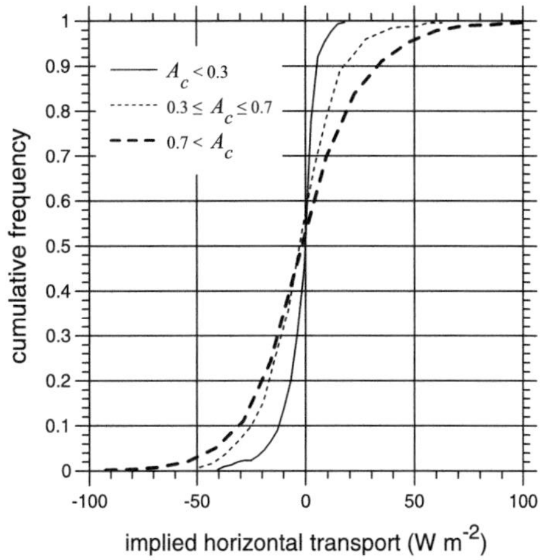

Fig. 12 As in Fig. 9 except these distributions are for implied horizontal transport of radiation as defined in (7)

the shadow of a cloud outside the sub-domain is cast into the sub-domain, but the majority of reflected photons come from, for example, a large planar cloud aloft (so that $R_{1D} \simeq R_{3D}$), $H > 0$ which might seem odd given that transport through the column's sides is less than it would have been had the external cloud been absent. If the external cloud is now moved so that direct-beam radiation undercuts the layer cloud and some radiation gets scattered by the external cloud back into the sub-domain, $H < 0$. Since non-zero H can arise through differences between 1D and 3D values of R, T or A as well as actual horizontal flux through vertical boundaries, perhaps it is best to think of H as implied horizontal flux.

Figure 12 shows histograms of H for the three classes of sub-domains using data from the five orbits considered here. Values of $|H| \gtrsim 50$ W m^{-2} occur for $\sim$10% of the scenes with $A_c > 0.7$ and point directly to the potential importance of 3D effects on the radiation budget of atmospheric volumes resembling those of global model grid-cells. Although the scale at which H was evaluated here (50×50 km domains) and the nature of the domains differ much from those assessed by Marshak and Davis (2005b), values are similar to those they arrive at (cf. their Fig. 12.4). When sub-domain size is reduced to that of a CERES footprint, distributions of H for cases with $A_c \in [0.3,0.7]$ and $A_c > 0.7$ resemble one another closely but the fractions of occurrences of $|H| \gtrsim 50$ W m^{-2} increase to roughly 20%. Again then, it appears to be crucial that 3D solar transfer be utilized by the EarthCARE mission if it is to make a serious go at both estimating surface-atmosphere radiation budgets and assessing the quality of active-passive retrievals by demanding that simulated TOA fluxes be within $\pm$10 W m^{-2} of values inferred from its broadband radiometer measurements for (10 km^2) scenes (Domenech et al. 2011), especially for broken cloud scenarios.

Finally, Fig. 13 shows sub-domain average heating rates computed by the 3D model for a 13,000 km stretch of the orbit on 5 July 2006. It also shows percentage differences between 3D and 1D transfer. The thing to point out is that differences rarely exceed $\pm$20%. When they do, however, the bias persists from the surface to the base of those clouds that are responsible for maximum heating. Once a larger number of orbits get processed the intention is to examine heating rates for scenes sorted according to various cloud properties including cloud-type (e.g., Xu et al. 2005; Bacmeister and Stephens 2010).

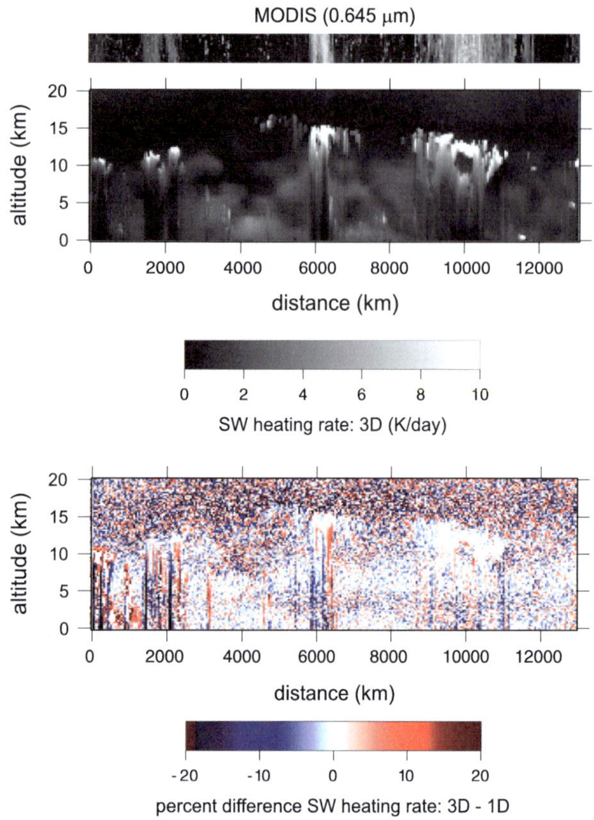

Fig. 13 Beneath the 60-km-wide visible MODIS image are solar heating rate profiles computed by the 3D Monte Carlo model using 50 × 50 km sections of the 13,000-km-long constructed domain for the orbit shown in Fig. 4. The *lower panel* shows percentage differences between solar heating rates predicted by the 3D and 1D transport models

7 Conclusions and Recommendations

The primary purpose of this study was to begin to get a global picture of errors in estimates of solar fluxes, for cloudy domains roughly the size of those found in global models, that are incurred by applying the 1D independent column approximation (ICA). Both the ICA and 3D benchmark values were obtained by applying a Monte Carlo photon transport model to cloud fields constructed from A-train satellite data. The results are of relevance to both global models that use ICA-based methods to compute mean radiative fluxes (e.g., Randall et al. 2003; Morcrette et al. 2008; Shonk and Hogan 2008) and to radiative closure studies.

3D cloud scenes were created from 2D A-train satellite data via application of the Barker et al. (2011) 3D scene construction algorithm. Their algorithm was developed for the EarthCARE mission which intends to perform continuous radiative closure experiments to assess the quality of its active-passive retrievals. As many A-train cloud profiles consisted of a cloud mask only, profiles of cloud properties had to be allocated subject to constraint by MODIS retrievals. Based on preliminary results using cloud system-resolving model (CSRM) data, it appears that the simple technique used here slightly overestimates differences between 1D and 3D transfer models. Hence, one way view of the results is that they represent an approximate upper-bound for differences between 1D and 3D solutions.

Other limitations of this study were its: inability to address impacts of horizontal fluctuations in cloud extinction at scales smaller than ~ 1 km; treatment of scattering by ice crystals as though they were liquid spheres; and neglect of aerosols. It did, however, include important fluctuations at scales larger than 1 km (e.g., Zuidema and Evans 1998), and according to simulations by Cole et al. (2005b), cloud albedo depends weakly on horizontal resolution for grid-spacings less than ~ 2 km. The general impression, therefore, was that 3D cloud structural features, plus some information pertaining to microphysical variations, were inferred well enough from A-train data to begin setting the stage for a global statistical analysis of differences between 1D and 3D solar transfer. The quality of this comparison is expected to improve once greater attention is given to ice cloud, aerosol and surface optical properties, and synergistic active-passive retrieval methods mature (e.g., Hogan et al. 2006).

A Monte Carlo transport algorithm was applied to constructed scenes using A-train time-of-crossing values of cosine of solar zenith angle μ_0. Calculations were performed on domains measuring 61 $\times$ 14,000 km and results were partitioned into 1303 50 $\times$ 50 km sub-domains which nominally represent the size of cells found in conventional global models. In corroboration with previous studies, 3D–1D differences are largest for sub-domains with intermediate values of total cloud fractions A_c, 3D transfer leading to larger reflectances at TOA and atmospheric absorption, and smaller surface irradiances. Results were, however, somewhat dependent on μ_0 with mean differences being almost zero at large μ_0 and about ~ 10 W m^{-2} for TOA reflectance, -10 W m^{-2} for surface irradiance, and ~ 5 W m^{-2} for atmospheric absorption for $\mu_0 \simeq 0.2$. Similarly, as shown for a 13,000 km stretch of a single orbit, CERES-inferred and model-predicted solar TOA fluxes agreed to with ± 10 W m^{-2} about 30% of the time. Similar values were found for other orbits.

More so than biases the most defining characteristic of 3D–1D flux differences, regardless of both μ_0 and A_c, was large *random* variations. For instance, for $A_c > 0.3$ it was not unusual to find surface irradiance differences exceeding ± 30 W m^{-2}. This scatter of results even occurs for near-overcast conditions. The reason for this may be related to multiple layers of clouds that lead to larger photon pathlengths, and thus slightly larger atmospheric absorption, but can sway reflectance and surface irradiance deviations one way or another.

It is difficult to say whether deviations of these magnitudes will have significant ramifications on simulations of clouds, weather and climate. The few experiments with CSRMs that have attempted to explore deviations like these suggest not (Mechem et al. 2008; Cole et al. 2005a; Pincus and Stevens 2009). But those tests were far from exhaustive. More telling, perhaps, are the Monte Carlo Independent Column Approximation (McICA) experiments with large-scale models (Pincus et al. 2003; Barker et al. 2008; Morcrette et al. 2008). In those studies the amount of radiative noise associated with random sampling of clouds far exceeded that shown in Fig. 8. Nevertheless, there was very little impact on the evolution of simulations.

What these results do suggest is that in order to limit errors in radiative closure studies, regardless of whether TOA fluxes or radiances get used, one should at least conditionally apply a 3D solar radiative transfer algorithm to retrieved/constructed scenes that exhibit highly variable cloud. This is especially so if one seeks to assess retrieval performance for small individual scenes using solar measurements. This is indeed the plan for the

EarthCARE mission which, as far as the authors are aware, will represent the first attempt in the atmospheric sciences to employ a 3D radiative transfer model operationally.

Acknowledgments This study was supported by a contract issued by the European Space Agency (ESA). Part of the work in this study used the facilities of the Shared Hierarchical Academic Research Computing Network (SHARCNET:http://www.sharcnet.ca). Thanks are extended to two anonymous reviewers for helpful and constructive comments.

References

Bacmeister JT, Stephens GL (2011) Spatial statistics of likely convective clouds in CloudSat data. J Geophys Res 116:D04104. doi:10.1029/2010JD014444

Barker HW, Davies JA (1992) Solar radiative fluxes for broken cloud fields above reflecting surfaces. J Atmos Sci 49:749–761

Barker HW (1992) Solar radiative transfer for clouds possessing isotropic variable extinction coefficient. Q J R Meteorol Soc 118:1145–1162

Barker HW et al (2003) Assessing 1D atmospheric solar radiative transfer models: interpretation and handling of unresolved clouds. J Clim 16:2676–2699

Barker HW, Morcrette J-J, Alexander GD (1998) Broadband solar fluxes and heating rates for atmospheres with 3D broken clouds. Q J R Meteorol Soc 124:1245–1271

Barker HW, Stephens GL, Fu Q (1999) The sensitivity of domain-averaged solar fluxes to assumptions about cloud geometry. Q J R Meteorol Soc 125:2127–2152

Barker HW, Cole JNS, Morcrette J-J, Pincus R, Räisänen P, von Salzen K, Vaillancourt PA (2008) The Monte Carlo independent column approximation: an assessment using several global atmospheric models. Q J R Meteorol Soc 134:1463–1478

Barker HW, Jerg MP, Wehr T, Kato S, Donovan DP, Hogan RJ (2011) A 3D cloud-construction algorithm for the EarthCARE satellite mission. Q J R Meteorol Soc, doi:10.1002/qj.824

Bloom SA et al (2005) Documentation and validation of the Goddard earth observing system (GEOS) data assimilation system version-4. NASA Technical Report. Series on global modeling and data assimilation NASA/TM-2005-104606, vol 26, 181 pp

Cahalan RF, Ridgway W, Wiscombe WJ, Bell TL, Snider JB (1994) The albedo of fractal stratocumulus clouds. J Atmos Sci 51:2434–2460

Cahalan RF, 26 others (2005) The I3RC: bringing together the most advanced radiative transfer tools for cloudy atmospheres. Bull Am Met Soc 86:1275–1293

Chandrasekhar S (1950) Radiative transfer. Reprinted by Dover Publications (1960). Oxford University Press, New York

Cole JNS, Barker HW, Randall DA, Khairoutdinov MF, Clothiaux EE (2005) Global consequences of interactions between clouds and radiation at scales unresolved by global climate models. Geophys Res Lett 32:L06703. doi:10.1029/2004GL020945

Cole JNS, Barker HW, O'Hirok W, Clothiaux EE, Khairoutdinov MF, Randall DA (2005) Atmospheric radiative transfer through global arrays of 2D clouds. Geophys Res Lett 32:L19817, doi: 10.1029/2005GL023329

Cox C, Munk W (1956) Slopes of the sea surface deduced from photographs of the sun glitter. Bull Scripps Inst Oceanog 6:401–488

Davies R (1978) The effect of finite geometry on the three-dimensional transfer of solar irradiance in clouds. J Atmos Sci 35:1712–1725

Davis A, Gabriel P, Lovejoy S, Schertzer D, Austin G (1990) Discrete angle radiative transfer—Part III: Numerical results and applications. J Geophys Res 95:11729–11742

Dobbie JS, Li J, Chylek P (1999) Two and four stream optical properties for water clouds at solar wavelenghts. J Geophys Res 104:2067–2079

Domenech C, Lopez-Baeza E, Donovan DP, Wehr T (2011) Radiative flux estimation from a broadband radiometer using synthetic angular models in the EarthCARE Mission framework. Part I: Methodology. J Appl Met Clim 50:974–993

Dupont J-C, Haeffelin M, Morille Y, Comstock JM, Flynn C, Long CN, Sivaraman C, Newson RK (2011) Cloud properties derived from two lidars over the ARM SGP site. Geophys Res Lett 38, doi: 10.1029/2010GL046274
Ellingson RG (1982) On the effects of cumulus dimensions on longwave irradiance and heating rate calculations. J Atmos Sci 39:886–896
ESA (2001) The five candidate earth explorer missions: EarthCARE—earth clouds, aerosols and radiation explorer, ESA SP-1257(1), September 2001, 130 pp. Prepared by J. P. V. Poiares-Baptista
Fu Q (1996) An accurate parameterization of the solar radiative properties of cirrus clouds for climate models. J Clim 9:2058–2082
Fu Q, Yang P, Sun WB (1998) An accurate parameterization of the infrared radiative properties of cirrus clouds for climate Models. J Clim 11:2223–2237
Hinkelman LM, Stevens B, Evans KF (2005) A large-eddy simulation study of anisotropy in fair-weather cumulus cloud fields. J Atmos Sci 62:2155–2171
Hogan RJ, Donovan DP, Tinel C, Brooks MA, Illingworth AJ, Poiares Baptista JPV (2006) Independent evaluation of the ability of spaceborne radar and lidar to retrieve the microphysical and radiative properties of ice clouds. J Atmos Oceanic Technol 23:211–227
Kobayashi T (1988) Parameterization of reflectivity for broken cloud fields. J Atmos Sci 45:3034–3045
Kato S, Rose FG, Charlock TP (2005) Computation of domain-averaged irradiance using satellite-derived cloud properties. J Atmos Oceanic Technol 22:146–164
Kato S, Sun-Mack S, Miller WF, Rose FG, Chen Y, Minnis P, Wielicki BA (2010) Relationships among cloud occurrence frequency, overlap, and effective thickness derived from CALIPSO and CloudSat merged cloud vertical profiles. J Geophys Res 115:D00H28. doi:10.1029/2009JD012277
Lindner TH, Li J (2000) Parameterization of the optical properties for water clouds in the infrared. J Clim 13:1797–1805
Loeb NG, Kato S, Loukachine K, Smith NM (2005) Angular distribution models for top-of-atmosphere radiative flux estimation from the clouds and the earth's radiant energy system instrument on the Terra satellite. Part I: Methodology. J Atmos Oceanic Technol 22:338–351
Loeb NG, Kato S, Loukachine K, Smith NM (2006) Angular distribution models for top-of-atmosphere radiative flux estimation from the clouds and the earth's radiant energy system instrument on the Terra satellite. Part II: validation. J Atmos Oceanic Technol 24:564–584
Marshak A, Davis AB, Wiscombe WJ, Cahalan RF (1995) Radiative smoothing in fractal clouds. J Geophys Res 100:26247–26261
Marshak A, Davis AB (2005) Scale-by-scale analysis and fractal cloud models. In: Marshak A, Davis AB (eds) 3D radiative transfer in cloudy atmospheres. Springer, Heidelberg
Marshak A, Davis AB (2005) Horizontal fluxes and radiative smoothing. In: Marshak A, Davis AB (eds) 3D radiative transfer in cloudy atmospheres. Springer, Heidelberg
McKee TB, Cox SK (1974) Scattering of visible radiation by finite clouds. J Atmos Sci 31:1885–1892
Minnis P, Trepte QZ, Sun-Mack S, Chen Y, Doelling DR, Young DF, Spangenberg DA, Miller WF, Wielicki BA, Brown RR, Gibson SC, Geier EB (2008) Cloud detection in non-polar regions for CERES using TRMM VIRS and Terra and Aqua MODIS data. IEEE Trans Geosci Remote Sens 46:3857–3884
Minnis P, Sun-Mack S, Young DF, Heck PW, Garber DP, Chen Y, Spangenberg DA, Arduini RF, Trepte QZ, Smith WL Jr., Ayers JK, Gibson SC, Miller WF, Chakrapani V, Takano Y, Liou K-N, Xie Y (2010a) CERES Edition-2 cloud property retrievals using TRMM VIRS and Terra and Aqua MODIS data, Part I: Algorithms. Submitted to IEEE Trans Geosci Remote Sens
Minnis P, Sun-Mack S, Chen Y, Khaiyer MM, Yi Y, Ayers JK, Brown RR, Dong X, Gibson SC, Heck PW, Lin B, Nordeen ML, Nguyen L, Palikonda R, Smith WL Jr, Spangenberg DA, Trepte QZ, Xi B (2010b) CERES Edition-2 cloud property retrievals using TRMM VIRS and Terra and Aqua MODIS data, Part II: Examples of average results and comparisons with other data. Submitted to IEEE Trans Geosci Remote Sens
Morcrette J-J, Barker HW, Cole JNS, Iacono MJ, Pincus R (2008) Impact of a new radiation package, McRad, in the ECMWF integrated forecasting system. Mon Wea Rev 136:4773–4798
O'Hirok W, Gautier C (1998) A three-dimensional radiative transfer model to investigate the solar radiation within a cloudy atmosphere. Part I: Spatial effects. J Atmos Sci 55:2162–2179
Pincus R, Barker HW, Morcrette J-J (2003) A new radiative transfer model for use in GCMs. J Geophys Res 108(D13):4376, doi:10.1029/2002JD003322
Pincus R, Stevens B (2009) Monte Carlo spectral integration: a consistent approximation for radiative transfer in large eddy simulations. J Adv Mod Earth Syst 1, doi:10.3894/JAMES.2009.1.1
Protat A, Delanoë J, Bouniol D, Heymsfield AJ, Bansemer A, Brown P (2007) Evaluation of ice water content retrievals from cloud radar reflectivity and temperature using a large airborne in situ microphysical database. J Appl Met Clim 46:557–572

Räisänen P, Isaac GA, Barker HW, Gultepe I (2003) Solar radiative transfer for stratiform clouds with horizontal variations in liquid water path and droplet effective radius. Q J R Meteorol Soc 129:2135–2149
Randall D, Khairoutdinov M, Arakawa A, Grabowski W (2003) Breaking the cloud parameterization deadlock. Bull Am Met Soc 84:1547–1564
Rutan D, Rose F, Roman M, Manalo-Smith N, Schaaf C, Charlock T (2009) Development and assessment of broadband surface albedo from clouds and the earth's radiant energy system clouds and radiation swath data product. J Geophys Res 114:D08125. doi:10.1029/2008JD010669
Sassen K, Wang Z (2008) Classifying clouds around the globe with the CloudSat radar: 1-year of results. Geophys Res Lett 35, doi:10.1029/2007GL032591.
Scinocca JF, McFarlane NA, Lazare M, Li J (2008) The CCCma third generation AGCM and its extension into the middle atmosphere. Atmos Chem Phys 8:7055–7074
Shonk JKP, Hogan RJ (2008) Tripleclouds: an efficient method for representing horizontal cloud inhomogeneity in 1D radiation schemes by using three regions at each height. J Clim 21:2352–2370
Stephens GL (1988) Radiative transfer through arbitrary shaped optical media. I: A general method of solution. J Atmos Sci 45:1818–1836
Stephens GL, Austin RT, Benedetti A, Boain R, Durden SL, Illingworth A, Mace G, Miller SD, Mitrescu C, O'Conner E, Rossow WB, Sassen K, Vane DG, Wang Z (2002) The CloudSat mission and the EPS constellation: a new dimension of space-based observations of clouds and precipitation. Bull Am Met Soc 83:1771–1790
Welch RM, Wielicki BA (1984) Stratocumulus cloud field reflected fluxes: the effect of cloud shape. J Atmos Sci 41:3085–3103
Welch RM, Wielicki BA (1985) A radiative parameterization of stratocumulus cloud fields. J Atmos Sci 42:2888–2897
Wielicki BA, Barkstrom BR, Harrison EF, Lee RB III, Smith LG, Cooper JE (1996) Mission to planet earth: role of clouds and radiation in climate. Bull Am Meteorol Soc 77:853–868
Winker DM, Hunt WH, McGill MJ (2007) Initial performance assessment of CALIOP. Geophys Res Lett 34, doi:10.1029/2007GL030135
Xu K-M, Wong T, Wielicki BA, Parker L, Eitzen ZA (2005) Statistical analyses of satellite cloud object data from CERES. Part I: Methodology and preliminary results of 1998 El Niño/2000 La Niña. J Clim 18:2497–2514
Yang P, Zhang L, Hong G, Nasiri SL, Baum BA, Huang H-L, King MD, Platnick S (2007) Differences between collection 4 and 5 MODIS ice cloud optical/microphysical products and their impact on radiative forcing simulations. IEEE Trans Geosci Remote Sens 45:2886–2899
Zuidema P, Evans KF (1998) On the validity of the independent pixel approximation for the boundary layer clouds observed during ASTEX. J Geophys Res 103:6059–6074

Surv Geophys (2012) 33:677–691
DOI 10.1007/s10712-011-9163-x

The Clouds of the Middle Troposphere: Composition, Radiative Impact, and Global Distribution

Kenneth Sassen · Zhien Wang

Received: 29 June 2011 / Accepted: 14 November 2011 / Published online: 8 December 2011

Abstract The clouds of the middle troposphere span the temperature range where both ice and liquid water in a supercooled state can exist. However, because one phase tends to dominate, of the two midlevel cloud types, altostratus are deep ice-dominated, while altocumulus are shallow water-dominated, mixed-phase clouds with ice crystal virga typically trailing below. Multiple remote sensor examples of these cloud types are given to illustrate their main features, and the radiative consequences of the different cloud microphysical compositions are discussed. Spaceborne radar and lidar measurements using the CloudSat and CALIPSO satellites are analyzed to determine the global distributions of cloud frequencies and heights of these clouds. It is found that together these little-studied clouds cover ~25% of the Earth's surface, which is about one-third of the total cloud cover, and thus represent a significant contribution to the planet's energy balance.

Keywords Tropospheric clouds · CloudSat · CALIPSO · Lidar · Radar

1 Introduction

Clouds play a critical role in maintaining the energy balance of the Earth via their redistribution of the incoming solar radiation and outgoing infrared radiation (Stephens et al. 1990; Wielicki et al. 1995; Stephens 2005). The factors that affect the flow of radiation through clouds are controlled by a number of things, mainly their geographical locations and heights, and their microphysical properties including cloud particle phase,

ISSI Workshop on *Observing and Modeling Earth's Energy Flows*, January 10–14, 2011, Bern, Switzerland.

K. Sassen (✉)
Geophysical Institute, University of Alaska Fairbanks, Fairbanks, AK, USA
e-mail: ksassen@gi.alaska.edu

Z. Wang
Department of Atmospheric Science, University of Wyoming, Laramie, WY, USA
e-mail: ZWang@uwyo.edu

shape, and number concentration. These latter properties determine their basic radiative consequences as can be expressed by their liquid or ice water path, mean particle size, and visible and infrared optical depths. It is apparent then that the different types of clouds inhabiting the Earth's atmosphere can result in quite different radiative forcings (Hartmann et al. 1992).

The clouds that are the subject of this report are those of the middle troposphere, which because of their intermediate heights in the atmosphere are typically formed at subfreezing temperatures. However, liquid water drops can exist in a supercooled state down to the homogenous freezing point of pure water, or to temperatures approaching about −40°C, while ice crystals can form in these clouds either homogeneously or heterogeneously through the action of special suspended particles, ice nuclei (Pruppacher and Klett 1997). Some ice nuclei, like desert dust particles, start to become activated within a few to several degrees below the freezing point, and the numbers activated increase strongly with decreasing temperature. Due to the thermodynamic advantage of the growth of the ice phase over that of supercooled water at the same temperature (due to the difference in their saturation vapor pressures), the strong competition between the two phases tends to result in the depletion of cloud droplets by ice before the homogeneous freezing point is reached. Thus, in practice, the coldest temperatures at which supercooled cloud droplets have been measured with in situ probes are between −35° and −36°C (e.g., Sassen et al. 1985). Middle-level clouds are therefore the only (nondeep convective) clouds that can be dominated by either the ice or water phase because of their strategic location in the atmosphere and are commonly of mixed-phase composition (Zhang et al. 2010). Partly as a result, they are poorly represented in climate models (Zhang 2005).

In response to the hypothesized global warming trend and other human-induced atmospheric alterations, what would be the roles played by the Earth's middle clouds? Certainly, the fact that they are often of mixed-phase composition means that any cloud layer temperature change will affect the balance of the two phases with a potentially large radiative impact (Sassen and Khvorostyanov 2007). Indirect aerosol effects on cloud content may be even more important because of potential ice nuclei concentration changes on midlevel cloud phase. An example of this is that increasing desertification and atmospheric dust content could lead to increased ice crystal numbers and glaciations of supercooled clouds.

In this paper, we first discuss and define the nature of the recognized types of middle-level clouds in our atmosphere, including by providing modern multiple remote sensor studies of the two basic midlevel cloud types, altostratus and altocumulus. Their basic radiative properties are then summarized. Next, we provide the latest data derived from the CALIPSO and CloudSat satellites using active remote sensors (lidar and radar) to characterize the global cloud frequencies and heights of these ubiquitous clouds, followed by our Conclusions.

2 Types of Middle-level Clouds

Classifications for the clouds of the Earth's atmosphere date back about two centuries, when natural philosophers trained mainly in classifying flora and fauna turned their attention to clouds (WMO 1987). For example, fibrous *cirrus* clouds were named after the fan-like appendages of barnacles used in food gathering. Because the family of cirrus clouds was acknowledged to occur high in the atmosphere, they were cold enough to be dominated by ice crystals. They often have a thin, bluish appearance through which the sun or moon is visible, produce halos, and comprise the leading edge of thickening cloud systems developing as part of cyclonic storm systems. Thus, the visual appearance and relative heights of clouds were important criteria in identifying the basic cloud types of the upper, middle, and lower atmosphere. Surface weather observers learn these fundamental distinctions aided by cloud atlas books (WMO 1987), which facilitated the first global climatologies of cloud type frequencies (Warren et al. 1986; Hahn and Warren 1999). Middle level means that these clouds correspond to the heights of the midtroposphere, which depend on which latitude belt (i.e. the polar regions, midlatitudes, and tropics) they are found in, because the average height of the tropopause progressively decreases with increasing latitude (WMO 1988). For example, as a rule of thumb, midlevel cloud bases span the altitude ranges of ~2.0 km to 7.0, 8.0, and 10 km above ground level in these belts, respectively. Figure 1 shows the breakdown of low, middle, and high cloud base heights for the three belts used in our current algorithm. Surface weather observers categorize cloud types according to their visually estimated cloud base heights, because cloud layers typically block viewing of the higher cloud tops.

These early attempts to classify clouds have generally proved to be well founded and retain their value in terms of modern concerns such as judging their radiative impacts according to cloud height, optical depth, and particle size, phase, and shape. More recently, satellite remote sensors illustrated the value of continuous global measurements for distinguishing general cloud types and predicting weather. The International Satellite Cloud Climatology Project (ISCCP) adopted a scheme using cloud categories divided into apparent height and cloud optical depth intervals derived from multispectral radiance data (Rossow and Schiffer 1999). Most recently, the availability of active remote sensors in Earth orbit, namely the polarization lidar on CALIPSO (Winker et al. 2003) and the W-band (3.2 mm) cloud radar on CloudSat that are part of the A-train formation of satellites (Stephens et al. 2003), has generated interest in improving the global characterization of clouds from space (Sassen and Wang 2008).

There are two categories of middle-level clouds that are both visually and radiatively distinct. Fundamentally, it is a question of cloud thermodynamic phase: altocumulus (Ac)

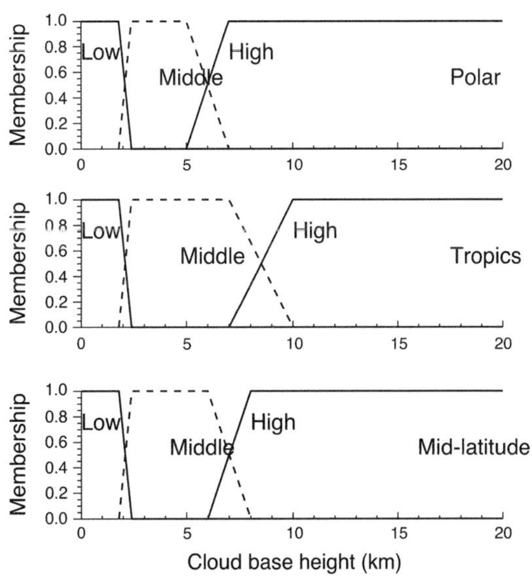

Fig. 1 The cloud base height (above ground level) distributions, or memberships, for low, middle, and high clouds in the three major latitude belts, used in the algorithm for classifying cloud types

are water-dominated, and altostratus (As) are ice-dominated. That is, although mixed water- and ice-phase conditions are common, the scattering of light by Ac is dominated by small spherical cloud droplets, while that by As is dominated by relatively large non-spherical ice crystals. Naturally, these scattering differences are reflected in the visual appearance of these cloud types. Ac tend to be cellular and relatively thin, so that they appear translucent (bluish) to gray to dark, while As are thick and dark enough to partially or totally block the disk of the sun or moon.

Below, we provide examples of the nature of these clouds as revealed by their scattering/emission properties measured in the visible, infrared, and microwave spectral regions.

2.1 Altostratus

Two recent examples of multiple remote sensor studies of As clouds collected at the Arctic Facility for Atmospheric Remote Sensing (AFARS) in the interior of Alaska (Sassen and Khvorostyanov 2008) are presented in Figs. 2 and 3 to illustrate their basic properties. From top to bottom, shown in Fig. 2 are a time series of the column infrared (IR) brightness temperature (T_{IR}) measured by a narrow-beam IR-window (9.5–11.5 μm) radiometer coaligned with the lidar; a time versus height display of Cloud Polarization Lidar (CPL, with 0.693 μm wavelength and 10-s time and 6-m height resolutions) relative attenuated backscattering (plotted using a logarithmic gray scale), with the local Fairbanks, Alaska, temperature (solid line) and relative humidity profiles shown at right; and the corresponding W-band (3.2 mm wavelength) Doppler radar displays of water-equivalent radar reflectivity factor (in dBZ) and Doppler velocity (where negative values represent downward motions).

The quintessential character of As is revealed by the T_{IR} trends and the differences in the lidar/radar backscatter displays in Fig. 2. Radar reflectivity peaks at about 10 dBZ near the middle of the display corresponding to where the lidar returns are so heavily attenuated that data from the upper cloud region are lost. Comparing the radar and lidar cloud heights (cloud base heights are characteristically very similar), prior to 2,300 UTC, the top heights are similar; at 2,330 UTC, the lidar shows the effects of range-limiting attenuation; and after 2,345 UTC, the lidar reveals two cloud layers that go undetected by the radar. These clouds are a cirrostratus (Cs) between 8.0 and 10.0 km MSL (above mean sea level) and a supercooled ($\sim -10°$) Ac cloud with ice virga just above 4.0 km MSL. Both of these cloud layers contain particles too small to be detected by the radar, although typically with time the ice virga will grow to sizes easily seen by a millimeter-wave cloud radar. The downwelling infrared cloud signal (top panel) reveals a gradual increase as the Ac cloud thickens both physically and optically. These are basic characteristics of As when probed in the visible, infrared, and millimeter-wave regions. Finally, note that the maximum Doppler radar vertical velocities are in the -1.5 to -2.0 m/s range in connection with the lower ice crystal fallstreaks, and this is a signature of the fall velocities of ice crystal aggregates. Also note that, when observed with lidar (as in Fig. 2), As frequently develop a strongly scattering and highly attenuating cloud base layer also indicative of the presence of large aggregates that form in the lower (warmer) cloud region.

Another basic feature of As is illustrated in Fig. 3, which shows an additional time series of middle and high cloud data collected at AFARS depicting in this case infrared and CPL returned energy and depolarization data. In this instance, the cloud layer initially appeared to be a Cs between $\sim$5.5 and 10.0 km MSL, with typical linear depolarization ratios (δ, the ratio of the returned laser energies in the orthogonal and parallel polarization planes relative to that

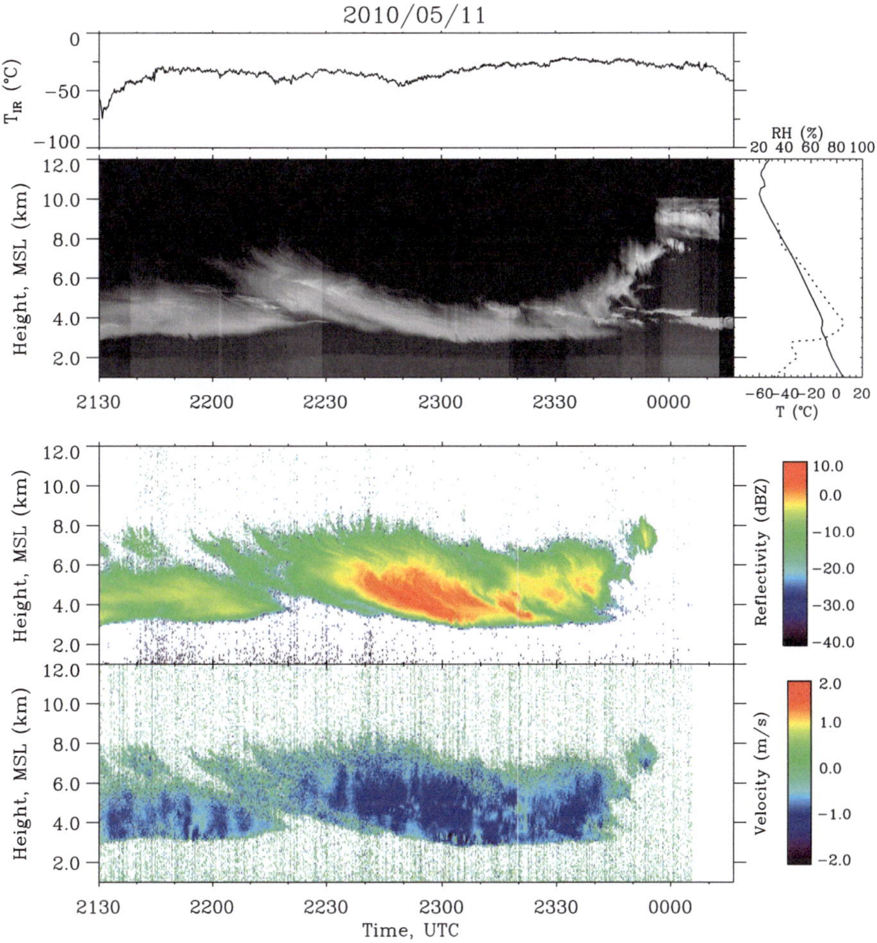

Fig. 2 An example from the Arctic Facility for Atmospheric Remote Sensing (AFARS, at a height of 218 m MSL) of an As cloud layer studied on May 11, 2010, showing, from *top* to *bottom*, a time series of midinfrared radiometer downwelling temperatures, and height versus time displays of CPL ruby (0.693 μm) lidar attenuated backscatter (with solid temperature and dashed relative humidity profiles to *right*), and W-band (3.2 mm) radar reflectivity factor and Doppler velocity (with color keys to *right*)

transmitted) of ~0.35. Following, lidar cloud top information is lost at around 21.10 UTC and from ~22.00 to 23.00 UTC, corresponding to the peak infrared signals. Another infrared maximum occurs at the end of the data record, when a lidar-blocking (from strong attenuation) Ac layer appears at about 3.0 km MSL and −5.0°C. Interestingly, a similar supercooled layer appeared briefly at ~22.10 UTC embedded in the base of the As. These liquid layers produce near-zero δ at the cloud base, but the δ then tends to increase with height because of photon multiple scattering among the dense assembly of cloud droplets, depending on lidar design (Sassen 2005a). Depolarization in the lower As, however, is also very low at times at temperature between −10° and −20°C, as a result of the non-depolarizing specular reflections from horizontally oriented plate crystals when probed by a zenith lidar. This highly anisotropic scattering medium is common in midlevel ice clouds because of the favorable temperatures for rapid planar ice crystal growth (Pruppacher and Klett 1997).

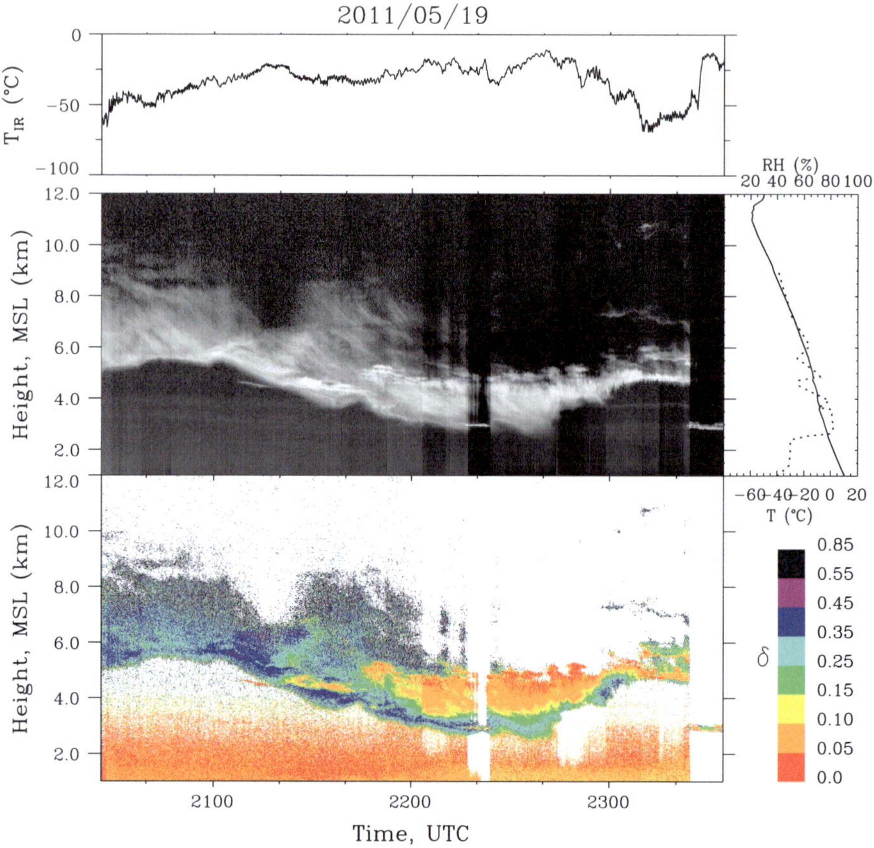

Fig. 3 As in Fig. 2, except for another As cloud studied on May 19, 2011, and showing lidar linear depolarization ratios at bottom (with color δ key to *right*)

2.2 Altocumulus

Figures 4 and 5 illustrate Ac lidar examples from two widely different parts of the globe, namely the Arctic of northern Alaska and the near tropics of southern Florida, respectively. The lidar returned energy height–time display in Fig. 4 comes from the MPACE field campaign at Barrow, Alaska (Verlinde et al. 2007), and was collected with the infrared channel (1.06 μm, at 1.0-s and 6-m resolutions) of the mobile Polarization Diversity Lidar (PDL) system. The strongly scattering cellular Ac layer (shown in white) has a −11.1°C cloud top temperature corresponding to the top of a temperature inversion. Considerably more noticeable, however, is the ice crystal virga below that is composed of numerous fallstreaks and extends a maximum of ~1.5 km below the supercooled liquid layer. The crystal fallstreaks show very little wind shear effects, and the entire cloud is about 45 km in length based on the 12.8 m/s average wind speed at this level. This is an Arctic mesoscale Ac cloud system. Data from a nearby cloud radar (not shown) had no difficulty in detecting the ice virga, but failed to see the uppermost cloud layer dominated by small cloud droplets and just-nucleated ice crystals.

The lidar images in Fig. 5 were collected during the CRYSTAL-FACE field campaign from a site in the Florida Everglades (Sassen et al. 2003). Both linear depolarization

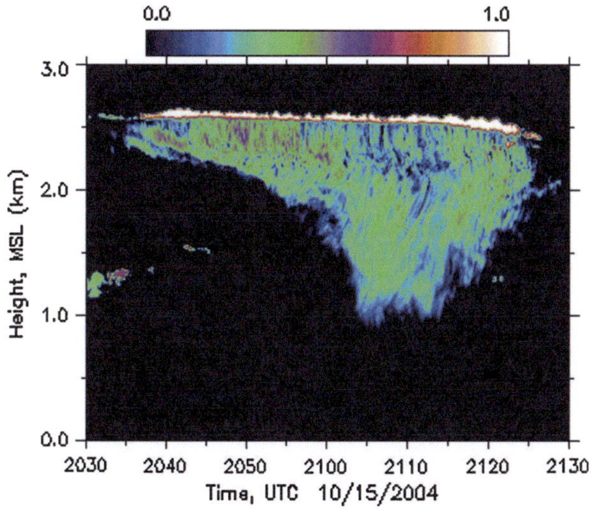

Fig. 4 An example of a relative returned energy 1.06 μm PDL lidar time–height display (note color relative returned energy scale at *top*) of a supercooled altocumulus cloud at 2.6 km MSL altitude with ice virga trailing below, sampled in the Arctic at Barrow, Alaska, on October 15, 2004

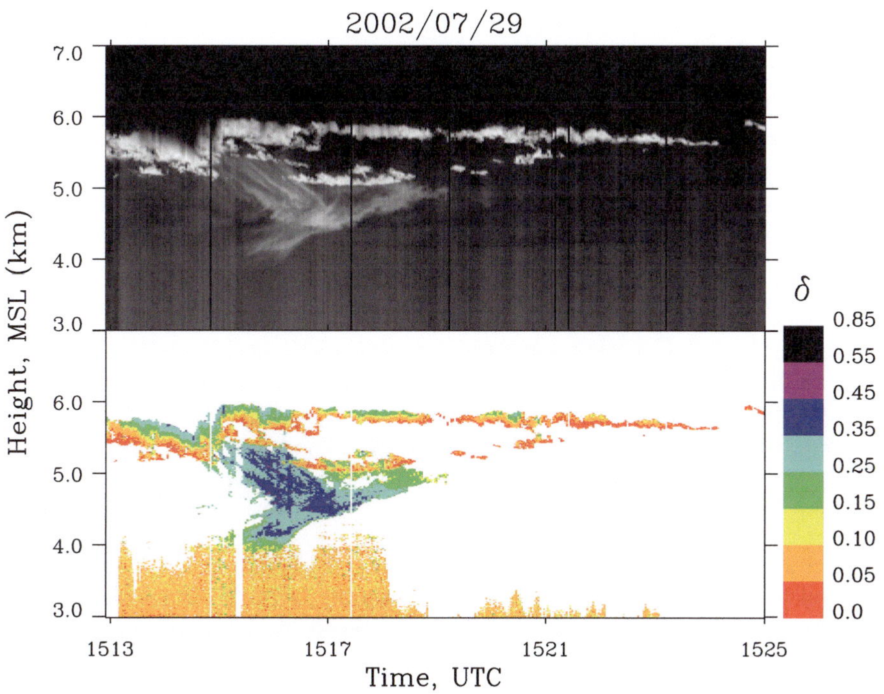

Fig. 5 An example of 0.532 μm PDL lidar relative returned energy (*top*) and depolarization (*bottom*, *δ* key to *right*) of a mildly supercooled Ac cloud, with a brief episode of depolarizing ice virga, sampled in the subtropics from a site in the Everglades in southern Florida on July 29, 2002

(bottom, note δ color scale at right) and relative returned energy plots are shown from the PDL using the 0.532-μm channel with 0.1-s and 1.5-m resolutions. The Ac layer is identified by the low δ at the ~5.5 km MSL cloud base, following by increasing δ aloft from multiple scattering in the optically thicker liquid cloud parts. The temperatures in this relatively warm supercooled cloud range from ~−5.0° to −9.0°C, but the fallstreak of depolarizing ice crystals around 1517 UTC (δ ~0.3–0.4) indicates ice nucleation as the Ac cloud base descended into a layer containing Saharan dust transported across the Atlantic Ocean. Because many desert dust particles are effective ice nuclei, these data suggest a brief supercooled cloud glaciation caused by the dust particles (Sassen et al. 2003). In general, it appears that Ac are high (i.e. cold) enough in the atmosphere to be supercooled and hence are typically mixed phase with ice virga trailing well below the liquid cloud top layer. According to the extended polarization lidar study reported in Sassen and Khvorostyanov (2007) from a midlatitude location, the mean observed Ac cloud base temperature was −14.7°C, ranging from −2.3° to −34.6°C. The last value is close to the temperature where pure supercooled water drops will freeze homogeneously (Pruppacher and Klett 1997), showing in this case an extreme (lack of ice nuclei at such high altitudes). The average Ac cloud base height in this study was 5.2 km MSL, which is close to the mean value given in Fig. 1 for the midlatitudes.

3 Radiative Properties

The microphysical properties of mixed-phase clouds, on which their radiative characteristics rely, depend on the dominant thermodynamic phase of the cloud because it determines the shapes, sizes, and concentrations of the constituent cloud particles (Khvorostyanov et al. 2001). Oddly, although proper As model simulations have not to our knowledge been undertaken, model and experimental studies of deep "cirrostratus" clouds that surely resemble As are available because of the attractiveness of treating "healthy" ice clouds and probing them with millimeter-wave radars, which may have difficulties with proper cirrus (e.g., Starr and Cox 1985; Mace et al. 2001; Khvorostyanov and Sassen 2002). Microphysical/radiative models of Ac clouds have often been based on observational studies (Heymsfield et al. 1991; Liu and Krueger 1998; Fleishauer et al. 2002; Sassen and Khvorostyanov 2007).

In terms of the fundamental scattering properties used in models, much is known from previous research dealing with water, ice, and, to some extent, mixed-phase clouds. Sassen et al. (2001), for example, employed a combined lidar/infrared radiometer dataset and a water cloud microphysical model to illuminate the dependencies of cirrus/As and Ac infrared emittances on cloud thickness and temperature, and the relationship between visible optical depth and cloud emittance in comparison with available radiative theory (Fu and Liou 1993). To separate Cs from classic As, a visible optical depth of ~ 3.0 and an infrared emissivity of 0.85 were inferred.

In all radiative calculations, the form of the particle size distribution, generally of a lognormal type, must be assumed and treated statistically, so that a mean particle size has meaning in the radiative computations. Parameterizations of visible and infrared optical properties, and their relationship, can then be made in terms of integrated ice/water paths (g/m^2) and an effective particle size. Water clouds are treated with Mie theory, while ice clouds must rely on approximate theories because of the consequences of the various possible nonspherical crystal shapes (Liou et al. 2002). Mixed-phase clouds must utilize a combination of these approaches for improved accuracy. It should also be reiterated that both As and Ac frequently produce horizontally oriented planar ice crystals, leading to a highly anisotropic scattering medium, which require more sophistication in modeling them.

4 Global Distribution of As and Ac Clouds

The algorithm used to process CloudSat radar and CALIPSO lidar data by this group has been developed over the past several years, evolving from ground-based remote sensing studies (Wang and Sassen 2001; Wang et al. 2004; Sassen and Wang 2008; Zhang et al. 2010). The classification of the midlevel Ac or As cloud types begins with the identification of a cloud cluster in the backscatter data, that is, a series of horizontally connected and similar cloud profiles at the appropriate altitudes for its latitude belt (see Fig. 1). In keeping with the above discussions, As (Ac) are not strongly precipitating (i.e. CloudSat radar signals are unlikely to be strong enough to be measured at the Earth's surface), thick to thin (moderate to thin) clouds of ice (water) or mixed phase. Zhang et al. (2010) provide many details but, basically, the algorithm matches the fundamental microphysical properties of As and Ac with the likely responses to radar and lidar signals, including signal slopes, apparent attenuation effects, and basic remote sensor sensitivities to characteristic particle sizes (i.e. essentially particle phase). The cloud classifier uses fuzzy logic in cloud type decision making.

Given in Figs. 6, 7, 8, 9 are the global CloudSat/CALIPSO midlevel cloud findings averaged over the years of 2007 and 2008 in terms of seasonally dependent (for four three-month periods) geographical and height (in 5.0° zonal latitude belts) frequency of occurrence displays. The geographical distribution of the As clouds identified by the algorithm (Fig. 6) displays a relatively low occurrence in the tropics and subtropics and a high occurrence in the midlatitude and polar regions, particularly in the winter hemispheres. Midlatitude storm tracks are apparent, but the frequencies are not as strong as in many polar areas. There is evidence for considerable orographic As enhancements as, for example, over northwestern North America, Greenland, and Antarctic coastal regions. Slight regional enhancements are associated with intense, seasonal deep convective cloud outbreaks (e.g., the Indian Monsoon in JJA and south-central Africa in DFJ). Zonal cloud height dependencies (Fig. 8) show peak occurrences at mid- to high latitudes in the winter middle troposphere and reflect a general dependence on planetary-scale (e.g., Hadley cell) circulation patterns that cause deep ascending and descending air motions, similar to the basic patterns for cirrus clouds (Sassen et al. 2008).

The identified Ac clouds indicate a quite different geographical occurrence (Fig. 7), where strong maxima occur in association with deep convection in the tropics. Generation of Ac along the midlatitude storm tracks is apparent, but with a strong seasonal asymmetry in the summer hemispheres extending to high latitudes, especially JJA in the northern hemisphere. This probably reflects the effects of deep convection in Ac generation, and the highest frequencies are indeed linked to tropical thunderstorms (Sassen et al. 2009). Apparently, this is a type of Ac termed altocumulus *cumulogenitus*, which is produced by midlevel cloud detrainment during the pulsewise development of deep convective towers at the heights of relatively stable layers. The vertical distribution of Ac in Fig. 9 again reveals the cloud minima in descending air surrounding the subtropics, but also depicts the relatively restricted heights of midlevel supercooled water-topped clouds due to the minimum temperature associated with water clouds: The algorithm uses a minimum temperature of −38.5°C.

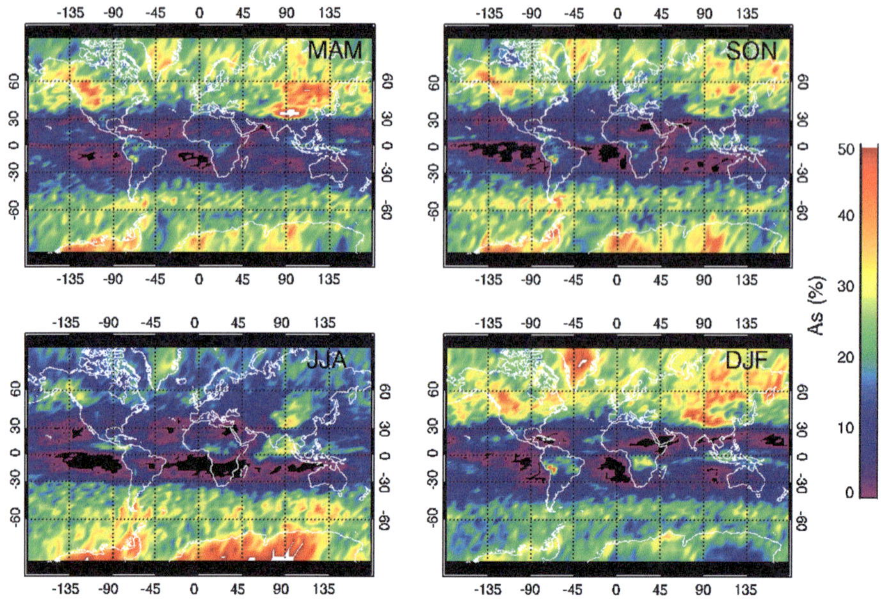

Fig. 6 The global distribution of As clouds derived from a 2-year average of CloudSat radar and CALIPSO lidar data using our cloud type identification algorithm (note color frequency scale to *right*)

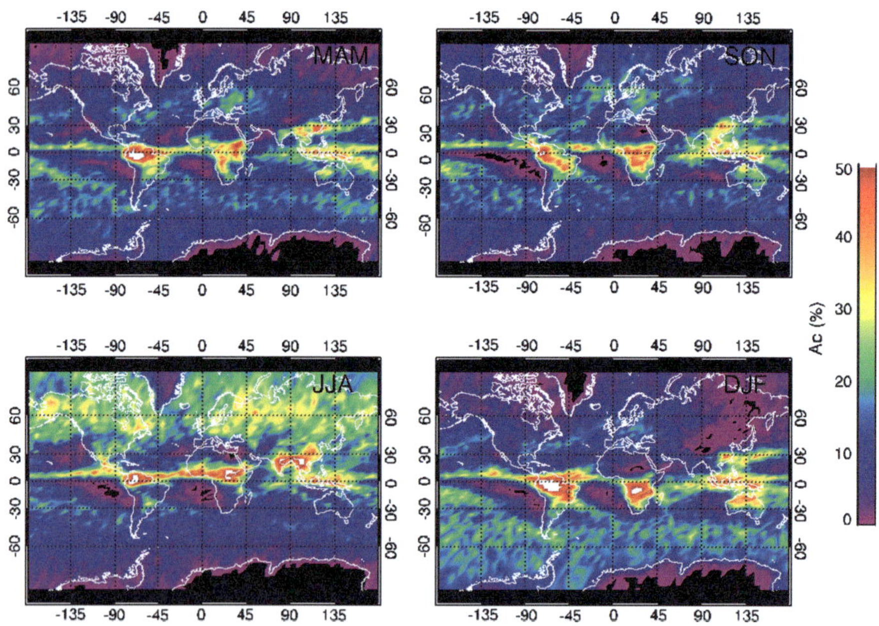

Fig. 7 As in Fig. 6 except for the global distribution of Ac clouds

Finally, the global two-year average frequencies of As and Ac are contrasted in each season to the total cloud cover determined from the CloudSat/CALISPO algorithm (Table 1). Compared to our total cloud cover average of nearly 75% of the Earth's surface,

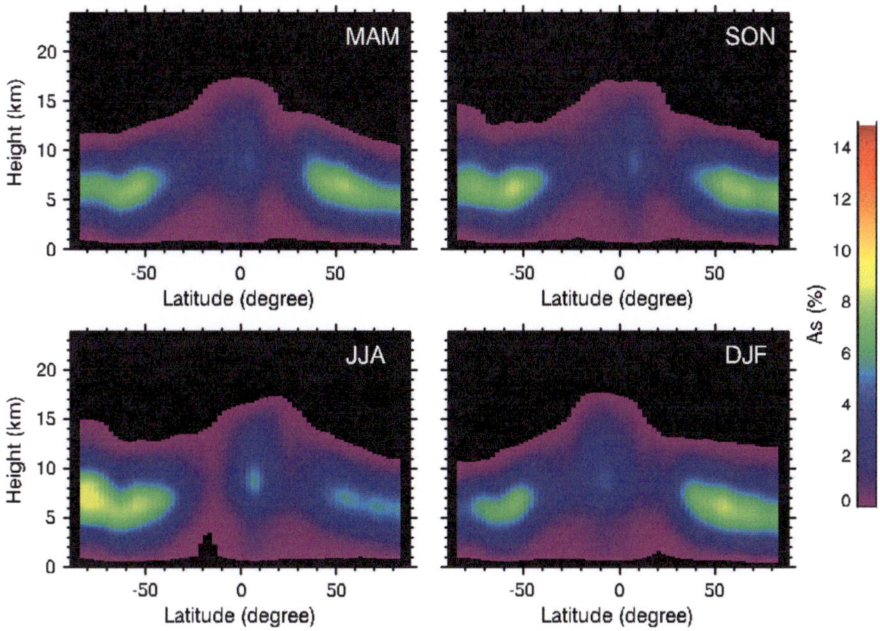

Fig. 8 The height versus zonally averaged latitude frequencies (note color scale at *right*) for As clouds derived from a 2-year average of CloudSat and CALIPSO data

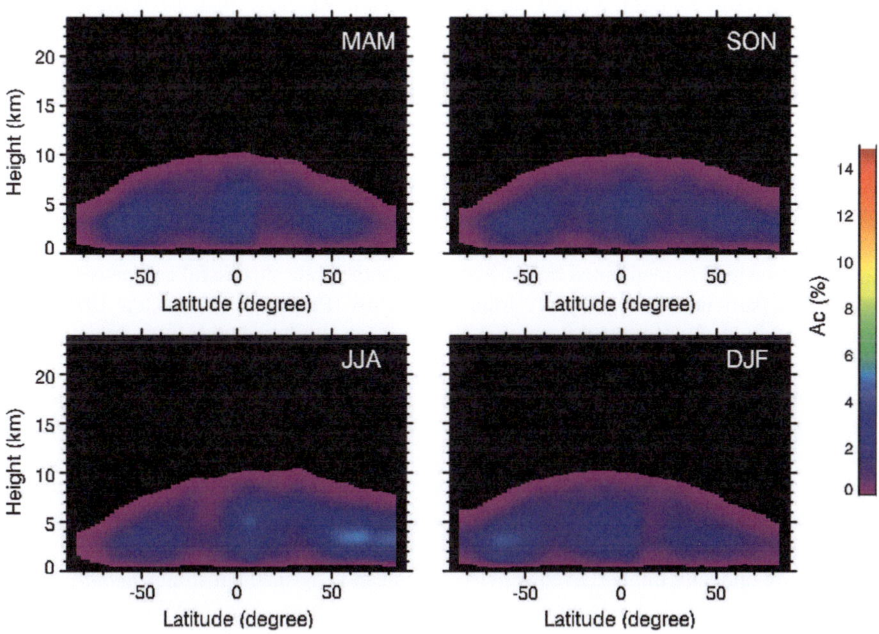

Fig. 9 As in Fig. 8, except for Ac clouds

Table 1 Midlevel and total cloud frequencies derived from the algorithm using the global CloudSat/CALIPSO dataset, in percent

	MAM	JJA	SON	DJF
All clouds	74.4	74.0	74.5	74.5
As	16.4	14.8	15.9	15.6
Ac	9.9	12.7	10.1	10.8

The frequencies are calculated as the ratio of vertical profiles containing only As or Ac clouds to the total profiles. Clouds in each height group are only counted once for each data profile even if multiple layers of each type is present. MAM is for March, April and May, JJA June, July and August, etc

As (15.7%) and Ac (10.9%) together comprise about 25%, or roughly one-third of the total clouds in the low, middle, and upper troposphere.

5 Discussion and Conclusions

The clouds of the middle troposphere are unglamorous to all but a few researchers (Gedzelman 1988, 2011). They do not generate significant amounts of rain or snow and have often been neglected in modern research: They are "forgotten clouds" (Vonder Haar et al. 1997). Nonetheless, As and Ac comprise a significant fraction (about one-third) of the total cloud cover of the Earth, and because of their widespread distribution and varied basic radiative effects, they must be considered an integral part of the planet's energy balance.

The data presented here reveal, not surprisingly, a basic dependence of midlevel clouds on planetary-scale circulation patterns and features of synoptic meteorology. Cloud frequency maxima appear to be found for Ac in association with tropical and midlatitude continental deep convection (variety *cumulogenitus*), while in contrast, As tend to avoid the tropics and are apparently most strongly influenced by the effects of orography and cyclonic development in midlatitude storm tracks. Their vertical distributions show quite deep As and relatively shallow Ac clouds, the latter a consequence of limits on the survival of supercooled water in the atmosphere. Nonetheless, it is interesting to note that the geographical frequency distributions of As and Ac clouds indicate that they inhabit reciprocal cloud niches in accordance with the action of the Wegener–Bergeron–Findeisen mechanism, which as mentioned before accounts for the quite successful competition of ice over water particles where they are found to coexist (Pruppacher and Klett 1997). The differences in their dominant formation mechanisms discussed here also may lead to different geographical locations for As and Ac.

An interesting feature of midlevel clouds that should be investigated further is appreciating other forms of As clouds that differ from the classic dense As layer commonly associated with midlatitude cyclonic activity. Although the current CloudSat/CALIPSO analysis has not directly addressed this issue, it is clear from surface-based lidar/radar cloud studies from AFARS and elsewhere that unrecognized subtypes of As differing from the classical layer occur and have probably strongly influenced the current results. These subtypes are optically thinner, even subvisual, and can be termed altostratus *nebulosus* and *altocumulogenitus* based on standard cloud typing terminology (WMO 1987). These midtropospheric ice clouds can resemble cirrus *fibratus* or Cs, but typically have a diffuse appearance with rounded forms and indistinct cloud edges.

Specifically, As that are derived from glaciated Ac (Sassen and Khvorostyanov 2007), heterogeneous depositional ice nucleation without the presence of the liquid phase (Sassen 2005b), or other cloud formation processes typically leave that behind relatively thin ice clouds; such clouds are no doubt part of the current satellite-based As sample. Orographic uplift may often be the cause of the required atmospheric humidification, and temperatures in the $-10°$ to $-20°C$ range particularly favorable for the planar ice crystal habits mean that the horizontally oriented plates would produce a highly anisotropic scattering medium. For example, in lidar studies of midlatitude cirrus clouds, Sassen and Campbell (2001) reported that oriented plate crystal layers present below the cirrus under study were present during 4.5% of the total observation periods. These were often subvisual clouds and thought to be associated with glaciated upwind Ac clouds formed by local mountain ranges. In any case, ice clouds that occupy the correct heights for As, but are not dense enough to block the sun or moon, may be commonplace, as our recent experience at AFARS in the Arctic would suggest. Since such clouds would have different radiative properties than the classic As, more study is needed to assess their climatic impact.

Finally, comparison with the height and geographical distributions of high-level cirrus clouds derived from the same satellite sources (Sassen et al. 2008, 2009), which should share some characteristics with As because of the common cirrostratus-to-altostratus transition (Sassen 2002), yields some surprising results. It was found that cirrus derived from spreading deep convective cloud anvils in the tropics was the dominant source of cirrus over the surface of the Earth, with those created in midlatitude storm tracks also prominent. A comparison with Fig. 6 shows poor agreement with this, in that the tropics lack much As while the polar regions have relatively high frequencies of occurrence. Although the basic height versus latitude As distributions in Fig. 8 are similar to cirrus, As lack the high-altitude tropical maximum and also have some of their highest frequencies extending to high latitudes. It is probable that the kinds of optically thin As mentioned above have some impact on these polar findings, as regular AFARS research would support (as currently under study). The Ac in Fig. 7, on the other hand, share the location of the maximum cirrus frequencies in the tropics, since both appear to be the result of mass detrainment in deep convection, although the effects of cirrus generation in midlatitude storm tracks are much more apparent. This indicates a dominance of mainly continental Ac formation mechanisms due to convection or orography.

Such unexpected findings open up new avenues for researching the clouds of the middle troposphere to improve our understanding of their effects on the planet's energy balance, thanks to the unique active remote sensing capabilities of the CloudSat and CALIPSO satellites.

Acknowledgments This research has been supported by NSF grants ATM-0645644 and AGS-037000, NASA grant NNX10AN18G, and contracts NAS-7-1407 and 1270909 from the Jet Propulsion laboratory.

References

Fleishauer RP, Larson VE, Vonder Haar TH (2002) Observed microphysical structure of midlevel, mixed-phase clouds. J Atmos Sci 59:1779–1804

Fu Q, Liou KN (1993) Parameterization of the radiative properties of cirrus clouds. J Atmos Sci 50:2008–2025

Gedzelman SD (1988) In praise of altocumulus. Weatherwise 41:143–149

Gedzelman SD (2011) Beautifying altostratus. Weatherwise 64:32–41

Hahn CJ, Warren SG (1999) Extended edited cloud reports from ships and land stations over the globe, 1952–1996. Numerical data package NDP-026C, Carbon Dioxide Information Analysis Center (CDIAC), Department of energy, Oak Ridge, Tennessee (Documentation, pp 79)
Hartmann DL, Ockert-bell ME, Michelsen ML (1992) The effect of cloud type on earth's energy balance: global analysis. J Clim 5:1281–1304
Heymsfield AJ, Miloshevich LM, Slingo A, Sassen K, DO'C Starr (1991) An observational and theoretical study of highly supercooled altocumulus. J Atmos Sci 48:923–945
Khvorostyanov VI, Sassen K (2002) Microphysical processes in cirrus and their impact on radiation: a mesoscale modeling perspective. In: Lynch D, Sassen K, DO'C Starr, Stephens GL (eds) cirrus. Oxford University Press, New York
Khvorostyanov VI, Curry JA, Pinto JO, Shupe M, Baker BA, Sassen K (2001) Modeling with explicit spectral water and ice microphysics of a two-layer cloud system of altostratus and cirrus observed during the FIRE arctic clouds experiment. J Geophys Res 106:15099–15112
Liou KN, Takano Y, Yang P, Gu Y (2002) Radiative transfer in cirrus clouds: light scattering and spectral information. In: Lynch D, Sassen K, Starr DO'C, Stephens GL (eds) Cirrus. Oxford University Press, New York
Liu S, Krueger SK (1998) Numerical simulation of altocumulus using a cloud model and a mixed-layer model. Atmos Res 47–48:461–474
Mace GG, Ackerman TA, Minnis P, Young DF (2001) The composite characteristics of cirrus clouds: bulk properties revealed by one year of continuous cloud radar data. J Clim 14:2185–2203
Pruppacher HR, Klett JD (1997) Microphysics of clouds and precipitation, 2nd edn. Kluwer, New York
Rossow WB, Schiffer RA (1999) Advances in understanding clouds from ISCCP. Bull Amer Meteor Soc 80:2261–2286
Sassen K (2002) Cirrus clouds: a modern perspective. In: Lynch D, Sassen K, DO'C Starr, Stephens GL (eds) Cirrus. Oxford University Press, New York
Sassen K (2005a) Polarization in lidar. In: Weitkamp C (ed) Lidar. Springer Press, New York
Sassen K (2005b) Dusty ice clouds over Alaska. Nature 434:456
Sassen K, Campbell JR (2001) A midlatitude cirrus cloud climatology from the facility for atmospheric remote sensing: I. macrophysical and synoptic properties. J Atmos Sci 58:481–496
Sassen K, Khvorostyanov VI (2007) Microphysical and radiative properties of mixed phase altocumulus: a model evaluation of glaciation effects. Atmos Res 84:390–398
Sassen K, Khvorostyanov VI (2008) Cloud effects from boreal forest fire smoke: evidence for ice nucleation from polarization lidar data and cloud model simulations. Environ Res Lett 3:025006. doi: 10.1088/1748-9326/3/2/025006
Sassen K, Wang Z (2008) Classifying clouds around the globe with the CloudSat radar: 1-year of results. Geophys Res Lett 35:L04805. doi:10.1029/2007GL032591
Sassen K, Liou KN, Kinne S, Griffin M (1985) Highly super-cooled cirrus cloud water: confirmation and climatic implications. Science 227:411–413
Sassen K, Comstock JM, Wang Z (2001) Parameterization of the radiative properties of midlatitude high and middle level clouds. Geophys Res Lett 28:729–732
Sassen K, DeMott PJ, Prospero JM, Poellot MR (2003) Saharan dust storms and indirect aerosol effects on clouds: CRYSTAL-FACE results. Geophys Res Lett 30:1633. doi:10.1029/2003GL017371
Sassen K, Wang Z, Liu D (2008) The global distribution of cirrus clouds from CloudSat/Cloud-Aerosol Lidar and Infrared Pathfinder Satellite Observations (CALIPSO) measurements. J Geophys Res 113:DOO12. doi:10.129/2008JD009972
Sassen K, Wang Z, Liu D (2009) Cirrus clouds and deep convection in the tropics: insights from CALIPSO and CloudSat. J Geophys Res 114:DOOH06. doi:10.1029/2009JD011916
Starr DO'C, Cox SK (1985) Cirrus clouds. II: numerical experiments on the formation and maintenance of cirrus clouds. J Atmos Sci 42:2682–2694
Stephens GL (2005) Cloud feedbacks in the climate system: a critical review. J Clim 18:237–273
Stephens GL, Tsay S, Stackhouse PW, Flatau PJ (1990) The relevance of the microphysical and radiative properties of cirrus clouds to climate and climate feedback. J Atmos Sci 47:1742–1753
Stephens GL, Vane DG, Boain R, Mace GG, Sassen K, Wang Z, Illingworth A, O'Conner E, Rossow WB, Durden SL, Miller S, Austin R, Benedetti A, Mitrescu C (2003) The CloudSat mission and the A-train: a new dimension of space-based observations of clouds and precipitation. Bull Amer Meteor Soc 83:1771–1790
Verlinde J, Harrington JY, McFarquhar GM, Yannuzzi VT, Avramov A, Greenberg S, Johnson N, Zhang G, Poellot MR, Mather JH, Turner DD, Eloranta EW, Zak BD, Prenni AJ, Daniel JS, Kok GL, Tobin DC, Holz R, Sassen K, Spangenberg D, Minnis P, Tooman TP, Ivey MD, Richardson SJ, Bahrmann CP,

Shupe M, Demott PJ, Heymsfield AJ, Schofield R (2007) The mixed-phase arctic cloud experiment. Bull Amer Meteor Soc 88:205–221
Vonder Haar TH et al (1997) Overview and objectives of the DOD Center for Geosciences sponsored ''Complex Layered-Cloud Experiment',' paper presented at Cloud Impacts on DOD Operations and Systems Conference. Philips Lab, Newport
Wang Z, Sassen K (2001) Cloud type and property retrieval using multiple remote sensors. J Appl Meteor 40:1665–1682
Wang Z, Sassen K, Whiteman D, Demoz B (2004) Studying altocumulus with ice virga using ground-based active and passive remote sensors. J Appl Meteor 43:449–460
Warren SG, Hahn CJ, London J, Chervin RM, Jenne R (1986) Global distribution of total cloud cover and cloud type amount over land. NCAR Tech. Note TN-273 STR. Natl. Cent. Atmos. Res., Boulder, Colo. USA, 229 pp
Wielicki BA, Cess RD, King MD, Randle DA, Harrison EF (1995) Mission to planet Earth: role of clouds and radiation in climate. Bull Am Meteor Soc 76:2125–2153
Winker DM, Pelon J, McCormick MP (2003) The CALIPSO mission: spaceborne lidar for observation of aerosols and clouds. Proc SPIE Int Soc Opt Eng 4893:1–11
WMO (1987) International cloud atlas, vol II. World Meteorological Organization, Geneva
WMO (1988) Manual on codes, vol. 1, WMO Publ vol 306. World Meteorological Organization, Geneva
Zhang MH (2005) Comparing clouds and their seasonal variations in 10 atmospheric general circulation models with satellite measurements. J Geophys Res 110:D15S02. doi:10.1029/2004JD005021
Zhang D, Wang Z, Liu D (2010) A global view of midlevel liquid-layer topped stratiform cloud distribution and phase partition from CALIPSO and CloudSat measurements. J Geophys Res 115:DOOH13. doi: 10.1029/2009JD012143

Surv Geophys (2012) 33:693–700
DOI 10.1007/s10712-011-9160-0

Aerosol Forcing: Rapporteur's Report and Summary

Frida A-M. Bender

Received: 16 September 2011 / Accepted: 2 November 2011 / Published online: 6 December 2011

Abstract This paper summarizes the discussions held during the session dedicated to *Aerosol forcing* at the Workshop *Observing and Modelling Earth's Energy Flows*. The session *Aerosol forcing* was convened by P. Ingmann and J. Heintzenberg and included 10 presentations given by R. Kahn, D. Winker, U. Baltensperger, J. Haywood, S. Schwartz, J. Heintzenberg, H. Le Treut, U. Lohmann, R. Wood, and E. Philipona. The presentations given ranged from overviews of current observational capabilities to analyses of aerosol–cloud interactions in observations and models of varying complexity. This paper is organized around a few key points, summarizing the major points of agreement, disagreement, and discussion that the presentations gave rise to. The focus is largely on the uncertainties that remain with regard to aerosol forcing, particularly related to aerosol-cloud interactions and indirect aerosol effects on climate.

Keywords Aerosol forcing · Aerosols and climate · Aerosol indirect effects · Aerosol-cloud interactions

1 Introduction

Aerosol forcing, and particularly the increase in aerosol loading due to anthropogenic activity, has long been recognized as a significant contributor to global and regional climate change (see, e.g., Bryson and Wendland 1970; Rasool and Schneider 1971; Bolin and Charlson 1976; Charlson et al. 1991, 1992; Ramanathan et al. 2001). However, the total magnitude of this forcing remains uncertain, with implications for attribution of observed temperature changes and constraint of climate sensitivity (see, e.g., Hansen et al. 1997a; Anderson et al. 2003; Andreae et al. 2005; Knutti 2008; Schwartz et al. 2010).

A brief introduction to the concept of atmospheric aerosol and aerosol forcing is warranted. Aerosols are liquid or solid particles suspended in the atmosphere, as a result of

F. A-M.Bender (✉)
Center for Clouds, Chemistry and Climate (C4), Scripps Institution of Oceanography, University of California, San Diego, La Jolla, CA, USA
e-mail: fbender@ucsd.edu; frida@misu.su.se

condensation of precursor gases or of direct emission or wind action on the Earth surface. The aerosols and precursors can be natural or anthropogenic in origin; for example wind-blown desert dust and sea salt, biogenic production, forest fires, volcanic eruptions, as well as biomass and fossil fuel burning and land use changes all contribute to the aerosol loading of the atmosphere.

The lifetime of aerosols against deposition and rainout is of the order of days to weeks (see, e.g., Textor et al. 2007). The short residence time of the tropospheric aerosol is reflected in the geographical distribution of aerosol optical depth (AOD) that mirrors the natural and anthropogenic source regions (Kinne et al. 2006; Ramanathan et al. 2007).

Aerosol impact on the radiation budget and energy flows of the Earth system is manifold. First, aerosols depending on their chemical and physical composition can scatter, reflect, and absorb incoming sunlight. This is referred to as the direct effect of aerosols on climate. In a simplified manner, aerosols can be said to be either white or black, meaning that they either reflect or absorb radiation, and thereby have either a cooling or a warming effect on climate.

Sulfate, which is strongly reflecting, and black carbon, which is strongly absorbing, are two aerosol species largely dependent on anthropogenic activity, originating primarily from fossil fuel and biomass burning. Consequently, their burdens are highly elevated as compared to pre-industrial levels. The global anthropogenic sulfur emissions decreased during the end of the 20th century, with decreases in the Northern Hemisphere (e.g., Europe and North America) and increases in the Southern Hemisphere (e.g., Asia and Africa), but recent estimates again point at rising levels, primarily due to coal combustion in Asia (Stern 2004; Smith et al. 2011; Kaufmann et al. 2011). The distribution of black carbon emissions has strong local maxima over South America and Africa, due to open burning, and over Southeast Asia, due to other combustion, with an increasing estimated global burden (Bond et al. 2004; Skeie et al. 2011).

Second, aerosols in various ways affect cloud properties and thereby have indirect effects on climate. These cloud-mediated aerosol effects can either be caused by redistribution of heat in the atmospheric column due to aerosol absorption and scattering or by changes in cloud microphysical structure. The latter is motivated by the fact that aerosols are essential in cloud formation, acting as cloud condensation and ice nuclei. Hence, changes in concentration and composition of the atmospheric aerosol can affect the number and size of cloud droplets and cloud ice particles, which may in turn cause changes in the radiative properties of the clouds and their tendency to precipitate, their lifetime and extent and thereby their radiative effects on climate.

By convention (see, e.g., IPCC 2007), distinctions are often made between (1) the first indirect effect, or the Twomey or cloud albedo effect (Twomey 1974, 1977), by which an increase in aerosol concentration under the assumption of constant liquid water content in the cloud gives rise to smaller and more numerous droplets, yielding more reflective clouds, (2) the second indirect effect, or the Albrecht or cloud lifetime effect (Albrecht 1989), by which an increase in aerosol concentration increases the cloud lifetime through reduction of precipitation efficiency due to smaller droplets, leading to increased cloudiness and (3) the semi-direct effect by which aerosol absorption in a cloud layer causes clouds to burn off, thereby decreasing their lifetime and reducing cloudiness (Hansen et al. 1997b; Ackerman et al. 2000). In addition, thermodynamic, glaciation, and riming effects, all affecting the precipitation efficiency of mixed-phase clouds, appear in the literature (see, e.g., Lohmann and Feichter 2005 and references therein). However, the magnitudes and signs of the indirect aerosol effects are not unequivocal, as exemplified, e.g., by the results of Feingold et al. (2005), Wood (2007), Xue et al. (2008), Sandu et al. (2008),

Wilcox (2010). Particularly, the second indirect or cloud lifetime effect has been criticized as representing a too simplified picture of the relationship between aerosol, clouds, and precipitation (e.g., Stevens and Feingold 2009), and the distinction between aerosol effects as described above may be criticized (see also Sect. 2.6).

In addition to their direct and indirect effects on climate, aerosols also have a negative impact on human health (WHO 2003; Nel 2005). This motivates reduction of aerosol emissions, regardless of their climatic impact.

2 Summary of Discussions

2.1 Aerosol Matters

There is a general agreement that aerosol forcing is an important factor in the present radiative budget and no one argues against a significant direct radiative effect of anthropogenic aerosol. The net global mean anthropogenic direct aerosol radiative forcing (DARF; estimated to -0.5 ± 0.4 Wm^{-2} in the Fourth Assessment Report (AR4) of the IPCC (Intergovernmental Panel on Climate Change); IPCC 2007) is potentially offsetting part of the positive forcing from greenhouse gases (GHG), on a global mean scale. It should be noted here that this DARF is the sum of the competing effects of reflecting and absorbing aerosol, which is not clear from the canonical diagram showing radiative forcing components in the IPCC AR4 summary for policy makers (http://www.ipcc.ch/publications_and_data/ar4/wg1/en/figure-2-20.html). It should also be noted that aerosols due to their shorter lifetime have a different geographical distribution than GHG prohibiting uniform offsetting or partial cancelation of their positive forcing.

Examples are also given of regional effects on the radiative budget, like solar dimming and brightening at the surface, due to changes in aerosol optical depth (Philipona et al. 2009).

2.2 Aerosol Forcing is Highly Uncertain

Aerosol forcing is still highly uncertain, and, to use IPCC terminology, a low level of scientific understanding remains, not least with regard to aerosol-cloud interaction and indirect aerosol forcing. Whereas the direct aerosol forcing is reasonably well constrained, some indirect effects are not even quantified, and new possible aerosol-induced alterations of cloud radiative properties continue to be discovered and hypothesized. AR4 quantifies the cloud albedo radiative forcing (the first indirect effect) to between -0.3 and -1.8 Wm^{-2} (-0.7 Wm^{-2} central value), based on a number of climate model simulations, but does not quantify other possible indirect forcings. Even for DARF, the uncertainty range (e.g., given by AR4 as -0.5 ± 0.4 Wm^{-2}) is very large compared with that in long-lived GHG forcing (in AR4 estimated to 2.63 ± 0.26 Wm^{-2} due to increases in CO_2, CH_4, N_2O, and halocarbons).

The uncertainty is associated with variability in aerosol forcing in models, and several Workshop presentations refer to the relation found among current-generation GCMs (general circulation models) by which models with low climate sensitivity have large total forcing due to small aerosol cooling, over the 20th century, and vice versa (Kiehl 2007). Taking this further, aerosol forcing can be said to determine the allowable CO_2 emissions for a certain temperature change.

2.3 Can Uncertainty be Reduced?

First of all, uncertainty needs to be properly quantified. The error estimates on aerosol radiative forcing (ARF) in IPCC AR4 reflect spread among models, not necessarily true uncertainty. The two are not equivalent and must not be confused with each other. We are increasing our knowledge about aerosols and aerosol forcing, but that does not imply we can reduce the error bars based on model diversity. If we were to change the definition of these error bars, so that they reflected uncertainty rather than model spread they would likely increase. However, maybe the measure based on model spread must be used (for IPCC purposes) for consistency with estimates for other forcing agents. Even if the definition is kept as is, the inclusion of more aerosol-cloud interactions in more models is still likely to increase the spread rather than reduce it. Either way, it is pointed out as a difficult but important task to communicate this to the public and policy makers. At the same time, the benefit of too much focus in the community on satisfying external constraints, like reducing error bars, is questioned.

Sensitivities in AOD from passive satellite measurements (for clear sky, i.e. direct forcing) are presently around 0.05 (Kahn et al. 2005, 2010), whereas a desired value would be ~ 0.02, which is the upper bound for TOA DARF accuracy of 1 Wm^{-2}. It is noted, however, that even this target value would corresponds to radiative flux uncertainties that are not negligible.

A key to better constrained AOD from space is determination of aerosol type on a global scale. There is, however, great complexity in aerosol chemical composition, making this difficult. Improved understanding of aerosol processes, sources and sinks, including emissions and wet removal of aerosol, is also needed.

2.4 Integrating Different Types of Observations

A need and a will to integrate different types of observations, like passive and active remote sensing, in situ and satellite observations, etc., is expressed. Different measurements complement each other. For instance, CALIPSO (cloud-aerosol lidar and infrared pathfinder satellite observations; Winker et al. 2009) can give vertical distribution of thin cloud and aerosol in the same column, whereas MODIS (MODerate resolution Imaging Spectroradiometer; King et al. 2003) and MISR (Multi-angle Imaging SpectroRadiometer; Diner et al. 1988) give column integral values, for cloud-free sky, the definition and determination of which are not unambiguous (see, e.g., Zhao and Di Girolamo 2006; Charlson et al. 2007). But, on the other hand, CALIPSO has sparser spatial sampling than MODIS and MISR. In situ observations can be used to characterize microphysical properties of different air mass types, giving a distribution that can then be applied to other places and times, with similar ambient conditions.

The meaningfulness of comparing point observations with satellite is discussed, i.e. the problem of representativity, but examples are shown where it does appear useful and fruitful.

2.5 Combining Models and Observations

Integrating models and observations is also put forward as important. We are presented with examples of creative thinking in model-data comparison, for instance combining model output, and various observational data sets to study an individual aerosol event (a volcanic eruption), day by day at high spatial resolution (rather than comparing values averaged over large temporal and/or spatial scales; Haywood et al. 2010).

Models can also be useful in linking aerosols to their sources, to deduce the anthropogenic contribution to AOD and ARF.

The question of how much detail in aerosol is meaningful to include in global climate models is discussed. Many aerosol processes are not well understood but there is also a multitude of processes that are not yet included in models. In a way models are lagging the research, as process study findings are not included in models but, in another way, models are leading research, featuring processes that have actually not been observed. Despite such lack of precision, climate models show a broad agreement on key parameters like global surface temperature change (Kiehl 2007; Schwartz et al. 2007; Knutti 2008), which may be seen as indicative of compensating errors in models.

2.6 Indirect Aerosol Effects on Climate

Interaction between aerosols and clouds and possible indirect aerosol radiative effects are very uncertain. Although necessary for understanding of aerosol-cloud interactions at a process level, it may not be meaningful to distinguish between different pathways for aerosol indirect effects on climate. Neither should the first indirect (Twomey) effect be singled out, for several reasons; (1) it is based on an assumption about constant cloud water content that often does not hold, (2) observational evidence for it is poor, (3) it may be counteracted by competing effects. Indirect aerosol effects on climate may preferably be bundled together, rather than being estimated separately. Another option suggested is to take out aerosol indirect forcing of climate models, and/or the IPCC report altogether, but this line of discussion is not pursued.

The question is also raised of how meaningful it is to separate direct and indirect effects at all, primarily for two reasons. First, the distinction between clear sky and clouds is not obvious and, regardless of the definition of a cloud, there will be a region of uncertainty around the defined threshold. Second, aerosols affect clouds and clouds affect aerosols and clear-sky aerosol forcing may well be influenced by the cloud-processing history of the aerosol.

2.7 The Importance of Clouds

The importance of understanding clouds better is heavily stressed. The boldest phrasing of this emphasis is that there is a misconception that aerosol effects on clouds are an aerosol problem rather than a cloud problem. We need to ask ourselves whether we understand clouds and their responses to changes in meteorology well enough for it to even make sense to try to model aerosol indirect effects on climate. Results shown indicate that the relative importance and even the sign of indirect aerosol effects can vary depending on cloud distribution and, hence, a correct cloud distribution is crucial (Wood 2007). Again, active remote sensing tools, like CALIPSO and CloudSat (Stephens et al. 2002), are mentioned as being useful.

The counter-question is whether the aerosol community can wait for full understanding of clouds before investigating possible aerosol-cloud interaction pathways and their effects on cloud radiative properties. The discussions end in a consensus that we should study and work on these issues in parallel and there should be better communication between the aerosol and the cloud communities.

2.8 Forcing Versus Feedback

The question whether aerosols should be seen as a forcing or a feedback is raised. Here, it is useful to distinguish between direct and indirect aerosol effects. Whereas the direct radiative effects of anthropogenic aerosols must reasonably be seen as a forcing, the radiative effects of aerosol perturbations to clouds may be seen as a feedback to this forcing. It is argued that the time scale of this adjustment is so short that it should be seen as part of the forcing, but it is also argued that definitions of a significantly short time scale for adjustment, or processes that should be allowed to adjust, are arbitrary.

2.9 The Future

There is a general spirit of "bringing it all together"—aerosol and cloud community, different observational approaches, models and observations, and so on. The meeting invited just such interaction, which is a step in the right direction, but more such efforts are clearly needed.

It is pointed out that, as the GHG forcing increases, the relative importance of aerosol forcing will decrease. However, such statements cannot be made without reservations. Better understanding, quantification and characterization of aerosol and aerosol forcing is necessary to understand the past, as well as the near future, and to answer general questions like how the Earth's temperature will respond if aerosol emissions decrease, and how this will depend upon what type of aerosol emission is decreased.

The poor understanding and quantification of aerosol forcing is also relevant to any discussion of geo-engineering. The evaluation of aerosol modeling, specifically that of stratospheric aerosol, was mentioned as providing insight into stratospheric geo-engineering proposals. But the complexity and great uncertainty involved in aerosol modeling, as revealed by the presentations and emphasized in the discussions, gave rise to skeptical comments regarding such proposed "armchair engineering".

The uncertainty in aerosol forcing is pointed out by several speakers as an important contributor to total forcing and thereby to estimated climate sensitivity. There can be said to be two primary approaches toward reduced uncertainty in climate sensitivity; either estimates are based on better constrained observations of total forcing (with significant contribution from aerosol forcing) and temperature change or estimates are based on increasingly realistic model predictions. Ideally, progress will be made along both lines. Regardless of which pathway is chosen, direct and indirect aerosol effects will play significant roles, as significant contributors to total forcing. And either way there are difficulties to overcome, related to the complexity of aerosol effects on climate and the prematurity of the aerosol-cloud interaction modeling compared with the understanding and modeling of cloud processes.

The situation at hand—aerosol constituting a substantial forcing agent, and substantial uncertainty remaining, with major implications for interpretation of past and projected future climate change—calls for continued and enhanced effort in quantifying aerosol influences on radiation, cloud, and precipitation processes.

In summary, the importance of a better communication between the cloud and aerosol communities is noted, and more joint activities and research are encouraged.

Reference

Ackerman AS, Toon OB, Stevens DE, Heymsfield AJ, Ramanathan V, Welton EJ (2000) Reduction of tropical cloudiness by soot. Science 288:1042–1047

Albrecht BA (1989) Aerosols, cloud microphysics and fractional cloudiness. Science 245:1227–1230
Anderson TL, Charlson RJ, Schwartz SE, Knutti R, Boucher O, Rodhe H, Heintzenberg J (2003) Climate forcing by aerosols—a hazy picture. Science 300:1103–1104
Andreae MO, Jones CD, Cox PM (2005) Strong present-day aerosol cooling implies a hot future. Nature 435:1187–1190
Bolin B, Charlson RJ (1976) On the role of the tropospheric sulfur cycle in the shortwave radiative climate of the earth. Ambio 5(2):47–54
Bond TC, Streets DG, Yarber KF, Nelson SM, Woo J-H, Klimont Z (2004) A technology-based inventory of black carbon and organic carbon emissions from combustion. J Geophys Res 109:D14203
Bryson RA, Wendland WM (1970) Climatic effects of atmospheric pollution. In: Singer SF (eds) Global effects of environmental pollution. Springer, New York, pp 130–138
Charlson RJ, Langner J, Rodhe H, Levoy CB, Warren SG (1991) Perturbation of the northern hemisphere radiative balance by backscattering from anthropogenic sulfate aerosols. Tellus 43 AB:152–163
Charlson RJ, Schwartz SE, Hales JM, Cess RD, Coakley JA, Hansen JE, Hohmann DJ (1992) Climate forcing by anthropogenic aerosols. Science 255:423–430
Charlson RJ, Ackerman AS, Bender FA-M, Anderson TL, Liu Z (2007) On the climate forcing consequences of the albedo continuum between cloudy and clear air. Tellus 59:715–727
Diner DJ, Beckert JC, Reilly TH, Bruegge CJ, Conel JE, Kahn R, Martonchik JV, Ackerman TP, Davies R, Gerstl SAW, Gordon HR, Muller J-P, Myneni RB, Sellers RJ, Pinty B, Verstraete MM (1988) Multiangle imaging spectroRadiometer (MISR) description and experiment overview. IEEE Trans Geosci Rem Sens 36:1072–1087
Feingold G, Jiang H, Harrington JY (2005) On smoke suppression of clouds in Amazonia. Geophys Res Lett 32:L02804
Hansen J, Sato M, Ruedy R (1997) The missing climate forcing. Phil Trans R Soc Lond 352:231–240
Hansen J, Sato M, Ruedy R (1997) Radiative forcing and climate response. J Geophys Res 102(D6): 6831–6864
Haywood JM, Jones A, Clarisse L, Bourassa A, Barnes J, Telford P, Bellouin N, Boucher O, Agnew P, Clerbaux C, Coheur P, Degenstein D, Braesicke P (2010) Observations of the eruption of the Sarychev volcano and simulations using the HadGEM2 climate model. J Geophys Res 115:D21212
IPCC (2007) In: Solomon S, Qin D, Manning M, Chen Z, Marquis M, Averyt KB, Tignor M, Miller HL (eds) Climate change 2007: the physical science basis. Contribution of Working Group I to the Fourth Assessment Report of the Intergovernmental panel on climate change. Cambridge University Press, Cambridge, p 996
Kahn RA, Gaitley BJ, Martonchik JV, Diner DJ, Crean KA, Holben B (2005) MISR global aerosol optical depth validation based on two years of coincident AERONET observations. J Geophys Res 110: D10S04
Kahn RA, Gaitley BJ, Garay MJ, Diner DJ, Eck TF, Smirnov A, Holben BN (2010) Multiangle Imaging SpectroRadiometer global aerosol product assessment by comparison with the aerosol robotic network. J Geophys Res 115:D23209
Kaufmann RK, Kauppi H, Mann ML, Stock JH (2011) Reconciling anthropogenic climate change with observed temperature 1998–2008. PNAS 108(29):11790–11793
Kiehl JT (2007) Twentieth century climate model response and climate sensitivity. Geophys Res Lett 34:L22710
King MD, Menzel WP, Kaufman YJ, Tanre D, Gao B-C, Platnick S, Ackerman SA, Remer LA, Pincus R, Hubanks PA (2003) Cloud and aerosol properties, precipitable water, and profiles of temperature and water vapor from MODIS. IEEE Trans Geosci Remote Sens 41:442–458
Kinne S et al (2006) An AeroCom initial assessment optical properties in aerosol component modules of global models. Atmos Chem Phys 6:1815–1834
Knutti R (2008) Why are climate models reproducing the observed global mean temperature so well? Geophys Res Lett 35:L18704
Lohmann U, Feichter J (2005) Global indirect aerosol effects: a review. Atmos Chem Phys 5:715–737
Nel A (2005) Air pollution-related illness: effects of particles. Science 308:804–806
Philipona R, Behrens K, Ruckstuhl C (2009) How declining aerosols and rising greenhouse gases forced rapid warming in Europe since the 1980s. Geophys Res Lett 36:L02806
Ramanathan V et al (2001) The Indian Ocean experiment: an integrated assessment of the climate forcing and effects of the great Indo-Asian Haze. J Geophys Res 106(D22):28371–28399
Ramanathan V, Li F, Ramana MV, Praveen PS, Kim D, Corrigan CE, Nguyen H, Stone EA, Schauer JJ, Carmichael GR, Adhikary B, Yoon SC (2007) Atmospheric brown clouds: hemispherical and regional variations in long-range transport, absorption, and radiative forcing. J Geophys Res 112:D22S21

Rasool SI, Schneider SH (1971) Atmospheric carbon dioxide and aerosols: effects of large increases on global climate. Science 173:138–141
Sandu I, Brenguier J-L, Geoffroy O (2008) Aerosol impacts on the diurnal cycle of marine stratocumulus. J Atmos Sci 65:2705–2718
Schwartz SE, Charlson RJ, Rodhe H (2007) Quantifying climate change—too rosy a picture? Nat Rep Clim Chang 1:23–24
Schwartz SE, Charlson RE, Kahn RA, Ogren JA, Rodhe H (2010) Why hasn't Earth warmed as much as expected? J Clim 23:2453–2464
Skeie RB, Berntsen T, Myhre G, Pedersen CA, Ström J, Gerland S, Ogren JA (2011) Black carbon in the atmosphere and snow, from pre-industrial times until present. Atmos Chem Phys 11:6809–6836
Smith SJ, van Aardenne J, Klimont Z, Andres RJ, Volke A, Delgado Arias S (2011) Anthropogenic sulfur dioxide emissions: 1850–2005. Atmos Chem Phys 11:1101–1116
Stephens GL, Vane DG, Boain RJ, Mace GG, Sassen K, Wang Z, Illingworth AJ, O'Connor EJ, Rossow WB, Durden SL, Miller SD, Austin RT, Benedetti A, Mitrescu C, the CloudSat Science Team (2002) The CloudSat mission and the A-TRAIN: a new dimension to space-based observations of clouds and precipitation. Bull Am Met Soc 83:1771–1790
Stern DI (2004) Global sulfur emissions from 1850 to 2000. Chemosphere 58:163–175
Stevens B, Feingold G (2009) Untangling aerosol effects on clouds and precipitation in a buffered system. Nature 461:607–613
Textor C et al (2007) The effect of harmonized emissions on aerosol properties in global models an AeroCom experiment. Atmos Chem Phys 7:4489–4501
Twomey SA (1974) Pollution and the planetary albedo. Atmos Environ 8:1251–1256
Twomey SA (1977) The influence of pollution on the shortwave albedo of clouds. J Atmos Sci 34: 1149–1152
WHO (2003) Health aspects of air pollution with particulate matter, ozone and nitrogen dioxide. Report on a WHO working group. WHO, Bonn, Germany, EUR/03/5042688. http://www.euro.who.int/document/e79097.pdf
Wilcox EM (2010) Stratocumulus cloud thickening beneath layers of absorbing smoke aerosol. Atmos Chem Phys 10:11769–11777
Winker DM, Vaughan MA, Omar A, Hu Y, Powell KA, Liu Z, Hunt WH, Young SA (2009) Overview of the CALIPSO mission and CALIOP data processing algorithms. J Atmos Ocean Technol 26: 2310–2323
Wood R (2007) Cancellation of aerosol indirect effects in marine stratocumulus through cloud thinning. J Atmos Sci 64:2657–2669
Xue H, Feingold G, Stevens B (2008) Aerosol effects on clouds, precipitation, and the organization of shallow cumulus convection. J Atmos Sci 65:392–406
Zhao G, Di Girolamo L (2006) Cloud fraction errors for trade wind cumuli from EOS-Terra instruments. Geophys Res Lett 33:L20802

Surv Geophys (2012) 33:701–721
DOI 10.1007/s10712-011-9153-z

Reducing the Uncertainties in Direct Aerosol Radiative Forcing

Ralph A. Kahn

Received: 26 July 2011 / Accepted: 11 October 2011 / Published online: 27 October 2011

Abstract Direct aerosol radiative forcing (DARF) remains a leading contributor to climate prediction uncertainty. To monitor the spatially and temporally varying global atmospheric aerosol load, satellite remote sensing is required. Despite major advances in observing aerosol amount, type, and distribution from space, satellite data alone cannot provide enough quantitative detail, especially about aerosol microphysical properties, to effect the required improvement in estimates of DARF and the anthropogenic component of DARF. However, the combination of space-based and targeted suborbital measurements, when used to constrain climate models, represents an achievable next step likely to provide the needed advancement.

Keywords Aerosol remote sensing · Direct aerosol radiative forcing

1 Introduction

Airborne particles, which include desert and soil dust, wildfire smoke, sea salt, volcanic ash, black carbon, natural and anthropogenic sulfate, nitrate, and organic aerosol, affect Earth's energy budget both directly, by reflecting and absorbing sunlight, and indirectly, by altering cloud microphysical processes. This paper reviews the current status, and evaluates future prospects for quantifying a major component of the aerosol influence on Earth's climate system, through the direct radiative impact on incoming solar radiation. Aerosol-cloud interactions, and the indirect aerosol radiative effects that these produce, are beyond the scope of the current paper. Global-scale observational constraints on the indirect processes are much less well established and, as such, the continuum between "clear-sky" conditions, which are the subject of this paper, and "all-sky" conditions (e.g., Várnai and Marshak 2011), is considered here only in the context of the cloud masking techniques applied to aerosol remote-sensing observations.

R. A. Kahn (✉)
Laboratory for Atmospheres, NASA Goddard Space Flight Center, Greenbelt, MD 20771, USA
e-mail: ralph.kahn@nasa.gov

The Intergovernmental Panel on Climate Change (IPCC) assessment of direct aerosol radiative forcing (DARF) uncertainty changed from "Very Low" to "Medium–Low" between their Third Assessment Report (IPCC 2001), completed as data from NASA's earth observing system (EOS) satellites were just becoming available, and their Fourth Assessment 6 years later. The IPCC uncertainties are estimated in light of the diversity in performance among a collection of benchmark climate models. Their DARF uncertainty designations represent the first and third rungs, respectively, of a five-step scale that reports consensus judgment, based upon available physical and chemical evidence. Between the third and fourth IPCC assessment reports, a climatology of global, monthly aerosol optical depth (AOD) was produced by combining data from the EOS Multi-angle imaging spectro-radiometer (MISR) and MODerate resolution imaging spectro-radiometer (MODIS) instruments, with ground-based values obtained from the Aerosol Robotic Network (AERONET; Kinne et al. 2006). The resulting AOD constraints made a key contribution toward reducing the reported model uncertainty (Haywood and Schulz 2007).

However, DARF uncertainty remains a limiting factor for reducing global average radiative forcing assessment, as the model-based estimated value is about -0.5 ± 0.4 Wm^{-2} (i.e., net cooling) for the global anthropogenic, all-sky, top-of-atmosphere component, assessed since the pre-industrial era (IPCC 2007), and is about -1.1 ± 0.4 Wm^{-2} for clear-sky conditions over ocean (CCSP 2009); for comparison, the CO_2 forcing is given as 1.66 ± 0.17 Wm^{-2} (IPCC 2007). Calculations suggest that instantaneous, mid-visible AOD measurement accuracy of about 0.02 is typically required under cloud-free conditions to constrain DARF to approximately 1 Wm^{-2} (McComiskey et al. 2008; CCSP 2009), whereas the corresponding uncertainties in the current global AOD products from MISR and MODIS are 0.03 or larger over dark water, and 0.05 or larger over land (Kahn et al. 2010; Levy et al. 2010; Remer et al. 2005). Theoretical DARF sensitivity analysis identified particle single-scattering albedo (SSA) as the other leading factor in most situations, especially important for determining radiative forcing over land surfaces, and requiring an instantaneous constraint of about 0.02, though varying with other factors, particularly AOD and surface albedo (McComiskey et al. 2008).

Perturbation studies based on one-dimensional radiative transfer calculations suggest that the quantitative uncertainties reported by the IPCC (2007), which were derived only from model diversity rather than aggregated error, are actually underestimated by a factor between two and four (Loeb and Su 2010). This study also concludes that, along with AOD, SSA uncertainty makes a leading contribution to overall DARF uncertainty, especially when clouds are present. A separate assessment of DARF uncertainty was performed by comparing forward simulations of DARF over the industrial era, with the same quantity calculated from the difference between (1) overall climate model forcing required to reproduce the temperature record, and (2) all forcing elements other than aerosol (Anderson et al. 2003). For the anthropogenic aerosol component, the differences between these two estimates were significant: the estimate was approximately -1.0 Wm^{-2} for the inverse calculations, and a statistically larger net cooling of roughly -1.5 Wm^{-2} for the forward approach, with scatter of order ± 0.5 Wm^{-2} for both methods.

One consequence of the current DARF uncertainty is the limitation it places on constraining climate model predictive ability (Anderson et al. 2003, 2005; Kiehl 2007; Schwartz et al. 2010). In its simplest form, a model's climate sensitivity, λ (K/(Wm^{-2})) is often represented as a factor linearly relating the model-predicted change in global-mean surface temperature (GMST), ΔT, to an increase in effective global-mean forcing, ΔF:

$$\Delta F = \lambda^{-1} \Delta T + N \tag{1}$$

Here, N is the global heat content rate-of-change, associated primarily with warming of the ocean, that approaches zero as equilibrium relative to a perturbed forcing is reached. The climate sensitivity term includes the modeled climate feedbacks; complicating factors include the time-scale over which the forcing acts, and that required for the system to relax to equilibrium, as well as the need to account for regional differences in assessing the global aggregate, nonlinear aspects of the system, and whether total or only anthropogenic aerosols are considered. (Note that, although DARF typically refers to direct radiative forcing by all aerosols, the term "Direct Aerosol Climate Forcing" is sometimes used to mean the direct forcing due to anthropogenic aerosols alone. To avoid confusion, the term "anthropogenic component of DARF" is used here, when that is intended.)

A key method of evaluating climate model performance overall is to compare the simulated global mean surface temperature (GMST) during the twentieth century with the observed temperature record for the same period, when the model forcing includes best-estimate greenhouse gas accumulation, solar variability, and aerosol effects, so uncertainty in DARF translates directly into model climate sensitivity uncertainty. As such, DARF is a leading factor limiting the confidence with which current climate models can predict changes in GMST (IPCC 2007), and among 11 widely cited climate models that reproduce the twentieth-century temperature anomaly, climate sensitivity is inversely related to the assumed anthropogenic forcing, of which aerosol forcing contributes the largest uncertainty (Kiehl 2007). The IPCC (2007) best estimate for λ is 0.8 K/(Wm^{-2}), with a one-σ range of about 0.54–1.2 K/(Wm^{-2}) based again on model diversity, so loosely translating DARF uncertainty into modeled equilibrium GMST uncertainty, DARF uncertainty of 1 Wm^{-2} yields a best-estimate GMST uncertainty of 0.8 K, ranging to values in excess of 1 K.

This paper provides a brief review the strengths and limitations of aerosol measurements that bear upon current, short-wave DARF under cloud-free conditions, and discusses how the contributions of satellite and suborbital measurements, when combined with aerosol transport modeling, can reduce the uncertainties in DARF.

2 What Satellites Offer

Aerosols are especially challenging to study because they originate from many, diverse sources, both natural and anthropogenic, and exhibit an enormous range of chemical compositions and physical properties. Unlike long-lived atmospheric gases, airborne particles are typically removed from the troposphere by precipitation or gravitational settling within about a week, so aerosol amount and type vary dramatically on many spatial and temporal scales. For this reason, the frequent, global coverage provided by space-based remote-sensing instruments has played a central role in the study of aerosol impacts on the global energy budget (e.g., Bellouin et al. 2008; Kim and Ramanathan 2008; Loeb and Manalo-Smith 2005; Myhre et al. 2009; Quaas et al. 2008: Yu et al. 2006).

This section gives an overview of passive aerosol remote-sensing capabilities from space; in addition to the references cited below, further detail about many of the retrieval algorithms involved can be found in Kokhanovsky et al. (2009) and de Leeuw et al. (2011).

2.1 Early Aerosol Optical Depth (AOD) Measurements

The aerosol parameter most commonly derived from spacecraft remote-sensing data, and used in global energy budget calculations, is total-column, mid-visible, aerosol extinction optical depth. It is a measure of aerosol amount based on the fraction of incident light that is either scattered or absorbed by particles, integrated over the vertical column through the depth of the atmosphere. Formally, AOD is a dimensionless quantity, the product of the particle number concentration, the particle-effective extinction cross-section (which accounts for particle scattering + absorption), and the path length through the atmosphere, assessed along an effective vertical path.

To infer AOD from the upwelling radiances observed by space-based passive imagers, retrieval algorithms must sort out surface from atmospheric contributions, must account for particle absorption and the angular variation of particle light-scattering intensity, as well as scene heterogeneity, cloud contamination, and any gaseous absorption in the spectral bands used. The problem is underdetermined, and assumptions are made, as needed, depending on the retrieval approach adopted. Most retrieval algorithms to date assume the aerosol is horizontally homogeneous over the retrieval region, which can be comprised of one or more pixels. Different approaches are often applied depending on surface type; it is generally easier to perform aerosol retrievals with these data over dark, uniform ocean or dark land than brighter surfaces or terrain that is topographically complex or variable on retrieval-region length scales. Much creative effort during the past 30 years has focused on developing reliable overland remote-sensing retrieval methods. Aerosol vertical distribution is usually either assumed or adopted from external models or measurements (e.g., Sect. 2.4 below). The required particle optical properties are either assumed based on climatological considerations, or constrained as part of self-consistent interpretation of the observed radiances.

The advanced very high resolution radiometer (AVHRR) imagers have been collecting daily, global, multispectral data from polar orbit since late 1978 (Table 1). Although these instruments were designed primarily to observe Earth's surface, over-ocean, total-column AOD has been derived, initially from single-channel observations, assuming a completely dark ocean surface in the red band (0.63 micrometers effective wavelength) along with medium-sized, purely scattering particles (Stowe et al. 1997). Subsequently, two-channel retrievals were performed (Mishchenko et al. 1999; Ignatov et al. 2004), adding the 0.85 micrometer band, refining the assumed ocean surface reflectance model, and also retrieving the Ångström exponent (ANG, related to the spectral slope of AOD), a quantity that can provide some indication of particle size. Similarly, AOD retrieval algorithms have been developed for imagers aboard geostationary satellites, such as NOAA's GOES series (Knapp 2002) and European Space Agency's Spinning Enhanced Visible and Infrared Imager (SEVIRI; Popp et al. 2007; Wagner et al. 2010), providing high-temporal-resolution products as frequently as every 15 or 30 min, making it possible to at least qualitatively map the development of aerosol plumes. Although broad spectral bands and poor radiometric calibration limit the quantitative application of these data, early results demonstrated the long-range, over-ocean transport of dust, smoke, and pollution particles from terrestrial sources (e.g., Husar et al. 1997), highlighting the need to consider aerosol effects on the global energy budget.

Other remote sensing approaches are required to separate the surface from atmospheric contributions to the observed scene brightness, making it possible to retrieve AOD over more reflective coastal regions and most land. Due to the intense scattering of ultraviolet (UV) light by atmospheric molecules, the surface is obscured when viewed from space in the wavelength range 0.33–0.38 micrometers, and many types of surfaces themselves are relatively dark in the UV. In the late 1990s, it was realized that the total ozone mapping spectrometers (TOMS), versions of which had already been orbiting for nearly 20 years, contained the spectral channels needed to detect aerosols over land as well as water, based

Table 1 Major, currently available, satellite tropospheric aerosol data products

Sensor/platform	Spatial coverage	Temporal coverage	Retrieved aerosol parameters	Leading sources of AOD uncertainty
AVHRR/NOAA-series	~Daily, global ocean	1981–present	AOD, ANG	CAL, CLD, TYP
TOMS/Nimbus, ADEOS1, EP	~Daily global, land & ocean	1979–2001	AOD, AAI, SSA, AAOD	HT, CLD, CAL, TYP, RES
ATSR-2/ERS-2; Advanced-ATSR/ENVISAT	~Weekly global, land & ocean	1995–2003; 2003–present	AOD, ANG	TYP, CLD, CAL
GOME/ERS-2; GOME-2/METOP	~Weekly global, land & ocean; ~daily global, land & ocean	1995–2003; 2007–present	AAI, AOD, TYP in synergy with ATSR-2/AVHRR	RES, SURF, HT, CLD, CAL
POLDER-1, -2, PARASOL	~Daily global, land & ocean	1997–present	AOD, ANG, NSPH (fine-mode over land)	CAL, CLD, TYP
MODIS/Terra, Aqua	~Daily global, land & ocean	2000–present (Terra) 2002–present (Aqua)	AOD, ANG (ocean), FMF (ocean)	CAL, CLD, TYP, SURF (over land)
MISR/Terra	~Weekly global, land & ocean + bright desert & nadir sun-glint	2000–present	AOD, ANG, SML, NSPH, SSA (2–4 bins), PH	CLD, CAL, TYP
GLAS/ICESat	Global curtain, land & ocean, 16-day repeat cycle	2002–present (~3 months/year)	Extinction/backscatter VP	–
SEVIRI/MSG	15 min intervals, Europe, Africa, Atlantic	2003–present	AOD	CAL, TYP, SURF
SCIAMACHY/ENVISAT	~Weekly global, land & ocean	2003–present	AAI; AOD, TYP in synergy with ATSR	CAL, SURF, CLD, TYP
MERIS/ENVISAT	~3-day global, land & ocean	2003–present	AOD	CLD, TYP, SURF
OMI/Aura	~Daily global, land & ocean	2004–present	AOD, AAI, SSA, AAOD	HT, CLD, CAL, TYP
CALIOP/CALIPSO	Global curtain, land & ocean, 16-day repeat cycle	2006–present	VP: extinction/backscatter, color ratio, depolarization ratio, AOD	EBR, CAL, CLD, D_SNR

Sources of uncertainty vary with conditions; they are listed in approximate order of significance

AOD aerosol optical depth, *AI* aerosol index, *ANG* Ångström exponent, *TYP* aerosol type, *AAI* absorbing aerosol index, *FMF* fine-mode fraction, *SML* AOD fraction small, medium, large, *SSA* single-scattering albedo, *AAOD* absorbing AOD, *NSPH* fraction AOD nonspherical, *PH* near-source plume height, *VP* vertical profile, *CAL* radiometric calibration, *CLD* cloud contamination, *TYP* aerosol-type assumption or constraint, *HT* aerosol layer height, *SURF* bright surface reflectance, *RES* spatial resolution, *EBR* extinction/backscatter ratio (a function of aerosol type), *D_SNR* signal/noise ratio for daytime observations (e.g., Kacenelenbogen et al. 2011) on aerosol absorption of the upwelling background UV light (Herman et al. 1997). This resulted in the Aerosol Index or Absorbing Aerosol Index, a qualitative measure of UV-absorbing aerosols such as dust and smoke, subsequently converted to quantitative estimates of AOD using a combination of available constraints and assumptions about aerosol type (Torres et al. 1998). A similar approach has been used to derive Aerosol Index from the GOME series of satellites, over coarse surface sampling regions averaging about 40×320 km (de Graaf et al. 2005). These retrievals have limited sensitivity to near-surface aerosols, and depend sensitively on the altitude of the particles and their optical properties, but the TOMS maps provided the first comprehensive, long-term record of aerosol source regions and overland transports (e.g., Torres et al. 2002; Prospero et al. 2002). A next-generation ozone monitoring instrument (OMI), with additional spectral channels in the visible, and 13×24 km sampling resolution at nadir (compared to 39×39 km for the Earth Probe TOMS, the latest in the series), has been flying aboard the EOS Aura spacecraft since July 2004. AOD at UV and visible wavelengths, as well as estimates of column-effective SSA, are routinely derived from OMI observations (Torres et al. 2007).

One of the first and most widely used aerosol remote-sensing techniques is surface-based sun photometry, a multi-angle approach that involves measuring the instantaneous intensity of the solar disk in narrow (~ 10 nm wide) bands in spectral regions of minimal gaseous absorption. The method predates satellite observations (e.g., Flowers et al. (1969), who present the results of a remarkable network of about 40 Voltz sun photometers across the United States), and relies on highly accurate radiometric calibration. Calibration is achieved for reference instruments by the Langley method: the systematic increase in atmospheric opacity and the corresponding decrease in solar brightness, are observed as the sun is viewed through longer atmospheric slant paths at stable, pristine mountain top observatories. Subsequently, field instruments are inter-calibrated with these references. AOD is measured, typically at multiple wavelengths, allowing some information about particle size to be derived. Though they offer only limited spatial coverage, networks of autonomous surface-based sun photometers have been deployed, such as AERONET (Holben et al. 1998), a global federation of several hundred stations, and SkyNet in East Asia (Kim et al. 2004); ships-of-opportunity provide additional AOD validation measurements in the generally under-sampled ocean regions (Smirnov et al. 2009). When pre- and post-observation instrument calibration is performed, the measured mid-visible AOD accuracy is estimated at about 0.01 for the AERONET CIMEL instruments (Eck et al. 1999); the results are used in part to validate the assumptions needed to retrieve AOD and ANG from space, by both statistical and event-by-event comparisons with the satellite retrievals (e.g., Remer et al. 2005; Levy et al. 2010; Kahn et al. 2010).

Multi-angle approaches have also been applied to space-based observations, beginning in 1995 with the European Space Agency (ESA) two-view-angle Along-Track Scanning Radiometer-2 (ATSR-2) imager (Veefkind et al. 1998, 2000; North 2002; Table 1). The satellite technique measures light scattered by the scene below, so unlike the situation for surface-based sun photometry, additional assumptions about aerosol and surface optical properties are required to retrieve column-integrated AOD. However, the atmospheric contribution to the signal, relative to that of the surface, increases systematically as the slant path increases, making surface-atmosphere separation possible. The steeper slant paths also improve sensitivity to thinner aerosol layers.

A third approach, in addition to UV and multi-angle, for separating surface from atmospheric signals in top-of-atmosphere observations was pioneered by the French Space Agency (CNES), which began flying their POLarization and Directionality of the Earth's

Reflectances (POLDER) series of imagers in 1996 (Deschamps et al. 1994). The POLDER instruments collected multi-angle, multispectral polarization data from orbit, at a pixel resolution of about 6 km at the sub-spacecraft point. Over most common land surface types, except bright desert, the polarization effects are fairly independent of wavelength (Waquet et al. 2009), making it possible to separate the more constant surface polarization contribution to the satellite signal from the spectrally varying atmospheric contribution.

Aerosols are sometimes divided into two size groups, depending on whether their effective diameter is greater or less than a certain size, usually taken to be around 1 micrometer for satellite and some field observations, and 2.5 micrometers for air quality applications. These distinctions are somewhat arbitrary, but are based on loose physical considerations (e.g., Whitby 1978); the majority of smoke and aerosol pollution particles, the products of combustion and chemical processing, tend to fall into the smaller "accumulation" or "fine" mode. Mechanically produced desert dust and sea salt particles tend to be weighted toward the "coarse" mode. And, if inhaled, particles smaller than about 2.5 micrometers in diameter are more likely to penetrate the lungs. From its combination of optical measurements, POLDER maps column-integrated fine-mode and total aerosol optical depth over water, as well as fine-mode optical depth over land (Deuzé et al. 2001; Herman et al. 2005).

2.2 AOD Measurements in the EOS Era

The MISR and MODIS instruments were launched into a sun-synchronous polar orbit with a 10:30 local equator-crossing time, descending on the dayside, in December 1999 aboard the EOS Terra satellite (Table 1). They were designed with enhanced capabilities for monitoring aerosols from space, compared to earlier efforts. A second MODIS instrument was deployed aboard the EOS Aqua satellite in May 2002, in a polar orbit having a 1:30 local equator-crossing time, ascending on the day side. MODIS follows the multispectral approach of the AVHRR instruments, but with higher spatial resolution (a maximum of 250 m, compared with 1 km for AVHRR), 36 spectral channels, and much higher radiometric calibration accuracy and stability. Global, total-column AOD over cloud- and glint-free ocean, as well as darker land surfaces, is produced routinely every 2 days, along with fine-mode fraction over ocean (Kaufman et al. 1997; Tanré et al. 1997; Remer et al. 2005; 2008; Levy et al. 2010).

Over ocean, standard glint and whitecap models are applied, and the surface is assumed dark at red and near-infrared wavelengths; using a look-up table approach, the algorithm selects the best-matching simulated radiances derived from assumed combinations of one fine-mode and one coarse-mode component (Remer et al. 2005). Several new techniques were developed to separate surface from atmospheric signals for multispectral, single-view imagers observing over land. For example, the top-of-atmosphere radiance in the MODIS 2.1 micrometer channel, for which the aerosol and atmospheric gas opacities (except water vapor in some places) are much lower than in the visible bands, with AOD often negligible, is used to determine the surface reflectance at this wavelength. The 2.1 micrometer reflectance is then used to evaluate the 0.47 and 0.67 micrometer surface reflectances, based on empirically derived relationships that depend on viewing geometry and a surface greenness index (Kaufman et al. 1997; Levy et al. 2007). Lacking sensitivity to particle type over land, the MODIS algorithm assumes particle types based on a climatology derived from AERONET observations (Levy et al. 2007). A similar approach is planned for the next-generation of U.S. geostationary satellites (Laszlo et al. 2008).

A related method has also been applied to data from the medium resolution imaging spectrometer (MERIS) aboard ESA's polar-orbiting ENVISAT spacecraft. However, the 15 programmable MERIS spectral bands do not extend as far into the near-infrared as those of MODIS, so a vegetation index calculated from 0.675 and 0.870 micrometer radiances is used to constrain the surface reflectance (von Hoyningen-Huene et al. 2007).

Yet another technique for retrieving AOD over land is based on looking at the time-dependent variation in measured radiance. By carefully selecting dark pixels in a time-series of images, representing the surface reflectance under cloud-free conditions and having the lowest available AOD, the AOD on hazier days can be estimated; this method has been applied to aggregates of geostationary satellite data about 2 weeks in length, taken at the same time-of-day to assure similar solar geometry (Knapp et al. 2007). For MODIS, up to 16 days of observations, capturing a range of view angles and atmospheric conditions, have been used in an automated algorithm to determine the 2.1–0.47 micrometer surface reflectance ratio dynamically, and to retrieve column-effective AOD, ANG, and surface bidirectional reflectance properties simultaneously at 1 km resolution (Lyapustin and Wang 2009; Lyapustin et al. 2011). Efforts are also being made, using the 0.412 micrometer "deep blue" channel, to extend the MODIS aerosol retrieval over land surfaces such as desert that are relatively dark in this band, but brighter at longer wavelengths (Hsu et al. 2004, 2006). In addition, MODIS thermal infrared channels are used to detect thermal hot spots, which can help locate fires and identify the associated smoke plumes (Giglio et al. 2006; Ichoku et al. 2008).

MISR complements MODIS, acquiring four-spectral-channel, near-simultaneous views of Earth at nine angles, spatial sampling up to 275 m/pixel, and similarly high radiometric calibration accuracy ($\sim$3% absolute, 1.5% band-to-band) and stability (Diner et al. 1998). Having a narrower swath than MODIS, MISR takes about a week to image the entire planet. With the multi-angle coverage, AOD and surface reflectance are retrieved self-consistently over land, including bright desert surfaces, providing more accurate AOD overland results, compared to other techniques (Martonchik et al. 2002, 2009). The nine views also sample light scattered in different directions, which yields some information about column-averaged particle size, shape, and SSA under favorable retrieval conditions (Kahn et al. 2010; Sect. 2.3 below).

For aerosol retrievals aimed at clear-sky conditions, cloud screening is an essential component of the algorithms. Brightness thresholds are routinely used, in the thermal infrared, when available, to identify cold, high clouds, and in the visible and near-infrared, for clouds having sufficient optical depth at these wavelengths (e.g., Ackerman et al. 1998; Remer et al. 2005; Stowe et al. 1998; Zhao and Di Girolamo 2004). Scene variability at high spatial resolution is applied in the MODIS algorithm over water to identify heterogeneity likely to be due to cloud (Martins et al. 2002), and the MODIS 1.38 micrometer channel is used to provide greater sensitivity to high, thin cirrus in the aerosol algorithm (Gao et al. 2002). For MISR, the angular information is applied in several ways, including changes in the reflected-light spectral ratio with angle (Di Girolamo and Wilson 2003) and stereo-derived heights (Moroney et al. 2002) to identify reflecting layers above the nominal surface, as well as angular smoothness and angle-to-angle spatial correlation tests to detect cloud edges, changes in the scene during multi-angle data acquisition, and sub-pixel scene heterogeneity (Martonchik et al. 2002, 2009).

However, these cloud-screening techniques sometimes mask thick aerosol plumes and, more generally, they offer limited sensitivity to clouds having visible optical depth below a few tenths. Systematic increases in MODIS-retrieved AOD with cloud fraction are observed (e.g., Loeb and Manalo-Smith 2005; Zhang and Reid 2006), and effects including aerosol humidification near clouds (Twohy et al. 2009; Tackett and Di Girolamo 2009), sub-pixel cloud contamination (Zhao et al. 2009), and the three-dimensional effects of cloud-scattered light (Marshak et al. 2008), have been identified as likely contributors under different conditions. Efforts at quantifying the contribution of clouds to the MODIS AOD identified a net bias of about 0.02 (Kaufman et al. 2005a), but the transition regions between cloudy and cloud-free atmosphere are not well resolved in the satellite observations, and validation data are scarce, so considerable uncertainty remains in this aspect of retrieval quality.

To evaluate confidence over the entire range of retrieved AOD values, an absolute criterion is needed to take account of the lower limit of AOD measurement sensitivity (e.g., 0.03 or 0.05), and in addition, a relative criterion (e.g., percent of AOD) is required to adequately represent AOD variability and other factors that scale as AOD itself (Kahn et al. 2011). Based on statistical comparisons with large numbers of near-coincident AERONET surface sun photometer observations, the Collection 5 MODIS AOD product confidence envelope (1σ) is $\pm$(0.03 +5% of AOD) over dark water (Remer et al. 2005), and $\pm$ (0.05 + 15% of AOD) over land (Levy et al. 2010). The Version 22 MISR overall confidence envelope, assessed more conservatively than the MODIS studies, amounts to the larger of 0.05 or 20% of AOD, with higher confidence over water (Kahn et al. 2010).

These AOD assessments do have limitations. For example, there are significant gaps in the global distribution of sun photometer sites, and apparent AOD discrepancies are often caused by actual AOD variability on kilometer scales, as sampled differently by the satellite vs. the ground-based instruments. And these comparisons can be made only when the satellite and surface instrument cloud masks both report cloud-free conditions; this precludes evaluating the satellite cloud-masking methods with these data.

Nevertheless, the validation results show that although the required instantaneous AOD accuracy of about 0.02 has not yet been achieved, it is within reach. The contribution already made by the combined MISR, MODIS, and AERONET data toward constraining climate model AOD and the resulting DARF estimates was discussed in Sect. 1. The same data combination has also made it possible to identify regional, and to some extent global, over-ocean AOD trends for the decade beginning in 2000 (Zhang and Reid 2010). Efforts to extend the record back through the early years of AVHRR have met with success in documenting regional trends over ocean, providing results consistent with regional patterns of observed solar dimming (Mishchenko and Geogdzhayev 2007), but the advances offered by the EOS-generation instruments are apparently required to isolate global-average behavior over ocean (Zhang and Reid 2010). AOD trends are especially difficult to assess, in part because they represent only one possible component of AOD variability, and individual outlier events can sometimes confuse a simple trend analysis (see, e.g., Kahn et al. 2011). As longer time-series become available, the analysis of AOD trends over ocean will become more tractable, though the greater challenge of identifying and quantifying trends over land awaits additional retrieval algorithm refinement and/or new instrumentation.

Improvements to the present MISR and MODIS AOD products rest in part on reducing radiometric calibration uncertainties, but primarily on upgrading the assumptions used to help constrain aerosol type, that are an essential component of the AOD retrieval process (Kahn et al. 2010; Levy et al. 2010). The assumed aerosol vertical distribution and surface reflectance must also be considered when interpreting observed top-of-atmosphere radiance signals in terms of AOD. We turn to these factors in the next two subsections.

2.3 Aerosol Type

The spectral dependence of AOD contains some information about particle size (King et al. 1978), and this is the basis for reporting ANG and fine-mode-fraction in the AVHRR and MODIS over-ocean products. Desert dust and sea salt tend to have higher fractions of coarse-mode particles, whereas smoke, secondary organic aerosol, and pollution particles tend to be fine-mode dominated. One caveat with this aerosol-type identification approach is that interpretation of ANG is ambiguous, especially when multiple aerosol modes are present in the atmospheric column (Schuster et al. 2006), which is a common occurrence. Yet, the fine-to-coarse AOD ratio from these instruments exhibits the generally expected regional and seasonal patterns, especially for desert dust from the Sahara and smoke originating from sub-Saharan grassland or tropical forest fires that is subsequently advected over the Atlantic, lending confidence to the results (Mishchenko et al. 1999; Remer et al. 2008).

Kaufman et al. (2005b) took the further step of interpreting the combination of MODIS over-ocean AOD and fine-mode fraction in terms of "anthropogenic" AOD. To accomplish this, they assume that a fixed fraction of fine-mode aerosol is smoke from natural sources, and that the tails of the dust and sea spray size distributions contribute fixed fractions to the fine-mode AOD, based on observations under relatively well-constrained conditions. This approach again reproduces generally expected patterns, but the results are easily over-interpreted, as the information content of the MODIS data is limited to a qualitative size constraint, and the actual size-resolved fractional contributions of different aerosol components vary immensely with conditions.

By combining the column AOD from MODIS with collocated UV radiances from OMI under cloud-free conditions, ambiguities in the OMI retrieval associated with aerosol altitude, total aerosol amount, and sub-pixel cloud contamination can be reduced, so tighter constraints on both aerosol layer height and UV absorption can be obtained than is possible from OMI alone (Satheesh et al. 2009). A paucity of coincident ground-truth data precludes quantitative validation of the retrieved UV SSA, but the relative values give an indication of the regionally and seasonally varying dust and smoke distributions.

Multi-angle views and polarized channels yield additional constraints on particle type. For example, the polarized multi-angle, multispectral radiance measurements from POLDER contain information on particle shape as well as size. Depolarizing desert dust can be distinguished from spherical particle types (Deuzé et al. (2001), and over ocean, fine- and coarse-mode AOD values can be derived, along with the coarse-mode nonspherical component AOD (Herman et al. 2005).

A classification of aerosol type, based on column-averaged particle size, shape, and SSA constraints, can be derived from the MISR multi-angle, multispectral observations (Fig. 1). The instrument can distinguish about three-to-five groupings based on particle size, two-to-four groupings in SSA, and spherical vs. nonspherical dust and cirrus particles (Chen et al. 2008; Kahn et al. 2001, 2010; Kalashnikova and Kahn 2006; Pierce et al. 2010). Although AOD retrievals tend to be robust under different scene conditions due to the systematic variation in atmospheric slant path observed by the MISR cameras, aerosol-type information varies considerably with the range of scattering angles sampled, with surface reflectance properties, and with AOD itself. As such, MISR aerosol-type information is essentially categorical, making it possible to map out about a dozen aerosol air-mass types globally when the AOD is greater than about 0.15, observations at an adequate range of scattering angles are obtained, and the surface is sufficiently uniform or well defined

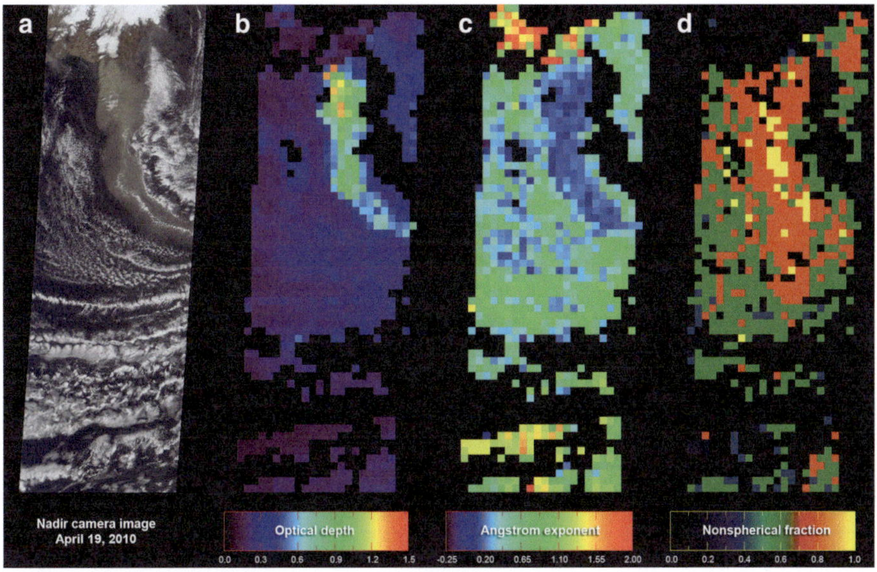

Fig. 1 An example of aerosol amount and type mapping from MISR. **a** This true-color, nadir view shows the (*brown*) ash plume emanating from Iceland's *Eyjafjalljökull* volcano, at about 12:51 UTC on April, 19 2010. The volcano itself is at the top-center of the image. The plume stands out as having higher mid-visible AOD (**b**) and being dominated by larger particles (lower ANG, **c**), than the background. The ash plume appears, but is less distinct, in the fraction AOD nonspherical map (**d**), as the algorithm also picks up some adjacent nonspherical cirrus. The aerosol properties are reported on 17.6 km retrieval regions. Ash from several eruptions of this volcano during Spring 2010 curtailed air traffic over much of Europe [*MISR Team, Jet Propulsion Lab/Caltech and NASA Goddard Space Flight Center*]

(Kahn et al. 2010). However, this information alone is not enough to provide the detailed particle microphysical properties required to fully constrain DARF.

A multispectral method has also been applied to synergetically exploit data from two instruments aboard ESA's ENVISAT spacecraft: the 1 km pixel-resolution AATSR radiometer for cloud screening and AOD retrieval, and the SCanning Imaging Absorption spectroMeter for Atmospheric CHartographY (SCIAMACHY) spectrometer to constrain aerosol type. Only the nadir view is used for the aerosol-type classification, which relies upon fitting spectral reflectances at 10 visible wavelengths from 405 to 670 nm; with the coarse SCIAMACHY spatial resolution of at about 30×60 km^2, four broad aerosol categories, water-soluble, soot, mineral dust, and sea salt, can be differentiated (Holzer-Popp et al. 2008).

Aerosol type has also been retrieved from active sensor data. Specifically, a layer-resolved aerosol-type classification has been produced from the combination of spectral backscatter, extinction, and polarization measurements made by the lidar aboard the Cloud-Aerosol Lidar and Infrared Pathfinder Satellite Observations (CALIPSO) spacecraft, flying since 2006 in formation with the EOS Aqua satellite carrying MODIS (Cattrall et al. 2005). Based on a clustering of optical properties measured by the active lidar sensor, qualitative distinctions among typical maritime, desert dust, biomass burning, and urban/industrial particles can be made, along with a mixture of types common in SE Asia.

One factor, in addition to the information content of the satellite remote-sensing data, limiting the quantitative assessment of aerosol type from spacecraft is a lack of validation data. By scanning across the almucantar in addition to making direct-sun measurements, the AERONET surface stations collect the data needed to retrieve several modes of column-integrated particle size distribution, along with constraints on column-effective (though not component-specific) SSA and particle sphericity (Dubovik and King 2000; Dubovik et al. 2006). Despite limited sampling, and significantly more assumptions than are required for the direct-sun spectral AOD measurements, these observations represent among the most extensive aerosol-type data available, and they have been used to derive general-purpose aerosol-type climatologies (Dubovik et al. 2002; Eck et al. 2008, 2010). AERONET-based climatologies have also been developed specifically for use in the CALIPSO (Omar et al. 2009) and the MODIS overland (Levy et al. 2007) retrievals.

In addition to sampling issues, the surface-based sun-sky radiometer constraints on SSA require AOD $> \sim 0.4$ at 0.44 micrometers, a criterion achieved infrequently in many parts of the world, though not in some regions dominated by major anthropogenic emissions, including major urban areas of North America and Europe, and parts of China and India. And AERONET-retrieved column-effective SSA does not distinguish between different aerosol modes or layers in the atmosphere. However, species-specific aerosol chemical and microphysical property validation data can be obtained with much greater detail from surface and aircraft direct sampling. Such measurements can provide the most accurate size distribution, SSA values, and chemical composition currently available, but sampling is extremely limited; coincident aircraft and satellite observations, obtained on field-campaign "Golden Days," can critically test aerosol-type retrievals, but only on these very special occasions (e.g., Kahn et al. 2009).

In summary, qualitative aerosol-type classifications have been produced from both passive and active EOS-era remote sensors, yielding broad-swath, column-effective, and layer-resolved curtain coverage, respectively. This represents a major advance in capability but, in itself, does not provide sufficient microphysical detail to resolve DARF uncertainties, nor does it provide the compositional constraints required to identify the anthropogenic aerosol component.

2.4 Aerosol Vertical Distribution and Surface Reflectance

Aerosol vertical distribution is a contributing factor when calculating DARF, especially for absorbing species, and it is also of primary importance for assessing aerosol transports and material fluxes (e.g., Yu et al. 2008). The highest-accuracy space-based measurements of aerosol vertical distribution are obtained from lidar, which can detect multiple aerosol layers day and night, as optically thin as 0.005 at mid-visible wavelengths, though signal-to-noise is lower under daylight conditions (Winker et al. 2009). Space-based lidar makes it possible to create a global, climatological picture of transported aerosols, but the lidar swath is the width of the laser beam, about 70 meters in size for CALIPSO, so coverage of specific sites or events is serendipitous (e.g., Kahn et al. 2008). The technique was demonstrated in 1994, with the Lidar In-space Technology Experiment (LITE) aboard NASA's Space Shuttle Discovery, and has been followed by the geoscience laser altimeter system (GLAS) and CALIPSO instruments on polar-orbiting EOS satellites (Table 1). The vertical resolution of the CALIPSO observations is as high as 30 m, and the horizontal resolution of the CALIPSO profile product is as high as 330 m, though it can be reduced when greater horizontal spatial averaging is applied to enhance signal-to-noise (Winker et al. 2009).

As discussed in Sect. 2.3 above, the TOMS/OMI passive aerosol retrieval technique is sensitive to layer height for UV-absorbing aerosols, and by combining these data with a constraint on total-column AOD, elevation can be obtained to within about a kilometer under favorable retrieval conditions, e.g., when there is sufficient AOD (Satheesh et al. 2009). The UV-absorption method also makes it possible to detect aerosols when they reside over cloud, and can take advantage of the broad-swath, once-daily global coverage of the TOMS and OMI instruments (Torres et al. 2007). Because the radiative impact of absorbing species such as wildfire smoke and black carbon is enhanced when the aerosols reside above highly reflective clouds (e.g., Zarzycki and Bond 2010; Loeb and Su 2010), the ability to map these situations can make an important contribution toward reducing DARF uncertainty.

Another passive radiance technique, often called "split-window," is based on measuring top-of-atmosphere radiance differences between two channels in the infrared spectral region, usually at 11 and 12 micrometers. This approach is commonly used to obtain the elevation of condensate clouds, and has been applied to constraining the heights of fairly optically thick volcanic ash plumes, dominated by coarse-mode particles, with AVHRR as well as SEVIRI data (Prata 1989; Pavolonis et al. 2006). The vertical atmospheric weighting functions for the split-window channels typically sample within and sometimes below aerosol plumes, depending on plume characteristics, creating some ambiguity in the result.

From stereo imaging, the heights of clouds and aerosol plumes can also be derived geometrically (Moroney et al. 2002; Muller et al. 2002). This technique is not sensitive to radiometric calibration accuracy, but is effective only near aerosol sources, where plumes exhibit sufficient spatial contrast to be matched in multiple angular views; the resulting maps capture the injection heights of wildfire smoke, desert dust, and volcanic emission plumes (Kahn et al. 2007; Scollo et al. 2010; Val Martin et al. 2010). In situations where the contrast elements seen from space come from different vertical locations within the plume, a histogram loosely representing the aerosol vertical distribution is obtained. As such, the stereo-derived and lidar aerosol elevation products tend to be complementary, one providing sufficient coverage to map aerosol injection height and near-source plume behavior, and the other having the sensitivity to characterize extended aerosol layers downwind (Kahn et al. 2008), a combination that offers powerful constraints on aerosol transport models.

Satellites also retrieve surface albedo and bidirectional reflectance distribution function (BRDF), the viewing- and solar-geometry dependent surface reflection property needed to specify the lower boundary condition for DARF calculations. Several of the aerosol retrieval approaches self-consistently derive surface reflectance parameters (e.g., Martonchik et al. 2009; Lyapustin and Wang 2009; Pinty et al. 2000; Govaerts et al. 2010), and the standard MODIS BRDF product (Schaaf et al. 2002) is widely used. Limited validation data are available for these products (e.g., Román et al. 2009), but it is beyond the scope of the current paper to review this subject in greater depth.

3 A Way Forward

Enormous advances have been made in space-based aerosol remote sensing. Yet some measurements needed to constrain DARF globally at the required level-of-accuracy elude these techniques. AOD sensitivity seems to be at least within reach, if not yet within grasp, but particle size is only loosely constrained, and SSA even less so, especially when multiple aerosol components are present in the atmospheric column. Detailed SSA, particle shape, indices of refraction, and compositional information needed to identify the anthropogenic fraction are obtained primarily by collecting samples of the particles themselves, in situ. Further, the observations from those space-based instruments yielding the most particle-microphysical-property detail fly in polar orbit, providing once-daily global snapshots at best, and cannot retrieve under clouds and in other regions where algorithmic issues arise, such as over most mountainous, snow, and ice-covered surfaces, so significant gaps remain in spatial and diurnal coverage.

Surface stations are deployed in places to sample near-surface aerosols such as pollution particles in urban settings. Such stations are often equipped to measure particle size and mass, as well as to collect samples for chemical analysis. Intensive field campaigns can provide more complete atmospheric characterization, at least on a regional basis; these involve aircraft, satellites, and ground stations making coordinated observations, often with the help of predictions from aerosol transport models.

A next-generation satellite instrument suite will likely encompass all the capabilities of previous efforts, including the wide-swath, sub-kilometer-pixel-resolution imaging of a MODIS-like instrument, with multiple viewing angles, spectral channels from the near-UV to the infrared and polarized bands, along with advanced lidar systems. High-precision multi-angular, multispectral polarization measurements offer the promise of adding important aerosol-type discrimination ability to a space-based remote-sensing instrument, retrieving in the future at least one additional moment in the particle size distribution and tighter constraints on real part of the refractive index, and offering greater flexibility in performing aerosol retrievals (Mishchenko and Travis 1997; Hasekamp and Landgraf 2007). Yet, it is not clear that the aerosol microphysical property information retrieved from even next-generation space-based remote sensing alone can provide enough quantitative detail to achieve the required improvement in global aerosol direct radiative forcing estimates.

One important simplifying factor for global-scale aerosol studies is that aerosols tend to be transported long distances in relatively thin atmospheric layers, so near-source plume height mapping combined with downwind lidar curtains offer significant constraints on vertical distribution. Another simplifying factor is that for most aerosol sources and specified seasons, emitted and evolved particle microphysical and chemical properties tend to be repeatable, even if the aerosol amounts vary considerably from day-to-day and year-to-year. This means it is at least feasible for an airborne observing program, aimed at routinely measuring particle properties in situ, to capture probability distribution functions (PDFs) of particle size, spectral absorption and scattering, shape, chemical composition, hygroscopic properties, and mass extinction efficiency, thereby characterizing the major aerosol-air-mass types in enough detail to adequately address the global aerosol direct radiative forcing problem.

This approach amounts to: (1) using satellite data to provide spatially extensive snapshots of AOD and aerosol type, as well as vertical distribution, with the coverage and at the level-of-detail possible, (2) relying on the accuracy of direct, suborbital measurements, targeted with the help of satellite aerosol-air-mass-type mapping, to assign detailed particle microphysical properties to the major aerosol-air-mass types, and (3) depending upon models, constrained by the aggregate of observational data, to calculate the regional and global radiation fields and material fluxes with adequate space–time resolution to produce the best result we can achieve.

These points are illustrated in Fig. 2. Here, satellites provide global maps of AOD and aerosol air-mass type, along with vertical distribution. Suborbital platforms perform the traditional role of validating and helping refine space-based remote-sensing retrieval algorithms, as well as providing detail unobtainable from space. Satellites also offer regional context, needed to target and appropriately interpret the in situ aerosol-air-mass-

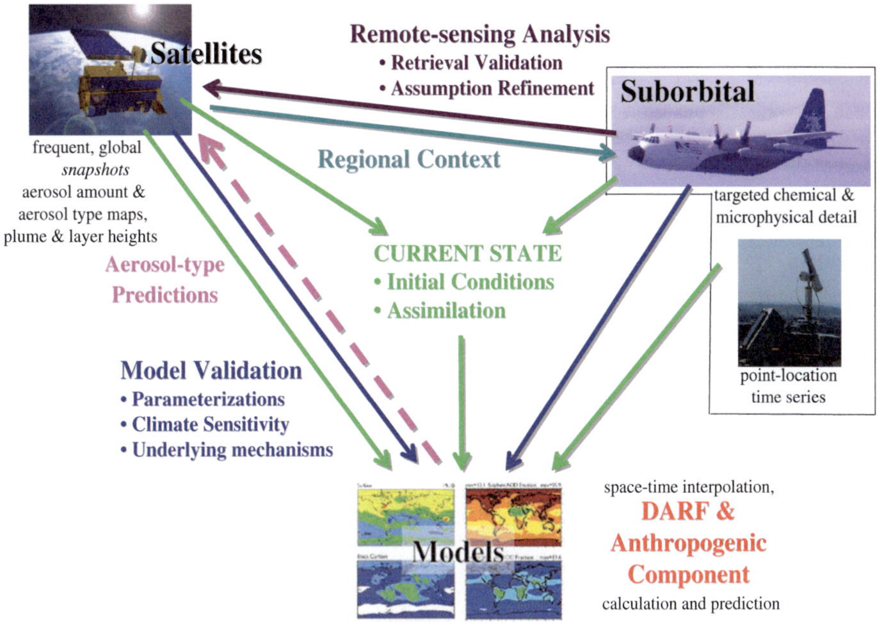

Fig. 2 Schematic representation of the relationships between satellite and suborbital measurements and models for constraining direct aerosol radiative forcing

type measurements. Taken together, these observations are used to constrain (green arrows) and validate (blue arrows) models that can calculate DARF and its anthropogenic component. The dashed pink arrow suggests the possibility of using aerosol transport model simulations to help constrain the choice of aerosol type within satellite retrieval algorithms, in situations where aerosol-type information in the satellite radiance measurements is limited.

Much work remains to reach this goal. Aerosol-type mapping from space-based remote sensing must be improved, in part by refining current retrieval algorithms, by combining data from multiple sources, and eventually by flying a next-generation instrument suite. Adequate in situ sampling of aerosol microphysical property PDFs for the major aerosol-air-mass types is needed as well. This would probably require developing and deploying a compact, low-weight instrument package that can be flown routinely and economically on a single, relatively small aircraft at least several times per week, making the requisite measurements. [A program of this type, though with somewhat different objectives, has already been successfully demonstrated (Andrews et al. 2004).] Finally, better ways of constraining aerosol transport models with the aggregate of measurements are also needed, taking account of spotty observational coverage and varying information content when establishing model result uncertainties. Yet, the goal of adequately constraining DARF is achievable, provided the aerosol measurement and modeling communities can obtain the resources, and retain the inclination, to put all the pieces together.

Acknowledgments I thank the International Space Science Institute (ISSI), and the organizers of the ISSI workshop on Observing and Modeling Earth's Energy Flows, for stimulating discussion of the material described in this paper. I also thank Mian Chin, Pete Colarco, Gerrit de Leeuw, Tom Eck, Michael Garay, Thomas Holzer-Popp, Rob Levy, Alexei Lyapustin, Omar Torres, and two anonymous reviewers for comments on an early version of the manuscript. The work of R. Kahn is supported in part by NASA's Climate and Radiation Research and Analysis Program, under H. Maring, NASA's Atmospheric Composition Program, and the EOS-MISR project.

References

Ackerman SA, Strabala KI, Menzel WP, Frey RA, Moeller CC, Gumley LE (1998) Discriminating clear sky from clouds with MODIS. J Geophys Res 103:32141–32157

Anderson TL, Charlson RJ, Schwartz SE, Knutti R, Bucher O, Rodhe H, Heintzenberg J (2003) Climate forcing by aerosols—a hazy picture. Science 300:1103–1104

Andreae MO, Jones CD, Cox PM (2005) Strong present-day aerosol cooling implies a hot future. Nature 435:1187–1190. doi:10.1038/nature03671

Andrews E, Sheridan PJ, Ogren JA, Ferrare R (2004) In situ aerosol profiles over the Southern Great Plains cloud and radiation test bed site: 1. Aerosol optical properties. J Geophys Res 109:D06208. doi:10.1029/2003JD004025

Bellouin N, Jones A, Haywood J, Christopher SA (2008) Updated estimate of aerosol direct radiative forcing from satellite observations and comparison against the Hadley Centre climate model. J Geophys Res 113:D10205

Cattrall C, Reagan J, Thome K, Dubovik O (2005) Variability of aerosol and spectral lidar and backscatter and extinction ratios of key aerosol types derived from selected Aerosol Robotic Network locations. J Geophys Res 110:D10S11. doi:10.1029/2004JD005124

Chen W-T, Kahn R, Nelson D, Yau K, Seinfeld J (2008) Sensitivity of multi-angle imaging to optical and microphysical properties of biomass burning aerosols. J Geophys Res 113:D10203. doi:10.1029/2007JD009414

Chin M, Kahn R, Schwartz S (eds) (2009) CCSP (U.S. Climate Change Science Program) synthesis and assessment product 2.3. Atmospheric aerosol properties and climate impacts, p 116

de Graaf M, Stammes P, Torres O, Koelemeijer RBA (2005) Absorbing aerosol index: sensitivity analysis, application to GOME and comparison with TOMS. J Geophys Res 110:D01201. doi:10.1029/2004JD005178

de Leeuw G, Kinne S, Leon JF, Pelon J, Rosenfeld D, Schaap M, Veefkind PJ, Veihelmann B, Winker DM, von Hoyningen-Huene W (2011) Retrieval of aerosol properties. In: Burrows JP, Platt U, Borrell P (eds) The remote sensing of tropospheric composition from space. Springer, Berlin, pp 359–414. doi:10.1007/978-3-642-14791-3

Deschamps PY, Bréon F-M, Leroy M, Podaire A, Bricaud A, Buriez J-C, Sèze G (1994) The POLDER mission: instrument characteristics and scientific objectives. IEEE Trans Geosci Remote Sens 32:3598–3615

Deuzé JL, Bréon FM, Devaux C, Goloub P, Herman M, Lafrance B, Maignan F, Marchand A, Nadal F, Perry G, Tanré D (2001) Remote sensing of aerosols over land surfaces from POLDER-ADEOS-1 polarized measurements. J Geophys Res 106:4913–4926

Di Girolamo L, Wilson MJ (2003) A first look at band-differenced angular signatures for cloud detection from MISR. IEEE Trans Geosci Remote Sens 41:1730–1734

Diner DJ, Beckert JC, Reilly TH, Bruegge CJ, Conel JE, Kahn R, Martonchik JV, Ackerman TP, Davies R, Gerstl SAW, Gordon HR, Muller J-P, Myneni R, Sellers RJ, Pinty B, Verstraete MM (1998) Multi-angle imaging spectroradiometer (MISR) instrument description and experiment overview. IEEE Trans Geosci Remote Sens 36:1072–1087

Dubovik O, King MD (2000). A flexible inversion algorithm for retrieval of aerosol optical properties from Sun and sky radiance measurements. J Geophys Res 105(D16):20673–20696

Dubovik O, Holben BN, Eck TF, Smirnov A, Kaufman YJ, King MD, Tanré D, Slutsker I (2002) Variability of absorption and optical properties of key aerosol types observed in worldwide locations. J Atmos Sci 59:590–608

Dubovik O et al (2006) Application of spheroid models to account for aerosol particle nonsphericity in remote sensing of desert dust. J Geophys Res 111:D11208. doi:10.1029/2005JD006619

Eck TF, Holben BN, Reid JS, Dubovik O, Smirnov A, O'Neill NT, Slutsker I, Kinne S (1999) Wavelength dependence of the optical depth of biomass burning, urban, and desert dust aerosols. J Geophys Res 104:31333–31349
Eck TF et al (2008) Spatial and temporal variability of column-integrated aerosol optical properties in the southern Arabian Gulf and United Arab Emirates in summer. J Geophys Res 113:D01204. doi:10.1029/2007JD008944
Eck TF et al (2010) Climatological aspects of the optical properties of fine/coarse mode aerosol mixtures. J Geophys Res 115:D19205. doi:10.1029/2010JD014002
Flowers EC, McCormick RA, Kurfis KR (1969) Atmospheric turbidity over the United States, 1961–1966. J Appl Meteorol 8:955–962
Gao B, Kaufman YJ, Tanre D, Li R (2002) Distinguishing tropospheric aerosols from thin cirrus clouds for improved aerosol retrievals using the ratio of 1.38 μm and 1.24 μm channels. Geophys Res Lett 29. doi:10.1029/2002GL015475
Giglio L, Csiszar I, Justice CO (2006) Global distribution and seasonality of active fires as observed with the Terra and Aqua Moderate Resolution Imaging Spectroradiometer (MODIS) sensors. J Geophys Res 111. doi:10.1029/2005JG000142
Govaerts YM, Wagner SC, Lattanzio A, Watts P (2010) Joint retrieval of surface reflectance and aerosol optical depth from MSG/SEVIRI observations with an optimal estimation approach: 1. Theory. J Geophys Res 115:D02203. doi:10.1029/2009JD011779
Hasekamp OP, Landgraf J (2007) Retrieval of aerosol properties over land surfaces: capabilities of multiple-viewing-angle intensity and polarization measurements. Appl Opt 46:3332–3344
Haywood J, Schulz M (2007) Causes of the reduction in uncertainty in the anthropogenic radiative forcing of climate between IPCC (2001) and IPCC (2007). Geophys Res Lett 34:L20701. doi:10.1029/2007GL030749
Herman JR, Bhartia PK, Torres O, Hsu C, Seftor C, Celarier E (1997) Global distribution of UV-absorbing aerosols from Nimbus-7/TOMS data. J Geophys Res 102:16911–16922
Herman M, Deuzé J-L, Marchand A, Roger B, Lallart P (2005) Aerosol remote sensing from POLDER/ADEOS over the ocean: Improved retrieval using nonspherical particle model. J Geophys Res 110. doi:10.1029/2004JD004798
Holben BN et al (1998) AERONET—a federated instrument network and data archive for aerosol characterization. Remote Sens Environ 66:1–16
Holzer-Popp T, Schroedter-Homscheidt M, Breitkreuz MH, Klüser L, Martynenko D (2008) Improvements of synergetic aerosol retrieval for ENVISAT. Atmos Chem Phys 8:7651–7672
Hsu N, Tsay S, King M, Herman J (2004) Aerosol properties over bright-reflecting source regions. IEEE Trans Geosci Remote Sens 42:557–569
Hsu NC, Tsay SC, King MD, Herman JR (2006) Deep Blue retrievals of Asian aerosol properties during ACE-Asia. IEEE Trans Geosci Remote Sens 44:3180–3195
Husar RB, Prospero JM, Stowe LL (1997) Characterization of troposhperic aerosols over the oceans with the NOAA advanced very high resolution radiometer optical thickness operational product. J Geophys Res 102:16889–16909
Ichoku C, Giglio L, Wooster MJ, Remer LA (2008) Global characterization of biomass-burning patterns using satellite measurements of Fire Radiative Energy. Remote Sens Environ 112:2950–2962
Ignatov A, Sapper J, Cox S, Laszlo I, Nalli NR, Kidwell KB (2004) Operational aerosol observations (AEROBS) from AVHRR/3 on board NOAA-KLM satellites, *J*. Atmos Oceanic Technol 21:3–26. doi:10.1175/1520-0426(2004)021
IPCC (Intergovernmental Panel on Climate Change) (2001) Radiative forcing of climate change, in Climate Change 2001. Cambridge University Press, New York
IPCC (Inter-governmental Panel on Climate Change) (2007) The physical science basis. In: Solomon S, Qin D, Manning H, Chen Z, Marquis M, Averyt K, Tignor M, Miller H (eds) Contribution of working group I to the fourth assessment report of the intergovernmental panel on climate change. Cambridge University press
Kacenelenbogen M, Vaughan MA, Redemann J, Hoff RM, Rogers RR, Ferrare RA, Russell PB, Hostetler CA, Hair JW, Holben BN (2011) An accuracy assessment on fthe CALIOP/CALIPSO version 2/version 3 daytime aerosol extinction product based on a detailed multi-sensor, multi-platform case study. Atmos Chem Phys 11:3981–4000
Kahn R, Banerjee P, McDonald D (2001) The sensitivity of multiangle imaging to natural mixtures of aerosols over ocean. J Geophys Res 106:18219–18238
Kahn RA, Li W-H, Moroney C, Diner DJ, Martonchik JV, Fishbein E (2007) Aerosol source plume physical characteristics from space-based multiangle imaging. J Geophys Res 112:D11205. doi:10.1029/2006JD007647

Kahn RA, Chen Y, Nelson DL, Leung F-Y, Li Q, Diner DJ, Logan JA (2008) Wildfire smoke injection heights: two perspectives from space. Geophys Res Lett 35:L04809. doi:10.1029/2007GL032165
Kahn R, Petzold A, Wendisch M, Bierwirth E, Dinter T, Esselborn M, Fiebig M, Heese B, Knippertz P, Muller D, Schladitz A, von Hoyningen-Huene W (2009) Desert dust aerosol air mass mapping in the Western Sahara, using particle properties derived from space-based multi-angle imaging. Tellus 61B:239–251. doi:10.1111/j.1600-0889.2008.00398.x
Kahn RA, Gaitley BJ, Garay MJ, Diner DJ, Eck T, Smirnov A, Holben BN (2010) Multiangle Imaging SpectroRadiometer global aerosol product assessment by comparison with the Aerosol Robotic Network. J Geophys Res 115:D23209. doi:10.1029/2010JD014601
Kahn RA, Garay MJ, Nelson DL, Levy RC, Bull MA, Diner DJ, Martonchik JV, Hansen EG, Remer LA, Tanré D (2011) Response to "Toward unified satellite climatology of aerosol properties. 3. MODIS versus MISR versus AERONET." J Quant Spectro Rad Transf 112:901–909. doi:10.1016/j.jqsrt.2009.11.003
Kalashnikova OV, Kahn R (2006) Ability of multiangle remote sensing observations to identify and distinguish mineral dust types: part 2. Sensitivity over dark water. J Geophys Res 111:D11207. doi:10.1029/2005JD006756
Kaufman YJ, Tanré D, Remer LA, Vermote EF, Chu DA, Holben BN (1997) Operations remote sensing of tropospheric aerosol over land from EOS moderate resolution imaging spectroradiometer. J Geophys Res 102:17051–17067
Kaufman YJ, Remer L, Tanré D, Li R, Kleidman R, Mattoo S et al (2005a) A critical examination of the residual cloud contamination and diurnal sampling effects on MODIS estimates of aerosol over ocean. IEEE Trans Geosci Remote Sens 43:2886–2897
Kaufman YJ, Boucher O, Tanré D, Chin M, Remer LA, Takemura T (2005b) Aerosol anthropogenic component estimated from satellite data. Geophys Res Lett 32. doi:10.1029/2005/GL023125
Kiehl JT (2007) Twentieth century climate model response and climate sensitivity. Geophys Res Lett 34. doi:10.1029/2007/GL031383
Kim D, Ramanathan V (2008) Solar radiation budget and radiative forcing due to aerosols and clouds. J Geophys Res 113. doi:10.1029/2007JD008434
Kim D-H, Sohn B-J, Nakajima T, Takamura T, Takemura T, Choi B-C, Yoon S-C (2004) Aerosol optical properties over East Asia determined from ground-based sky radiation measurements. J Geophys Res 109:D02209. doi:10.1029/2003JD003387
King MD, Byrne DM, Herman BM, Reagan JA (1978) Aerosol size distributions obtained by inversion of spectral optical depth measurements. J Atmosph Sci 35:2153–2167
Kinne S, Schulz M, Textor C et al (2006) An AeroCom initial assessment—optical properties in aerosol component modules of global models. Atmos Chem Phys 6:1815–1834
Knapp KR (2002) Quantification of aerosol signal in the GOES 8 visible imagery over the United States. J Geophys Res 107. doi:10.1029/2001JD002001
Knapp KR, Frouin R, Kondradunta S, Prados A (2007) Toward aerosol optical depth retrievals over land from GOES visible radiances: determining surface reflectance. Int J Remote Sens 26:4097–4116
Kokhanovsky A, de Leeuw G (eds) (2009) Satellite aerosol remote sensing over land. Springer, Berlin, p 388
Laszlo I, Ciren P, Liu H, Kondragunta S, Tarpley JD, Glodberg MD (2008) Remote sensing of aerosol and radiation from geostationary satellites. Adv Space Res 41:1882–1893
Levy RC, Remer LA, Dubovik O (2007) Global aerosol optical properties and application to Moderate Resolution Imaging Spectroradiometer aerosol retrieval over land. J Geophys Res 112:D13210. doi:10.1029/2006JD007815
Levy RC, Remer LA, Kleidman RG, Mattoo S, Ichoku C, Kahn R, Eck TF (2010) Global evaluation of the Collection 5 MODIS dark-target aerosol products over land. Atmos Chem Phys 10:10399–10420. doi:10.5194/acp-10-10399-2010
Loeb NG, Manalo-Smith N (2005) Top-of-atmosphere direct radiative effect of aerosol over global oceans from merged CERES and MODIS observations. J Climate 18:3506–3526
Loeb NG, Su W (2010) Direct aerosol radiative forcing uncertainty based on a radiative perturbation analysis. J Climate 23:5288–5293
Lyapustin A, Wang Y (2009) The time series technique of aerosol retrievals over land from MODIS. In: Kokhanovsky A, de Leeuw G (eds) Satellite aerosol remote sensing over land. Springer, Berlin, pp 69–99
Lyapustin A, Wang Y, Laszlo I, Kahn R, Korkin S, Remer L, Levy R, Reid JS (2011) Multiangle implementation of atmospheric correction (MAIAC): 2. Aerosol algorithm. J Geophys Res 116:D03211. doi:10.1029/2010JD014986

Marshak A, Wen G, Coakley JA, Remer LA, Loeb NG, Cahalan RF (2008) A simple model for the cloud adjacency effect and the apparent bluing of aerosols near clouds. J Geophys Res 113. doi:10.1029/2007JD009196

Martins JV, Tanre D, Remer L, Kaufman Y, Mattoo S, Levy R (2002) MODIS cloud screening for remote sensing of aerosol over oceans using spatial variability. Geophys Res Lett 29. doi:10.1029/2001GL013252

Martonchik JV, Diner DJ, Crean KA, Bull MA (2002) Regional aerosol retrieval results from MISR. IEEE Trans Geosci Remote Sens 40:1520–1531

Martonchik JV, Kahn RA, Diner DJ (2009) Retrieval of aerosol properties over land using MISR observations. In: Kokhanovsky A, de Leeuw G (eds) Satellite aerosol remote sensing over land. Springer, Berlin, pp 267–293

McComiskey A, Schwartz SE, Schmid B, Guan H, Lewis ER, Ricchiazzi P, Ogren JA (2008) Direct aerosol forcing: calculation from observables and sensitivities to inputs. J Geophys Res 113:D09202. doi:10.1029/2007JD009170

Mishchenko MI, Geogdzhayev IV (2007) Satellite remote sensing reveals regional tropospheric aerosol trends. Opt Exp 15:7423–7438

Mishchenko MI, Travis LD (1997) Satellite retrieval of aerosol properties over the ocean using polarization as well as intensity of reflected sunlight. J Geophys Res 102:16989–17012

Mishchenko MI, Geogdzhayev IV, Cairns B, Rossow WB, Lacis AA (1999) Aerosol retrievals over the ocean by use of channels 1 and 2 AVHRR data: sensitivity analysis and preliminary results. Appl Opt 38:7325–7341

Moroney C, Davies R, Muller J-P (2002) Operational retrieval of cloud-top heights using MISR data. IEEE Trans Geosci Remote Sens 40:1532–1540

Muller J-P, Mandanayake A, Moroney C, Davies R, Diner DJ, Paradise S (2002) Operational retrieval of cloud-top heights using MISR data. IEEE Trans Geosci Remote Sens 40:1532–1546

Myhre G et al (2009) Modelled radiative forcing of the direct aerosol effect with multi-observation evaluation. Atmos Chem Phys 9:1365–1392

North PRJ (2002) Estimation of aerosol opacity and land surface bidirectional reflectance from ATSR-2 dual-angle imagery: operational method and validation. J Geophys Res 107. doi:10.1029/2000JD000207

Omar AH, Winker DM, Kittaka C et al (2009) The CALIPSO automated aerosol classification and lidar ratio selection algorithm. J Atm Ocean Tech 26:1994–2014

Pavolonis MJ, Feltz WF, Heidinger AK, Gallina GM (2006) A daytime complement to the reverse absorption technique for improved automated detection of volcanic ash. J Atmos Oceanic Tech 23:1422–1444

Pierce JR, Kahn RA, Davis MR, Comstock JM (2010) Detecting thin cirrus in multiangle imaging spectroradiometer aerosol retrievals. J Geophys Res 115:D08201. doi:10.1029/2009JD013019

Pinty B, Roveda F, Verstraete MM, Gobron N, Govaerts Y, Martonchik JV, Diner DJ, Kahn RA (2000) Surface albedo retrieval from Meteosat: 2. Applications J Geophys Res 105:18113–18134

Popp C, Hauser A, Foppa N, Wunderle S (2007) Remote sensing of aerosol optical depth over central Europe from MSG-SEVIRI data and accuracy assessment with ground-based AERONET measurements. J Geophys Res 112:D24S11. doi:10.1029/2007JD008423

Prata AJ (1989) Observations of volcanic ash clouds in the 10–12 micrometer window using AVHRR/2 data. Int J Remote Sens 10:751–761

Prospero JM, Ginoux P, Torres O, Nicholson SE, Gill TE (2002) Environmental characterization of global sources of atmospheric soil dust identified with the Nimbus 7 Total Ozone Mapping Spectrometer (TOMS) absorbing aerosol product. Rev Geophys 40. doi:10.1029/2000RG000095

Quaas J, Boucher O, Bellouin N, Kinne S (2008) Satellite-based estimate of the direct and indirect aerosol climate forcing. J Geophys Res 113. doi:10.1029/2007JD008962

Remer LA, Kaufman YJ, Tanre D, Mattoo S, Chu DA, Martins JV, Li R–R, Ichoku C, Levy RC, Kleidman RG, Eck TF, Vermote E, Holben BN (2005) The MODIS aerosol algorithm, products, and validation. J Atmos Sci 62:947–973

Remer LA, Kleidman RG, Levy RC, Kaufman YJ, Tanre D, Mattoo S, Martins JV, Ichoku C, Koren I, Yu H, Holben BN (2008) Global aerosol climatology from the MODIS satellite sensors. J Geophys Res113. doi:10.1029/2007JD009661

Román MO, Schaaf CB, Yang X, Woodcock CE, Strahler AH, Braswell RH et al (2009) The MODIS (Collection V005) BRDF/albedo product: assessment of spatial representativeness over forested landscapes. Remote Sens Environ 113:2476–2498. doi:10.1016/j.rse.2009.07.009

Satheesh SK, Torres O, Remer LA, Babu SS, Vinoj V, Eck TF, Kleidman RG, Holben BN (2009) Improved assessment of aerosol absorption using OMI-MODIS joint retrieval. J Geophys Res 114. doi:10.1029/2008JD011024

Schaaf CB et al (2002) First operational BRDF, Albedo and Nadir reflectance products from MODIS. Remote Sens Environ 83:135–148

Schuster GL, Dubovik O, Holben BN (2006) Angstrom exponent and bimodal aerosol size distributions. J Geophys Res 111:D07207. doi:10.1029/2005JD006328

Schwartz SE, Charlson RJ, Kahn RA, Ogren JA, Rodhe H (2010) Why hasn't Earth warmed as much as expected? J Clim 23:2453–2464

Scollo S, Folch A, Coltelli M, Realmuto VJ (2010) Three-dimensional volcanic aerosol dispersal: a comparison between Multiangle Imaging Spectroradiometer (MISR) data and numerical simulations. J Geophys Res 115:D24210. doi:10.1029/2009JD013162

Smirnov A, Holben BN, Slutsker I, Giles DM, McClain CR, Eck TF, Sakerin SM, Macke A, Croot P, Zibordi G, Quinn PK, Sciare J, Kinne S, Harvey M, Smyth TJ, Pikety S, Zielinski T, Proshutinsky A, Goes JI, Nelson NB, Larouche P, Radionov VF, Goloub P, Krishnamoorthy K, Matarrese R, Robertson EJ, Jourdin F (2009) Maritime aerosol network as a component of aerosol robotic network. J Geophys Res 114:D06204. doi:10.1029/2008JD011257

Stowe LL, Ignatov AM, Singh RR (1997) Development, validation, and potential enhancements to the second-generation operational aerosol product at the National Environmental Satellite, Data, and Information Service of the National Oceanic and Atmospheric Administration. J Geophys Res 102:16923–16934

Stowe LL, Wellemeyer CG, Eck TF, Yeh HYM, Nimbus-7 cloud processing team (1998) Nimbus-7 global cloud climatology. Part I: Algorithms and validation. J Clim 5:445–470

Tackett JL, Di Girolamo L (2009) Enhanced aerosol backscatter adjacent to torpical trade wind clouds revealed by satellite-based lidar. Geophys Res Lett 36. doi:10.1029/2009GL039264

Tanré D, Kaufman YJ, Herman M, Mattoo S (1997) Remote sensing of aerosol properties over oceans using the MODIS/EOS spectral radiances. J Geophys Res 102:16971–16988

Torres O, Bhartia PK, Herman JR, Ahmad Z, Gleason J (1998) Derivation of aerosol properties from satellite measurements of backscattered ultraviolet radiation: theoretical basis. J Geophys Res 103:17099–17110

Torres O, Bhartia PK, Herman JR, Syniuk A, Ginoux P, Holben B (2002) A long-term record of aerosol optical depth from TOMS observations and comparison to AERONET measurements. J Atmos Sci 59:398–413

Torres O, Tanskanen A, Veihelmann B, Ahn C, Braak R, Bhartia PK, Veefkind P, Levelt P (2007) Aerosols and surface UV products from Ozone Monitoring Instrument observations: an overview. J Geophys Res 112:D24S47. doi:10.1029/2007JD008809

Twohy CH, Coakley JA, Tahnk WR (2009) Effect of changes in relative humidity on aerosol scattering near clouds. J Geophys Res 114:D05205. doi:10.1029/2008JD010991

Val Martin M, Logan JA, Kahn R, Leung F-Y, Nelson D, Diner D (2010) Fire smoke injection heights over North America constrained from the Terra Multi-angle Imaging SpectroRadiometer. Atm Chem Phys 10:1491–1510

Várnai T, Marshak A (2011) Global CALIPSO observations of aerosol changes near clouds. IEEE Geosci Remote Sens Lett 8:19–23

Veefkind JP, de Leeuw G, Durkee PA (1998) Retrieval of aerosol optical depth over land using two-angle view satellite radiometry during TARFOX. Geophys Res Lett 25:3135–3138

Veefkind JP, de Leeuw G, Stammes P, Koelemeijer RBA (2000) Regional distribution of aerosol over land, derived from ATSR-2 and GOME. Remote Sens Environ 74:377–386

von Hoyningen-Huene W, Kokhanovsky AA, Wuttke MW, Buchwitz M, Noel S, Gerilowski K, Burrows JP, Latter B, Siddans R, Kerridge BJ (2007) Validation of SCIAMACHY top-of-atmosphere reflectance for aerosol remote sensing using MERIS L1 data. Atmos Chem Phys 7:97–106

Wagner SC, Govaerts YM, Lattanzio A (2010) Joint retrieval of surface reflectance and aerosol optical depth from MSG/SEVIRI observations with an optimal estimation approach: 2. Implementation and evaluation. J Geophys Res 115:D02204. doi:10.1029/2009JD011780

Waquet F, Léon J-F, Cairns B, Goloub P, Deuzé J-L, Auriol F (2009) Analysis of the spectral and angular response of the vegetated surface polarization for the purpose of aerosol remost esesing over land. Appl Opt 48:1228–1236

Whitby KT (1978) The physical characteristics of sulfur aerosols. Atmos Environ 12:135–159

Winker DM, Vaughan MA, Omar A, Hu Y, Powell KA, Liu Z, Hunt WH, Young SA (2009) Overview of the CALIPSO mission and CALIOP data processing algorithms. J Atmos Ocean Tech 26:2310–2323

Yu H, Kaufman YJ, Chin M, Feingold G, Remer LA, Anderson TL, Balkanski Y, Bellouin N, Boucher O, Christopher S, DeCola P, Kahn R, Koch D, Loeb N, Reddy MS, Schulz M, Takemura T, Zhou M (2006) A review of measurement-based assessment of aerosol direct radiative effect and forcing. Atmos Chem Phys 6:613–666

Yu H, Remer LA, Chin M, Bian H, Kleidman RG, Diehl T (2008) A satellite-based assessment of trans-pacific transport of pollution aerosol. J Geophys Res 113:D14S12. doi:10.1029/2007JD009349

Zarzycki CM, Bond TM (2010) How much can the vertical distribution of black carbon affect its global direct radiative forcing? Geophys Res Lett 37:L20807. doi:10.1029/2010GL044555

Zhang J, Reid JS (2006) MODIS aerosol product analysis for data assimilation: assessment of over-ocean level 2 aerosol optical thickness retrievals. J Geophys Res 111:D22207. doi:10.1029/2005JD006898

Zhang J, Reid JS (2010) A decadal regional and global trend analysis of the aerosol optical depth using a data-assimilation grade over-water MODIS and Level 2 MISR aerosol products. Atmos Chem Phys 10. doi:10.5194/acp-10-1-2010

Zhao G, Di Girolamo L (2004) A cloud fraction versus view angle technique for automatic in scene evaluation of the MISR cloud mask. J Appl Meteorol 43:860–869. doi:10.1175/1520-0450

Zhao G, Di Girolamo L, Dey S, Jones AL, Bull M (2009) Examination of direct cumulus contamination on MISR-retrieved aerosol optical depth and angstrom coefficient over ocean. Geophys Res Lett 36. doi:10.1029/2009GL038549

Surv Geophys (2012) 33:723–731
DOI 10.1007/s10712-012-9190-2

Greenhouse Gases, Aerosols and Reducing Future Climate Uncertainties

Hervé Le Treut

Received: 23 January 2012 / Accepted: 15 March 2012 / Published online: 11 May 2012

Abstract The past record of global earth surface temperatures is unable to offer quantitative evidence about the amplitude of climate sensitivity, due to the competing effects of long-lived greenhouse gases and short-lived aerosols. This factor constitutes one of the reasons why uncertainties about climate sensitivity have remained almost unchanged for more than 30 years, and it is also limiting our current capacity to propose reliable climate projections for the coming century. This paper offers a short review of the studies that have dealt with this issue. A number of approaches aim at a process-oriented diagnostic of current models. These studies are in constant progress since the launch of remote-sensing instruments, such as those from the A-train satellite constellation. Past climate fluctuations may also offer some limited possibilities to discriminate the effects of greenhouse gases and aerosols. There is therefore a real hope that climate projections may eventually become more accurate, which would be extremely useful in monitoring global warming during the next decades.

Keywords Greenhouse gases · Aerosols · Climate sensitivity · Climate feebacks

1 Introduction

Providing a quantitative comparison between the anthropogenic warming by greenhouse gases and the cooling effects that result from the increasing atmospheric aerosol content constitutes a scientific challenge of huge importance, because it would help reduce the current uncertainties affecting the quantification of future anthropogenic climate change. The two effects are asymmetric.

The greenhouse gases, which have increased due to human emissions, are mostly long-lived, with an estimated atmospheric residence time of more than a century for carbon dioxide (CO_2) or nitrous oxide (N_2O), and about 10 years for methane (CH_4). In consequence, the respective atmospheric concentrations of these gases are mostly uniform over

H. Le Treut (✉)
Laboratoire de Météorologie Dynamique/Institut Pierre Simon Laplace, Université Pierre et Marie Curie, Paris, France
e-mail: letreut@ipsl.jussieu.fr

the globe and easy to document. Their radiative impact can be computed precisely for given atmospheric conditions. The aerosols have almost opposite properties: they are short-lived within the atmosphere, with a tropospheric residence time of about a week, and their geographical distribution is very contrasted.

The radiative impact of both the CO_2 and the aerosols is often characterized by a quantity called radiative forcing. It is defined as a modification of net radiative fluxes at the tropopause, in response to a prescribed change of the atmospheric composition. It is computed in conditions when only the stratospheric temperature has adjusted and is in balance with the changed composition. This rather complex definition may create an added uncertainty, which will not be discussed in this paper.

Because greenhouse gases accumulate within the atmosphere, they are bound to be the first determinants of climate evolution over the twenty-first century (Dufresne et al. 2005)—and more so because policies to improve air quality have managed to reduce aerosol emissions in different parts of the world. However, since aerosols may have masked part of the initial greenhouse effect (Schwartz et al. 2010), using the observed initial and transient phase of the Earth warming as a quantitative indication of what may happen in the future is extremely difficult.

The objective of this paper is to review some of the elements, which explain why these uncertainties have not been significantly reduced over the last decades. This analysis comes at a specific moment, since the modeling community is carrying out a major collective exercise, the Coupled Model Intercomparison Program 5 (CMIP5) (Taylor et al. 2012), whose results will form part of the upcoming IPCC fifth assessment report. CMIP5 will also benefit from the information brought up by new observing tools (such as the satellites of the A-train satellite constellation (L'Ecuyer and Jiang 2010), when it comes to better constraining aerosol forcing, cloud feedbacks and climate sensitivity). While the conclusions from the AR5 are not yet available, the various assessments carried out by past IPCC reports and, in particular, the AR4 (IPCC 2007) provide a few key figures, which have not strongly evolved since then, and are sufficient to illustrate the problem.

2 The Permanence of Uncertainties

Since the preindustrial era, the Earth has experienced a global surface warming of approximately 0.7° (the largest uncertainty being related to the determination of the pre-industrial temperature level). The anthropogenic radiative forcing that results from the combined increase of greenhouse gases and aerosols (with smaller contributions from ground albedo modifications, short-lived gases or stratospheric modifications), as reviewed by the IPCC AR4 (Forster et al. 2007) and by Haywood and Schulz (2007), lies within a very large range extending from 0.6 W m^{-2} to 2.4 W m^{-2}. The largest part of this uncertainty is due the aerosol contribution of about 2 W m^{-2} corresponding to an uncertainty that is roughly 5 times larger than that of the anthropogenic greenhouse gases. Some very rough estimates may be used to characterize the implications of this. Climate sensitivity is defined as the equilibrium temperature response to a CO_2 doubling, for example, to a mostly uniform radiative forcing of 3.8 W m^{-2}. If, for the sake of simplicity, we consider that, within an evolving transient, roughly half of the radiative imbalance is being stored into the ocean, the climate sensitivity would then be approximately equivalent to the transient temperature response of the climate system to a forcing twice as large. For a 2.4 W m^{-2} radiative forcing, assuming an overall linear response of the surface temperatures to the radiative forcing would then associate the current transient warming with a climate sensitivity of about 2 °C, whereas a stronger role of the aerosols leading to a 0.6 W m^{-2} forcing would imply a climate sensitivity of almost 9 °C.

These estimates correspond to the range derived by Huber and Knutti (2011), using the Bern climate model, a model of intermediate complexity (Claussen et al. 2002) in the context of attribution studies. The IPCC AR4 reviews similar studies that use the observed historical climate record as a way to estimate Probability Density Functions (PDFs) of the climate sensitivity and provide similar uncertainty ranges (Meehl et al. 2007).

Climate sensitivity may also be diagnosed from model simulations. In this case, its determination reflects exclusively the parameterization of key model processes, with no explicit contribution of the observed trends. Such model-based estimates of the climate sensitivity currently range from approximately 1.5 to 4.5 °C, which was also the range quoted in the "Charney report" (Charney et al. 1979). Sensitivities lower than 1.5 °C are not likely with present models, whereas there is a larger probability of sensitivities above 4.5 °C (Meehl et al. 2007).

These two different approaches to estimate climate sensitivity therefore provide overlapping ranges of results. This is a key source of uncertainty, which affects our assessment of future climate evolution, because it means that one approach will not substantially constrain the other one. Another implication of this large overlap is that current climate evolution may always be interpreted as resulting from anthropogenic forcings: whatever the value of the climate sensitivity within the currently accepted range, the observed warming can be explained as a response to an anthropogenic radiative forcing which also falls within the current uncertainty range. Kiehl (2007) showed that this does explain why the different models have the capacity to replicate the recent warming, in spite of marked differences in climate sensitivity, because they tend to use an adapted amplitude of the aerosol forcing.

The aim of the present paper is to provide a brief review of some of the past studies that have brought some insight into this competition between greenhouse gases and aerosols, and also to describe some of the scientific approaches that may potentially help us to provide a better diagnosis in the future.

3 Comparing the Greenhouse and Aerosol Effects: Process-Oriented Studies

3.1 A Brief History

The recognition that carbon dioxide resulting from anthropogenic emissions may be captured by the ocean on long timescales, and consequently may also be stored for a long period of time in the atmosphere, was first suggested by Revelle and Suess (1957) and first quantified by Bolin and Eriksson (1959). It was then confirmed by the analysis of continuous atmospheric measurements carried out at Mauna Loa since 1957, showing the systematic growth of CO_2. The "Charney Report," compiled under the auspices of the US National Academy of Sciences, provided an early estimate of the consequences of a continuing growth. As mentioned above, this estimate is essentially unchanged after more than 30 years (Charney et al. 1979).

The potential cooling effect of aerosols was also recognized very early, mostly in relation with volcanic eruptions. Peterson et al. (2008) have presented a review of "warming" and "cooling" papers published during the 1960s and 1970s. Their review shows that the often advocated myth, stating that most scientists were fearing at that time an exceptional and catastrophic cooling event, cannot be traced back to the published literature. It also shows that the scientific community had identified quite early the fact that aerosol loading could delay the greenhouse warning, but was not likely to reverse it.

A quantitative estimate of these effects depends on two series of uncertain processes:

1. Climate feedbacks determining the overall climate sensitivity. When considering the added effect of the greenhouse gases and aerosols, a key problem is also to know whether the same feedbacks are triggered by a horizontally homogeneous longwave forcing (such as the greenhouse effect) and a combined forcing with much stronger geographical contrasts (backscatter, absorption in both the longwave and shortwave domains, such as the aerosol radiative impact).
2. Processes affecting the aerosol radiative forcing. They include the direct backscatter to space, the absorption of solar or terrestrial radiation, as mentioned above, but also the multiple indirect effects by which aerosols modify cloud microphysical and possibly macrophysical properties and therefore their contribution to the radiation budget.

The main approach to reduce those uncertainties has been to develop process studies.

3.2 Climate Sensitivity and Feedbacks: Process-Oriented Approaches

It is beyond the scope of this paper to review the different climate feedbacks and their present assessment. Off-line linear decomposition of the changes in radiative fluxes associated with a greenhouse warming (Hansen et al. 1984; Cess et al. 1989) as well as further nonlinear approaches (Colman et al. 1997) have managed to distinguish the role and amplitude of water vapour, cloud, lapse-rate and surface albedo feedbacks. In the tropics, those feedbacks manifest themselves quite distinctly in ascending and descending regions. This provides a way to separate dynamic and thermodynamic feedbacks, and therefore to use observed short-time climate fluctuations to assess the latter. This has prompted the use of two-column models (Pierrehumbert 1995) or the use of the 500 hPa vertical velocity to stratify the relations between the different radiative parameters and sea surface temperatures. As shown by Bony et al. (1995, 2004, 2006), these methods make it possible to use observed variations of the radiative forcing associated with the diurnal and interannual scales to evaluate and constrain the models. They have also been extended to areas outside the tropics using other types descriptors for the atmospheric dynamics (Tselioudis and Jakob 2002). More conceptual models can also strengthen our understanding of some of the feedbacks, which have been identified in climate models and increase their robustness. These methods, associated with new satellite data, are expected to provide new and powerful constraints on climate models. Whether this will be enough to provide a better estimate of the overall climate sensitivity is still uncertain, as the dynamic feedbacks that operate at different time scales are not exactly similar.

3.3 Aerosol Radiative Effect: Direct Determinations from Process Studies

It is even more difficult to evaluate quantitatively the aerosol radiative forcing and again it will not be reviewed in detail here. The different assessments (Haywood and Boucher 2000; IPCC 2001, 2007) show a regular progress in the identification of the various aerosol types, their distribution and optical properties, and their indirect effects, which amplify their impacts. The emergence of new data such as those from the different instruments of the A-train satellite constellation has helped a better identification of the associated uncertainties. However, there is an increasing realization of the complexity of the aerosol effects, especially through their impact on cloud microphysics, and as a result the uncertainty which affect their quantification has not been reduced decisively. It is unlikely that, in the near future, it should be reduced in a way that would significantly improve our estimate of climate sensitivity.

The identification of the aerosol forcing component will have to rely on some of the characteristics of the aerosol impact that best distinguish it from the greenhouse effect, such as:

1. Their different geographical distributions: this reflects not only the different forcing distributions—and, in particular, the North–South asymmetry but also the feedback amplification which may differ from model to model, being more local or more global (Le Treut et al. 1994; Roeckner et al. 1999, Jones et al. 2007).
2. The different impact on the atmospheric vertical stratification.

We know that greenhouse gases tend to warm the lower atmospheric layers, cool the stratosphere and elevate the tropopause, but this effect is not easy to use quantitatively, because of the water vapour feedback, which adds a greenhouse component to any initial forcing. Aerosols probably have a different vertical signature (and a different one for scattering and absorbing aerosols).

3. The different impact on the diurnal temperature range (Hansen et al. 1995) and other variables of the climate system, for example, evaporation and runoff (Gedney et al. 2006).

In the next paragraph, we review some of the observed past climate changes, which may be used to discriminate those effects.

4 Perspectives in Using Past Observed Climate Changes to Discriminate Aerosol and Greenhouse Impacts

4.1 The Last Glacial Maximum

Climate changes over the Quaternary era (Lorius et al. 1990) or climate changes at the last glacial maximum (Schmittner et al. 2011) have been used as a test to evaluate climate sensitivity. However, many aspects of climate forcings during these periods are imperfectly known. In the case of the last glacial maximum, in particular, the dust load of the atmosphere is certainly much larger than in the present conditions, but is difficult to quantify. Ramstein et al. (1998) showed that the cloud feedbacks may have occured differently in glacial times, as the changes in the Hadley cells may have a stronger effect on cloud distribution than the direct greenhouse effect, with its general symmetry between both hemispheres. Past data participate to better constrain the models, but have limitations. The same can be said of climate fluctuations that are caused by volcanic eruptions (Wigley et al. 2005).

4.2 Using the Records of Climate Evolution During the Last Century

A first important difference between the aerosol and greenhouse gas radiative forcings is that the former presents a much stronger asymmetry between the two hemispheres. The North–South asymmetry of climate evolution has been studied in that perspective for a long time. It presents two main characteristics: (1) the rate of the warming is much stronger in the Northern than in the Southern polar regions—but this first of all reflects a different strength in the feedbacks leading to a stronger polar amplification in the Northern hemisphere. (2) The Northern hemisphere has experienced a transient cooling phase in the 50s and 60s, which is not matched by a similar evolution in the Southern hemisphere.

Several interpretations of this asymmetric cooling can be found in the literature. The effect of an inaccurate correction of sea surface temperatures was shown by Thompson et al. (2008). Many papers—an early example is that of Schlesinger and Ramankutty (1995)—have also emphasized the potential importance of fluctuations in the Atlantic circulation, whether oscillatory or chaotic. But many elements also hint at the important role that may be played by the combined effect of aerosols and greenhouse gas increases. Quaas et al. (2004) have provided an estimate of the radiative anthropogenic forcing of the Earth over the period 1930–1989, which shows a negative (cooling) phase developing strongly immediately after World War II, followed by a positive (warming) phase beginning one decade afterwards. This may be interpreted very simply as resulting from an initially dominating aerosol effect, soon overbalanced by the accumulation of the long-lived greenhouse gases in the atmosphere.

This evolution of the climate system over the last decades is part of the experiments, which were carried out by all models in the framework of CMIP5 and, in spite of the uncertainties affecting SST observations, some progress may be expected from these simulations.

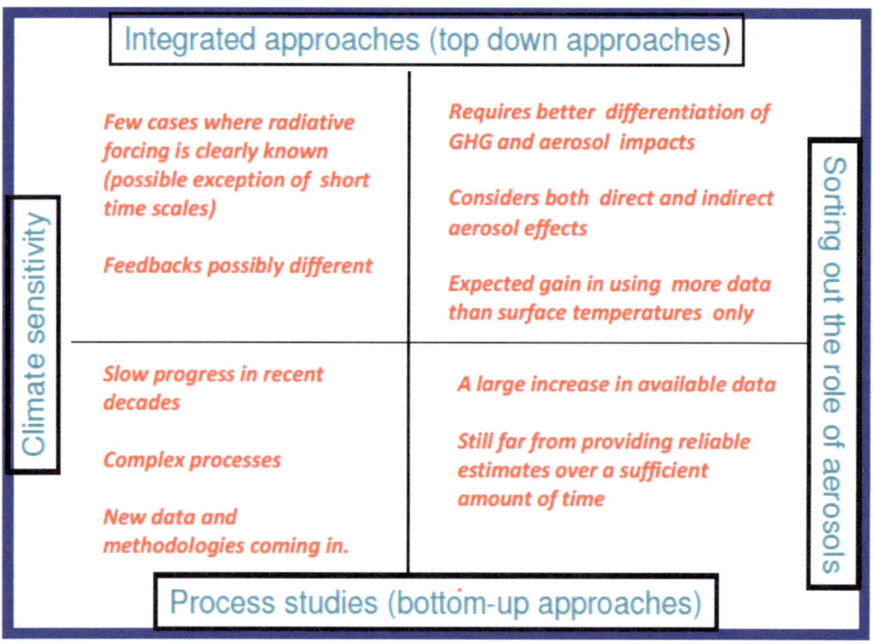

Fig. 1 A simplified diagram showing the different approaches toward a better determination of climate sensitivity, distinguishing on the one hand bottom-up approaches (process-oriented methodologies) and on the other hand top-down approaches (using observed climate fluctuations to diagnose the role of feedbacks). These latter studies may also give a different weight to the aerosol contribution, depending on the time scales under consideration. GHG refers to Greenhouse Gases

Methods to better distinguish the climate response to increasing concentrations of greenhouse gases, and to aerosol forcing, should have an impact on detection and attribution studies, although evaluating the geographical patterns of natural variability is key for those studies and is difficult to assess in a context where the Northern hemisphere has been strongly affected by aerosols during the last decades. However, this should be improved by getting access to a longer climate record, together with a more systematic search for the fingerprints of the greenhouse gas and aerosol forcing, using more parameters than surface temperature only.

5 Conclusion

Providing a better quantification of climate sensitivity would in turn affect many aspects of future projections, with important consequences in terms of policy planning. Four crucial issues are: (1) the "ultimate" warming of the planet, which will depend on the concentration of greenhouse gases in the atmosphere at which the anthropocene era finally stabilizes, (2) the date at which key climate thresholds may be reached (in particular, the possible timing of a 2 °C warming, which has also become an important political marker), (3) the climate impact of air quality policies, and (4) the potential relevance of geoengineering proposals, using the dissemination of aerosols as a way to limit future climate evolution.

This brief review shows that we are at a key moment where there has been a large increase in our qualitative understanding of the processes constraining climate sensitivity, with a remaining difficulty to translate it into a better quantitative assessment. As we attempt to summarize in Fig. 1, no single method alone is expected to bring new accurate estimates of climate sensitivity, but progress will benefit from the combination of a wide range of complementary approaches. In this context, it is not only important to know how much we may reduce uncertainties, but also when, information that also has a strong potential importance for decision-making in relation to climate.

Acknowledgments The author thanks Olivier Boucher for useful comments and discussions.

References

Bolin B, Eriksson E (1959) Changes in the carbon dioxide content of the atmosphere and sea due to fossil fuel combustion. Rockefeller Institute Press, New York, pp 130–142

Bony S, Duvel JP, Le Treut H (1995) Observed dependence of the water vapor and clear-sky greenhouse effect on sea surface temperature: comparison with climate warming experiments. Clim Dyn 11:307–320

Bony SJ-L, Dufresne H Le, Treut J–J Morcrette, Senior C (2004) On dynamic and thermodynamic components of cloud changes. Clim Dyn 22(2–3):71–86

Bony S et al (2006) How well do we understand and evaluate climate change feedback processes? J Clim 19:3445–3482. doi:10.1175/JCLI3819.1

Cess Robert D et al (1989) Interpretation of cloud-climate feedback as produced by 14 atmospheric general circulation models. Science 245:513–516

Charney JG et al (1979) Carbon dioxide and climate: a scientific assessment. National Academy of Sciences, Washington, DC

Clausen M, Mysak LA, Weaver AJ, Crucifix M, Fichefet T, Loutre MF, Weber SL, Alcamo J, Alexeev VA, Berger A, Calov R, Ganapolski A, Goosse H, Lohmann G et al (2002) Earth system models of intermediate complexity: closing the gap in the spectrum of climate system models. Clim Dyn 18:579–586. doi:10.1007/s00382-001-0200-1

Colman RA, Power SB, McAvaney BJ (1997) Non-linear climate feedback analysis in an atmospheric GCM. Clim Dyn 13:717–731

Dufresne J-L , Quaas J, Boucher O, Denvil S, Fairhead L (2005) Contrast in the effects on climate of anrgropogenic sulfate aerosols between the 20th and the 21st century. Geophys Res Lett 32:L21703. doi:10.1029/2005GL023619

Forster P, Ramaswamy V, Artaxo P, Berntsen T, Betts R, Fahey DW, Haywood J, Lean J, Lowe DC, Myhre G, Nganga J, Prinn R, Raga G, Schulz M, Van Dorland R (2007) Changes in atmospheric constituents and in radiative forcing. In: Solomon S, Qin D, Manning M, Chen Z, Marquis M, Averyt KB, Tignor M, Miller HL (eds) Climate change 2007: the physical science basis. Contribution of working group I to the fourth assessment report of the Intergovernmental Panel on Climate Change. Cambridge University Press, Cambridge

Gedney N, Cox PM, Betts RA, Boucher O, Huntingford C, Stott PA (2006) Detection of a direct carbon dioxide effect in continental river runoff records. Nature 439:835–838. doi:10.1038/nature04504

Hansen J, Lacis A, Rind D, Russell G, Stone P, Fung I, Ruedy R, Lerner J (1984) Climate sensitivity: analysis of feedback mechanisms. In: Hansen JE, Takahashi T (eds) Climate processes and climate sensitivity, AGU geophysical monograph 29, Maurice Ewing vol 5. American Geophysical Union, Washington, DC, pp 130–163

Hansen JM, Sato R Ruedy (1995) Long-term changes of the diurnal temperature cycle: implications about the mechanisms of global climate change. Atmos Res 37:175–209

Haywood J, Boucher O (2000) Estimates of the direct and indirect radiative forcing due to tropospheric aerosols: a review. Rev Geophys 38(4):513–543

Haywood J, Schulz M (2007) Causes of the reduction in uncertainty in the anthropogenic radiative forcing of climate between IPCC (2001) and IPCC (2007). Geophys Res Lett 34:L20701. doi:10.1029/2007GL030749

Huber M, Knutti R (2011) Anthropogenic and natural warming inferred from changes in Earth's energy balance. Nat Geosci. doi:10.1038/ngeo1327

IPCC (2001) Climate change 2001: the scientific basis. In: Houghton JT, Ding Y, Griggs DJ, Noguer M, van der Linden PJ, Dai X, Maskell K, Johnson CA (eds) Contribution of working group I to the third assessment report of the intergovernmental panel on climate change. Cambridge University Press, Cambridge, United Kingdom, 881pp

IPCC (2007) Contribution of working group I to the fourth assessment report of the intergovernmental panel on climate change. In: Solomon S, Qin D, Manning M, Chen Z, Marquis M, Averyt KB, Tignor M, Miller HL (eds) Cambridge University Press, Cambridge

Jones A, Haywood JM, Boucher O (2007) Aerosol forcing, climate response and climate sensitivity in the Hadley Centre climate model. J Geophys Res 112:D20211. doi:10.1029/2007JD008688

Kiehl JT (2007) Twentieth century climate model response and climate sensitivity. J Geophys Res Lett 34:L22710. doi:10.1029/2007GL031383

L'Ecuyer TS, Jiang JH (2010) Touring the atmosphere aboard the A-train. Phys Today 63(7):36. doi:10.1063/1.3463626

Le Treut H, Li ZX, Forichon M (1994) Sensitivity of the LMD general circulation model to greenhouse forcing associated with two different cloud water parametrizations. J Clim 7(12):1827–1841

Lorius C, Jouzel J, Raynaud D, Hansen J, Le Treut H (1990) Greenhouse warming, climate sensitivity and ice core data. Nature 347:139–145

Meehl GA, Stocker TF, Collins WD, Friedlingstein P, Gaye AT, Gregory JM, Kitoh A, Knutti R, Murphy JM, Noda A, Raper SCB, Watterson IG, Weaver AJ, Zhao Z-C (2007) Global climate projections. In: Solomon S, Qin D, Manning M, Chen Z, Marquis M, Averyt KB, Tignor M, Miller HL (eds) Climate change 2007: the physical science basis. Contribution of working group I to the fourth assessment report of the Intergovernmental Panel on Climate Change. Cambridge University Press, Cambridge

Peterson TC, Connolley WM, Fleck J (2008) The myth of the 1970s global cooling scientific consensus. Bull Am Meteorol Soc 89(9):1325–1337. doi:10.1175/2008BAMS2370.1

Pierrehumbert RT (1995) Thermostats, radiator fins, and the local runaway greenhouse. J Atmos Sci 52:1784–1806

Quaas J, Boucher O, Dufresne J-L, Le Treut H (2004) Impacts of greenhouse gases and aerosol direct and indirect effects on clouds and radiation in atmospheric GCM simulations of the 1930–1989 period. Clim Dyn 23(7–8):779–789. doi:10.1007/s00382-004-0475-0

Ramstein G, Serafini V, Le Treut H, Forichon M, Joussaume S (1998) Cloud processes associated with past and future climate changes. Clim Dyn 14:233–247

Revelle R, Suess H (1957) Carbon dioxide exchange between atmosphere and ocean and the question of an increase of atmospheric CO_2 during the past decades. Tellus 9:18–27

Roeckner E, Bengtsson L, Feichter J, Lelieveld J, Rodhe H (1999) Transient climate change simulations with a coupled atmosphere–ocean GCM including the tropospheric sulfur cycle. J Clim 12:3004–3032

Schlesinger ME, Ramankutty N (1995) Is the recently reported 65–70 year surface-temperature oscillation the result of climatic noise? J Geophys Res 100(13):767–13774

Schmittner A, Urban NM, Shakun JD, Mahowald NM, Clark PU, Bartlein PJ, Mix AC, Rosell-Melé A (2011) Climate sensitivity estimated from temperature reconstructions of the last glacial maximum. Science. doi:10.1126/science.1203513

Schwartz SE, Charlson RJ, Kahn RA, Ogren JA, Rodhe H (2010) Why hasn't Earth warmed as much as expected? J Clim 23:2453–2464. doi:10.1175/2009JCLI3461.1

Taylor KE, Stouffer RJ, Meehl GA (2012) An overview of CMIP5 and the experiment design. Bull Am Meteor Soc 93. doi:10.1175/BAMS-D-11-00094.1

Thompson DWJ, Kennedy JJ, Wallace JM, Jones PD (2008) A large discontinuity in the mid-twentieth century in observed global-mean surface temperature. Nature 453:646–649. doi:10.1038/nature06982

Tselioudis G, Jakob C (2002) Evaluation of midlatitude cloud properties in a weather and a climate model: dependence on dynamic regime and spatial resolution. J Geophys Res 107(D24):4781. doi:10.1029/2002JD002259

Wigley TML, Ammann CM, Santer BD, Raper SCB (2005) Effect of climate sensitivity on the response to volcanic forcing. J Geophys Res 110:D09107. doi:10.1029/2004JD005557

Surv Geophys (2012) 33:733–744
DOI 10.1007/s10712-012-9187-x

Diagnosing Climate Feedbacks in Coupled Ocean–Atmosphere Models

Eui-Seok Chung · Brian J. Soden · Amy C. Clement

Received: 22 November 2011 / Accepted: 21 February 2012 / Published online: 15 March 2012

Abstract We review the methodologies used to quantify climate feedbacks in coupled models. The method of radiative kernels is outlined and used to illustrate the dependence of lapse rate, water vapor, surface albedo, and cloud feedbacks on (1) the length of the time average used to define two projected climate states and (2) the time separation between the two climate states. Except for the shortwave component of water vapor feedback, all feedback processes exhibit significant high-frequency variations and intermodel variability of feedback strengths for sub-decadal time averages. It is also found that the uncertainty of lapse rate, water vapor, and cloud feedback decreases with the increase in the time separation. The results suggest that one can substantially reduce the uncertainty of cloud and other feedbacks with the accumulation of accurate, long-term records of satellite observations; however, several decades may be required.

Keywords Climate feedbacks · Kernels · Time dependence

1 Background

Climate models employed to predict future climate in response to the projected increase in anthropogenic greenhouse gases exhibit substantial intermodel variations. Considering that the response of climate system to a given radiative forcing is determined by a number of feedback processes that amplify or dampen the initial radiative perturbation, the intermodel spread is attributed to the differences in climate feedbacks depicted in climate models and thus requires better understanding of climate feedbacks in order to reduce the uncertainty in climate sensitivity.

Climate feedbacks have been quantified by a number of different approaches. At first, offline processing approaches have been applied to CO_2 doubling experiments of climate models with mixed-layer oceans. Hansen et al. (1984) conducted feedback calculations for water vapor, snow/ice albedo, and clouds based on one-dimensional radiative convective

E.-S. Chung (✉) · B. J. Soden · A. C. Clement
Rosenstiel School of Marine and Atmospheric Science, University of Miami, Miami, FL, USA
e-mail: e.chung@miami.edu

model (RCM) and suggested that cloud feedback amplifies global warming. However, it was pointed out that spatiotemporal averages of nonlinear variables required for RCM methods may cause uncertainties in global feedback values (e.g., Wetherald and Manabe 1988). By contrast, Wetherald and Manabe (1988) introduced an offline partial radiative perturbation (PRP) method in which the change in top-of-atmosphere (TOA) radiative fluxes is computed by substituting one variable at a time from the perturbed climate state into the control climate. This method can be complicated and computationally expensive due to repeated offline three-dimensional radiative transfer calculations. Colman (2003) intercompared the published strengths for water vapor, lapse rate, surface albedo, and cloud feedbacks determined from the offline PRP methods for CO_2 doubling experiments and showed that the intermodel differences are substantial for all climate feedbacks. However, Colman (2003) emphasized that the substantial intermodel differences could be caused by the absence of a consistent comparison methodology.

The CO_2 doubling experiments require substantial amounts of computer time for the ocean to equilibrate, and differences in control climates could cause spurious differences in climate sensitivity between climate models (e.g., Cess et al. 1990). In order to avoid such difficulties, uniform perturbations in sea surface temperature were prescribed for a fixed season to induce a change in TOA radiative fluxes, and the resulting changes in clear-sky fluxes and cloud radiative forcing were used to infer clear-sky sensitivity and cloud feedback (Cess et al. 1990). Although this approach is much simpler compared to the offline approaches, uniform perturbations in sea surface temperature for a fixed season may not represent realistic climate change conditions (e.g., Colman 2003; Soden and Held 2006). In addition, feedbacks other than clouds cannot be quantified from this approach. Furthermore, it was pointed out that non-cloud feedbacks can be included in the cloud-forcing term (e.g., Zhang et al. 1994; Colman 2003; Soden et al. 2004; Soden and Held 2006).

A radiative kernel approach was proposed as an alternative method to facilitate the analysis of climate feedbacks (Soden and Held 2006; Soden et al. 2008; Shell et al. 2008). Radiative kernels describe the differential response of TOA radiative fluxes to incremental changes in the feedback variables. The use of radiative kernels enables one to decompose a feedback into two parts: one that depends on radiative transfer and the unperturbed climate state, and a second factor that arises from the climate response of the feedback variable. This approach avoids the effects of field decorrelation which introduce substantial biases in traditional feedback analyses that assume all feedback variables to be temporally uncorrelated (Colman 2003; Colman and McAvaney 2009). Furthermore, the kernel technique requires far fewer computations and is easier to implement than the traditional offline PRP methods (e.g., Wetherald and Manabe 1988). Thus, this approach facilitates an understanding of the causes and implications of differences among models, or between models and observations. The radiative kernel method showed that previous surveys of climate feedbacks exaggerated the intermodel differences in water vapor, lapse rate, and surface albedo feedbacks due to inconsistent methodology and that cloud feedback ranges from nearly neutral to positive in all climate models (Soden and Held 2006; Soden et al. 2008). In addition, the robustness of water vapor feedback was confirmed through the radiative kernel approach (John and Soden 2007).

Feedback processes affect internal variations (i.e., seasonal to decadal time scales) in surface temperature as well as externally forced changes, implying that the calculation of climate feedbacks between two climate states involves a combination of both internally and externally driven responses. Unfortunately, the length of satellite observations is not long enough to examine the time dependence of climate feedbacks (e.g., Dessler et al.

2008; Murphy et al. 2009; Chung et al. 2010a, b; Dessler 2010). Furthermore, the impact of internal variability on the climate feedbacks has not been fully investigated even in the cases of climate model simulations. Thus, it is important to gain insight on the time scales of climate feedbacks and the implications for detecting them from satellite observations. In doing so, the time dependence of feedback processes is examined using climate model simulations of the Intergovernmental Panel on Climate Change (IPCC) Fourth Assessment Report (AR4).

2 Data and Methodology

Climate change simulations for the IPCC AR4 Special Report on Emissions Scenarios (SRES) A1B scenario, which approximately corresponds to a doubling of carbon dioxide concentration between 2000 and 2100, are used to quantify the strength of feedbacks for temperature, water vapor, surface albedo, and clouds for 15 coupled ocean–atmosphere models (CCCMA CGCM3.1, CCCMA CGCM3.1-T63, CNRM CM3, CSIRO MK3.0, GFDL CM2.0, GFDL CM2.1, GISS ER, IAP FGOALS1.0.G, INMCM3.0, IPSL CM4, MIROC3.2-MEDRES, MPI ECHAM5, NCAR CCSM3.0, NCAR PCM1, and UKMO HADGEM1).

The feedback strength (λ_x) for variable x is computed following Soden and Held (2006), that is,

$$\lambda_x = (\partial R/\partial x)\left(\mathrm{d}x/\mathrm{d}\overline{T_s}\right)$$

where R and T_s denote the top-of-atmosphere net incoming radiative flux and surface temperature, respectively, and the overbar indicates the global averaging. The first term of λ_x, which depends on the radiative transfer and the unperturbed climate state, has been calculated by Soden et al. (2008) for temperature, water vapor (shortwave and longwave, separately), and surface albedo. We employed their values (termed radiative kernel) given as functions of latitude, longitude, altitude, and monthly resolved season. The second term represents the model response for each variable x and can be computed from model outputs by dividing the difference of variable x between two climate states A and B by the change of global mean surface temperature, that is, $\mathrm{d}x/\mathrm{d}\overline{T_s} = (x^B - x^A)/(\overline{T_s}^B - \overline{T_s}^A)$. The product of the model-computed climatic responses for each variable x with the radiative kernel corresponding to latitude, longitude, altitude, and month is vertically integrated from the surface to the tropopause. Then, we average the vertically integrated feedback parameter over the entire globe to provide global-average estimates of each feedback parameter.

It is noted that the temperature feedback is further subdivided into the response that results from a uniform temperature change throughout the troposphere (Planck feedback) and the modification due to the variation of atmospheric temperature profile (lapse rate feedback). Meanwhile, because of cloud masking effects (Zhang et al. 1994; Colman 2003; Soden et al. 2004, 2008; Soden and Held 2006), cloud feedback is computed by adjusting model-simulated change in cloud radiative forcing between two climate states following Soden et al. (2008).

The magnitude of climate feedbacks can also be estimated from short-term climate fluctuations such as ENSO or volcanic eruptions by regressing changes in the radiative fluxes at the top-of-atmosphere with surface temperature changes (e.g., Dessler and Wong 2009; Dessler 2010). However, the characteristics of radiative forcings associated with short-term climate fluctuations and resultant meridional pattern of surface temperature changes are different from those corresponding to changes in greenhouse gas concentrations. As a result, it is unlikely that climate feedbacks estimated from short-term climate variations are directly comparable to climate feedbacks due to changes in greenhouse gases.

The subsequent sections examine the variations of each feedback parameter as a function of (1) the length of time average of the climate states A and B and (2) the time separation between climate states used to compute climate feedbacks. Both factors are critical for the design of observational systems intended to reduce uncertainties in climate feedbacks and climate sensitivity (National Research Council 2007).

3 Influences of the Length of Time Average to Define Climate States

Feedback strengths are computed for each model of the IPCC AR4 SRES A1B scenario by changing the length of climate states A and B, which ranges 1–20 years, with the fixed time separation between two climate states of 50 years. Considering the variation of feedback strengths in time (e.g., Murphy 1995; Senior and Mitchell 2000), feedback computations are repeated 50 times by moving the time location of climate states A and B by 1 year. Figures 1 and 2 display the computed feedback strengths as functions of time location as well as the length of time average used to define climate states. For example, the feedback strengths of year 2050 are derived by scaling the radiative kernels with the differences of projected climate of years (2050–2050) [(2050–2069)] from that of (2100–2100) [(2100–2119)] for the 1-year [20-year] time average case. In addition, standard deviation of a given feedback parameter is computed for each climate model, and the numeral in the upper-right corner of each panel denotes the average of the standard deviations of the individual models.

Figure 3 displays the average of the standard deviations of the individual models (closed circles) as a function of the length of the time average chosen for feedback computation, and the associated error bars (in blue) denote the range of standard deviations. Because the temperature increase is greater in the troposphere compared to the surface, mainly over the tropics, the reduction in lapse rate with the surface warming leads to an increase in the top-of-atmosphere outgoing longwave radiation (Fig. 1a) with a large intermodel variability due to a difference in the meridional patterns of surface warming (Soden and Held 2006). The magnitude of lapse rate feedback exhibits a smaller variation for the longer time average cases (Fig. 3a). Mean standard deviation decreases from 0.235 $W\ m^{-2}\ K^{-1}$ (1-year case) to 0.071 $W\ m^{-2}\ K^{-1}$ (20-year case) with a substantial reduction occurring for the shorter-term averaging cases ($\leq$5 years). Moreover, a couple of models produce even positive values for 1-year averaging case (Fig. 1a). Considering that the reduction in standard deviation and intermodel variability is not distinct for the averaging time scale longer than 10 years, a filtering of the sub-decadal timescale temperature fluctuations (primarily ENSO) is preferred to reduce the uncertainty in computed lapse rate feedback strengths. Although feedback strength is approximately constant (about $-3.2\ W\ m^{-2}\ K^{-1}$) for Planck feedback, the year-to-year variation is not trivial for the shorter-term averaging cases (not shown).

The increase in water vapor with the surface warming results in a positive feedback with larger magnitudes for longwave spectrum regardless of the length of climate states (Fig. 1b, c). Although a couple of models show noticeable interannual variations in the case of the shortwave component, standard deviations and intermodel variability are quite small across different averaging timescales (Fig. 3b). On the other hand, the longwave

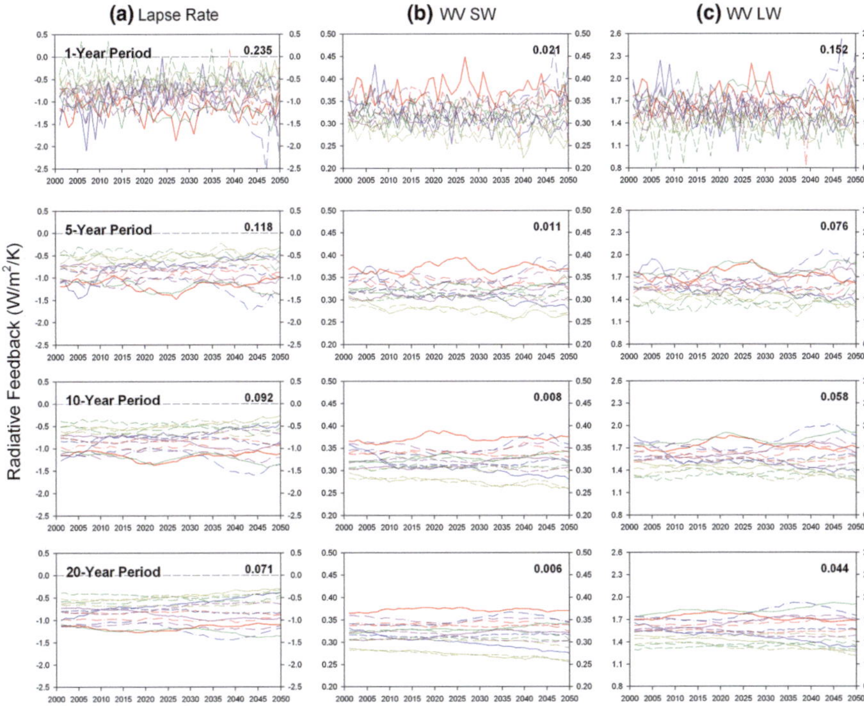

Fig. 1 Temporal variations of the feedback strengths computed from the IPCC AR4 SRES A1B simulations with the time separation between climate states A and B of 50 years: **a** lapse rate, **b** water vapor (shortwave), and **c** water vapor (longwave). The *numeral* in the *upper-right* corner of each panel denotes the average of the standard deviations of 15 climate models for a given feedback parameter and the length of time average used to define climate states

component of water vapor feedback exhibits distinctly larger temporal variations that are comparable to those of lapse rate feedback. Figure 3c shows the decrease of standard deviation from 0.152 W m^{-2} K^{-1} to 0.044 W m^{-2} K^{-1} with the increase in the length of time average from 1 to 20 years. Note that most of this decrease occurs for the sub-decadal time averages. Comparisons between lapse rate and water vapor feedbacks in Fig. 1 indicate that the larger feedback strength of water vapor feedback relative to lapse rate feedback renders the sum of these two feedbacks as being positive. In addition, the temporal variations are negatively correlated with each other, consistent with the strong coupling between the vertical structure of temperature and water vapor changes (e.g., Colman 2003; Allan 2006; Soden and Held 2006; John and Soden 2007; Soden et al. 2008; Colman and McAvaney 2009; Chung et al. 2010a). Therefore, the sum of lapse rate and water vapor feedbacks shows reduced temporal variations and intermodel variability, and less sensitivity to the length of the time average employed for feedback computations.

The lapse rate and water vapor feedbacks are augmented by the feedback process between surface temperature and surface albedo (Fig. 2a). In spite of a much smaller strength compared to lapse rate and water vapor feedbacks, the accuracy of computed surface albedo feedback is also influenced by the length of the time average of climate states (Fig. 3d). A large intermodel variability is likely to stem from differences in the

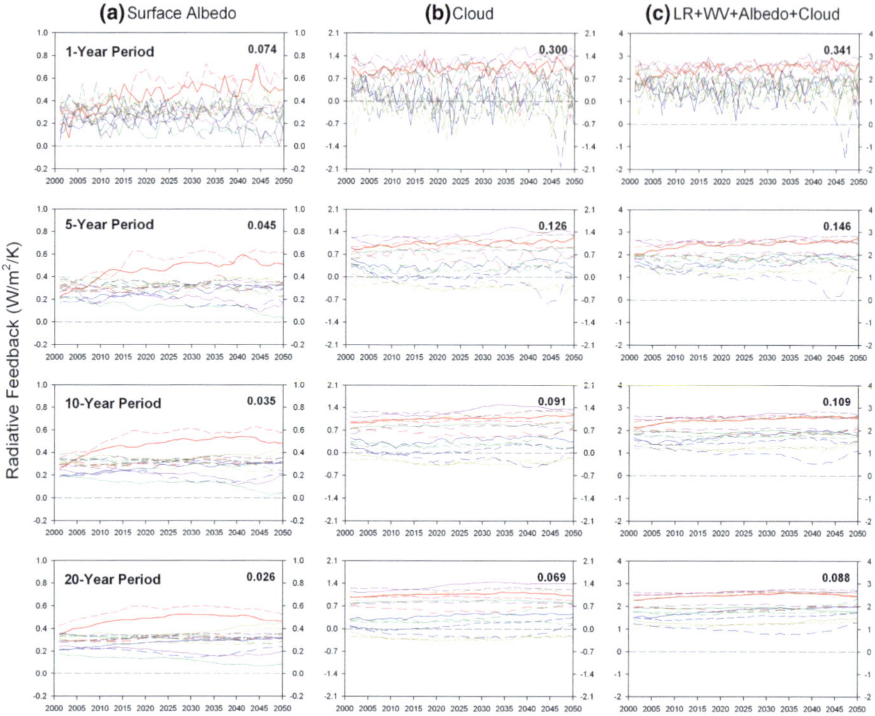

Fig. 2 Same as in Fig. 1, but for: **a** surface albedo, **b** cloud, and **c** summation except Planck feedback

depiction of snow/sea ice coverage changes and the meridional distributions of the changes (e.g., Colman and McAvaney 2009).

The cloud feedback was calculated from the change of the top-of-atmosphere net cloud radiative forcing and the difference between the all-sky and clear-sky radiative kernels following Soden et al. (2008). It is noted that changes in aerosol emission and thus aerosol indirect effect may affect the changes in cloud distribution in the model simulations of SRES A1B scenario. The temporal variations and intermodel differences are larger than any other feedback processes for all time averages (Fig. 2b). However, like other feedback processes, the use of longer time average reduces the uncertainty of inferred feedback strengths with an approximately exponential decrease for the sub-decadal time average cases (Fig. 3e). Given the offsetting influences between lapse rate and water vapor feedbacks arising from the strong coupling between the vertical structure of temperature and water vapor changes and smaller magnitude of surface albedo feedback, the uncertainty of combined feedback (Fig. 3f) is determined primarily by the cloud feedback (e.g., Colman 2003; Bony and Dufresne 2005; Bony et al. 2006; Soden et al. 2008; Colman and McAvaney 2009; Chung et al. 2010b; Dessler 2010).

Figure 3 also displays the dependence of the difference between the maximum and minimum value of the computed feedback strengths for a particular averaging interval with respect to the length of the time average used to define two climate states (open triangles). As shown in the cases of standard deviation, the ranges of computed feedback strengths manifest the influences of averaging time scale on lapse rate, water vapor (longwave component), and cloud feedbacks. Sub-decadal time average cases are characterized by

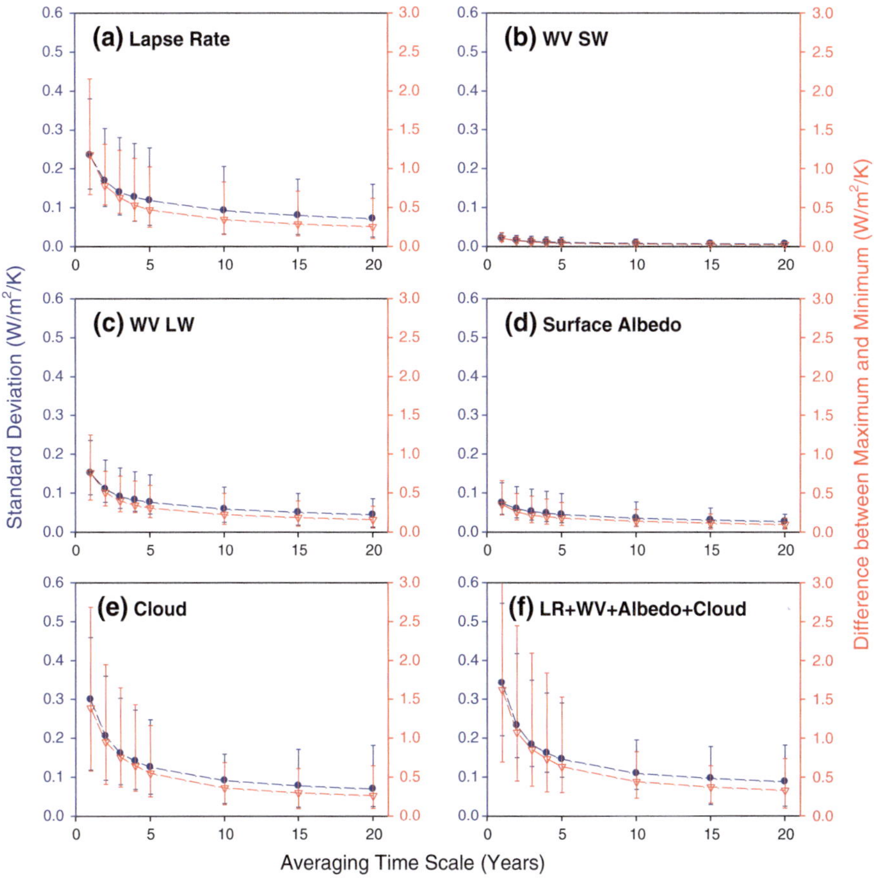

Fig. 3 Variations of standard deviation (*circles*) and the difference between the maximum and minimum value (*triangles*) of the feedback strengths computed, by setting the time separation between climate states A and B to 50 years, from the IPCC AR4 SRES A1B simulations as a function of the length of time average: **a** lapse rate, **b** water vapor (shortwave), **c** water vapor (longwave), **d** surface albedo, **e** cloud, and **f** summation except Planck feedback. *Symbols* (*circles* and *triangles*) and error bars denote the average and range of 15 climate models, respectively. Note that the axis and corresponding labels for the difference between the maximum and minimum value of the feedback strengths are placed on the right of the graphs

larger uncertainties due to smaller value of signal-to-noise ratio associated with seasonal to decadal scale natural variability. In addition, the intermodel variability denoted by the error bars (in red) decreases significantly with the increase in averaging time scale. Thus, the use of time average longer than sub-decadal time scale is required to filter high-frequency internal variations and thereby enhance the stability and robustness in the feedback computations.

Because of the difference in magnitude of each feedback process as shown in Figs. 1 and 2, the same value of standard deviation or the difference between the maximum and minimum could produce substantially different feedback uncertainty in terms of a percent error. Thus, the fractional differences are computed with respect to the feedback strengths of the 20-year averaging case (Table 1). All feedback processes produce the largest fractional difference for the 1-year averaging case, and the fractional differences reduce

Table 1 Fractional differences of feedback strengths relative to the values of 20-year averaging case

	Lapse rate	WV SW	WV LW	Albedo	Cloud	All
1-Year	19.3 ± 46.7	3.8 ± 11.0	4.4 ± 16.1	−5.8 ± 21.2	−64.0 ± 205.3	−18.7 ± 34.3
5-Year	18.2 ± 24.2	3.3 ± 3.8	4.2 ± 5.7	−10.8 ± 19.0	−15.6 ± 41.8	−6.7 ± 10.6
10-Year	7.3 ± 9.8	1.4 ± 1.4	1.8 ± 2.6	−4.9 ± 13.2	−11.7 ± 25.5	−1.6 ± 6.3
15-Year	4.8 ± 7.8	0.7 ± 0.9	1.0 ± 1.8	−2.4 ± 6.0	−2.8 ± 10.4	−0.2 ± 2.5

Numerals denote the average and standard deviation of 15 climate models. The unit is %

with the increase in averaging time scale. As expected, the largest percent error comes from cloud feedback, affecting the accuracy of the sum of all feedback processes. In addition, Table 1 exhibits that the fractional errors are larger for lapse rate and albedo feedback compared to both the components of water vapor feedback and that the strength of water vapor feedback can be confidently determined from even short-term observations (e.g., Allan 2006; John and Soden 2007; Dessler et al. 2008; Chung et al. 2010a).

4 Influences of the Time Separation of Two Climate States

With the operation of space-based hyper-spectral observations (e.g., AIRS and IASI), profiles of temperature and water vapor can be retrieved with sufficient vertical resolution to enable the computation of observational feedbacks using radiative kernels. While previous studies have estimated feedback strengths using those products (e.g., Dessler et al. 2008), short observational records (<10 years) have been an obstacle to investigating the time dependence of feedback parameters.

This limitation raises the following question. If one could perfectly measure temperature and humidity profiles and other climate variables from those satellite instruments, how many years of observations are required to determine climate feedbacks within a certain level of uncertainty? The problem is not strictly an observational one. The same question can be posed to the analysis of climate models, for example, How many years of model simulations are required to ensure the robustness of a model-diagnosed feedback strength? To address these questions, we examine the sensitivity of feedback parameters to the time separation between two climate states A and B.

Feedback computations are conducted with a time separation ranging from 20 to 50 years for three cases of time averaging (5, 10, and 20 years). For example, the feedback strengths of year 2001 for the 10-year averaging case are derived from the radiative kernels in conjunction with the differences of projected climate of years (2001–2010) from that of (2021–2030) [(2051–2060)] for the 20-year [50-year] time separation. Then, the temporal variations and intermodel variability of model-inferred feedback strengths are examined as in the previous section. On account of considerable high-frequency variations and intermodel discrepancies, the computation has not been conducted for the time averaging of 1–4 years. Figure 4 shows the average of the models' standard deviations of a given feedback parameter as a function of the time separation for three cases of time average length. All feedback processes manifest larger standard deviations for the shorter time average case, indicating that filtering of high-frequency internal fluctuations is required to reduce uncertainties in the feedback computations. The influence of the length of time average diminishes with the increase in the time separation between climate states A and B. This is a relevant consideration as a continuous record of high spectral resolution

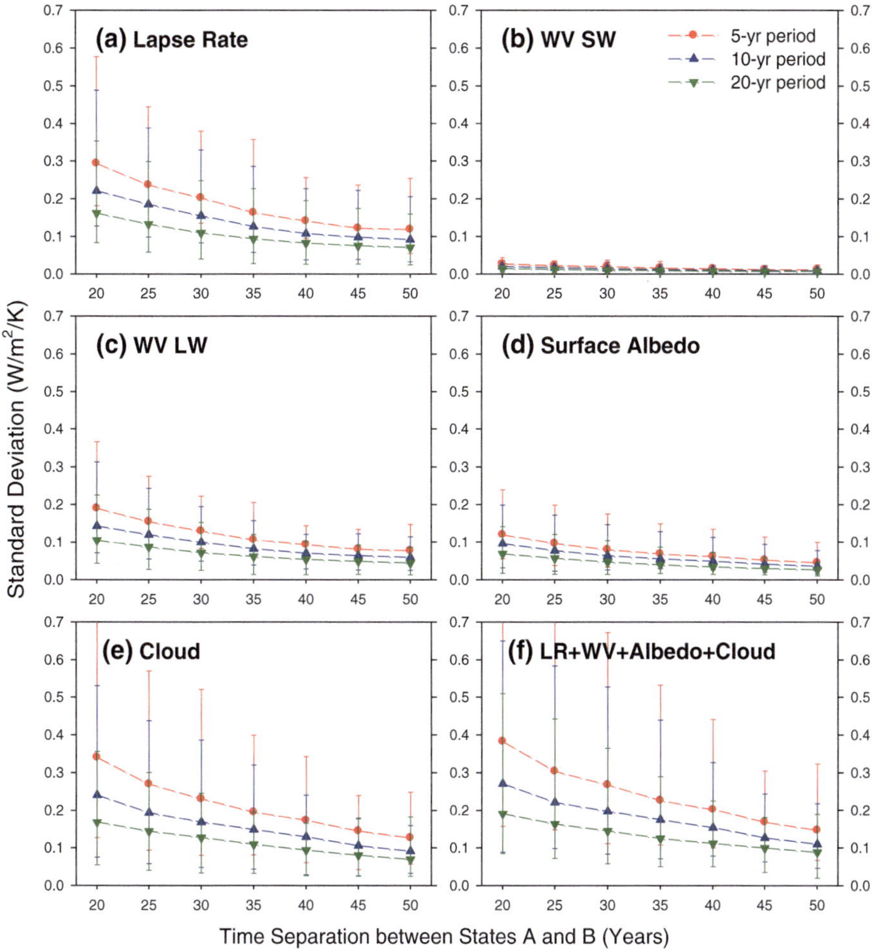

Fig. 4 Variations of standard deviation of the feedback strengths computed from the IPCC AR4 SRES A1B simulations as a function of the time separation between climate states A and B: **a** lapse rate, **b** water vapor (shortwave), **c** water vapor (longwave), **d** surface albedo, **e** cloud, and **f** summation except Planck feedback. *Symbols* and *error bars* denote the average and range of 15 climate models, respectively. *Different symbols* represent different length of the time average used to define climate states

measurements may not be needed if future observing systems have sufficiently accurate absolute calibration.

The temporal variations and intermodel variability of feedback strengths gradually decrease as the climate state B is located far from the climate state A. The extent of decrease is listed in the descending order as follows: cloud, lapse rate, water vapor (longwave component), surface albedo, and water vapor (shortwave component), indicating the differences in temporal variability between feedback processes. For example, weak feedbacks, such as the shortwave component of water vapor feedback, can be accurately quantified from a short period of observations. On the other hand, we need considerably longer satellite observations/model simulations to reduce uncertainties for other feedback processes that are stronger or more variable, for example, cloud feedback

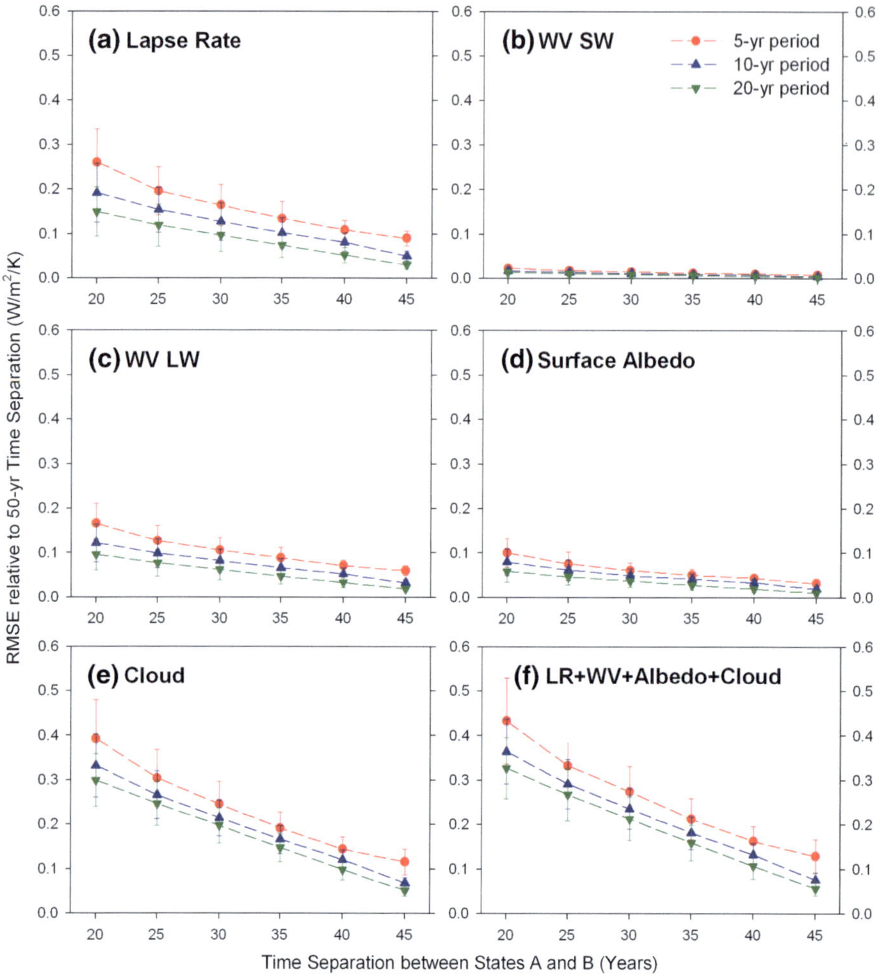

Fig. 5 The root-mean-square-errors of the feedback strengths computed from the IPCC AR4 SRES A1B simulations relative to the 50-year time separation cases: **a** lapse rate, **b** water vapor (shortwave), **c** water vapor (longwave), **d** surface albedo, **e** cloud, and **f** summation except Planck feedback. *Symbols* and *error bars* denote the average and standard deviation of 15 climate models, respectively. *Different symbols* represent different length of the time average used to define climate states

(Senior and Mitchell 2000; Chung et al. 2010b; Dessler 2010). This time dependence of feedback strengths is mainly attributable to the inter-hemispheric difference of temperature change associated with the large thermal inertia of Southern Ocean (Senior and Mitchell 2000).

Finally, we examine the relative accuracy of computed feedback strengths in terms of root-mean-square-error assuming that the 50-year time separation case produces true values. While the use of more long-term model simulations/satellite observations does not substantially improve the accuracy of inferred feedback strengths in the cases of the shortwave component of water vapor feedback and surface albedo feedback, Fig. 5 clearly exhibits a significant reduction in the uncertainty for cloud feedback and the combined feedback with the increase in the time separation between two climate states. These results suggest that we can keep reducing the uncertainty of cloud feedback and the combined feedback (or climate sensitivity) with the accumulation of satellite observations under the assumption that the satellite measurements are perfect and thus reduce the present inter-model discrepancy of climate sensitivity.

In this study, the time dependence of climate feedbacks has been examined using a radiative kernel approach. However, due to the insufficient record of observations, it may not be practical to derive feedback strengths from observations using a radiative kernel method. Instead, another approach such as regressing radiative fluxes at the top-of-atmosphere with surface temperature can be utilized for observation-based estimation of climate feedbacks (e.g., Dessler and Wong 2009; Dessler 2010). Therefore, better understanding of the time dependence of climate feedbacks may be obtained for shorter time scales by applying the regression approach to both observations and climate model simulations.

Acknowledgments We would like to thank two anonymous reviewers for their constructive and valuable comments which led to an improved version of the manuscript. We also thank Bruce Wielicki and David Young of NASA Langley Research Center for valuable discussion. This research was supported by a grant from the NASA/CLARREO Program and the NOAA Climate Program Office.

References

Allan RP (2006) Variability in clear-sky longwave radiative cooling of the atmosphere. J Geophys Res 111:D22105. doi:10.1029/2006JD007304

Bony S, Dufresne J-L (2005) Marine boundary layer clouds at the heart of tropical cloud feedback uncertainties in climate models. Geophys Res Lett 32:L20806. doi:10.1029/2005GL023851

Bony S et al (2006) How well do we understand and evaluate climate change feedback processes? J Clim 19:3445–3482

Cess RD et al (1990) Intercomparison and interpretation of climate feedback processes in 19 atmospheric general circulation models. J Geophys Res 95(D10):16601–16615

Chung E-S, Yeomans D, Soden BJ (2010a) An assessment of climate feedback processes using satellite observations of clear-sky OLR. Geophys Res Lett 37:L02702. doi:10.1029/2009GL041889

Chung E-S, Soden BJ, Sohn B-J (2010b) Revisiting the determination of climate sensitivity from relationships between surface temperature and radiative fluxes. Geophys Res Lett 37:L10703. doi: 10.1029/2010GL043051

Colman R (2003) A comparison of climate feedbacks in general circulation models. Clim Dyn 20:865–873

Colman R, McAvaney B (2009) Climate feedbacks under a very broad range of forcing. Geophys Res Lett 36:L01702. doi:10.1029/2008GL036268

Dessler AE (2010) A determination of the cloud feedback from climate variations over the past decade. Science 330:1523–1527

Dessler AE, Wong S (2009) Estimates of the water vapor climate feedback during El Niño-Southern Oscillation. J Clim 22:6404–6412

Dessler AE, Zhang Z, Yang P (2008) Water-vapor climate feedback inferred from climate fluctuations, 2003–2008. Geophys Res Lett 35:L20704. doi:10.1029/2008GL035333

Hansen J, Lacis A, Rind D, Russell G, Stone P, Fung I, Ruedy R, Lerner J (1984) Climate sensitivity: analysis of feedback mechanisms, in climate processes and climate sensitivity. AGU, Washington, pp 130–163

John VO, Soden BJ (2007) Temperature and humidity biases in global climate models and their impact on climate feedbacks. Geophys Res Lett 34:L18704. doi:10.1029/2007GL030429

Murphy JM (1995) Transient response of the Hadley Centre coupled ocean-atmosphere model to increasing carbon dioxide. Part III: analysis of global-mean response using simple models. J Clim 8:496–514

Murphy DM, Solomon S, Portmann RW, Rosenlof KH, Forster PM, Wong T (2009) An observationally based energy balance for the Earth since 1950. J Geophys Res 114:D17107. doi:10.1029/2009JD012105

National Research Council (2007) Earth science and applications from space: national imperatives for the next decade and beyond. The National Academies Press, Washington

Senior CA, Mitchell JFB (2000) The time-dependence of climate sensitivity. Geophys Res Lett 27:2685–2688
Shell KM, Kiehl JT, Shields CA (2008) Using the radiative kernel technique to calculate climate feedbacks in NCAR's Community Atmospheric Model. J Clim 21:2269–2282
Soden BJ, Held IM (2006) An assessment of climate feedbacks in coupled ocean-atmosphere models. J Clim 19:3354–3360
Soden BJ, Broccoli AJ, Hemler RS (2004) On the use of cloud forcing to estimate cloud feedback. J Clim 17:3661–3665
Soden BJ, Held IM, Colman R, Shell KM, Kiehl JT, Shields CA (2008) Quantifying climate feedbacks using radiative kernels. J Clim 21:3504–3520
Wetherald RT, Manabe S (1988) Cloud feedback processes in a general circulation model. J Atmos Sci 45:1397–1415
Zhang MH, Hack JJ, Kiehl JT, Cess RD (1994) Diagnostic study of climate feedback processes in atmospheric circulation models. J Geophys Res 99:5525–5537

Surv Geophys (2012) 33:745–777
DOI 10.1007/s10712-012-9180-4

Determination of Earth's Transient and Equilibrium Climate Sensitivities from Observations Over the Twentieth Century: Strong Dependence on Assumed Forcing

Stephen E. Schwartz

Received: 20 July 2011 / Accepted: 20 January 2012 / Published online: 4 May 2012

Abstract Relations among observed changes in global mean surface temperature, ocean heat content, ocean heating rate, and calculated radiative forcing, all as a function of time over the twentieth century, that are based on a two-compartment energy balance model, are used to determine key properties of Earth's climate system. The increase in heat content of the world ocean, obtained as the average of several recent compilations, is found to be linearly related to the increase in global temperature over the period 1965–2009; the slope, augmented to account for additional heat sinks, which is an effective heat capacity of the climate system, is 21.8 ± 2.1 W year m^{-2} K^{-1} (one sigma), equivalent to the heat capacity of 170 m of seawater (for the entire planet) or 240 m for the world ocean. The rate of planetary heat uptake, determined from the time derivative of ocean heat content, is found to be proportional to the increase in global temperature relative to the beginning of the twentieth century with proportionality coefficient 1.05 ± 0.06 W m^{-2} K^{-1}. Transient and equilibrium climate sensitivities were evaluated for six published data sets of forcing mainly by incremental greenhouse gases and aerosols over the twentieth century as calculated by radiation transfer models; these forcings ranged from 1.1 to 2.1 W m^{-2}, spanning much of the range encompassed by the 2007 assessment of the Intergovernmental Panel on Climate Change (IPCC). For five of the six forcing data sets, a rather robust linear proportionality obtains between the observed increase in global temperature and the forcing, allowing transient sensitivity to be determined as the slope. Equilibrium sensitivities determined by two methods that account for the rate of planetary heat uptake range from 0.31 ± 0.02 to 1.32 ± 0.31 K (W m^{-2})$^{-1}$ (CO_2 doubling temperature 1.16 ± 0.09–4.9 ± 1.2 K), more than spanning the IPCC estimated "likely" uncertainty range, and strongly anticorrelated with the forcing used to determine the sensitivities. Transient sensitivities, relevant to climate change on the multidecadal time scale, are considerably lower, 0.23 ± 0.01 to 0.51 ± 0.04 K (W m^{-2})$^{-1}$. The time constant characterizing the response of the upper ocean compartment of the climate system to perturbations is

S. E. Schwartz (✉)
Atmospheric Sciences Division, Brookhaven National Laboratory, Upton, NY 11973, USA
e-mail: ses@bnl.gov
URL: http://www.ecd.bnl.gov/steve

estimated as about 5 years, in broad agreement with other recent estimates, and much shorter than the time constant for thermal equilibration of the deep ocean, about 500 years.

Keywords Climate sensitivity · Forcing · Global mean surface temperature · Heat capacity · Time constant

1 Introduction

Quantifying the response of Earth's climate system to radiative forcings, changes in the Earth radiation budget that are imposed externally to the climate system, is central to understanding prior climate change over the industrial period and to planning mitigation of and/or adaptation to future climate change. Imposition of a forcing, which induces an immediate imbalance in the top of the atmosphere (TOA) radiation budget, induces changes in the climate system, including change in global mean near-surface air temperature (GMST) and changes in the long- and short-wave components of the TOA radiation budget that would ultimately lead to restoration of the energy balance. A widely employed measure of climate system response to imposed perturbations is the so-called equilibrium climate sensitivity, conventionally defined as the long-term steady-state change GMST that would result from a sustained externally imposed change in global net absorbed radiation (forcing), normalized to this forcing.[1] Numerous climate model studies have indicated that such a change in GMST would be proportional to the magnitude of the imposed forcing, but relatively insensitive to nature of the forcing, and thus that the equilibrium sensitivity is an intrinsic property of Earth's climate system. Such studies indicate as well that other changes in climate scale with changes in GMST. Consequently, the equilibrium sensitivity is widely viewed as essential to assessing the magnitude of climate change, generally, that would result from a given forcing. Determining the equilibrium sensitivity has thus been the objective of much of the research endeavor directed to understanding Earth's climate and its response to perturbations.

Frequently, the equilibrium climate sensitivity is expressed as a "CO_2 doubling temperature" $\Delta T_{2\times}$, the amount by which GMST would ultimately increase in response to a sustained doubling of atmospheric CO_2. $\Delta T_{2\times}$ is related to the equilibrium sensitivity S_{eq} by $\Delta T_{2\times} = F_{2\times} S_{eq}$, where $F_{2\times}$ is the forcing that would result from a doubling of CO_2. The equilibrium sensitivity (or equivalently $\Delta T_{2\times}$) is quite uncertain; the best estimate for $\Delta T_{2\times}$ given by the 2007 assessment report of the Intergovernmental Panel on Climate Change (IPCC 2007) is 3 K, with an uncertainty range (central 66% of the probability distribution function) of 2–4.5 K (relative range 83%). This uncertainty greatly limits confidence in the interpretation of climate change over the industrial period and precludes effective planning of energy futures (Schwartz et al. 2010).

Broadly speaking, approaches to determining the climate sensitivity can be distinguished as model-based and observation-based. Model-based determination is generally taken to mean through the use of general circulation models (GCMs) of the Earth climate system. Such models are capable of imposing a forcing of known magnitude on the climate system

[1] For reasons having to do with stratospheric adjustment that occurs rapidly (months) following an increase CO_2, which has traditionally been used as a benchmark forcing in model studies of climate sensitivity, the forcing pertinent to climate change and to determination of climate sensitivity has long been considered to be the change in net absorbed radiation at the tropopause. Increasingly, however, it is becoming recognized (e.g., Gregory and Forster 2008) that the measure of forcing pertinent to the global energy balance is the change in net radiation at the top of the atmosphere, again following such rapid adjustment.

and determining the sensitivity from the climate system response, accounting for the departure of the system from equilibrium through the net heating rate of the planet (Forster and Taylor 2006). Observationally based determination generally requires knowledge of both the forcing that is thought to have induced a change in GMST over a given period of time and the resultant temperature change attributable to that forcing, necessitating confident attribution of the response to the forcing. In principle, this approach might be based on equilibrium change in radiation and GMST over a long period of time, up to and including differences between glacial ice ages or the mid-Cretaceous and the present temperate period. Alternatively, the empirical approach might be based on shorter-term forcing and response, e.g., volcanic aerosols, or over the industrial period with time-dependent forcing, with some means of accounting for the observed response being only a fraction of the equilibrium response (Gregory et al. 2002; Forster and Gregory 2006; Murphy et al. 2009).

An intrinsic concern with determining the equilibrium climate sensitivity from long-time climate response to a perturbation is the long time required to approach a new steady state following imposition of a forcing, in the real world or in coupled atmosphere–ocean climate models, of the order of 1,000 years (Held et al. 2010; Hansen et al. 2011; Jarvis and Li 2011). Such a long response time makes it impractical to determine the equilibrium sensitivity by this long-time response, empirically or in climate models. The long time required to reach a new "equilibrium" also raises the question of the utility of the equilibrium sensitivity to societal decision-making about reducing carbon dioxide emissions to limit near-term global warming (Allen and Frame 2007).

Several investigators (Gregory 2000; Gregory and Forster 2008; Baker and Roe 2009; Held et al. 2010) have examined the response of Earth's climate system to perturbations using two-compartment models that exhibit a rapidly achieved (decadal) near–steady-state response to a perturbation by a fast-responding compartment, which comprises the atmosphere, surface, and upper ocean and a much more slowly responding (multiple centuries) deep ocean compartment that is responsible for the slow approach to the ultimate steady state following imposition of a sustained forcing. In such models, in response to a (positive) forcing, there results, in addition to changes in the net irradiance at the top of the atmosphere (TOA), a flow of heat energy from the fast-responding compartment to the deep ocean, which has large heat capacity and long response time, that diminishes climate system response relative to the long-time response, in which the imposed forcing is offset only by change in net TOA irradiance. If the heat capacity and time constant of the deep ocean compartment are large compared to the upper compartment, the flow of heat energy into this compartment would be expected to be proportional to the change in global mean temperature. This situation leads in turn to an expected proportionality between the increase in GMST and imposed forcing that is achieved on decadal time scales. This proportionality is denoted here and elsewhere (Held et al. 2010; Padilla et al. 2011) as the *transient climate sensitivity*, although other terminology, e.g., "transient climate response," (Dufresne and Bony 2008) is used.

Here, in order to determine the dependence of transient and equilibrium sensitivities on assumed forcing over twentieth century, I examine relations among observed changes in GMST, ocean heat content, and ocean heating rate, together with several published model-based estimates of forcing, all as a function of time. Interpretation of these relations within the two-compartment model yields quantities pertinent to climate system response to perturbations: the effective heat capacity of the climate system pertinent to climate change over this period, the heat uptake coefficient relating the rate of increase of planetary heat content and the increase in GMST, the coefficient of proportionality between increase in GMST and forcing (transient climate sensitivity), and the equilibrium climate sensitivity.

Estimates are also provided of the time constants for response of the two compartments to perturbations. Within the two-compartment model, all of these quantities are intrinsic properties of Earth's climate system. The effective heat capacity and heat uptake coefficient adduced by the present analysis are independent of assumptions about radiative forcing over this period, but the sensitivities are rather strongly dependent on the radiative forcing employed in the analysis.

The outline of this paper is as follows. Section 2 sets out the two-compartment model. Section 3 examines relations among the observables (GMST anomaly, ocean heat content anomaly, and global heating rate) and model-based estimates of global mean forcing and presents results for the climate system properties determined by this analysis. Section 4 places the findings in context with other work and examines concerns with the present analysis. Summary and conclusions are given in Sect. 5.

2 Theory

The energy conservation equation for Earth's climate system is

$$\frac{\mathrm{d}H}{\mathrm{d}t} \equiv N = Q - E \tag{1}$$

Here, H is the global heat content anomaly (relative to an arbitrary year or period); $N \equiv \mathrm{d}H/\mathrm{d}t$ is the net heat flux into the planet; Q is the absorbed solar energy; and E is the emitted long-wave flux. Other sources or sinks of energy are negligible (Schwartz 2008b; Pilewskie 2011). In considering the consequences of a forcing, i.e., a radiative flux perturbation imposed on the climate system, it is useful to consider a situation in which a forcing is imposed on a system that is initially at radiative steady state (commonly denoted "equilibrium") in which the net flux is equal to zero: $N_0 = 0$. Whether such a situation is ever achieved is perhaps a question of threshold and averaging time. For example, fluctuations in the solar constant on an 11-year cycle would set a floor on the steadiness of the initial state; similarly, there are inevitably fluctuations arising from internal variability such as ENSO and from occasional forcing by aerosols from eruptive volcanos. In this steady-state situation, the emitted long-wave radiation, averaged over the planet and over a sufficiently long time (greater than a year, to average over seasonal variation), is very closely equal to the absorbed solar radiation: $E_0 = Q_0$.

Upon application of a perturbation in radiative flux, or forcing external to the climate system, F, the net flux into the climate system, N, is altered, a positive forcing corresponding to an increase in heat content of the system (positive N),

$$N(t) = F(t) + Q(t) - E(t), \tag{2}$$

and the system responds by an increase in GMST; this increase in temperature and other changes in the climate system induced by the forcing result in changes in the absorbed and/or emitted power at the TOA. These changes are represented by an expression that denotes the dependence on GMST as a leading term but allows for dependence on higher order terms as well:

$$N = F + Q_0 - E_0 + \frac{\partial(Q - E)}{\partial T}\Delta T + \text{higher order terms}, \tag{3}$$

where the partial derivative denotes the changes in Q and E due to the response of the climate system to the change in temperature (but explicitly not including the change in net flux that is the forcing itself). The first-order term would include changes in the climate system that scale with change in GMST, for example, surface long-wave irradiance, water vapor amount, and cloud amount. The higher order terms would include terms that are second order in ΔT and any dependence on geographical or temporal distribution of change in surface temperature. As is conventional, the effects of these higher order terms are omitted in the analysis that leads to a climate sensitivity. In the present analysis, I restrict consideration to the first-order term, but I return to consideration of the higher order terms and their implications in Sect. 4.5.

Conventionally, the partial derivative in Eq. 3 is expressed as a *climate response coefficient*,

$$\lambda \equiv -\frac{\partial(Q-E)}{\partial T}, \qquad (4)$$

the sign of which is chosen to make λ a positive quantity. For a positive forcing, initially N is increased by the magnitude of the forcing; as the temperature increases in response to this forcing, the value of N decreases as the system approaches a new steady state. For this constant forcing, N again approaches zero, and, under the assumption that the properties of the climate system affecting the climate system response to the forcing are unchanged as the new equilibrium is reached, ΔT would be equal to $\lambda^{-1}F$. The inverse of λ, $S_{eq} \equiv \lambda^{-1}$ is denoted the *equilibrium climate sensitivity*, the amount by which GMST would ultimately change in response to a sustained forcing, normalized to the value of the forcing, for properties of the climate system affecting the partial derivative in Eq. 4 unchanged. By this definition,

$$S_{eq} = \lambda^{-1} = \left[\left(\frac{\partial(Q-E)}{\partial T}\right)_0\right]^{-1},$$

where the subscript 0 denotes a given unperturbed climate state, the equilibrium sensitivity is seen to be a property of Earth's climate system and thus of intrinsic, as well as practical, interest.

Commonly, especially in the context of examining the consequences of alternative scenarios of future CO_2 emissions, the equilibrium sensitivity is expressed as the amount by which the global mean surface temperature would ultimately increase in response to a sustained doubling of atmospheric CO_2, CO_2 *doubling temperature*, $\Delta T_{2\times}$; within the linear range of the relation between increase in GMST and forcing, $\Delta T_{2\times}$ is equal to $F_{2\times}S_{eq}$, where $F_{2\times}$ is the forcing corresponding to a doubling of CO_2, which is about 3–4 W m^{-2}. $F_{2\times}$ is commonly (Myhre et al. 1998; IPCC 2007) given, or readily inferable, as 3.71 W m^{-2}, although that precision is hardly justified by the accuracy with which $F_{2\times}$ is known for the actual climate system (Stevens and Schwartz 2012) or represented in current climate models (Webb et al. 2006; Forster and Taylor 2006) because, *inter alia*, of issues involving radiation transfer, especially in cloudy atmospheres (Collins et al. 2006), and short-term responses of clouds to changes in CO_2 (Andrews et al. 2009). In this paper, the primary results for climate sensitivity are presented in systematic units, K (W m^{-2})$^{-1}$, with values of $\Delta T_{2\times}$, evaluated with $F_{2\times}$ taken as 3.71 W m^{-2}, presented for convenience.

Equation 3 (*cf.* also, Gregory and Forster 2008) suggests a direct proportionality (constant ratio) between $\Delta T(t)$ and $F(t)$–$N(t)$, where the dependence of all quantities on secular time t, as would result from temporally variable forcing, is explicitly noted,

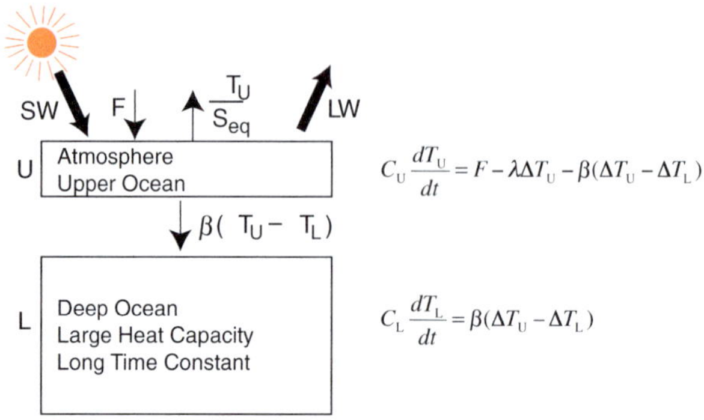

Fig. 1 Two-compartment model for Earth's climate system, consisting of *upper* compartment U with small heat capacity C_U and short time constant for reaching steady state following a perturbation, and *lower* compartment L with large heat capacity C_L and long time constant for reaching steady state. *Thick arrows* denote initial short-wave (SW) and long-wave (LW) fluxes, and *thin arrows* denote perturbations: forcing F and resultant heat flows, a change in net top-of-atmosphere irradiance given by the time-dependent forcing minus the time-dependent change in temperature ΔT_U of the *upper* compartment upon the equilibrium sensitivity S_{eq}, and a heat flux from the *upper* compartment to the *lower* compartment L given by the heat exchange coefficient β times the difference in temperature changes between the *upper* (ΔT_U) and *lower* (ΔT_L) compartments

$$F(t) - N(t) = \lambda \Delta T(t), \tag{5}$$

or, equivalently, in terms of the equilibrium sensitivity

$$\Delta T(t) = S_{eq}(F(t) - N(t)). \tag{6}$$

Equation 6 leads to the hypothesis that a plot of $\Delta T(t)$ versus $F(t)$–$N(t)$, for both $F(t)$ and $\Delta T(t)$ measured from the same initial steady-state condition, would be expected to be linear through the origin with slope S_{eq} and thus that S_{eq} can be determined as the slope. To distinguish this method of determining S_{eq} from a second method described below, I denote this approach to determining the equilibrium sensitivity the "*F*–*N* method".

A second approach to determining climate sensitivity is based a simple two-compartment model of the climate system illustrated in Fig. 1 (*cf.* also Gregory 2000; Held et al. 2010). The upper compartment, which consists principally of the atmosphere and the upper ocean, is coupled radiatively to the incoming solar irradiance and the outgoing thermal infrared radiation at the TOA and thermally to the deep ocean compartment. A positive forcing applied to the climate system induces an increase in the temperature of the upper compartment that induces not only a radiative response identical to that described above (Eq. 3) but also a heat flow into the larger–heat-capacity deep-ocean compartment. It is posited that the time required for the lower compartment to respond to any forcing is much greater than that for the upper compartment; Held et al. (2010) refer to this compartment as "recalcitrant" to changes in heat content. For an initial estimate of the heat capacities of the upper compartment from the depth of the mixed layer (ca., 100 m) and of the lower compartment from the average depth of the world ocean (ca 3,800 m), the ratio of the heat capacities would be approximately 40. Hence, following imposition of an external forcing, it would be expected that the temperature change of the lower compartment would be much less than that of the low–heat-capacity, rapidly accommodating upper compartment. This leads to the hypothesis that the rate of heat transport from the upper compartment to the lower compartment would be proportional to the increase in temperature of the upper compartment ΔT; *cf.* also Forster and Taylor (2006), Boer et al. (2007), Gregory and Forster (2008). Observationally, as shown below, the rate of increase of heat content of the climate system is found to be proportional to ΔT,

$$N(t) = \kappa \Delta T(t). \tag{7}$$

The *heat uptake coefficient* κ, like the equilibrium climate sensitivity, is posited to be an intrinsic property of Earth's climate system.

The two-compartment model (Gregory 2000; Knutti et al. 2008, Schwartz 2008a; Held et al. 2010) exhibits two time constants (inverses of the eigenvalues) that characterize the rate of relaxation of the system to a perturbation. Specializing to the situation in which the lower compartment is much larger than the upper compartment, and has much greater heat capacity C_L than that of the upper compartment C_U, yields (Held et al. 2010) for the two time constants (to first order in C_U/C_L)

$$\tau_s = \frac{C_U}{\beta + \lambda} \quad \tau_l = C_L\left(\frac{1}{\lambda} + \frac{1}{\beta}\right), \tag{8}$$

where the subscripts s and l denote short and long time constants characterizing the response of the upper and lower compartments, respectively. Although it does not appear that these time constants can be exactly determined by the present analysis, as shown below, they can be estimated based on observations and reasonable assumptions.

The two-compartment model suggests that the heat capacities of the two compartments of the climate system may be identified as follows. The heat capacity of the large compartment is that of the global ocean, which may be evaluated as the volume of the global ocean times the volumetric heat capacity seawater. For the fractional area of Earth covered by ocean as 0.71 and the average depth of the world ocean taken as 3,800 m, the average depth per area of the entire planet is 2,700 m. For the volumetric heat capacity of seawater taken as 4.0×10^6 J m^{-3} K^{-1}, the areal heat capacity (per area of the planet) $C_L = 1.1 \times 10^{10}$ J m^{-2} K^{-1}. For climate change considerations, it is convenient to express this heat capacity in the unit W year m^{-2} K^{-1}, yielding $C_L = 340$ W year m^{-2} K^{-1}. An effective heat capacity of that part of the climate system that is coupled to the change in global temperature may be obtained (Schwartz 2007) from the relation between the observed change in global heat content H and the observed change in GMST. As shown previously (Schwartz 2007) and updated below for the further available measurements, a linear relation is obtained between ocean heat content anomaly and global temperature anomaly,

$$H_{ocn} = H_0 + C_{ocn}\Delta T, \tag{9}$$

allowing an effective ocean heat capacity to be evaluated as the slope of a graph of the two quantities. The heat capacity thus determined is then augmented to account for other heat sinks (heating of the atmosphere and upper solid earth; melting of ice) to yield an estimate for the effective heat capacity of the climate system, C_{eff}. As determined below, this quantity is more than an order of magnitude less than C_L, consistent with the two-compartment model outlined above and with the expressions for the time constants given in Eq. 8.

A proportionality between global heating rate and the increase in GMST would result in a further proportionality, between the time-dependent increase in GMST $\Delta T(t)$ and the time-dependent forcing $F(t)$: from Eqs. 5 and 7,

$$F(t) = (\kappa + \lambda)\Delta T(t). \quad (10)$$

Equation 10 suggests a so-called *transient climate sensitivity* analogous to the equilibrium sensitivity,

$$S_{tr} \equiv (\kappa + \lambda)^{-1} \quad (11)$$

that relates the time-dependent increase in surface temperature to the time-dependent forcing,

$$\Delta T(t) = S_{tr}F(t), \quad (12)$$

which implies that a graph of $\Delta T(t)$ versus $F(t)$ would be linear through the origin, with slope S_{tr}. This transient climate sensitivity is the same quantity as that determined from observed temperature change and modeled forcing by Gregory and Forster (2008), except that their quantity is referred to the forcing of doubled CO_2 and is thus greater by a factor of 3.7.

The temperature response to a step-function forcing of the two-compartment system with a given transient sensitivity and upper compartment time constant is initially virtually identical to that of a single compartment system having the same time constant and equilibrium sensitivity, Fig. 2a; the influence of the long–time-constant, large–heat-capacity, lower compartment initially is to provide a further heat sink, in addition to the decrease in net radiation absorbed at the TOA (Eq. 3), that diminishes the initial temperature response. Only at long times, appreciable relative to the time constant of the lower compartment, when the increase in temperature of the lower compartment ΔT_L becomes appreciable relative to the increase in temperature of the upper compartment ΔT_U (Fig. 1), does the heat transport to the lower compartment become appreciably reduced, and the temperature of the upper compartment approaches its "equilibrium" value, Fig. 2b. The plateau in the temperature change of the upper compartment shown in Fig. 2a is the basis of the transient sensitivity concept.

In order to determine the transient and equilibrium sensitivities and their dependence on assumed forcing over twentieth century, the present study examines the several proportionalities between observations of temperature change, planetary heat content, and heating rate over the twentieth century, $\Delta T(t)$, $\Delta H(t)$, and $N(t)$, and between these observables and estimates of the forcing over this period, $F(t)$, as follows: $H(t)$ versus $\Delta T(t)$, slope, C_{eff},

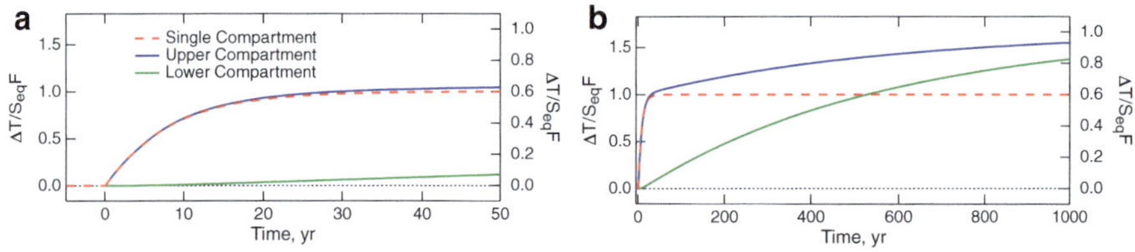

Fig. 2 Normalized temperature response of one- (*left* axis) and two-compartment (*right* axis) models to step-function forcing initiated at time = 0. Equilibrium sensitivity of one-compartment model and transient sensitivity of two-compartment model are 0.4 K (W m^{-2})$^{-1}$ ($\Delta T_{2\times} = 1.5$ K); equilibrium sensitivity of two-compartment model is 0.67 K (W m^{-2})$^{-1}$ ($\Delta T_{2\times} = 2.5$ K). Heat exchange coefficient in two-compartment model is 1 W m^{-2} K^{-1}. Time constant of one-compartment model and of *upper* compartment of two-compartment model is 8 years; time constant of *lower* compartment is 570 years. Heat capacity of upper compartment and *lower* compartment are 20 and 340 W year m^{-2} K^{-1}, respectively. *Vertical* axes are scaled by the ratio of the equilibrium sensitivities of the two models

(Eq. 9); $N(t)$ versus $\Delta T(t)$, slope κ, Eq. 7; $\Delta T(t)$ versus $F(t)$, slope S_{tr}, Eq. 12; $\Delta T(t)$ versus $F(t)-N(t)$, slope S_{eq}, Eq. 6. Note that (except for the relation between H and ΔT) each of these several relations is a *proportionality*, that is, a linear relation with zero intercept. The order of the relation indicates the dependent and independent variables, respectively, under the expectation that the temperature change depends on the forcing but that the heat content and heating rate depend on the temperature change. The relation between S_{eq} and S_{tr},

$$S_{eq} = \frac{1}{S_{tr}^{-1} - \kappa}, \tag{13}$$

leads to a second method of estimating of S_{eq} based on the heat uptake coefficient κ rather than the individual measurements of heating rate $N(t)$; I denote this method the "κ method".

The estimates of climate sensitivities obtained in this way are not wholly observationally based but are hybrids between observationally and model-based estimates of these quantities, as the time-dependent forcing is not directly observed but is based on radiative transfer calculations for measured or modeled changes in atmospheric composition. Examining these quantities for a suite of forcing estimates allows assessing the consistency of the model and the estimates of forcing, which would be manifested by a linear proportionality of temperature change and forcing, and determination of the dependence of the inferred climate sensitivities on the forcing.

This entire analysis is rooted in the forcing-response model of climate change, that is, that the change in global mean temperature is a consequence only of forcing under the assumption that forcings are fungible and that GMST response, normalized to forcing, is independent of the nature and geographical distribution of the forcing. I return to this point in Sect. 4.5.

3 Results and Interpretation

3.1 Global Heat Content and Heating Rate

The major accessible reservoir for storing planetary heat energy is the world ocean on account of the high specific heat of water and the relatively rapid rate of heat exchange within the ocean by virtue of circulations and turbulent mixing on a variety of scales. The basis for determination of the global heating rate rests on the data base of historical measurements of ocean temperature as a function of location, depth, and time. These temperature measurements are converted into heat content anomaly as a function of time (relative to a specified base period) via the heat capacity of ocean water and integration over the volume of the world ocean. The measurement data base extends back to about 1950, but the early measurements are sparse, and there remain questions about the accuracy of the primary measurements and the coherence of measurements by different types of sounding instruments and platforms, which have changed over time (Gouretski and Reseghetti 2010). The measurements have been analyzed by several groups, with broad agreement but significant differences based on differences of approach and assumptions about measurement techniques; for reviews, see Palmer et al. (2010) and Lyman (2011). A composite of the data compilations for anomaly of heat content of the world ocean from the surface to 700 m is shown in Fig. 3; most of the increase in ocean heat content is due to

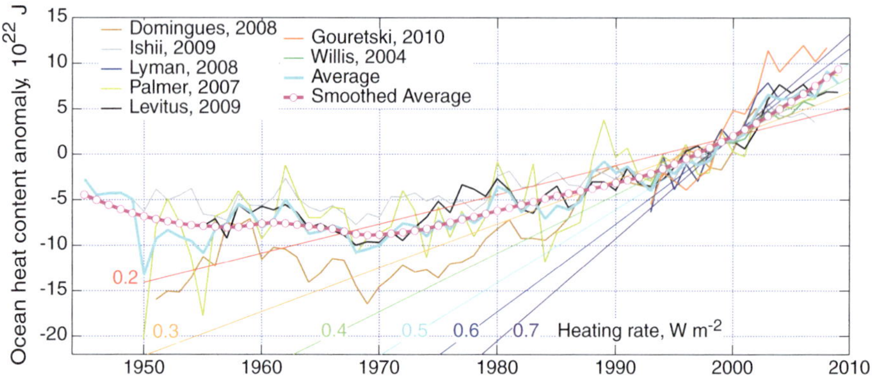

Fig. 3 Recent evaluations of ocean heat content anomaly (0–700 m), relative to 1993–2002. Also shown are the average of the several estimates and a LOWESS smoothed curve based on the average. Slopes of *lines* show heating rate referred to Earth surface area. Data from http://www.ncdc.noaa.gov/bams-state-of-the-climate/2009-time-series/ohc (retrieved November 5, 2010) which gives the citations to the publications; data have been updated from the indicated publications by Palmer et al. (2010)

warming within the top 700 m, and most of that within the top 300 m (Levitus et al. 2005). Also shown in the figure is the average of the data obtained by the several groups; in order not to bias the average at times where different data sets contributed to the average, all anomalies were computed relative to the years 1993–2002, for which period all data sets are represented.

The effective areal heat capacity of the portion of the climate system that is actively coupled to the change in global temperature C_{eff} is evaluated from the dependence of the change in ocean heat content on the change in global temperature. As seen in Fig. 4, a linear relation is exhibited between the areal heat content anomaly of the ocean (to 700 m) and the surface temperature anomaly; the fit was restricted to the data subsequent to 1965, consistent with the analysis for heating rate, but the slope for the entire data set (subsequent to 1945) differed negligibly. The slope of this plot, 14.1 ± 1.0 W year m^{-2} K^{-1}, represents an effective areal heat capacity of the upper 700 m of the world ocean. This effective heat capacity is well less the actual areal heat capacity of seawater to this depth, evaluated as the fractional area of the world ocean, 0.71, times the volumetric heat capacity of seawater, 4.0×10^6 J m^{-3} K^{-1}, times the depth, 700 m, and expressed in the same units, 63 W year m^{-2} K^{-1}, indicative of the fact that the entire global amount of seawater to this depth is not in thermal equilibrium with the increasing global temperature but is substantially lagging the increase in temperature. Accounting for heating of the ocean deeper than 700 m adds another 30% to this effective heat capacity, yielding 18.3 W year m^{-2} K^{-1}. An alternative way of looking at this result is that this heat capacity corresponds to a hypothetical depth of seawater in thermal equilibrium with the surface temperature, 145 m (for the entire planet) or 200 m for ocean fractional area of 0.71. Finally, accounting for heat sinks in the climate system (air, solid earth, melting of ice) adds a further 19% to the effective heat capacity, yielding 21.8 ± 2.1 W year m^{-2} K^{-1}, where the uncertainty is a one-sigma estimate that takes into account the uncertainty in the slope together with estimated uncertainties (25%) in the two augmentations.

Determining the heating rate of the global ocean requires taking a derivative of the heat content; this is complicated by the temporal variability of the data. This variability may be noise associated with the measurements and/or may be a manifestation of actual variability

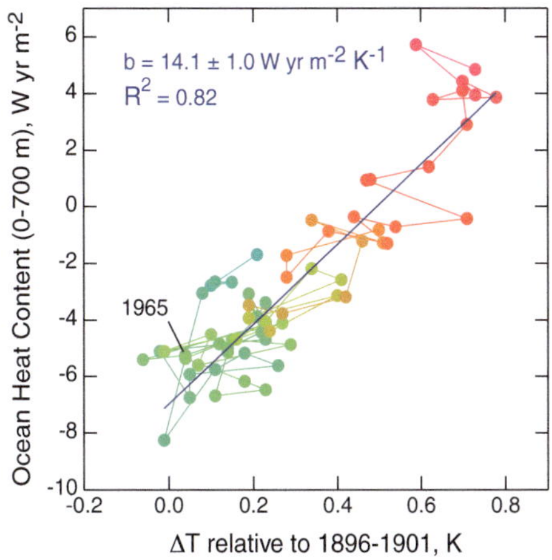

Fig. 4 Examination of relation between global ocean heat content anomaly and temperature anomaly (Eq. 9). Heat content anomaly of the world ocean, relative to 1993–2002, from surface to 700 m, expressed per area of the planet (Fig. 3) is plotted against global mean surface temperature anomaly relative to 1896–1901 ΔT_{1900}. Fit to data in the form $\Delta H = a + b\Delta T$ (Eq. 9) is evaluated for data from 1965 through 2009; uncertainty in regression slope is standard error, neglecting autocorrelation. Also indicated as R^2 is fraction of variance in the original data accounted for by the regression. *Colors* of data points denote date, from 1945 (*light blue*) to 2009 (*red*)

of ocean heat content, a consequence perhaps of internal climate variability such as ENSO or of change in heating rate due to variability in forcing, such as by volcanic aerosol forcing. In order to smooth the data to permit taking the time derivative, the LOWESS algorithm (locally weighted scatterplot smoothing; Cleveland and Devlin 1988) was used to construct a smooth curve that retained the slow temporal variability of the data (magenta curve in Fig. 3); the areal heating rate of the ocean was obtained as the derivative of this quantity. A sense of the magnitude of the slope is provided by auxiliary lines drawn on the figure whose slopes correspond to the heating rates indicated, in units of watts per square meter, expressed relative to the total Earth surface area (5.1×10^{14} m^2), not just the area of the world ocean. There is indication that since 1970, the slope has been increasing from perhaps 0.3 W m^{-2} to perhaps 0.6 W m^{-2}.

Accounting for additional ocean heat uptake and other heat sinks, as with ocean heat content, yields the rate of global heating shown as a function of time in Fig. 5. Prior to about 1970, the heating rate exhibited fairly large fluctuations and indication even of negative values (net cooling of the planet); these fluctuations may be due to sampling or measurement issues or may reflect actual changes in global heat content. Subsequent to about 1970, the rate of change of global heat content has been consistently positive, albeit not monotonically increasing until after about 1992. Although this variable pattern may again be a consequence of sampling or measurement issues, that possibility seems increasingly unlikely in the later time frame on account of the increase in number and quality of the data and the general tightness of the several compilations of the ocean heat content data. In further support of this argument is the general agreement of the increase in heating rate N with the increase in GMST over the entire time period from about 1960 to the present shown in the time series in Fig. 5.

The relation between global heating rate and global mean surface temperature anomaly (Eq. 7) is examined further in Fig. 6, in which the heating rate is plotted against global temperature anomaly relative to 1900, ΔT_{1900}. For the fit restricted to the data from 1965 to

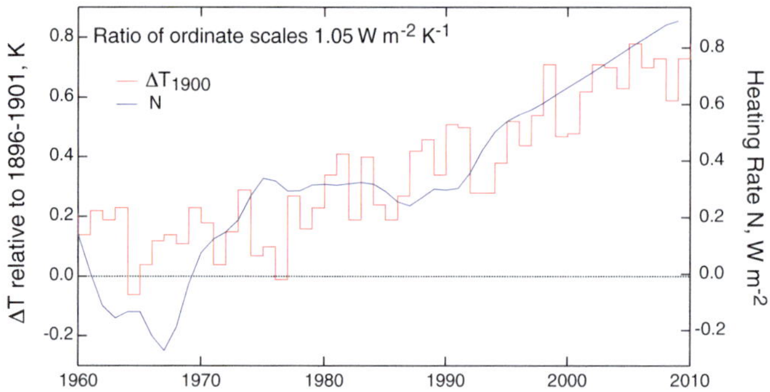

Fig. 5 Global mean surface temperature anomaly relative to 1896–1901 ΔT_{1900} (*left* axis), evaluated from the Goddard Institute for Space Studies (GISS) Combined Land-Surface Air and Sea-Surface Water Temperature Anomalies (Land–Ocean Temperature Index, LOTI) http://data.giss.nasa.gov/gistemp/tabledata/GLB.Ts+dSST.txt, (Hansen et al. 2010) and global heating rate N (*right* axis) evaluated as derivative of smoothed average ocean heat content anomaly shown in Fig. 3, augmented to account for deep ocean heating and other heat sinks as described in text, and expressed per area of the planet. Ordinate axes are scaled by the slope of regression forced through the origin (Fig. 6)

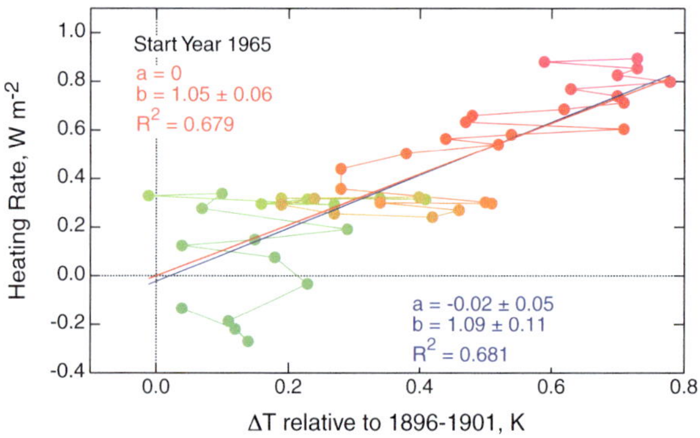

Fig. 6 Examination of proportionality between global heating rate and temperature anomaly (Eq. 7). Global heating rate (evaluated as derivative of smoothed average ocean heat content anomaly shown in Fig. 3, augmented to account for deep ocean heating and other heat sinks as described in text and expressed per area of planet) is plotted against global mean surface temperature anomaly relative to 1896–1901 ΔT_{1900}. Fits to data in the form $dH/dt = a + b\Delta T$ are evaluated for data from 1965 through 2009, with and without constraining regression line to pass through origin; uncertainties in regression coefficients are standard error, calculated under assumption of no autocorrelation. Also indicated as R^2 is fraction of variance in the original data accounted for by the regression. *Colors* of data points denote date from 1965 (*green*) to 2009 (*red*)

the present, the regression line has an intercept quite close to zero, -0.02 ± 0.05 W m^{-2} (1 sigma, calculated under assumption of zero autocorrelation), and the regression fit constrained to pass through the origin accounts for virtually the same fraction of the variance (68%) as the two parameter fit, that is, the heating rate is proportional to the increase in surface temperature relative to 1896–1901. This finding is relatively insensitive to the start date of the correlation, with intercept ranging from -0.06 ± 0.05 W m^{-2} (start date 1960) to $+0.10 \pm 0.04$ W m^{-2} (start date 1970), so the finding of linear proportionality (zero intercept) would appear to be robust. The corresponding values for the slope of the regression forced through the origin are 1.02 ± 0.06, 1.05 ± 0.06, and 1.07 ± 0.05 W m^{-2} K^{-1} for start date 1960, 1965, and 1970, respectively. The rather tight correlation between global heating rate (as inferred from ocean heating rate) and the increase in global temperature relative to the beginning of the twentieth century supports the hypothesis that this heating rate is proportional to global mean surface temperature anomaly, Eq. 7, with heat uptake coefficient $\kappa = 1.05 \pm 0.06$ W m^{-2} K^{-1}. The proportionality of heating rate and global temperature, independently determined, is consistent with the two-compartment model with a much greater heat capacity in the lower compartment, as described above and supports the utility of using the time-dependent heating rate and/or the regression slope in examination of the relation between global temperature change and various estimates of forcing over the twentieth century.

3.2 Forcing

The observationally based determination of climate sensitivity rests on knowledge of climate forcing over the instrumental record. Knowledge of this time-dependent forcing is required also in climate model calculations over the twentieth century. Two approaches have been taken in climate modeling studies to determine the forcing. The forcing may be specified as an input to the model based on measured or modeled changes in atmospheric composition and other radiation influencing quantities. Alternatively, the atmospheric substances influencing radiation in the model are themselves modeled within the climate model, and the resulting changes in radiation are calculated within the climate model. The latter approach is increasingly being taken by various modeling groups (Lohmann et al. 2010). As a consequence, it seems to be difficult to extract the forcing from climate model runs of the twentieth century. An examination of the literature revealed only a limited number of forcing data sets that were suitable for the present study. In addition to anthropogenic gases and aerosols, whose forcings are to be calculated, modeling the twentieth century requires representation of natural forcings by changes in solar irradiance and, importantly, forcing by stratospheric aerosols that result from eruptive volcanoes. Volcanic eruptions in the twentieth century have resulted in several instances of short-duration forcing, the magnitudes of which are substantial in the context of anthropogenic forcing, and which are manifested in the record of GMST. The several explicitly calculated time series of forcing over the twentieth century available from the literature are summarized in Table 1. In addition to the data sets examined here, Forster and Taylor (2006) have presented forcing data sets inferred from the increase in GMST in AR4 model runs over the twentieth century together with the transient sensitivity of these models determined from model runs with 1% per year increment CO_2, for which the forcing is known. The approach appears to yield a fairly accurate estimate of the time-dependent forcing as shown by comparison with forcing data sets employed in the GISS (Goddard Institute for Space Studies; Hansen et al. 2005) and MIROC (Model for Interdisciplinary Research On Climate; Takemura et al. 2006) model studies for which forcings were explicitly calculated. This method was subsequently applied to determine the forcing time series over the twentieth century in the Hadley Centre model (Jones et al. 2011). In view of the indirect means of inferring the forcing, these forcing data sets were not examined in the present study, although doing so might provide further insight.

The forcing data set employed in climate model studies by the investigators at the Goddard Institute for Space Studies (GISS; Hansen et al. 2005), Fig. 7, is illustrative of current forcing data sets. The positive (warming) forcing by the well mixed greenhouse gases (GHGs) and the negative (cooling) forcings due to scattering and cloud-brightening effects of aerosols exhibit relatively smooth increases in magnitude with similar temporal trend. Negative forcing by volcanic stratospheric aerosols in contrast is highly irregular,

Table 1 Forcing data sets examined in this study

Data Set	Reference and data source	Forcing, 1900–1990, W m^{-2}
PCM, parallel climate model, national center for atmospheric research	Meehl et al. (2003), Fig. 1; numerical file provided by G. Strand, 2011	2.1
GFDL, geophysical fluid dynamics laboratory	Held et al. (2010), Fig. 2; numerical file provided by I. Held, 2011	1.9
GISS, goddard institute for space studies	Hansen et al. (2005), Fig. 1a; http://data.giss.nasa.gov/modelforce/RadF.txt	1.6
RCP—representative concentration pathways	Meinshausen et al. (2011); http://www.pik-potsdam.de/~mmalte/rcps/data/20THCENTURY_MIDYEAR_RADFORCING.xls	1.6
MIROC, model for interdisciplinary research on climate	Takemura et al. (2006), Fig. 2a; numerical file provided by T. Takemura, 2011.	1.1
Myhre	Myhre et al. (2001), Fig. 1; http://folk.uio.no/gunnarmy/data/rf_time/rf_time.dat	1.0

Column 3 gives forcing in 1990 (prior to the 1991 Pinatubo eruption) relative to the average for 1896–1901 (also relatively free of influences of volcanic aerosols) used to relate inferred climate properties to twentieth century forcing

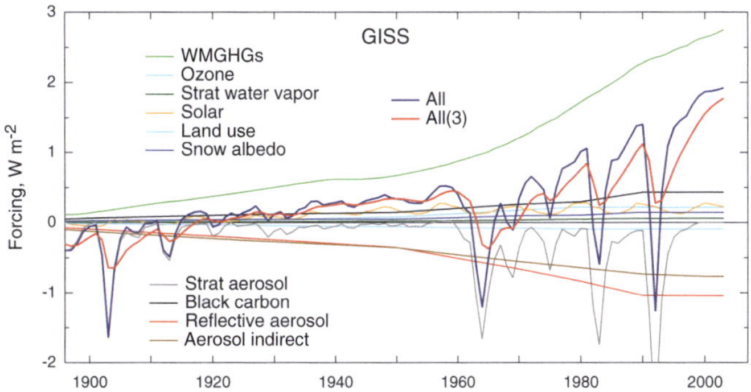

Fig. 7 Global mean forcings (relative to 1880) as employed in the Goddard Institute for Space Studies (GISS) climate model (Hansen et al. 2005; http://data.giss.nasa.gov/modelforce/RadF.txt). WMGHGs, well mixed greenhouse gases. "All" denotes the total forcing, the sum of the individual forcings, and "All(3)" denotes that forcing convolved with the function $\exp(-t/3)\ [1-\exp(-1/3)]$

exhibiting large values immediately following volcanic eruptions, but decaying to near zero over the course of a few years. Other forcings are relatively small, even in the aggregate. Solar forcing exhibits a cyclical behavior with a roughly 11-year period, indicative of the change in the solar constant associated with the sunspot cycle. Forcings by black carbon from incomplete combustion offset part of the negative aerosol cooling forcing. There are minor contributions from the increase in tropospheric and stratospheric ozone, and the increase in stratospheric water vapor resulting from the increase in tropospheric methane, some of which makes its way into the stratosphere and is oxidized there. Changes in surface albedo from black carbon on snow and from land use changes also make minor contributions. The total forcing, evaluated as the sum of the several components, exhibits a general positive trend over the period but is punctuated by the large, short-duration negative forcings by volcanic stratospheric aerosols, such that the total forcing can, even in the latter part of the twentieth century, exhibit brief excursions into negative values (relative to 1880) before returning to the gradual positive forcing.

In order to relate forcing to temperature change, it is necessary to take into account the damping effect of global surface temperature on the rapid impulses in forcing associated with volcanic eruptions. The approach I have taken is to impose a damping on the forcing by convolving the total forcing from the several data sets with a decaying exponential function having a 3-year time constant, $\exp(-t/3)\ [1-\exp(-1/3)]$, where t is time in years and $[1-\exp(-1/3)]$ is a normalization factor. As expected, this damped forcing, also shown in Fig. 7, exhibits reduced magnitude but extended duration of the stratospheric volcanic aerosol forcing. To assess the suitability of this approach, I compared the time series of the damped forcing with that of observed global mean surface temperature anomaly, Fig. 8. It is seen that the two time series exhibit qualitatively similar behavior in the response to the impulses due to volcanic aerosols, suggesting that the 3-year exponential function is doing a reasonable job of accounting for the time lag of global temperature to the volcanic forcing, especially during the latter half of the twentieth century during which the volcanic aerosol forcings are perhaps better characterized. The 3-year time constant is similar to that found in analysis of GCM data by Held et al. (2010), 4 ± 1 years, but is somewhat shorter than that determined from analysis of autocorrelation of global mean temperature data, 6–11 years (Scafetta 2008; Schwartz 2008a). A fairly short damping time constant is desired so as to minimize any artificial lag of the more

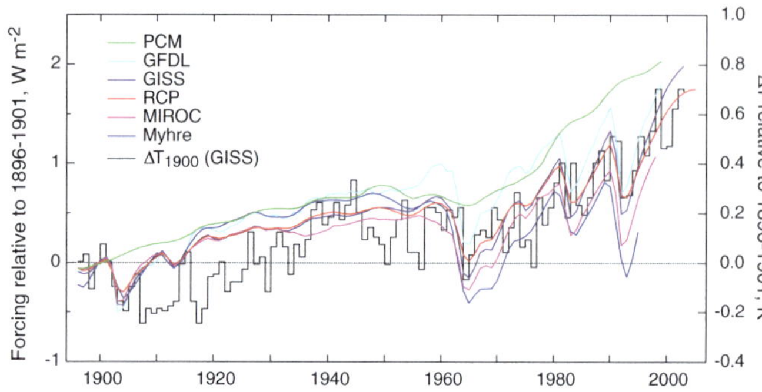

Fig. 8 Total forcings from the several forcing data series examined in this study, convolved with the function $\exp(-t/3)$ $[1-\exp(-1/3)]$. Also shown for comparison is the time series of global mean surface temperature anomaly as tabulated by the Goddard Institute for Space Studies (GISS; Combined Land-Surface Air and Sea-Surface Water Temperature Anomalies, LOTI; http://data.giss.nasa.gov/gistemp/tabledata/GLB.Ts+dSST.txt)

slowly varying forcings arising from the gradual increases in greenhouse gases and aerosols.

Also shown in Fig. 8 are the time series for the several other forcing data sets examined in this study, similarly convolved with the 3-year exponential decay function. With the exception of the PCM data series, which did not include volcanic aerosol forcing, all the data series exhibit qualitatively similar behavior over the time period, and all exhibit an increase in forcing over the twentieth century, reflecting the dominant contribution of forcing by GHGs. However, closer comparison of the several data sets, Fig. 9, shows considerable differences among them. Examination of the scatter plots shows strong correlation of the GISS, RCP, GFDL, and MIROC forcings but with offsets from the diagonal, which would indicate perfect agreement. Not surprisingly, the PCM forcing data set exhibits poorer correlation because of volcanic aerosol forcing not being included in that data set; this is manifested by downward departure of the other data sets from the PCM data set in “streamers” associated with periods of volcanic aerosol forcing. The data set of Myhre et al. (2001) exhibits rather poor correlation with the other forcing data sets.

3.3 Correlation of Temperature Anomaly and Forcing

As the intent of this study is to examine the relation of the change in GMST to radiative forcing over the twentieth century, for which the forcing and temperature data are available, it was considered essential that the forcing and temperature anomaly time series be defined relative to values of these quantities in a common time period at the beginning of the century. As the variation in the several data sets over the base period 1896–1901 is quite small relative to variation over the data set as a whole (Fig. 8), the average over this time period was taken as the reference value for both forcing and temperature change.

Attention is called also to the large range of forcing over the full span of the time series. To some extent, this is a consequence of the differing end dates of the time series; it might be argued that the time series of Myhre et al., which terminates in 1995, is not fully recovered from the Pinatubo event. However, even so, in 1990, prior to the Pinatubo eruption, the forcing in the six data sets, Table 1, exhibits a span of a factor of 2 and a

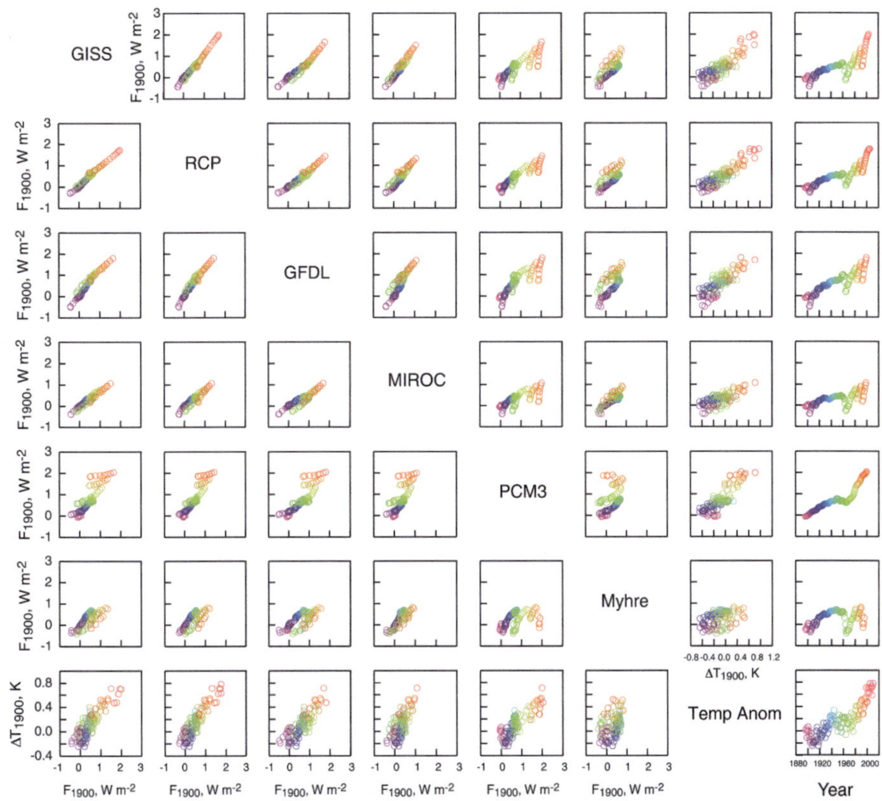

Fig. 9 Scatterplot matrix of the several forcings (relative to 1896–1901) employed in this study and also global mean surface temperature anomaly (also relative to 1896–1901). The column and row headings of the several data sets are given along the trace of the matrix, and each graph represents a scatter plot of the quantity identified by the row heading on the *y* axis versus the quantity identified by the column heading on the *x* axis. All of the forcings are shown on the same scale; perfect agreement between two different forcing data sets would be manifested by all data lying along the diagonal. Similarly a linear dependence of temperature anomaly on forcing would be manifested by a linear array of the data points. The *right* hand column in the matrix of graphs gives the time series of each of the forcings and temperature anomaly. The *color* coding reflects the date (*violet*, 1896 to *red*, 2009)

relative range (range divided by median) of 69%. This range is characteristic of, but well less than, the uncertainty in total forcing over the twentieth century, as given in the IPCC Fourth Assessment Report (2007), best estimate 1.6 W m^{-2}; 90% confidence range (0.6–2.4 W m^{-2}), or relative range 113%. A range in forcing will inevitably lead to a corresponding range in any assessment of climate sensitivity that is obtained using the several values of forcing together with the time series of observed surface temperature. A key objective of this study is to assess the range in transient and equilibrium sensitivities that results from the range of forcing estimates.

The relation between surface temperature anomaly and forcing was examined by means of graphs of surface temperature anomaly relative to the 1896–1901 base period versus forcing relative to the same period, Fig. 10; these graphs are the same as in the bottom row of Fig. 9. The *y*-data of all graphs are the observed temperature change; the *x*-data, the

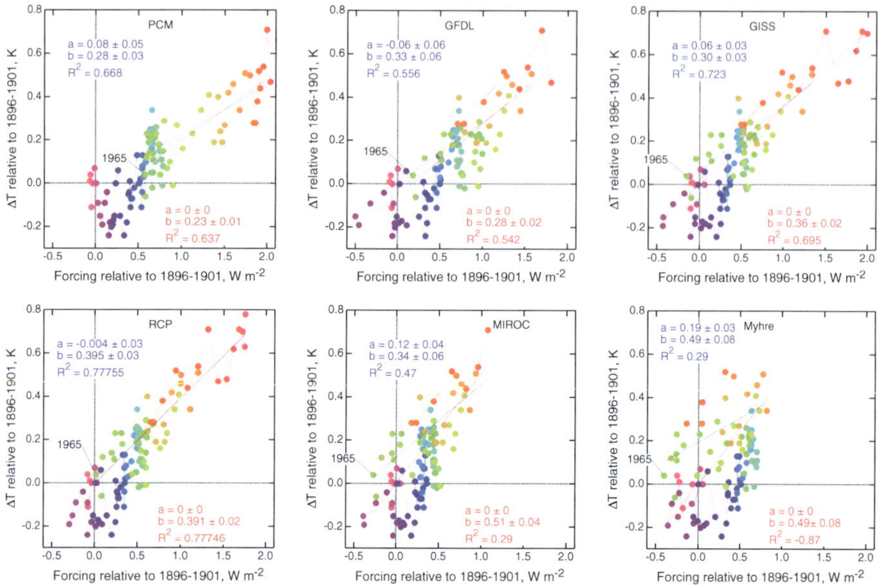

Fig. 10 Examination of proportionality of surface temperature anomaly and forcing (Eq. 12). Surface temperature anomaly (relative to 1896–1901) is plotted against total forcing from the several forcing data series examined in this study, as convolved with the function $\exp(-t/3)\,[1-\exp(-1/3)]$. Fits to data in the form $\Delta T = a + bF$ from 1965 (indicated) to the ends of the several forcing data series are ordinary linear least squares, *blue*, and least squares but with a held equal to 0, *red*. Also indicated as R^2 is fraction of variance in the original data accounted for by the regression; negative value indicates that variance in the residuals about the regression exceeds that in the temperature data themselves. The *color* coding reflects the date (*violet*, 1896 to *red*, 2009)

forcings adduced by the several modeling groups, differ among the figures. As the hypothesis under examination is that surface temperature has increased in response to the applied forcing, the forcing is taken as the independent (known) variable in computing the regressions. The regressions are restricted to the forcing and temperature data for 1965 and beyond, consistent with the start date of the regression for heating rate, but both the forcing and temperature anomaly data are referenced to the common base period, 1896–1901. However, the entire data sets for forcing and temperature from 1896 to the end of the several forcing data sets are shown, and it would seem from inspection of the graphs that, for most of the forcing data sets, the entire time series is well represented by a linearly proportional relation. In the regressions, a rather robust linear proportionality is exhibited for most of the forcing data sets between surface temperature and forcing, but with different slopes. The fraction of the variance in the temperature data accounted for by the regression forced through the origin is over 50% for four of the six forcing data sets. For most of the data sets, the intercept is near zero; constraining the regression line to pass through the origin results in little decrease in the fraction of the variance in the data accounted for by the regression, denoted R^2 in the figures and in Table 2. (For the 2-parameter fits R^2 is equal to the square of the Pearson product-moment regression coefficient, as usual; for the 1-parameter fits the quantity denoted R^2 is explicitly evaluated from the residuals; the value of R^2 for the 2-parameter fit necessarily exceeds that for the 1-parameter fit.) A high correlation with zero intercept, that is, temperature anomaly proportional to forcing (Eq. 12), is consistent with a planetary heating rate N that is

Table 2 Equilibrium climate sensitivity S_{eq} and transient sensitivity S_{eq} and other quantities determined for the several forcing data sets examined in this study

Method	Quantity	Unit	Forcing data set						Relative range
			PCM	GFDL	GISS	RCP	MIROC	Myhre	
	F(1900–1990)	W m^{-2}	2.1	1.9	1.6	1.6	1.1	1.0	0.69
F–N	S_{eq}	K (W m^{-2})$^{-1}$	0.30	0.39	0.59	0.70	–	–	0.82
	$\sigma(S_{eq})$	K (W m^{-2})$^{-1}$	0.02	0.03	0.04	0.02	–	–	
	R^2		0.52	0.25	0.50	0.58	−0.97	−2.9	
	$\Delta T_{2\times}$	K	1.10	1.45	2.19	2.60	–	–	0.82
	$\sigma(\Delta T_{2\times})$	K	0.07	0.12	0.14	0.15	–	–	
κ	S_{tr}	K (W m^{-2})$^{-1}$	0.23	0.28	0.36	0.39	0.51	–	0.79
	$\sigma(S_{tr})$	K (W m^{-2})$^{-1}$	0.01	0.02	0.02	0.02	0.04	–	
	R^2		0.64	0.54	0.69	0.78	0.29	−0.87	
	S_{eq}	K (W m^{-2})$^{-1}$	0.31	0.43	0.63	0.74	1.32	–	1.61
	$\sigma(S_{eq})$	K (W m^{-2})$^{-1}$	0.02	0.04	0.06	0.07	0.31	–	
	$\Delta T_{2\times}$	K	1.16	1.60	2.32	2.75	4.90	–	1.61
	$\sigma(\Delta T_{2\times})$	K	0.09	0.16	0.22	0.25	1.15	–	

S_{tr} was evaluated as coefficient of proportionality between temperature anomaly and forcing. S_{eq} was determined by two methods: directly as coefficient of proportionality between temperature anomaly and forcing minus planetary heating rate (*F–N* method), and via S_{tr}, accounting for the planetary heating rate, as $S_{eq} = (S_{tr}^{-1} - \kappa)^{-1}$ (Eq. 10) where κ is the coefficient of proportionality between heating rate and temperature anomaly (κ method). R^2 denotes fraction of variance in temperature anomaly data set accounted for by the linear regression with zero intercept. Uncertainties denote 1-sigma estimates inferred from least-squares fits (neglecting autocorrelation), suitably propagated for derived quantities. Relative range denotes range divided by median

likewise proportional to the temperature increase (Eq. 7). The sole exception among the forcing data sets examined to the linear proportionality between surface temperature and forcing is for the forcing data set of Myhre et al. (2001), which exhibits poor correlation and for which constraining the regression line to pass through the origin results in a variance about the regression line that actually exceeds the variance in the temperature data themselves. Thus, the Myhre forcing data set would seem entirely inconsistent with a linear proportionality between observed temperature change and forcing.

3.4 Determination of Equilibrium Climate Sensitivity by the κ Method

As described above, if the energy imbalance of the planet (net energy flow into the planet) is linearly proportional to the change in temperature, the transient sensitivity (slope of a graph of temperature change versus forcing, Fig. 10) would be related (Eq. 11) to the geophysical quantities κ and λ as $S_{tr} \equiv (\kappa + \lambda)^{-1}$. Thus, within this model, for κ independently determined from observations of the change in GMST and rate of change of global heat content (Fig. 6), it is possible to determine the equilibrium sensitivity as $S_{eq} = (S_{tr}^{-1} - \kappa)^{-1}$, Eq. 13. These relations are used to evaluate S_{eq} from the transient sensitivities determined for the several forcing data sets; the results are given in Table 2 as are the corresponding values of the CO_2 doubling temperature. As is the case with the values of the transient sensitivity, the differences in the values of the equilibrium sensitivity determined for the several forcing data sets are due entirely to differences in the forcing data sets. The relative range of the values of the equilibrium climate sensitivity,

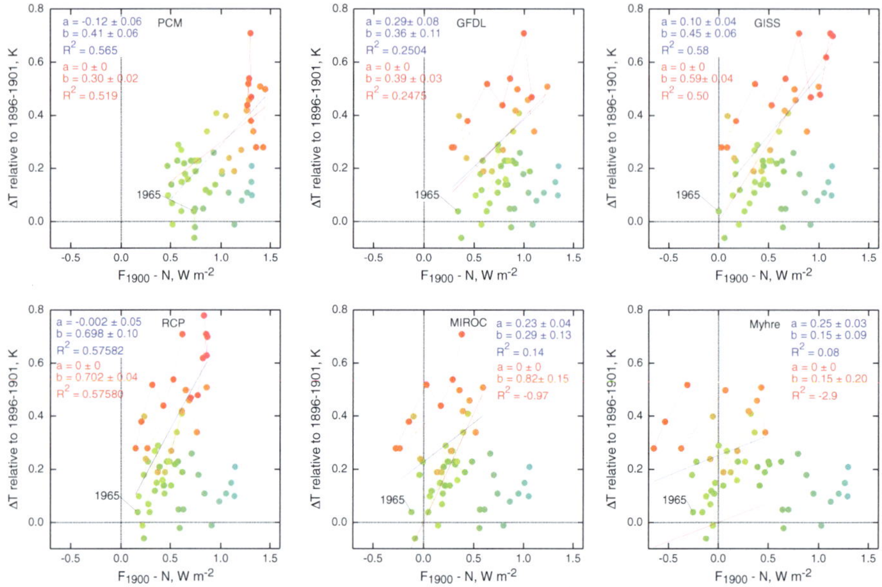

Fig. 11 Examination of proportionality between temperature anomaly and forcing minus heating rate (Eq. 6). Surface temperature anomaly (relative to 1896–1901) is plotted against $F–N$, where F is total forcing from the several forcing data series examined in this study, as convolved with the function $\exp(-t/3)$ $[1-\exp(-1/3)]$, and N is planetary heating rate shown in Fig. 4. Data are shown for the entire time period for which ocean heat content data are available, extending to 1949, but regression fits to data in the form $\Delta T = a + b(F–N)$ are limited to the time period from 1965 (indicated) to the ends of the individual forcing data series. Fits are ordinary linear least squares, *blue*, and least squares but with the intercept a held equal to 0 K, *red*. Also indicated as R^2 is fraction of variance in the original data accounted for by the regression; negative value indicates that variance in the residuals about the regression exceeds the variance in the temperature data themselves. The color coding reflecting the date (*light blue*, 1949, to *red*, 2009), is consistent with the color coding of Fig. 9)

123%, is substantially greater than that in the forcings themselves or in the transient sensitivities because of the subtraction involved in the calculation of S_{eq}.

3.5 Determination of Equilibrium Climate Sensitivity from Correlation of Temperature Change with $F–N$

According to Eq. 6, surface temperature anomaly as a function of time would be expected to exhibit a linear proportionality to forcing F minus planetary heating rate N, also both functions of time, with slope equal to the equilibrium sensitivity S_{eq}. This relation was examined by means of graphs of surface temperature anomaly relative to the 1896–1901 base period versus forcing relative to the same period minus heating rate N for the several forcing data sets, Fig. 11. In these graphs, the time period represented by the data is limited to the period subsequent to 1949, for which heating rate data are available. As the accuracy and representativeness of the ocean heat content data prior to 1965 are questionable, fits to obtain linear regressions were carried out only for the data from 1965 through the ends of the several forcing data sets. Differences in the slopes among the several graphs are entirely a consequence of the differences in the forcing data sets. As the hypothesis is that surface temperature increases in response to the forcing minus the planetary heating rate, the latter is taken as the independent (known) variable in computing the regressions.

Correlations between ΔT and $F–N$ (see also Table 2) are fairly robust, $R^2 > 0.5$, for three (GISS, RCP, and PCM) of the six forcing data sets, with intercepts, corresponding to zero forcing, close to 0 K, and with little decrease in the fraction of the variance accounted for by the regression for the regression line constrained through the origin, indicative of a linearly proportional dependence of ΔT on $F–N$ expected for the energy balance model. In all cases, the R^2 values are less than that exhibited in the regression of heating rate on surface temperature anomaly, 0.68 (Fig. 5), which would seem to indicate that the forcing data introduce additional variance into the relation. For the MIROC and Myhre forcing data sets, only a rather small fraction of the variance is accounted for by the regression, and constraining the regression line to pass through the origin at $F = 0$ results in the variance about the regression line exceeding the variance characterizing the temperature anomaly data about their mean value; this situation is revealed by a negative value of R^2, evaluated as the fraction of the variance in the data accounted for by the regression. The departure from linear proportionality should not necessarily be taken as implying that the forcing data do not provide an accurate picture of actual forcing of Earth's radiation balance over this time period, but it would certainly seem that these forcing data, together with the observations of increase in temperature and planetary heating rate, are inconsistent with an energy balance model for which the change in net emitted irradiance at the top of the atmosphere is proportional to the increase in surface temperature.

The values of equilibrium sensitivity determined by the $F–N$ method would appear to exhibit a much smaller relative range than those determined by the κ method. However, the difference is due entirely to inclusion of the sensitivity determined with the MIROC forcing data in the set of values of S_{eq} obtained by the κ method, whereas this data set is excluded from the values of S_{eq} obtained by the $F–N$ method because of failure to exhibit a linear regression. If that value of S_{eq} is removed from the set of values obtained by the κ method, the relative range is virtually identical and, indeed, the values of S_{eq} by the two methods agree within their one-sigma uncertainties. The identity of the results obtained by the two different methods lends support to the two methods and values obtained.

3.6 Estimates of Upper- and Lower-Compartment Time Constants

As the expressions for the time constants of the upper and lower compartments (Eq. 8) involve quantities that are not determined directly in this analysis, it is not possible to explicitly determine values for these time constants. However, algebraic manipulation of the expression for the time constant of the upper compartment yields an expression for τ_s as the product of scoping value, τ'_s, times two correction factors both of which are near unity, and of opposite sense,

$$\tau_s = \frac{C_U}{\beta + \lambda} = \tau'_s \cdot \frac{C_U}{C_{eff}} \cdot \frac{\lambda + \kappa}{\lambda + \beta}; \ \tau'_s = C_{eff} S_{tr}.$$

Values of τ'_s for the several forcing data sets range from 4 to 9 years, from which it may be concluded that τ_s is comparable. By a similar argument,

$$\tau_1 = C_L\left(\frac{1}{\lambda}+\frac{1}{\beta}\right)_1 = \tau_1'\cdot\left(\frac{\lambda}{\kappa+\lambda}\right)\cdot\left(\frac{\kappa}{\lambda}+\frac{\kappa}{\beta}\right);\ \tau_1' = \frac{C_L}{(\kappa - S_{tr}\kappa^2)}.$$

Values of τ' for the several data sets range from 400 to 580 years, establishing the long time constant of the lower compartment. These estimates are consistent with the assumed large separation of time constants for the two compartments that is a premise of the model.

4 Discussion

4.1 Effective Heat Capacity of the Climate System

The graph of ocean heat content anomaly versus temperature anomaly (Fig. 4) gives rise, as a slope, to the effective heat capacity of the ocean that is coupled to the climate system, from which the total effective heat capacity of the climate system was obtained by taking into account other sinks of heat in the climate system. The resulting effective heat capacity 21.8 ± 2.1 W year m^{-2} K^{-1} is somewhat greater, and is determined with considerably less uncertainty, than the value given by a similar approach by Schwartz (2007), 14 ± 6 W year m^{-2} K^{-1}, a consequence mainly of improved estimates of ocean heat content anomaly subsequent to the study of Levitus et al. (2005) on which the earlier determination was based. However, this uncertainty reflects only scatter about the regression line and does not encompass systematic uncertainty that would result from inadequate sampling and methodological artifacts in the heat content measurements. As determination of this effective heat capacity does not rely on modeled forcings, it is not subject to uncertainties in the forcings and therefore is entirely observationally determined. In an analysis of ocean uptake in climate model runs with the Hadley Centre coupled atmosphere–ocean general circulation model (AOGCM) in which the CO_2 mixing ratio was increased by 1% year^{-1}, compounded, Gregory (2000) found an increase in ocean heat content of 7.5×10^{23} J for a temperature increase of 1.8 K, from which a heat capacity of 25.9 W year m^{-2} K^{-1} may be inferred. Based on climate model studies in which model parameters were allowed to vary, Frame et al. (2005) presented values (0.1–2.05) GJ m^{-2} K^{-1} (5–95% confidence), equivalent to (3.2–65) W year m^{-2} K^{-1}, a range that encompasses the value obtained here but provides little constraint or insight. Andrews and Allen (2008) present an analysis of this quantity as inferred from studies with coupled AOGCMs yielding a probability distribution function (PDF) for this quantity that peaks at about 0.7 GJ m^{-2} K^{-1} (principal contribution to the PDF between 0.4 and 1.4 GJ m^{-2} K^{-1}; equivalently, 22 W year m^{-2} K^{-1}; range 13–44 W year m^{-2} K^{-1}), broadly consistent with the observationally determined heat capacity found here.

Although the analysis of the ocean heat content observations presented here provides a fairly tight constraint on the effective heat capacity, interpretation of the physical meaning of this quantity remains problematic. As the determination is based on the increase in heat content of the entire ocean (and other tightly coupled components of the climate system), the resulting heat capacity is not the heat capacity just of the upper compartment of the climate system but accounts for the increase in heat content in the deeper compartment as well; the penetration of heat into the deep ocean occurs in conjunction with downwelling ocean circulations (Levitus et al. 2005). Still, this effective heat capacity is certainly well less than the heat capacity of the deep ocean, which is an order of magnitude greater. Insight into the multiple response times of the climate system from a two-compartment model is provided by Held et al. (2010), and a more thorough interpretation through an upwelling diffusion model is provided by Hoffert et al. (1980). An approach to separating the shallow ocean compartment from the deep ocean compartment was advanced by Gregory (2000), who examined temporal autocorrelation of temperature as a function of depth with that in the top ocean layer in output of the Hadley Centre AOGCM, finding a sharp break at about 100 m, for which the heat capacity, expressed per area of planet would be 0.28 GJ m^{-2} K^{-1} (9.0 W year m^{-2} K^{-1}). It would seem that a similar, observationally based approach might usefully distinguish the upper and lower oceanic compartments of the actual climate system. In this respect, attention is called to a time series of measurements of ocean temperature as a function of depth (surface to 800 m) over a 15-year period (Sutton and Roemmich 2001) that shows the damping of the amplitude of the seasonal cycle of temperature with depth from about 6 K peak-to-peak at the surface to about 1 K at 150 m, suggesting an observational approach to determining the time-dependent penetration of heat due to a secular change in surface temperature induced by changing atmospheric composition.

4.2 Heat Uptake Coefficient

The heat uptake coefficient κ, the rate of heat uptake by the planet normalized to the global temperature anomaly was determined as the slope of the linear proportionality between the rate of increase of ocean heat content, evaluated as the time derivative of the ocean heat content, augmented to include estimates of additional heat sinks, and the surface temperature anomaly. This heat uptake coefficient is 1.05 ± 0.06 W m^{-2} K^{-1}, where again the uncertainty characterizes the regression but not systematic errors in the heat content data. For the observed increase in global temperature relative to 1900 $\Delta T_{obs} = 0.78$ K in 2010 (Hansen et al. 2010, as extended at http://data.giss.nasa.gov/gistemp/), this value of κ would indicate a heat flux into the climate system, which is equal to the energy imbalance of the climate system, of 0.82 W m^{-2}. Such an energy imbalance is consistent with the present heating rate shown in Fig. 5 (from which it is derived) and with other current observationally based estimates of this quantity (Palmer et al. 2010; Lyman 2011).

As with the effective heat capacity, the heat uptake coefficient is wholly observationally determined. Further, it would seem that this heat uptake coefficient is an intrinsic property of Earth's climate system, rather than a property that is dependent on the nature and/or magnitude of recent forcings. Certainly, a proportionality between the rate of heat input into the climate system and the increase in global temperature following imposition of a radiative forcing would not be maintained indefinitely, as the system approaches its ultimate new steady state. Nonetheless at least in the early years following the onset of forcing, and even more, in situations of continuously increasing forcing, the linear proportionality between the rate of increase of heat content and the temperature anomaly would seem to be a useful means of quantifying the heat input into the climate system and the consequences of the departure from steady state following imposition of the forcing.

Despite considerable interest in the heat uptake coefficient in interpreting climate model calculations (Gregory and Forster 2008; Dufresne and Bony 2008), there does not seem to be prior observationally based determination of this quantity. Gregory and Forster determined κ as the slope of a regression of net heat flux into the planet, relative to control runs, against GMST anomaly in the output of 16 AOGCMs that participated in the intercomparison of models over the twentieth century carried out by the IPCC (2007) Fourth Assessment, the same approach as employed here with observation-derived data. For the 16 models examined, the mean value of κ was 0.62 ± 0.13 (1 sigma); maximum 0.83; minimum 0.41. It would thus seem that the heat uptake coefficient may be somewhat underestimated in current climate models. For a given equilibrium sensitivity, an underestimate would result in the rate of increase in GMST being overestimated in climate model calculations. As the net heat flux is subtractive from the forcing in determining the rate of temperature increase, the magnitude of the overestimate would depend on the forcing. Alternatively, knowledge of the heat transfer coefficient might be used to infer the equilibrium sensitivity of a climate model from S_{tr} obtained from the dependence of modeled temperature anomaly on forcing (κ method, Eq. 11); as κ is subtractive from S_{tr}^{-1}, an erroneously low value of κ would result in an overestimate of S_{eq}, the magnitude of which would depend on the value ofS_{tr}^{-1}.

As with the effective heat capacity, it would seem that questions remain regarding the interpretation of this heat flux. Both observationally and from the climate model output, the quantity that is determined is the total net heat flux into the planet, normalized to the global temperature anomaly, whereas the heat flux that is calculated in two-compartment models is the heat flux from the upper, short–time-constant compartment to the lower, long–time-constant compartment. As the upper compartment undoubtedly comprises a substantial fraction of the effective heat capacity of the system, it would seem fruitful to more explicitly distinguish between the two compartments in refining these concepts in future work. Nonetheless, it is clear that the net heat flux into the climate system that is subtractive from the applied forcing to yield the equilibrium sensitivity of the climate system, Eq. 6, is the heat flux into the entire climate system. The observationally based finding of a linear proportionality between heat flux and temperature anomaly supports the relation between transient and equilibrium sensitivities, Eq. 13 that can be used to infer the equilibrium sensitivity from the transient sensitivity based on observations (and assumed forcings) or to infer the transient sensitivity from climate model runs that yield the equilibrium sensitivity with slab-ocean models.

4.3 Transient and Equilibrium Climate Sensitivities

This study has examined the relation between observed increase in GMST and forcing, as calculated by several groups, in terms of a two-compartment energy balance model. According to this model, a linear proportionality would be expected between the two quantities, the slope of which would be interpreted as a transient climate sensitivity S_{tr}. This expectation is borne out for five of the six forcing data sets examined, for forcing and temperature anomaly over the twentieth century. However, the regression slopes (limited to measurements subsequent to 1965) differ for the several forcing data sets by amounts that substantially exceed the uncertainties in the regression slopes (Table 2) and by amounts that are significant in the context of interpretation of climate change over the twentieth century. Recognition that the planetary heating rate N is subtractive from the forcing F to yield the equilibrium sensitivity S_{eq} suggests a further linear proportionality between GMST anomaly and F–N, the proportionality constant being S_{eq}, leading to determination of S_{eq} as this slope, the F–N method. A linear proportionality was found for four of the six data sets examined; again the sensitivities so determined differ by amounts that substantially exceed the uncertainties in the regressions and that are important in the context of understanding climate change. The finding of a linear proportionality between the heating rate and GMST anomaly yielding the heat uptake coefficient κ permits determination of S_{eq} from S_{tr}, the κ method. This method yielded values of S_{eq} for the five forcing data sets for which it was possible to determine S_{tr}. The values of S_{eq} so determined agreed closely with the values of S_{eq} determined by the F–N method. Here, it should be

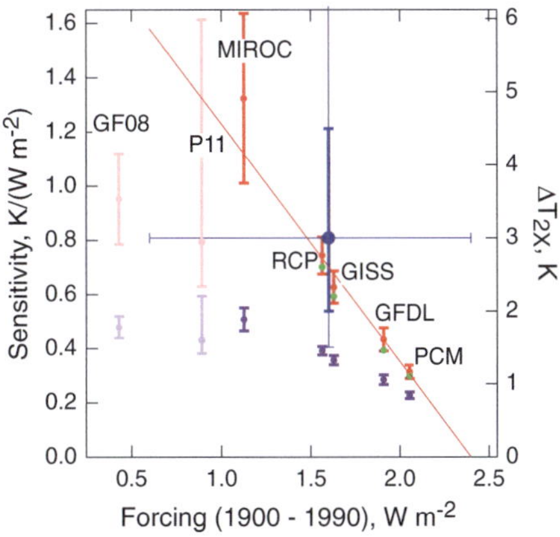

Fig. 12 Dependence of transient climate sensitivity S_{tr}, Eq. 10, (*purple*) and equilibrium climate sensitivity S_{eq} inferred by the *F–N* method, Eq. 6, (*green*) and by the κ method, Eq. 11, (*red*) on forcing between 1900 and 1990. Uncertainties in equilibrium sensitivity (shown for the κ method) represent one sigma, estimated by error propagation from the uncertainties in S_{tr} (also shown, one sigma) and κ. Right axis gives equivalent CO_2 doubling temperature evaluated from S_{tr}, or S_{eq}. Also shown are S_{tr}, determined by Gregory and Forster (2008) and Padilla et al. (2011) and S_{eq} evaluated from those values of S_{tr} using the value of κ determined here, with associated one-sigma uncertainties. Shown in *blue* are best estimates of equilibrium climate sensitivity and anthropogenic forcing (relative to preindustrial) and associated uncertainties as given by the IPCC Fourth Assessment Report (2007); *thick* uncertainty *lines* correspond to central 66% of the likelihood function (roughly equivalent to one sigma); *thin* uncertainty *lines* denote "very likely" range corresponding to the central 90% of the likelihood function. *Red line* denotes extrapolation of equilibrium sensitivities to the full "very likely" range of forcing given by the 2007 IPCC assessment

emphasized that the absence of linear proportionality between GMST anomaly and forcing for a single forcing data set of the six examined (that of Myhre et al. 2001) does not demonstrate those forcing data are an inaccurate representation of forcing over the latter part of the twentieth century but only that these forcing data are inconsistent with the energy balance model considered. This inconsistency in a single forcing data set, and more broadly the differences in the sensitivities determined from the several forcing data sets examined, underscore the importance of accurate determination of climate forcing over the twentieth century, especially the latter part of the twentieth century for which ocean heat content data are available, to observational determination of Earth's climate sensitivities.

Values of the equilibrium sensitivities were determined for five of the six forcing data sets examined, all of which are within the "very likely" range[2] given for this forcing by the IPCC (2007) Assessment. These sensitivities range from 0.31–1.32 K $(\mathrm{W\ m^{-2}})^{-1}$, corresponding to $\Delta T_{2\times}$ 1.2–4.9 K, and, with the exception of the MIROC forcing data set, are less than the best estimate, 3 K given for this quantity by the 2007 IPCC Assessment, Fig. 12. Two of the five forcing data sets yield sensitivities within the IPCC "likely" range (see Footnote 2), 2–4.5 K, with two below and one (MIROC) above this range, although the large uncertainty attached to the latter extends well into this range. For the PCM and GFDL forcing data sets, the equilibrium sensitivities are below the lower bound of the "likely" range for this quantity, and indeed are nearly at, or below, the limit of the "very likely" range for this quantity, $\Delta T_{2\times} = 1.5$ K.

[2] Here the term "very likely" is used in the sense of the 2007 IPCC Assessment Report; that is, corresponding to the estimate of the central 90% of the PDF for the quantity. Likewise the term "likely" is used to denote the estimate of the central 66% of the PDF.

Examination of the relation between the values of S_{tr} and S_{eq} determined by this analysis and the twentieth century climate forcing used to infer the sensitivity from the observed increase in GMST (Fig. 12) shows distinct anticorrelation, that is, a low forcing yields a high sensitivity, and vice versa. An anticorrelation between forcing and sensitivity, which would be expected for a given increase in GMST has been noted previously in both empirical inference (Gregory et al. 2002; Schwartz 2004) and in analysis of the equilibrium sensitivity of climate models (Kiehl 2007; Knutti 2008). Of course the equilibrium climate sensitivity, which is a property of Earth's climate system, cannot depend on the forcing. Rather it is the equilibrium climate sensitivities that are inferred from estimates of the forcing that exhibit such dependence. The anticorrelation between inferred equilibrium sensitivity and forcing found here indicates that the only way that Earth's equilibrium climate sensitivity could be as great as the central value of the IPCC estimate, $\Delta T_{2\times} = 3$ K, would be for the total forcing (recall that the forcing corresponds to the period 1900–1990) to be about 0.8 W m^{-2}. Such a low forcing, which is at the low end of the IPCC "very likely" range, would require a rather large negative aerosol forcing to offset the forcing, by the well mixed greenhouse gases, about 2.3 W m^{-2} in 1990; here, it must be emphasized that this is not an "inverse" calculation of aerosol forcing, as would be obtained by using a modeled sensitivity, but rather an observational constraint on this forcing together with the best estimate of the equilibrium climate sensitivity given by the IPCC assessment. Extrapolation of the anticorrelation between equilibrium sensitivity and forcing to the entire span of the "very likely" range for the total forcing given by the IPCC 2007 assessment (0.6–2.4 W m^{-2}) yields a range in equilibrium sensitivity from near zero to 1.6 K (W $m^{-2})^{-1}$ ($\Delta T_{2\times}$ 6 K). This wide range of equilibrium climate sensitivity underscores the importance of constraining the forcing if the climate sensitivity is to be determined with accuracy, either observationally or in climate model studies.

Relevant prior studies examining the relation of observed temperature change to forcing are of Gregory and Forster (2008) and Padilla et al. (2011). Gregory and Forster presented, for forcing determined by the Hadley Centre climate model, graphs of observed temperature change versus anthropogenic forcing or versus total forcing; years strongly influenced by volcanic emissions were excluded from determination of the regression slope. The transient sensitivity found in that study as the slope of a regression of ΔT observed over the years 1970–2006 against "median" estimates of total forcing by anthropogenic greenhouse gases and aerosols (years strongly affected by volcanic aerosols excluded) was 0.48 ± 0.04 K (W $m^{-2})^{-1}$; the corresponding value of equilibrium sensitivity, evaluated with the value of κ determined here is 0.95 ± 0.18 K (W $m^{-2})^{-1}$ ($\Delta T_{2\times} = 3.5$ K). Padilla et al. used a statistical approach to infer the quantity denoted here as $\kappa + \lambda$ from the observed temperature record together with an composite forcing based on the forcings from the GISS, GFDL, and Gregory-Forster forcing data sets. The resulting transient sensitivity was 0.43 (+0.16; −0.05) K (W $m^{-2})^{-1}$. For the value of κ determined in this study, the corresponding equilibrium sensitivity is 0.79 (+0.82; −0.16) K (W $m^{-2})^{-1}$. The sensitivities determined in those studies are somewhat to substantially greater than the values determined for the forcing data sets examined here, Fig. 12. Correspondingly, the total forcings over the twentieth century employed in these analyses were lower to considerably lower, 0.89 and 0.43 W m^{-2}, than those obtained with the forcings from the studies examined here; the forcing data set employed by Gregory and Forster is less even than the lower bound of the "very likely" range for forcing up to 2005 as given by the IPCC, although that gap is closed by the incremental forcing between 1990 and 2005. As also found here, the Padilla et al. analysis indicates that the transient sensitivity differed for the several forcings; however, for the three forcings examined by Padilla et al., the range of the central values of the transient sensitivities, 0.43–0.53 K (W m^{-2})$^{-1}$, was considerably narrower than found in the present study and not monotonic with the forcings.

4.4 Climate System Time Constants

Although the estimates of time constants for relaxation of perturbations in heat content of the two compartments of the climate system determined by this analysis depend on the magnitudes of the forcings of the several data sets examined, for each of the forcing data sets, the values associated with the two compartments are clearly separated into a short time constant (median 6.3 year, range 4.1–9.2 years) and a long time constant (median 470 year, range 400–580 year). A rapid initial response to a step-function forcing that exhibits most of the climate system response followed by a further response of lower magnitude and long duration is widely exhibited in coupled climate model runs (Brasseur and Roeckner 2005; Matthews and Caldeira 2007; Knutti et al. 2008; Held et al. 2010; Hansen et al. 2011; Knutti and Plattner 2012). Examining a two-compartment model having parameters chosen to match experiments with the Hadley Centre climate model, Gregory (2000) found that the response could be characterized by a short time constant of about 12 year and as second time constant at least an order of magnitude greater. Boucher and Reddy (2008) found that the response of GMST in the HadCM3 climate model to a forcing that was rapidly ramped up and then held constant was accurately approximated by a two-exponential function with time constants 8.4 and 410 years. A large gap in time constants was noted also by Held et al. (2010), in an analysis of the response of the GFDL climate model to step-function changes in forcings, with a short time constant characterizing the transient response of the climate system to an applied forcing and a much longer response time, which they characterize as "centuries or longer" associated with the second, large-heat-capacity compartment of the climate system. Based on an analysis of model runs of two AOGCMs and a model of intermediate complexity run out to equilibrium Jarvis and Li (2011) likewise find two discrete time scales, one of about 20 years and a second of about 700 years. Based on an examination of transient runs across the climate models that participated in the intercomparison reported in the 2007 IPCC Assessment, Lucarini and Ragone (2011) characterized the longer relaxation time as of order 500–700 years.

A response of Earth's climate system to perturbations that is characterized by two such widely separated time constants would have important implications for the interpretation of climate change over the past half century during which forcing has likely been systematically increasing. Specifically, for such a situation, the response of GMST to the forcing would be in rather close steady state to the forcing, lagging the response given by the transient sensitivity by about one time constant, that is, just a few years. Thus, Held et al. (2010) found that the response of the GFDL model to twentieth century forcing is accurately matched by a single-box model with a time constant of 4 year and a transient sensitivity of 0.43 K (W m^{-2})$^{-1}$. Ultimately, Held et al. found that the heat input into what they term the "recalcitrant" compartment of the climate system becomes appreciable, but under a situation of continuously increasing forcing, the increase in temperature of this compartment remains a small fraction of that of the short–time-constant compartment. On the time scale certainly of several decades, it would seem that the increase in temperature of the deep ocean is insufficient to appreciably diminish the proportionality between temperature anomaly and heat flux. In other words, the response of the climate system to forcing would continue to be given by the transient sensitivity, not the equilibrium sensitivity. This situation would likely continue to hold as temperature anomaly continues to increase. A departure from proportionality between temperature anomaly and heat flux to the long–time-constant compartment might be expected only after some decades of constant forcing, at which point the back flux of heat from the deep ocean compartment to the shallow ocean compartment would begin to become appreciable. Based on these considerations it would seem that the most relevant measure of climate sensitivity for policy purposes is the transient sensitivity, as it is this sensitivity that will govern climate system response to changes in forcing, given that the time constant for the deep ocean compartment to respond to forcings is several centuries.

4.5 Concerns with this Analysis

This analysis consists of three distinguishable components, listed here in order of increasing dependence on assumptions:

1. Determination of C_{eff} and κ. Determination of these quantities rests entirely on measurements (temperature, ocean heat content) and is independent of details of the two-compartment model or of assumptions about the forcings over the twentieth century.
2. Determination of S_{tr} and S_{eq}. Determination of these quantities rests on the assumed first-order response of GMST to forcing, or equivalently, on a linear response of net irradiance at the TOA to change in GMST. The existence of two distinguishable sensitivities rests on the climate system being characterized by two (or more) distinguishable compartments with widely separated response times, but is independent of the details of the two-compartment model. Determination of both sensitivities depends also on choice of the forcing data set.
3. Estimation of τ_{s}, τ_{l}. In addition to dependence on the above assumptions and input data, determination of these quantities depends explicitly also on the climate system being represented by a two-compartment model.

A source of possible error, affecting determination of C_{eff} and κ, and hence of all the quantities determined here, is error in the increase in planetary heat content inferred here from the increase of heat content of the world ocean, and in turn in the planetary energy imbalance, the time derivative of the heat content. Ocean heat content is subject to uncertainty in the measurements, arising from changes in methods and limited and nonuniform sampling (geographically and as a function of depth) over the period of record, as discussed by Lyman (2011). A greater rate of heat uptake would result in a greater inferred equilibrium climate sensitivity than that determined here, and conversely a smaller rate of heat uptake would result in an even lower equilibrium sensitivity, demonstrating the importance of this quantity also in determining equilibrium climate sensitivity. An inherent limitation to this approach is that it requires determining the rate of heat uptake as a time derivative of the ocean heat content with the attendant requirements on precision and the associated requirement of measurements over an extended time period before the derivative can be inferred with confidence. An alternative approach to determining heat imbalance is through satellite measurements of the net flux at the top of the atmosphere. As discussed by Stevens and Schwartz (2012) an improvement in measurement accuracy of more than an order of magnitude is required to yield an accuracy that is comparable to that of the heat content measurements.

Determination of Earth's climate sensitivities and compartment time constants is a hybrid based on largely modeled forcings in addition to observations. Determination of the transient sensitivity depends on GMST anomaly, which is known with considerable confidence from multiple independent determinations (NASA GISS, Hansen et al. 2010; NOAA, Smith et al. 2008; Climatic Research Unit, UK, Brohan et al. 2006). The equilibrium sensitivity depends also on κ, as discussed above. Forcings are obtained by model-based calculations of the radiative influence of measured or modeled perturbations of atmospheric composition (gases and aerosols). The dependence on choice of forcing data set is explicitly examined. Clearly, the several forcings employed in this analysis, which are characteristic of present understanding and uncertainty, cannot all be correct, and the strong dependence of the inferred climate sensitivity to the forcing employed in the analysis underscores the importance of determining forcing with greater accuracy if climate sensitivity is to be determined with confidence (Schwartz 2004; Schwartz et al. 2010).

As the analysis presented here rests entirely on the assumed independence of normalized climate system response, as manifested in change in net TOA irradiance per forcing, to the nature and geographical distribution of the forcing and to the constancy of this response at least over the time period for which the data are available. Clearly, such constancy of climate system response to forcing must at best be an approximation to the actual climate system response; any departure from this assumption would weaken the conclusions drawn here. Support for this approach comes from linear dependence of change in GMST on forcing, with sensitivity independent of the nature of the forcing, which has been found in numerous model studies, with efficiencies (response per forcing, relative, for example to response to CO_2 forcing) typically within 20% (e.g. Hansen et al. 1997; Joshi et al. 2003; Kloster et al. 2010).

A potential basis for non-constancy of climate sensitivity in a given model (or in the real world) would be a dependence of feedback processes on locus of temperature change, under the expectation that certain geographical regions would exhibit greater or lesser feedbacks than others. Ultimately, such a model fails if, for example, there are local feedbacks such as melting of surface ice that amplify the effects of forcings in one location relative to another. This hypothesis was examined in a coupled climate model study by Boer and Yu (2002), who found that the spatial distribution of temperature response generally exhibited what the investigators denoted a "generic" pattern independent of spatial distribution of the forcing rather than being characteristic of the forcing patterns themselves and leading therefore to only weak dependence of the forcing on the nature or spatial distribution of the forcing. That study noted as well that the patterns of temperature change exhibited "a remarkable linearity or additivity" whereby the sum of response patterns for different forcings was very similar to that for the sum of the forcings. In contrast, also in calculations with coupled models, Watterson and Dix (2005) found that the equilibrium sensitivity, evaluated by the equivalent of the present Eq. 6, varied by as much as -20 to $+70\%$ compared with a standard doubled CO_2 case, depending on whether forcing was confined to either low- or high-latitude regions, raising caution over the assumption inherent in the sensitivity concept that an increase in GMST depends only on global mean forcing independent of geographical distribution.

Finally, although it might be granted that the analysis such as the present one that is based on measurements of the transient response of the climate system to imposed forcing is capable of yielding information about the transient sensitivity of Earth's climate system, the question might nonetheless be raised whether such an analysis is capable of yielding information pertinent to the equilibrium sensitivity, which would be approached only after 1,000 years or more. Here, it should be recalled that the equilibrium sensitivity is as defined here, and generally, is a property of Earth's present climate system in the sense that a derivative of a mathematical function is a temporally local property of that function. This temporally local approach to the determination of the equilibrium sensitivity is rooted in Eq. 6 which displays the relation of equilibrium sensitivity to values of forcing, net energy imbalance, and temperature change at any given time.

5 Summary and Conclusions

The present analysis has examined several relations among global mean quantities pertinent to climate change: ocean heat content with temperature anomaly (slope, effective heat capacity of the climate system); global heating rate with surface temperature anomaly (slope, heat uptake coefficient); surface temperature anomaly with forcing over the twentieth century (slope, transient climate sensitivity); and surface temperature anomaly with forcing minus heating rate (slope, equilibrium climate sensitivity). Six published estimates of forcing over the twentieth century were examined. The first two relations, yielding effective heat capacity and heat uptake coefficient, are independent of the forcings and thus do not encompass any model-based uncertainty associated with the forcings, allowing determination of the effective planetary heat capacity and the heat uptake coefficient from measurements of global mean surface temperature anomaly and ocean heat content anomaly over the second half of the twentieth century. All of these relations, and for the relations involving forcings, for most of the forcings examined, exhibited robust linear proportionality (i.e., zero intercept in linear regression; in the case of heat capacity, linear relation), consistent with expectation based on an energy balance model of the climate system Additionally, a second method was applied for evaluating equilibrium climate sensitivity using the heat uptake coefficient that yielded values essentially identical to those of the direct regression.

The results of these correlations are interpreted in terms of a two-compartment energy balance model of Earth's climate system that is characterized by an upper, small–heat-capacity, short–time-constant compartment that corresponds to the atmosphere and upper ocean and a lower, large–heat-capacity, long–time-constant compartment that corresponds to the deep ocean, which is only weakly coupled to the upper compartment. This model leads to a distinction between a transient climate sensitivity, which pertains to the upper compartment, and an equilibrium climate sensitivity, which pertains to the entire climate system including the deep ocean. The analysis leads to estimates of the time constants of the two compartments, which are dependent on the forcing data sets employed, but which are quite separated from one another, about 5 years for the upper compartment, and about 500 years for the lower compartment. In this situation, the response of the climate system to forcings over the twentieth century is governed by the short time constant of the upper compartment, which is the time constant that governs the relation between forcing and surface temperature anomaly. The proportionality between forcing and temperature anomaly exhibited for all but one of the forcing data sets examined, together with the large separation in time constants, supports the transient climate sensitivity, evaluated as the proportionality coefficient of these two quantities, as the measure of Earth's climate sensitivity pertinent to interpreting climate change over the twentieth century and to informing policy decisions about future emissions.

The analysis presented here, although focusing on observational data, nonetheless rests heavily on the forcings over the twentieth century as calculated by several modeling groups based, ultimately, on measured or modeled changes in atmospheric composition. Of these, the forcing due to anthropogenic aerosols is the source of the greatest uncertainty, and it this uncertainty that is mainly responsible for the differences in forcings over the twentieth century. Confident determination of Earth's climate sensitivities thus remains hostage to accurate determination of these forcings.

Acknowledgments I thank the several modeling groups for providing the forcing data sets employed in this analysis. An earlier version of this paper was presented at the Workshop on Observing and Modelling

Earth's Energy Flows organized and sponsored by the International Space Science Institute in Bern Switzerland, January, 10–14, 2011, and I thank Lennart Bengtsson for his encouragement of this study. This study benefited from comments by Bjorn Stevens and a second, anonymous reviewer. This work was supported by the U.S. Department of Energy's Atmospheric System Research Program (Office of Science, OBER) under Contract No. DE-AC02-98CH10886.

References

Allen MR, Frame DJ (2007) Call off the quest. Science 318:582–583. doi:10.1126/science.1149988

Andrews DG, Allen MR (2008) Diagnosis of climate models in terms of transient climate response and feedback response time. Atmos Sci Lett 9:7–12. doi:10.1002/asl.163

Andrews T, Forster PM, Gregory JM (2009) A surface energy perspective on climate change. J Clim 22(10):2557–2570. doi:10.1175/2008JCLI2759.1

Baker MB, Roe GH (2009) The shape of things to come: why is climate change so predictable? J Clim 22:4574–4589. doi:10.1175/2009JCLI2647.1

Boer GJ, Yu B (2002) Climate sensitivity and climate state. Clim Dyn 21:167–176. doi:10.1007/s00382-003-0323-7

Boer GJ, Stowasser M, Hamilton K (2007) Inferring climate sensitivity from volcanic events. Clim Dyn 28:481–502. doi:10.1007/s00382-006-0193-x

Boucher O, Reddy MS (2008) Climate trade-off between black carbon and carbon dioxide emissions. Energy Policy 36:193–200. doi:10.1016/j.enpol.2007.08.039

Brasseur GP, Roeckner E (2005) Impact of improved air quality on the future evolution of climate. Geophys Res Lett 32:L23704. doi:10.1029/2005GL023902

Brohan P, Kennedy JJ, Harris I, Tett SFB, Jones PD (2006) Uncertainty estimates in regional and global observed temperature changes: a new dataset from 1850. J Geophys Res 111:D12106. doi:10.1029/2005JD006548

Cleveland WS, Devlin SJ (1988) Locally-weighted regression: an approach to regression analysis by local fitting. J Am Stat Assoc 83(403):596–610. doi:10.2307/2289282

Collins WD, Ramaswamy V, Schwarzkopf MD, Sun Y, Portmann RW, Fu Q, Casanova SEB, Defresne J-L, Fillmore DW, Forster PMD, Galin VY, Gohar LK, Ingram WJ, Kratz DP, Lefebvre M-P, Li J, Marquet P, Oinas V, Tsushima T, Uchiyama T, Zhong WY (2006) Radiative forcing by well-mixed greenhouse gases: estimates from climate models in the IPCC AR4. J Geophys Res 111:D14317. doi:10.1029/2005JD006713

Domingues CM, Church JA, White NJ, Gleckler PJ, Wijffels SE, Barker PM, Dunn JR (2008) Improved estimates of upper-ocean warming and multi-decadal sea-level rise. Nature 453:1090–1093. doi:10.1038/nature07080

Dufresne J-L, Bony S (2008) An assessment of the primary sources of spread of global warming estimates from coupled atmosphere-ocean models. J Clim 21:5135–5144. doi:10.1175/2008JCLI2239.1

Forster PMF, Gregory JM (2006) The climate sensitivity and its components diagnosed from earth radiation budget data. J Clim 19:39–52. doi:10.1175/JCLI3611.1

Forster PMD, Taylor KE (2006) Climate forcings and climate sensitivities diagnosed from coupled climate model integrations. J Clim 19:6181–6194. doi:10.1175/JCLI3974.1

Frame DJ, Booth BBB, Kettleborough JA, Stainforth DA, Gregory JM, Collins M, Allen MR (2005) Constraining climate forecasts: the role of prior assumptions. Geophys Res Lett 32:L09702. doi:09710.01029/02004GL022241

Gouretski V, Reseghetti F (2010) On depth and temperature biases in bathythermograph data: development of a new correction scheme based on the analysis of global ocean data. Deep-Sea Res I 57:812–833. doi:10.1016/j.dsr.2010.03.011

Gregory JM (2000) Vertical heat transports in the ocean and their effect on time-dependent climate change. Clim Dyn 16:501–515. doi:10.1007/s003820000059

Gregory JM, Forster PM (2008) Transient climate response estimated from radiative forcing and observed temperature change. J Geophys Res 113:D23105. doi:10.1029/2008JD010405

Gregory JM, Stouffer RJ, Raper SCB, Stott PA, Rayner NA (2002) An observationally based estimate of the climate sensitivity. J Clim 15(22):3117–3121. doi:10.1175/1520-0442(2002)015<3117:AOBEOT>2.0.CO;2

Hansen J, Sato M, Ruedy R (1997) Radiative forcing and climate response. J Geophys Res 102:6831–6864. doi:10.1029/96JD03436

Hansen J, Nazarenko L, Ruedy R, Sato M, Willis J, DelGenio A, Koch D, Lacis A, Lo K, Menon S, Tovakov T, Perlwitz J, Russell G, Schmidt GA, Tausnev N (2005) Earth's energy imbalance: confirmation and implications. Science 308:1431–1435. doi:1410.1126/science.1110252

Hansen J, Ruedy R, Sato M, Lo K (2010) Global surface temperature change. Rev Geophys 48:RG4004. doi: 10.1029/2010RG000345

Hansen J, Sato M, Kharecha P, von Schuckmann K (2011) Earth's energy imbalance and implications. Atmos Chem Phys 11:13421–13449. doi:10.5194/acp-11-13421-2011

Held IM, Winton M, Takahashi K, Delworth T, Zeng F, Vallis GK (2010) Probing the fast and slow components of global warming by returning abruptly to preindustrial forcing. J Clim 23:2418–2427. doi:10.1175/2009JCLI3466.1

Hoffert MI, Callegari AJ, Hsieh CT (1980) The role of deep sea heat storage in the secular response to climate forcing. J Geophys Res 85:6667–6679. doi:10.1029/JC085iC11p06667

IPCC (2007) Climate change 2007: the physical science basis. In: Solomon S, Qin D, Manning M (eds) Intergovernmental panel on climate change, Geneva. http://www.ipcc.ch/publications_and_data/ar4/wg1/en/contents.html

Ishii M, Kimoto M (2009) Reevaluation of historical ocean heat content variations with time-varying XBT and MBT depth bias corrections. J Oceanogr 65:287–299. doi:10.1007/s10872-009-0027-7

Jarvis A, Li S (2011) The contribution of timescales to the temperature response of climate models. Clim Dyn 36:523–531. doi:10.1007/s00382-010-0753-y

Jones GS, Christidis N, Stott PA (2011) Detecting the influence of fossil fuel and bio-fuel black carbon aerosols on near surface temperature changes. Atmos Chem Phys 11:799–816. doi:10.5194/acp-11-799-2011

Joshi M, Shine K, Ponater M, Stuber N, Sausen R, Li L (2003) A comparison of climate response to different radiative forcing in three general circulation models: towards and improved metric of climate change. Clim Dyn 20:843–854. doi:10.1007/s00382-003-0305-9

Kiehl JT (2007) Twentieth century climate model response and climate sensitivity. Geophys Res Lett 34:L22710. doi:10.1029/2007GL031383

Kloster S, Dentener F, Feichter J, Raes F, Lohmann U, Roeckner E, Fischer-Bruns I (2010) A GCM study of future climate response to aerosol pollution reductions. Clim Dyn 34:1177–1194. doi:10.1007/s00382-009-0573-0

Knutti R (2008) Why are climate models reproducing the observed global surface warming so well? Geophys Res Lett 35:L18704. doi:10.1029/2008GL034932

Knutti R, Plattner G-K, (2012) Comment on "Why Hasn't earth warmed as much as expected?" by Schwartz et al. 2010. J Clim 25:2192–2199. doi:10.1175/2011JCLI4038.1

Knutti R, Krähenmann S, Frame DJ, Allen MR (2008) Comment on ''Heat capacity, time constant, and sensitivity of Earth's climate system'' by S. E. Schwartz. J Geophys Res 113:D15103. doi:10.1029/2007JD009473

Levitus S, Antonov J, Boyer T (2005) Warming of the world ocean, 1955–2003. Geophys Res Lett 32: L02604. doi:10.1029/2004GL021592

Levitus S, Antonov JI, Boyer TP, Locarnini RA, Garcia HE, Mishonov AV (2009) Global ocean heat content 1955–2008 in light of recently revealed instrumentation problems. Geophys Res Lett 36:L07608. doi:10.1029/2008GL037155

Lohmann U, Rotstayn L, Storelvmo T, Jones A, Menon S, Quaas J, Ekman AML, Koch D, Ruedy R (2010) Total aerosol effect: radiative forcing or radiative flux perturbation? Atmos Chem Phys 10(7): 3235–3246. doi:10.5194/acp-10-3235-2010

Lucarini V, Ragone F (2011) Energetics of climate models: net energy balance and meridional enthalpy transport. Rev Geophys 49:RG1001. doi:10.1029/2009RG000323

Lyman JM (2011) Estimating global energy flow from the global upper ocean. Surv Geophys. doi: 10.1007/s10712-011-9167-6

Lyman JM, Johnson GC (2008) Estimating annual global upper ocean heat content anomalies despite irregular in situ ocean sampling. J Clim 21:5629–5641. doi:10.1175/2008JCLI2259.1

Matthews HD, Caldeira K (2007) Transient climate-carbon simulations of planetary geoengineering. Proc Natl Acad Sci USA 104:9949–9954. doi:10.1073/pnas.0700419104

Meehl GA, Washington WM, Wigley TML, Arblaster JM, Dai A (2003) Solar and greenhouse gas forcing and climate response in the twentieth century. JClim 16:426–444. doi:10.1175/1520-0442(2003)016<0426:SAGGFA>2.0.CO;2

Meinshausen M, Smith S, Calvin K, Daniel JS, Kainuma M, Lamarque J-F, Matsumoto K, Montzka SA, Raper SCB, Riahi K, Thomson AM, Velders GJM, van Vuuren D (2011) The RCP greenhouse gas concentrations and their extension from 1765 to 2300. Clim Chang. doi:10.1007/s10584-011-0156-z

Murphy DM, Solomon S, Portmann RW, Rosenlof KH, Forster PM, Wong T (2009) An observationally based energy balance for the Earth since 1950. J Geophys Res 114:D17107. doi:10.1029/2009 JD012105

Myhre G, Highwood EJ, Shine KP, Stordal F (1998) New estimates of radiative forcing due to well mixed greenhouse gases. Geophys Res Lett 25:2715–2718. doi:10.1029/98GL01908

Myhre G, Myhre A, Stordal F (2001) Historical evolution of radiative forcing of climate. Atmos Environ 35:2361–2373. doi:10.1016/S1352-2310(00)00531-8

Padilla LE, Vallis GK, Rowley CW (2011) Probabilistic estimates of transient climate sensitivity subject to uncertainty in forcing and natural variability. J Clim 24:5521–5537. doi:10.1175/2011JCLI3989.1

Palmer MD, Haines K, Tett SFB, Ansell TJ (2007) Isolating the signal of ocean global warming. Geophys Res Lett 34:L23610. doi:10.1029/2007GL031712

Palmer M, Antonov J, Barker P, Bindoff N, Boyer T, Carson M, Domingues C, C. SG, Gleckler P, Good S, Gouretski V, Guinehut S, Haines K, Harrison DE, Ishii M, Johnson G, Levitus S, Lozier S, Lyman J, Meijers A, Schuckmann Kv, Smith D, Wijffels S, Willis J (2010) Future observations for monitoring global ocean heat content. In: Hall J, Harrison DE, Stammer D (eds) Proceedings of the "OceanObs' 09: Sustained ocean observations and information for society" Conference (vol 2), Venice, Italy, 21–25 September 2009. ESA Publication WPP-306. doi:10.5270/OceanObs09.cwp.68

Pilewskie P (2011) Measurements of solar spectral irradiance. ISSI Workshop on Observing and Modeling Earth's Energy Flows. Bern, 10–14 January 2011

Scafetta N (2008) Comment on "Heat capacity, time constant, and sensitivity of Earth's climate system" by S. E. Schwartz. J Geophys Res 113:D15104. doi:10.1029/2007JD009586

Schwartz SE (2004) Uncertainty requirements in radiative forcing of climate change. J Air Waste Manag Assoc 54:1351–1359. doi:10.1080/10473289.2004.10471006

Schwartz SE (2007) Heat capacity, time constant, and sensitivity of Earth's climate system. J Geophys Res 112(D24):D24S05. doi:10.1029/2007JD008746

Schwartz SE (2008a) Reply to comments by G. Foster et al., R. Knutti et al., and N. Scafetta on "Heat capacity, time constant, and sensitivity of Earth's climate system". J Geophys Res 113:D15105. doi: 10.1029/2008JD009872

Schwartz SE (2008b) Uncertainty in climate sensitivity: causes, consequences, challenges. Energy Environ Sci 1:430–453. doi:10.1039/B810350J

Schwartz SE, Charlson RJ, Kahn RA, Ogren JA, Rodhe H (2010) Why hasn't Earth warmed as much as expected? J Clim 23:2453–2464. doi:10.1175/2009JCLI3461.1

Smith TM, Reynolds RW, Peterson TC, Lawrimore J (2008) Improvements to NOAA's historical merged land–ocean surface temperature analysis (1880–2006). J Clim 21:2283–2296. doi:10.1175/2007 JCLI2100.1

Stevens B, Schwartz SE (2012) Observing and modeling Earth's energy flows. Surv Geophys. doi: 10.1007/s10712-012-9184-0

Sutton P, Roemmich D (2001) Ocean temperature climate off north-east New Zealand. N Z J Marine Freshw Res 35:553–565. doi:10.1080/00288330.2001.9517022

Takemura T, Tsushima Y, Yokohata T, Nozawa T, Nagashima T, Nakajima T (2006) Time evolutions of various radiative forcings for the past 150 years estimated by a general circulation model. Geophys Res Lett 33:L19705. doi:10.1029/2006GL026666

Watterson G, Dix MR (2005) Effective sensitivity and heat capacity in the response of climate models to greenhouse gas and aerosol forcings. Q J R Meteorol Soc 131:259–279. doi:10.1256/qj.03.232

Webb MJ, Senior CA, Sexton DMH, Ingram WJ, Williams KD, Ringer MA, McAvaney BJ, Colman R, Soden BJ, Gudgel R, Knutson T, Emori S, Ogura T, Tsushima Y, Andronova N, Li B, Musat I, Bony S, Taylor KE (2006) On the contribution of local feedback mechanisms to the range of climate sensitivity in two GCM ensembles. Clim Dyn 27:17–38. doi:10.1007/s00382-006-0111-2

Willis JK, Roemmich D, Cornuelle B (2004) Interannual variability in upper ocean heat content, temperature, and thermosteric expansion on global scales. J Geophys Res 109:C12036. doi:10.1029/2003 JC002260

Surv Geophys (2012) 33:779–816
DOI 10.1007/s10712-012-9184-0

Observing and Modeling Earth's Energy Flows

Bjorn Stevens · Stephen E. Schwartz

Received: 29 June 2011 / Accepted: 17 February 2012 / Published online: 11 May 2012

Abstract This article reviews, from the authors' perspective, progress in observing and modeling energy flows in Earth's climate system. Emphasis is placed on the state of understanding of Earth's energy flows and their susceptibility to perturbations, with particular emphasis on the roles of clouds and aerosols. More accurate measurements of the total solar irradiance and the rate of change of ocean enthalpy help constrain individual components of the energy budget at the top of the atmosphere to within ± 2 W m^{-2}. The measurements demonstrate that Earth reflects substantially less solar radiation and emits more terrestrial radiation than was believed even a decade ago. Active remote sensing is helping to constrain the surface energy budget, but new estimates of downwelling surface irradiance that benefit from such methods are proving difficult to reconcile with existing precipitation climatologies. Overall, the energy budget at the surface is much more uncertain than at the top of the atmosphere. A decade of high-precision measurements of the energy budget at the top of the atmosphere is providing new opportunities to track Earth's energy flows on timescales ranging from days to years, and at very high spatial resolution. The measurements show that the principal limitation in the estimate of secular trends now lies in the natural variability of the Earth system itself. The forcing-feedback-response framework, which has developed to understand how changes in Earth's energy flows affect surface temperature, is reviewed in light of recent work that shows fast responses (adjustments) of the system are central to the definition of the effective forcing that results from a change in atmospheric composition. In many cases, the adjustment, rather than the characterization of the compositional perturbation (associated, for instance, with changing greenhouse gas concentrations, or aerosol burdens), limits accurate determination of the radiative forcing. Changes in clouds contribute importantly to this adjustment and thus contribute both to uncertainty in estimates of radiative forcing and to uncertainty in the response. Models are indispensable to calculation of the adjustment of the system to a compositional change but are known to be flawed in their representation of

B. Stevens (✉)
Max-Planck-Institüt for Meteorology, KlimaCampus, 20146 Hamburg, Germany
e-mail: bjorn.stevens@zmaw.de

S. E. Schwartz
Atmospheric Sciences Division, Brookhaven National Laboratory, Upton, NY 11973, USA

clouds. Advances in tracking Earth's energy flows and compositional changes on daily through decadal timescales are shown to provide both a critical and constructive framework for advancing model development and evaluation.

Keywords Climate change · Cloud radiative effects · Aerosol · Energy budget · Climate sensitivity · Radiative forcing

1 Introduction

Modern climate science, wherein descriptive pictures of the climate system began to be complemented by quantitative theory, is only about a hundred years old. In the late nineteenth Century understanding of radiative transfer, particularly at the infrared wavelengths associated with terrestrial radiation, was developing rapidly, and it became possible to formulate quantitative descriptions of the relationship between the flux of energy through the Earth system and quantities like the average surface temperature. Exemplary in this respect are Arrhenius' 1896 calculations suggesting that changes in carbon dioxide would induce changes in the surface temperature. Arrhenius' study was essentially an exercise in radiative transfer in which he quantified the flow of solar and terrestrial radiation through the Earth system and the roles of the various processes that influenced these transports. Greenhouse gases such as carbon dioxide and water vapor played a key role in Arrhenius' calculations, and other factors influencing solar radiation such as clouds and surface properties were also accounted for. Although prescient in many respects, for instance with respect to the role of the carbon cycle and feedbacks associated with water vapor and surface albedo, Arrhenius did not touch on a number of issues that have come to dominate the discourse with respect to climate change. Among these are rate of increase in ocean enthalpy and changes in the ocean circulation; changes in patterns of precipitation; the role of aerosols, both in influencing clouds and in the energy budget as a whole; and also the possibility that changes in cloudiness may enhance or offset other changes in the climate system.

An increased emphasis on precisely those issues largely left out of the early studies has marked a fundamental shift in climate science over the last forty or so years. Through the latter part of the 1960s, climate science was preoccupied with resolving controversies related to the radiative transfer that was the basis for the analysis of Arrhenius and his followers. Confusion about the nature and importance of the details of the spectroscopy of CO_2 and H_2O in the thermal infrared (e.g., Plass 1956; Kaplan 1960) was resolved only by the calculations of Manabe and Wetherald (1967), which showed that earlier controversies lost relevance when the vertical structure of the atmosphere is properly accounted for (cf., Pierrehumbert 2011). Their research propelled the field into the current era, wherein qualitatively new questions, in particular the role of clouds and aerosol particles, demanded a more detailed understanding and accounting of the energy flows through the climate system.

Arrhenius was perhaps the first to appreciate how small changes in flows of energy through the system can have a large impact on Earth's climate. A doubling of the mixing ratio (commonly "concentration") of atmospheric CO_2, which is the paradigmatic example of a forcing of climate change, gives rise to a radiative perturbation of 3–4 W m^{-2}, which is about 1 % of the solar radiation incident at the top-of-atmosphere;[1] the total radiative

[1] Unless otherwise noted the radiative forcing estimates we cite are for the radiative perturbation that arises at the TOA after the stratosphere has adjusted to the presence of compositional change, but with the tropospheric temperatures held constant.

forcing attributed to long-lived greenhouse gases introduced through human activities over the industrial era thus far is about 3 W m^{-2}. By way of comparison, in the tropics, the diurnal variation of incident solar radiation is more than 1,000 W m^{-2}. The presence of a high cloud can change the outgoing long-wave radiation by 100–200 W m^{-2}, comparable in magnitude to seasonal changes in radiative fluxes, albeit more short lived. From the perspective of climate system response, the central value of current estimates of the increase in global mean surface temperature that would result from a doubling of CO_2, 3 K, is about 1 % of the global mean surface temperature, 288 K, and again, much less than geographical and temporal variability. The rather large consequences of such small changes in Earth's energy flows, and the complexity of the system that mediates these flows, make determining the effects of changing atmospheric composition on Earth's energy flows, through measurement or modeling, a challenging scientific problem.

In the following, we reflect on this challenge as framed by discussions at a recent ISSI workshop titled "Observing and Modelling Earth's Energy Flows". Our presentation is organized around three basic issues: (1) what is the status of present understanding of Earth's energy budget; (2) how does the composition of the atmosphere, particularly clouds and aerosols, influence this budget; and (3) how can modeling help constrain a description of processes regulating the flow of energy through the climate system. These issues, particularly those aspects central to advancing understanding of the climate system, are discussed in turn below. Although the discussion presented herein benefitted greatly from presentations and discussions at the Workshop, responsibility for the material presented rests with the present authors and should not be taken as representing the views of participants in the workshop or as a workshop consensus.

2 Present Understanding of Earth's Energy Budget

2.1 Global Energy Balance

Measurements have been indispensable to the advancing understanding of energy flows through the climate system. Current understanding of these flows is summarized in Fig. 1, which we have constructed based on the available literature (especially Trenberth et al. 2009; Kato et al. 2012; Stephens et al. 2012). The energy flows are more certain at the top of the atmosphere than at the surface, as measurements at the top of the atmosphere have benefitted greatly from advances in satellite remote sensing. Measurements from the Earth Radiation Budget Experiment (ERBE) (Ramanathan 1987) and now those from the Clouds and Earth's Radiant Energy System (CERES) mission (Wielicki et al. 1996) have convincingly shown that Earth reflects less short-wave radiation and emits more long-wave radiation than previously thought (cf., Table 1) and have contributed to closure of the top-of-atmosphere energy budget to within a few watts per square meter.

Also contributing to the reduction of uncertainty in the balance of energy flows at the top of the atmosphere are new measurements of total solar irradiance and ocean heat uptake. The total solar irradiance determined by the Solar Radiance and Climate Experiment, 1,360.8 ± 0.5 W m^{-2} at solar minimum (Kopp and Lean 2011), is well below the range of previous estimates as summarized in Table 1. Confidence in these lower estimates and their associated assessment of uncertainty is gained through the identification of artifacts in the older measurements that lends enhanced credence to the newer lower estimates. Because the atmosphere has a relatively small heat capacity, an imbalance of energy flows at the TOA can be sustained only an increase in ocean enthalpy, augmented to

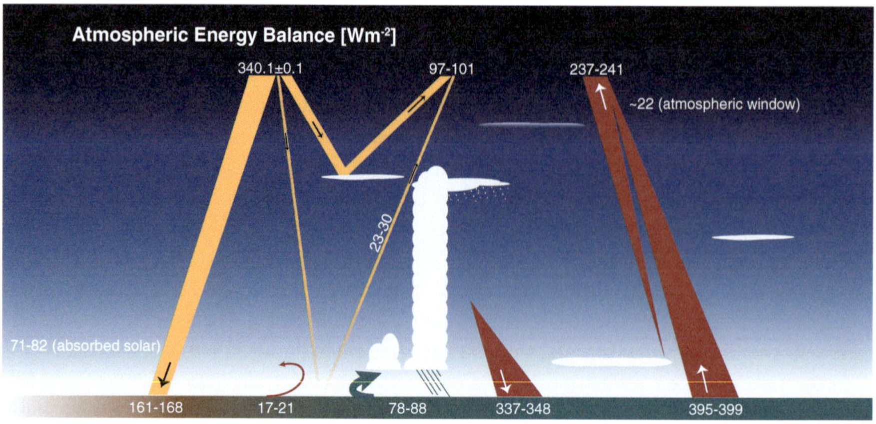

Fig. 1 Earth's global and annual mean top-of-atmosphere (TOA) and surface energy budget. Values are presented as a two-sigma range (which is to say the authors assign a roughly 68 % likelihood that the actual values fall within the stated range). These values are subjectively determined based on a review of the literature, and complemented by global simulations, as discussed in the Appendix

Table 1 Estimates of Earth's energy budget, subjectively determined based on a review of the existing literature and the best estimates of the net imbalance at the surface and top-of-atmosphere

Source	H-1954	L-1957	R-1987	K-1997	Z-2005	T-2009	S-2012	This study
Top of atmosphere								
Incident SW	338	349	343	342	342	341	340	340
Reflected SW	115	122	106	107	106	102	98	100
Outgoing LW	223	227	237	235	233	239	239	239
Surface								
Absorbed SW	160	167	169	168	165	161	168	162
Downward LW	352	334	327	324	345	333	347	342
Upward LW	398	395	390	390	396	396	398	397
Latent heat flux	82	68	90	78	n/a	80	88	86
Sensible heat flux	30	33	16	24	n/a	17	24	20

References to prior estimates are: Houghton (1954), London (1957), Ramanathan (1987), Kiehl and Trenberth (1997), Trenberth et al (2009), Stephens et al. (2012). Because both the H-1954 and L-1957 estimates were for the northern hemisphere only, these estimates have been rescaled using the ratio of the global versus the northern hemisphere average from a high-resolution AMIP simulation using the ECHAM6 model. For basis of present estimates see Appendix

lesser extent by melting of the cryosphere and warming of the land surface and the atmosphere. As discussed by Lyman (2011), the rate of heating of the top 700 m of the world ocean over the period from 1993 to 2008, as inferred from temperature measurements and expressed per the area of the entire planet, is 0.64 ± 0.11 W m^{-2} (90 % confidence interval). Sparser measurements extending to ocean depths of 3 km reported by Levitus et al. (2005) suggest that the upper ocean takes up about three-quarters of the ocean heating. Levitus et al. (2005) also estimate that the contributions of other enthalpy sinks, including the atmosphere, the land, and the melting of ice, can account for an additional 0.04] W m^{-2}; based on these estimates the flux imbalance at the TOA is estimated to be 0.9 ± 0.3 W m^{-2}. The uncertainty is based on the 90 % confidence interval given by Lyman and the assumption that the relative uncertainty in the deep ocean enthalpy uptake estimates and in the estimates of heating by other components of the Earth system are about 50 %.[2] For reference, in constructing the energy-balanced version of the CERES data, Loeb et al. (2009) estimated the surface enthalpy uptake to be 0.85 W m^{-2} similar to the 0.9 W m^{-2} estimated here and employed by Trenberth et al. (2009) based on a somewhat different line of reasoning. An analysis based on more recent ARGO data and a revised analysis of the CERES measurements Loeb et al. (2012) suggest a somewhat lower central value of 0.50 ± 0.43 W m^{-2} for the rate of increase in Earth system enthalpy, per unit area of Earth's surface. In summary, although there is a rather larger, 6.5 W m^{-2}, inherent uncertainty in measurements of the reflected short-wave and emitted long-wave radiation at the TOA, associated principally with uncertainty in the absolute calibration of the CERES instruments (Loeb et al. 2009), the net irradiance at the TOA is constrained to within 1 W m^{-2} by improved measurements of increases in ocean enthalpy.

The surface energy budget is distributed over several terms, each of which exhibits uncertainty that is several-fold greater than the uncertainty in the net budget at TOA. As pointed out by Trenberth et al. (2009), if each of the terms in the surface energy budget is estimated individually, in isolation of the others, an imbalance can arise in the net surface flux that is as much as 20 W m^{-2}; this is more than an order of magnitude greater than current measurement-based estimates of the rate of increase in the enthalpy of the ocean, cryosphere, and land. In absolute terms, the uncertainty is largest for the long-wave irradiance downwelling at the surface and for the latent heat flux (precipitation). Here too, new measurements and improved modeling of energy flows are beginning to improve understanding. Active remote sensors such as the cloud profiling radar flown as part of the CloudSat mission (Stephens et al. 2008) and the Cloud-Aerosol Lidar with Orthogonal Polarization flown as part of the Cloud-Aerosol Lidar and Infrared Pathfinder Satellite Observation mission (Winker et al. 2010) are providing improved estimates of the vertical distribution of clouds, particularly cloud base. These measurements are crucial for estimates of the downward long-wave irradiance, which are based on radiative transfer modeling given a realistic atmospheric state. Using these measurements, Kato et al. (2012) estimate a downward long-wave irradiance at the surface of 345.4 ± 6.9 W m^{-2}, which is considerably larger than the value derived by Trenberth et al (2009) from the residual of the surface energy balance given existing precipitation climatologies. CloudSat and surface-based measurements also suggest that existing precipitation climatologies underestimate light rain from warm clouds, which is common over the tropical ocean (Nuijens et al. 2009; Stephens et al. 2012), and which may help reconcile the differences in the energy-based versus precipitation-based estimates of the strength of the hydrological cycle. Although still large, overall the uncertainty in the surface energy budget, as presented in Fig. 1 is somewhat smaller than that presented by Stephens et al. (2012), as global modeling (including reanalysis) estimates constrained by observations are given somewhat more weight in the present analysis, particularly for the upward turbulent fluxes and the long-wave irradiance.

Advances in understanding of Earth's energy flows become most apparent when looking beyond the globally and annually averaged estimates of these flows. Measurements by CERES instruments track day-to-day variations in Earth's energy flows on regional scales.

[2] Other sources of energy—geothermal, combustion of fossil fuel and nuclear production—are yet an order of magnitude smaller (P. Pilewskie, ISSI Workshop on Observing and modeling Earth's energy flows, Bern, 10–14 Jan., 2011).

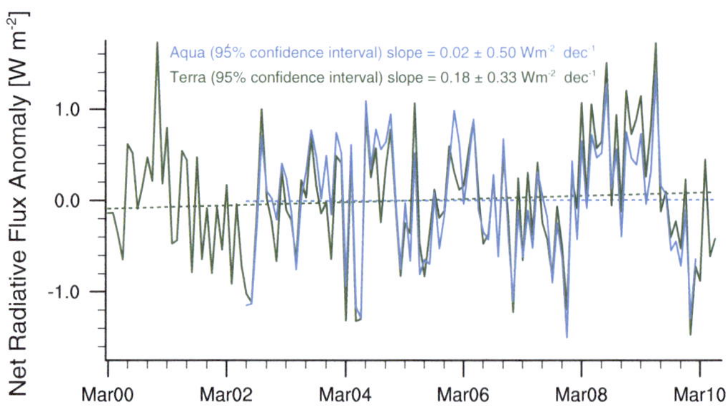

Fig. 2 Anomalies in the monthly and globally averaged top of the atmosphere net radiation from the CERES SSF1deg product. Anomalies are calculated by subtracting the CERES monthly climatology from the monthly fluxes. Regression line uncertainty is estimated assuming that monthly anomalies are independent. The standard deviation of the anomalies is about 0.65 W m^{-2}. Aqua, pm; Terra, am

Measurements extending over more than a decade have made it possible to explore how these energy flows fluctuate on intradecadal time-scales. As an example, Fig. 2 shows the existing record of monthly and globally averaged radiation anomalies. The radiation anomalies are calculated using the SSF1deg (single scanner footprint at one degree) data set, which includes information only from the twice daily measurements of the CERES platform on the polar orbiting Terra (ascending node equator crossing time 10:30 am) and Aqua (ascending node equator crossing time 1:30 pm). Unlike the CERES SYN or EBAF products, the SSF1deg data set does not incorporate measurements from a shifting complement of geostationary satellites, and thus more fully represents the inherent stability of the CERES measurements themselves. Monthly anomalies were constructed by subtracting the monthly climatology of the same data set from the global and monthly averaged values. The remarkable agreement between the Aqua and Terra measurements, over the period of the record where they coincide, suggests that the observed variability is an intrinsic property of Earth's climate system, rather than an artifact due to instrument precision or insufficient sampling. This inherent variability limits quantification of secular trends in the flow of energy through the TOA.

This variability in Earth's energy flows, manifest as fluctuations in the anomaly of net irradiance as large as 3 W m^{-2} over annual to inter-annual timescales, e.g., Fig. 2, is expected to be mirrored in other components of the climate system. By combining these trends with concurrent measurements of ocean enthalpy, Wong et al. (2006) and Trenberth and Fasullo (2010b) have begun to explore this question. By drawing attention to the what ended up being insignificant (and spurious) trends over the length of the record, Trenberth and Fasullo (2010b) have given what turns out to be a false impression of irreconcilable trends in the relation between ocean enthalpy increases and the TOA energy imbalance.[3] These results unfortunately overshadowed the real revolution in such approaches, namely

[3] The analysis by Trenberth and Fasullo (2010b) suggested that the radiative imbalance at the top of the atmosphere measured by CERES has been increasing by as much as 1 W m^{-2} dec $^{-1}$. Because of the limited heat capacity of the atmosphere such an imbalance would imply a large change in ocean enthalpy and/or surface ice amount, neither of which is observed. However, Trenberth and Fasullo (2010b) made use of preliminary CERES data for the time-period 2005–2010, over which the striking trend in radiative imbalance was noted, and failed to account for uncertainty in their estimated trend.

the recognition that the present observing system shows evidence of large inter-annual fluctuations in the flow of energy through the Earth system and that these might be trackable as they work their way through the system— for instance from anomalies in the TOA net radiation, to changes in ocean heat uptake.

We disagree with the assertion of Stephens et al. (2012) that the satellite measurements such as CERES have "not significantly changed our understanding of the TOA energy balance" relative to what was known based on the ERBE data. Although we acknowledge that estimates of the surface radiation budget do not significantly differ from those of Houghton (1954) more than a half century ago, we would argue in both cases that, when viewed from the perspective of initial attempts to infer the planetary energy balance from measurements looking down and up from Mt Wilson (Abbot and Fowle 1908), or on the basis of a network of pyrheliometers over North America (e.g., Houghton 1954), progress has been tremendous, even if the small magnitude of changes in estimates of the globally and annually averaged budgets (Table 1) would seem to belie this progress. The present era of global sensing is unprecedented: high-precision instruments are tracking Earth's energy flows on decadal timescales and regional spatial scales; an absolute accuracy of better than one part per thousand has been achieved in measurements of solar irradiance and is helping to close the TOA energy budget; networks of ocean-going floats are quantifying ocean heat uptake; and surface-based networks and profiling instruments are beginning to advance understanding of the surface energy budget. As should become increasingly clear through the course of this article, sustaining and expanding such measurements will be key to further advances in understanding of Earth's climate and its susceptibility to change.

2.2 Radiative Forcing, Response and Climate Sensitivity

In addition to providing understanding of energy flows in the current climate, consideration of the energy budget also provides a framework for understanding climate change. This framework, which has developed over the last 30 years, rests on the assumption that changes in the globally averaged surface temperature can be linearly related to a radiative forcing, F. The constant of proportionality between the forcing and the response is called the equilibrium climate sensitivity, S_{eq}, which can be formally defined as the steady-state change in T_s (the globally averaged near-surface air temperature) that would result from a sustained change in a radiative flux component of the Earth energy budget at the TOA (forcing), normalized to that flux change, with unit: K (W m^{-2})$^{-1}$. That is, the equilibrium sensitivity is the proportionality constant between the steady-state change in surface temperature and the applied forcing,

$$\Delta T_s = S_{eq}F. \tag{1}$$

Because it codifies the sensitivity of such an essential feature of the climate state, the globally averaged surface temperature, Eq. 1 provides a powerful description of past and prospective future climate change. To the extent that other properties of the climate system scale with its value, S_{eq} assumes an even broader significance. For these reasons, determination of S_{eq} has evolved into a central focus of climate science (e.g., Knutti and Hegerl 2008).

Equation 1 is well suited to the interpretation of models, for which the changes in the equilibrium state can be determined on the basis of numerical simulations. However, by extending the framework that leads to Eq. 1, a related framework has been developed which allows for the determination of S_{eq} from observations. Assuming that the forcing,

F were known, it could be related to the rate of change in the enthalpy of the Earth system, $\dot{H}$, and the temperature change between that at $t = 0$, defined to correspond to a time when the system is in stationarity and the forcing is applied, and that at some later time, t,

$$\dot{H} = F - S_{eq}^{-1}(T_s(t) - T_s(0)). \tag{2}$$

If both the forcing and the rate of change of the planetary enthalpy are known, the equilibrium sensitivity might be determined (Gregory et al. 2002) as

$$S_{eq} = \frac{T_s(t) - T_s(0)}{F - \dot{H}} \tag{3}$$

where the forcing and the change in global temperature are relative to a prior unperturbed state (at time, $t = 0$) and the rate of heating of the planet $\dot{H}$ can be inferred either from satellite measurements at the top of the atmosphere or from the change in enthalpy of the planet inferred from ocean calorimetry as discussed by Lyman (2011). However, application of Eq. 3 is limited by the fact that changes in the enthalpy and surface temperature must be measured relative to an equilibrium state that is not generally known. To sidestep this issue Forster and Gregory (2006) and Murphy et al. (2009) proposed to determine S_{eq} over shorter time-periods, not starting with the unperturbed state, as

$$S_{eq} = \frac{T_s'}{F' - \dot{H}'} \tag{4}$$

where primes denote changes in the quantities over a given time-period. The utility of any such determination of sensitivity rests on the uncertainties in the several quantities on the right-hand side of the equation.

Central to the conceptual framework surrounding the use of Eq. (1), or its time-dependent variants, are the assumptions: (1) that S_{eq} does not depend on the nature or magnitude of the forcing; and (2) that the forcing resulting from a given change in atmospheric composition or surface properties can be determined unambiguously. Both assumptions become increasingly problematic if the forcing-sensitivity framework is pushed too far, for instance by demanding ever more certain estimates of S_{eq}. Models have long suggested that S_{eq} depends on both the temperature and the nature of the forcing (e.g., Hansen et al. 1997; Colman and McAvaney 2009). For instance, Hansen et al. (1997) showed that if the TOA radiative forcing is equated with the initial radiative perturbation associated with a change in CO_2, the change in T_s will differ by as much as 25 % as compared to what arises for an equal radiative perturbation arising from a change in the solar irradiance. To get around this issue, the concept of an effective, or adjusted, forcing has been developed. The adjusted radiative forcing attempts to account for a rapid, and perturbation-dependent, adjustment of the climate system to a compositional change. The familiar example, and the one that motivated the original concept of an adjusted radiative forcing, is the adjustment of the stratosphere which proceeds differently depending on whether the initial radiative perturbation is caused by changes in well-mixed greenhouse gases as compared to changes in the solar irradiance, or stratospheric aerosols (e.g., Hansen et al. 1997). However, several studies have begun to show that, in addition to the stratosphere, changes in other components of the atmosphere, for instance clouds, also adjust rapidly in ways that depend on the nature of the external perturbation and thus must be factored into the calculation of an adjusted radiative forcing. An elegant framework for dealing with these adjustments has been proposed by Gregory et al. (2004). In their approach, the adjusted radiative forcing is defined as the *y*-intercept of the regression of the TOA radiative imbalance against T_s. This definition formalizes the idea of a rapid adjustment of the system to a perturbation which determines the effective, or adjusted radiative forcing, and which is temporally well separated from the slow response of the surface temperature to this adjusted radiative forcing. Of course this definition means that the forcing is no longer independent, or external to the system and thus cannot be determined *a priori*, for instance, through complete knowledge of radiative transfer and compositional changes alone. And as a result, internal physical processes, such as the nature of cloud adjustment, contribute to uncertainty in estimates of the effective radiative forcing driving changes in Earth's climate.

The dependence of the forcing on the system to which the perturbation is applied makes the framework surrounding Eq. (1) less useful, as it implies that both F and S_{eq} depend on the system. The extra uncertainty, which the fast response of the system introduces in the determination of the forcing, poses a particular challenge to observationally based attempts to estimate S_{eq}, for example through the use of either Eqs. (3) or (4). By focusing on periods over which the observational record is better constrained the numerator and denominator of Eqs. (3) and (4) become small; but in this case, because the denominator is the difference of two rather large terms, the answer becomes proportionally more sensitive to errors in either $F, \dot{H}$ (or their associated changes in time). So, although in the past, estimates of the forcing F have generally been thought to be limited by understanding to changes in the aerosol over the period of industrialization (e.g., Gregory et al. 2002), recent work suggests that factors such as clouds, which contribute to the fast response of the system, (Hansen et al. 1997; Gregory and Webb 2008; Andrews et al. 2009 2011), may place a more fundamental limit on the quantification of the effective radiative forcing that arises as a result of compositional changes to the atmosphere.

3 Clouds

3.1 Radiative Effects, Feedbacks, and Forcing

A framework for considering cloud feedbacks in the climate system, first formulated by Schneider (1972), introduced the idea of cloud-forcing (see also Cess 1976; Ramanathan 1987). Identifying the subset of short- and long-wave radiative fluxes associated with cloud-free scenes, and assuming that three-dimensional effects associated with neighboring clouds can be neglected, permits determining "cloud radiative forcing" as the difference between the all-sky or actual radiative flux and the contribution from scenes having skies identified as cloud-free. Here, as before, the term "cloud-forcing" is used to denote a change in radiative flux due to a change in atmospheric composition, namely clouds. However, because this terminology does not allow one to distinguish the radiative effect of the totality of clouds from the radiative perturbation that would accompany an external perturbation to cloud properties, it proves useful to distinguish between the cloud radiative effect, CRE, as the radiative effect of the background state of cloudiness, and reserve the phrase "radiative forcing" for radiative perturbations driven by externally imposed changes.

The definition of the cloud radiative effect can be made precise as follows. Let Q denote the rate of absorption of solar energy (short-wave irradiance) and E denote the rate of emission of infrared energy (long-wave irradiance), both at TOA. If an inward directed flux is defined to be positive, E at the TOA must be negative. With this sign convention and letting H denote the enthalpy of the Earth system, including components such as oceans, land and surface ice, then the rate of change of H with time,[4] $\dot{H}$ is given by $\dot{H} = Q + E$. Under the assumption that three-dimensional radiative effects are negligible the two radiative flux terms may be conceptually distinguished into components pertaining to contributions from cloudy and cloud-free regions of the planet:

$$Q = Q_\star[1 - \alpha_0(1 - A_c) - \alpha_c A_c] \tag{5}$$

$$E = -E_0(1 - A_c) - E_c A_c \tag{6}$$

Here A_c denotes the fraction of the area of the planet that is, on average, cloudy; $Q_\star$ denotes the average solar irradiance incident at the TOA ($Q_\star \approx 340$ W m^{-2}, e.g., Fig. 1); α_0 and α_c denote the effective TOA albedo of the cloud-free and cloudy scenes, respectively; and E_0, and E_c denote the emitted long-wave irradiance obtained by compositing over the cloud-free and cloudy scenes, respectively. The albedos are designated as "effective" quantities because Eq. (5) depends non-linearly on cloudiness and insolation and, thus, the effective albedos must account for the co-variability between these two quantities. So for instance, α_c is not the average cloud albedo, but rather the cloud albedo that the average cloudiness requires so that the planetary albedo, $\alpha = \alpha_0(1 - A_c) - \alpha_c A_c$ matches that observed, likewise for the α_0, the effective albedo of the cloud-free scenes. In general each of the terms on the right-hand side of Eqs. (5) and (6) except for $Q_\star$, is a function of the state of the system, importantly the vertical distribution of temperature, T_0, the composition of the atmosphere and the surface properties. E_c is the effective emitted irradiance of cloudy scenes that is calculated by compositing over all columns not identified as being cloud-free. It depends on the distribution of clouds, but because clouds have some transparency and the atmosphere has some opacity, it also depends on the composition and temperature of the atmosphere, and the co-variability among the two. The short- and long-wave components of the CRE, denoted by superscripts (Q) and (E) respectively, follow naturally as the difference between the all-sky radiative flux and the fluxes which would be manifest in the absence of clouds, i.e.,

$$F_c^{(Q)} = -Q_\star(\alpha_c - \alpha_0)A_c \tag{7}$$

$$F_c^{(E)} = -(E_c - E_0)A_c \tag{8}$$

Given our sign convention, and because α_c is generally greater than α_0 whereas E_c is generally less than E_0, the short-wave CRE is negative and the long-wave CRE is positive. Both quantities increase in magnitude with cloud amount, A_c. The net CRE, F_c, is given by $F_c^{(Q)} + F_c^{(E)}$. A secular increase in the magnitude of $F_c^{(Q)}$ would exert a cooling influence on the Earth system, whereas an increase in the magnitude of $F_c^{(E)}$ would exert a warming influence. It should be emphasized that the CRE depends not just on the properties of the cloudy fraction of the planet but on the *differences* between the cloudy and cloud-free portions of the planet. As reviewed by Loeb et al. (2009) the application of this concept to various data sets shows the short-wave CRE to range from −45.4 to −53.3 W m^{-2} and the

[4] By conservation of energy, and as other sources of energy in the Earth system are negligible (P. Pilewskie ISSI Workshop on Observing and modeling Earth's energy flows, Bern, 10–14 Jan., 2011) this time-rate of change in Earth's enthalpy is equal to the net flux at the top of the atmosphere, N. Because the atmospheric enthalpy changes so little, a consequence of the low heat capacity of the atmosphere, N at the top of the atmosphere is usually taken as identical with N at the surface. The enthalpy notation is adopted because it emphasizes what is being measured, namely the change in the heat content, or enthalpy of the Earth system.

long-wave CRE to range from 26.5 to 30.6 W m^{-2}; the associated net CRE from these prior estimates ranges from -16.7 to -24.5 W m^{-2}. The CERES EBAF data, upon which the TOA estimates in Fig. 1 are largely based, give a short-wave cloud radiative effect of -47.1 W m^{-2} and a long-wave CRE of $+26.5$ W m^{-2}; overall clouds, more precisely cloudy scenes, exert a net cooling influence on the Earth system of about 20 W m^{-2}.

The CRE concept has seen considerable use in the interpretation of feedbacks in the climate system. From the perspective of Eq. (1), a feedback is a change in a radiative flux that results from a change in global temperature; such a further change in radiative flux in addition to that caused by an initial forcing imposed on the climate system can enhance or diminish the temperature change induced by a given forcing (positive or negative feedback, respectively). The feedback concept has been quite useful in interpreting the contributions of different components of the climate system to S_{eq}. It is quite straightforward to demonstrate that changes in cloud radiative effects are not the same as cloud feedbacks. To appreciate this point note that, in the limit of small changes, the strength of the short-wave cloud feedback can be derived formally from (5) and (6) as (e.g., Soden et al. 2008; Schwartz 2011)

$$\lambda_c^{(Q)} = \frac{\partial Q}{\partial A_c}\frac{\partial A_c}{\partial T_s} + \frac{\partial Q}{\partial \alpha_c}\frac{\partial \alpha_c}{\partial T_s} \tag{9}$$

$$= -Q_\star\left[(\alpha_c - \alpha_0)\frac{\delta A_c}{\delta T_s} + A_c\frac{\delta \alpha_c}{\delta T_s}\right]. \tag{10}$$

A change in the CRE from surface temperature induced changes in cloud amount is, however, not simply equal to the cloud feedback times the change in surface temperature; but rather includes an additional term that accounts for the change in the albedo of the cloud-free scenes with the change in surface temperature:

$$\delta F_c^{(Q)} = \lambda_c^{(Q)}\delta T_s + Q_\star A_c\frac{\partial \alpha_0}{\partial T_s}\delta T_s. \tag{11}$$

The relationship between the cloud feedback and the change in the CRE follows similarly for the long-wave part of the spectrum. Equation (11) explicitly includes the dependence of the CRE on factors other than cloudiness, in the present example also the cloud-free sky albedo and long-wave emission from cloud-free scenes. Thus if, for example, in a changing climate, the surface albedo or cloud-free-sky aerosol changed, the CRE, as conventionally defined, would change for reasons that have nothing to do with changes in cloud properties. If the cloud properties remained fixed, the actual cloud feedback would be zero. This situation is described by Soden et al (2008) as a masking effect. This example is readily extended to the long-wave, where the change in the cloud-free-sky emission, E_0 can cause a change in the CRE. Because δE_0 and δa_0 are in principle observable, the net effect of changing clouds on the response of the system to an external perturbation can, in principle, be determined from measurements. Soden et al. (2008), illustrate how radiative kernels can be used to diagnose cloud feedbacks from changes in CRE, although the kernel methods conflate cloud feedbacks with cloud mediated CO_2 indirect forcing (adjustment), as discussed below. Irrespective of how a feedback is calculated, care must be taken in its interpretation, as the definition of a feedback is dependent on how one defines their system. Choosing relative humidity instead of absolute humidity as a thermodynamic coordinate, or potential temperature instead of height as a vertical coordinate, can give very different pictures of the feedbacks in the system for reasons that have nothing to do with the processes taking place in the atmosphere.

A further subtlety of the cloud-forcing concept is that the CRE, as defined above, depends not only on the properties of the cloud-free scenes, but also on other radiation-influencing constituents of the climate system that might be correlated with the presence of clouds. Hence, it is really a cloudy scene radiative effect, i.e., the difference between the total radiative flux and that which would obtain if cloud-free conditions always prevailed. The emphasis on cloudy scenes rather than clouds is because the former admits the possibility that the atmosphere in cloudy scenes is systematically different from the atmosphere in cloud-free scenes. For instance, if the atmosphere in cloudy scenes tends to be more humid than in cloud-free scenes then this difference in humidity will, through the definition of cloud radiative effect Eq. (8), be interpreted as an effect of clouds. This distinction has important consequences for how cloud radiative effects are calculated in models versus in measurements. In models, the radiative transfer calculation is usually performed twice, the second time with the clouds removed from the input. The first call to the radiation defines the all-sky radiative flux, the second call defines the clear-sky radiative flux and their difference is called the cloud radiative effect. In observations one cannot remove clouds[5], so the clear-sky radiative flux is estimated based on scenes where no clouds are identified in the first place, rather than based on all scenes but with removal only of the clouds from those scenes in which they occur. If the atmosphere around the clouds is different (for instance more humid) in the cloudy scenes, as one might expect, this leads to differences between the two estimates of CRE. Such effects were shown by Sohn et al. (2010) to lead to systematic discrepancies between the observed long-wave CRE and that calculated by models. Such discrepancies can, however, be readily overcome by calculating the CRE in models in the same way as is done in observations, by compositing over cloud-free scenes when determining the clear-sky flux.

In Sect. 2.2, it was pointed out that rapid adjustments by clouds can contribute to the adjusted radiative forcing associated with an external perturbation to the atmospheric composition. It proves instructive to illustrate this idea using the conceptual framework developed above. To do so, consider the simplified case in which only the long-wave emission of the atmosphere depends on the greenhouse gas concentration, so that $E = E(\chi, T_s)$ where χ denotes the greenhouse gas concentration. In this case, the net radiative forcing that results from a perturbation in a greenhouse gas concentration, $\delta\chi$, can be expressed as follows

$$F_\chi = F_\chi^{(E)} = \frac{\partial E}{\partial \chi}\delta\chi. \tag{12}$$

To the extent that the perturbation in greenhouse gas concentrations also influences cloud amount (for reasons that will become clear shortly), there is a resulting further contribution to the radiative forcing of the greenhouse gas perturbation, so that the effective radiative forcing due to the perturbation in concentration becomes

$$F_\chi = \left[\left(\frac{\partial Q}{\partial A_c} + \frac{\partial E}{\partial A_c}\right)\frac{\partial A_c}{\partial \chi} + \frac{\partial E}{\partial \chi}\right]\delta\chi. \tag{13}$$

The first two terms on the right-hand side introduce the idea of an indirect forcing of greenhouse gases that is mediated by clouds. The use of the word "indirect" signifies that

[5] It might appear that there are exceptions. Intrieri et al. (2002) and Mauritsen et al (2011) calculate the clear-sky radiative fluxes using measurements of the atmosphere in cloudy conditions, but do not include the clouds in the radiative transfer calculations. However, this is not an actual measurement of radiative effects, but rather a calculation of radiative effects given measured atmospheric properties.

the change in the TOA irradiance is not a direct consequence of the greenhouse gas concentration on the clear-sky emissivity, but rather results from the sensitivity of cloud amount to the long-wave emissivity of the atmosphere, i.e., a cloud adjustment. Given our description of the system through Eqs. (5) and (6), an indirect CO_2 forcing follows as soon as one admits that cloudiness may depend on the concentration of atmospheric CO_2 (Forster and Gregory 2006; Gregory and Webb 2008; Andrews et al. 2009). The idea that clouds may be sensitive to the concentration of greenhouse gases and thus may rapidly adjust in ways that change the initial forcing, actually predates the idea that clouds may depend on surface temperature and hence act as a feedback (Plass 1956). This is not just a mathematical abstraction. Because the radiative cooling at the top of stratiform cloud layers, which is important to their sustenance, is sensitive to the downwelling long-wave radiative flux, which in turn depends on the long-wave opacity of the overlying atmosphere (Caldwell and Bretherton 2009; Stevens et al. 2003) it is likely that clouds respond rapidly (adjust) to perturbations in atmospheric CO_2, thereby providing at least one justification for the *ansatz* that $A_c = A_c(\chi)$.

3.2 Cloud Amount

A determination of whether changes in cloudiness are causing changes in Earth's energy flows depends on the ability to identify clouds, and cloud changes, unambiguously. This proves to be a challenge, which stems in part from the very nature of clouds. Cloud are, in a word, nebulous. A cloud, like an aerosol more generally, is a dispersion of particulate matter in an often turbulent flow. But, additionally and in contrast to clear-air aerosols, clouds are inherently ephemeral, as they contain a substantial amount of condensed (liquid or solid) water, the presence and amount of which are maintained by local supersaturation, and which can quickly dissipate by evaporation, converting a cloudy scene to a cloud-free scene, making it difficult to determine the boundaries or even the presence of a cloud. This situation leads inevitably to a certain arbitrariness in whether a cloud is present at a given location. And this arbitrariness can lead to large differences in quantities, such as cloud fraction, which are central to interpreting Earth's energy flows (Stephens 1988).

It might be argued that a clear basis for defining a cloud is provided by Köhler theory, namely as the set of particles that exist in an environment that is supersaturated relative to the equilibrium supersaturation over the particle surface and for which the equilibrium state is unstable. However, such a definition is not very useful in practice as the theory applies at best only to liquid clouds, and deliquesced aerosol in humid environments, or evaporating hydrometers in subsaturated environments are often optically indistinguishable from clouds defined on the basis of Köhler theory, the transition region extending over distances up to kilometers or more (Koren et al. 2007; Tackett and Di Girolamo 2009; Twohy et al. 2009; Bar-Or et al. 2010).

As a consequence of such concerns, the presence of clouds is often determined based on their radiative properties. This approach is advantageous from a practical perspective, as satellite borne instruments can be built to be sensitive to such properties, and thus afford the opportunity for reproducible measurements with global coverage and high spatial and temporal resolution. But it also introduces the possibility that quantities such as cloud fraction will be determined, in part, by the characteristics of the instruments with which they are measured. The seriousness of this issue is illustrated with the help of Figs. 3, 4 and 5, which show that co-located measurements by multiple approaches yield results that can differ profoundly at a single time, in monthly averages, and in the seasonal pattern. These issues can be ameliorated by focusing on anomalies over longer time-periods, which as

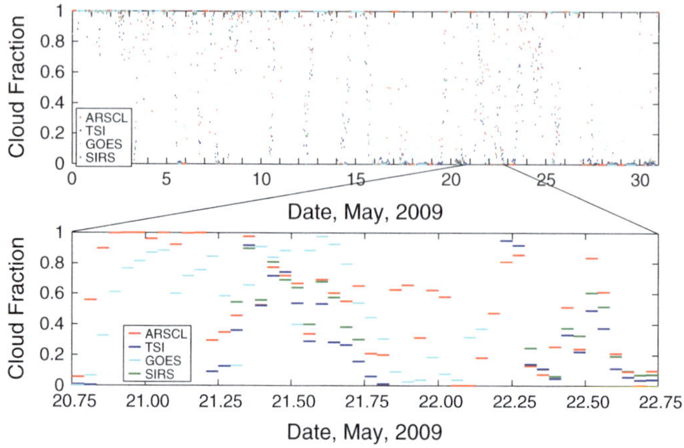

Fig. 3 Cloud Fraction in north central Oklahoma determined by several techniques. Both *panels* show 3 h mean cloud fraction as a function of local standard time; *upper panel* gives 1-month overview; *lower panel* expands a 2 days period. ARSCL (active remote aensing of clouds) product of the Department of Energy Atmospheric Radiation Measurement (ARM) Program (Clothiaux et al. 2000), is time-average based on vertically pointing lidars and millimeter cloud radars; Solar infrared radiation system (SIRS) product of the ARM program (Long et al. 2006) is time-average based on downwelling short-wave irradiance within nominal 160° field of view. TSI (Total Sky Imager) is based on fraction of cloudy pixels within hemispheric field of view geostationary operational environmental satellite (GOES) is based on average of all pixels (4 km pixel size) within 20 km of the surface measurement site (Genkova et al. 2004). Figure provided by W. Wu (Brookhaven National Laboratory 2011)

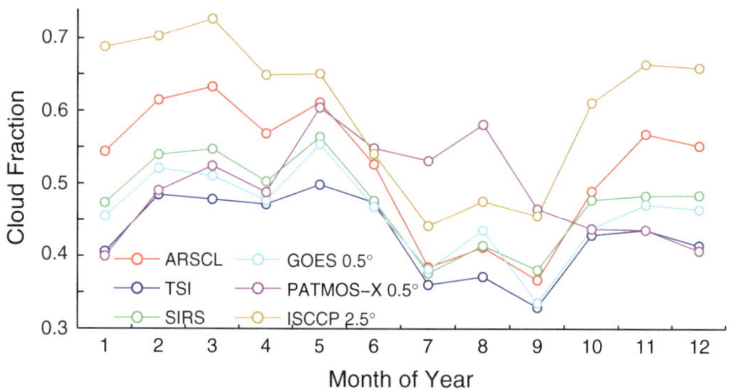

Fig. 4 Monthly mean cloud fraction in north central Oklahoma determined by several techniques, as in Fig. 3. ISSCP reference imaging radiometers on operational weather satellites (ISCCP); PATMOS-X Pathfinder Atmospheres Extended Product (Heidinger and Pavolonis 2009). Figure provided by W. Wu (Brookhaven National Laboratory 2011)

show in Fig. 5 are less instrument dependent. When a comparison to models is of interest, instrument simulators can also help address these issues. Even so, differences among measurement techniques make the unambiguous determination of long-term trends more difficult, and potentially more sensitive to inhomogeneities in the observational record.

Natural variability in cloudiness compounds these issues. CERES SSF1deg short-wave CRE data analyzed and presented in Fig. 6 show that the CERES instruments on the different satellites are in relatively good agreement in their quantification of monthly

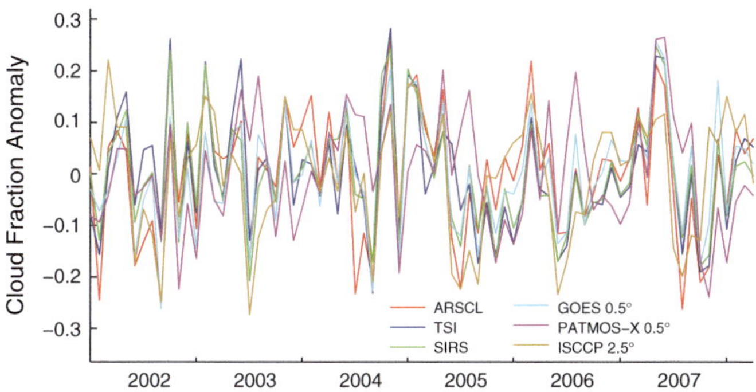

Fig. 5 Anomalies in monthly mean cloud fraction in north central Oklahoma determined by several techniques, as in Fig. 3. *Trend lines* (not shown) are all zero within approximately one sigma, as evaluated assuming the absence of autocorrelation. Figure provided by W. Wu (Brookhaven National Laboratory 2011)

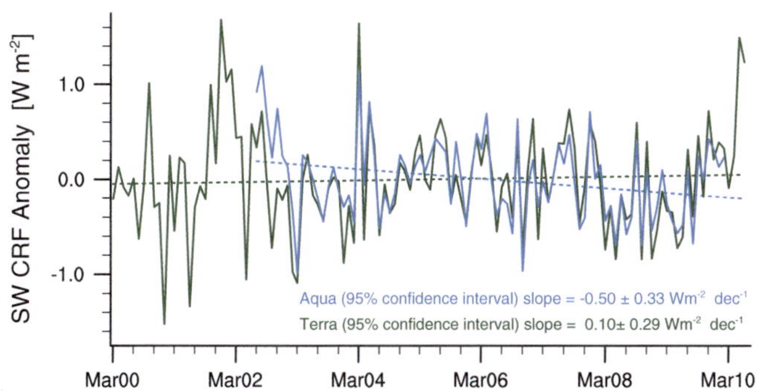

Fig. 6 Anomalies in the monthly and globally averaged top-of-atmosphere short-wave cloud radiative effect (CRE) from the CERES SSF1deg product. Cloud short-wave forcing, difference between all-sky and cloud-free downwelling irradiance at the TOA, e.g., Eq. (7), is negative. Anomalies are calculated by subtracting the CERES monthly mean cloud-forcing (average over the data set) from the individual monthly forcing. Negative slope, i.e., decrease in cloud-forcing with time, corresponds to increase in cloud reflectivity climatology from the monthly fluxes. Aqua, pm; Terra, am

anomalies during their period of overlap. Given that these estimates are based on similar instruments whose measurements are processed in a similar way, albeit for different samples of the planetary cloudiness, it is not surprising that the correlation between the two time-series is markedly stronger than what one deduces with different instruments, even if they sample the same cloud field, as in Fig. 5. The agreement between the CERES measurements aboard the TERRA and AQUA satellites suggests that natural variability is the main cause of month-to-month fluctuations, which can be as large 1–2 W m^{-2}. These fluctuations are responsible for the stated uncertainty of about 0.3 W m^{-2} dec^{-1} in the 95 % significance ranges in monthly global anomalies of Q_c, an uncertainty range which is just large enough to explain a nearly 0.6 W m^{-2} difference in the decadal trends estimated from the two instruments.

Such a large uncertainty, associated with natural variability alone, suggests that even very high quality, dedicated measurements, such as those provided by CERES will not be able to document potentially substantial changes in global cloud radiative effects on timescales shorter than half a century. From the perspective of climate change, a change in the irradiance associated with cloud-covered scenes that is comparable to a given radiative forcing over the time-period of interest can be considered substantial. Over the period 1960–2005, CO_2 increased at an average rate of 1.4 ppm yr^{-1} (Forster et al. 2007). A 3.7 W m^{-2} forcing associated with a doubling of CO_2 (Forster et al. 2007) would result in a trend in the energy budget of about 0.25 W m^{-2} dec^{-1}, in the absence of any feedbacks (and assuming a linear increase in the forcing with time. In this context, a change in cloud properties in response to this forcing that resulted in a further change in the radiation budget $|\dot{Q}| > 0.05$ W m^{-2} dec^{-1} would constitute an appreciable change. Based on this we argue that observationally constraining the response of the climate system to such a perturbation, and to some meaningful degree, requires an ability to detect changes in the radiation budget $|\dot{Q}|$ of 0.05 W m^{-2} dec^{-1}. The analysis of Loeb et al. (2007) suggests detection of such a trend with 90 % confidence would require 50 years of data, and this analysis is optimistic as it assumes a perfect instrument and insignificant decadal variability. A similar conclusion has been reached by Dessler (2010). Moreover, as the number of years of data required to establish a trend is proportional to $|\dot{Q}|^{2/3}$, where $|\dot{Q}|$ is the magnitude of the trend (Weatherhead et al. 2000), establishing a trend even twice as large, which becomes interesting as the rate of forcing increases super linearly with time, would still require 30 years—a timeframe which, if anything is a lower bound as it does not factor in the effects of instrumental limitations, or issues related to the ambiguity of cloud identification, as discussed above.

4 Aerosol Radiative Effects and Forcing

4.1 Aerosol Radiative Effects

The radiative properties of cloud-free skies depend on, in addition to surface reflectivity, the gaseous composition of the atmosphere and on the spatial and temporal distribution and chemical and microphysical properties of atmospheric aerosols. Atmospheric aerosol particles also serve as cloud condensation nuclei and ice-forming nuclei that are the seed particles for the formation of cloud liquid drops and ice crystals and, thus, changes in the aerosol are expected to change cloud properties, thereby contributing to an indirect aerosol radiative forcing (Twomey 1974; Charlson et al. 1992; Albrecht 1989; Stevens and Feingold 2009). Hence, atmospheric aerosols influence, to varying degrees, the flows of energy in almost all of the constituent parts of Fig. 1. Whereas aerosol effects on the short-wave radiative properties of the atmosphere are usually emphasized (Charlson et al. 1992), aerosol effects on the emissivity of the atmosphere (Markowicz et al. 2003) or on the downwelling long-wave irradiance of thin clouds (Lubin and Vogelmann 2006) may also be important. Because human activity has long been recognized as contributing to increasing aerosol burdens, aerosols also contribute to the net radiative forcing that appears in Eq. (3). The radiative forcing of aerosol is generally appreciated to be negative. As a consequence, taking this forcing into account in estimating climate sensitivity Eq. (3) results in a greater sensitivity than would be inferred from consideration only of the greenhouse gas forcing (Gregory et al. 2002; Schwartz et al. 2010). This situation necessitates accurate knowledge of the radiative forcing associated with changes in the aerosol burden. Not only does this require an understanding of how the aerosol burden has changed as a result of human activity, but also an understanding of how Earth's climate system adjusts to such changing aerosol burdens, thereby determining the effective radiative forcing. However, in both instances, this understanding is poor, and hence the magnitude of the aerosol radiative forcing is highly uncertain.

The complexity of aerosols is one reason it has proven difficult to understand how human activity has contributed to a changing aerosol burden. This complexity is expressed in terms of the heterogeneous chemical and microphysical properties and also the highly variable spatial and temporal distribution of the aerosol. The heterogeneous composition, which contrasts with the well defined molecular properties of the greenhouse gases, is a consequence of the numerous contributions to atmospheric aerosols: Primary emissions from natural and anthropogenic sources and gas-to-particle conversion resulting from atmospheric reactions of precursor gases, importantly sulfur and nitrogen oxides from combustion sources, ammonia from agriculture and animal husbandry, organics from anthropogenic sources and vegetation, and numerous other sources. Gas-to-particle conversion processes lead both to new particle formation and to growth of pre-existing particles. The resulting aerosols undergo further evolution in the atmosphere through condensation and coagulation and in cloud processing. Ultimately, the aerosol particles are removed from the atmosphere, importantly by precipitation. The optical and cloud-nucleating properties of aerosols, and thus their influences on climate and climate change, are strongly dependent on the size and chemical composition of the particles comprising the aerosol. For example, growth of particles with increasing relative humidity, which greatly increases their ability to scatter visible light, is dependent on composition.

The complexity in the spatio-temporal distribution of aerosols is hinted at even upon inspection of their long-term average global distribution undifferentiated by aerosol type as seen in the large spatial variability of aerosol optical depth (AOD) at 550 nm averaged over the available 11 years of multi-angle imaging spectroRadiometer (MISR) data, Fig. 7. Major contributions arise from windblown dust (e.g., Northern Africa, Arabian peninsula, western China) and biomass burning (e.g., central Africa, Amazonia, Indonesia). Substantial contributions from human activity, mainly combustion related, can be inferred over

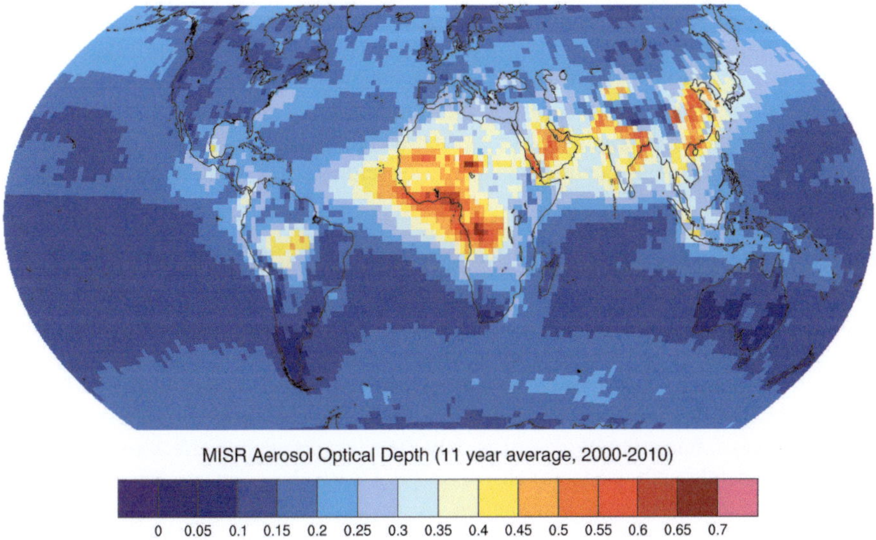

Fig. 7 Annually averaged aerosol optical depth climatology inferred from eleven (2000–2010) years of multi-angle imaging spectro radiometer (MISR) measurements

southeast and eastern Asia and extending into the western North Pacific. The highly industrialized regions of Europe and North America (extending to the North Atlantic) also exhibit noticeable enhancement of AOD relative to pristine continental regions and major portions of the Southern Hemisphere Ocean. The spatial heterogeneity of the distribution of these aerosols is a consequence of the heterogeneous distribution of sources together with the short atmospheric residence times of these aerosols, about a week, together with the intermittent removal by precipitation. Because aerosol sources have pronounced seasonality, and because sink and transport processes of all aerosols are heavily dependent on variable meteorological conditions, the distribution of aerosols shown in Fig. 7, being a long-term average, considerably understates the complexity of the spatial distribution of atmospheric aerosols. This points out that the complexity of aerosols is manifested not only by their varied chemical and microphysical properties, but also by their heterogeneous spatial distribution; for example an aerosol particle above a bright surface has a different radiative effect compared to even the same particle over a darker surface. All of these considerations make it much more difficult to quantify aerosol forcing than is the case with the incremental greenhouse gases.

The short residence time of aerosol particles in the troposphere not only complicates characterization of their radiative influences, but also has implications on climate change that would result from future changes in emissions, especially as most of the incremental aerosol arises from emissions associated with fossil fuel combustion. If at some point in the future, emissions of CO_2 from combustion are substantially reduced and if this were accompanied by reduction of associated emissions of sulfur and nitrogen oxides, major precursors of light-scattering tropospheric aerosols, the result would likely be for temperatures to initially increase because of the reduction of aerosol forcing. An initially abrupt increase in temperature following an abrupt cessation of aerosol forcing has been shown in climate model studies (e.g., Brasseur and Roeckner 2005; Matthews and Caldeira 2007).

As a consequence of all these considerations, understanding of energy flows in the Earth system and changes in these flows over the industrial era is challenged by poor understanding of the effect of aerosols on cloud-free skies where one can speak of the direct aerosol radiative forcing. This challenge is even greater with respect to aerosol effects on clouds (Lohmann and Feichter 2005; Stevens and Feingold 2009) where it has become common to speak of the indirect aerosol radiative forcing, which results from modification of the radiative influences of clouds that result from changes in the aerosol environment in which they form. The discrimination of aerosol forcing into direct and indirect components structures thinking about aerosol influences on climate and is used to structure the discussion below.

4.2 Aerosol Direct Forcing

Analogous to the way in which a CRE is calculated, the direct aerosol radiative effect, DARE, is the change in the irradiance at the top of the atmosphere that results from the total aerosol present. This can be contrasted with the direct aerosol radiative forcing, DARF, which is the quantity pertinent to forcing of climate change. As is the case with greenhouse gas forcing, DARF describes only the aerosol radiative effect associated with the secular (mainly anthropogenic) change in the DARE due to incremental aerosols, as that is the externally forced contribution. A negative forcing denotes a decrease in absorbed short-wave irradiance; opposite in sign to the positive greenhouse gas forcing, and, within the framework of the forcing-response paradigm, would offset some fraction of the greenhouse gas forcing.

The DARF depends on the difference between the present day and pre-industrial aerosol loading. This difference must account for the co-variability of the aerosol with surface and circulation features, as changes in the aerosol loading over bright surfaces, or above clouds, will have a different impact, and may even have a different sign, than changes over dark surfaces, or in clear air. And because the radiative effect of aerosols depends on their intensive properties, this difference must include a characterization of aerosol properties and how they have changed between the pre-industrial versus the present day. Because the aerosol loading is heterogeneous, the radiative effect of a change in the aerosol will also be heterogeneous and thus likely generate responses in the circulation as the atmosphere adjusts to different regional heating patterns. These changes can be thought of as adjustments, which further complicate an estimate of the adjusted forcing associated with the direct effect of the anthropogenic aerosol. Finally, because there are few reliable proxies for the pre-industrial aerosol loading, this must be determined using models, although attributing accumulation mode aerosols, aerosols which are formed by gas-to particle conversion, and which operationally have diameter less than about 1 μm, to the anthropogenic perturbation, provides some observational constraints. Still this approach has its limitations; for example, it would misattribute dust aerosol from tilled soils (which have particle diameters larger than 1 μm), and likewise would misattribute background sulfate and smoke, or natural organic haze such as gave rise to the name of the Great Smoky Mountains in the eastern United States. For all of these reasons, the direct aerosol radiative forcing is a much more challenging quantity to estimate than greenhouse gas forcing.

A starting point for quantifying the DARF is an estimate of the magnitude of the present day aerosol loading. This quantity is commonly expressed by the aerosol optical depth, AOD, the vertical integral of the aerosol extinction coefficient, typically given in the mid-visible, 500 or 550 nm. AOD can, with diligence, be measured by Sun photometry during the daytime and in the absence of clouds in the path to the Sun to an accuracy of 0.01 (e.g., Holben et al. 2001; Kim et al. 2008; Michalsky et al. 2010); the aerosol contribution to path extinction is determined by accounting for the extinction due to Rayleigh scattering and to atmospheric gases (importantly ozone). However, such measurements provide no information about the vertical distribution, are limited in their spatial coverage and to cloud-free skies during the day, and are especially lacking over the oceans (Smirnov et al. 2009). Hence, the approach taken to characterizing the distribution of aerosols globally has been to determine AOD by satellite, as for example illustrated by Fig. 7. It should be stressed, however, that the satellite measurement of AOD is rather indirect and also provides little information about the vertical distribution. The AOD is inferred from the enhancement of the path radiance over that which would be obtained in the absence of the aerosol. Rigorous cloud screening, which recalls the discussion (Sect. 3.2) of what is a cloud, is also required. By consistently accounting for contributions to path radiance from the surface and from Rayleigh scattering the contribution to path radiance attributable to light scattering by the aerosol can be determined. Converting from radiance to optical depth rests on assumptions about the scattering phase function (angular distribution of light scattering) and the fraction of the extinction by the aerosol that is due to scattering versus absorption. Typically look-up tables are used, with identification of aerosol type informed by the wavelength dependence of the aerosol light scattering and climatology of aerosol types. The MISR (Multi-angle Imaging SpectroRadiometer) instrument, upon which the measurements illustrated in Fig. 7 are based, takes advantage of measurements at multiple scattering angles to better constrain the phase function and in turn the AOD (Martonchik and Kahn 2009; Kahn 2011). This is particularly helpful over bright and complex surfaces, because the inferred AOD is highly sensitive to errors in estimates of surface-leaving radiance. Measurements at multiple angles also constrains inferences of aerosol type, and hence absorption, which is another large source of error in inferring AOD. Further detail on determination of AOD by satellite and associated uncertainties is provided by Kahn et al. (2010) and Kahn (2011).

Importantly, in determining the DARE, and how it has changed, it is necessary to account for the effects of clouds. The effect of cloud contamination of pixels used to determine aerosol optical depth as already been noted; such contamination would result in a gross overestimation of aerosol optical depth, and it is thus necessary to apply stringent cloud screening (Mishchenko et al. 1999). However, being overly stringent runs the risk that high-humidity regions will be misclassified as being cloudy, with the resultant effect that the swelling of particles at high relative humidity and the attendant increase in light-scattering cross-section, optical depth, and forcing is excluded from the measurements. These effects can be substantial; for sulfate aerosols, the scattering cross-section increases fourfold between 90 and 97 % relative humidity (Nemesure et al. 1995). Surface-based measurements are also subject to a similar concern as the technique requires a direct path to the Sun. More intrinsic to the radiative forcing issue, the direct radiative effect of light scattering by aerosols is greatly diminished in the presence of clouds, which prevents solar irradiance from reaching the aerosol if clouds are above the aerosol, or by providing a bright underlying albedo, minimizing the effect of aerosol scattering if clouds are below the aerosol. Clouds beneath an absorbing aerosol greatly increase the amount of solar absorbed irradiance relative to the cloud-free situation. Because it is difficult to retrieve aerosol amounts in the presence of clouds, it is has been common to estimate the DARE on a global basis simply by multiplying the cloud-free DARE, determined by measurement or by modeling of the amount and optical properties of the aerosol, by the cloud-free sky fraction. Such an approach has been shown in model calculations to yield a value of DARF whose magnitude is erroneously large (Bellouin et al. 2008) by a factor of two.

Increases in the amount of aerosol loading that can confidently be ascribed to anthropogenic emissions have been thought to give rise to changes in global average atmospheric radiative fluxes (aerosol forcing) that are a substantial fraction of the greenhouse gas forcing over the industrial era. As part of the fourth assessment report of the IPCC the DARF was estimated to be -0.5 ± 0.4 W m^{-2} (Forster et al. 2007, 90 % confidence limits). Although considerably reduced in magnitude as compared to earlier estimates, such as the value of -1.3 W m^{-2} initially estimated for the sulfate aerosol alone (Charlson et al. 1992), a value of DARF at the high magnitude end of the range cited by IPCC would still offset a substantial fraction of greenhouse gas warming over the industrial era (about 3 W m^{-2}), and thereby imply a much larger climate sensitivity as inferred on the basis of Eq. (3) with F denoting the sum of greenhouse gas and aerosol forcing. To the extent that the DARF is large, advances in satellite remote sensing make it conceivable to look for a direct signal, by looking for signatures of aerosol trends in radiative fluxes measured by passive satellite borne instrumentation over the past decade of intensive Earth observations. Eleven years of MISR measurements (upper panel of Fig. 8) of AOD show large-scale shifts in specific regions. Large increases in AOD over northern India southeast and eastern Asia and around the Persian Gulf likely reflects the economic development of these regions. Over the region ranging from southwest of North America, across the Atlantic and into North Africa, the Mediterranean and central Europe, and over the maritime continent, AODs have decreased sharply, by as much as 0.1. A modest decrease is evident across the southern ocean, and there is an apparent increase over western Canada. The uncertainties associated with the retrievals of AOD from space (Kahn et al. 2010), and the susceptibility of decadal trends to the effects of inter-annual modes of variability, such as El Nino,

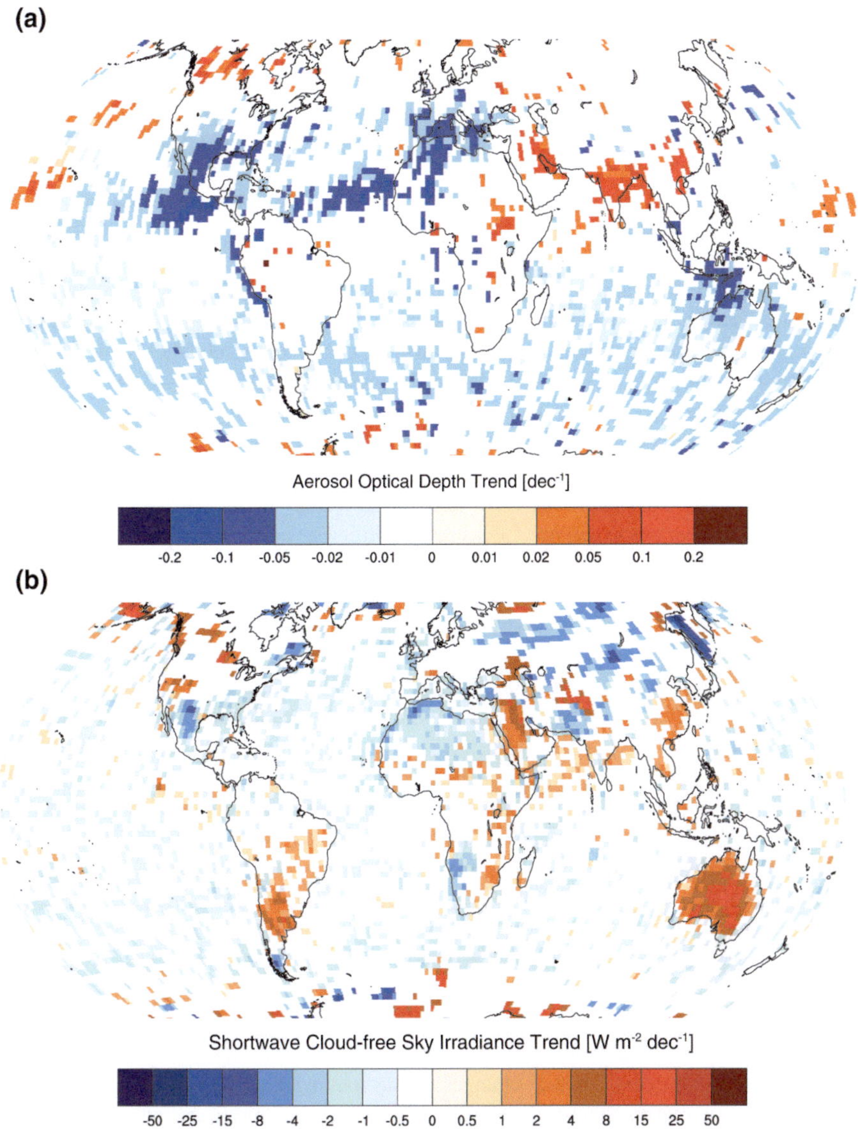

Fig. 8 Decadal trend in the annually averaged aerosol optical depth at (550 nm) inferred from eleven years (2000–2010) of multi-angle imaging spectroRadiometer (MISR) measurements (*upper*) and in cloud-free scene (TOA) radiative fluxes taken from CERES measurements (*lower*). Both the CERES and MISR instruments are aboard the Terra satellite. Only trends that are significantly different from zero at the 95 % confidence level (as estimated based on the ratio of the residual variance and mean trend assuming no auto correlation in the yearly data) are shown. An increase in reflected short-wave irradiance, as indicated by the redder colors, is indicative of a brightening of the cloud-free skies

preclude drawing confident conclusions from Fig. 8; however, the broad conclusions drawn by the figure are also supported by a more systematic analysis based on multiple platforms (Zhang and Reid 2010).

A concern over the interpretation of these measurements is that the marked changes in aerosol optical depth inferred from the MISR and multi-instrument aerosol measurements are not mirrored in trends in the outgoing short-wave irradiance in cloud-free scenes over the same time-period as measured by the highly calibrated CERES radiometer on the same platform, lower panel of Fig. 8. Although the irradiance in cloud-free circumstances shows trends in some regions where aerosol optical depth retrieved by MISR has been increasing, for instance over south Asia, particularly the middle east and southeast China, and to some extent over western Canada, the relationship between changes in AOD and changes in clear-sky short-wave radiative fluxes is not striking. The largest changes (irrespective of sign) in cloud-free-sky radiation appear over land in regions where there are no discernible trends in AOD. In almost every case statistically significant trends in CERES absorbed short-wave radiative irradiance over land are also evident in changes in land-surface properties measured over effectively the same time-period: the brightening, at TOA, of Australia; the brightening in southeastern Asia; the brightening/dimming patterns over the southwestern United States and northern Mexico as well as the brightening/dimming patterns over Argentina and South America, and the dimming of the very northern tip of Africa are all associated with consistent changes in soil moisture, whereby an increase in the outgoing short-wave irradiance for scenes identified as being cloud-free corresponds to a decrease in soil moisture and evapotranspiration (Jung et al. 2010). Notwithstanding the merit of exploring why such large apparent trends in AOD over the ocean are not seen in the CERES clear-sky short-wave irradiance, this analysis provides little support for a strong DARF.

In summary, the evolving understanding of radiative forcing, one in which the process of adjustment plays an important role; the complexity of the aerosol and its co-variability with clouds and surface features; and the role of even small amounts of absorption of aerosols over bright surfaces, suggests that back-of-the-envelope estimates of the DARF can easily be misleading. Based on this reasoning, and in consideration of more detailed calculations that suggest previous work underestimates the uncertainty in aerosol forcing (Loeb and Su 2010), it would not be surprising if the sum of the various contributions to the DARF is much closer to zero than previously thought.

4.3 Aerosol Indirect Forcing (Cloud Adjustments)

It has long been appreciated that clouds adjust to changes in tropospheric aerosols and this adjustment affects their albedo and precipitation development. Based on this understanding, it has been hypothesized that changes in the loading and properties of tropospheric aerosols may indirectly affect the radiative influences of clouds (i.e., alter the CRE) by modifying cloud properties and/or amount. These changes are referred to as aerosol indirect effects, or the indirect radiative forcing resulting from anthropogenic (or secular) changes to tropospheric aerosol loading and properties. They are analogous to the CO_2 indirect forcing discussed in the context of Eq. (13) and, like the CO_2 indirect forcing, are more usefully thought of as an adjustment to a compositional change of the atmosphere. Because such adjustments convolve spatially heterogeneous changes in the aerosol with changes in cloudiness, they tend to be complex and uncertain, and are only briefly touched upon them here.

Broadly speaking cloud adjustments to aerosol perturbations can be classified into adjustments of cloud intensive properties and adjustments in cloud extensive properties. Examples of an adjustment in cloud intensive properties would be an increase in the cloud droplet number concentration (Twomey 1974) and/or cloud droplet size distribution (Liu and Daum 2002), and in turn in changes in the cloud radiative influence, that result from an adjustment to increased concentrations of cloud active aerosol. Adjustments of cloud extensive properties are often called cloud lifetime effects, such as hypothesized by Albrecht (1989) and Pincus and Baker (1994), wherein changes to cloud droplet number concentrations modify the precipitation efficiency of clouds, resulting in modifications of cloud amount (cf., Stevens and Feingold 2009; Khain 2009). Research continues to explore how clouds adjust to aerosol perturbations, with interest increasingly focusing on changes to ice-forming nuclei and hence adjustments of ice or mixed phase clouds. Because clouds interact so potently with radiation, a slight, but systematic, change in the amount or albedo of clouds globally or over large regions of the Northern Hemisphere could exert a change in the global radiation budget that is comparable in magnitude to greenhouse gas forcing. For example, in a back-of-the-envelope calculation, Charlson et al. (1992) showed that a 30 % increase in the number concentration of cloud drops in marine stratus clouds globally would, other things being equal, exert a global radiative forcing of about -1 W m^{-2}. In the intervening years, there have been abundant climate model calculations of the global magnitude of the cloud adjustment to aerosol forcing over the industrial era, with estimates ranging from near-zero to -3 W m^{-2} or more (Lohmann et al. 2010), based on differing assumptions regarding the relation between aerosols and cloud properties and various other controlling factors.

To be sure, a wealth of observational support exists for Twomey's hypothesis that the cloud drop number concentration increases, cloud drop radius decreases, and cloud albedo increases with increasing aerosol particle concentration. However, quantification of the net radiative forcing that can be attributable to the indirect effects of aerosols on clouds has proven elusive. That cloud drop radii are reduced and cloud albedo is increased by aerosols is clearly shown in ship tracks (e.g., Segrin et al. 2007), but the net effect of such changes are partially offset by changes in cloud water content. Regionally, reduction in cloud drop effective radius is associated with enhanced concentration of anthropogenic aerosol (e.g., Schwartz et al. 2002), but the expected increase in cloud albedo is often absent. Such lack of enhancement of cloud albedo is also likely due to a decrease in cloud liquid water path with increasing aerosol concentration, at variance with Twomey's *ansatz* of other things, especially cloud water content, remaining equal. A global survey using satellite observations showed roughly equal likelihood of negative, near-zero, or positive correlation of column liquid water and column drop concentrations in liquid water clouds (Han et al. 2002). Although several studies show strong correlations between cloud amount and aerosol optical depth (e.g., Nakajima et al. 2001; Koren et al. 2010), the interpretation of such correlations is difficult, as a variety of processes (both physical and retrieval artifacts) can be expected to produce such correlations, quite independently of whether or not the aerosol is interacting with the cloud (e.g., Loeb and Schuster 2008). For example, both aerosol optical depth and cloudiness increase with humidity and, thus, it is not surprising that modeling studies might overestimate the tendency of clouds to adjust to aerosol perturbations. By regressing the logarithm of the retrieved aerosol optical depth against the logarithm of the retrieved cloud droplet concentrations over a number of geographic regions, Quaas et al. (2005) found statistically significant slopes that range from 0.1 to 0.3 depending on location, with values over the ocean three times greater than those over land and with a global mean value of just under 0.2 (see also Quaas et al. 2009). Based on this analysis, Quaas et al. (2009) estimate quite a low aerosol indirect forcing, -0.2 ± 0.1 W m^{-2}. As noted by Quaas et al. (2005), this uncertainty is parametric, and contributions to the uncertainty from structural effects can be expected to be substantial; hence, observational estimates cannot, on their own, establish with confidence even the sign of the effect hypothesized by Twomey, despite arguments based on simple physical considerations that it is negative. The structural uncertainty that frustrates attempts to quantify the Twomey, or Twomey-like, effects also makes it more difficult to test cloud lifetime hypotheses and all the more to quantify the resultant forcing. Moreover, to the extent that precipitation processes become involved the difficulties are compounded, in no small part because of the sensitivity of aerosol amount to wet scavenging by precipitation.

In summary, although there is little doubt of the importance of aerosol-cloud interactions in influencing the amount of atmospheric aerosol, as well as cloud properties, the variety of ways in which clouds adjust to aerosol perturbations (Stevens and Feingold. 2009), many of which are not possible to account for given the relatively crude description of cloud processes in climate models, lends weight to the argument that, after a full accounting, the radiative forcing attributable to cloud adjustments to aerosol perturbations is likely to be small, at least on a global scale.

5 Modeling Earth's Energy Flows

The preceding discussion demonstrates that, irrespective of the sophistication of observing systems, models are indispensable for estimates of fundamental properties of the climate system. The forcing-feedback-response framework is useful only in so far as compositional perturbations can be associated with a radiative forcing. And if it is to have any general meaning, this radiative forcing must incorporate the perturbation-dependent fast response of the system, the adjustment. Complications posed by adjustment have long been appreciated for inhomogeneous perturbations, such as those due to aerosols, and in special cases for homogeneous forcings such as those due to greenhouse gases, i.e., stratospheric adjustment. Adjustments that involve changes in cloudiness, whether due to aerosol or greenhouse gas forcing, can be large, are often model dependent, and add significant uncertainty to ultimate estimates of the response of the climate system to compositional changes.

Uncertainties in the modeling arise because models are imperfect. And, given the singular nature of the task to which climate models are to be applied, these imperfections are difficult to quantify. Independent realizations of perturbations to the energy flows in Earth-like planets, which could be used to evaluate the reliability of climate models, do not exist. So the empiricism through which the adequacy of models could be assessed does not exist, and critical tests are invariably indirect. As an example, Hall and Qu (2006) showed that in models the surface albedo feedback, which is not observable, correlates with the relationship between seasonal variations of surface temperature and surface albedo, which is observable. The implication is that models which capture the observed relationship between seasonal changes in surface temperature and surface albedo more reliably represent the surface albedo feedback. Of course, the relationship that models show between their seasonal cycle and their response to a secular perturbation might simply be an artifact of how the models are constructed, and indeed there is evidence that relationships between quantities one desires from a model, and quantities that one can measure, may say more about the models from which such relationships are derived than they do about the physical system (e.g., Klocke et al. 2011). Nonetheless, the Hall and Qu (2006) example shows how focusing on perturbations to Earth's energy flows might lead to the development of a critical framework for assessing the reliability of models. This idea hinges on advancements in the ability to measure Earth's energy flows, as developed further below.

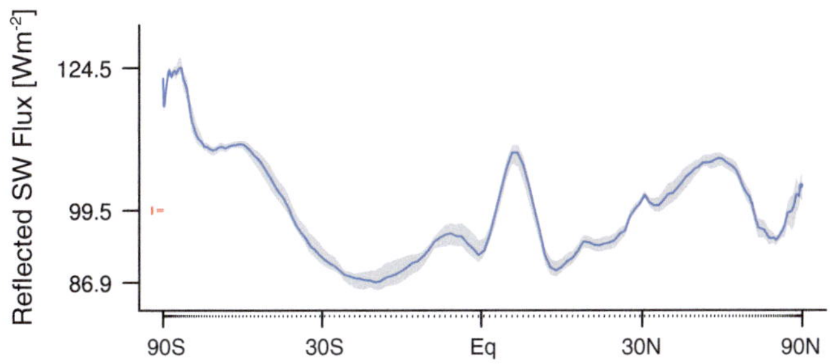

Fig. 9 Zonally averaged reflected short-wave radiation from ten years of CERES EBAF (energy balanced and filled) data. The figure shows the upward-directed short-wave irradiance ($Q^{\uparrow}(\varphi)$, where φ denotes latitude) plotted versus the sine of latitude. The year-to-year variability is indicated by the gray shading (which spans the range of yearly values). The *blue line* shows the 10-year mean. The *red vertical bar* near the *vertical axis* denotes the range in the global and annual means. The *two small horizontal bars* near the vertical axis show the 10-year means averaged over the northern and southern hemisphere separately

One of the key energy flows illustrated by Fig. 1, is that of short-wave radiation, which powers the climate system. The near constancy of the total solar irradiance is thought to be a prerequisite for the development of life, but this would come to nought were it not for a commensurate constancy in the planetary albedo. The constancy of the planetary albedo is evident not only on globally and annually averaged scales, but also within latitude zones. This point is illustrated by the latitudinal dependence of zonal averages from a decade of CERES measurements of the reflected short-wave radiation at the TOA, Fig. 9. Also shown are the range of annual averages of the reflected short-wave radiation, evaluated as

$$\langle Q_n^{\uparrow} \rangle = \frac{1}{2} \int Q_n^{\uparrow} \cos(\varphi) \mathrm{d}\varphi \qquad (14)$$

where $Q_n^{\uparrow}$ denotes the reflected short-wave radiation as a function of latitude φ and year n; and the values averaged over the northern and southern hemisphere separately. The surprising feature of Earth's climate system that is revealed in these measurements is the small inter-annual variability (the range in the yearly averages is 1.16 W m^{-2} and the standard deviation is 0.36 W m^{-2}), despite the zonal average spanning more than 40 W m^{-2}, with a root-mean squared variability (weighted by area) of nearly 9 W m^{-2}. Also the difference between the two hemispherically averaged values is very small, only 0.35 W m^{-2}.

Because the globally and annually averaged albedo can be, and is, readily tuned by adjusting global parameters in models, the fidelity with which it is represented is not a critical test of models. However, the ability of models to represent natural fluctuations in the energy flows about the parameter constrained global values begins to provide a much more critical test. The simplest example of such a fluctuation is that embodied by the climatology of zonally averaged anomalies, Fig. 10. These zonal patterns help to regulate the meridional heat transport and are not directly specified through the adjustment of global parameters. Broadly speaking, climate models, here represented by a subjective selection of five climate models, are skillful in representing zonally averaged anomalies; correlations between the observed and simulated latitudinal anomaly range between 0.60 and 0.94 for the models shown. However, the departures from the observations, which are due mainly to treatment of clouds in the models, are substantial; particularly when viewed in the context of radiative forcing over the industrial era. The root-mean square error in the zonal

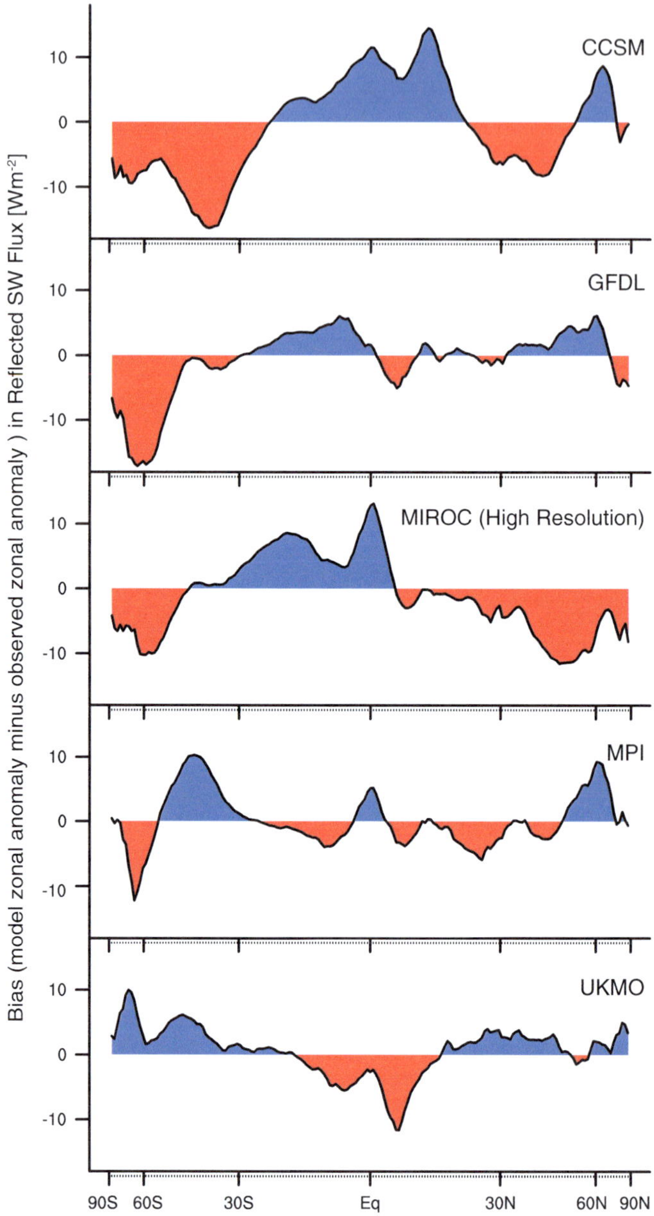

Fig. 10 Biases in zonal anomalies in annually and zonally averaged reflected short-wave radiation. Here the biases are the differences between modeled and observed estimates, so that a negative (*red*) anomaly corresponds to too much absorption in the model. The model estimates are taken from five CMIP3 models, each processed over the last 10 years of a slab-control simulation, the observations are taken from ten years of CERES data

residuals ranges from 4 to 8 W m^{-2}, comparable to the departure of the zonal mean from the global mean and considerably greater than a short-wave cloud feedback that would be important in the context of understanding climate feedbacks, 1–2 W m^{-2}. The ways in which the models are wrong also varies, although some patterns emerge. In a comparison of zonal monthly mean albedo calculated with twenty climate models and ERBE observations over 1985–1990 Bender et al. (2006) found substantial positive and negative departures, not infrequently as great as 0.1, that were reproducible from year-to-year in a given model but differed substantially from model to model in space and time. Trenberth and Fasullo (2010a) show that most models reflect too little solar radiation in the southern storm tracks and misrepresent in one fashion or another the structure of the tropical convergence zones (cf., Lin. 2007). Most models reasonably represent the poleward increase of reflected short-wave radiation in the mid-latitudes of the northern hemisphere, presumably because part of this is carried by the influence of specified surface features such as the Saharan desert and Tibetan Plateau. The models also tend to accurately represent the remarkable constancy in the globally averaged values, and even the year-to-year variability within latitude bands, but almost all fail to properly capture the near equality of the hemispherically averaged values.

Given a globally and annually averaged flow of energy into the Earth system, models would ideally also produce a model state that is consistent with what is observed. In this respect, there is also room for improvement. The pre-industrial control climate can differ substantially among models, even if the flow of energy into the system is prescribed, or tuned to match the best estimate of the observations. This is evident in Fig. 11, which shows simulated global mean temperature over the twentieth century taken from all of the relevant simulations, 58 in total, in the CMIP3 archive. Simulated temperatures at the end of the twentieth century exhibit a range of nearly 3 °C (from 12.8 to 15.5 °C). Most models are biased cold, despite being forced with a total solar irradiance that is now thought to be too large. The multi-model mean temperature is more than 0.5 °C lower than measured, an offset that is comparable to the temperature change observed over the twentieth century. From a certain perspective, the agreement is excellent; errors in temperature of 1 K out of 288 K corresponds to an error of only 0.35 %, albeit somewhat larger (5.5 W m^{-2} or nearly 1.4 % when translated into an energy flux). However, even such a small temperature error can alter the modeled climate in ways that are as great as the climate change that has occurred over the twentieth century or are projected for the twenty-first century. Such an error would seem to have implications for model projections of climate change; so it is surprising that, despite these differences the models, individually and collectively, still represent the trend in twentieth century temperatures as accurately as they do. This surprise is tempered by a realization that the agreement in the twentieth century temperature trend may also be a reflection of the model development process and the considerable latitude that uncertainty in the aerosol forcing gives model developers in matching the observed temperature trend (Kiehl 2007).

Although tests like the ones just discussed are neither exhaustive, nor absolute, it seems reasonable to assume that the ability of a model to represent perturbations to Earth's energy flows that arise from natural forcings should be indicative of their ability to represent perturbations stemming from human sources (cf., Lucarini and Ragone 2011). Viewed comprehensively, an improvement in an ability to model these natural perturbations in Earth's energy flows, could reasonably be associated with more reliable projections of perturbations in Earth's energy flows stemming from prospective future changes in atmospheric composition. Viewed over many model generations, substantial improvements in these respects, should correlate with a reduction of model spread in the representation of perturbations to Earth's energy flows that are otherwise not observable. To test the conjecture that models which better represent natural perturbations in Earth's energy flows (for instance associated with seasons, decadal variations in solar forcings, volcanos, or different

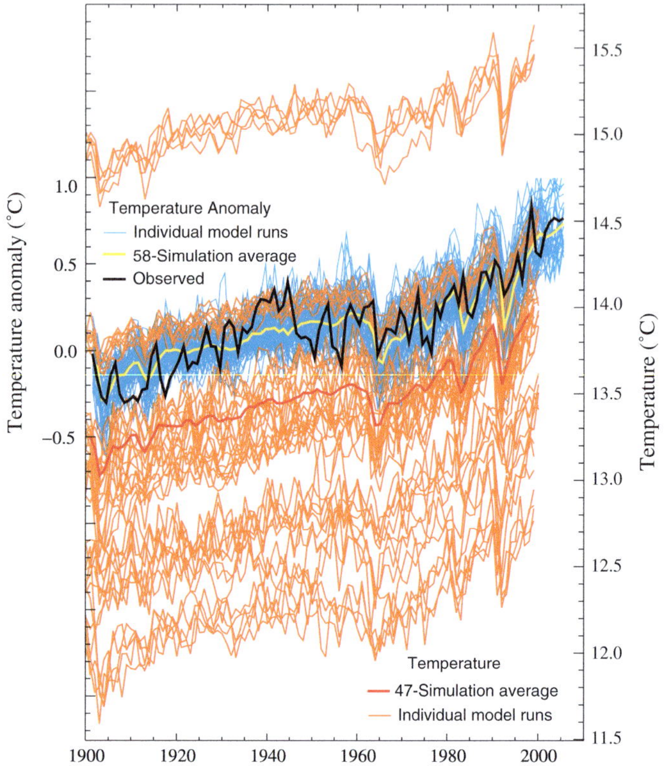

Fig. 11 Change in global mean temperature anomaly relative to 1900–1950 (*left axis*) and global mean temperature (*right axis*) over the twentieth century as evaluated with climate models that participated in the 2007 IPCC Assessment. The observed change in temperature anomaly (HadCRUT3) is plotted in *black*. The 13.75 °C offset in the vertical scales corresponds to the 1900–1950 global mean surface temperature (Jones et al. 1999, HadCRUT3). Adapted from Tredger (2009) and IPCC AR4 Figure 9.5

surface boundary conditions) are also better at representing forced perturbations requires both sustaining and advancing observations of Earth's energy flows, but also standardized experiments within evolving intercomparison protocols, such as the coupled model intercomparison project (CMIP).

6 Concluding Remarks

Tremendous advances in observing systems now make it possible to track energy flows through the Earth system with high precision, and on temporal and spatial scales that would have been difficult to imagine just a few decades ago. More accurate measurements of the total solar irradiance and the increase in ocean enthalpy are helping to constrain the balance between net solar irradiance and outgoing long-wave irradiance to less than ± 2 W m^{-2} at the top of the atmosphere. Precise measurements of Earth's energy flows, as part of NASA' s CERES program are providing new insights into how the energy flows at the top of the atmosphere vary spatially on timescales ranging from days to now more than a decade. And although, with the exception of the planetary albedo, which the satellite record has convincingly shown to be much smaller than pre-satellite estimates, the several terms in Earth's energy budget do not appear to have changed markedly as a result of the satellite record, the confidence in this budget has been greatly enhanced by the accuracy and precision of the satellite measurements. Measurements have advanced to the point where estimates of long-term trends in the top-of-atmosphere energy budget are now limited principally by natural variability, rather than instrument precision. Active remote sensing, from both surface- and space-based platforms is helping to constrain the surface energy budget. And seasonally resolved climatologies of aerosol optical depth are available both from surface- and space-based networks.

These measurement advances are not without their attendant controversies, nor do they yet provide a complete picture of Earth's energy flows. In particular, satellite-based estimates of precipitation remain difficult to reconcile with estimates of downwelling surface irradiance in the long-wave, given uncertainties in other terms in the energy budget. Also, the absolute uncertainty of global climatologies of energy related quantities remains comparable to, or larger than, perturbations expected from changes in atmospheric composition resulting from human activities, precluding measurement-based quantification of changes in flux terms resulting from the anthropogenic perturbation. Furthermore, distinguishing aerosols from clouds, or complex surfaces, and quantifying aerosol forcings remain challenging, even for a well characterized aerosol perturbation. These challenges limit confidence in estimates of total secular forcing. Nonetheless, the comprehensiveness of the measurements is making it possible to track energy flows through the Earth system on space and time scales that are unprecedented, thereby offering the possibility to pose critical tests of both models and understanding of how energy flows through the Earth system.

Climate change, as quantified, for instance, by changes to the globally averaged surface temperature, can be expected to result from compositional changes to Earth's atmosphere which perturb the flow of energy through the Earth system. The forcing-response-feedback framework, which has developed to understand such changes, posits that compositional changes in the Earth system can be associated with an effective radiative forcing, so that distinct perturbations that result in quantitatively the same effective radiative forcing will elicit the same response, as measured by globally averaged surface temperature. The magnitude of the response to a sustained forcing (per unit forcing), the climate sensitivity, is thought to be an intrinsic property of the Earth system. Notwithstanding objections that climate change can be manifest in changes that do not scale with changes in the globally averaged surface temperature, this framework relies on an ability to determine this effective radiative forcing, given some externally imposed change to the system. However, because this effective forcing depends on the fast response (or adjustments) of the system, to a given perturbation, it cannot be determined from first principles. And so climate models, which previously were thought to be indispensable to an assessment of the climate sensitivity, have become equally indispensable to the estimation of the effective radiative forcing. Consequently those poorly modeled elements of the climate system that affect the flow of energy through the system, for instance clouds and convection, have become the principal limit in quantification of not only the climate sensitivity, but also the effective forcing – especially for complex perturbations such as those associated with aerosols.

Although climate models have become indispensable, they also lack critical tests. Independent realizations of perturbations to Earth-like planets do not exist and, hence, model-based estimates of the effective radiative forcing that accompanies a compositional perturbation, or the climate sensitivity that emerges as a result of feedbacks in the Earth system, can not be independently assessed. However, by focusing on the ability of models to represent the ways in which energy is observed to flow through the Earth system, particularly for repeatable events such as seasons, ENSOs, solar cycles, and perhaps volcanic perturbations, and as a function of surface state or latitude, it might be possible to develop a measure of reliability of climate models, somewhat analogous to what is done for weather forecasting. So doing would complement the heuristic use of models that currently proves essential to development of understanding. However, this will required sustained and expanded efforts to measure the flows of energy through the Earth system and international programs dedicated to documenting the evolving skill of models to represent these flows.

Acknowledgments We thank the organizers of the workshop for giving us the opportunity to present our views. In addition through the development of this manuscript exchanges with Andrea Brose, Robert Cahalan, Robert Cess, Ralph Kahn, Seiji Kato, Norman Loeb, John Lyman, Jochem Marotzke, David Randall, and Graeme Stephens, were a great resource in the development of our ideas and analysis methods. MISR and CERES data were obtained from the NASA Langley Research Center Atmospheric Sciences Data Center. We thank Wei Wu (BNL) for providing cloud cover data for north Central Oklahoma as well as Stefan Kinne and Thorsten Mauritsen for critical and substantive comments on an initial draft version of this manuscript. Several critical reviews of the manuscript, two of which were anonymous, also resulted in major changes to the manuscript that we believe improved the presentation of our ideas, and these reviewers are thanked for their frankness, and their attention to our ideas. We acknowledge the modeling groups, the Program for Climate Model Diagnosis and Intercomparison (PCMDI) and the WCRP's Working Group on Coupled Modeling (WGCM) for their roles in making available the WCRP CMIP3 multi-model dataset. Support of this dataset is provided by the Office of Science, U.S. Department of Energy. Work by S. Schwartz was supported by the United States Department of Energy (Office of Science, OBER) under Contract No. DE-AC02-09CH10886. B. Stevens was supported by the Max Planck Society for the Advancement of Science.

Appendix: A Energy Budget Justification

This appendix provides a brief, term-by-term, justification, based on observations and simulations with the ECHAM6 climate model, for energy flows at the top of the atmosphere and at the surface. The best estimate for each term is provided in Table 1, and the uncertainty is provided as a "likely" range (which we interpret as the central 68 %, or two-sigma, of the distribution) in Fig. 1. Estimates were constrained by measurements of the increase in ocean enthalpy, from which the energy flux imbalance at the TOA and surface, both taken as about 1 W m^{-2}, can be inferred. ECHAM6 simulations were performed using prescribed sea-surface temperatures for the period 1977–1986 with T255 spectral resolution, or roughly 0.5°; these simulations benefitted from modifications to ECHAM6, relative to ECHAM5, designed to include more advanced treatments of clear-sky radiative transfer and surface albedo, as well as tuning that better reflects present understanding of the radiative imbalance at TOA. The syntheses of the energy budgets presented here were greatly informed by those of Trenberth et al (2009) and Stephens et al. (2012), denoted below as T-2009 and S-2012 respectively.

Total Solar Irradiance

Kopp and Lean (2011) estimate a total solar irradiance of 1360.8 ± 0.5 W m^{-2} at the last solar minimum. This estimate gains credibility because of laboratory measurements that

Table 2 Model-based estimates of non-radiative heat fluxes as constrained by observations for the reanalysis products, and sea-surface temperature and top-of-atmosphere balance in the case of ECHAM6

Source	NRA1	NRA2	JRA1	JRA2	ERA	ERA-interim	ECHAM6 (1980–1990)
Latent heat flux	80.2	83.1	85.1	90.2	82.3	81.6	87.2
Sensible heat flux	15.3	15.6	18.8	19.4	15.3	17.7	19.0

NRA, JRA and ERA refer to the reanalysis products of the National (US) Center for Environmental Prediction, the Japanese Meteorological Agency, and the European Center Reanalysis Products, ERA40 and ERA-Interim. The ERA-Interim data are for the period from January 1989 to December 2007. The different eras, e.g., NRA1 and NRA2 refer to estimates over the ERBE epoch (lasting from February 1985 to March 1989) versus the CERES period which T-2009 define as lasting from March 2000 to April 2004 (cf., Trenberth et al. 2009)

show that the measurement geometry of earlier generations of satellites that had reported higher values of solar irradiance were biased high as a result of scattered light. Accounting for the total solar maximum being roughly 1.6 W m^{-2} greater (Kopp and Lean 2011) yields an average over the solar cycle of 1361.6 W m^{-2}. For the average ratio of Earth's surface area to its projected area, 4.0034, the average TOA irradiance is 340.1 $\pm$ 0.1 W m^{-2} (Table 2).

Reflected Short-wave Irradiance (TOA)

Based largely on the CERES data S-2012 estimate this term to be 97.7 $\pm$ 2 W m^{-2}. We maintain their estimate of the uncertainty at $\pm$2 W m^{-2}, in recognition of constraints posed by estimates of the rate of increase in ocean enthalpy and associated uncertainty in the outgoing long-wave irradiance at TOA. However, we center our estimate at a slightly greater value (99 W m^{-2}) and give a best estimate of 100 W m^{-2}, in the light of the estimates of T-2009 and previous assessments of the ERB data which gave much higher values of reflected short-wave irradiance at TOA.

Outgoing Long-wave Irradiance (TOA)

Both S-2012 and T-2009 estimate a value of 239 W m^{-2}, which we adopt as a central value. S-2012 estimate the uncertainty in measurements of this term to be $\pm$3 W m^{-2}. We adopt a somewhat smaller estimate of the uncertainty, as we believe this better represents the constraint posed by measurements of the other terms at TOA, together with the rate of change of ocean enthalpy.

Absorbed Solar Radiation (Surface)

In the absence of notable advancements in measurements of this quantity we let the values cited by T-2009, 161 W m^{-2}, and S-2012, 168 W m^{-2}, span the range of our estimate. The resultant range encompasses nearly all past estimates, as well as global climate modeling results using ECHAM6. Larger values, as suggested by S-2012 were not adopted because we could not reconcile them with other measurements and the need to maintain surface energy balance. For reference, Stackhouse et al. (2011) and Kato et al (2012) estimate this term to be 167 and 169 W m^{-2}, respectively: these values are at the upper end, or even just beyond, the range of values we present as most likely.

Downwelling Long-wave Irradiance (Surface)

In estimating this term, we discount the much lower estimates by T-2009 and adopt values closer to the central estimates of S-2012, which largely summarizes results from earlier studies by Kato et al (2012) and Stackhouse et al. (2011). However, because the central estimate of S-2012 is at the upper end of the range of earlier studies upon which it is based, we do not extend our "likely" range to include values at the upper end of their uncertainty range.

Upwelling Long-wave Irradiance (Surface)

The two-sigma range presented in Fig. 1 is slightly larger than the range spanned by the estimates of S-2012 and T-2009. Our ECHAM6 estimates, constrained by observed sea-surface temperatures yield a value of 397.7 W m^{-2}. Although one would expect this quantity to be well constrained by surface temperature, uncertainties in the surface emissivity, as reviewed by T-2009 and references therein, contribute to uncertainty. We note that if the upwelling surface infrared flux is evaluated according to the Stefan-Boltzmann radiation law with the emissivity of the surface taken as unity. Under this assumption, the upwelling thermal infrared flux given in Fig. 1 as 395–399 W m^{-2} corresponds to Ts of 288.9–289.6 K. Likewise the downwelling flux from the atmosphere to the surface is evaluated under the assumption of unity absorptivity. A departure from unity in emissivity and absorptivity would result in substantial reduction in both fluxes. Measurement-based emissivities in the thermal infrared range from 0.89 to 0.99 for terrestrial surfaces (Sutherland 1986); and 0.96–0.99 for the ocean surface (Hanafin and Minnett 2005). Other things being equal, for an upwelling flux of 400 W m^{-2}, a decrease in emissivity by 0.1 would result in a reduction in flux of 40 W m^{-2}. However, as a reduction in emissivity is accompanied by equal reduction in absorptivity, reduction in both fluxes from those calculated for unity emissivity and absorptivity would be locally compensated to great extent. Zhou et al. (2002) present calculations with a radiation transfer model coupled to a land-surface model that shows that a reduction in emissivity/absorptivity over terrestrial surfaces results in an increase in surface temperature and near-surface air temperature; a decrease in the emissivity by 0.1 was found to result in a decrease in net and upward long-wave radiation by about 6.6 and 8.1 W m^{-2}, respectively. The decrease in radiative emission was found to be further offset by an increase in sensible heat flux. In view of these considerations, it would seem that the uncertainty associated with the rates of bidirectional radiative transfer at the surface may be somewhat greater than has previously been appreciated; however, the net effect is likely smaller than might be assumed based on emissivity considerations alone because of compensating effects. In addition to considerations of emissivity/absorptivity, insufficient resolution of the spatio-temporal variability tends to bias estimates low. For model-based estimates the latter are somewhat ameliorated by the ECHAM6 resolution in which radiative fluxes are calculated at 20 min intervals on a T255 grid; however, the ECHAM6 calculations are based on a constant surface emissivity of 0.996 which is likely too large; hence we take the ECHAM6 estimates as an upper bound.

Latent Heat Flux

Our "likely" range of 78–88 W m^{-2} attempts to bridge the gap between estimates derived from the climatology of precipitation versus those inferred from a consideration of the surface energy budget. The GEWEX radiation panel specifies a best estimate of globally and annually averaged precipitation of 2.61 ± 0.233 mm d^{-1} (Gruber and Levizzani 2008). Assuming an effective evaporation/precipitation temperature of 23 °C yields a conversion factor of 28.25 W m^{-2}(mm $d^{-1})^{-1}$ so the global precipitation estimate become 73.7 ± 6.6 W m^{-2} . Assuming a density of water of 1 g cm^{-3} and an enthalpy of evaporation of 2.501 kJ kg^{-1}, which would correspond to an effective evaporation/precipitation temperature of 4 °C, leads to a conversion factor of 28.83 W m^{-2} (mm $d^{-1})^{-1}$, which is about 2 % or 1.5 W m^{-2} larger than the value at 23 °C. Kiehl and Trenberth (1997) appear to use this larger factor, whereas T-2009 assume a conversion factor of 80.0/ 2.76 = 29.0 W m^{-2} (mm $d^{-1})^{-1}$, which is yet larger. T-2009 cite a value of 2.63 mm d^{-1} for the global and annual averaged precipitation rate from the global precipitation climatology project; thus their best estimate (assuming an effective precipitation/evaporation temperature of 23 °C) is nearly 8 %, not 5 %, larger than that measured by GPCP. These calculations suggest that because the precipitation does not fall at the same temperature, or in the same phase, as it evaporated, differences in the contribution of precipitation, versus evaporation/sublimation, to the surface energy budget may be as large as 1 W m^{-2}; this difference is not accounted for in Fig. 1. S-2012 estimate a yet larger, 88 ± 10 W m^{-2}, value for globally and annually averaged precipitation, justified in part by a consideration of larger estimates of downwelling long-wave irradiance at the surface and the constraints of surface energy imbalance. Our estimated likely range overlaps with the upper part of the likely range given by Gruber and Levizzani (2008) and incorporates the central value of S-2012, but gives less weight to the upper range of estimates by S-2012. Values larger than 88 W m^{-2} are necessary only if values in the upper end of the range for the downwelling irradiance estimated by S-2012 are adopted but would require a bias in the precipitation climatologies more than twice as large as the stated (1-σ) uncertainty.

Surface Sensible Heat Flux

Our "likely" range of 17–21 W m^{-2} attempts to bridge the gap between the estimate by T-2009 and the much larger central value estimated by S-2012. We adopt somewhat smaller values than estimated by the latter investigators because the larger values are difficult to reconcile with the reanalysis of meteorological observations and because high-resolution climate models constrained by sea-surface temperatures robustly give estimates toward the lower end of our stated range. Here we note that estimates of surface fluxes from the ERA-Interim reanalysis are nearly 2 W m^{-2} larger than those by the ERA40. The ERA40 reanalysis helped motivate the lower estimate of surface sensible heat flux by T-2009, but the physics of the model used for the ERA40 is generally assumed to be inferior to what is used in the later (ERA-Interim) reanalysis product.

Directly Transmitted Surface Radiation

Recent spectroscopic and theoretical work suggests that the amount of directly transmitted surface radiation through the atmospheric window is markedly less than previously realized. The estimate of 22 W m^{-2} provided in Fig. 1 for this quantity can be compared to the estimate of 40 W m^{-2}, as for instance given by Trenberth et al. (2009). The present estimate adopts the smaller numbers based on new measurements of the water vapor continuum, (e.g., see the review article in this volume by Shine et al. 2012) that have been incorporated into revised global estimates by Costa and Shine (2012), and as reported in the workshop. These revised estimates are also the source for Stephens et al. (2012) who adopt the same value of directly transmitted surface emission as we do.

Overall we note that our estimate of the net surface irradiance (long-wave and short-wave summed) varies between 99 and 121 W m^{-2}. This differs considerably from the range 86–148 W m^{-2} cited by S-2012, the much higher values of net surface irradiance included in the S-2012 estimate being too difficult to reconcile with understanding of turbulent exchange processes at the surface. These considerations motivate a somewhat lower upper bound in the downwelling long-wave irradiance and absorbed solar irradiance at the surface given in Fig. 1.

References

Abbot CG, Fowle FE (1908) Radiation and terrestrial temperature. Ann Astrophys Obs Smithson Inst 2(2):125–224

Albrecht BA (1989) Aerosols, cloud microphysics, and fractional cloudiness. Science 245(4923):1227–1230

Andrews T, Forster PM, Gregory JM (2009) A surface energy perspective on climate change. J Climate 22(10):2557–2570

Andrews T, Gregory JM, Forster PM, Webb MJ (2011) Cloud adjustment and its role in CO_2 radiative forcing and climate sensitivity: a review. Surv Geophys. doi:10.1007/s10712-011-9152-0

Bar-Or RZ, Koren I, Altaratz O (2010) Estimating cloud field coverage using morphological analysis. Environ Res Lett 5(1):014,022

Bellouin N, Jones A, Haywood J, Christopher SA (2008) Updated estimate of aerosol direct radiative forcing from satellite observations and comparison against the Hadley Centre climate model. J Geophys Res 113(D10), D10205

Bender FAM, Rodhe H, Charlson RJ, Ekman AML, Loeb NG (2006) 22 views of the global albedo-comparison between 20 GCMs and two satellites. Tellus A 58(3):320–330

Brasseur GP, Roeckner E (2005) Impact of improved air quality on the future evolution of climate. Geophys Res Lett 32(23), L23704

Caldwell P, Bretherton CS (2009) Response of a subtropical stratocumulus-capped mixed layer to climate and aerosol changes. J Climate 22(1):20–38

Cess RD (1976) Climate change: an appraisal of atmospheric feedback mechanisms employing zonal climatology. J Atmos Sci 33(10):1831–1843

Charlson RJ, Schwartz SE, Hales JM, Cess RD, Coakley JA, Hansen JE, Hofmann DJ (1992) Climate forcing by anthropogenic aerosols. Science 255(5043):423–430

Clothiaux EE, Ackerman TP, Mace GG, Moran KP, Marchand RT, Miller MA, Martner BE (2000) Objective determination of cloud heights and radar reflectivities using a combination of active remote sensors at the ARM CART sites. J Appl Meteorol 39(5):645–665

Colman R, McAvaney B (2009) Climate feedbacks under a very broad range of forcing. Geophys Res Lett 36(1), L01702

Costa S, Shine KP (2012) Outgoing long-wave radiation due to directly-transmitted surface emission. J Atmos Sci 69 (to appear)

Dessler AE (2010) A determination of cloud feedback from climate variations over the past decade. Science 330:1523–1527

Forster P, Ramaswamy V, Artaxo P, Berntsen T, Betts R, Fahey D, Haywood J, Lean J, Lowe D, Myhre G, Nganga J, Prinn R, Raga G, Schulz M, van Dorland R (2007) Changes in atmospheric constituents and in radiative forcing. In: Solomon S, Qin D, Manning M, Chen Z, Marquis M, Averyt K, Tignor M (eds) Climate change 2007: the physical science basis. Contribution of working group I to the fourth assessment report of the intergovernmental panel on climate change, Cambridge University Press, Cambridge, pp 1–106

Forster PMF, Gregory JM (2006) The climate sensitivity and its components diagnosed from Earth radiation budget data. J Clim 19(1):39–52

Genkova I, Long C, Minnis P, Heck P (2004) Point-to-point comparison of satellite and ground-based cloud properties at the ARM southern great plains central facility. In: Fourteenth ARM science team meeting proceedings, albuquerque, New Mexico, March 22–26, 2004, pp 1–18

Gregory J, Webb M (2008) Tropospheric adjustment induces a cloud component in CO_2 forcing. J Clim 21(1):58–71

Gregory JM, Stouffer RJ, Raper SCB, Stott PAA, Rayner NA (2002) An observationally based estimate of the climate sensitivity. J Clim 15(22):3117–3121
Gregory JM, Ingram WJ, Palmer MA, Jones GS, Stott PAA, Thorpe RB, Lowe JA, Johns TC, Williams KD (2004) A new method for diagnosing radiative forcing and climate sensitivity. Geophys Res Lett 31(3):L03,205
Gruber A, Levizzani V (2008) Assessment of global precipitation prodcuts. Tech. rep
Hall A, Qu X (2006) Using the current seasonal cycle to constrain snow albedo feedback in future climate change. Geophys Res Lett 33(3):L03,502–L03,504
Han Q, Rossow WB, Zeng J, Welch R (2002) Three different behaviors of liquid water path of water clouds in aerosol—cloud interactions. J Atmos Sci 59:726–735
Hanafin JA, Minnett PJ (2005) Measurements of the infrared emissivity of a wind-roughened sea surface. Appl Opt 44:398–411
Hansen J, Sato M, Ruedy R (1997) Radiative forcing and climate response. J Geophys Res 102(D6):6831–6864
Heidinger AK, Pavolonis MJ (2009) Gazing at cirrus clouds for 25 years through a split window. Part I: methodology. J Appl Meteor Climatol 48(6):1100–1116
Holben BN, Tanré D, Smirnov A, Eck TF, Slutsker I, Abuhassan N, Newcomb W, Schafer J, Catenet B, Lavenu F, Kaufman Y, Vande Castle J, Setzer A, Markham B, Clark D, Frouin R, Halthore R, Karneli A, O'Neill N, Pietras C, Pinker R, Voss K, Zibordi G (2001) An emerging ground-based aerosol climatology: aerosol optical depth from AERONET. J Geophys Res pp 1–31
Houghton HG (1954) On the annual heat balance of the northern hemisphere. J Meteor 11(1):1–9
Intrieri J, Shupe M, Uttal T, McCarty B (2002) An annual cycle of Arctic cloud characteristics observed by radar and lidar at SHEBA. J Geophys Res C 10:1–15
Jones P, New M, Parker DE, Martin S, Rigor IG (1999) Surface air temperature and its changes over the past 150 years. Rev Geophys 37(2):173–199
Jung M, Reichstein M, Ciais P, Seneviratne SI, Sheffield J, Goulden ML, Bonan G, Cescatti A, Chen J, de Jeu R, Dolman AJ, Eugster W, Gerten D, Gianelle D, Gobron N, Heinke J, Kimball J, Law BE, Montagnani L, Mu Q, Mueller B, Oleson K, Papale D, Richardson AD, Roupsard O, Running S, Tomelleri E, Viovy N, Weber U, Williams C, Wood E, Zaehle S, Zhang K (2010) Recent decline in the global land evapotranspiration trend due to limited moisture supply. Nature 467(7318):951–954
Kahn RA (2011) Reducing the uncertainties in direct aerosol radiative forcing. Surv Geophys
Kahn RA, Gaitley BJ, Garay MJ, Diner DJ, Eck TF, Smirnov A, Holben BN (2010) Multiangle imaging spectroRadiometer global aerosol product assessment by comparison with the aerosol robotic network. J Geophys Res 115:D23209
Kaplan LD (1960) The influence of carbon dioxide variations on the atmospheric heat balance. Tellus 12:204–208
Kato S, Rose FG, Sun-Mack S, Miller WF, Chen Y, Rutan DA, Stephens GL, Loeb NG, Minnis P, Wielicki BA, Winker DM, Charlock TP, Stackhouse PW, Kuan-man X, Collins WD (2012) Improvement of top-of-atmosphere and surface irradiance computations with CALIPSO, CloudSAT, and MODIS-derived cloud and aerosol properties. J Climate 25:1–63
Khain A (2009) Notes on state-of-the-art investigations of aerosol effects on precipitation: a critical review. Environ Res Lett 4:015004 (20 pp)
Kiehl JT (2007) Twentieth century climate model response and climate sensitivity. Geophys Res Lett 34(22), L22710
Kiehl JT, Trenberth KE (1997) Earth's annual global mean energy budget. Bull Am Meteorol Soc 78:197–208
Kim SW, Yoon SC, Dutton E, Kim J, Wehrli C, Holben B (2008) Global surface-based Sun photometer network for long-term observations of column aerosol optical properties: intercomparison of aerosol optical depth. Aerosol Sci Tech 42(1):1–9
Klocke D, Pincus R, Quaas J (2011) On constraining estimates of climate sensitivity. J Climate 24:6092–6099
Knutti R, Hegerl GC (2008) The equilibrium sensitivity of the Earth's temperature to radiation changes. Nat Geosci 1(11):735–743
Kopp G, Lean JL (2011) A new, lower value of total solar irradiance: evidence and climate significance. Geophys Res Lett 38(1), L01706
Koren I, Remer LA, Kaufman YJ, Rudich Y, Martins JV (2007) On the twilight zone between clouds and aerosols. Geophys Res Lett 34(8):L08,805
Koren I, Feingold G, Remer LA (2010) The invigoration of deep convective clouds over the Atlantic: aerosol effect, meteorology or retrieval artifact. Atmos Chem Phys 10(18):8855–8872

Levitus S, Antonov J, Boyer T (2005) Warming of the world ocean, 1955–2003. Geophys Res Lett 32(2):L02,604
Lin JL (2007) The double-ITCZ problem in IPCC AR4 coupled GCMs: ocean—atmosphere feedback analysis. J Climate 20(18):4497–4525
Liu Y, Daum PH (2002) Anthropogenic aerosols: Indirect warming effect from dispersion forcing. Nature 419(6907):580–581
Loeb NG, Schuster GL (2008) An observational study of the relationship between cloud, aerosol and meteorology in broken low-level cloud conditions. J Geophys Res 113(D14), D14214
Loeb NG, Su W (2010) Direct aerosol radiative forcing uncertainty based on a radiative perturbation analysis. J Climate 23:5288–5293
Loeb NG, Wielicki BA, Su W, Loukachine K, Sun W, Wong T, Priestley KJ, Matthews G, Miller WF, Davies R (2007) Multi-instrument comparison of top-of-atmosphere reflected solar radiation. J Climate 20(3):575–591
Loeb NG, Wielicki BA, Doelling DR, Smith GL, Keyes DF, Kato S, Manalo-Smith N, Wong T (2009) Toward optimal closure of the Earth's top-of-atmosphere radiation budget. J Climate 22(3):748–766
Loeb NG, Lyman JM, Johnson GC, Allan RP, Doelling DR, Wong T, Soden BJ, Stephens GL (2012) Observed changes in top-of-the-atmosphere radiation and upper-ocean heating consistent within uncertainty. Nat Geosci 5(2):110–113
Lohmann U, Feichter J (2005) Global indirect aerosol effects: a review. Atmos Chem Phys 5:715–737
Lohmann U, Rotstayn L, Storelvmo T, Jones A, Menon S, Quaas J, Ekman AML, Koch D, Ruedy R (2010) Total aerosol effect: radiative forcing or radiative flux perturbation. Atmos Chem Phys 10(7):3235–3246
London JA (1957) A study of the atmospheric heat balance. Tech. rep., Department of Meteorology and Oceanography, New York University
Long CN, Ackerman TP, Gaustad KL, Cole JNS (2006) Estimation of fractional sky cover from broadband shortwave radiometer measurements. J Geophys Res 111(D11):D11,204
Lubin D, Vogelmann AM (2006) A climatologically significant aerosol longwave indirect effect in the Arctic. Nature 439:241–242
Lucarini V, Ragone F (2011) Energetics of climate models: net energy balance and meridional enthalpy transport. Rev Geophys 49:RG1001
Lyman JM (2011) Estimating global energy flow from the global upper ocean. Surv Geophys. doi: 10.1007/s10712-011-9167-6
Manabe S, Wetherald RT (1967) Thermal equilibrium of atmosphere with a given distribution of relative humidity. J Atmos Sci 24(3):241
Markowicz KM, Flatau PJ, Vogelmann AM, Quinn PK, Welton EJ (2003) Clear-sky infrared aerosol radiative forcing at the surface and the top of the atmosphere. QJR Meteorol Soc 129(594):2927–2947
Martonchik JV, Kahn RA (2009) Retrieval of aerosol properties over land using MISR observations. In: KA A (ed) Satellite aerosol remote sensing over land, Springer, New York, pp 267–293
Matthews HD, Caldeira K (2007) Transient climate—carbon simulations of planetary geoengineering. Proc Natl Acad Sci 104(24):9949–9954
Mauritsen T, Sedlar J, Tjernström M, Leck C, Martin M, Shupe M, Sjogren S, Sierau B, Persson POG, Brooks IM, Swietlicki E (2011) An arctic CCN-limited cloud-aerosol regime. Atmos Chem Phys 11(1):165–173
Michalsky J, Denn F, Flynn C, Hodges G, Kiedron P, Koontz A, Schlemmer J, Schwartz SE (2010) Climatology of aerosol optical depth in north-central Oklahoma: 1992–2008. J Geophys Res 115(D7): D07203
Mishchenko MI, Geogdzhayev IV, Cairns B, Rossow WB, Lacis AA (1999) Aerosol retrievals over the ocean by use of channels 1 and 2 AVHRR data: sensitivity analysis and preliminary results. Appl Opt 38(36):7325–7341
Murphy DM, Solomon S, Portmann RW, Rosenlof KH, Forster PM, Wong T (2009) An observationally based energy balance for the Earth since 1950. J Geophys Res-Atmos 114
Nakajima T, Higurashi A, Kawamoto K, Penner JE (2001) A possible correlation between satellite-derived cloud and aerosol microphysical parameters. Geophys Res Lett 28(7):1171–1174, D17107
Nemesure S, Wagener R, Schwartz SE (1995) Direct shortwave forcing of climate by the anthropogenic sulfate aerosol: Sensitivity to particle size, composition, and relative humidity. J Geophys Res 100(D12):26,105–26,116
Nuijens L, Stevens B, Siebesma AP (2009) The environment of precipitating shallow cumulus convection. J Atmos Sci 66(7):1962–1979
Pierrehumbert R (2011) Infrared radiation and planetary temperature. Physics Today 64:33–38

Pincus R, Baker MB (1994) Effect of precipitation on the albedo susceptibility of clouds in the marine boundary layer. Nature 372(6503):250–252
Plass GN (1956) The carbon dioxide theory of climatic change. Tellus 8(2):140–154
Quaas J, Boucher O, Bellouin N, Kinne S (2005) Satellite-based estimate of the direct and indirect aerosol climate forcing. J Geophys Res 113:9
Quaas J, Ming Y, Menon S, Takemura T, Wang M, Penner JE, Gettelman A, Lohmann U, Bellouin N, Boucher O, Sayer AM, Thomas GE, McComiskey A, Feingold G, Hoose C, Kristjansson JE, Liu X, Balkanski Y, Donner LJ, Ginoux PA, Stier P, Grandey B, Feichter J, Sednev I, Bauer SE, Koch D, Grainger RG, Kirkevaring A, Iversen T, Seland O, Easter R, Ghan SJ, Rasch PJ, Morrison H, Lamarque JF, Iacono MJ, Kinne S, Schulz M (2009) Aerosol indirect effects—general circulation model intercomparison and evaluation with satellite data. Atmos Chem Phys 9(22):8697–8717
Ramanathan V (1987) The role of Earth radiation budget studies in climate and general circulation research. J Geophys Res 92(D4):4075–4095
Schneider SH (1972) Cloudiness as a global climatic feedback mechanism: the effects on the radiation balance and surface temperature of variations in cloudiness. J Atmos Sci 29(8):1413–1422
Schwartz SE (2011) Feedback and sensitivity in an electrical circuit: an analog for climate models. Climatic Change 106(2):315–326
Schwartz SE, Harshvardhan, Benkovitz CM (2002) Influence of anthropogenic aerosol on cloud optical depth and albedo shown by satellite measurements and chemical transport modeling. Proc Natl Acad Sci 99(4):1784–1789
Schwartz SE, Charlson RJ, Kahn RA, Ogren JA, Rodhe H (2010) Why hasn't Earth warmed as much as expected. J Climate 23(10):2453–2464
Segrin MS, Coakley JAJ, Tahnk WR (2007) MODIS observations of ship tracks in summertime stratus off the west coast of the United States. J Atmos Sci 64(12):4330–4345
Shine KP, Ptashnik IV, Rädel G (2012) The water vapour continuum: brief history and recent developments. Surv Geophys. doi:10.1007/s10712-011-9170-y
Smirnov A, Holben BN, Slutsker I, Giles DM, McClain CR, Eck TF, Sakerin SM, Macke A, Croot P, Zibordi G, Quinn PK, Sciare J, Kinne S, Harvey M, Smyth TJ, Piketh S, Zielinski T, Proshutinsky A, Goes JI, Nelson NB, Larouche P, Radionov VF, Goloub P, Krishna Moorthy K, Matarrese R, Robertson EJ, Jourdin F (2009) Maritime aerosol network as a component of aerosol robotic network. J Geophys Res 114(D6):D06,204
Soden BJ, Held IM, Colman R, Shell KM, Kiehl JT, Shields CA (2008) Quantifying climate feedbacks using radiative kernels. J Climate 21(14):3504–3520
Sohn BJ, Nakajima T, Satoh M, Jang HS (2010) Impact of different definitions of clear-sky flux on the determination of longwave cloud radiative forcing: NICAM simulation results. Atmos Chem Phys 10(23):11,641–11,646
Stackhouse PW, Gupta SK, Cox SJ, Zhang T, Mikovitz JC, Hinkelman LM (2011) 24.5 year surface radiation budget data set relaseased. GEWEX News 21(1):1–4
Stephens G (1988) Radiative transfer through arbitrarily shaped optical media. Part II: group theory and simple closures. J Atmos Sci 45:1837–1848
Stephens GL, Vane DG, Tanelli S, Im E, Durden S, Rokey M, Reinke D, Partain P, Mace GG, Austin R, L'Ecuyer T, Haynes J, Lebsock M, Suzuki K, Waliser D, Wu D, Kay J, Gettelman A, Wang Z, Marchand R (2008) CloudSat mission: performance and early science after the first year of operation. J Geophys Res 113, D00A18
Stephens GL, Juilin L, Wild M, Clayson CA, Loeb NG, Kato S, L'Ecuyer T, Stackhouse Jr PW, Andrews T (2012) The energy blance of the Earth's climate system. Nat Geosci 1–13
Stevens B, Feingold G (2009) Untangling aerosol effects on clouds and precipitation in a buffered system. Nature 461(7264):607–613
Stevens B, Lenschow D, Faloona I, Moeng C, Lilly D, Blomquist B, Vali G, Bandy AR, Campos T, Gerber H, Haimov S, Morley B, Thornton D (2003) On entrainment rates in nocturnal marine stratocumulus. QJR Meteorol Soc 129(595):3469–3493
Sutherland RA (1986) Broadband and spectral emissivities (2–19 μm) of some natural soils and vegetation. J Atmos Oceanic Technol 3:199–202
Tackett JL, Di Girolamo L (2009) Enhanced aerosol backscatter adjacent to tropical trade wind clouds revealed by satellite-based lidar. Geophys Res Lett 36(14):L14,804
Tredger E (2009) On the evaluation of uncertainty in climate models. PhD thesis, London School of Economics, London
Trenberth KE, Fasullo JT (2010a) Simulation of present-day and twenty-first-century energy budgets of the Southern Oceans. J Climate 23(2):440–454
Trenberth KE, Fasullo JT (2010b) Tracking Earth's energy. Science 328(5976):316–317

Trenberth KE, Fasullo JT, Kiehl J (2009) Earth's global energy budget. Bull Am Meteor Soc 90(3):311–323
Twohy CH, Coakley JA Jr, Tahnk WR (2009) Effect of changes in relative humidity on aerosol scattering near clouds. J Geophys Res 114(D5):D05,205
Twomey S (1974) Pollution and planetary albedo. Atmos Environ 8:1251–1256
Weatherhead EC, Reinsel GC, Tiao GC, Jackman CH, Bishop L, Frith SMH, De Luisi J, Keller T, Oltmans SJ, Fleming EL, Wuebbles DJ, Kerr JB, Miller AJ, Herman J, McPeters R, Nagatani RM, Frederick JE (2000) Detecting the recovery of total column ozone. J Geophys Res 105(D17):22,201–22,210
Wielicki BA, Barkstrom BR, Harrison EF, Lee RB, Louis Smith G, Cooper JE (1996) Clouds and the Earth's radiant energy system (CERES): an Earth observing system experiment. Bull Am Meteorol Soc 77(5):853–868
Winker DM, Pelon J, Coakley JAJ, Ackerman SA, Charlson RJ, Colarco PR, Flamant P, Fu Q, Hoff RM, Kittaka C, Kubar TL, Le Treut H, McCormick MP, Megie G, Poole L, Powell K, Trepte C, Vaughan MA, Wielicki BA (2010) THE CALIPSO MISSION A Global 3D view of aerosols and clouds. Bull Am Meteorol Soc 91(9):1211–1229
Wong T, Wielicki BA, Lee RB III, Smith GL, Bush KA, Willis JK (2006) Reexamination of the observed decadal variability of the Earth radiation budget using altitude-corrected ERBE/ERBS nonscanner WFOV data. J Clim 19(16):4028–4040
Zhang J, Reid JS (2010) A decadal regional and global trend analysis of the aerosol optical depth using a data-assimilation grade over-water MODIS and Level 2 MISR aerosol products. Atmos Chem Phys 10(22):10,949–10,963
Zhou L, Dickinson RE, Tian Y, Jin M, Ogawa K, Yu H, T S (2002) A sensitivity study of climate and energy balance simulations with use of satellite-derived emissivity data over Northern Africa and the Arabian Peninsula. J Geophys Res 108:4795

MIX
Papier aus verantwortungsvollen Quellen
Paper from responsible sources
FSC® C105338

If you have any concerns about our products,
you can contact us on
ProductSafety@springernature.com

In case Publisher is established outside the EU,
the EU authorized representative is:
Springer Nature Customer Service Center GmbH
Europaplatz 3, 69115 Heidelberg, Germany

Printed by Libri Plureos GmbH
in Hamburg, Germany